JACARANDA
MATHS QUEST 9

AUSTRALIAN CURRICULUM | FOURTH EDITION

JACARANDA MATHS QUEST 9

AUSTRALIAN CURRICULUM | FOURTH EDITION

CATHERINE SMITH

JAMES SMART

GEETHA JAMES

CAITLIN MAHONY

BEVERLY LANGSFORD WILLING

CONTRIBUTING AUTHORS

Michael Sheedy | Kahni Burrows | Paul Menta

jacaranda
A Wiley Brand

Fourth edition published 2022 by
John Wiley & Sons Australia, Ltd
42 McDougall Street, Milton, Qld 4064

First edition published 2011
Second edition published 2015
Third edition published 2018

Typeset in 10.5/14 pt Times LT Std

ISBN: 978-0-7303-9364-1

Front cover image: © Perepadia Y/Shutterstock

Illustrated by diacriTech and Wiley Composition Services

Typeset in India by diacriTech

 A catalogue record for this book is available from the National Library of Australia

Printed in Singapore
M11634R1_130522

Contents

NAPLAN practice   online only

Set A Calculator allowed
Set B Non-calculator
Set C Calculator allowed
Set D Non-calculator
Set E Calculator allowed
Set F Non-calculator

NAPLAN practice

Go online to complete practice NAPLAN tests. There are 6 NAPLAN-style question sets available to help you prepare for this important event. They are also useful for practising your Mathematics skills in general.

Also available online is a video that provides strategies and tips to help with your preparation.

SET A
Calculator allowed

SET B
Non-calculator

SET C
Calculator allowed

SET D
Non-calculator

SET E
Calculator allowed

SET F
Non-calculator

Everything you need (and *want*) at your fingertips

A full lesson on one screen in your online course. Trusted, curriculum-aligned theory. Engaging, rich multimedia. All the teacher support resources you need. Deep insights into progress. Immediate feedback for students. Create custom assignments in just a few clicks.

Practical teaching advice and ideas

Teaching videos for all lessons

Reading content and rich media

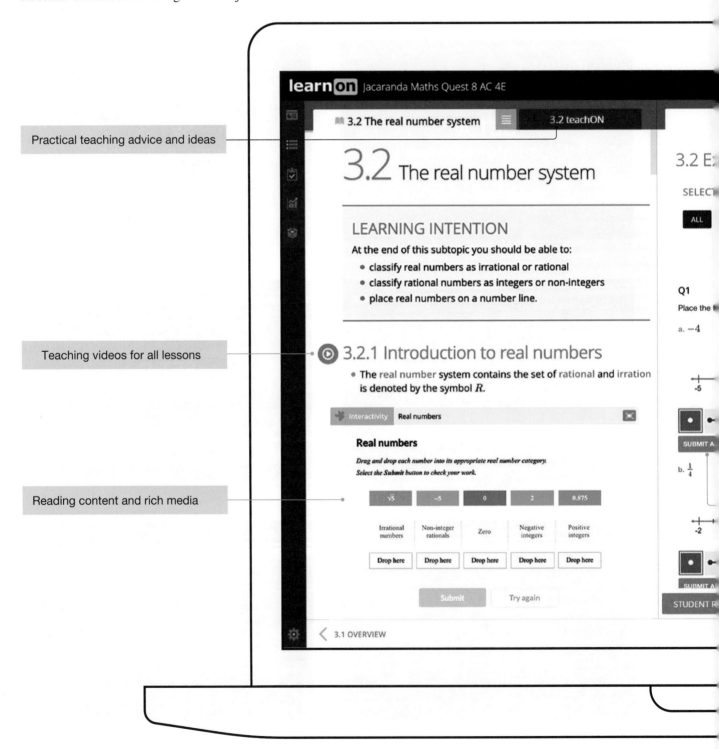

Differentiated question sets

Teacher and student views

Textbook questions

eWorkbook

Solutions

Digital documents

Video eLessons

Interactivities

ProjectsPLUS

Extra teaching resources

Questions with immediate feedback

| Teacher | Student | Year7A-Jacaranda-VC-... | Help | Samwell Tarly |

.2 Ex 1   3.2 TBQ   1

se

PATHWAY

ACTISE	CONSOLIDATE	MASTER
estions: 8, 11, 13, 9, 20, 27	Questions: 1, 6, 9, 12, 14, 17, 18, 21, 24, 26	Questions: 2, 7, 10, 15, 22, 23, 25, 28

numbers on a number line.

1 mark(s)

1 mark(s)

& MARKING

3.3 ADDING AND SUBTRACTING FRACTIONS >

RESOURCES

Topic PDF	1
eWorkbook	29
Solutions	1
Digital documents	1
Video eLessons	8
Interactivities	17
ProjectsPLUS	1
Weblinks	6
TEACHER eWorkbook	26
TEACHER Digital documents	9
TEACHER Weblinks	2

About this resource

Online, these new editions are the complete package — with trusted Jacaranda theory plus tools to support teaching and make learning more engaging, personalised and visible.

Available in learnON, our most powerful online learning platform

Instant reports give teachers and students deep insights into progress, including mapping of results against the cognitive processes

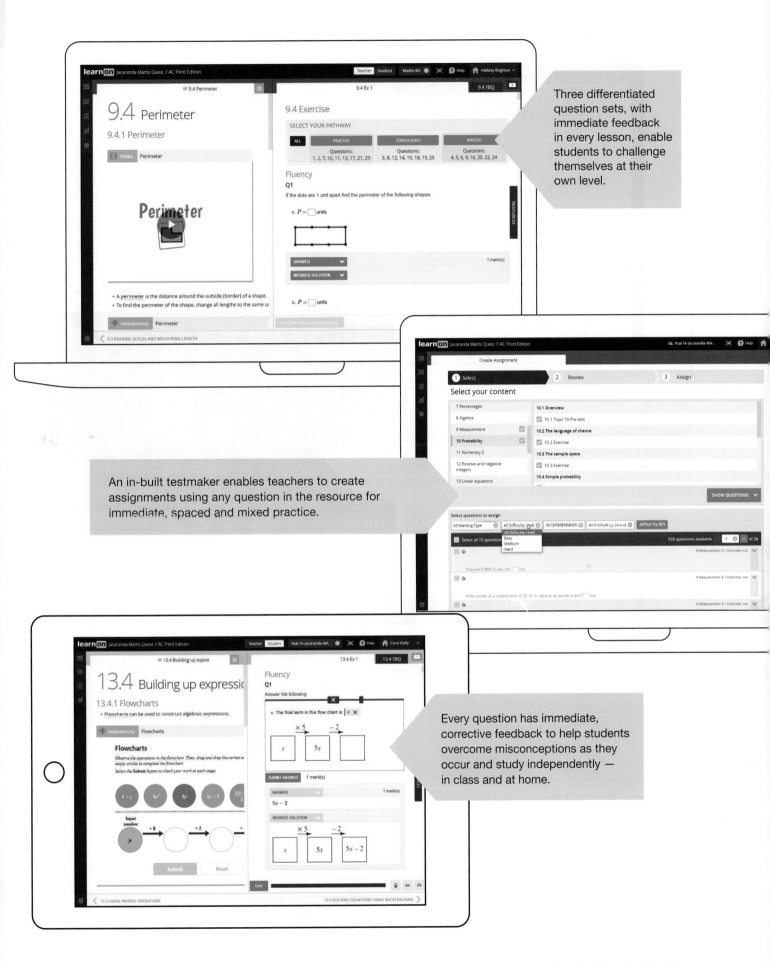

Three differentiated question sets, with immediate feedback in every lesson, enable students to challenge themselves at their own level.

An in-built testmaker enables teachers to create assignments using any question in the resource for immediate, spaced and mixed practice.

Every question has immediate, corrective feedback to help students overcome misconceptions as they occur and study independently — in class and at home.

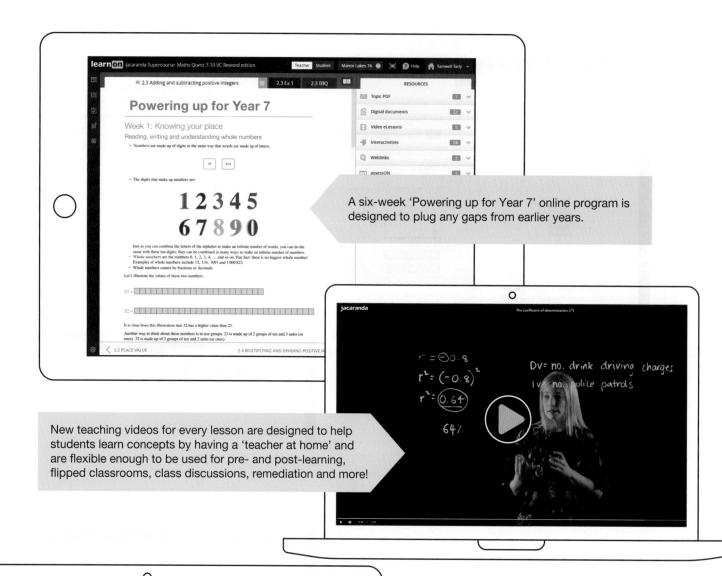

A six-week 'Powering up for Year 7' online program is designed to plug any gaps from earlier years.

New teaching videos for every lesson are designed to help students learn concepts by having a 'teacher at home' and are flexible enough to be used for pre- and post-learning, flipped classrooms, class discussions, remediation and more!

For teachers, the online teachON section contains practical teaching advice including learning intentions and three levels of differentiated teaching programs.

The eWorkbook enables teachers and students to download additional activities to support deeper learning.

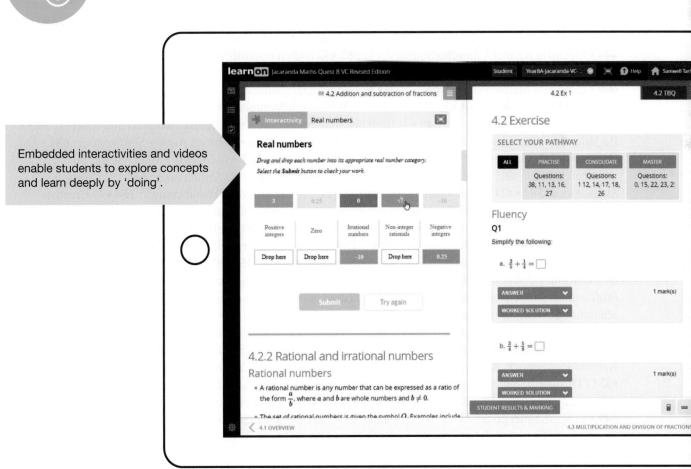

Embedded interactivities and videos enable students to explore concepts and learn deeply by 'doing'.

Acknowledgements

The authors and publisher would like to thank the following copyright holders, organisations and individuals for their assistance and for permission to reproduce copyright material in this book.

Images

• a. © PeopleImages/Getty Images; b. Dean Drobot/Shutterstock: **435** • © a. Vitalii Vodolazskyi/Shutterstock; b. RTimages/Shutterstock: **440** • © a. Yellow duck/Shutterstock; b. pattang/Shutterstock; c. Denis Tabler/Shutterstock; d. Nattika/Shutterstock; e. rangizzz/Shutterstock: **608** • © Corbis Corporation: **737** • © Desmos: **725** • © alexsl/ Getty Images: **272** • © SensorSpot/Getty Images: **634** • © Stephen Studd/Getty Images: **297** • © imageaddict.com.au: **529** • © John Wiley & Sons Australia/Photo by Jennifer Wright: **510** • © John Wiley & Sons Australia/Photo by Renee Bryon: **90** • © John Wiley & Sons Australia/Lyn Elms: **297** • © John Wiley & Sons Australia/Renee Bryon: **83** • © John Wiley & Sons Australia/Kari-Ann Tapp: **625** • © John Wiley & Sons Australia/Photo by Renee Bryon: **236, 288** • © John Wiley & Sons: Australia/Photo taken by Renee Bryon: **552, 578** • © Screenshot reprinted by permission from Microsoft Corporation: **612** • © NASA/Barney Magrath: **37** • © PhotoDisc: **519, 521** • © Photodisc: **210, 280** • © Photodisc Inc.: **276** • © Public Domain: **83** • © Adisa/Shutterstock: **615** • © Africa Studio/Shutterstock: **556** • © Aila Images/Shutterstock: **9** • © airdynamic/Shutterstock: **185** • © Alan Poulson Photography/Shutterstock: **590** • © Albachiaraa/Shutterstock: **495** • © Alex James Bramwell/Shutterstock: **591** • © ALPA PROD/Shutterstock: **444** • © Analia Aguilar Camacho/Shutterstock: **507** • © andrea crisante/Shutterstock: **505** • © Andresr/Shutterstock: **347, 571** • © Andrii Yalanskyi/Shutterstock: **409** • © anek.soowannaphoom/Shutterstock: **609** • © Angela Matthews/Shutterstock: **vii** • © antoniodiaz/Shutterstock: **437** • © Aquarius Studio/Shutterstock: **638** • © ARTvektor/Shutterstock: **587** • © Ashley Whitworth/Shutterstock: **493** • © Bankoo/Shutterstock: **68** • © bikeriderlondon/Shutterstock: **548** • © Binikins/Shutterstock: **464** • © BlueBarronPhoto/Shutterstock: **59** • © Bo Valentino/Shutterstock: **565** • © Brian A Jackson/Shutterstock: **4** • © Bruce Aspley/Shutterstock: **681** • © Business stock/Shutterstock: **445** • © cassiede alain/Shutterstock: **717** • © CGN089/Shutterstock: **50** • © Chatchai Somwat/Shutterstock: **619** • © CHRISTOPHE ROLLAND/Shutterstock: **648** • © chrupka/ Shutterstock: **584** • © Ciprian Vladut/Shutterstock: **312** • © ClaudioValdes/Shutterstock: **154** • © Creativa Images/Shutterstock: **434** • © David H.Seymour/Shutterstock: **520** • © Daxiao Productions/Shutterstock: **167** • © Dean Drobot/Shutterstock: **86** • © defpicture/Shutterstock: **494** • © del-Mar/Shutterstock: **500** • © DenisProduction.com/Shutterstock: **172** • © Dima Moroz/Shutterstock: **253** • © Dimedrol68/Shutterstock: **408** • © Diyana Dimitrova/Shutterstock: **59** • © djgis/Shutterstock: **358** • © Dmitry Natashin/Shutterstock: **50** • © dotshock/Shutterstock: **738** • © Edw/Shutterstock: **175** • © Eric Gevaert/Shutterstock: **416** • © ESB Professional/Shutterstock: **403** • © Fablok/Shutterstock: **36** • © fizkes/Shutterstock: **461** • © foto infot/ Shutterstock: **164** • © freesoulproduction/Shutterstock: **26** • © Galina Barskaya/Shutterstock: **73** • © Gordon Bell/Shutterstock: **351** • © Hallgerd/Shutterstock: **230** • © Harvepino/Shutterstock: **446** • © hddigital/Shutterstock: **463** • © Henri et George/Shutterstock: **175** • © hidesy/Shutterstock: **432** • © Iakov Kalinin/Shutterstock: **451** • © ILYA GENKIN/Shutterstock: **352** • © Inspiredbyart/Shutterstock: **407** • © ITTIGallery/Shutterstock: **453** • © J.D.S/Shutterstock: **120** • © Janaka Dharmasena/Shutterstock: **398** • © jessicakirshcreative/Shutterstock: **590** • © Jesus Cervantes/Shutterstock: **649** • © Jiripravda/Shutterstock: **549** • © jiris/Shutterstock: **22** • © JOAT/Shutterstock: **397** • © jocic/Shutterstock: **14** • © Juergen Faelchle/

Shutterstock: **484** • © Kahan Shan/Shutterstock: **401** • © Kaspars Grinvalds/Shutterstock: **402**• © Khakimullin Aleksandr/Shutterstock: **174** • © Kitch Bain/Shutterstock: **577** • © Krisana Tongnantree/Shutterstock.com: **197** • © ksy9/Shutterstock: **739** • © Kuki Ladron de Guevara/Shutterstock: **14** • © kurhan/Shutterstock: **171** • © Laitr Keiows/Shutterstock: **483** • © larryrains/Shutterstock: **124** • © leolintang/Shutterstock: **417** • © Lerner Vadim/Shutterstock: **359** • © Lightspring/Shutterstock: **301** • © Liudmila Pleshkun/Shutterstock: **218** • © lucadp/Shutterstock: **506** • © LuckyImages/Shutterstock: **632** • © Luna Vandoorne/Shutterstock: **167** • © Magdalena Kucova/Shutterstock: **472** • © Marian Weyo/Shutterstock: **77** • © Marques/Shutterstock: **48** • © Martin Charles Hatch/Shutterstock: **660** • © matteo_it/Shutterstock: **471** • © Maxisport/Shutterstock: **219** • © metamorworks/Shutterstock: **538** • © Michael William/Shutterstock: **164** • © michaeljung/Shutterstock: **444** • © Mila Supinskaya Glashchenko/Shutterstock: **161** • © MilanB/Shutterstock: **211** • © Mmaxer/Shutterstock: **78** • © Monkey Business Images/Shutterstock: **51, 81, 438, 452** • © MVolodymyr/Shutterstock: **287** • © Nata-Lia/Shutterstock: **521** • © Naypong/Shutterstock: **197** • © NDAB Creativity/Shutterstock: **748** • © Neale Cousland/Shutterstock: **82** • © New Africa/Shutterstock: **448** • © Nikodash/Shutterstock: **482** • © NikoNomad/Shutterstock: **36, 58** • © Nilobon Sweeney/Shutterstock: **14** • © Norman Pogson/Shutterstock: **246** • © Olaf Speier/Shutterstock: **546** • © OSTILL is Franck Camhi/Shutterstock: **748** • © OZBEACHES/Shutterstock: **682** • © PavleMarjanovic/Shutterstock: **747** • © Petr Toman/Shutterstock: **411** • © Philip Willcocks/Shutterstock: **143** • © Philip Yuan/Shutterstock: **196** • © photka/Shutterstock: **400** • © Phovoir/ Shutterstock: **292** • © pics4sale/Shutterstock: **495** • © pikselstock/Shutterstock: **442** • © PinkBlue/Shutterstock: **506** • © PlusONE/Shutterstock.com: **227** • © Potstock/Shutterstock: 157 • © Pressmaster/Shutterstock: **742** • © pryzmat/Shutterstock: **351** • © puhhha/Shutterstock: **2** • © quka/Shutterstock: **191** • © rachel ko/Shutterstock: **162** • © Radu Bercan/Shutterstock: **734** • © Ramon grosso dolarea/Shutterstock: **82** • © Rawpixel.com/Shutterstock: **604** • © Rawpixel.com/Shutterstock: **132, 177** • © REDPIXEL.PL/Shutterstock: **452** • © Redshinestudio/ Shutterstock: **162** • © Rido/Shutterstock: **440** • © Robert Kneschke/Shutterstock: **20** • © Robyn Mackenzie/Shutterstock: **406, 577** • © Roma Koshel/Shutterstock: **545** • © s74/Shutterstock: **742** • © Sara Julin Ingelmark/Shutterstock: **537** • © SERGEI PRIMAKOV/Shutterstock: **409** • © Sergey Nivens/Shutterstock: **33** • © Shenjun Zhang/Shutterstock: **729** • © Sinisa Botas/Shutterstock: **557** • © Skylines/Shutterstock: **120** • © Sonja Foos/Shutterstock.com: **636** • © Stefan Schurr/Shutterstock: **420** • © StepStock/Shutterstock: **693** • © StockPhotosLV/Shutterstock. **319** • © Stuart Monk/Shutterstock: **158** • © Sunflowerey/Shutterstock: **431** • © sunsetman/Shutterstock: **412** • © Suzanne Tucker/Shutterstock: **172, 586** • © Syda Productions/Shutterstock: **168, 603** • © szefei/Shutterstock: **460** • © Taki O/Shutterstock: **311** • © Taras Vyshnya/Shutterstock: **409** • © ThomasLENNE/Shutterstock: **117** • © Tom Hirtreiter/Shutterstock.com: **425** • © Torderiul/Shutterstock: **500** • © trexdruid/Shutterstock: **206** • © Triff/Shutterstock: **1** • © Tyler Olson/Shutterstock: **327** • © UfaBizPhoto/Shutterstock: **453** • © Val Thoermer/Shutterstock: **557** • © Vitalinka/Shutterstock: **584** • © Vladi333/Shutterstock: **67** • © Vladimir Borozenets/Shutterstock: **172** • © Vladir09/Shutterstock: **522** • © Volt Collection/Shutterstock: **254** • © Walter Cicchetti/Shutterstock: **132** • © wavebreakmedia/Shutterstock: **551** • © wavebreakmedia/Shutterstock: **38, 178** • © weknow/Shutterstock: **128** • © WilleeCole Photography/Shutterstock: **231** • © WitR/Shutterstock: **186** • © Worytko Pawel/ Shutterstock: **590** • © XiXinXing/Shutterstock: **443** • © Yermek/Shutterstock: **662** • © Yuriy Belmesov/ Shutterstock: **245** • © Sylex Ergonomics: **263** • © John Wiley & Sons Australia: **494, 505, 507, 549, 550, 708** • © Avesun/Shutterstock: **262** • © ChiccoDodiFC/Shutterstock: **281** • © Christina Richards/Shutterstock: **288** • © Haali/Shutterstock: **53** • © John Wiley & Sons Australia/Photo by Renee Bryon: **262, 281** • © John Wiley & Sons Australia/ Renee Bryon: **281** • © Photodisc Inc.: **288** • © Polka Dot Images: **288**

Every effort has been made to trace the ownership of copyright material. Information that will enable the publisher to rectify any error or omission in subsequent reprints will be welcome. In such cases, please contact the Permissions Section of John Wiley & Sons Australia, Ltd.

1 Number skills and index laws

1.1 Overview

Why learn this?

The earliest mathematics that humans used involved simple counting. Knowing that two goats could be swapped for six chickens was an important part of life twelve thousand years ago, so being able to count was a big deal.

Of course, being able to count is still a big deal today, but over thousands of years we have moved on from counting sheep. Ancient Egypt was the first recorded civilisation to think about fractions. Later on, ancient Greek mathematicians became the first people we know of to think about numbers that can't be expressed as fractions — these thoughts gave rise to the idea of irrational numbers. The value of π (pi) — which is the ratio of the circumference of a circle to its diameter — is one such irrational number. The value of π can only be approximated — if you were to try to say all of the digits in the value of π (that we presently know of) at a rate of 10 digits a second it would take you more than 6 million years to say them!

As new kinds of numbers have been discovered, new ways of writing them have been developed. These new ways of writing numbers can make it easier to work with numbers like the massively large ones that describe the distances between stars, or the incredibly small ones that describe the size of subatomic particles.

In this topic we will continue our mathematical journey by exploring a range of different types of numbers and the ways they can be written.

Where to get help

Go to your learnON title at **www.jacplus.com.au** to access the following digital resources. The Online Resources Summary at the end of this topic provides a full list of what's available to help you learn the concepts covered in this topic.

Video eLessons

Interactivities

Fully worked solutions to every question

Digital documents

eWorkbook

Complete this pre-test in your learnON title at www.jacplus.com.au and receive **automatic marks,** **immediate corrective feedback** and **fully worked solutions**.

1. State whether the following statement is True or False.
   Decimals that do not terminate and also do not recur are irrational numbers.

2. State the kind of number that $7.\overline{324}$ is an example of.

3. State whether the following statements are True or False.
   a. π is an irrational number.
   b. Every surd is an irrational number.

4. Calculate $\sqrt[3]{27}$.

5. **MC** Select the simplest form of $\sqrt{50}$ from the following.
   A. $5\sqrt{10}$
   B. $2\sqrt{5}$
   C. $5\sqrt{2}$
   D. $10\sqrt{5}$
   E. $25\sqrt{2}$

6. **MC** Select the correct answer for the expression $\sqrt{5} - 2\sqrt{45} + 3\sqrt{20}$.
   A. $7\sqrt{5}$
   B. $\sqrt{5}$
   C. $2\sqrt{70}$
   D. $19\sqrt{5}$
   E. $18\sqrt{5}$

7. The length of the hypotenuse of a right-angled triangle uses the formula $c^2 = a^2 + b^2$.
   Calculate the length of the hypotenuse for a triangle with side lengths of $\sqrt{2}$ and $\sqrt{3}$. Leave your answer as a surd.

8. **MC** Select how many significant figures there are in 0.0305.
   A. 1
   B. 2
   C. 3
   D. 4
   E. 5

9. Round 7.43861 to 3 decimal places.

10. Round 7.43861 to 3 significant figures.

11. **MC** Select the correct scientific notation to 3 significant figures for 0.000 000 021 37.
    A. 2.137×10^8
    B. 2.14×10^8
    C. 2.14×10^{-8}
    D. 2.137×10^{-8}
    E. 21.4×10^{-9}

12. **MC** Select the series of numbers that is correctly ordered from smallest to largest.
    A. $-\sqrt[3]{8}, -\sqrt[3]{27}, \sqrt[4]{81}$
    B. $\sqrt{3}, e, 5 \times 10^2$
    C. $2.1 \times 10^{-5}, 2.13 \times 10^2, 2.75 \times 10^{-1}$
    D. $\left(\dfrac{2}{5}\right)^2, \sqrt{0.16}, 0.\dot{3}$
    E. $\sqrt{3}, e, 5 \times 10^2, 3.\overline{14}, \pi, \dfrac{22}{7}$

13. An electron has a mass of $9.1093819 \times 10^{-31}$ kg, correct to 8 significant figures. A neutron is 1836 times the size of an electron. Evaluate the mass of a neutron correct to 4 significant figures.

14. Rationalise $\dfrac{4}{\sqrt{3}}$.

15. Simplify $\dfrac{5}{2}\sqrt{\dfrac{10}{5}}$.

1.2 Real numbers

▶ 1.2.1 Natural numbers

eles-4374

• The earliest use of numbers by humans was for counting things. The numbers we used were the positive whole numbers: 1, 2, 3, 4...
• These counting numbers are now known as the **natural numbers**.
• We use the symbol N to represent the set of natural numbers.

Integers

• **Integers** include the entire set of whole numbers. This means that the integer group contains:
  • the positive whole numbers (i.e. the natural numbers) 1, 2, 3, 4...
  • the negative whole numbers −1, −2, −3, −4...
  • 0 (which is neither positive nor negative).
• We use the symbol Z to represent the full set of integers. You may also see Z^- used to represent only the group of negative integers and Z^+ used for only the group of positive integers. This means that Z^+ and N represent the same group of numbers.

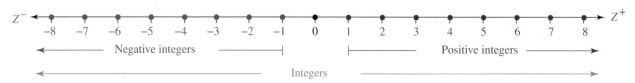

Rational numbers

• Integers, along with recurring decimals, are part of a larger group called the **rational numbers**.
• Rational numbers are numbers that can be expressed as a fraction with whole numbers both above and below the dividing sign.
• Examples of rational numbers include $\frac{3}{7}$, -5 (which can be written as $\frac{-5}{1}$), 3.2 (which can be written as $\frac{16}{5}$) and -0.036 (which can be written as $\frac{-36}{1000}$).
• We use the symbol Q to represent the set of rational numbers.

The denominator in a fraction

The denominator (bottom number) in a fraction cannot be 0. The quotient $9 \div 0$ has no answer. Because of this, numbers such as $\dfrac{9}{0}$ are said to be **undefined**. Because they do not exist, they are not rational numbers!

WORKED EXAMPLE 1 Writing rational numbers as fractions

By writing each of the following in fraction form, show that these numbers are rational.

a. 7   b. −11   c. 0   d. $-4\dfrac{2}{5}$ e. 1.2

THINK	WRITE
Each number must be written as a fraction using integers.	
a. The number 7 has to be written in fraction form. To write a number as a fraction, it must be written with a numerator and denominator. For a whole number like 7, the denominator is always 1.	a. $7 = \dfrac{7}{1}$. This number is rational.
b. The number −11 has to be written in fraction form. To write a number as a fraction, it must be written with a numerator and denominator. Because −11 is a whole number, the denominator is 1.	b. $-11 = \dfrac{-11}{1}$. This number is rational.
c. The number 0 has to be written in fraction form. To write a number as a fraction, it must be written with a numerator and a denominator. Because 0 is a whole number, the denominator is 1.	c. $0 = \dfrac{0}{1}$. This number is rational.
d. Write $-4\dfrac{2}{5}$ as an improper fraction.	d. $-4\dfrac{2}{5} = \dfrac{-22}{5}$. This number is rational.
e. The number 1.2 can be expressed as $1 + 0.2$. This can then be expressed as $1 + \dfrac{2}{10}$. Write this as an improper fraction.	e. $1.2 = \dfrac{12}{10}$. This number is rational.

⊙ 1.2.2 Recurring decimals

eles-4377

- When a rational number is written in decimal form, there are two possibilities:
  1. The decimal will come to an end (or **terminate**). For example, $\dfrac{5}{4} = 1.25$, which is a decimal that terminates (ends) after two decimal places.
  2. The decimal will repeat itself endlessly. For example, $\dfrac{7}{6} = 1.16666\ldots$ The 6 in this number is a repeating digit, which means the decimal does not terminate – it has no end.
- Decimals that endlessly repeat digits (or blocks of digits) are called **recurring decimals**.
- Recurring decimals are written by placing dots or a line above the repeating digits.

Writing recurring decimals

Number of recurring digits	Example	How to write the decimal
1 recurring digit	1.615 555 555 ... (Only the digit 5 repeats.)	$1.61\dot{5}$
2 recurring digits	1.615 151 515 ... (The digits 1 and 5 repeat.)	$1.6\dot{1}\dot{5}$ or $1.6\overline{15}$
3 or more recurring digits	1.615 615 615 ... (The digits 6, 1 and 5 repeat.)	$1.\dot{6}1\dot{5}$ or $1.\overline{615}$

WORKED EXAMPLE 2 Writing recurring decimals in extended form

Write the first 8 digits of each of the following recurring decimals.

a. $3.0\dot{2}$ b. $47.\dot{1}$ c. $11.5\dot{4}\dot{9}$

THINK

a. In $3.0\dot{2}$, the digits 0 and 2 repeat.

b. In $47.\dot{1}$, the digit 1 repeats.

c. In $11.5\dot{4}\dot{9}$, the digits 4 and 9 repeat.

WRITE

a. 3.020 202 0

b. 47.111 111

c. 11.549 494

WORKED EXAMPLE 3 Writing fractions as recurring decimals

Write each fraction as a recurring decimal using dot notation. Use a calculator for the division.

a. $\dfrac{5}{6}$ b. $\dfrac{57}{99}$ c. $\dfrac{25}{11}$ d. $\dfrac{4}{7}$

THINK

a. $5 \div 6 = 0.833\,333\,3$
 3 repeats, so a dot goes above the 3.

b. $57 \div 99 = 0.575\,757\,575\,7$
 5 and 7 repeat, so dots go above the 5 and the 7.

c. $25 \div 11 = 2.272\,727\,27$
 2 and 7 repeat, so dots go above the 2 and the 7.

d. $4 \div 7 = 0.571\,428\,571$
 It looks as though 571 428 will recur, so dots go above the 5 and the 8.

WRITE

a. $\dfrac{5}{6} = 0.8\dot{3}$

b. $\dfrac{57}{99} = 0.\dot{5}\dot{7}$

c. $\dfrac{25}{11} = 2.\dot{2}\dot{7}$

d. $\dfrac{4}{7} = 0.\dot{5}71\,42\dot{8}$

▶ 1.2.3 Irrational numbers

eles-4378

- A number is **irrational** if it cannot be written either as a fraction or as a terminating or recurring decimal.
- Irrational numbers may be expressed as decimals. In irrational numbers the digits do not repeat themselves in any particular order. For example:

$$\sqrt{18} = 4.242\ 640\ 687\ 12\ ...$$

$$\sqrt{0.03} = 0.173\ 205\ 080\ 757\ ...$$

- We use the symbol **I** to represent the set of irrational numbers.

Identifying irrational numbers

To classify a number as either rational or irrational:
1. Determine whether it can be expressed as a whole number, a fraction or a terminating or recurring decimal.
2. If the answer to step 1 is yes, then the number is rational. If the answer is no, then the number is irrational.

Real numbers

- The set of **real numbers** is made up of the group of rational numbers and the group of irrational numbers. These groups can, in turn, be broken up into subsets of other number groups.
- A real number is any number that lies on the number line.
- We use the symbol **R** to represent the set of real numbers.

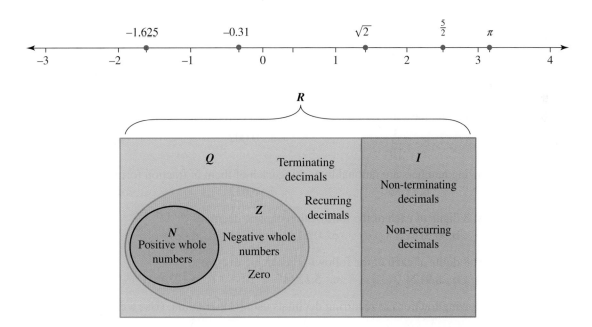

DISCUSSION

Determine how many recurring digits there will be for $\dfrac{1}{13}$.

π (pi) — a special number

The symbol π **(pi)** is used for a particular number. That number is the circumference of a circle whose diameter length is 1 unit. Pi can be approximated as a decimal that is non-terminating and non-recurring. Therefore, π is classified as an irrational number. It is also called a transcendental number.

In decimal form, $\pi = 3.14159265358979323\ldots$ It has been calculated to $29\,000\,000$ (29 million) decimal places with the aid of a computer.

on Resources

eWorkbook   Topic 1 Workbook (worksheets, code puzzle and project) (ewbk-2001)

Digital document SkillSHEET Operations with directed numbers (doc-6100)

Interactivity   Individual pathway interactivity: Rational numbers (int-4476)
                     The number system (int-6027)

Exercise 1.2 Real numbers

learnon

Individual pathways

■ PRACTISE	■ CONSOLIDATE	■ MASTER
1, 4, 9, 11, 14	2, 5, 6, 8, 12, 15	3, 7, 10, 13, 16, 17

To answer questions online and to receive **immediate corrective feedback** and **fully worked solutions** for all questions, go to your learnON title at www.jacplus.com.au.

Fluency

1. **WE1** Show that the following numbers are rational by writing each of them in fraction form.

   a. 15 b. -8 c. $2\dfrac{2}{3}$ d. $-5\dfrac{1}{8}$

2. Show that the following numbers are rational by writing each of them in fraction form.

   a. $\sqrt{16}$ b. $7\dfrac{3}{10}$ c. 0.002 d. 87.2

3. Show that the following numbers are rational by writing each of them in fraction form.

   a. 0 b. 1.56 c. 3.612 d. -0.08

4. **WE2** Write the first 8 digits of each of the following recurring decimals.

   a. $0.\dot{5}$ b. $0.5\dot{1}$ c. $0.\dot{5}\dot{1}$ d. $6.0\dot{3}\dot{1}$ e. $5.\dot{1}8\dot{3}$

5. **WE2** Write the first 8 digits of each of the following recurring decimals.

   a. $-7.02\dot{4}$ b. $8.\dot{9}12\dot{4}$ c. $5.123\dot{4}$ d. $5.\dot{1}23\dot{4}$ e. $3.0\dot{0}\dot{2}$

6. **WE3** Write the following fractions as recurring decimals using dot notation. (Use a calculator.)

   a. $\dfrac{5}{9}$ b. $\dfrac{3}{11}$ c. $2\dfrac{2}{11}$ d. $\dfrac{3}{7}$ e. $\dfrac{-173}{99}$

7. Write the following fractions as recurring decimals using dot notation. (Use a calculator.)

   a. $\dfrac{73}{990}$ b. $-3\dfrac{5}{6}$ c. $\dfrac{7}{15}$ d. $\dfrac{46}{99}$ e. $\dfrac{46}{990}$

Understanding

8. From the following list of numbers:

$$-3, \frac{-3}{7}, 0, 2.\dot{3}, 2\frac{3}{5}, 15$$

a. write down the natural numbers
b. write down the integers
c. write down the rational numbers
d. write down the recurring decimals.

9. Write these numbers in order from smallest to largest.

$2.\dot{1}, 2.\dot{1}\dot{2}, 2.\dot{1}2\dot{1}, 2.1\dot{2}\dot{1}, 2.1\dot{2}$

10. State which of the following statements is true.

a. All natural numbers are integers.
b. All integers are rational numbers.
c. All rational numbers have recurring decimals.
d. Natural numbers can be negative integers.
e. All numbers with decimals are rational numbers.
f. Irrational numbers are not real numbers.

Reasoning

11. Explain why all integers are rational numbers.

12. a. If positive integers are represented by $\mathbf{Z}^+$, explain what symbol we would use to represent the set of negative rational numbers.
    b. Explain why we could not write N^-.

13. Are all fractions in which both the numerator and denominator are integers rational? Explain why or why not.

Problem solving

14. a. Using the fraction $\frac{a}{b}$, where a and b are natural numbers, write 3 recurring decimals in fractional form using the smallest natural numbers possible.
    b. Determine the largest natural number you used.

15. Write 2 fractions that have the following number of repeating digits in their decimal forms.
    a. 1 repeating digit
    b. 2 repeating digits
    c. 3 repeating digits
    d. 4 repeating digits

16. If a \$2 coin weighs 6 g, a \$1 coin weighs 9 g, a 50 c coin weighs 15 g, a 20 c coin weighs 12g, a 10 c coin weighs 5 g and a 5 c coin weighs 3 g, determine what the maximum value of the coins would be if a bundle of them weighed exactly 10 kg.

17. A Year 9 student consumes two bottles of water every day, except on every fifth day when they only have one. Evaluate their annual bottled-water consumption.

1.3 Rounding numbers and significant figures

LEARNING INTENTION

At the end of this subtopic, you should be able to:
- round decimal numbers to a given number of decimal places
- identify the significant figures in a decimal number
- round decimal numbers to a given number of significant figures.

▶ 1.3.1 Rounding decimals

eles-4380

- If a number can't be written as an exact decimal, or if it has a lot of digits after the decimal point, we can use a shorter, approximate value.
- A number's approximate value can be found by using a calculator and then rounding it off to the desired level of accuracy.
- When we round numbers, we write them to either a certain number of decimal places or a certain number of significant figures.

Rounding to a given number of decimal places

1. Decide how many decimal places you need to round the number to.
2. Identify the digit that is located at the position you need to round to. This digit is called the rounding digit.
3. Look at the digit that comes after the rounding digit.
   - If the digit's value is 0, 1, 2, 3 or 4, leave the rounding digit as it is.
   - If the digit's value is 5, 6, 7, 8 or 9, then increase the rounding digit by adding 1 to its value.
4. Remove all of the numbers after the rounding digit.

WORKED EXAMPLE 4 Rounding a decimal number to three decimal places

Round 3.456 734 correct to three decimal places (3 d.p.).

THINK	WRITE
1. Identify the rounding digit located at the third decimal place. In this case it is 6.	3.456⃝734
2. Look at the digit that comes after the rounding digit.	↓ 3.456⃝734
3. The number after the rounding digit is a 7, so increase the value of the rounding digit by adding 1. Next, remove the digits after it.	3.457

WORKED EXAMPLE 5 Writing a real number to three decimal places

Write these numbers correct to three decimal places (3 d.p.).

a. $\sqrt{3}$ b. π c. $5.1\dot{9}$ d. $\dfrac{2}{3}$ e. **7.123 456**

THINK

a. $\sqrt{3} = 1.7320\ldots$ The rounding digit is 2. The next digit is 0, so leave 2 as it is.

b. $\pi = 3.1415\ldots$ The rounding digit is 1. The next digit is 5, so add 1 to the rounding digit.

c. $5.1\dot{9} = 5.1999\ldots$ The rounding digit is 9. The next digit is 9, so add 1 to the rounding digit.

d. $\dfrac{2}{3} = 0.6666\ldots$ The rounding digit is 6. The next digit is 6, so add 1 to the rounding digit.

e. $7.123\,456$. The rounding digit is 3. The next digit is 4, so leave 3 as is.

WRITE

a. $\sqrt{3} \approx 1.732$

b. $\pi \approx 3.142$

c. $5.1\dot{9} \approx 5.200$

d. $\dfrac{2}{3} \approx 0.667$

e. $7.1234 \approx 7.123$

⏵ 1.3.2 Rounding to significant figures

- Another method of rounding decimals is to write them so that they are correct to a certain number of significant figures.
- When decimals are rounded, they don't always have the same number of decimal places as significant figures. For example, the number 1.425 has three decimal places but four significant figures.

Counting significant figures in a decimal number

1. Identify the first non-zero digit in the number. This digit may be located either before or after the decimal point. This digit is counted as the first significant figure.
2. Continue to count all digits in the number to the end of the number as it is written. This will give you the total number of significant figures.

WORKED EXAMPLE 6 Counting significant figures

State the number of significant figures in each of the following numbers.

a. **25** b. **0.04** c. **3.02** d. **0.100**

THINK

a. In the number 25, the first significant figure is 2. There is 1 more significant figure after the first one.

b. In the number 0.04, the first significant figure is 4. There are no more significant figures after the first one.

c. In the number 3.02, the first significant figure is 3. There are 2 more significant figures after the first one.

d. In the number 0.100, the first significant figure is 1. There are 2 more significant figures after the first one.

WRITE

a. 2 significant figures

b. 1 significant figure

c. 3 significant figures

d. 3 significant figures

TOPIC 1 Number skills and index laws **11**

WORKED EXAMPLE 7 Rounding real numbers to five significant figures

Round these numbers so that they are correct to five significant figures.

a. π b. $\sqrt{200}$ c. $0.0\dot{3}$ d. **2530.166**

THINK	WRITE
a. $\pi \approx 3.14159\ldots$ The first significant figure is 3. Starting with 3, write down the next 4 digits. The 6th digit is 9, so add 1 to the value of 5, the 5th digit.	a. 3.1416
b. $\sqrt{200} \approx 14.1421\ldots$ The first significant figure is 1. Starting with 1, write down the next 4 digits. The 6th digit is 1, so leave 2, the 5th digit, as it is.	b. 14.142
c. $0.0\dot{3} = 0.030\,303\,0\ldots$ The first significant figure is 3. Starting with 3, write down the next 4 digits. The 6th digit is 0, so leave 3, the 5th digit, as it is.	c. $0.030\,303$
d. $2530.16\ldots$ The first significant figure is 2. Starting with 2, write the next 4 digits. The 6th digit is 6, so add 1 to the value of 1, the 5th digit.	d. 2530.2

DISCUSSION

Is there an equal number of rational numbers and irrational numbers? Explain.

on Resources

Exercise 1.3 Rounding numbers and significant figures   **learn**on

Individual pathways

■ PRACTISE	■ CONSOLIDATE	■ MASTER
1, 4, 7, 10, 15, 18	2, 5, 8, 11, 16, 19	3, 6, 9, 12, 13, 14, 17, 20

To answer questions online and to receive **immediate corrective feedback** and **fully worked solutions** for all questions, go to your learnON title at www.jacplus.com.au.

Fluency

1. **WE5** Write each of the following correct to 3 decimal places.

     a. $\dfrac{\pi}{2}$ b. $\sqrt{5}$ c. $\sqrt{15}$ d. 5.12×3.21 e. $5.\dot{1}$ f. $5.1\dot{5}$

2. Write each of the following correct to 3 decimal places.

     a. $5.\dot{1}5$ b. 11.72^2 c. $\dfrac{3}{7}$ d. $\dfrac{1}{13}$ e. $2\dfrac{3}{7}$ f. $0.999\,999$

3. Write each of the following correct to three decimal places.
    a. 4.000 01   b. 2.79 ÷ 11   c. 0.0254   d. 0.000 913 6   e. 5.000 01   f. 2342.156

4. **WE6** State how many significant figures there are in each of the following numbers.
    a. 36   b. 207   c. 1631   d. 5.04

5. State how many significant figures there are in each of the following numbers.
    a. 176.2   b. 95.00   c. 0.21   d. 0.01

6. State how many significant figures there are in each of the following numbers.
    a. 0.000 316   b. 0.1007   c. 0.010   d. 0.0512

7. **WE7** Write each number so that it is correct to five significant figures.
    a. $\dfrac{\pi}{2}$ b. $\sqrt{5}$ c. $\sqrt{15}$ d. 5.12×3.21 e. $5.\dot{1}$ f. $5.1\dot{5}$

8. Write each number so that it is correct to five significant figures.
    a. $5.\dot{1}\dot{5}$ b. 11.72^2 c. $\dfrac{3}{7}$
    d. $\dfrac{1}{13}$ e. $2\dfrac{3}{7}$ f. 0.999 999

9. Write each number so that it is correct to 5 significant figures.
    a. 6.581 29   b. 4.000 01   c. 2.79 ÷ 11   d. 0.0254   e. 0.000 913 6   f. 5.000 01

Understanding

10. Write the value of π correct to 4, 5, 6 and 7 decimal places.

11. State whether each of the following numbers has more significant figures than decimal places.
    a. 17.26   b. 0.0032   c. 1.06   d. 0.010 005

12. State whether each statement is True or False.
    a. Every surd is a rational number.
    b. Every surd is an irrational number.
    c. Every irrational number is a surd.
    d. Every surd is a real number.

13. State whether each statement is True or False.
    a. π is a rational number.
    b. π is an irrational number.
    c. π is a surd.
    d. π is a real number.

14. State whether each statement is True or False.
    a. $1.\dot{3}\dot{1}$ is a rational number.
    b. $1.\dot{3}\dot{1}$ is an irrational number.
    c. $1.\dot{3}\dot{1}$ is a surd.
    d. $1.\dot{3}\dot{1}$ is a real number.

Reasoning

15. Write a decimal number that has 3 decimal places, but 4 significant figures. Show your working.

16. Explain why the number 12.995 412 3 becomes 13.00 when rounded to 2 decimal places.

17. The rational numbers 3.1416 and $\frac{22}{7}$ are both used as approximations to π.

   a. Determine the largest number of decimal places these numbers can be rounded to in order to give the same value.
   b. Explain which of these two numbers gives the best approximation to π.

Problem solving

18. The area of a circle is calculated using the formula $A = \pi \times r^2$, where r is the radius of the circle. Pi (π) is sometimes rounded to 2 decimal places to become 3.14. A particular circle has a radius of 7 cm.

   a. Use $\pi = 3.14$ to calculate the area of the circle to 2 decimal places.
   b. Use the π key on your calculator to calculate the area of the circle to 4 decimal places.
   c. Round your answer for part b to 2 decimal places.
   d. Are your answers for parts a and c different? Discuss why or why not.

19. The volume of a sphere (a ball shape) is calculated using the formula $v = \frac{4}{3} \times \pi \times r^3$, where r is the radius of the sphere. A beach ball with a radius of 25 cm is bouncing around the crowd at the MCG during the Boxing Day Test.

   a. Calculate the volume of the beach ball to 4 decimal places.
   b. Calculate the volume to 4 decimal places and determine how many significant figures the result has.
   c. Explain whether the calculated volume is a rational number.

20. In a large sample of written English there are about 7 vowels for every 11 consonants. The letter e accounts for about one-third of the occurrence of vowels. Explain how many times you would expect the letter e to occur in a passage of 100 000 letters. Round your answer to the nearest 100.

1.4 Review of index laws

▶ 1.4.1 Index notation

eles-4382

- The product of factors can be written in a shorter form called **index notation** (also known as exponent notation).

$$\text{Base} \longrightarrow 6^{\overset{\displaystyle \text{Index (exponent)}}{4}} = \underbrace{6 \times 6 \times 6 \times 6}_{} \longleftarrow \text{Factor form}$$

$$= 1296$$

- Any composite number can be written as a product of powers of prime factors using a factor tree, or by other methods, such as repeated division.

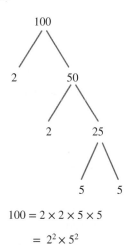

$$100 = 2 \times 2 \times 5 \times 5$$
$$= 2^2 \times 5^2$$

WORKED EXAMPLE 8 Using index notation to express a product of prime factors

Express 360 as a product of powers of prime factors using index notation.

THINK	WRITE
1. Express 360 as a product of a factor pair.	$360 = 6 \times 60$
2. Further factorise 6 and 60.	$= 2 \times 3 \times 4 \times 15$
3. Further factorise 4 and 15. They are both composite numbers.	$= 2 \times 3 \times 2 \times 2 \times 3 \times 5$
4. There are no more composite numbers.	$= 2 \times 2 \times 2 \times 3 \times 3 \times 5$
5. Write the answer using index notation. *Note:* The factors are generally expressed with their bases written in ascending order.	$360 = 2^3 \times 3^2 \times 5$

⏵ 1.4.2 Multiplication using indices

- When multiplying **terms** with the same bases, add the indices. This is the **First Index Law**.
- When more than one base is involved, apply the First Index Law to each base separately.

The First Index Law

$$a^m \times a^n = a^{m+n}$$

For example, $2x^2 \times x^3 = 2x^{2+3} = 2x^5$

WORKED EXAMPLE 9 Simplifying an expression using the First Index Law

Simplify $5e^{10} \times 2e^3$.

THINK	WRITE
1. The order is not important when multiplying, so place the coefficients first.	$5e^{10} \times 2e^3$ $= 5 \times 2 \times e^{10} \times e^3$
2. Simplify by multiplying the coefficients and then applying the First Index Law (add the indices). Write the answer.	$= 10e^{13}$

WORKED EXAMPLE 10 Simplifying using the First Index Law on multiple bases

Simplify $7m^3 \times 3n^5 \times 2m^8 n^4$.

THINK	WRITE
1. The order is not important when multiplying, so place the coefficients first and group all similar pronumerals together.	$7m^3 \times 3n^5 \times 2m^8 n^4$ $= 7 \times 3 \times 2 \times m^3 \times m^8 \times n^5 \times n^4$
2. Simplify by multiplying the coefficients and applying the First Index Law (add the indices). Write the answer.	$= 42m^{11} n^9$

⏵ 1.4.3 Division using indices

- When dividing terms with the same bases, subtract the indices. This is the **Second Index Law**.
- When the coefficients do not divide evenly, simplify by cancelling.

The Second Index Law

$$a^m \div a^n = a^{m-n} \text{ or } \frac{a^m}{a^n} = a^{m-n}$$

For example, $2x^5 \div x^3 = \frac{2x^5}{x^3} = 2x^{5-3} = 2x^2$

Simplify $\dfrac{25v^6 \times 8w^9}{10v^4 \times 4w^5}$.

THINK	WRITE
1. Simplify the numerator and the denominator by multiplying the coefficients.	$\dfrac{25v^6 \times 8w^9}{10v^4 \times 4w^5}$ $= \dfrac{200v^6w^9}{40v^4w^5}$
2. Simplify further by dividing the coefficients and applying the Second Index Law (subtract the indices). Write the answer.	$= \dfrac{\overset{5}{\cancel{200}}}{\underset{1}{\cancel{40}}} \times \dfrac{v^6}{v^4} \times \dfrac{w^9}{w^5}$ $= 5v^2w^4$

Simplify $\dfrac{7t^3 \times 4t^8}{12t^4}$.

THINK	WRITE
1. Simplify the numerator by multiplying the coefficients.	$\dfrac{7t^3 \times 4t^8}{12t^4}$ $= \dfrac{28t^{11}}{12t^4}$
2. Simplify further by dividing the coefficients by the highest common factor. Then apply the Second Index Law (subtract the indices). Write the answer.	$= \dfrac{28}{12} \times \dfrac{t^{11}}{t^4}$ $= \dfrac{7t^7}{3}$

1.4.4 Zero index

eles-4385

- Any number divided by itself is 1.
- This means that $\dfrac{a^m}{a^m} = 1$. So, $a^{m-m} = a^0 = 1$.
- In general, any number (except) to the power of 0 is equal to 1. This is the **Third Index Law**.

> ### The Third Index Law
>
> $$a^0 = 1, \text{where } a \neq 0$$
>
> For example, $2^0 = 1$

WORKED EXAMPLE 13 Evaluating using the Third Index Law

Evaluate the following.

a. t^0 b. $(xy)^0$ c. 17^0 d. $5x^0$ e. $(5x)^0 + 2$ f. $5^0 + 3^0$

THINK	WRITE
a. Apply the Third Index Law.	a. $t^0 = 1$
b. Apply the Third Index Law.	b. $(xy)^0 = 1$
c. Apply the Third Index Law.	c. $17^0 = 1$
d. Apply the Third Index Law.	d. $5x^0 = 5 \times x^0$ $\quad = 5 \times 1$ $\quad = 5$
e. Apply the Third Index Law.	$(5x)^0 = 1 + 2$ $\quad = 3$
f. Apply the Third Index Law.	f. $5^0 + 3^0 = 1 + 1$ $\quad = 2$

WORKED EXAMPLE 14 Simplifying using the First and Third index laws

Simplify $\dfrac{9g^7 \times 4g^4}{6g^3 \times 2g^8}$.

THINK	WRITE
1. Simplify the numerator and the denominator by applying the First Index Law.	$\dfrac{9g^7 \times 4g^4}{6g^3 \times 2g^8}$
2. Simplify the fraction further by applying the Second Index Law.	$= \dfrac{36g^{11}}{12g^{11}}$ $= \dfrac{^3\cancel{36}g^{11}}{^1\cancel{12}g^{11}}$
3. Simplify by applying the Third Index Law.	$= 3g^0$ $= 3 \times 1$
4. Write the answer.	$= 3$

▶ 1.4.5 Cancelling fractions

eles-4386

- Consider the fraction $\dfrac{x^3}{x^7}$. This fraction can be cancelled by dividing the denominator and the numerator by the highest common factor (HCF), which is x^3, so $\dfrac{x^3}{x^7} = \dfrac{1}{x^4}$.

  Note: $\dfrac{x^3}{x^7} = x^{-4}$ when we apply the Second Index Law. We will study negative indices in a later section.

DISCUSSION

How do the First, Second and Third index laws help with calculations?

WORKED EXAMPLE 15 Using the first three index laws to simplify expressions

Simplify these fractions by cancelling.

a. $\dfrac{x^5}{x^7}$

b. $\dfrac{6x}{12x^8}$

c. $\dfrac{30x^5y^6}{10x^7y^3}$

THINK	WRITE
a. Divide the numerator and denominator by the highest common factor (HCF), which is x^5.	a. $\dfrac{x^5}{x^7} = \dfrac{1}{x^2}$
b. 1. Divide the numerator and denominator by the HCF, which is $6x$.	b. $\dfrac{6x}{12x^8} = \dfrac{6}{12} \times \dfrac{x}{x^8}$ $= \dfrac{1}{2} \times \dfrac{1}{x^7}$
2. Simplify and write the answer.	$= \dfrac{1}{2x^7}$
c. 1. Divide the numerator and denominator by the HCF, which is $10x^5y^3$.	c. $\dfrac{30x^5y^6}{10x^7y^3} = \dfrac{30}{10} \times \dfrac{x^5}{x^7} \times \dfrac{y^6}{y^3}$ $= \dfrac{3}{1} \times \dfrac{1}{x^2} \times \dfrac{y^3}{1}$
2. Simplify and write the answer.	$= \dfrac{3y^3}{x^2}$

 Resources

📋 **eWorkbook**	Topic 1 Workbook (worksheets, code puzzle and project) (ewbk-2001)
📄 **Digital documents**	SkillSHEET Index form (doc-6225)
	SkillSHEET Using a calculator to evaluate numbers in index form (doc-6226)
▶ **Video eLesson**	Index notation (eles-1903)
🔁 **Interactivities**	Individual pathway interactivity: Review of index laws (int-4516)
	Review of index form (int-3708)
	First Index Law (int-3709)
	Second Index Law (int-3711)
	Third Index Law (int-3713)

Exercise 1.4 Review of index laws

learn on

Individual pathways

■ PRACTISE	■ CONSOLIDATE	■ MASTER
1, 3, 6, 10, 13, 15, 19, 20, 21, 26	2, 4, 7, 8, 11, 18, 22, 23, 27, 28	5, 9, 12, 14, 16, 17, 24, 25, 29, 30

To answer questions online and to receive **immediate corrective feedback** and **fully worked solutions** for all questions, go to your learnON title at www.jacplus.com.au.

Fluency

1. **WE8** Use index notation to express each of the following as a product of powers of prime factors.

   a. 12   b. 72   c. 75   d. 240   e. 640   f. 9800

2. **WE9** Simplify each of the following.

    a. $4p^7 \times 5p^4$ b. $2x^2 \times 3x^6$ c. $8y^6 \times 7y^4$ d. $3p \times 7p^7$ e. $12t^3 \times t^2 \times 7t$ f. $6q^2 \times q^5 \times 5q^8$

3. **WE10** Simplify each of the following.

    a. $2a^2 \times 3a^4 \times e^3 \times e^4$ b. $4p^3 \times 2h^7 \times h^5 \times p^3$ c. $3k^2 \times 2k^{11} \times k$ d. $3^4 \times 3^4 \times 3^4$

4. Simplify each of the following.

    a. $2m^3 \times 5m^2 \times 8m^4$ b. $2gh \times 3g^2\,h^5$

    c. $5p^4q^2 \times 6p^2q^7$ d. $8u^3w \times 3uw^2 \times 2u^5w^4$

5. Simplify each of the following.

    a. $9y^8d \times y^5d^3 \times 3y^4d^7$

    b. $7b^3c^2 \times 2b^6c^4 \times 3b^5c^3$

    c. $4r^2s^2 \times 3r^6s^{12} \times 2r^8s^4$

    d. $10h^{10}v^2 \times 2h^8v^6 \times 3h^{20}v^{12}$

6. **WE11** Simplify each of the following.

    a. $\dfrac{15p^{12}}{5p^8}$ b. $\dfrac{18r^6}{3r^2}$ c. $\dfrac{45a^5}{5a^2}$ d. $\dfrac{60b^7}{20b}$

7. Simplify each of the following.

    a. $\dfrac{100r^{10}}{5r^6}$ b. $\dfrac{9q^2}{q}$ c. $\dfrac{130d^3}{d^5}$ d. $\dfrac{21t^{11}}{3t^7}$

8. **WE12** Simplify each of the following.

    a. $\dfrac{8p^6 \times 3p^4}{16p^5}$ b. $\dfrac{12b^5 \times 4b^2}{18b^2}$ c. $\dfrac{25m^{12} \times 4n^7}{15m^2 \times 8n}$ d. $\dfrac{27x^9y^3}{12xy^2}$

9. Simplify each of the following.

    a. $\dfrac{12j^8 \times 6f^5}{8j^3 \times 3f^2}$ b. $\dfrac{8p^3 \times 7r^2 \times 2s}{6p \times 14r}$ c. $\dfrac{27a^9 \times 18b^5 \times 4c^2}{18a^4 \times 12b^2 \times 2c}$ d. $\dfrac{81f^{15} \times 25g^{12} \times 16h^{34}}{27f^9 \times 15g^{10} \times 12h^{30}}$

10. **WE13** Evaluate the following.

    a. m^0 b. $6m^0$ c. $(6m)^0$ d. $(ab)^0$ e. $5(ab)^0$

11. Evaluate the following.

    a. w^0x^0 b. 85^0 c. $85^0 + 15^0$ d. $x^0 + 1$ e. $5x^0 - 2$

12. Evaluate the following.

    a. $\dfrac{x^0}{y^0}$ b. $x^0 - y^0$ c. $3x^0 + 11$ d. $3a^0 + 3b^0$ e. $3\left(a^0 + b^0\right)$

13. **WE14** Simplify each of the following.

    a. $\dfrac{2a^3 \times 6a^2}{12a^5}$ b. $\dfrac{3c^6 \times 6c^3}{9c^9}$ c. $\dfrac{5b^7 \times 10b^5}{25b^{12}}$ d. $\dfrac{8f^3 \times 3f^7}{4f^5 \times 3f^5}$ e. $\dfrac{9k^{12} \times 4k^{10}}{18k^4 \times k^{18}}$

14. Simplify each of the following.

    a. $\dfrac{2h^4 \times 5k^2}{20h^2 \times k^2}$ b. $\dfrac{p^3 \times q^4}{5p^3}$ c. $\dfrac{m^7 \times n^3}{5m^3 \times m^4}$ d. $\dfrac{8u^9 \times v^2}{2u^5 \times 4u^4}$ e. $\dfrac{9x^6 \times 2y^{12}}{3y^{10} \times 3y^2}$

Understanding

15. **WE15** Simplify each of the following by cancelling.

    a. $\dfrac{x^7}{x^{10}}$ b. $\dfrac{m}{m^9}$ c. $\dfrac{m^3}{4m^9}$ d. $\dfrac{12x^6}{6x^8}$

16. Simplify each of the following by cancelling.

a. $\dfrac{12x^8}{6x^6}$ b. $\dfrac{24t^{10}}{t^4}$ c. $\dfrac{5y^5}{10y^{10}}$ d. $\dfrac{35x^2y^{10}}{20x^7y^7}$

17. Simplify each of the following by cancelling.

a. $\dfrac{12m^2n^4}{30m^5n^8}$ b. $\dfrac{16m^5n^{10}}{8m^5n^{12}}$ c. $\dfrac{20x^4y^5}{10x^5y^4}$ d. $\dfrac{a^2b^4c^6}{a^6b^4c^2}$

18. Calculate the value of each of the following expressions if $a = 3$.

a. $2a$ b. a^2 c. $2a^2$ d. $a^2 + 2$ e. $a^2 + 2a$ f. $a^2 - a$

Reasoning

19. Explain why x^2 and $2x$ are not the same number. Use an example to illustrate your reasoning.

20. **MC**

a. $12a^8b^2c^4(de)^0f$, when simplified, is equal to:

   A. $12a^8b^2c^4$ **B.** $12a^8b^2c^4f$ **C.** $12a^8b^2f$ **D.** $12a^8b^2$ **E.** $12a^8b^2c^4f^2$

b. $\left(\dfrac{6}{11}a^2b^7\right)^0 \times - (3a^2b^{11})^0 + 7a^0b$, when simplified, is equal to:

   A. $7b$ **B.** $1 + 7b$ **C.** $-1 + 7ab$ **D.** $-1 + 7b$ **E.** $-1 - 7b$

c. You are told that there is an error in the statement $3p^7q^3r^5s^6 = 3p^7s^6$. Explain what the left-hand side should be for this statement to be correct.

   A. $(3p^7q^3r^5s^6)^0$ **B.** $(3p^7)^0q^3r^5s^6$ **C.** $3p^7(q^3r^5s^6)^0$ **D.** $3p^7(q^3r^5)^0s^6$ **E.** $6p^7(q^3r^5)^0s^6$

21. a. You are told that there is an error in the statement $\dfrac{8f^6g^7h^3}{6f^4g^2h} = \dfrac{8f^2}{g^2}$. Explain what the left-hand side should be for this statement to be correct.

   A. $\dfrac{8f^6(g^7h^3)^0}{(6)^0f^4g^2(h)^0}$ **B.** $\dfrac{8(f^6g^7h^3)^0}{(6f^4g^2h)^0}$ **C.** $\dfrac{8(f^6g^7)^0h^3}{(6f^4)^0g^2h}$ **D.** $\dfrac{8f^6g^7h^3}{(6f^4g^2h)^0}$ **E.** $\dfrac{8f^8(g^7h^3)^0}{(6)^0f^4g(h)^0}$

b. Select what $\dfrac{6k^7m^2n^8}{4k^7(m^6n)^0}$ is equal to.

   A. $\dfrac{6}{4}$ **B.** $\dfrac{3}{2}$ **C.** $\dfrac{3n^8}{2}$ **D.** $\dfrac{3m^2n^8}{2}$ **E.** $\dfrac{3m^2n^6}{2}$

22. Explain why $5x^5 \times 3x^3$ is not equal to $15x^{15}$.

23. A multiple-choice question requires a student to multiply 5^6 by 5^3. The student is having trouble deciding which of these four answers is correct: 5^{18}, 5^9, 25^{18} or 25^9.

a. State the correct answer.

b. Explain your answer by using another example to explain the First Index Law.

24. A multiple-choice question requires a student to divide 5^{24} by 5^8. The student is having trouble deciding which of these four answers is correct: 5^{16}, 5^3, 1^{16} or 1^3.

a. Determine the correct answer.

b. Explain your answer by using another example to explain the Second Index Law.

25. a. Calculate the value of $\dfrac{5^7}{5^7}$.

b. Determine the value of any number divided by itself.

c. According to the Second Index Law, which deals with exponents and division, $\dfrac{5^7}{5^7}$ should equal 5 raised to what index?

d. Use an example to explain the Third Index Law.

Problem solving

26. Answer the following questions.

    a. For $x^2 x^\triangle = x^{16}$ to be true, determine what number must replace the triangle.

    b. For $x^\triangle x^\circ x^\diamond = x^{12}$ to be true, there are 55 ways of assigning positive whole numbers to the triangle, circle, and diamond. Give at least four of these.

27. Answer the following questions.

    a. Determine a pattern in the units digit for powers of 3.

    b. The units digit of 3^6 is 9. Determine the units digit of 3^{2001}.

28. Answer the following questions.

    a. Determine a pattern in the units digit for powers of 4.

    b. Determine the units digit of 4^{105}.

29. Answer the following questions.

    a. Investigate the patterns in the units digit for powers of 2 to 9.

    b. Predict the units digit for the following.

        i. 2^{35} **ii.** 3^{16} **iii.** 8^{51}

30. Write $4^{n+1} + 4^{n+1}$ as a single power of 2.

1.5 Raising powers

LEARNING INTENTION

At the end of this subtopic you should be able to:
- apply the Fourth, Fifth and Sixth index laws to raise a power to another power
- apply the first six index laws to simplify expressions involving indices.

1.5.1 More index laws

eles-4387

- When a power is raised to another power the indices are multiplied. This is the **Fourth Index Law**.
- The Fourth Index Law can then be applied to products and quotients of mixed bases to develop the Fifth and Sixth index laws.

The Fourth Index Law

$$(a^m)^n = a^{m \times n}$$

For example, $(x^2)^3 = x^{2 \times 3} = x^6$

The Fifth Index Law

$$(a \times b)^m = a^m \times b^m$$

For example, $(2x)^3 = 2^3 \times x^3 = 8x^3$

The Sixth Index Law

$$\left(\frac{a}{b}\right)^m = \frac{a^m}{b^m}$$

For example, $\left(\dfrac{4x}{2}\right)^2 = \dfrac{4^2 x^2}{2^2} = 4x^2$

WORKED EXAMPLE 16 Simplifying using the Fourth Index Law

Simplify the following.

a. $\left(7^4\right)^8$

b. $\left(3a^2 b^5\right)^3$

THINK	WRITE
a. Simplify by applying the Fourth Index Law (multiply the indices).	**a.** $\left(7^4\right)^8$ $= 7^4 \times 8$ $= 7^{32}$
b. 1. Write the expression, including all indices.	**b.** $\left(3^1 a^2 b^5\right)^3$
2. Simplify by applying the Fourth Index Law (multiply the indices) for each term inside the brackets.	$= 3^{1 \times 3} a^{2 \times 3} b^{5 \times 3}$ $= 3^3 a^6 b^{15}$
3. Write the answer.	$= 27 a^6 b^{15}$

WORKED EXAMPLE 17 Simplifying using the Fifth Index Law

Simplify $\left(2b^5\right)^2 \times (5b)^3$.

THINK	WRITE
1. Write the expression, including all indices.	$\left(2^1 b^5\right)^2 \times \left(5^1 b^1\right)^3$
2. Simplify by applying the Fifth Index Law.	$= 2^{1 \times 2} b^{5 \times 2} \times 5^{1 \times 3} b^{1 \times 3}$ $= 2^2 \times b^{10} \times 5^3 \times b^3$
3. Calculate the number values and bring them to the front. Then use the First Index Law to simplify the powers of b.	$= 4 \times 125 \times b^{10+3}$
4. Write the answer.	$= 500 b^{13}$

WORKED EXAMPLE 18 Simplifying using the Sixth Index Law

Simplify $\left(\dfrac{2a^5}{d^2}\right)^3$.

THINK	WRITE
1. Write the expression, including all indices.	$\left(\dfrac{2^1 a^5}{d^2}\right)^3$

2. Simplify by applying the Sixth Index Law for each term inside the brackets.

$$= \frac{2^{1\times3} a^{5\times3}}{d^{2\times3}}$$

$$= \frac{2^3 a^{15}}{d^6}$$

3. Simplify and then write the answer.

$$= \frac{8 a^{15}}{d^6}$$

TI | THINK	WRITE
In a new problem, on a Calculator page, complete the entry line as: $\left(\dfrac{2a^5}{d^2}\right)^3$ Press ENTER.	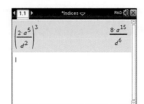 $$\left(\frac{2a^5}{d^2}\right)^3 = \frac{8a^{15}}{d^6}$$

CASIO | THINK	WRITE
On the Main screen complete the entry line as: $\left(\dfrac{2a^5}{d^2}\right)^3$ Press EXE.	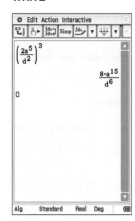 $$\left(\frac{2a^5}{d^2}\right)^3 = \frac{8a^{15}}{d^6}$$

DISCUSSION

What difference, if any, is there between the operation of the index laws on numeric terms compared with similar operations on algebraic terms?

---

on Resources

 eWorkbook Topic 1 Workbook (worksheets, code puzzle and project) (ewbk-2001)

 Interactivity Individual pathway interactivity: Raising a power to another power (int-4517)

Fourth Index Law (int-3716)

Fifth and sixth index laws (int-6063)

Exercise 1.5 Raising powers

learnon

Individual pathways

■ PRACTISE	■ CONSOLIDATE	■ MASTER
1, 3, 7, 9, 14, 15, 20	2, 5, 8, 10, 12, 16, 17, 21	4, 6, 11, 13, 18, 19, 22, 23

To answer questions online and to receive **immediate corrective feedback** and **fully worked solutions** for all questions, go to your learnON title at www.jacplus.com.au.

Fluency

1. **WE16** Simplify each of the following.
   a. $(e^2)^3$ b. $(f^8)^{10}$ c. $(p^{25})^4$ d. $(r^{12})^{12}$ e. $(2a^2)^3$

2. Simplify each of the following.
   a. $(a^2b^3)^4$ b. $(pq^3)^5$ c. $(g^3h^2)^{10}$ d. $(3w^9q^2)^4$ e. $(7e^5r^2q^4)^2$

3. **WE17** Simplify each of the following.
   a. $(p^4)^2 \times (q^3)^2$ b. $(r^5)^3 \times (w^3)^3$ c. $(b^5)^2 \times (n^3)^6$ d. $(j^6)^3 \times (g^4)^3$ e. $(2a)^3 \times (b^2)^2$

4. Simplify each of the following.
   a. $(q^2)^2 \times (r^4)^5$ b. $(h^3)^8 \times (j^2)^8$ c. $(f^4)^4 \times (a^7)^3$ d. $(t^5)^2 \times (u^4)^2$ e. $(i^3)^5 \times (j^2)^6$

5. **WE18** Simplify each of the following.
   a. $\left(\dfrac{3b^4}{d^3}\right)^2$ b. $\left(\dfrac{5h^{10}}{2j^2}\right)^2$ c. $\left(\dfrac{2k^5}{3t^8}\right)^3$ d. $\left(\dfrac{7p^9}{8q^{22}}\right)^2$

6. Simplify each of the following.
   a. $\left(\dfrac{5y^7}{3z^{13}}\right)^3$ b. $\left(\dfrac{4a^3}{7c^5}\right)^4$ c. $\left(\dfrac{-4k^2}{7m^6}\right)^3$ d. $\left(\dfrac{-2g^7}{3h^{11}}\right)^4$

Understanding

7. Simplify each of the following.
   a. $(2^3)^4 \times (2^4)^2$ b. $(t^7)^3 \times (t^3)^4$ c. $(a^4)^0 \times (a^3)^7$ d. $(e^7)^8 \times (e^5)^2$

8. Simplify each of the following.
   a. $(g^7)^3 \times (g^9)^2$ b. $(3a^2) \times (2a^6)^2$ c. $(2d^7)^3 \times (3d^2)^3$ d. $(10r^2)^4 \times (2r^3)^2$

9. **MC** $(p^7)^2 \div p^2$ is equal to:
   A. p^7 **B.** p^{12} **C.** p^{16} **D.** $p^{4.5}$ **E.** p^{13}

10. **MC** $\dfrac{(w^5)^2 \times (p^7)^3}{(w^2)^2 \times (p^3)^5}$ is equal to:
    A. w^2p^6 **B.** $(wp)^6$ **C.** $w^{14}p^{36}$ **D.** w^2p^2 **E.** $(wp)^7$

11. **MC** $(r^6)^3 \div (r^4)^2$ is equal to:
    A. r^3 **B.** r^4 **C.** r^8 **D.** r^{10} **E.** r^{12}

12. Simplify each of the following.
    a. $(a^3)^4 \div (a^2)^3$ b. $(m^8)^2 \div (m^3)^4$ c. $(n^5)^3 \div (n^6)^2$
    d. $(b^4)^5 \div (b^6)^2$ e. $(f^7)^3 \div (f^2)^2$ f. $(g^8)^2 \div (g^5)^2$

13. Simplify each of the following.
    a. $(p^9)^3 \div (p^6)^3$ b. $(y^4)^4 \div (y^7)^2$ c. $\dfrac{(c^6)^5}{(c^5)^2}$ d. $\dfrac{(f^5)^3}{(f^2)^4}$ e. $\dfrac{(k^3)^{10}}{(k^2)^8}$ f. $\dfrac{(p^{12})^3}{(p^{10})^2}$

Reasoning

14. a. Replace the triangle with the correct index for the equation $4^7 \times 4^7 \times 4^7 \times 4^7 \times 4^7 = (4^7)^\triangle$.
    b. The expression $(p^5)^6$ means to write p^5 as a factor how many times?

c. If you rewrote the expression from part **b** without any exponents, in the format $p \times p \times p...$, determine how many factors you would need.

d. Explain the Fourth Index Law.

15. a. Simplify each of the following.

  i. $(-1)^{10}$ ii. $(-1)^{7}$ iii. $(-1)^{15}$ iv. $(-1)^{6}$

b. Write a general rule for the result obtained when -1 is raised to a positive power. Explain your answer.

16. Jo and Danni are having an algebra argument. Jo is sure that $-x^2$ is equivalent to $(-x^2)$, but Danni thinks otherwise. Explain who is correct and justify your answer.

17. A multiple-choice question requires a student to calculate $(5^4)^3$. The student is having trouble deciding which of these three answers is correct: 5^{64}, 5^{12} or 5^7.

  a. Determine the correct answer.

  b. Explain your answer by using another example to illustrate the Fourth Index Law.

18. a. Without using your calculator, simplify each side of the following equations to the same base and then solve each of them.

  i. $8^x = 32$

  ii. $27^x = 243$

  iii. $1000^x = 100\,000$

  b. Explain why all 3 equations have the same solution.

19. Consider the expression 4^{3^2}. Identify which is the correct answer, 4096 or 262 144. Justify your choice.

Problem solving

20. The diameter of a typical atom is so small that it would take about 10^8 atoms arranged in a line to reach a length of just 1 centimetre. Estimate how many atoms are contained in a cubic centimetre. Write this number as a power of 10.

21. Writing a base as a power itself can be used as a way to simplify an expression. Copy and complete the following calculations.

  a. $16^{\frac{1}{2}} = (4^2)^{\frac{1}{2}} = \underline{\quad\quad}$

  b. $343^{\frac{2}{3}} = (7^3)^{\frac{2}{3}} = \underline{\quad\quad}$

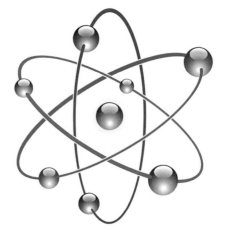

22. Simplify the following using index laws.

  a. $8^{\frac{1}{3}}$ b. $27^{\frac{4}{3}}$ c. $125^{-\frac{2}{3}}$ d. $512^{\frac{2}{9}}$

  e. $16^{-\frac{1}{2}}$ f. $4^{-\frac{1}{2}}$ g. $32^{-\frac{1}{5}}$ h. $49^{-\frac{1}{2}}$

23. a. Use the index laws to simplify the following.

  i. $(3^2)^{\frac{1}{2}}$ ii. $(4^2)^{\frac{1}{2}}$ iii. $(8^2)^{\frac{1}{2}}$ iv. $(11^2)^{\frac{1}{2}}$

  b. Use your answers from part **a** to calculate the value of the following.

  i. $9^{\frac{1}{2}}$ ii. $16^{\frac{1}{2}}$ iii. $64^{\frac{1}{2}}$ iv. $121^{\frac{1}{2}}$

  c. Use your answers to parts **a** and **b** to write a sentence describing what happens when you raise a number to a power of one-half.

1.6 Negative indices

LEARNING INTENTION

At the end of this subtopic you should be able to:
- apply the Seventh Index Law to evaluate expressions using negative indices
- apply the first seven index laws to simplify expressions involving indices.

▶ 1.6.1 The Seventh Index Law

eles-4388

- Negative indices occur when the power (or exponent) is a negative number.
  For example, 3^{-2}.
- To explain the meaning or value of negative indices it is useful to consider patterns of numbers written in index form.
  For example, in the sequence $3^4 = 81$, $3^3 = 27$, $3^2 = 9$, $3^1 = 3$, $3^0 = 1$ each number is $\dfrac{1}{3}$ of the number before it.

Powers decrease by 1

3^4	3^3	3^2	3^1	3^0
81	27	9	3	1

$\times \dfrac{1}{3}$ $\quad$ $\times \dfrac{1}{3}$ $\quad$ $\times \dfrac{1}{3}$ $\quad$ $\times \dfrac{1}{3}$

- It is logical, then, that the next numbers in this sequence are:

$$3^{-1} = \frac{1}{3},\ 3^{-2} = \frac{1}{9},\ 3^{-3} = \frac{1}{27},\ 3^{-4} = \frac{1}{81}$$

Powers decrease by 1

3^0	3^{-1}	3^{-2}	3^{-3}	3^{-4}
1	$\dfrac{1}{3}$	$\dfrac{1}{9}$	$\dfrac{1}{27}$	$\dfrac{1}{81}$

$\times \dfrac{1}{3}$ $\quad$ $\times \dfrac{1}{3}$ $\quad$ $\times \dfrac{1}{3}$ $\quad$ $\times \dfrac{1}{3}$

The Seventh Index Law

$$a^{-n} = \frac{1}{a^n},\ \text{where } a \neq 0$$

For example, $2^{-3} = \dfrac{1}{2^3} = \dfrac{1}{8}$

WORKED EXAMPLE 19 Evaluating using the Seventh Index Law

Evaluate the following.

a. 5^{-2}

b. 7^{-1}

c. $\left(\dfrac{3}{5}\right)^{-1}$

THINK	WRITE
a. 1. Apply the rule $a^{-n} = \dfrac{1}{a^n}$.	a. $5^{-2} = \dfrac{1}{5^2}$
2. Simplify and write the answer.	$= \dfrac{1}{25}$
b. Apply the rule $a^{-n} = \dfrac{1}{a^n}$.	b. $7^{-1} = \dfrac{1}{7^1}$
	$= \dfrac{1}{7}$
c. 1. Apply the Sixth Index Law, $\left(\dfrac{a}{b}\right)^m = \dfrac{a^m}{b^m}$.	c. $\left(\dfrac{3^1}{5^1}\right) = \dfrac{3^{-1}}{5^{-1}}$
2. Apply the Seventh Index Law, $a^{-n} = \dfrac{1}{a^n}$, to the numerator and denominator.	$= \dfrac{1}{3} \div \dfrac{1}{5}$
3. Simplify and write the answer.	$= \dfrac{1}{3} \times \dfrac{5}{1}$
	$= \dfrac{5}{3}$

WORKED EXAMPLE 20 Writing with positive indices

Write the following with positive indices.

a. x^{-3}

b. $5x^{-6}$

c. $\dfrac{x^{-3}}{y^{-2}}$

THINK	WRITE
a. Apply the Seventh Index Law, $a^{-n} = \dfrac{1}{a^n}$.	a. $x^{-3} = \dfrac{1}{x^3}$
b. 1. Write in expanded form and then apply the Seventh Index Law, $a^{-n} = \dfrac{1}{a^n}$.	b. $5x^{-6} = 5 \times x^{-6}$
	$= 5 \times \dfrac{1}{x^6}$
2. Simplify and write the answer.	$= \dfrac{5}{x^6}$

c. 1. Write the fraction using division.

c. $\dfrac{x^{-3}}{y^{-2}}$

$= x^{-3} \div y^{-2}$

2. Apply the Seventh Index Law, $a^{-n} = \dfrac{1}{a^n}$.

$= \dfrac{1}{x^3} \div \dfrac{1}{y^2}$

3. Simplify and write the answer.

$= \dfrac{1}{x^3} \times \dfrac{y^2}{1}$

$= \dfrac{y^2}{x^3}$

WORKED EXAMPLE 21 Simplifying products of powers with positive and negative indices

Simplify the following expressions, writing your answers with positive indices.

a. $x^3 \times x^{-8}$

b. $3x^{-2}y^{-3} \times 5xy^{-4}$

THINK

WRITE

a. 1. Apply the First Index Law $a^n \times a^m = a^{m+n}$.

a. $x^3 \times x^{-8}$

$= x^{3+-8}$

$= x^{-5}$

2. Write the answer with a positive index.

$= \dfrac{1}{x^5}$

b. 1. Write in expanded form and apply the First Index Law.

b. $3x^{-2}y^{-3} \times 5xy^{-4}$

$= 3 \times 5 \times x^{-2} \times x^1 \times y^{-3} \times y^{-4}$

2. Apply the rule $a^{-n} = \dfrac{1}{a^n}$.

$= 15x^{-1}y^{-7}$

$= \dfrac{15}{1} \times \dfrac{1}{x} \times \dfrac{1}{y^7}$

3. Simplify and write the answer.

$= \dfrac{15}{xy^7}$

WORKED EXAMPLE 22 Simplifying quotients of powers with positive and negative indices

Simplify the following expressions, writing your answers with positive indices.

a. $\dfrac{t^2}{t^{-5}}$

b. $\dfrac{15m^{-5}}{10m^{-2}}$

THINK

WRITE

a. Apply the Second Index Law, $\dfrac{a^n}{a^m} = a^{n-m}$.

a. $\dfrac{t^2}{t^{-5}}$

$= t^{2-(-5)}$

$= t^{2+5}$

$= t^7$

b. **1.** Apply the Second Index Law and simplify.

2. Write the answer with positive indices.

$$\text{b. } \frac{15m^{-5}}{10m^{-2}} = \frac{15}{10} \times \frac{m^{-5}}{m^{-2}}$$

$$= \frac{3}{2} \times m^{-5-(-2)}$$

$$= \frac{3}{2} \times m^{-3}$$

$$= \frac{3}{2} \times \frac{1}{m^3}$$

$$= \frac{3}{2m^3}$$

DISCUSSION

What strategy will you use to remember the index laws?

 Resources

 eWorkbook   Topic 1 Workbook (worksheets, code puzzle and project) (ewbk-2001)

 Video eLesson Negative indices (eles-1910)

 Interactivity   Individual pathway interactivity: Negative indices (int-4518)

Negative indices (int-6064)

Exercise 1.6 Negative indices

learnon

Individual pathways

■ PRACTISE	■ CONSOLIDATE	■ MASTER
1, 2, 5, 8, 11, 14, 17, 21	3, 6, 9, 12, 15, 18, 19, 22	4, 7, 10, 13, 16, 20, 23, 24

To answer questions online and to receive **immediate corrective feedback** and **fully worked solutions** for all questions, go to your learnON title at www.jacplus.com.au.

Fluency

1. Copy and complete the following patterns.

a. $3^5 = 243$

$3^4 = 81$

$3^3 = 27$

$3^2 =$

$3^1 =$

$3^0 =$

$3^{-1} = \frac{1}{3}$

$3^{-2} = \frac{1}{9}$

$3^{-3} =$

$3^{-4} =$

$3^{-5} =$

b. $5^4 = 625$

$5^3 =$

$5^2 =$

$5^1 =$

$5^0 =$

$5^{-1} =$

$5^{-2} =$

$5^{-3} =$

$5^{-4} =$

c. $10^4 = 10\ 000$

$10^3 =$

$10^2 =$

$10^1 =$

$10^0 =$

$10^{-1} =$

$10^{-2} =$

$10^{-3} =$

$10^{-4} =$

2. **WE19** Evaluate each of the following expressions.

    a. 2^{-5} **b.** 3^{-3} **c.** 4^{-1} **d.** 10^{-2}

3. **WE19** Evaluate each of the following expressions.

    a. 5^{-3} **b.** $\left(\dfrac{1}{7}\right)^{-1}$ **c.** $\left(\dfrac{3}{4}\right)^{-1}$ **d.** $\left(\dfrac{3}{4}\right)^{-2}$

4. **WE19** Evaluate each of the following expressions.

    a. $\left(\dfrac{1}{3}\right)^{-3}$ **b.** $\left(\dfrac{3}{2}\right)^{-1}$ **c.** $\left(2\dfrac{1}{4}\right)^{-2}$ **d.** $\left(\dfrac{2}{7}\right)^{-2}$

5. **WE20** Write each expression with positive indices.

    a. x^{-4} **b.** y^{-5} **c.** z^{-1} **d.** a^2b^{-3} **e.** $m^{-2}n^{-3}$

6. Write each expression with positive indices.

    a. $\left(m^2n^3\right)^{-1}$ **b.** $\dfrac{x^2}{y^{-2}}$ **c.** $\dfrac{5}{x^{-3}}$ **d.** $\dfrac{x^{-2}}{w^{-5}}$ **e.** $\dfrac{1}{x^{-2}y^{-2}}$

7. Write each expression with positive indices.

    a. $a^2b^{-3}cd^{-4}$ **b.** $\dfrac{a^2b^{-2}}{c^2d^{-3}}$ **c.** $10x^{-2}y$ **d.** $3^{-1}x$ **e.** $\dfrac{m^{-3}}{x^2}$

Understanding

8. **WE 21** Simplify the following expressions, writing your answers with positive indices.

    a. $a^3 \times a^{-8}$ **b.** $m^7 \times m^{-2}$ **c.** $m^{-3} \times m^{-4}$ **d.** $2x^{-2} \times 7x$ **e.** $2b^{-2} \times 3b^2$ **f.** $ab^{-5} \times 5b^2$

9. Simplify the following expressions, writing your answers with positive indices.

    a. $x^5 \times x^{-8}$ **b.** $3x^2y^{-4} \times 2x^{-7}y$ **c.** $10x^5 \times 5x^{-2}$

    d. $x^5 \times x^{-5}$ **e.** $10a^2 \times 5a^{-1}$ **f.** $10a^{10} \times a^{-6}$

10. Simplify the following expressions, writing your answers with positive indices.

    a. $16w^2 \times -2w^{-5}$ **b.** $4m^{-2} \times 4m^{-2}$ **c.** $\left(3m^2n^{-4}\right)^3$

    d. $\left(a^2b^5\right)^{-3}$ **e.** $\left(a^{-1}b^{-3}\right)^{-2}$ **f.** $\left(5a^{-1}\right)^2$

11. **WE22** Simplify the following expressions, writing your answers with positive indices.

    a. $\dfrac{x^3}{x^8}$ **b.** $\dfrac{x^{-3}}{x^8}$ **c.** $\dfrac{x^3}{x^{-8}}$ **d.** $\dfrac{x^{-3}}{x^{-8}}$ **e.** $\dfrac{6a^2c^5}{a^4c}$

12. Simplify the following expressions, writing your answers with positive indices.

    a. $10a^2 \div 5a^8$ **b.** $5m^7 \div m^8$ **c.** $\dfrac{a^5b^6}{a^5b^7}$ **d.** $\dfrac{a^2b^8}{a^5b^{10}}$ **e.** $\dfrac{a^{-3}bc^3}{abc}$

13. Simplify the following expressions, writing your answers with positive indices.

    a. $\dfrac{4^{-2}ab}{a^2b}$ **b.** $\dfrac{m^{-3} \times m^{-5}}{m^{-5}}$ **c.** $\dfrac{2t^2 \times 3t^{-5}}{4t^6}$ **d.** $\dfrac{t^3 \times t^{-5}}{t^{-2} \times t^{-3}}$ **e.** $\dfrac{\left(m^2n^{-3}\right)^{-1}}{\left(m^{-2}n^3\right)^2}$

14. Write the following numbers as powers of 2.

    a. 1 **b.** 8 **c.** 32 **d.** 64 **e.** $\dfrac{1}{8}$ **f.** $\dfrac{1}{32}$

15. Write the following numbers as powers of 4.

    a. 1 **b.** 4 **c.** 64 **d.** $\dfrac{1}{4}$ **e.** $\dfrac{1}{16}$ **f.** $\dfrac{1}{64}$

16. Write the following numbers as powers of 10.

    a. 1   b. 10   c. 10 000   d. 0.1   e. 0.01   f. 0.000 01

Reasoning

17. Answer the following questions.

    a. The result of dividing 3^7 by 3^3 is 3^4. Determine the result of dividing 3^3 by 3^7.
    b. Explain what it means to have a negative index.
    c. Explain how you write a negative index as a positive index.

18. Indices are encountered in science, where they help to deal with very small and large numbers. The diameter of a proton is 0.000 000 000 000 3 cm.
    Explain why it is logical to express this number in scientific notation as 3×10^{-13}.

19. Answer the following questions.

    a. When asked to write an expression with positive indices that is equivalent to $x^3 + x^{-3}$, a student gave the answer x^0. Is this answer correct? Explain why or why not.
    b. When asked to write an expression with positive indices that is equivalent to $\left(x^{-1} + y^{-1}\right)^{-2}$, a student gave the answer $x^2 + y^2$. Is this answer correct? Explain why or why not.

20. Answer the following questions.

    a. When asked to write an expression with positive indices that is equivalent to $x^8 - x^{-5}$, a student gave the answer x^3. Is this answer correct? Explain why or why not.
    b. Another student said that $\dfrac{x^2}{x^8 - x^5}$ is equivalent to $\dfrac{1}{x^6} - \dfrac{1}{x^3}$. Is this answer correct? Explain why or why not.

Problem solving

21. Write the following numbers as basic numerals.

    a. 4.8×10^{-2} b. 7.6×10^3 c. 2.9×10^{-4} d. 8.1×10^0

22. Determine the value of n in the following expressions.

    a. $4793 = 4.793 \times 10^n$ b. $0.631 = 6.31 \times 10^n$ c. $134 = 1.34 \times 10^n$ d. $0.000 56 = 5.6 \times 10^n$

23. Evaluate the following.

    a. Half of 2^{20} b. One-third of 3^{21}

24. Simplify the following expressions.

    a. $\left(2^{-1} + 3^{-1}\right)^{-1}$ b. $\dfrac{3^{400}}{6^{200}}$

1.7 Scientific notation

> **LEARNING INTENTION**
>
> At the end of this subtopic you should be able to:
> • identify numbers that are written in scientific notation
> • write numbers in scientific notation as decimal numbers
> • write decimal numbers using scientific notation.

▶ 1.7.1 Using scientific notation

eles-4389

- Scientists often work with extremely large and extremely small numbers. These can range from numbers as large as 1 000 000 000 000 000 000 km (the approximate diameter of our galaxy) down to numbers as small as 0.000 000 000 06 mm (the diameter of an atom).

- Doing calculations with numbers written this way can be challenging — it can be easy to lose track of all of those zeros.
- It is more useful in these situations to use a notation system based on powers of 10. This system is called **scientific notation** (or **standard form**).
- A number written in scientific notation looks like this:

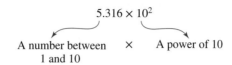

$$5.316 \times 10^2$$

A number between 1 and 10   ×   A power of 10

Using a calculator for scientific notation

- Numbers written in scientific notation can be entered into a calculator using its scientific notation buttons.
- Some calculators have different ways of displaying numbers written in scientific notation. For example, some calculators display 5.71×10^4 as 5.71E4. When you are writing down scientific notation yourself, you should always show this number as 5.71×10^4.

Writing any number in scientific notation

1. Move the decimal point so that it is between the first and second significant figures.
2. Consider how many places the decimal point has moved — this number of places corresponds to the appropriate power of 10.
3. If the decimal point has moved to the left, the power of 10 will have a positive index (this happens with very large numbers).
4. If the decimal point has moved to the right, the power of 10 will have a negative index (this happens with very small numbers).

WORKED EXAMPLE 23 Writing decimal numbers in scientific notation

Write each of these numbers in scientific notation.
a. 827.2   **b.** 51 920 000 000   **c.** 0.0051   **d.** 0.000 000 007 648

THINK

a. 1. The numerical part must be written so it is a number between 1 and 10. This means we must move the decimal point 2 steps to the left.

   2. Moving the decimal point 2 steps left corresponds to the power 10^2. Note that the number of steps moved is equal to the index.

b. 1. Even though the decimal point is not written, we know that it lies after the final zero.

   2. To get a number between 1 and 10 we move the decimal 10 steps to the left. The corresponding power will be 10^{10}.

WRITE

8.272
Decimal moves 2 steps left
8.272×10^2

5.192×10^{10}

c. To get a number between 1 and 10 we must move the decimal point 3 steps to the right. This corresponds to the power 10^{-3}. Note that moving to the right gives a negative index.

5.1×10^{-3}

d. To get a number between 1 and 10 we must move the decimal point 9 steps to the right. This corresponds to the power 10^{-9}.

7.648×10^{-9}

⏵ 1.7.2 Converting scientific notation to decimal notation

eles-4390

- Converting scientific notation to decimal notation is simply the reverse of the process outlined in section 1.7.1.

WORKED EXAMPLE 24 Converting scientific notation with positive indices to decimals

The following numbers are written in scientific notation. Write them in decimal notation.
a. 7.136×10^2 **b.** 5.017×10^5 **c.** 8×10^6

THINK

a. The index on the power of 10 is positive 2, therefore we move the decimal point 2 steps to the right.

b. The index on the power of 10 is positive 5, therefore we move the decimal point 5 steps to the right. To do this we will need to add extra zeros.

c. The index on the power of 10 is positive 6, therefore we move the decimal point 6 steps to the right. While the decimal point has not been written as part of the scientific notation, we know it lies after the 8. We will need to add extra zeros.

WRITE

a. $7.136 \times 10^2 = 713.6$

b. $5.017 \times 10^5 = 501\,700$

c. $8 \times 10^6 = 8\,000\,000$

WORKED EXAMPLE 25 Converting scientific notation with negative indices to decimals

Write these numbers in decimal notation.
a. 9.12×10^{-1} **b.** 7.385×10^{-2} **c.** 6.32×10^{-7}

THINK

a. The index on the power of 10 is negative 1, therefore we move the decimal point 1 step to the left.

b. The index on the power of 10 is negative 2, therefore we move the decimal point 2 steps to the left. To do this we will need to add extra zeros.

c. The index on the power of 10 is negative 7, therefore we move the decimal point 7 steps to the left. To do this we will need to add extra zeros.

WRITE

a. $9.12 \times 10^{-1} = 0.912$

b. $7.3857 \times 10^{-2} = 0.073\,857$

c. $6.32 \times 10^{-7} = 0.000\,000\,632$

DISCUSSION

What is the advantage of converting numbers into scientific notation?

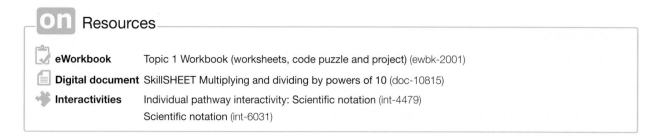
Exercise 1.7 Scientific notation

learnon

Individual pathways

■ PRACTISE	■ CONSOLIDATE	■ MASTER
1, 4, 6, 7, 10, 13, 14, 19, 23	2, 5, 8, 11, 15, 16, 20, 24	3, 9, 12, 17, 18, 21, 22, 25, 26

To answer questions online and to receive **immediate corrective feedback** and **fully worked solutions** for all questions, go to your learnON title at www.jacplus.com.au.

Fluency

1. **WE24** Write these numbers in decimal notation.
   a. 6.14×10^2 b. 6.14×10^3 c. 6.14×10^4 d. 3.518×10^2

2. Write these numbers in decimal notation.
   a. 1×10^9 b. $3.926\,73 \times 10^2$ c. 5.911×10^2 d. 5.1×10^3

3. Write these numbers in decimal notation.
   a. 7.34×10^5 b. 7.1414×10^6 c. 3.51×10 d. 8.05×10^4

4. **WE23** Write these numbers in scientific notation.
   a. 5000   b. 431   c. 38   d. 350 000

5. Write these numbers in scientific notation.
   a. 72.5   b. 725   c. 7250   d. 725 000 000

6. Write these numbers in scientific notation, correct to 4 significant figures.
   a. 43.792   b. 5317   c. 258.95   d. 110.11   e. 1 632 000   f. 1 million

7. **WE25** Write these numbers in decimal notation.
   a. 2×10^{-1} b. 4×10^{-3} c. 7×10^{-4} d. 3×10^{-2}

8. Write these numbers in decimal notation.
   a. 8.273×10^{-2} b. 7.295×10^{-2} c. 2.9142×10^{-3} d. 3.753×10^{-5}

9. Write these numbers in decimal notation.
   a. 5.29×10^{-4} b. 3.3333×10^{-5} c. 2.625×10^{-9} d. 1.273×10^{-15}

10. Write these numbers in scientific notation.
    a. 0.7   b. 0.005   c. 0.000 000 3   d. 0.000 000 000 01

11. Write these numbers in scientific notation.
    a. 0.231   b. 0.003 62   c. 0.000 731   d. 0.063

12. Write these numbers in scientific notation, correct to 3 significant figures.
    a. 0.006 731   b. 0.142 57   c. 0.000 068 3
    d. 0.000 000 005 12   e. 0.0509   f. 0.012 46

Understanding

13. Write each of the following sets of numbers in ascending order.
    a. 8.31×10^2, 3.27×10^3, 9.718×10^2, 5.27×10^2
    b. 7.95×10^2, 4.09×10^2, 7.943×10^2, 4.37×10^2
    c. 5.31×10^{-2}, 9.29×10^{-3}, 5.251×10^{-2}, 2.7×10^{-3}
    d. 8.31×10^2, 3.27×10^3, 7.13×10^{-2}, 2.7×10^{-3}

14. One carbon atom weighs 1.994×10^{-23} g.
    a. Write this weight as a decimal.
    b. Calculate how much one million carbon atoms will weigh.
    c. Calculate how many carbon atoms there are in 10 g of carbon. Give your answer correct to 4 significant figures.

15. The distance from Earth to the Moon is approximately 3.844×10^5 km. If you could drive there at a constant speed of 100 km/h, calculate how long it would take. Determine how long that is in days, correct to 2 decimal places.

16. Earth weighs 5.97×10^{24} kg and the Sun weighs 1.99×10^{30} kg. Calculate how many Earths it would take to balance the Sun's weight. Give your answer correct to 2 decimal places.

17. Inside the nucleus of an atom, a proton weighs 1.6703×10^{-28} kg and a neutron weighs 1.6726×10^{-27} kg. Determine which one is heavier and by how much.

18. Earth's orbit has a radius of 7.5×10^7 km and the orbit of Venus has a radius of 5.4×10^7 km. Calculate how far apart the planets are when:
    a. they are closest to each other
    b. they are furthest apart from each other.

Reasoning

19. A USB stick has 8 MB (megabytes) of storage.
    1 MB = 1 048 576 bytes.
    a. Determine the number of bytes in 8 MB of storage correct to the nearest 1000.
    b. Write the answer to part a in scientific notation.

20. The basic unit of electric current is the ampere. It is defined as the constant current flowing in 2 parallel conductors 1 metre apart in a vacuum, which produces a force between the conductors of 2×10^7 newtons (N) per metre.
    Complete the following statement. Write the answer as a decimal.
    1 ampere $= 2 \times 10^7$ N/m = _____ N/m

21. a. Without performing the calculation, state the power(s) of 10 that you believe the following equations will have when solved.
    i. $5.36 \times 10^7 + 2.95 \times 10^3$ ii. $5.36 \times 10^7 - 2.95 \times 10^3$
    b. Evaluate equations i and ii, correct to 3 significant figures.
    c. Were your answers to part a correct? Why or why not?

22. Explain why $2.39 \times 10^{-3} + 8.75 \times 10^{-7} = 2.39 \times 10^{-3}$, correct to 3 significant figures.

Problem solving

23. Distance is equal to speed multiplied by time. If we travelled at $100\,\text{km/h}$ it would take us approximately 0.44 years to reach the Moon, 89.6 years to reach Mars, 1460 years to reach Saturn and 6590 years to reach Pluto.

   a. Assuming that there are 365 days in a year, calculate the distance (as a basic numeral) between Earth and:
      i. the Moon   ii. Mars   iii. Saturn   iv. Pluto.

   b. Write your answers to part a, correct to 3 significant figures.

   c. Write your answers to part a using scientific notation, correct to 3 significant figures.

24. A light-year is the distance that light travels in one year. Light travels at approximately $300\,000\,\text{km/s}$.
   a. i. Calculate the number of seconds in a year (assuming 1 year = 365 days).
      ii. Write your answer to part i using scientific notation.
   b. Calculate the distance travelled by light in one year. Express your answer:
      i. as a basic numeral
      ii. using scientific notation.

Alpha Centauri

   c. The closest star to Earth (other than the Sun) is in the star system Alpha Centauri, which is 4.3 light-years away.
      i. Determine how far this is in kilometres, correct to 4 significant figures.
      ii. If you were travelling at $100\,\text{km/h}$, determine how many years it would take to reach Alpha Centauri.

25. Scientists used Earth's gravitational pull on nearby celestial bodies (for example, the Moon) to calculate the mass of Earth. Their answer was that Earth weighs approximately 5.972 sextillion metric tonnes.

   a. Write 5.972 sextillion using scientific notation.
   b. Determine how many significant figures this number has.

26. Atoms are made up of smaller particles called protons, neutrons and electrons. Electrons have a mass of $9.109\,381\,88 \times 10^{-31}$ kilograms, correct to 9 significant figures.

   a. Write the mass of an electron correct to 5 significant figures.
   b. Protons and neutrons are the same size. They are both 1836 times the size of an electron. Use the mass of an electron (correct to 9 significant figures) and your calculator to evaluate the mass of a proton correct to 5 significant figures.
   c. Use the mass of an electron (correct to 3 significant figures) to calculate the mass of a proton, correct to 5 significant figures.
   d. Explain why it is important to work with the original amounts and then round to the specified number of significant figures at the end of a calculation.

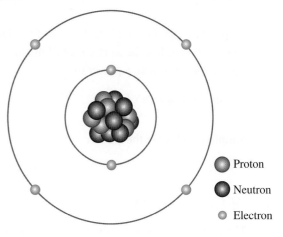
Proton
Neutron
Electron

1.8 Square roots and cube roots

▶ 1.8.1 Square roots

eles-4391

- The symbol $\sqrt{}$ (the radical symbol) means '**square root**' — a number that multiplies by itself to give the original number.
- Every number actually has a positive and negative square root. For example, $(2)^2 = 4$ and $(-2)^2 = 4$.
- This means that the square root of 4 is $+2$ or -2. For this topic, however, assume that $\sqrt{}$ is positive unless otherwise indicated.
- A square root is the inverse of squaring (raising to the power 2).
- Because of this, a square root of a number is equivalent to raising that number to an index of $\frac{1}{2}$.
- In general, $\sqrt{a} = a^{\frac{1}{2}}$.

WORKED EXAMPLE 26 Evaluating square roots

Evaluate $\sqrt{16p^2}$.

THINK	WRITE
1. We need to work out the square root of both 16 and p^2.	$\sqrt{16p^2} = \sqrt{16} \times \sqrt{p^2}$
2. Remember that 4 is multiplied by itself to give 16, so the square root of 16 is 4. For the square root of the pronumeral, replace the square root sign with a power of $\frac{1}{2}$.	$= 4 \times \left(p^2\right)^{\frac{1}{2}}$
3. Use the Fourth Index Law.	$= 4 \times p^{2 \times \frac{1}{2}}$ $= 4 \times p^1$
4. Simplify and write the answer.	$= 4p$

▶ 1.8.2 Cube roots

eles-4392

- The symbol $\sqrt[3]{}$ means '**cube root**' — a number that multiplies by itself 3 times to give the original number.
- The cube root is the inverse of cubing (raising to the power 3).
- Because of this, a cube root is equivalent to raising a number to an index of $\dfrac{1}{3}$.
- In general, $\sqrt[3]{a} = a^{\frac{1}{3}}$.
- In more general terms, $a^{\frac{n}{m}} = \sqrt[m]{a^n}$.

WORKED EXAMPLE 27 Evaluating cube roots

Evaluate $\sqrt[3]{8j^6}$

THINK	WRITE
1. We need to work out the cube root of both 8 and j^6.	$\sqrt[3]{8j^6} = \sqrt[3]{8} \times \sqrt[3]{j^6}$
2. Remember that 2 multiplied by itself 3 times gives 8, so the cube root of 8 is 2. For the cube root of the pronumeral, replace the cube root sign with a power of $\dfrac{1}{3}$.	$= 2 \times \left(j^6\right)^{\frac{1}{3}}$
3. Use the Fourth Index Law.	$= 2 \times j^{6 \times \frac{1}{3}}$ $= 2 \times j^2$
4. Simplify and write the answer.	$= 2j^2$

DISCUSSION

Why does the cube root of a number always have the same sign (positive or negative) as the number itself?

 Resources

Exercise 1.8 Square roots and cube roots

learnon

Individual pathways

■ PRACTISE	■ CONSOLIDATE	■ MASTER
1, 4, 7, 10, 14	2, 5, 8, 11, 15, 16	3, 6, 9, 12, 13, 17

To answer questions online and to receive **immediate corrective feedback** and **fully worked solutions** for all questions, go to your learnON title at www.jacplus.com.au.

Fluency

1. Write the following in index form.

   a. $\sqrt{15}$ b. $\sqrt{m}$ c. $\sqrt[3]{t}$ d. $\sqrt[3]{w^2}$

2. **WE26** Evaluate the following.

   a. $49^{\frac{1}{2}}$ b. $4^{\frac{1}{2}}$ c. $27^{\frac{1}{3}}$ d. $125^{\frac{1}{3}}$

3. Evaluate the following.

   a. $1000^{\frac{1}{3}}$ b. $64^{\frac{1}{2}}$ c. $64^{\frac{1}{3}}$ d. $1\,000\,000^{\frac{1}{2}}$ e. $1\,000\,000^{\frac{1}{3}}$ f. $\left(27^{\frac{1}{3}}\right)^2$

Understanding

4. Simplify the following expressions.

   a. $\sqrt{m^2}$ b. $\sqrt[3]{b^3}$ c. $\sqrt{36t^4}$ d. $\sqrt[3]{m^3 n^6}$

5. Simplify the following expressions.

   a. $\sqrt[3]{125t^6}$ b. $\sqrt[5]{x^5 y^{10}}$ c. $\sqrt[4]{a^8 m^{40}}$ d. $\sqrt[3]{216y^6}$

6. Simplify the following expressions.

   a. $\sqrt[3]{64x^6 y^6}$ b. $\sqrt{25a^2 b^4 c^6}$ c. $\sqrt[7]{b^{49}}$ d. $\sqrt[3]{b^3} \times \sqrt{b^4}$

7. **MC** $\sqrt[3]{8000 m^6 n^3 p^3 q^6}$ is equal to:

   A. $2666.6 m^2 n p q^2$ **B.** $20 m^2 n p q^2$ **C.** $20 m^3 n^0 p^0 q^3$ **D.** $7997 m^2 n p q^2$ **E.** $2 m^2 n p q^2$

8. **MC** $\sqrt[3]{3375 a^9 b^6 c^3}$ is equal to:

   A. $1125 a^3 b^2 c$ **B.** $1125 a^6 b^3 c^0$ **C.** $1123 a^6 b^3$ **D.** $15 a^3 b^2 c$ **E.** $5 a^3 b^2 c$

9. **MC** $\sqrt[3]{15\,625 f^3 g^6 h^9}$ is equal to:

   A. $25 f g^2 h^3$ **B.** $25 f^0 g^3 h^6$ **C.** $25 g^3 h^6$ **D.** $5208.3 f g^2 h^3$ **E.** $250 f g^2 h^3$

Reasoning

10. Answer the following questions.

    a. Using the First Index Law, explain how $3^{\frac{1}{2}} \times 3^{\frac{1}{2}} = 3$.

    b. Express $3^{\frac{1}{2}}$ in another way.

    c. Write $\sqrt[n]{a}$ in index form.

    d. Without a calculator, evaluate $8^{\frac{1}{3}}$.

11. Answer the following questions.

    a. Explain why calculating $z^{2.5}$ is a square root problem.
    b. Explain whether $z^{0.3}$ is a cube root problem.

12. Prithvi and Yasmin are having an algebra argument. Prithvi is sure that $\sqrt{x^2}$ is equivalent to x, but Yasmin thinks otherwise. State who is correct. Explain how you would resolve this disagreement.

13. Verify that $(-8)^{\frac{1}{3}}$ can be evaluated and explain why $(-8)^{\frac{1}{4}}$ cannot be evaluated.

Problem solving

14. The British mathematician Augustus de Morgan enjoyed telling his friends that he was x years old in the year x^2. Determine the year of Augustus de Morgan's birth, given that he died in 1871.

15. Kepler's Third Law describes the relationship between the distance of planets from the Sun and their orbital periods. It is represented by the equation $d^{\frac{1}{2}} = t^{\frac{1}{3}}$. Solve for:

    a. d in terms of t b. t in terms of d.

16. If $n^{\frac{3}{4}} = \dfrac{8}{27}$, evaluate the value of n.

17. An unknown number is multiplied by 4 and then has 5 subtracted from it. It is now equal to the square root of the original unknown number, squared.

    a. Determine how many solutions to this problem are possible. Explain why.
    b. Determine all possible values for the unknown number.

1.9 Surds

LEARNING INTENTION

At the end of this subtopic you should be able to:
- identify when a square root forms a surd
- place surds in their approximate position on a number line
- simplify expressions containing surds.

▶ 1.9.1 Identifying surds

eles-4393

- In many cases the square root or cube root of a number results in a rational number. For example, $\sqrt{25} = \pm 5$ and $\sqrt[3]{1.728} = 1.2$.
- When the square root of a number is an irrational number, it is called a **surd**.
- For example, $\sqrt{10} \approx 3.162\ 277\ 660\ 17 \ldots$ Since $\sqrt{10}$ cannot be written as a fraction, a recurring decimal or a **terminating decimal**, it is irrational and therefore it is a surd.
- The value of a surd can be approximated using a number line. For example, we know that $\sqrt{21}$ lies between 4 and 5, because it lies between $\sqrt{16}$ (which equals 4) and $\sqrt{25}$ (which equals 5). We can show its approximate position on the number line like this:

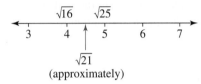

WORKED EXAMPLE 28 Placing surds on a number line

Place $\sqrt{34}$ on a number line.

THINK

1. The next lowest square number that is smaller than 34 is 25. The next highest square number that is larger than 34 is 36.

2. Draw a number line to show the approximate position of $\sqrt{34}$.

WRITE

$\sqrt{34}$ lies between $\sqrt{25}$ and $\sqrt{36}$.

Identify which of the following are surds.

a. $\sqrt{0}$ b. $\sqrt{20}$ c. $-\sqrt{9}$ d. $\sqrt[3]{6}$

THINK	WRITE
a. $\sqrt{0} = 0$. This is a rational number and therefore not a surd.	a. $\sqrt{0} = 0$. This is not a surd.
b. $\sqrt{20} \approx 4.472\,135\,955\ldots$ This is an irrational number and therefore a surd.	b. $\sqrt{20}$ is a surd.
c. $-\sqrt{9} = -3$. This is a rational number and therefore not a surd.	c. $-\sqrt{9} = -3$. This is not a surd.
d. $\sqrt[3]{6} \approx 1.817\,120\,592\,83\ldots$ This is an irrational number and therefore a surd.	d. $\sqrt[3]{6}$ is a surd.

1.9.2 Multiplying and dividing surds

eles-4394

- Multiplication of surds is done by multiplying the numerical parts of each surd under a single radical sign.
- Division of surds is done by placing the quotient of the numerical parts of each surd under a single radical sign.

Multiplying surds

$$\sqrt{a} \times \sqrt{b} = \sqrt{ab}$$

Dividing surds

$$\frac{\sqrt{a}}{\sqrt{b}} = \sqrt{\frac{a}{b}}$$

Evaluate the following, leaving your answer in surd form.

a. $\sqrt{7} \times \sqrt{2}$ b. $5 \times \sqrt{3}$ c. $\sqrt{5} \times \sqrt{5}$ d. $-2\sqrt{3} \times 4\sqrt{5}$

THINK	WRITE
a. Apply the rule $\sqrt{a} \times \sqrt{b} = \sqrt{ab}$.	a. $\sqrt{7} \times \sqrt{2} = \sqrt{14}$
b. Only $\sqrt{3}$ is a surd. It is multiplied by 5, which is not a surd.	b. $5 \times \sqrt{3} = 5\sqrt{3}$
c. Apply the rule $\sqrt{a} \times \sqrt{b} = \sqrt{ab}$.	c. $\sqrt{5 \times 5} = \sqrt{25}$ $= 5$
d. Multiply the whole numbers. Then multiply the surds.	d. $-2\sqrt{3} \times 4\sqrt{5} = -2 \times 4 \times \sqrt{3} \times \sqrt{5}$. $= -8 \times \sqrt{15}$ $= -8\sqrt{15}$

Evaluate the following, leaving your answer in surd form.

a. $\dfrac{\sqrt{10}}{\sqrt{5}}$
b. $\sqrt{\dfrac{10}{5}}$
c. $\dfrac{-6\sqrt{8}}{4\sqrt{4}}$
d. $\dfrac{\sqrt{20}}{\sqrt{5}}$
e. $\dfrac{5}{\sqrt{5}}$

THINK	WRITE
a. Apply the rule $\sqrt{a} \div \sqrt{b} = \sqrt{\dfrac{a}{b}}$.	a. $\dfrac{\sqrt{10}}{\sqrt{5}} = \sqrt{\dfrac{10}{5}}$ $= \sqrt{2}$
b. Simplify the fraction.	b. $\sqrt{\dfrac{10}{5}} = \sqrt{2}$
c. Simplify the whole numbers. Then apply the rule $\sqrt{a} \div \sqrt{b} = \sqrt{\dfrac{a}{b}}$.	c. $\dfrac{-6\sqrt{8}}{4\sqrt{4}} = \dfrac{-3\sqrt{2}}{2}$
d. Apply the rule $\sqrt{a} \div \sqrt{b} = \sqrt{\dfrac{a}{b}}$.	d. $\dfrac{\sqrt{20}}{\sqrt{5}} = \sqrt{\dfrac{20}{5}} = \sqrt{4}$ $= 2$
e. Rewrite the numerator as the product of two surds and then simplify.	e. $\dfrac{5}{\sqrt{5}} = \dfrac{\sqrt{5} \times \sqrt{5}}{\sqrt{5}}$ $= \sqrt{5}$

▶ 1.9.3 Simplifying surds

eles-4395

- Just as a rational number can be written many different ways (e.g. $\dfrac{1}{2} = \dfrac{5}{10} = \dfrac{7}{14}$), so can a surd. It is expected that surds should normally be written in their simplest form.
- A surd is in its simplest form when the number inside the radical sign has the smallest possible value.
- Note that $\sqrt{24}$ can be factorised several ways. For example:

$$\sqrt{24} = \sqrt{2} \times \sqrt{12}$$

$$\sqrt{24} = \sqrt{3} \times \sqrt{8}$$

$$\sqrt{24} = \sqrt{4} \times \sqrt{6}$$

In the last example, $\sqrt{4} = 2$, which means:

$$\sqrt{24} = 2 \times \sqrt{6}$$
$$= 2\sqrt{6}$$

$2\sqrt{6}$ is equal to $\sqrt{24}$, and $2\sqrt{6}$ is $\sqrt{24}$ written in its simplest form.

- To simplify a surd you must find a factor that is also a perfect square. For example, 4, 9, 16, 25, 36 or 49
- A surd like $\sqrt{22}$, for example, cannot be simplified because 22 has no perfect square factors.
- Surds can be simplified in more than one step.

$$
\begin{aligned}
\sqrt{72} &= \sqrt{4} \times \sqrt{18} \\
&= 2\sqrt{18} \\
&= 2 \times \sqrt{9} \times \sqrt{2} \\
&= 2 \times 3\sqrt{2} \\
&= 6\sqrt{2}
\end{aligned}
$$

WORKED EXAMPLE 32 Simplifying surds

Simplify the following surds.

a. $\sqrt{18}$

b. $6\sqrt{20}$

THINK	WRITE
a. 1. Rewrite 18 as the product of two numbers, one of which is a perfect square (9).	a. $\sqrt{18} = \sqrt{9} \times \sqrt{2}$
2. Simplify.	$= 3 \times \sqrt{2}$ $= 3\sqrt{2}$
b. 1. Rewrite 20 as the product of two numbers, one of which is square (4).	b. $6\sqrt{20} = 6 \times \sqrt{4} \times \sqrt{5}$
2. Simplify.	$= 6 \times 2 \times \sqrt{5}$ $= 12\sqrt{5}$

▶ 1.9.4 Entire surds

eles-4396

- The surd $\sqrt{45}$, when simplified, is written as $3\sqrt{5}$.
  The surd $\sqrt{45}$ is called an entire surd because it is written entirely inside the radical sign. The surd $3\sqrt{5}$, however, is not an entire surd.
- Writing a surd as an entire surd reverses the process of simplification.

WORKED EXAMPLE 33 Writing entire surds

Write $3\sqrt{7}$ as an entire surd.

THINK	WRITE
1. In order to place the 3 inside the radical sign it has to be written as $\sqrt{9}$.	$3\sqrt{7} = \sqrt{9} \times \sqrt{7}$
2. Apply the rule $\sqrt{a} \times \sqrt{b} = \sqrt{ab}$.	$= \sqrt{63}$

Determine which number is larger, $3\sqrt{5}$ or $5\sqrt{3}$.

THINK	WRITE
1. Write $3\sqrt{5}$ as an entire surd.	$3\sqrt{5} = \sqrt{9} \times \sqrt{5}$ $= \sqrt{45}$
2. Write $5\sqrt{3}$ as an entire surd.	$5\sqrt{3} = \sqrt{25} \times \sqrt{3}$ $= \sqrt{75}$
3. Compare the values of each surd.	$\sqrt{75} > \sqrt{45}$
4. Write your answer.	$5\sqrt{3}$ is the larger number.

1.9.5 Addition and subtraction of surds

eles-4397

- Surds can be added or subtracted if they have like terms.
- Surds should be simplified before adding or subtracting like terms.

Simplify each of the following.

a. $6\sqrt{3} + 2\sqrt{3} + 4\sqrt{5} - 5\sqrt{5}$

b. $3\sqrt{2} - 5 + 4\sqrt{2} + 9$

THINK	WRITE
a. Collect the like terms ($\sqrt{3}$ and $\sqrt{5}$).	a. $6\sqrt{3} + 2\sqrt{3} + 4\sqrt{5} - 5\sqrt{5}$ $= 8\sqrt{3} - \sqrt{5}$
b. Collect the like terms and simplify.	b. $3\sqrt{2} - 5 + 4\sqrt{2} + 9$ $= 3\sqrt{2} + 4\sqrt{2} - 5 + 9$ $= 7\sqrt{2} + 4$

Simplify $5\sqrt{75} - 6\sqrt{12} + \sqrt{8} - 4\sqrt{3}$.

THINK	WRITE
1. Simplify $5\sqrt{75}$.	$5\sqrt{75} = 5 \times \sqrt{25} \times \sqrt{3}$ $= 5 \times 5 \times \sqrt{3}$ $= 25\sqrt{3}$

2. Next, simplify $6\sqrt{12}$.

$$6\sqrt{12} = 6 \times \sqrt{4} \times \sqrt{3}$$
$$= 6 \times 2 \times \sqrt{3}$$
$$= 12\sqrt{3}$$

3. Lastly, simplify $\sqrt{8}$.

$$\sqrt{8} = \sqrt{4} \times \sqrt{2}$$
$$= 2\sqrt{2}$$

4. Rewrite the original expression and simplify by adding like terms.

$$5\sqrt{75} - 6\sqrt{12} + \sqrt{8} - 4\sqrt{3}$$
$$= 25\sqrt{3} - 12\sqrt{3} + 2\sqrt{2} - 4\sqrt{3}$$
$$= 9\sqrt{3} + 2\sqrt{2}$$

TI | THINK	WRITE	CASIO | THINK	WRITE
In a new problem, on a Calculator page, complete the entry line as: $5\sqrt{75} - 6\sqrt{12} + \sqrt{8} - 4\sqrt{3}$ Press ENTER.	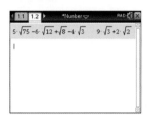 $$5\sqrt{75} - 6\sqrt{12} + \sqrt{8} - 4\sqrt{3}$$ $$9\sqrt{3} + 2\sqrt{2}$$	Make sure the calculator is set to Standard mode. On the Main screen, complete the entry line as: $5\sqrt{75} - 6\sqrt{12} + \sqrt{8} - 4\sqrt{3}$ Press EXE.	 $$5\sqrt{75} - 6\sqrt{12} + \sqrt{8} - 4\sqrt{3}$$ $$9\sqrt{3} + 2\sqrt{2}$$

DISCUSSION

Are all square root numbers surds?

 Resources

eWorkbook	Topic 1 Workbook (worksheets, code puzzle and project) (ewbk-2001)
Digital documents	SkillSHEET Calculating the square root of a number (doc-6101)
	SkillSHEET Using a calculator to evaluate numbers in index form (doc-6102)
	SkillSHEET Simplifying surds (doc-10813)
Video eLesson	Surds (eles-1906)
Interactivities	Individual pathway interactivity: Surds (int-4477)
	Simplifying surds (int-6028)
	Surds on the number line (int-6029)

Exercise 1.9 Surds

learnon

Individual pathways

■ PRACTISE	■ CONSOLIDATE	■ MASTER
1, 5, 7, 10, 13, 16, 19, 21, 23, 26, 29, 33, 35, 39	2, 4, 9, 11, 15, 17, 22, 24, 28, 30, 31, 36, 40	3, 6, 8, 12, 14, 18, 20, 25, 27, 32, 34, 37, 38, 41, 42

To answer questions online and to receive **immediate corrective feedback** and **fully worked solutions** for all questions, go to your learnON title at www.jacplus.com.au.

Fluency

1. Write down the square roots of each of the following.
   a. 1
   b. 4
   c. 0
   d. $\dfrac{1}{9}$
   e. $1\dfrac{9}{16}$

2. Write down the square roots of each of the following.
   a. 0.16
   b. 400
   c. 10 000
   d. $\dfrac{4}{25}$
   e. 1.44

3. Write down the square roots of each of the following.
   a. 20.25
   b. 1 000 000
   c. 0.0009
   d. 256
   e. $2\dfrac{23}{49}$

4. Write down the value of each of the following.
   a. $\sqrt{81}$
   b. $-\sqrt{81}$
   c. $\sqrt{121}$
   d. $-\sqrt{441}$

5. Write down the value of each of the following.
   a. $\sqrt[3]{8}$
   b. $\sqrt[3]{64}$
   c. $\sqrt[3]{343}$
   d. $\sqrt[4]{81}$

6. Write down the value of each of the following.
   a. $\sqrt[5]{1024}$
   b. $\sqrt[3]{125}$
   c. $-\sqrt{49}$
   d. $\sqrt[3]{-27}$

7. **WE28** Identify which of the following are surds.
   a. $\sqrt[3]{0}$
   b. $\sqrt{10}$
   c. $-\sqrt{36}$
   d. $\sqrt[3]{9}$

8. Identify which of the following are surds.
   a. $-\sqrt[3]{216}$
   b. $\sqrt[3]{-216}$
   c. $-\sqrt{2}$
   d. $1+\sqrt{2}$

9. Identify which of the following are surds.
   a. 1.32
   b. $1.\dot{3}\dot{2}$
   c. $\sqrt[4]{64}$
   d. 1.752 16

10. **WE29** Simplify each of the following.
    a. $\sqrt{3}\times\sqrt{7}$
    b. $-\sqrt{3}\times\sqrt{7}$
    c. $2\times\sqrt{6}$
    d. $2\times3\sqrt{7}$
    e. $2\sqrt{7}\times5\sqrt{2}$

11. Simplify each of the following.
    a. $3\sqrt{7}\times4$
    b. $\sqrt{7}\times9$
    c. $-2\sqrt{5}\times-11\sqrt{2}$
    d. $2\sqrt{3}\times11$
    e. $\sqrt{3}\times\sqrt{3}$

12. Simplify each of the following.
    a. $-3\sqrt{2}\times-5\sqrt{5}$
    b. $2\sqrt{3}\times4\sqrt{3}$
    c. $\sqrt{6}\times\sqrt{6}$
    d. $\sqrt{11}\times\sqrt{11}$
    e. $\sqrt{15}\times2\sqrt{15}$

13. **WE30** Simplify each of the following.

  a. $\sqrt{\dfrac{12}{4}}$ b. $-\sqrt{\dfrac{10}{5}}$

  c. $\dfrac{\sqrt{18}}{\sqrt{3}}$ d. $\dfrac{-\sqrt{15}}{-\sqrt{3}}$

14. Simplify each of the following.

  a. $\dfrac{15\sqrt{6}}{5\sqrt{2}}$ b. $\dfrac{15\sqrt{6}}{10}$

  c. $\dfrac{15\sqrt{6}}{\sqrt{3}}$ d. $\dfrac{5}{3}\sqrt{\dfrac{15}{3}}$

15. Simplify each of the following.

  a. $\dfrac{-10\sqrt{10}}{5\sqrt{2}}$ b. $\dfrac{\sqrt{9}}{\sqrt{3}}$ c. $\dfrac{3}{\sqrt{3}}$ d. $\dfrac{7}{\sqrt{7}}$

16. **WE31** Simplify each of the following.

  a. $\sqrt{20}$ b. $\sqrt{8}$ c. $\sqrt{18}$ d. $\sqrt{49}$

17. Simplify each of the following.

  a. $\sqrt{30}$ b. $\sqrt{50}$ c. $\sqrt{28}$ d. $\sqrt{108}$

18. Simplify each of the following

  a. $\sqrt{288}$ b. $\sqrt{48}$ c. $\sqrt{500}$ d. $\sqrt{162}$

19. **WE32** Simplify each of the following.

  a. $2\sqrt{8}$ b. $5\sqrt{27}$ c. $6\sqrt{64}$ d. $7\sqrt{50}$ e. $10\sqrt{24}$

20. Simplify each of the following.

  a. $5\sqrt{12}$ b. $4\sqrt{42}$ c. $12\sqrt{72}$ d. $9\sqrt{45}$ e. $12\sqrt{242}$

21. **WE33** Write each of the following in the form $\sqrt{a}$ (that is, as an entire surd).

  a. $2\sqrt{3}$ b. $5\sqrt{7}$ c. $6\sqrt{3}$ d. $4\sqrt{5}$ e. $8\sqrt{6}$

22. Write each of the following in the form $\sqrt{a}$ (that is, as an entire surd).

  a. $3\sqrt{10}$ b. $4\sqrt{2}$ c. $12\sqrt{5}$ d. $10\sqrt{6}$ e. $13\sqrt{2}$

23. **WE35** Simplify each of the following.

  a. $6\sqrt{2}+3\sqrt{2}-7\sqrt{2}$ b. $4\sqrt{5}-6\sqrt{5}-2\sqrt{5}$

  c. $-3\sqrt{3}-7\sqrt{3}+4\sqrt{3}$ d. $-9\sqrt{6}+6\sqrt{6}+3\sqrt{6}$

24. Simplify each of the following.

  a. $10\sqrt{11}-6\sqrt{11}+\sqrt{11}$ b. $\sqrt{7}+\sqrt{7}$

  c. $4\sqrt{2}+6\sqrt{2}+5\sqrt{3}+2\sqrt{3}$ d. $10\sqrt{5}-2\sqrt{5}+8\sqrt{6}-7\sqrt{6}$

25. Simplify each of the following.

  a. $5\sqrt{10}+2\sqrt{3}+3\sqrt{10}+5\sqrt{3}$ b. $12\sqrt{2}-3\sqrt{5}+4\sqrt{2}-8\sqrt{5}$

  c. $6\sqrt{6}+\sqrt{2}-4\sqrt{6}-\sqrt{2}$ d. $16\sqrt{5}+8+7-11\sqrt{5}$

26. **WE36** Simplify each of the following.
    a. $\sqrt{8} + \sqrt{18} - \sqrt{32}$
    b. $\sqrt{45} - \sqrt{80} + \sqrt{5}$
    c. $-\sqrt{12} + \sqrt{75} - \sqrt{192}$
    d. $\sqrt{7} + \sqrt{28} - \sqrt{343}$
    e. $\sqrt{24} + \sqrt{180} + \sqrt{54}$

27. Simplify each of the following.
    a. $\sqrt{12} + \sqrt{20} - \sqrt{125}$
    b. $2\sqrt{24} + 3\sqrt{20} - 7\sqrt{8}$
    c. $3\sqrt{45} + 2\sqrt{12} + 5\sqrt{80} + 3\sqrt{108}$
    d. $6\sqrt{44} + 4\sqrt{120} - \sqrt{99} - 3\sqrt{270}$
    e. $2\sqrt{32} - 5\sqrt{45} - 4\sqrt{180} + 10\sqrt{8}$

28. **MC** Choose the correct answer from the given options.
    a. $\sqrt{2} + 6\sqrt{3} - 5\sqrt{2} - 4\sqrt{3}$ is equal to:

    A. $-5\sqrt{2} + 2\sqrt{3}$ **B.** $-3\sqrt{2} + 23$ **C.** $6\sqrt{2} + 2\sqrt{3}$ **D.** $-4\sqrt{2} + 2\sqrt{3}$ **E.** $4\sqrt{2} + 2\sqrt{3}$

    b. $6 - 5\sqrt{6} + 4\sqrt{6} - 8$ is equal to:

    A. $-2 - \sqrt{6}$ **B.** $14 - \sqrt{6}$ **C.** $-2 + \sqrt{6}$ **D.** $-2 - 9\sqrt{6}$ **E.** $2 + 9\sqrt{6}$

    c. $4\sqrt{8} - 6\sqrt{12} - 7\sqrt{18} + 2\sqrt{27}$ is equal to:

    A. $-7\sqrt{5}$ **B.** $29\sqrt{2} - 18\sqrt{3}$ **C.** $-13\sqrt{2} - 6\sqrt{3}$ **D.** $-13\sqrt{2} + 6\sqrt{3}$ **E.** $7\sqrt{5}$

    d. $2\sqrt{20} + 5\sqrt{24} - \sqrt{54} + 5\sqrt{45}$ is equal to:

    A. $19\sqrt{5} + 7\sqrt{6}$ **B.** $9\sqrt{5} - 7\sqrt{6}$ **C.** $-11\sqrt{5} + 7\sqrt{6}$ **D.** $-11\sqrt{5} - 7\sqrt{6}$ **E.** $-9\sqrt{5} - 7\sqrt{6}$

Understanding

29. **MC** Choose the correct answer from the given options.
    a. $\sqrt{1000}$ is equal to:

    A. 31.6228 **B.** $50\sqrt{2}$ **C.** $50\sqrt{10}$ **D.** $10\sqrt{10}$ **E.** $100\sqrt{10}$

    b. $\sqrt{80}$ in simplest form is equal to:

    A. $4\sqrt{5}$ **B.** $2\sqrt{20}$ **C.** $8\sqrt{10}$ **D.** $5\sqrt{16}$ **E.** $2 + 16\sqrt{5}$

    c. Choose which of the following surds is in simplest form.

    A. $\sqrt{60}$ **B.** $\sqrt{147}$ **C.** $\sqrt{105}$ **D.** $\sqrt{117}$ **E.** $\sqrt{20}$

    d. Choose which of the following surds is **not** in simplest form.

    A. $\sqrt{102}$ **B.** $\sqrt{110}$ **C.** $\sqrt{116}$ **D.** $\sqrt{118}$ **E.** $\sqrt{105}$

30. **MC** Choose the correct answer from the given options.
    a. $6\sqrt{5}$ is equal to:

    A. $\sqrt{900}$ **B.** $\sqrt{30}$ **C.** $\sqrt{150}$ **D.** $\sqrt{180}$ **E.** $\sqrt{1800}$

    b. Choose the expression out of the following that is **not** equal to the rest.

    A. $\sqrt{128}$ **B.** $2\sqrt{32}$ **C.** $8\sqrt{2}$ **D.** $64\sqrt{2}$ **E.** $6\sqrt{2}$

    c. Choose the expression out of the following that is **not** equal to the rest.

    A. $4\sqrt{4}$ **B.** $2\sqrt{16}$ **C.** 8 **D.** 16 **E.** 32

    d. $5\sqrt{48}$ is equal to:

    A. $80\sqrt{3}$ **B.** $20\sqrt{3}$ **C.** $9\sqrt{3}$ **D.** $21\sqrt{3}$ **E.** $10\sqrt{3}$

31. Reduce each of the following to simplest form.
    a. $\sqrt{675}$ b. $\sqrt{1805}$ c. $\sqrt{1792}$ d. $\sqrt{578}$

32. Reduce each of the following to its simplest form.

a. $\sqrt{a^2c}$ **b.** $\sqrt{bd^4}$ **c.** $\sqrt{h^2jk^2}$ **d.** $\sqrt{f^3}$

33. **WE11** Determine which number in each of the following pairs of numbers is larger.

a. $\sqrt{10}$ or $2\sqrt{3}$ **b.** $3\sqrt{5}$ or $5\sqrt{2}$ **c.** $10\sqrt{2}$ or $4\sqrt{5}$ **d.** $2\sqrt{10}$ or $\sqrt{20}$

34. Write the following sequences of numbers in order from smallest to largest.

a. $6\sqrt{2}$, 8, $2\sqrt{7}$, $3\sqrt{6}$, $4\sqrt{2}$, $\sqrt{60}$
b. $\sqrt{6}$, $2\sqrt{2}$, $\sqrt{2}$, 3, $\sqrt{3}$, 2, $2\sqrt{3}$

Reasoning

35. The formula for calculating the speed of a car before it brakes in an emergency is $v = \sqrt{20d}$, where v is the speed in m/s and d is the braking distance in metres.
Calculate the speed of a car before braking if the braking distance is 32.50 m.
Write your answer as a surd in its simplest form.

36. A gardener wants to divide their vegetable garden into 10 equal squares, each of area exactly $2\,\text{m}^2$.

a. Calculate the exact side length of each square. Explain your reasoning.
b. Determine the side length of each square, correct to 2 decimal places.
c. The gardener wishes to group the squares in their vegetable garden next to each other so as to have the smallest perimeter possible for the whole garden.
   i. Determine how they should arrange the squares.
   ii. Explain why the exact perimeter of the vegetable patch is $14\sqrt{2}\,\text{m}$.
   iii. Calculate the perimeter, correct to 2 decimal places.

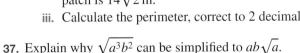

37. Explain why $\sqrt{a^3b^2}$ can be simplified to $ab\sqrt{a}$.

38. Determine the smallest values that a and b can have, given that they are both natural numbers, that $1 < a < b$, and also that $\sqrt{ab}$ is not a surd.

Problem solving

39. Kyle wanted a basketball court in his backyard, but he could not fit a full-sized court in his yard. He was able to get a rectangular court laid with a width of $6\sqrt{2}\,\text{m}$ and a length of $3\sqrt{10}\,\text{m}$. Calculate the area of the basketball court and represent it in its simplest surd form.

40. A netball team went on a pre-season training run. They completed 10 laps of a triangular course that had side lengths of $(200\sqrt{3} + 50)$ m, $(50\sqrt{2} + 75\sqrt{3})$ m and $(125\sqrt{2} - 18)$ m. Determine the distance they ran, in its simplest surd form.

41. The area of a square is x cm². Explain whether the side length of the square would be a rational number.

42. To calculate the length of the hypotenuse of a right-angled triangle, use the formula $c^2 = a^2 + b^2$.

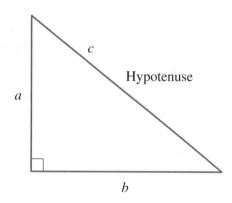

a. Calculate the length of the hypotenuse for triangles with other side lengths as given in the following pairs of values.

   i. $\sqrt{5}, \sqrt{8}$ **ii.** $\sqrt{2}, \sqrt{7}$ **iii.** $\sqrt{15}, \sqrt{23}$

b. Describe any pattern in your answers to part **a**.

c. Calculate the length of the hypotenuse (without calculations) for triangles with the following side lengths.

   i. $\sqrt{1000}, \sqrt{500}$ **ii.** $\sqrt{423}, \sqrt{33}$ **iii.** $\sqrt{124}, \sqrt{63}$

d. Your friend wrote down the following explanation.
$$\sqrt{b} + \sqrt{c} = \sqrt{a}$$
Is this answer correct? If not, explain what the correct answer should be.

1.10 Review

1.10.1 Topic summary

Natural numbers

- The natural numbers are the positive whole numbers, e.g. 1, 2, 3, 4 ... They were our first counting numbers.
- The set of natural numbers is represented by the symbol N.

Integers

- Integers are the whole number values on the number line. They include the positive and negative whole numbers as well as zero, e.g. $-2, -1, 0, 1, 2$.
- The set of integers is represented by the symbol Z.

Rational numbers

- Rational numbers are numbers that can be expressed as a whole number fraction, i.e. as $\frac{a}{b}$, where a and b are both integers and $b \neq 0$.
- The symbol Q is used for the set of rational numbers.
- The set of rational numbers includes the integers as well as terminating decimals and recurring decimals.
  e.g. $\frac{3}{7}$, $-5 \left(\frac{-5}{1}\right)$, $3.2 \left(\frac{16}{5}\right)$, $-0.036 \left(\frac{-36}{1000}\right)$

Rounding to significant figures

- In a decimal number, the significant figures start at the first non-zero digit.
- To round a decimal number to significant figures:
  1. Start at the first non-zero digit.
  2. Write down the digits that follow until you have reached the required number of significant figures. (You may need to round off the final digit.)
  3. Leave out the digits that come after the last significant figure that is required.
  e.g. 1.425 has 3 decimal places but 4 significant figures.

Recurring decimals

- A recurring decimal is a decimal number with digits or groups of digits that repeat themselves endlessly.
- When writing recurring decimals, we place dots above the first and last repeating digits.
  For example, 3.333333 … becomes $3.\dot{3}$ and 11.923 923 923 … becomes $11.\dot{9}2\dot{3}$.

Square and cube roots

- A square root is the inverse of squaring (raising to the power of 2).
- Every number has a positive and a negative square root.
  e.g. $2^2 = 4$ and $-2^2 = 4$
- The cube root is the inverse of cubing (raising to the power of 3).

Rounding to decimal places

- Identify the rounding digit.
- If the digit after the rounding digit is 0, 1, 2, 3 or 4, leave the rounding digit as it is.
- If the digit after the rounding digit is 5, 6, 7, 8 or 9, increase the rounding digit by 1.
- Leave out all digits that come after the rounding digit.

Rounding

- Rounding is used to write approximate values of irrational numbers, recurring decimals or long terminating decimals.
- Numbers can be rounded to a set number of decimal places or to a set number of significant figures.

Transcendental numbers

- Transcendental numbers are irrational numbers that have special mathematical significance.
- Pi (π) is an example of a transcendental number.

Real numbers

- Real numbers lie on the number line.
- The set of real numbers includes the irrational numbers and the rational numbers.
- The symbol R is used for real numbers.

Irrational numbers

- Irrational numbers are numbers that result in decimals that do not terminate and do not have repeated patterns of digits.
  e.g. $\sqrt{18} = 4.242\ 640\ 687\ 12...$
- The symbol I is used for the set of irrational numbers.

NUMBER SKILLS AND INDEX LAWS

Index laws

- There are a number of index laws:
  - $a^m \times a^n = a^{m+n}$
  - $a^m \div a^n = a^{m-n}$
  - $a^0 = 1$, where $a \neq 0$
  - $(a^m)^n = a^{m \times n}$
  - $(a \times b)^m = a^m \times b^m$
  - $\left(\frac{a}{b}\right)^m = \frac{a^m}{b^m}$
  - $a^{-n} = \frac{1}{a^n}$, where $a \neq 0$
  - $a^{\frac{1}{n}} = \sqrt[n]{a}$.

Scientific notation

- Scientific notation is used to write very large and very small numbers.
- A number written in scientific notation has the form: (number between 1 and 10) × (a power of 10).
- Large numbers have powers of 10 with positive indices, while small numbers have powers of 10 with negative indices. For example:
  - 8079 can be written as 8.079×10^3
  - 0.000 15 can be written as 1.5×10^{-4}.

Surds

- Surds result when the root of a number is an irrational number.
  e.g. $\sqrt{10}$
  $\approx 3.162\ 277\ 660\ 17 ...$
- They are written with a root (radical) sign in them.
  - $\sqrt{a} \times \sqrt{b} = \sqrt{ab}$
  - $\frac{\sqrt{a}}{\sqrt{b}} = \sqrt{\frac{a}{b}}$

1.10.2 Success criteria

Tick the column to indicate that you have completed the subtopic and how well you have understood it using the traffic light system.

(**Green:** I understand; **Yellow:** I can do it with help; **Red:** I do not understand)

Subtopic	Success criteria			
1.2	I can identify real numbers that are natural, integer, rational or irrational.			
	I can recall and use correct symbols for the sets of real, natural, integer, rational or irrational numbers.			
	I can write integers and decimals as rational number fractions.			
	I can write recurring decimals using dot notation.			
	I can explain the difference between recurring and terminating decimals.			
1.3	I can round decimals to a given number of decimal places.			
	I can identify how many significant figures a decimal number has.			
	I can round decimals to a given number of significant figures.			
1.4	I can identify the base and the index of a number written in index form.			
	I can recall the first three index laws.			
	I can use the first three index laws to simplify expressions involving indices.			
1.5	I can apply the Fourth, Fifth and Sixth index laws to raise a power to another power.			
	I can apply the first six index laws to simplify expressions involving indices.			
1.6	I can apply the Seventh Index Law to evaluate expressions using negative indices.			
	I can apply the first seven index laws to simplify expressions involving indices.			
1.7	I can identify numbers that are written in scientific notation.			
	I can write numbers in scientific notation as decimal numbers.			
	I can write decimal numbers using scientific notation.			
1.8	I can write square and cube roots using the radical symbol.			
	I can use fractional indices to represent square roots and cube roots.			
	I can simplify expressions that use square roots and cube roots.			

(continued)

(continued)

1.9	I can identify when a square root forms a surd.			
	I can place surds in their approximate position on a number line.			
	I can simplify expressions containing surds.			

1.10.3 Project

Concentric squares

Consider a set of squares drawn around a central point on a grid, as shown.

These squares can be regarded as concentric squares, because the central point of each square is the point labelled **X**. The diagram shows four labelled squares drawn on one-centimetre grid paper.

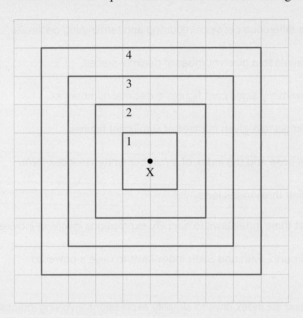

1. Use the diagram above to complete the following table, leaving your answers in simplest surd form (if necessary). The first square has been completed for you.

Square	1	2	3	4
Side length (cm)	2			
Diagonal length (cm)	$2\sqrt{2}$			
Perimeter (cm)	8			
Area (cm²)	4			

Observe the patterns in the table and answer the following questions.

2. What would be the side length of the tenth square of this pattern?
3. What would be the length of the diagonal of this tenth square?

Consider a different arrangement of these squares on one-centimetre grid paper, as shown.

These squares all still have a central point, labelled **Y**.

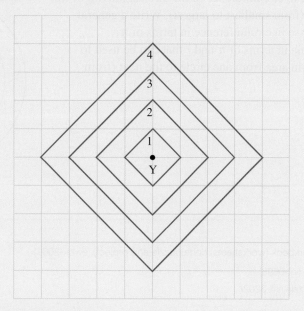

4. Use the diagram to complete the following table for these squares, leaving your answers in simplest surd form if necessary. The first square has been done for you.

Square	1	2	3	4
Side length (cm)	$\sqrt{2}$			
Diagonal length (cm)	2			
Perimeter (cm)	$4\sqrt{2}$			
Area (cm²)	2			

5. What would be the side length of the tenth square in this pattern of squares?
6. What would be the length of the diagonal of the tenth square?

Use the two diagrams above to answer the following questions.

7. In which of the two arrangements does the square have the greater side length?
8. Which of the two arrangements shows the squares with the greater diagonal length?
9. Compare the area of a square (your choice) in the first diagram with the area of the corresponding square in the second diagram.
10. Compare the perimeter of the same two corresponding squares in each of the two diagrams.
11. Examine the increase in area from one square to the next in these two diagrams. What increase in area would you expect from square 7 to square 8 in each of the two diagrams?

Consider a set of concentric circles around a centre labelled **Z**, drawn on one-centimetre grid paper, as shown.

12. Investigate the change in circumference of the circles, moving from one circle to the next (from smallest to largest). Write a general formula for this increase in circumference in terms of π.

13. Write a general formula in terms of π and r that can be used to calculate the increase in area from one circle to the next (from smallest to largest).

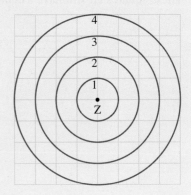

Exercise 1.10 Review questions

learnon

To answer questions online and to receive **immediate corrective feedback** and **fully worked solutions** for all questions, go to your learnON title at www.jacplus.com.au.

Fluency

1. **MC** Choose which of the given numbers are rational: $\sqrt{\dfrac{6}{12}}$, $\sqrt{0.81}$, 5, -3.26, $0.\dot{5}$, $\dfrac{\pi}{5}$, $\sqrt{\dfrac{3}{12}}$.

   A. $\sqrt{0.81}$, 5, -3.26, $0.\dot{5}$ and $\sqrt{\dfrac{3}{12}}$ B. $\sqrt{\dfrac{6}{12}}$ and $\dfrac{\pi}{5}$

   C. $\sqrt{\dfrac{6}{12}}$, $\sqrt{0.81}$ and $\sqrt{\dfrac{3}{12}}$ D. 5, -3.26 and $\sqrt{\dfrac{6}{12}}$

   E. $\sqrt{\dfrac{6}{12}}$, 5 and $\sqrt{0.81}$

2. For each of the following, state whether the number is rational or irrational and give the reason for your answer.
   a. $\sqrt{12}$ b. $\sqrt{121}$ c. $\dfrac{2}{9}$ d. $0.\dot{6}$ e. $\sqrt[3]{0.08}$

3. **MC** Choose which of the numbers in the set $\left\{3\sqrt{2},\ 5\sqrt{7},\ 9\sqrt{4},\ 6\sqrt{10},\ 7\sqrt{12},\ 12\sqrt{64}\right\}$ are surds.

   A. $9\sqrt{4}$, $12\sqrt{64}$ B. $3\sqrt{2}$ and $7\sqrt{12}$ only

   C. $3\sqrt{2}$, $5\sqrt{7}$ and $6\sqrt{10}$ only   D. $3\sqrt{2}$, $5\sqrt{7}$, $6\sqrt{10}$ and $7\sqrt{12}$

   E. $9\sqrt{4}$ and $7\sqrt{12}$

4. State which of the following are not irrational numbers.
   a. $\sqrt{7}$ b. $\sqrt{8}$ c. $\sqrt{81}$ d. $\sqrt{361}$

5. **MC** Choose which of the following represents the number 0.000 67 written using scientific notation.

   A. 67×10^{-5} B. 0.67×10^{-3} C. 6.7×10^{-4} D. 6.7×10^{-5} E. 6.7×10^{-3}

6. **MC** Choose which of the following is a simplification of the expression $\sqrt{250}$.

   A. $25\sqrt{10}$ B. $5\sqrt{10}$ C. $10\sqrt{5}$ D. $5\sqrt{50}$ E. $10\sqrt{25}$

7. Simplify each of the following.
   a. $b^7 \times b^3$
   b. $m^9 \times m^2$
   c. $k^3 \times k^5$
   d. $f^2 \times f^8 \times f^4$
   e. $h^4 \times h^5 \times h$

8. Simplify each of the following.
   a. $2q^5 \times 3q^2 \times q^{10}$
   b. $5w^3 \times 7w^{12} \times w^{14}$
   c. $2e^2 p^3 \times 6e^3 p^5$
   d. $5a^2 b^4 \times 3a^8 b^5 \times 7a^6 b^8$
   e. $a^5 \div a^2$

9. Simplify each of the following.
   a. $t^5 \div t$
   b. $r^{19} \div r^{12}$
   c. $p^8 \div p^5$
   d. $\dfrac{f^{17}}{f^{12}}$
   e. $\dfrac{y^{100}}{y^{10}}$

10. Simplify each of the following.
    a. $\dfrac{m^{24}}{m^{14}}$
    b. $\dfrac{g^4 \times g^5}{g^2}$
    c. $\dfrac{x^6 \times x^2 \times x}{x^8}$
    d. $\dfrac{d^6 \times d^7 \times d^2}{d^8}$
    e. $\dfrac{t^7 \times t \times t^3}{t^2 \times t^4}$

11. Simplify each of the following.
    a. $\dfrac{p^5 \times p^3 \times p \times p^4}{p^2 \times p^4 \times p^2}$
    b. $\dfrac{16k^{13}}{21} \div \dfrac{8k^9}{42}$
    c. $\dfrac{22b^{15}}{c} \div \dfrac{2b^8}{c^6}$
    d. $\dfrac{9d^8}{16e^{10}} \div \dfrac{2d^{10}}{e^{16}}$
    e. $\dfrac{2a^3 \times ab}{4 \times a \times b}$

12. Simplify each of the following.
    a. 5^0
    b. 12^0
    c. 345^0
    d. q^0
    e. r^0

13. Simplify each of the following.
    a. ab^0
    b. $3w^0$
    c. $5q^0 - 2q^0$
    d. $100s^0 + 99t^0$
    e. $a^7 b^0$

14. Simplify each of the following.
    a. $v^{10}w^0$
    b. prt^0
    c. $a^9 b^4 c^0$
    d. $j^8 k^0 m^3$
    e. $4e^2 f^0 - 36\left(a^2 b^3\right)^0$

15. Simplify each of the following.
    a. $2x^0 \times 3xy^2$
    b. $-8\left(18x^2 y^4 z^6\right)^0$
    c. $15 - 12x\left(\dfrac{3x}{8}\right)^0$
    d. $-4p^0 \times 6\left(q^2 r^3\right)^0 \div 8\left(-12q^2\right)^0$
    e. $3\left(6w^0\right)^2 \div 2\left(5w^5\right)^0$

16. Raise each of the following to the given power.
    a. $\left(b^4\right)^2$
    b. $\left(a^8\right)^3$
    c. $\left(k^7\right)^{10}$
    d. $\left(j^{100}\right)^2$

▶

17. **MC** $\left(\dfrac{4b^4}{d^2}\right)^3$ is equal to:

   A. $\dfrac{4b^3}{d^3}$
   B. $\dfrac{12b^{12}}{d^6}$
   C. $\dfrac{64b^{12}}{d^6}$
   D. $\dfrac{64b^7}{d^5}$
   E. $\dfrac{64b^{16}}{d^5}$

18. Raise each of the following to the given power.

   a. $(a^5b^2)^3$
   b. $(m^7n^{12})^2$
   c. $(st^6)^3$
   d. $(qp^{30})^{10}$

19. Write each of the following with positive indices.

   a. a^{-1}
   b. k^{-4}
   c. $4m^3 \div 2m^7$
   d. $7x^3y^{-4} \times 6x^{-3}y^{-1}$

20. Write each of the following using a negative index.

   a. $\dfrac{1}{x}$
   b. $\dfrac{2}{y^4}$
   c. $z \div z^4$
   d. $45p^2q^{-4} \times 3p^{-5}q$

21. Simplify each of the following.

   a. $\sqrt{100}$
   b. $\sqrt{36}$
   c. $\sqrt{a^2}$
   d. $\sqrt{b^2}$
   e. $\sqrt{49f^4}$

22. Simplify each of the following.

   a. $\sqrt[3]{27}$
   b. $\sqrt[3]{1000}$
   c. $\sqrt[3]{x^3}$
   d. $\sqrt[3]{8d^3}$
   e. $\sqrt[3]{64f^6g^3}$

Problem solving

23. If your body produces about 2.0×10^{11} red blood cells each day, determine how many red blood cells your body has produced after one week.

24. If your body produces about 1.0×10^{10} white blood cells each day, determine how many white blood cells your body produces each hour.

25. If Earth is approximately 1.496×10^8 km from the Sun, and Mercury is about 5.8×10^7 km from the Sun, evaluate the distance from Mercury to Earth.

26. Calculate the closest distance between Mars and Saturn if Mars is 2.279×10^8 km from the Sun and Saturn is 1.472×10^9 km from the Sun.

27. Consider the numbers 6×10^9, 3.4×10^7 and 1.5×10^9.
   a. Place these numbers in order from smallest to largest.
   b. If the population of Oceania is 3.4×10^7 and the world population is 6×10^9, determine the percentage of the world population that is represented by the population of Oceania. Give your answer to 2 decimal places.
   c. Explain what you need to multiply 1.5×10^9 by to get 6×10^9.
   d. Assume the surface area of Earth is approximately $1.5 \times 10^9 \text{km}^2$, and that the world's population is about 6×10^9. If we divided Earth up into equal portions, determine how much surface area each person in the world would get.

28. The total area of the state of Victoria is $2.27 \times 10^5 \text{ km}^2$. If the total area of Australia is $7.69 \times 10^6 \text{ km}^2$, calculate the percentage of the total area of Australia that is occupied by the state of Victoria. Write your answer correct to two decimal places.

29. A newly discovered colony of bees contains 2.05×10^8 bees. If 0.4% of these bees are estimated to be queen bees, calculate how many queen bees live in the colony.

30. Third-generation fibre optics that can carry up to 10 trillion (1×10^{12}) bits of data per second along a single strand of fibre have recently been tested. Based on the current rate of improvement in this technology, the amount of data that one strand of fibre can carry is tripling every 6 months. This trend is predicted to continue for the next 20 years.
Calculate how many years it will take for the transmission speed along a single strand of fibre to exceed 6×10^{16} bits per second.

31. Mach speed refers to the speed of sound, where Mach 1 is the speed of sound (about 343 metres per second (m/s)). Mach 2 is double the speed of sound, Mach 3 is triple the speed of sound, and so on. Earth has a circumference of about 4.0×10^4 km.

  a. The fastest recorded aircraft speed is 11 270 kilometres per hour (km/h). Express this as a Mach speed, correct to 1 decimal place.
  b. Create and complete a table for three different speeds (Mach 1, 2 and 3) with columns labeled 'Mach number', 'Speed in m/s' and 'Speed in km/h'.
  c. Calculate how long it would take something travelling at Mach 1, 2 and 3 to circle Earth. Give your answer in hours, correct to 2 decimal places.
  d. Evaluate how long it would take for the world's fastest aircraft to circle Earth. Give your answer in hours, correct to 2 decimal places.
  e. Evaluate how many times the world's fastest aircraft could circle Earth in the time it takes an aircraft travelling at Mach 1, 2 and 3 to circle Earth once. Give your answers correct to 2 decimal places.
  f. Explain the relationship between your answers in part **e** and their corresponding Mach value.

on To test your understanding and knowledge of this topic, go to your learnON title at www.jacplus.com.au and complete the **post-test**.

Online Resources

 Resources

Below is a full list of **rich resources** available online for this topic. These resources are designed to bring ideas to life, to promote deep and lasting learning and to support the different learning needs of each individual.

eWorkbook

Download the workbook for this topic, which includes worksheets, a code puzzle and a project (ewbk-2001) ☐

Solutions

Download a copy of the fully worked solutions to every question in this topic (sol-0721) ☐

Digital documents

1.2 SkillSHEET Operations with directed numbers (doc-6100) ☐
1.3 SkillSHEET Rounding to a given number of significant figures (doc-10814) ☐
1.4 SkillSHEET Index form (doc-6225) ☐
   SkillSHEET Using a calculator to evaluate numbers in index form (doc-6226) ☐
1.5 SkillSHEET Multiplying and dividing by powers of 10 (doc-10815) ☐
1.6 SkillSHEET Calculating the square root of a number (doc-6101) ☐
   SkillSHEET Using a calculator to evaluate numbers in index form (doc-6102) ☐
   SkillSHEET Simplifying surds (doc-10813) ☐

Video eLessons

1.2 Natural numbers (eles-4374) ☐
   Recurring decimals (eles-4377) ☐
   Irrational numbers (eles-4378) ☐
1.3 Rounding decimals (eles-4380) ☐
   Rounding to significant figures (eles-4381) ☐
1.4 Index notation (eles-4382) ☐
   Multiplication using indices (eles-4383) ☐
   Division using indices (eles-4384) ☐
   Zero index (eles-4385) ☐
   Cancelling fractions (eles-4386) ☐
   Index notation (eles-1903) ☐
1.5 More index laws (eles-4387) ☐
1.6 The Seventh Index Law (eles-4388) ☐
   Negative indices (eles-1910) ☐
1.7 Using scientific notation (eles-4389) ☐
   Converting scientific notation to decimal notation (eles-4390) ☐
1.8 Square root (eles-4391) ☐
   Cube root (eles-4392) ☐
1.9 Identifying surds (eles-4393) ☐
   Multiplying and dividing surds (eles-4394) ☐
   Simplifying surds (eles-4395) ☐
   Entire surds (eles-4396) ☐
   Addition and subtraction of surds (eles-4397) ☐
   Surds (eles-1906) ☐

Interactivities

1.2 Individual pathway interactivity: Rational numbers (int-4476) ☐
   The number system (int-6027) ☐
1.3 Individual pathway interactivity: Real numbers (int-4478) ☐
   Rounding to significant figures (int-6030) ☐
1.4 Individual pathway interactivity: Review of index laws (int-4516) ☐
   Review of index form (int-3708) ☐
   First Index Law (int-3709) ☐
   Second Index Law (int-3711) ☐
   Third Index Law (int-3713) ☐
1.5 Individual pathway interactivity: Raising a power to another power (int-4517) ☐
   Fourth Index Law (int-3716) ☐
   Fifth and sixth index laws (int-6063) ☐
1.6 Individual pathway interactivity: Negative indices (int-4518) ☐
   Negative indices (int-6064) ☐
1.7 Individual pathway interactivity: Scientific notation (int-4479) ☐
   Scientific notation (int-6031) ☐
1.8 Individual pathway interactivity: Square roots and cube roots (int-4519) ☐
   Cube roots (eles-6065) ☐
   Square roots (eles-6066) ☐
1.9 Individual pathway interactivity: Surds (int-4477) ☐
   Simplifying surds (int-6028) ☐
   Surds on the number line (int-6029) ☐
1.10 Crossword (int-0698) ☐
   Sudoku puzzle (int-3202) ☐

Teacher resources

There are many resources available exclusively for teachers online.

To access these online resources, log on to **www.jacplus.com.au**.

Answers

Topic 1 Number skills and index laws

Exercise 1.1 Pre-test

1. True

2. Recurring decimal

3. a. True   b. True

4. 3

5. C

6. B

7. $\sqrt{5}$

8. C

9. 7.439

10. 7.44

11. C

12. B

13. 1.672×10^{-27}

14. $\dfrac{4\sqrt{3}}{3}$

15. $\dfrac{5\sqrt{2}}{2}$

Exercise 1.2 Real numbers

1. a. $\dfrac{15}{1}$ b. $\dfrac{-8}{1}$ c. $\dfrac{8}{3}$ d. $\dfrac{-41}{8}$

2. a. $\dfrac{4}{1}$ b. $\dfrac{73}{10}$ c. $\dfrac{2}{1000}$ d. $\dfrac{872}{10}$

3. a. $\dfrac{0}{1}$ b. $\dfrac{156}{100}$ c. $\dfrac{3612}{1000}$ d. $\dfrac{-8}{100}$

4. a. 0.555 555 55   b. 0.515 151 51
   c. 0.511 111 11   d. 6.031 313 1
   e. 5.183 183 1

5. a. $-7.024\,444\,4$ b. 8.912 491 2
   c. 5.123 434 3   d. 5.123 412 3
   e. 3.002 020 2

6. a. $0.\dot{5}$ b. $0.\dot{2}\dot{7}$ c. $2.\dot{1}\dot{8}$
   d. $0.\dot{4}2857\dot{1}$ e. $-1.\dot{7}\dot{4}$

7. a. $0.0\dot{7}\dot{3}$ b. $-3.8\dot{3}$ c. $0.4\dot{6}$
   d. $0.4\dot{6}$ e. $0.0\dot{4}\dot{6}$

8. a. 15
   b. $-3, 0, 15$
   c. $-3, \dfrac{-3}{7}, 0, 2.\dot{3}, 2\dfrac{3}{5}, 15$
   d. $2.\dot{3}$ and $-\dfrac{3}{7}$

9. $2.\dot{1}, 2.\dot{1}\dot{2}\dot{1}, 2.1\dot{2}\dot{1}, 2.\dot{1}\dot{2}, 2.1\dot{2}$

10. Statements a and b are true.

11. Every integer a can be written as $\dfrac{a}{1}$ and therefore every integer is rational.

12. a. Q^-
   b. N^- implies a set of negative natural numbers. As the natural numbers include only positive numbers, this set would not exist.

13. Yes, because the division of an integer by another integer will always result in a rational number.

14. a. $\dfrac{1}{3}, \dfrac{2}{3}, \dfrac{1}{6}$ b. 6

15. Answers will vary. Example answers are shown.

   a. $\dfrac{1}{9}$ b. $\dfrac{10}{99}$ c. $\dfrac{100}{999}$ d. $\dfrac{1000}{9999}$

16. $3330.20

17. 657 bottles of water per year

Exercise 1.3 Rounding numbers and significant figures

1. a. 1.571   b. 2.236   c. 3.873
   d. 16.435   e. 5.111   f. 5.156

2. a. 5.152   b. 137.358   c. 0.429
   d. 0.077   e. 2.429   f. 1.000

3. a. 4.000   b. 0.254   c. 0.025
   d. 0.001   e. 5.000   f. 2342.156

4. a. 2   b. 3   c. 4   d. 3

5. a. 4   b. 4   c. 2   d. 1

6. a. 3   b. 4   c. 2   d. 3

7. a. 1.5708   b. 2.2361   c. 3.8730
   d. 16.435   e. 5.1111   f. 5.1556

8. a. 5.1515   b. 137.36   c. 0.428 57
   d. 0.076 923   e. 2.4286   f. 1.0000

9. a. 6.5813   b. 4.0000   c. 0.253 64
   d. 0.025 400   e. 0.000 913 60   f. 5.0000

10. 3.1416, 3.14159, 3.141593, 3.1415927

11. a. Yes   b. No   c. Yes   d. No

12. a. False   b. True   c. False   d. True

13. a. False   b. True   c. False   d. True

14. a. True   b. False   c. False   d. True

15. A number of answers are possible. Example answer: 2.004.

16. The 5th digit (5) causes the 4th digit to round up from 9 to 10, which causes the 3rd digit to round up from 9 to 10, which causes the 2nd digit to round up from 2 to 3.

17. a. 2 decimal places   b. 3.1416

18. a. $153.86\,\text{cm}^2$
   b. $153.9380\,\text{cm}^2$
   c. $153.94\,\text{cm}^2$
   d. Yes, because 3.14 is used as an estimate and is not the value of π as used on the calculator.

19. a. $65\,449.8470\,\text{cm}^3$
   b. 9
   c. Yes, any number with a finite number of decimal places is a rational number.

20. 13 000

Exercise 1.4 Review of index laws

1. a. $2^2 \times 3$ b. $2^3 \times 3^2$ c. 3×5^2
   d. $2^4 \times 3 \times 5$ e. $2^7 \times 5$ f. $2^3 \times 5^2 \times 7^2$

2. a. $20p^{11}$ b. $6x^8$ c. $56y^{10}$
   d. $21p^8$ e. $84t^6$ f. $30q^{15}$

3. a. $6a^6e^7$ b. $8p^6h^{12}$ c. $6k^{14}$ d. 3^{12}

4. a. $80m^9$ b. $6g^3h^6$ c. $30p^6q^9$ d. $48u^9w^7$

5. a. $27d^{11}y^{17}$ b. $42b^{14}c^9$ c. $24r^{16}s^{18}$ d. $60h^{38}v^{20}$

6. a. $3p^4$ b. $6r^4$ c. $9a^3$ d. $3b^6$

7. a. $20r^4$ b. $9q$ c. $\dfrac{130}{d^2}$ d. $7t^4$

8. a. $\dfrac{3p^5}{2}$ b. $\dfrac{8b^5}{3}$ c. $\dfrac{5m^{10}n^6}{6}$ d. $\dfrac{9x^8y}{4}$

9. a. $3j^5f^3$ b. $\dfrac{4p^2rs}{3}$ c. $\dfrac{9a^5b^3c}{2}$ d. $\dfrac{20f^6g^2h^4}{3}$

10. a. 1   b. 6   c. 1   d. 1   e. 5

11. a. 1   b. 1   c. 2   d. 2   e. 3

12. a. 1   b. 0   c. 14   d. 6   e. 6

13. a. 1   b. 2   c. 2   d. 2   e. 2

14. a. $\dfrac{h^2}{2}$ b. $\dfrac{q^4}{5}$ c. $\dfrac{n^3}{5}$ d. v^2 e. $2x^6$

15. a. $\dfrac{1}{x^3}$ b. $\dfrac{1}{m^8}$ c. $\dfrac{1}{4m^6}$ d. $\dfrac{2}{x^2}$

16. a. $2x^2$ b. $24t^6$ c. $\dfrac{1}{2y^5}$ d. $\dfrac{7y^3}{4x^5}$

17. a. $\dfrac{2}{5m^3n^4}$ b. $\dfrac{2}{n^2}$ c. $\dfrac{2y}{x}$ d. $\dfrac{c^4}{a^4}$

18. a. 6   b. 9   c. 18   d. 11   e. 15   f. 6

19. Sample responses can be found in the worked solutions in the online resources.

20. a. B   b. D   c. D

21. a. A   b. D

22. Sample responses can be found in the worked solutions in the online resources.

23. a. 5^9
   b. Sample responses can be found in the worked solutions in the online resources.

24. a. 5^{16}
   b. Sample responses can be found in the worked solutions in the online resources.

25. a. 1
   b. 1
   c. 0
   d. Sample responses can be found in the worked solutions in the online resources.

26. a. $\Delta = 14$
   b. Answers will vary, but $\Delta + O + \Diamond$ must sum to 12. Possible answers include: $\Delta = 3, O = 2, \Diamond = 7$; $\Delta = 1, O = 3, \Diamond = 8$; $\Delta = 4, O = 4, \Diamond = 4$; $\Delta = 5, O = 1, \Diamond = 6$.

27. a. The repeating pattern is 1, 3, 9, 7.
   b. 3

28. a. The repeating pattern is 4, 6.
   b. 4

29. a. Sample responses can be found in the worked solutions in the online resources.

b.i. 8   ii. 1   iii. 2

30. 2^{2n+3}

Exercise 1.5 Raising powers

1. a. e^6 b. f^{80} c. p^{100} d. r^{144} e. $8a^6$

2. a. a^8b^{12} b. p^5q^{15} c. $g^{30}h^{20}$
   d. $81w^{36}q^8$ e. $49e^{10}r^4q^8$

3. a. p^8q^6 b. $r^{15}w^9$ c. $b^{10}n^{18}$
   d. $j^{18}g^{12}$ e. $8a^3b^4$

4. a. q^4r^{20} b. $h^{24}j^{16}$ c. $a^{21}f^{16}$
   d. $t^{10}u^8$ e. $i^{15}j^{12}$

5. a. $\dfrac{9b^8}{d^6}$ b. $\dfrac{25h^{20}}{4j^4}$ c. $\dfrac{8k^{15}}{27t^{24}}$ d. $\dfrac{49p^{18}}{64q^{44}}$

6. a. $\dfrac{125y^{21}}{27z^{39}}$ b. $\dfrac{256a^{12}}{2401c^{20}}$ c. $\dfrac{-64k^6}{343m^{18}}$ d. $\dfrac{16g^{28}}{81h^{44}}$

7. a. 2^{20} b. t^{33} c. a^{21} d. e^{66}

8. a. g^{39} b. $12a^{14}$ c. $216d^{27}$ d. $40\,000r^{14}$

9. B

10. B

11. D

12. a. a^6 b. m^4 c. n^3
    d. b^8 e. f^{17} f. g^6

13. a. p^9 b. y^2 c. c^{20}
    d. f^7 e. k^{14} f. p^{16}

14. a. 5
    b. 6
    c. 30
    d. Sample responses can be found in the worked solutions in the online resources.

15. a.i. 1   ii. -1 iii. -1 iv. 1
    b. $(-1)^{even} = 1 (-1)^{odd} = -1$. Sample responses can be found in the worked solutions in the online resources.

16. Danni is correct. Explanations will vary but should involve $(-x)(-x) = (-x)^2 = x^2$ and $-x^2 = -1 \times x^2 = -x^2$.

17. a. 5^{12}
    b. Sample responses can be found in the worked solutions in the online resources.

18. a. i. $x = \dfrac{5}{3}$ ii. $x = \dfrac{5}{3}$ iii. $x = \dfrac{5}{3}$
    b. When equating the powers, $3x = 5$.

19. Answers will vary. Possible answers are 4096 and 262144.

20. $10^8 \times 10^8 \times 10^8 = (10^8)^3$ atoms

21. a. 4^1 b. 7^2

22. a. 2^1 b. 3^4 c. $\dfrac{1}{5^2}$ d. 2^2
    e. $\dfrac{1}{2^2}$ f. $\dfrac{1}{2}$ g. $\dfrac{1}{2}$ h. $\dfrac{1}{7}$

23. a.i. 3   ii. 4   iii. 8   iv. 11
    b.i. 3   ii. 4   iii. 8   iv. 11
    c. Raising a number to a power of one-half is the same as finding the square root of that number.

Exercise 1.6 Negative indices

1. a. $3^5 = 243$, $3^4 = 81$, $3^3 = 27$, $3^2 = 9$, $3^1 = 3$, $3^0 = 1$,
   $3^{-1} = \dfrac{1}{3}$, $3^{-2} = \dfrac{1}{9}$, $3^{-3} = \dfrac{1}{27}$, $3^{-4} = \dfrac{1}{81}$, $3^{-5} = \dfrac{1}{243}$

   b. $5^4 = 625$, $5^3 = 125$, $5^2 = 25$, $5^1 = 5$, $5^0 = 1$, $5^{-1} = \dfrac{1}{5}$,
   $5^{-2} = \dfrac{1}{25}$, $5^{-3} = \dfrac{1}{125}$, $5^{-4} = \dfrac{1}{625}$

   c. $10^4 = 10000$, $10^3 = 1000$, $10^2 = 100$, $10^1 = 10$, $10^0 = 1$,
   $10^{-1} = \dfrac{1}{10}$, $10^{-2} = \dfrac{1}{100}$, $10^{-3} = \dfrac{1}{1000}$, $10^{-4} = \dfrac{1}{10\,000}$

2. a. $\dfrac{1}{32}$ b. $\dfrac{1}{27}$ c. $\dfrac{1}{4}$ d. $\dfrac{1}{100}$

3. a. $\dfrac{1}{125}$ b. 7 c. $\dfrac{4}{3}$ d. $\dfrac{16}{9}$

4. a. 27 b. $\dfrac{2}{3}$ c. $\dfrac{16}{81}$ d. $\dfrac{49}{4}$

5. a. $\dfrac{1}{x^4}$ b. $\dfrac{1}{y^5}$ c. $\dfrac{1}{z}$
   d. $\dfrac{a^2}{b^3}$ e. $\dfrac{1}{m^2 n^3}$

6. a. $\dfrac{1}{m^2 n^3}$ b. $x^2 y^2$ c. $5x^3$
   d. $\dfrac{w^5}{x^2}$ e. $x^2 y^2$

7. a. $\dfrac{a^2 c}{b^3 d^4}$ b. $\dfrac{a^2 d^3}{b^2 c^2}$ c. $\dfrac{10y}{x^2}$
   d. $\dfrac{x}{3}$ e. $\dfrac{1}{m^3 x^2}$

8. a. $\dfrac{1}{a^5}$ b. m^5 c. $\dfrac{1}{m^7}$
   d. $\dfrac{14}{x}$ e. 6 f. $\dfrac{5a}{b^3}$

9. a. $\dfrac{1}{x^3}$ b. $\dfrac{6}{x^5 y^3}$ c. $50x^3$
   d. 1 e. $\dfrac{50}{a^5}$ f. $10a^4$

10. a. $\dfrac{-32}{w^3}$ b. $\dfrac{16}{m^4}$ c. $\dfrac{27m^6}{n^{12}}$
    d. $\dfrac{1}{a^6 b^{15}}$ e. $a^2 b^6$ f. $\dfrac{25}{a^2}$

11. a. $\dfrac{1}{x^5}$ b. $\dfrac{1}{x^{11}}$ c. x^{11}
    d. x^5 e. $\dfrac{6c^4}{a^2}$

12. a. $\dfrac{2}{a^6}$ b. $\dfrac{5}{m}$ c. $\dfrac{1}{b}$
    d. $\dfrac{1}{a^3 b^2}$ e. $\dfrac{c^2}{a^4}$

13. a. $\dfrac{1}{16a}$ b. $\dfrac{1}{m^3}$ c. $\dfrac{3}{2t^9}$
    d. t^3 e. $\dfrac{m^2}{n^3}$

14. a. 2^0 b. 2^3 c. 2^5 d. 2^6 e. 2^{-3} f. 2^{-5}

15. a. 4^0 b. 4^1 c. 4^3 d. 4^{-1} e. 4^{-2} f. 4^{-3}

16. a. 10^0 b. 10^1 c. 10^4 d. 10^{-1} e. 10^{-2} f. 10^{-5}

17. a. $3^{-4} = \dfrac{1}{3^4}$

    b. Answers will vary, but should mention that if you are dividing, the power in the numerator is lower than that in the denominator. Sample responses can be found in the worked solutions in the online resources.

    c. Sample responses can be found in the worked solutions in the online resources.

18. Answers will vary, but should mention that the negative 13 means the decimal point is moved 13 places to the left of 3. Using scientific notation allows the number to be expressed more concisely. Sample responses can be found in the worked solutions in the online resources.

19. a. No. The equivalent expression with positive indices is $\dfrac{x^6 + 1}{x^3}$.

    b. No. The equivalent expression with positive indices is $\dfrac{(xy)^2}{(x+y)^2}$.

20. a. No. The equivalent expression with positive indices is $\dfrac{x^{13} - 1}{x^5}$.

    b. No. The correct equivalent expression is $\dfrac{1}{x^6 - x^3}$.

21. a. 0.048 b. 7600 c. $0.000\,29$ d. 8.1

22. a. 3 b. -1 c. 2 d. -4

23. a. 2^{19} b. 3^{20}

24. a. $\dfrac{6}{5}$ b. $\left(\dfrac{3}{2}\right)^{200}$

Exercise 1.7 Scientific notation

1. a. 614 b. 6140
   c. $61\,400$ d. 351.8

2. a. $1\,000\,000\,000$ b. 392.673
   c. 591.1 d. 5100

3. a. $734\,000$ b. $7\,141\,400$
   c. 35.1 d. $80\,500$

4. a. 5.00×10^3 b. 4.31×10^2
   c. 3.8×10^1 d. 3.5×10^5

5. a. 7.25×10^1 b. 7.25×10^2
   c. 7.25×10^3 d. 7.25×10^8

6. a. 4.379×10^1 b. 5.317×10^3 c. 2.590×10^2
   d. 1.101×10^2 e. 1.632×10^6 f. 1.000×10^6

7. a. 0.2 b. 0.004 c. 0.0007 d. 0.03

8. a. $0.082\,73$ b. $0.072\,95$
   c. $0.002\,914\,2$ d. $0.000\,037\,53$

9. a. $0.000\,529$ b. $0.000\,033\,333$
   c. $0.000\,000\,002\,625$ d. $0.000\,000\,000\,000\,001\,273$

10. a. 7×10^{-1} b. 5×10^{-3}
    c. 3×10^{-7} d. 1×10^{-11}

11. a. 2.31×10^{-1} b. 3.62×10^{-3}
    c. 7.31×10^{-4} d. 6.3×10^{-2}

12. a. 6.73×10^{-3} b. 1.43×10^{-1} c. 6.83×10^{-5}
    d. 5.12×10^{-9} e. 5.09×10^{-2} f. 1.25×10^{-2}

13. a. $5.27 \times 10^{2}, 8.31 \times 10^{2}, 9.718 \times 10^{2}, 3.27 \times 10^{3}$
    b. $4.09 \times 10^{2}, 4.37 \times 10^{2}, 7.943 \times 10^{2}, 7.95 \times 10^{2}$
    c. $2.7 \times 10^{-3}, 9.29 \times 10^{-3}, 5.251 \times 10^{-2}, 5.31 \times 10^{-2}$
    d. $2.7 \times 10^{-3}, 7.13 \times 10^{-2}, 8.31 \times 10^{2}, 3.27 \times 10^{3}$

14. a. $0.000\,000\,000\,000\,000\,000\,019\,94$ g
    b. 1.994×10^{-17} g
    c. 5.015×10^{23} atoms

15. 3844 hours ≈ 160.17 days

16. 333 333.33

17. The neutron is heavier by 2.3×10^{-30} kg.

18. a. 2.1×10^{7} b. 1.29×10^{8}

19. a. 8 389 000 bytes b. 8.389×10^{6} bytes

20. $0.000\,000\,2$ N/m

21. a. i. Sample responses can be found in the worked solutions in the online resources.
    ii. Sample responses can be found in the worked solutions in the online resources.
    b. i. 5.36×10^{7} ii. 5.36×10^{7}
    c. The answers are the same because 2.95×10^{3} is very small compared with 5.36×10^{7}.

22. 8.75×10^{-7} is such a small amount (0.000 000 875) that, when it is added to 2.39×10^{-3}, it doesn't affect the value when given to 3 significant figures.

23. a. i. 385 440 km
    ii. 78 489 600 km
    iii. 1 278960 000 km
    iv. 5 772 840 000 km
    b. i. 385 000 km
    ii. 78 500 000 km
    iii. 1 280 000 000 km
    iv. 5 770 000 000 km
    c. i. 3.85×10^{5} km
    ii. 7.85×10^{7} km
    iii. 1.28×10^{9} km
    iv. 5.77×10^{9} km

24. a. i. 31 536 000 s   ii. 3.1536×10^{7} s
    b. i. 9 460 800 000 000 km   ii. 9.4608×10^{12} km
    c. i. 4.068×10^{13} km   ii. 46 440 000 years

25. a. 5.972×10^{21} b. 4

26. a. 9.1094×10^{-31}
    b. 1.6725×10^{-27}
    c. 1.6726×10^{-27}
    d. It is important to work with the original amounts and leave rounding until the end of a calculation so that the answer is accurate.

Exercise 1.8 Square roots and cube roots

1. a. $15^{\frac{1}{2}}$ b. $m^{\frac{1}{2}}$ c. $t^{\frac{1}{3}}$ d. $(w^{2})^{\frac{1}{3}}$

2. a. 7 b. 2 c. 3 d. 5

3. a. 10 b. 8 c. 4
   d. 1000 e. 100 f. 9

4. a. m b. b c. $6t^{2}$ d. mn^{2}

5. a. $5t^{2}$ b. xy^{2} c. $a^{2}m^{10}$ d. $6y^{2}$

6. a. $4x^{2}y^{2}$ b. $5ab^{2}c^{3}$ c. b^{7} d. b^{3}

7. B

8. D

9. A

10. a. $3^{\frac{1}{2}+\frac{1}{2}} = 3^{1}$ b. $\sqrt{3}$ c. $a^{\frac{1}{n}}$ d. 2

11. a. $z^{2.5} = z^{\frac{5}{2}} = (z^{5})^{\frac{1}{2}} = \sqrt{z^{5}}$
    b. No, it is the tenth root: $z^{0.3} = z^{\frac{3}{10}} = (z^{3})^{\frac{1}{10}} = \sqrt[10]{z^{3}}$.

12. Prithvi is correct: $\sqrt{x^{2}} = x^{\frac{1}{2}\times 2} = x^{1} = x$; x can be a positive or negative number.

13. $(-2^{3})^{\frac{1}{3}} = -2$; answers will vary, but should include that we cannot take the fourth root of a negative number. Sample responses can be found in the worked solutions in the online resources.

14. $\sqrt{1871} \approx 43.25$
    $42^{2} = 1764$
    $43^{2} = 1849$
    He was 43 years old in 1849. Therefore, he was born in $1849 - 43 = 1806$.

15. a. $d = t^{\frac{2}{3}}$ b. $t = d^{\frac{3}{2}}$

16. $\dfrac{16}{81}$

17. a. There are 2 possible solutions, because the square root of a number always has a positive and a negative answer.
    b. $\dfrac{5}{3}, 1$

Exercise 1.9 Surds

1. a. 1 b. 2 c. 0
   d. $\dfrac{1}{3}$ e. $\dfrac{5}{4}$

2. a. 0.4 b. 20 c. 100
   d. $\dfrac{2}{5}$ e. 1.2

3. a. 4.5 b. 1000 c. 0.03
   d. 16 e. $\dfrac{11}{7}$

4. a. 9 b. -9 c. 11 d. -21

5. a. 2 b. 4 c. 7 d. 3

6. a. 4 b. 5 c. -7 d. -3

7. b and d

8. c and d

9. c

10. a. $\sqrt{21}$ b. $-\sqrt{21}$ c. $2\sqrt{6}$
    d. $6\sqrt{7}$ e. $10\sqrt{14}$

11. a. $12\sqrt{7}$ b. $9\sqrt{7}$ c. $22\sqrt{10}$
    d. $22\sqrt{3}$ e. 3

12. a. $15\sqrt{10}$ b. 24 c. 6
    d. 11 e. 30

13. a. $\sqrt{3}$ b. $-\sqrt{2}$ c. $\sqrt{6}$ d. $\sqrt{5}$

14. a. $3\sqrt{3}$ b. $\dfrac{3\sqrt{6}}{2}$ c. $15\sqrt{2}$ d. $\dfrac{5\sqrt{5}}{3}$

15. a. $-2\sqrt{5}$ b. $\sqrt{3}$ c. $\sqrt{3}$ d. $\sqrt{7}$

16. a. $2\sqrt{5}$ b. $2\sqrt{2}$ c. $3\sqrt{2}$ d. 7

17. a. $\sqrt{30}$ b. $5\sqrt{2}$ c. $2\sqrt{7}$ d. $6\sqrt{3}$

18. a. $12\sqrt{2}$ b. $4\sqrt{3}$ c. $10\sqrt{5}$ d. $9\sqrt{2}$

19. a. $4\sqrt{2}$ b. $15\sqrt{3}$ c. 48 d. $35\sqrt{2}$ e. $20\sqrt{6}$

20. a. $10\sqrt{3}$ b. $4\sqrt{42}$ c. $72\sqrt{2}$ d. $27\sqrt{5}$ e. $132\sqrt{2}$

21. a. $\sqrt{12}$ b. $\sqrt{175}$ c. $\sqrt{108}$ d. $\sqrt{80}$ e. $\sqrt{384}$

22. a. $\sqrt{90}$ b. $\sqrt{32}$ c. $\sqrt{720}$ d. $\sqrt{600}$ e. $\sqrt{338}$

23. a. $2\sqrt{2}$ b. $-4\sqrt{5}$ c. $-6\sqrt{3}$ d. 0

24. a. $5\sqrt{11}$ b. $2\sqrt{7}$
    c. $10\sqrt{2}+7\sqrt{3}$ d. $8\sqrt{5}+\sqrt{6}$

25. a. $8\sqrt{10}+7\sqrt{3}$ b. $16\sqrt{2}-11\sqrt{5}$
    c. $2\sqrt{6}$ d. $5\sqrt{5}+15$

26. a. $\sqrt{2}$ b. 0 c. $-5\sqrt{3}$
    d. $-4\sqrt{7}$ e. $5\sqrt{6}+6\sqrt{5}$

27. a. $2\sqrt{3}-3\sqrt{5}$ b. $4\sqrt{6}+6\sqrt{5}-14\sqrt{2}$
    c. $29\sqrt{5}+22\sqrt{3}$ d. $9\sqrt{11}-\sqrt{30}$
    e. $28\sqrt{2}-39\sqrt{5}$

28. a. D b. A c. D d. A

29. a. D b. A c. C d. C

30. a. D b. D c. D d. B

31. a. $15\sqrt{3}$ b. $19\sqrt{5}$ c. $16\sqrt{7}$ d. $17\sqrt{2}$

32. a. $a\sqrt{c}$ b. $d^2\sqrt{b}$ c. $hk\sqrt{j}$ d. $f\sqrt{f}$

33. a. $2\sqrt{3}$ b. $5\sqrt{2}$ c. $10\sqrt{2}$ d. $2\sqrt{10}$

34. a. $2\sqrt{7}, 4\sqrt{2}, 3\sqrt{6}, \sqrt{60}, 8, 6\sqrt{2}$
    b. $\sqrt{2}, \sqrt{3}, 2, \sqrt{6}, 2\sqrt{2}, 3, 2\sqrt{3}$

35. $5\sqrt{26}$

36. a. $\sqrt{2}$ m
    b. 1.41 m
    c. i. Any of the arrangements shown would work.

    ii. The length of each square is $\sqrt{2}$ m and there are 14 lengths around each shape.
    iii. 19.80 m

37. $\sqrt{a^3b^2} = \sqrt{a^3} \times \sqrt{b^2}$
    $\quad = \sqrt{a^2} \times \sqrt{a} \times \sqrt{b^2}$
    $\quad = a \times \sqrt{a} \times b$
    $\quad = ab\sqrt{a}$

38. $a = 2$ and $b = 8$.

39. $36\sqrt{5}\,\text{m}^2$

40. $\left(1750\sqrt{2} + 2750\sqrt{3} + 320\right)$ m

41. The side length will be rational if x is a perfect square. If x is not a perfect square, the side length will be a surd.

42. a. i. $\sqrt{13}$ ii. 3 iii. $\sqrt{38}$
    b. For side lengths $\sqrt{a}$ and $\sqrt{b}$, the length of the hypotenuse is $\sqrt{a+b}$.
    c. i. $\sqrt{1500} = 10\sqrt{15}$
       ii. $2\sqrt{114}$
       iii. $\sqrt{187}$
    d. No. The correct answer is $\sqrt{b+a} = \sqrt{c}$.

Project

1.

Square	1	2	3	4
Side length (cm)	2	4	6	8
Diagonal length (cm)	$2\sqrt{2}$	$4\sqrt{2}$	$6\sqrt{2}$	$8\sqrt{2}$
Perimeter (cm)	8	16	24	32
Area (cm²)	4	16	36	64

2. 20 cm
3. $20\sqrt{2}$ cm

4.

Square	1	2	3	4
Side length (cm)	$\sqrt{2}$	$2\sqrt{2}$	$3\sqrt{2}$	$4\sqrt{2}$
Diagonal length (cm)	2	4	6	8
Perimeter (cm)	$4\sqrt{2}$	$8\sqrt{2}$	$12\sqrt{2}$	$16\sqrt{2}$
Area (cm²)	2	8	18	32

5. $10\sqrt{2}$ cm
6. 20 cm
7. First arrangement
8. First arrangement
9. The squares in the first diagram are twice the area of each corresponding square in the second diagram.
10. The squares in the first diagram have a perimeter that is $\sqrt{2}$ times the perimeter of each corresponding square in the second diagram.
11. An increase of 60 cm² in the first diagram and an increase of 30 cm² for the second diagram.
12. 2π
13. $\pi(2r+1)$

Exercise 1.10 Review questions

1. A
2. a. Irrational, because it is equal to a non-recurring and non-terminating decimal.
   b. Rational, because it can be expressed as a whole number.
   c. Rational, because it is given in a rational form.
   d. Rational, because it is a recurring decimal.
   e. Irrational, because it is equal to a non-recurring and non-terminating decimal.

3. D

4. c and d

5. C

6. B

7. a. b^{10} b. m^{11} c. k^8 d. f^{14} e. h^{10}

8. a. $6q^{17}$ b. $35w^{29}$ c. $12e^5p^8$
   d. $105a^{16}b^{17}$ e. a^3

9. a. t^4 b. r^7 c. p^3 d. f^5 e. y^{90}

10. a. m^{10} b. g^7 c. x d. d^7 e. t^5

11. a. p^5 b. $4k^4$ c. $11b^7c^5$ d. $\dfrac{9e^6}{32d^2}$ e. $\dfrac{a^3}{2}$

12. a. 1   b. 1   c. 1   d. 1   e. 1

13. a. a b. 3   c. 3   d. 199   e. a^7

14. a. v^{10} b. pr c. a^9b^4 d. j^8m^3 e. $4e^2 - 36$

15. a. $6xy^2$ b. -8 c. $15 - 12x$
    d. -3 e. 54

16. a. b^8 b. a^{24} c. k^{70} d. j^{200}

17. C

18. a. $a^{15}b^6$ b. $m^{14}n^{24}$
    c. s^3t^{18} d. $q^{10}p^{300}$

19. a. $\dfrac{1}{a}$ b. $\dfrac{1}{k^4}$ c. $\dfrac{2}{m^4}$ d. $\dfrac{42}{y^5}$

20. a. x^{-1} b. $2y^{-4}$
    c. z^{-3} d. $135p^{-3}q^{-3}$

21. a. 10   b. 6   c. a d. b e. $7f^2$

22. a. 3   b. 10   c. x d. $2d$ e. $4f^2g$

23. 1.4×10^{12}

24. 4.167×10^8

25. 9.16×10^7 km

26. 1.2441×10^9 km

27. a. $3.4 \times 10^7, 1.5 \times 10^9, 6 \times 10^9$
    b. 0.57%
    c. 4
    d. 0.25 km^2

28. 2.95%

29. 820 000 queen bees

30. At the beginning of the 5th year, or in 5.5 years.

31. a. Mach 9.1

    b.

Mach number	Speed in m/s	Speed in km/h
1	343	1234.8
2	686	2469.6
3	1029	3704.4

    c. At Mach 1 it would take 32.39 hours.
       At Mach 2 it would take 16.20 hours.
       At Mach 3 it would take 10.79 hours.

    d. 3.55 hours

e. The fastest aircraft could circle Earth 9.12 times in the time an aircraft travelling at Mach 1 could circle Earth once.
   The fastest aircraft could circle Earth 4.56 times in the time an aircraft travelling at Mach 2 could circle Earth once.
   The fastest aircraft could circle Earth 3.04 times in the time an aircraft travelling at Mach 3 could circle Earth once.

f. (Number of times to circle Earth) $\times$ (Mach value) ≈ 9.1

2 Algebra

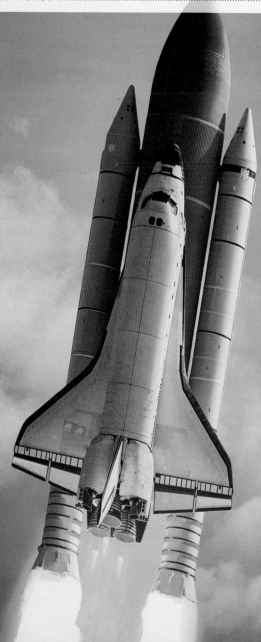

2.1 Overview

Why learn this?

Most people will tell you that if you want a career in fields like engineering, science, finance and software development, you need to study and do well in algebra at high school and university. Without algebra we wouldn't have landed on the Moon or be able to enjoy the technological marvels we do today, like our smartphones.

Learning algebra in school helps you to develop critical thinking skills. These skills can help with things like problem solving, logic, pattern recognition and reasoning.

Consider these situations:

- You want to work out the cost of a holiday to Japan at the current exchange rate.
- You're driving to a petrol station to refuel your car, but you only have $30 in your pocket. You want to know how much petrol you can afford.
- You're trying to work out whether a new phone plan that charges $5 per day over 24 months for the latest smartphone is worth it.

Working out the answers to these kinds of real-life questions is a direct application of algebraic thinking.

Understanding algebra can help you to make reasoned and well-considered financial and life decisions. It can help you make independent choices and assist you with many everyday tasks. Algebra is an important tool that helps us connect with — and make sense of — the world around us.

Where to get help

Go to your learnON title at www.jacplus.com.au to access the following digital resources. The Online Resources Summary at the end of this topic provides a full list of what's available to help you learn the concepts covered in this topic.

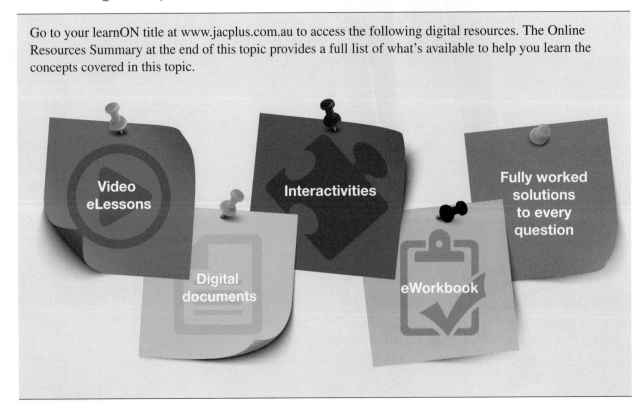

Exercise 2.1 Pre-test

Complete this pre-test in your learnON title at www.jacplus.com.au, and receive **automatic marks**, **immediate corrective feedback** and **fully worked solutions**.

1. **MC** Select the coefficient of x in the expression $\dfrac{-x^3}{3} + x^2 + 2ax + b$, where a and b are constants.

   A. 1   B. $-\dfrac{1}{3}$ C. 2   D. $2a$ E. b

2. Evaluate the following expression if $x = -3$, $y = 1$ and $z = -2$.

$$2x^2 - y^2 - xyz$$

3. Determine the value of m in the following equation if $E = 25$ and $v = 5$.

$$E = \frac{1}{2}mv^2$$

4. **MC** Choose the expression out of the following that is not equivalent.

   A. $x - y + z$ B. $z - y + x$ C. $x - (y - z)$ D. $z + x - y$ E. $z - (x + y)$

5. **MC** A bag of chips costs x cents. Select the correct expression for the cost of 6 bags of chips if each bag is discounted by 50 cents.

   A. $6(x - 50)$ B. $6(x - 0.5)$ C. $6x - 50$ D. $6x - 0.5$ E. $50 - 6x$

6. Expand and simplify the following expressions.
   a. $4(m + 3n) - (2m - n)$
   b. $-x(y + 3) + y(5 - x)$

7. Expand and simplify the expression $(3q - r)(2q + r)$.

8. **MC** Students were asked to expand $-2(3x - 4)$. They gave the following answers.
   Student 1: $-6x - 8$
   Student 2: $-6x + 6$
   Student 3: $6x - 8$
   Student 4: $-6x + 8$
   Student 5: $-5x + 6$
   Choose the student who has the correct answer.

   A. Student 1   B. Student 2   C. Student 3   D. Student 4   E. Student 5

9. State whether 23^2 can be expanded using the following method.

$$(20 + 3)^2 = 20^2 + 2 \times 20 \times 3 + 3^2$$

10. State whether the following statement is True or False.
    $(a + 3)^2$ expanded is equal to $a^2 + 9$

11. Expand and simplify the following expression.

$$(2b - 5)^2 - (b + 4)(b - 3)$$

12. Factorise the following expressions.

    a. $8p^2 + 20p$ **b.** $25x^2 - 10y$ **c.** $6xy + 8x^2y - 3xy^2$

13. Factorise the following expressions.

    a. $7a(b+3) - (b+3)$ **b.** $10xy + 5x - 4y - 2$

14. A box has the following dimensions: x, $2x - 1$ and $3x + 2$.

    Determine the correct expression for calculating the total surface area of the box.

15. Factorise the expression $x^2 + 3x - 4$.

2.2 Using pronumerals

LEARNING INTENTION

At the end of this subtopic you should be able to:
- identify the number of terms in an algebraic expression
- identify the variables and coefficients of the terms in an expression
- identify the constant terms in an expression
- evaluate an expression or formula by substituting values for each variable.

▶ 2.2.1 Expressions

eles-4588

- A **pronumeral** is a letter or symbol that stands for a number. Algebra is the branch of mathematics in which we use and manipulate pronumerals.
- In algebra an **expression** is a group of **terms** in which each term is separated by a $+$ sign or a $-$ sign .
- An expression with two terms is called a **binomial**. An expression with one term is called a **monomial**.
- Here are some important things to note about the expression $5x - 3y + 2$:
  - The expression has three terms and is called a **trinomial**.
  - The pronumerals x and y may take different values and are called **variables**.
  - The term 2 has only one value and is called a **constant**.
  - The **coefficient** (the number in front) of x is 5, and the coefficient of y is -3.

Algebraic expressions

An algebraic expression is a group of terms that are separated by $+$ or $-$ signs.

Consider the following expression: $5x^2 - 6y + 13x + 8 + 2y^3$.
- This expression is made up of **five terms**.
- The **pronumerals/variables** used in this expression are x and y .
- The **coefficients** of the terms are (in order): 5, -6, 8 and 2.
- The **coefficient of** x is 13.
- The **smallest coefficient** is -6.
- The **constant** is 8 — it does not have a pronumeral.

WORKED EXAMPLE 1 Identifying coefficients of terms in an expression

For the expression $6x - 3xy + z + 2 + x^2z + \dfrac{y^2}{7}$, determine:

a. the number of terms
b. the coefficient of the second term
c. the coefficient of the last term
d. the constant term
e. the term with the smallest coefficient
f. the coefficient of x^2z.

THINK

a. Count the number of terms.

b. The second term is $-3xy$. The number part is the coefficient.

c. The last term is $\dfrac{y^2}{7}$. This can be rewritten as $\dfrac{1}{7} \times y^2$.

d. The constant is the term with no pronumeral.

e. Identify the smallest coefficient and write the whole term to which it belongs.

f. x^2z can be written as $1x^2z$.

WRITE

a. There are six terms.

b. The coefficient of the second term is -3.

c. The coefficient of the last term is $\dfrac{1}{7}$.

d. The constant term is 2.

e. The term with the smallest coefficient is $-3xy$.

f. The coefficient of x^2z is 1.

2.2.2 Substitution

eles-4589

- We can evaluate (find the value of) an algebraic expression if we replace the pronumerals with their known values.
- This process is called substitution.
- Consider the expression $4x + 3y$. If we substitute the values $x = 2$ and $y = 5$, then the expression takes can be evaluated like this:

$$4 \times 2 + 3 \times 5 = 8 + 15$$
$$= 23$$

It is more common to use brackets for substitution instead of multiplication signs. For example:

$$4x + 3y = 4(2) + 3(5)$$
$$= 8 + 15$$
$$= 23$$

WORKED EXAMPLE 2 Substituting values into terms

Evaluate the following terms for the values $x = 4$ and $x = -4$.

a. $3x^2$
b. $-x^2$
c. $\sqrt{5 - x}$

THINK

a. 1. Substitute 4 for x.

   2. Only the x term is squared.

   3. Evaluate.

WRITE

a. if $x = 43$ $x^2 = 3(4)^2$

   $= 3 \times 16$

   $= 48$

4. Repeat for $x = -4$.

if $x = -4$: $3x^2 = 3(-4)^2$
$$= 3 \times 16$$
$$= 48$$

b. 1. Substitute 4 for x.

b. if $x = 4$: $-x^2 = -(4)^2$

2. Evaluate.

$$= -16$$

3. Repeat for $x = -4$.

if $x = -4$: $-x^2 = -(-4)^2$
$$= -16$$

c. 1. Substitute 4 for x.

c. if $x = 4$: $\sqrt{5-x} = \sqrt{5-4}$

2. Evaluate.

$$= \sqrt{1}$$
$$= 1$$

3. Repeat for $x = -4$.

if $x = -4$ $\sqrt{5-x} = \sqrt{5-(-4)}$
$$= \sqrt{5+4}$$
$$= \sqrt{9}$$
$$= 3$$

WORKED EXAMPLE 3 Evaluating expressions using substitution

If $x = 3$ and $y = -2$, evaluate the following expressions.
a. $3x + 2y$
b. $5xy - 3x + 1$
c. $2x^2 + y^2$

THINK

WRITE

a. 1. Write the expression.

a. $3x + 2y$

2. Substitute $x = 3$ and $y = -2$.

$$= 3(3) + 2(-2)$$

3. Evaluate.

$$= 9 - 4$$
$$= 5$$

b. 1. Write the expression.

b. $5xy - 3x + 1$

2. Substitute and $x = 3$ and $y = -2$.

$$= 5(3)(-2) - 3(3) + 1$$

3. Evaluate.

$$= -30 - 9 + 1$$
$$= -38$$

c. 1. Write the expression.

c. $2x^2 + y^2$

2. Substitute $x = 3$ and $y = -2$.

$$= 2(3)^2 + (-2)^2$$

3. Evaluate.

$$= 18 + 4$$
$$= 22$$

TI | THINK	WRITE	CASIO | THINK	WRITE
a–c. On a Calculator page, complete the entry lines as: $3x + 2y \| x = 3$ and $y = -2$ $5x \times y - 3x + 1 \|$ $x = 3$ and $y = -2$ $2x^2 + y^2 \| x = 3$ and $y = -2$ Press ENTER after each entry.	**a–c.** 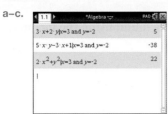 If $x = 3$ and $y = -2$, then **a.** $3x + 2y = 5$ **b.** $5xy - 3x + 1 = -38$ **c.** $2x^2 + y^2 = 22$.	**a–c.** On the Main screen, complete the entry lines as: $3x + 2y \| x = 3 \|$ $y = -2$ $5x \times y - 3x + 1 \|$ $x = 3 \| y = -2$ $2x^2 + y^2 \| x = 3 \|$ $y = -2$ Press EXE after each entry line.	**a–c.**  If $x = 3$ and $y = -2$, then **a.** $3x + 2y = 5$ **b.** $5xy - 3x + 1 = -38$ **c.** $2x^2 + y^2 = 22$.

▶ 2.2.3 Substitution into formulas

eles-4590

- A **formula** is a mathematical rule. Formulas are usually written using pronumerals.
- For example, the formula for the area of a rectangle is given by:

$$A = lw$$

where A represents the area of the rectangle, l represents the length and w represents the width.
- If a rectangular kitchen tile has length $l = 20$ cm and width $w = 15$ cm, we can substitute these values into the formula to find its area.

$$A = lw$$
$$= 20 \times 15$$
$$= 300 \, \text{cm}^2$$

- If the area of a kitchen tile is 400 cm^2 and its width is 55 cm, then we can substitute these values into the formula to find its length.

$$A = lw$$
$$400 = l \times 55$$
$$l = \frac{400}{55}$$
$$= \frac{80}{11}$$
$$\approx 7.3 \, \text{cm} \, (\text{to 1 d.p.})$$

The formula for the voltage in an electrical circuit can be found using the formula known as Ohm's Law:

$$V = IR$$

where I = current (in amperes)
R = resistance (in ohms)
V = voltage (in volts).

a. Calculate V when:
i. $I = 2$ amperes, $R = 10$ ohms
ii. $I = 20$ amperes, $R = 10$ ohms.
b. Calculate I when $V = 300$ volts and $R = 600$ ohms.

THINK	WRITE
a. i. 1. Write the formula.	a. i. $V = IR$
2. Substitute $I = 2$ and $R = 10$.	$= (2)(10)$
3. Evaluate and write the answer using the correct units.	$= 20$ The voltage is 20 volts.
ii. 1. Write the formula.	ii. $V = IR$
2. Substitute $I = 20$ and $R = 10$.	$= (20)(10)$
3. Evaluate and write the answer using the correct units.	$= 200$ The voltage is 200 volts.
b. 1. Write the formula.	b. $V = IR$
2. Substitute $V = 300$ and $R = 600$.	$300 = I(600)$
3. Evaluate and write the answer using the correct units.	$I = \dfrac{300}{600}$ $= 0.5$ The current is 0.5 amperes.

Note: Even if we know nothing about volts, amperes and ohms, we can still do this calculation.

DISCUSSION

Which letters (pronumerals) should you avoid using when writing algebraic expressions?

---

on Resources

eWorkbook   Topic 2 Workbook (worksheets, code puzzle and project) (ewbk-2002)

Digital document SkillSHEET Alternative expressions used for the four operations (doc-6122)

Video eLesson   Substitution 2 (eles-1892)

Interactivities   Individual pathway interactivity: Using pronumerals (int-4480)

                    Substituting positive and negative numbers (int-3765)

Exercise 2.2 Using pronumerals

Individual pathways

■ PRACTISE	■ CONSOLIDATE	■ MASTER
1, 4, 7, 10, 13, 14, 17, 18, 25	2, 5, 8, 11, 15, 19, 20, 21, 26	3, 6, 9, 12, 16, 22, 23, 24, 27

To answer questions online and to receive **immediate corrective feedback** and **fully worked solutions** for all questions, go to your learnON title at www.jacplus.com.au.

Fluency

1. Determine the coefficient of each of the following terms.

   a. $3x$
   b. $-2m$
   c. $\dfrac{x}{3}$
   d. $-\dfrac{at}{4}$

2. Evaluate the coefficient of each of the following terms.

   a. $5y$
   b. $-2pq$
   c. $\dfrac{x^2}{6}$
   d. $-\dfrac{2mn^2}{7}$

3. Determine the coefficient of each of the following terms.

   a. $15xy$
   b. $-\dfrac{2pq}{5}$
   c. $\dfrac{7w^2z^4}{2}$
   d. $-\dfrac{pq^3}{7}$

4. **WE1** For each expression below, determine:

   i. the number of terms
   ii. the coefficient of the first term
   iii. the constant term
   iv. the term with the smallest coefficient.

   a. $5x^2 + 7x + 8$
   b. $-9m^2 + 8m - 6$
   c. $5x^2y - 7x^2 + 8xy + 5$
   d. $9ab^2 - 8a - 9b^2 + 4$

5. For each expression below, determine:

   i. the number of terms
   ii. the coefficient of the first term
   iii. the constant term
   iv. the term with the smallest coefficient.

   a. $4a - 2 + 9a^2b^2 - 3ac$
   b. $5s + s^2t + 9 + 12t - 3u$
   c. $-m + 8 + 5n^2m + m^2 + 2n$
   d. $7c^2d + 5d^2 + 14 - 3cd^2 - 2e$

6. For each expression below, determine:

   i. the number of terms
   ii. the coefficient of the x term
   iii. the constant term
   iv. the term with the smallest coefficient.

   a. $-10x^2 + 3x - 7$
   b. $4y^2 + 6xy - 6x^2 + 7 - 3x$
   c. $10x^5 - 4x^{-3} + 2x^2 - 5x + 8$
   d. $6 - y^3 + 6x - 7y^5 + 10y - 15xy$

7. Solve the following expressions for $x = 0$, $x = 2$ and $x = -2$.

   a. $5x + 2$
   b. $x^2 - x$
   c. $x^2 + 3x - 1$
   d. $\dfrac{x}{2}$

8. Calculate the following expressions for $x = 0$, $x = 3$ and $x = -3$.

   a. $-5x + 4$
   b. $x^2 - 2x + 1$
   c. $x(x + 4)$
   d. $\dfrac{8}{x + 1}$

9. Determine the following expressions for $x = 0$, $x = 1$ and $x = -2$.

   a. $\sqrt{x + 3}$
   b. $\dfrac{2x^2 + 7x}{3}$
   c. $(x + 3)(x - 4)$
   d. $\dfrac{2x + 3}{5}$

10. **MC** Answer the following questions.

    a. If $x = -3$, then state the value of $-5x - 3$.
       A. -18
       B. 12
       C. 30
       D. 18
       E. -56

    b. Choose the expression that is a trinomial.
       A. $3x^2$
       B. $x + 5 + 2$
       C. $x^2 - 7$
       D. $x^2 + x + 2$
       E. $4x^3$

    c. Choose the value of x for which the expression $\sqrt{(x - 6)}$ cannot be evaluated.
       A. 2
       B. 6
       C. 7
       D. 10
       E. 16

    d. In the expression $5x^2 - 3xy + 0.5x - 0.3y - 5$, identify the smallest coefficient.
       A. 5
       B. -3
       C. 0.5
       D. -0.3
       E. -2

11. **WE3** Determine the value of the following expressions if $x = 2$, $y = -1$ and $z = 3$.

    a. $2x$
    b. $3xy$
    c. $2y^2z$
    d. $\dfrac{14}{x}$
    e. $6(2x + 3y - z)$
    f. $x^2 - y^2 + xyz$

12. If $x = 4$ and $y = -3$, evaluate the following expressions.

    a. $4x + 3y$
    b. $3xy - 2x + 4$
    c. $x^2 - y^2$

Understanding

13. **WE4** The change in the voltage in an electrical circuit can be found using the formula known as Ohm's Law $V = IR$, where $I = $ current (in amperes), $R = $ resistance (in ohms) and $V = $ voltage (in volts).

    a. Calculate V when:
       i. $I = 4$, $R = 8$
       ii. $I = 25$, $R = 10$.

    b. Calculate R when:
       i. $V = 100$, $I = 25$
       ii. $V = 90$, $I = 30$.

14. Evaluate each of the following by substituting the given values into each formula.

    a. If $A = bh$, calculate the value of A when $b = 5$ and $h = 3$.

    b. If $d = \dfrac{m}{v}$, determine the value of d when $m = 30$ and $v = 3$.

    c. If $A = \dfrac{1}{2}xy$, calculate the value of A when $x = 18$ and $y = 2$.

    d. If $A = \dfrac{1}{2}(a + b)h$, determine the value of A when $h = 10$, $a = 7$ and $b = 2$.

    e. If $V = \dfrac{AH}{3}$, determine the value of V when $A = 9$ and $H = 10$.

15. Calculate each of the following by substituting the given values into each formula.

    a. If $v = u + at$, calculate the value of v when $u = 4$, $a = 3.2$ and $t = 2.1$.

    b. If $t = a + (n - 1)d$, determine the value of t when $a = 3$, $n = 10$ and $d = 2$.

    c. If $A = \dfrac{1}{2}(x + y)h$, determine the value of A when $x = 5$, $y = 9$ and $h = 3.2$.

    d. If $A = 2b^2$, calculate the value of A when $b = 5$.

    e. If $y = 5x^2 - 9$, determine the value of y when $x = 6$.

16. Calculate each of the following by substituting the given values into each formula.

    a. If $y = x^2 - 2x + 4$, determine the value of y when $x = 2$.

    b. If $a = -3b^2 + 5b - 2$, calculate the value of a when $b = 4$.

    c. If $s = ut + \dfrac{1}{2}at^2$, determine the value of s when $u = 0.8$, $t = 5$ and $a = 2.3$.

    d. If $F = \dfrac{mp}{r^2}$, calculate the value of F correct to 2 decimal places, when $m = 6.9$, $p = 8$ and $r = 1.2$.

    e. If $C = \pi d$, determine the value of C correct to 2 decimal places if $d = 11$.

Reasoning

17. The area of a triangle is given by the formula $A = \dfrac{1}{2}bh$, where b is the length of the base and h is the perpendicular height of the triangle.

    a. Show that the area is $12\,\text{cm}^2$ when $b = 6\,\text{cm}$ and $h = 4\,\text{cm}$.

    b. Evaluate h if $A = 24\,\text{cm}^2$ and $b = 4\,\text{cm}$.

18. Using $E = F + V - 2$, where F is the number of faces on a prism, E is the number of edges and V is the number of vertices, calculate:

    a. E if $F = 5$ and $V = 7$ b. F if $E = 10$ and $V = 2$.

19. The formula to convert degrees Fahrenheit (F) to degrees Celsius (C) is $C = \dfrac{5}{9}(F - 32)$.

    a. Determine the value of C when $F = 59$.

    b. Show that when Celsius (C) is 15, Fahrenheit (F) is 59.

20. The length of the hypotenuse of a right-angled triangle (c) can be found using the formula $c = \sqrt{a^2 + b^2}$, where a and b are the lengths of the other 2 sides.

    a. Calculate c when $a = 3$ and $b = 4$.

    b. Show that when $a = 5$ and $c = 13$, $b = 12$.

21. The kinetic energy (E) of an object is found by using the formula $E = \dfrac{1}{2}mv^2$, where m is the mass and v is the velocity of the object.

    a. Calculate E when $m = 3$ and $v = 3.6$.

    b. Show that when $E = 25$ and $v = 5$, $m = 2$.

22. If the volume of a prism (V) is given by the formula $V = AH$, where A is the area of the cross-section and H is the height of the prism, calculate:

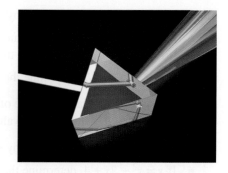

    a. V when $A = 7\,\text{cm}^2$ and $H = 9\,\text{cm}$

    b. H when $V = 120\,\text{cm}^3$ and $A = 30\,\text{cm}^2$.

23. The volume of a cylinder (v) is given by $v = \pi r^2 h$, where r is the radius in centimetres and h is the height of the cylinder in centimetres.

    a. Determine v correct to 2 decimal places if $r = 7$ and $h = 3$.

    b. Determine h correct to 2 decimal places if $v = 120$ and $r = 2$.

24. The surface area of a cylinder (S) is given by $S = 2\pi r (r + h)$, where r is the radius of the circular end and h is the height of the cylinder.

    a. Calculate S (to 2 decimal places) if $r = 14$ and $h = 5$.

    b. Show that for a cylinder of surface area 240 units^2 and radius 5 units, the height is 2.64 units, correct to 2 decimal places.

Problem solving

25. a. In a magic square, every row, column and diagonal adds to the same number. This number is called the magic number for that square. Complete the unfinished magic square by first finding an algebraic expression for the magic number.

    b. Make three different magic squares by substituting different values for g and d in the completed magic square from part **a.**

    c. Calculate the magic number for each magic square.

	$2d + 2g$	
	$d + g$	
0	$d + 2g$	

26. A flat, rectangular board is built by gluing together a number of square pieces of the same size. The board is m squares wide and n squares long. In terms of m and n, write expressions for:

    a. the total number of squares

    b. the number of completely surrounded squares

    c. the perimeter of the figure, given that each square is 1 unit by 1 unit.

27. Determine which of the following eight expressions is **not** equivalent to the others.

    $x - y + z$ $z - y + x$

    $z - (y - x)$ $-y + x + z$

    $z + x - y$ $x - (y - z)$

    $y - (z - x)$ $x + z - y$

2.3 Algebra in worded problems

▶ 2.3.1 Algebra in worded problems

eles-4591

- In order to make use of the skills you learn in algebra, it is useful to be able to take a real-life problem and convert it into an algebraic expression.
- The starting point for any worded problem is to define the unknown quantities in the question by assigning appropriate pronumerals to those unknown quantities.

Key language used in worded problems

In order to be able to turn a worded problem into an algebraic expression, it helps to look out for the following kinds of words.
- **Words for addition:** sum, altogether, add, more than, and, in total
- **Words for subtraction:** difference, less than, take away, take off, fewer than
- **Words for multiplication:** product, groups of, times, of, for each, double, triple
- **Words for division:** quotient, split into, halve, thirds

---

WORKED EXAMPLE 5 Converting worded problems into algebraic expressions

Write an algebraic expression for each of the following, choosing an appropriate pronumeral if necessary.
a. The number that is 6 more than Ben's age
b. The total value of a bundle of $10 notes
c. The total cost of 8 adults' and 3 children's train tickets
d. The product of a and w
e. w less than a

THINK	WRITE
a. 1. Ben's age is the unknown quantity. *Note*: You must not say, 'Let Ben $= b$', because b is a number.	a. Let Ben's age $= b$ years.
2. Write an expression for Ben's age plus 6.	$b + 6$
b. 1. The number of notes is the unknown quantity.	b. Let $n =$ the number of $10 notes
2. The value is 10 times the number of notes. Write the expression for the value of the notes.	Total value $= 10n$ dollars
c. 1. The cost of tickets is the unknown quantity.	c. Let an adult's ticket cost a dollars
2. The adults' tickets cost $8 \times a$ and the children's Tickets cost $3 \times c$. Write the expression for the total cost of the tickets.	and a child's ticket cost c dollars. Total cost $= 8a + 3c$ dollars.

▶

d. 'Product' means to multiply.
Write the expression.

d. $a \times w$

e. In this expression, w is subtracted from a.
Write the expression.

e. $a - w$

WORKED EXAMPLE 6 Solving a worded problem

Kot has a block of chocolate that is made up of 40 smaller pieces.
a. **If he breaks off x pieces to give to Akira, evaluate how much of the block remains.**
b. **Kot then eats $\dfrac{1}{5}$ of the remaining chocolate for dessert. Determine how much he ate.**
c. **Finally Kot gives a third of what remains to his sister Temar. Calculate how much Temar receives if the original amount Kot gave to Akira was 10 pieces.**

THINK	WRITE
a. 1. The unknown quantity is the amount passed to Akira, which is x.	**a.** $x =$ number of pieces given to Akira
2. Write an expression for x less than 40.	Amount remaining $= 40 - x$
b. 1. The unknown quantity is still x.	**b.** Amount eaten $= \dfrac{1}{5}$ of amount remaining
2. Using the answer to **a**, write an expression for $\dfrac{1}{5}$ of what is left. *Note:* 'of' is a word that means multiplication.	Amount eaten $= \dfrac{1}{5} \times (40 - x)$ Amount eaten $= \dfrac{1}{5}(40 - x)$
c. 1. The unknown quantity is still x from part **a**.	**c.** Amount left $= 40 - x - \dfrac{1}{5}(40 - x)$
2. Use your answers to **a** and **b** to write an expression for the amount left over after Kot eats $\dfrac{1}{5}$ of the remaining chocolate.	Amount passed to Temar is $\dfrac{1}{3}$ of what is left over after Kot's dessert.
3. Multiply that expression by $\dfrac{1}{3}$.	Temar's amount $= \dfrac{1}{3}\left(40 - x - \dfrac{1}{5}(40 - x)\right)$
4. Substitute the value of $x = 10$ into the expression to find the final answer. Write the answer in words.	When $x = 10$ $= \dfrac{1}{3}\left(40 - 10 - \dfrac{1}{5}(40 - 10)\right)$ $= \dfrac{1}{3}\left(30 - \dfrac{1}{5}(30)\right)$ $= \dfrac{1}{3}(30 - 6)$ $= \dfrac{1}{3}(24)$ $= 8$ Temar receives 8 pieces of chocolate from Kot.

Exercise 2.3 Algebra in worded problems

learn**on**

Individual pathways

■ **PRACTISE**	■ **CONSOLIDATE**	■ **MASTER**
1, 4, 5, 6, 10, 14	2, 7, 8, 11, 15, 16	3, 9, 12, 13, 17, 18

To answer questions online and to receive **immediate corrective feedback** and **fully worked solutions** for all questions, go to your learnON title at www.jacplus.com.au.

Fluency

1. Iam studies five more subjects than Xan. Determine how many subjects Iam studies if:

   a. Xan studies six subjects
   b. Xan studies x subjects
   c. Xan studies y subjects.

2. Eva and Juliette walk home from school together. Eva's home is 2 km further from school than Juliette's home. Calculate how far Eva walks if Juliette's home is:
   a. 1.5 km from school
   b. x km from school.

3. Lisa watched television for 2.5 hours today. Calculate how many hours she will watch tomorrow if she watches:

   a. 1.5 hours more than she watched today
   b. t hours more than she watched today
   c. y hours fewer than she watched today.

4. Samir ran d km today. Calculate the distance he will cover tomorrow if he runs:

   a. x km less than he ran today
   b. y km more than he ran today
   c. 4 km more than double the distance that he ran today.

Understanding

5. **WE5** Write an algebraic expression for each of the following questions.

   a. If it takes 10 minutes to iron a single shirt, calculate how long it would take to iron all of Anthony's shirts.
   b. Ross has 30 dollars more than Nick. If Nick has N dollars, calculate how much money Ross has.
   c. In a game of Aussie Rules football Luciano kicked 4 more goals than he kicked behinds.

      i. Determine how many behinds Luciano scored, if g is the number of goals kicked.
      ii. Calculate the number of points Luciano scored.
         (*Note:* 1 goal scores 6 points, 1 behind scores 1 point.)

6. Jeff and Chris play Aussie Rules football for opposing teams. Jeff's team won when the two teams played each other.

   a. Calculate how many points Jeff's team scored if they kicked:

      i. 14 goals and 10 behinds
      ii. x goals and y behinds.

   b. Calculate how many points Chris's team scored if his team kicked:

      i. 10 goals and 6 behinds
      ii. p goals and q behinds.

   c. Calculate how many points Jeff's team won by if:

      i. Chris's team scored 10 goals and 6 behinds, and Jeff's team scored 14 goals and 10 behinds
      ii. Chris's team scored p goals and q behinds, and Jeff's team scored x goals and y behinds.

   (*Note:* 1 goal scores 6 points, 1 behind scores 1 point.)

7. Yvonne's mother gives her x dollars for each school subject she passes. If she passes y subjects, determine how much money Yvonne receives.

8. Rosya buys a bag containing x Smarties.

   a. If they divide them equally among n people, calculate how many each person gets.
   b. If they keep half of the Smarties for themselves and divide the remaining Smarties equally among n people, calculate how many Smarties each person gets.

9. A piece of licorice is 30 cm long.

   a. If Ameer cuts d cm off, calculate how much licorice remains.
   b. If Ameer cuts off $\dfrac{1}{4}$ of the remaining licorice, calculate how much licorice has been cut off.
   c. Determine how much licorice remains after the two cuts that Ameer has made.

Reasoning

10. One-quarter of a class of x students plays tennis at the weekend. One-sixth of the class plays tennis and also swims at the weekend.

    a. Write an expression to represent the number of students playing tennis at the weekend.
    b. Write an expression to represent the number of students playing tennis and also swimming at the weekend.
    c. Show that the number of students playing only tennis at the weekend is $\dfrac{x}{12}$.

11. During a 24-hour period, Vanessa uses her computer for c hours. Her brother Darren uses it for $\dfrac{1}{7}$ of the remaining time.

    a. Determine the length of time for which Darren used the computer.
    b. Show that the total number of hours that Vanessa and Darren use the computer during a 24-hour period can be expressed as $\dfrac{6c + 24}{7}$.

12. Dizem had a birthday party last weekend and invited n friends. The table shown indicates the number of friends at Dizem's party at certain times during the evening. Everybody arrived by 8.30 pm and everyone left the party by 11 pm.

   a. Show that 1 person arrived between 7.00 pm and 7.30 pm.

   b. Determine the start and end times during which the most friends arrived.

   c. Calculate how many friends were invited but did not arrive.

   d. Show that 24 friends were invited in total.

   e. Determine the start and end times during which the most friends were present at the party.

Time	Number of friends
7.00 pm	$n - 24$
7.30 pm	$n - 23$
8.00 pm	$n - 8$
8.30 pm	$n - 5$
9.00 pm	$n - 5$
9.30 pm	$n - 7$
10.00 pm	$n - 12$
10.30 pm	$n - 18$
11.00 pm	$n - 24$

13. **WE6** Ty earns $\$y$ per week working at his part-time job.

   a. If he spends $30 a week on his phone, calculate how much money he has left.

   b. If he saves two-thirds of the remaining money for rent each week, calculate how much money is left.

   c. If, after paying for his phone, saving for rent and spending $100 on food, Ty still has $200 left over, determine how much he earns in a week.

Problem solving

14. A child builds a pyramid out of building blocks using the pattern shown.

   a. Determine a rule that gives the number of blocks on the bottom layer, b, for a tower that is h blocks high.

   b. If the child wants the tower to be 10 blocks high, determine how many blocks they should begin with on the bottom layer.

15. A shop owner is ordering vases from a stock catalogue. Red vases cost $4.20 and clear vases cost $5.70. If she buys 25 vases for $120, determine how many of each type of vase she bought.

16. A father is 4 times the age of his son. In 4 years time the father will be 3 times the age of his son. Determine how old the father is now.

17. If a circle has 3 equally spaced dots (A, B and C) on its circumference, only 1 triangle can be formed by joining the dots. Investigate to determine the number of triangles that could be formed by 4, 5 or more equally spaced dots on the circumference.

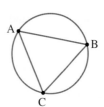
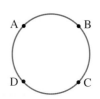

From your investigation, can you see a pattern in the number of triangles (t) formed for a circle with d equally spaced dots on its circumference? Determine whether it is possible to express this as a formula.

18. Answer the following questions, showing full working.

   a. Transpose the equation $T = \dfrac{x-3}{x-1}$ to make x the subject.

   b. Evaluate T when $x = 1$.

   c. Evaluate x when $T = 1$.

2.4 Simplification of algebraic expressions

> **LEARNING INTENTION**
>
> At the end of this subtopic you should be able to:
> - identify whether two terms are like terms
> - simplify algebraic expressions by adding and subtracting like terms
> - simplify algebraic expressions that involve multiplication and division of multiple algebraic terms.

2.4.1 Addition and subtraction of like terms

eles-4592

- Simplifying an expression involves writing it in a form with the least number of terms, without $\times$ or $\div$ signs, and with any fractions expressed in their simplest form.
- **Like terms** have identical pronumeral parts (including the power), but may have different coefficients. For example:
  - $5y$ and $10y$ are like terms, but $5y^2$ and $10y$ are not like terms.
  - $3mn^2$ and $-4mn^2$ and like terms, but $3m^2n$ and $-2mn^2$ are not like terms.
- If a term contains more than one pronumeral, the convention is to write the pronumerals in alphabetical order. This makes it easier to identify like terms. For example:
  - $3cb^2a$ and $-7b^2ca$ are like terms and should be written as $3ab^2c$ and $-7ab^2c$ to make this easier to identify.
- We can simplify like terms by adding (or subtracting) the coefficients to form a single term. We cannot simplify any terms that are not like terms. For example:
  - $10xy + 3xy - 7xy = 6xy$.

> ### WORKED EXAMPLE 7 Simplifying algebraic expressions
>
> **Simplify the following expressions.**
> **a.** $x + 2x + 3x - 5$ **b.** $9a^2b + 5ba^2 + ab^2$ **c.** $-12 - 4c^2 + 10 + 5c^2$
>
> | THINK | WRITE |
> |---|---|
> | **a. 1.** Write the expression. | **a.** $x + 2x + 3x - 5$ |
> | **2.** Collect the like terms x, $2x$ and $3x$. | $= 6x - 5$ |
> | **b. 1.** Write the expression. | **b.** $9a^2b + 5ba^2 + ab^2$ |
> | **2.** Collect the like terms $9a^2b$ and $5ba^2$. | $= 14a^2b + ab^2$ |
> | **c. 1.** Write the expression. | **c.** $-12 - 4c^2 + 10 + 5c^2$ |
> | **2.** Collect the like terms $4c^2$ and $5c^2$. | $= -12 + 10 - 4c^2 + 5c^2$ |
> | **3.** Simplify the expression. | $= -2 + c^2$ or $c^2 - 2$ |

2.4.2 Multiplication of algebraic terms

eles-4593

- Any number of algebraic terms can be multiplied together to produce a single term.
- Numbers can be multiplied in any order, so it is easiest to multiply the coefficients first, then multiply the pronumerals in alphabetical order.
- Remember that, when multiplying pronumerals, we also add their indices together.

Note: $x = x^1$

$$4ab \times 3a^2b \times 2a^3 = 4 \times a \times b \times 3 \times a^2 \times b \times 2 \times a^3$$
$$= 4 \times 3 \times 2 \times a \times a^2 \times a^3 \times b \times b$$
$$= 24 \times a^6 \times b^2$$
$$= 24a^6b^2$$

WORKED EXAMPLE 8 Multiplication of algebraic terms

Simplify the following expressions.

a. $4a \times 2b \times a$

b. $7ax \times -6bx \times -2abx$

THINK	WRITE
a. 1. Write the expression.	a. $4a \times 2b \times a$
2. Rearrange the expression, writing the coefficients first.	$= 4 \times 2 \times a \times a \times b$
3. Multiply the coefficients and pronumerals separately.	$= 8a^2b$
b. 1. Write the expression.	b. $7ax \times -6bx \times -2abx$
2. Rearrange the expression, writing the coefficients first.	$= 7 \times -6 \times -2 \times a \times a \times b \times b \times x \times x \times x \times x$
3. Multiply the coefficients and pronumerals separately.	$= 84a^2b^2x^3$

TI | THINK	WRITE	CASIO | THINK	WRITE
a–b. In a new problem, on a Calculator page, complete the entry lines as: $4a \times 2b \times a$ $7a \times x \times$ $-6b \times x \times -2a \times b \times x$ Press ENTER after each entry line.	a–b.  a. $4a \times 2b \times a = 8a^2b$ b. $7a \times x \times -6b \times x \times -2a \times b \times x = 84a^2b^2x^3$	a–b. On the Main screen, complete the entry lines as: $4a \times 2b \times a$ $7a \times x \times -6b \times$ $x \times -2a \times b \times x$ Press EXE after each entry line.	a–b. 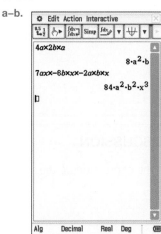 a. $4a \times 2b \times a = 8a^2b$ b. $7a \times x \times -6b \times x \times -2a \times b \times x = 84a^2b^2x^3$

▶ 2.4.3 Division of algebraic terms

eles-4594

- When dividing two algebraic terms, first rewrite the terms in fraction form.
- Simplify the fraction by dividing through by the highest common factor of the coefficients and cancelling down the pronumerals.
- Remember that when dividing pronumerals we subtract their indices.

$$36x^3yz^2 \div 40x^2y^2 = \frac{36x^3yz^2}{40x^2y^2}$$

$$= \frac{36^9 \times x \times \cancel{x} \times \cancel{x} \times \cancel{y} \times z \times z}{{}_{10}\cancel{40} \times \cancel{x} \times \cancel{x} \times y \times \cancel{y}}$$

$$= \frac{9xz^2}{10y}$$

WORKED EXAMPLE 9 Dividing algebraic terms

Simplify the following terms.

a. $\dfrac{12xy}{4xz}$

b. $-8ab^2 \div -16a^2b$

THINK		WRITE
a. 1. Write the term.	a.	$\dfrac{12xy}{4xz}$
2. Cancel the common factors.		$= \dfrac{{}^3\cancel{12} \times \cancel{x} \times y}{{}_1\cancel{4} \times \cancel{x} \times z}$
3. Write the answer.		$= \dfrac{3y}{z}$
b. 1. Write the term as a fraction.	b.	$\dfrac{-8ab^2}{-16a^2b}$
2. Cancel the common factors.		$= \dfrac{{}^1\cancel{-8} \times \cancel{a} \times \cancel{b}^2 b}{{}_2\cancel{-16} \times_a \cancel{a}^2 \times \cancel{b}_1}$
3. Simplify and write the answer.		$= \dfrac{1}{2} \times \dfrac{1}{a} \times \dfrac{b}{1}$
		$= \dfrac{b}{2a}$

DISCUSSION

Is the expression *ab* the same as *ba*? Explain your answer.

Exercise 2.4 Simplification of algebraic expressions

learn

Individual pathways

■ PRACTISE	■ CONSOLIDATE	■ MASTER
1, 3, 9, 10, 11, 14, 19, 22, 25, 26	2, 4, 5, 7, 12, 15, 17, 20, 23, 27, 28	6, 8, 13, 16, 18, 21, 24, 29, 30

To answer questions online and to receive **immediate corrective feedback** and **fully worked solutions** for all questions, go to your learnON title at www.jacplus.com.au.

Fluency

1. For each of the following terms, select the terms listed in brackets that are like terms.

   a. $6ab$ $\quad\left(7a, 8b, 9ab, -ab, 4a^2b^2\right)$
   b. $-x$ $\quad\left(3xy, -xy, 4x, 4y, -yx\right)$
   c. $3az$ $\quad\left(3ay, -3za, -az, 3z^2a, 3a^2z\right)$
   d. x^2 $\quad\left(2x, 2x^2, 2x^3, -2x, -x^2\right)$

2. For each of the following terms, select the terms listed in brackets that are like terms.

   a. $2x^2y$ $\quad\left(xy, -2xy, -2xy^2, -2x^2y, -2x^2y^2\right)$
   b. $3x^2y^5$ $\quad\left(3xy, 3x^5y^2, 3x^4y^3, -x^2y^5, -3x^2y^5\right)$
   c. $5x^2\,p^3w^5$ $\quad\left(-5x^3w^5\,p^3, p^3\,x^2\,w^5, 5xp^3w^5, -5x^2\,p^3w^5, w^5\,p^2\,x^3\right)$
   d. $-x^2y^5z^4$ $\quad\left(-xy^5, -y^2z^5x^4, -x+y+z, 4y^5z^4x^2, -2x^2z^4y^5\right)$

3. **WE7** Simplify the following expressions.

   a. $5x + 2x$
   b. $3y + 8y$
   c. $7m + 12m$
   d. $13q - 2q$
   e. $17r - 9r$
   f. $-x + 4x$

4. Simplify the following expressions.

   a. $5a + 2a + a$
   b. $9y + 2y - 3y$
   c. $7x - 2x + 8x$
   d. $14p - 3p + 5p$
   e. $2q^2 + 7q^2$
   f. $5x^2 - 2x^2$

5. Simplify the following expressions.

   a. $6x^2 + 2x^2 - 3y$
   b. $3m^2 + 2n - m^2$
   c. $-2g^2 - 4g + 5g - 12$
   d. $-5m^2 + 5m - 4m + 15$
   e. $12a^2 + 3b + 4b^2 - 2b$
   f. $6m + 2n^2 - 3m + 5n^2$

6. Simplify the following expressions.

   a. $3xy + 2y^2 + 9yx$
   b. $3ab + 3a^2b + 2a^2b - ab$
   c. $9x^2y - 3xy + 7yx^2$
   d. $4m^2n + 3n - 3m^2n + 8n$
   e. $-3x^2 - 4yx^2 - 4x^2 + 6x^2y$
   f. $4 - 2a^2b - ba^2 + 5b - 9a^2$

7. **MC** Choose which of the following is a simplification of the expression $18p - 19p$.

    A. p **B.** $-p$ **C.** p^2 **D.** -1 **E.** $p - 1$

8. **MC** Choose which of the following is a simplification of the expression $5x^2 - 8x + 6x - 9$.

    A. $3x - 9$ **B.** $3x^2 - 9$ **C.** $5x^2 + 2x - 9$

    D. $5x^2 - 2x - 9$ **E.** $-3x^2 + 6x - 9$

9. **MC** Choose which of the following is a simplification of the expression $12a - a + 15b - 14b$.

    A. $11a + b$ **B.** 12 **C.** $11a - b$

    D. $13a + b$ **E.** $-12a - b$

10. **MC** Choose which of the following is a simplification of the expression $-7m^2n + 5m^2 + 3 - m^2 + 2m^2n$.

    A. $-9m^2n + 4m^2 + 3$ **B.** $-9m^2n + 8$ **C.** $-5m^2n - 4m^2 + 3$

    D. $-5m^2n + 4m^2 + 3$ **E.** $5m^2n + 4m^2 + 3$

11. **WE8** Simplify the following.

    a. $3m \times 2n$ **b.** $4x \times 5y$ **c.** $2p \times 4q$ **d.** $5x \times -2y$ **e.** $3y \times -4x$

12. Simplify the following.

    a. $-3m \times -5n$ **b.** $5a \times 2a$ **c.** $3mn \times 2p$ **d.** $-6ab \times b$ **e.** $-5m \times -2mn$

13. Simplify the following.

    a. $-6a \times 3ab$ **b.** $-3xy \times -5xy \times 2x$ **c.** $4pq \times -p \times 3q^2$

    d. $4c \times -7cd \times 2c$ **e.** $-3a^2 \times -5ab^3 \times 2ab^4$

14. **WE9** Simplify the following.

    a. $\dfrac{6x}{2}$ **b.** $\dfrac{9m}{3}$ **c.** $\dfrac{12y}{6}$ **d.** $\dfrac{8m}{2}$ **e.** $12m \div 3$

15. Simplify the following.

    a. $14x \div 7$ **b.** $-21x \div 3$ **c.** $-32m \div 8$ **d.** $\dfrac{4m}{8}$ **e.** $\dfrac{6x}{18}$

16. Simplify the following.

    a. $\dfrac{8mn}{18n}$ **b.** $\dfrac{6ab}{12a^2b}$ **c.** $\dfrac{28xyz}{14x}$

    d. $\dfrac{2x^2yz}{8xz}$ **e.** $-7xy^2z^2 \div 11xyz$

Understanding

17. Simplify the following.

    a. $5x \times 4y \times 2xy$ **b.** $7xy \times 4ax \times 2y$ **c.** $\dfrac{6x^2y}{12y^2}$ **d.** $\dfrac{-15x^2ab}{12b^2x^2}$

18. Simplify the following.

    a. $\dfrac{2p^3q^2}{p^3q^2}$ **b.** $-4a \times -5ab^2 \times 2a$ **c.** $-a \times 4ab \times 2ba \times b$ **d.** $2a \times 2a \times 2a \times 2a$

19. Jim buys m pens at p cents each and n books at q dollars each.

    a. Calculate how much Jim spends in:

        i. dollars   **ii.** cents.

    b. Calculate how much change Jim will have if he starts with $20.

20. At a local discount clothing store 4 shirts and 3 pairs of shorts cost $138 in total. If a pair of shorts costs 2.5 times as much as a shirt, calculate the cost of each kind of clothing.

21. Anthony and Jamila are taking their 3 children to see a movie at the cinema. The total cost for the 2 adults and 3 children is $108. If an adult's ticket is 1.5 times the cost of a child's, calculate the cost of 1 child's ticket plus 1 adult's ticket.

Reasoning

22. Class 9A were given an algebra test. One of the questions is shown below.

    Simplify the following expression: $\dfrac{3ab}{2} \times \dfrac{4ac}{6b} \times 7c$.

    Sean, who is a student in class 9A, wrote his answer as $\dfrac{12aabc}{12b} \times 7c$. Explain why Sean's answer is incorrect, and write the correct answer.

23. Using the appropriate method to divide fractions, simplify the following expression.

$$\frac{15a^2b^2}{16c^4} \div \left(\frac{5ac^2}{4b^3} \times \frac{a}{2b^4} \right)$$

24. Answer the following questions.

    a. Simplify the expression $\dfrac{n}{(n+1)} \times \dfrac{(n+1)}{(n+2)}$.

    b. Simplify the expression $\dfrac{n}{(n+1)} \times \dfrac{(n+1)}{(n+2)} \times \dfrac{(n+2)}{(n+3)}$.

    c. Use the results from parts **a** and **b** to evaluate $\dfrac{1}{2} \times \dfrac{2}{3} \times \dfrac{3}{4} \times \ldots \times \dfrac{99}{100}$.

Problem solving

25. a. Fill in the empty bricks in the following pyramids. The expression in each brick is obtained by adding the expressions in the two bricks below it.

    i.
    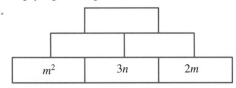

    ii.
    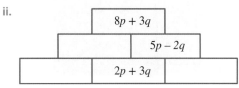

    b. Fill in the empty bricks in the following pyramids. The expression in each brick is obtained by multiplying the expressions in the two bricks below it.

    i.
    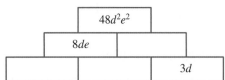

    ii.
    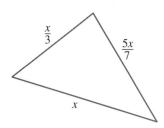

26. For the triangle shown:

    a. write an expression for the perimeter of the triangle

    b. show that the triangle's perimeter can be simplified to $\dfrac{43}{21}x$

    c. calculate the cost (to the nearest dollar) to frame the triangle using timber that costs $4.50 per metre, if x equals 2 metres.

27. For the rectangle shown:

   a. write an expression for the perimeter of the rectangle

   b. show that the expression for the perimeter can be simplified to $6\frac{4}{7}w$

   c. calculate the cost (to the nearest cent) to create a wire frame for the rectangle using wire that costs $1.57 per metre, if w equals seven metres.

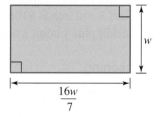

28. A rectangular chocolate block has dimensions x and $(4x - 7)$.

   a. Write an expression for the perimeter (P) in the form $P = -x + -(4x - 7)$.
   b. Expand and simplify this expression.
   c. Calculate the perimeter of the chocolate block when $x = 3$.
   d. Explain why x cannot equal 1.

29. A doghouse in the shape of a rectangular prism is to be reinforced with steel edging along its outer edges including the base. The cost of the steel edging is $1.50 per metre. All measurements are in centimetres.

   a. Write an expression for the total length of the straight edges of the frame.
   b. Expand the expression.
   c. Evaluate the cost of steel needed when $x = 80$ cm.

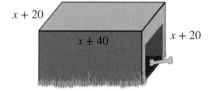

30. A tile manufacturer produces tiles that have the side lengths shown. All measurements are in centimetres.

   a. Write an expression for the perimeter of each shape.
   b. Evaluate the value of x for which the perimeter of the triangular tile is the same as the perimeter of the square tile.

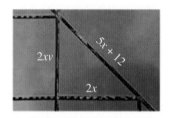

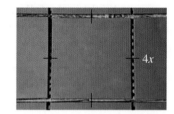

2.5 Expanding brackets

LEARNING INTENTION

At the end of this subtopic you should be able to:
• apply the distributive law to expand and simplify an expression with a single set of brackets
• apply the distributive law to expand and simplify expressions containing two or more sets of brackets.

▶ 2.5.1 The Distributive Law

eles-4595

• There are two ways of calculating the area of the rectangle shown.

1. The rectangle can be treated as a single shape with a length of 4 cm and a width of $5 + 3$ cm.

$$A = l \times w$$
$$= 4(5 + 3)$$
$$= 4 \times 8$$
$$= 32 \text{ cm}^2$$

2. The areas of the two smaller rectangles can be added together.

$$A = 4 \times 5 + 4 \times 3$$
$$= 20 + 12$$
$$= 32 \text{ cm}^2$$

- The length of the rectangle shown is a and its width is $b + c$.

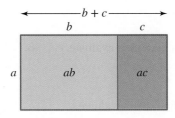

- This rectangle's area can be found in two different ways.

   1. $A = l \times w$
      $= a(b + c)$

   This expression is described as **factorised** because it shows one number (or factor) multiplied by another. The two factors are a and $(b + c)$.

   2. The areas of the two small rectangles can be added together.

$$A = ab + ac$$

This expression is described as **expanded**, which means that it is written without brackets.

- $a(b + c) = ab + ac$ is called the **Distributive Law**.

The Distributive Law

In order to expand a single set of brackets we apply the distributive law, which states that:

$$a(b + c) = a \times b + a \times c = ab + ac$$

- A helpful way to expand brackets is to draw arrows between factors as you work out each multiplication.

$$x(x - y) = x^2 \ldots$$
$$x(x - y) = x^2 - xy$$

WORKED EXAMPLE 10 Expanding a single set of brackets

Use two different methods to calculate the value of the following.
a. $7(5 + 15)$ b. $10(9 - 1)$

THINK	WRITE
a. 1. Method 1: Work out the brackets first, then evaluate.	a. $7(5 + 15)$ $= 7 \times 20$ $= 140$
2. Method 2: Expand the brackets first, then evaluate.	$7(5 + 15)$ $= 7 \times 5 + 7 \times 15$ $= 35 + 105$ $= 140$
b. 1. Method 1: Work out the brackets first, then evaluate.	b. $10(9 - 1)$ $= 10 \times 8$ $= 80$
2. Method 2: Expand the brackets first, then evaluate	$10(9 - 1)$ $= 10 \times 9 + 10 \times -1$ $= 90 - 10$ $= 80$

WORKED EXAMPLE 11 Expanding using arrows

Expand the following expressions.
a. $5(x + 3)$ b. $-4y(2x - w)$ c. $3 \times (5 - 6y + 2y)$

THINK	WRITE
a. 1. Draw arrows to help with the expansion.	a. $5(x + 3)$
2. Simplify and write the answer.	$= 5 \times x + 5 \times 3$ $= 5x + 15$
b. 1. Draw arrows to help with the expansion.	b. $-4y(2x - w)$
2. Simplify and write the answer.	$= -4y \times 2x - 4y \times -w$ $= -8xy + 4wy$
c. 1. Draw arrows to help with the expansion.	c. $3x(5 - 6x + 2y)$
2. Simplify and write the answer.	$= 3x \times 5 + 3x \times -6x + 3x \times 2y$ $= 15x - 18x^2 + 6xy$

▶ 2.5.2 Expanding and simplifying

- When a problem involves expanding that is more complicated, it is likely that the like terms will need to be collected and simplified after you have expanded the brackets.
- When solving these more complicated problems, expand all sets of brackets first, then simplify any like terms that result from the expansion.

WORKED EXAMPLE 12 Expanding and simplifying

Expand and simplify the following expressions by collecting like terms.

a. $4(x-4)+5$ **b.** $x(y-2)+5x$ **c.** $-x(y-2)+5x$ **d.** $7x-6(y-2x)$

THINK

a. 1. Expand the brackets.

   2. Simplify and write the answer.

b. 1. Expand the brackets.

   2. Simplify and write the answer.

c. 1. Expand the brackets.

   2. Simplify and write the answer. (*Note:* There are no like terms.)

d. 1. Expand the brackets.

   2. Simplify and write the answer.

WRITE

a.

$$4(x-4)+5$$

$$= 4x-16+5$$
$$= 4x-11$$

b.

$$x(y-2)+5x$$

$$= xy-2x+5x$$
$$= xy+3x$$

c.

$$-x(y-z)+5x$$

$$-xy+xz+5x$$

d.

$$7x-6(y-2x)$$

$$= 7x-6y+12x$$
$$= 19x-6y$$

WORKED EXAMPLE 13 Expanding two sets of brackets and simplifying

Expand and simplify the following expressions.

a. $5(x+2y)+6(x-3y)$ **b.** $5x(y-2)-y(x+3)$

THINK

a. 1. Expand each set of brackets.

   2. Simplify and write the answer.

b. 1. Expand each set of brackets.

   2. Simplify and write the answer.

WRITE

a.

$$5(x+2y)+6(x-3y)$$

$$= 5x+10y+6x-18y$$
$$= 11x-8y$$

b.

$$5x(y-2)-y(x+3)$$

$$= 5xy-10x-xy-3y$$
$$= 4xy-10x-3y$$

2.5.3 Expanding binomial factors

eles-4597

- Remember that a **binomial** is an expression containing two terms, for example $x + 3$ or $2y - z^2$. In this section we will look at how two binomials can be multiplied together.
- The rectangle shown has length $a + b$ and width $c + d$.

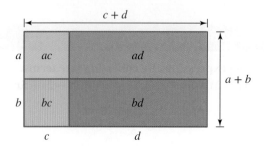

- There are two ways of finding the area of the large rectangle.
  1. $A = l \times w$

     $= (a + b) \times (c + d)$

     $= (a + b)(c + d)$

     This is a factorised expression in which the two factors are $(a + b)$ and $(c + d)$.
  2. The areas of the four small rectangles can be added together.

     $A = ac + ad + bc + bd$

     So $(a + b)(c + d) = ac + ad + bc + bd$
- There are several methods that can be helpful when remembering how to expand binomial factors. One commonly used method is **FOIL**.
- Each of the letters in FOIL stand for:

  First — multiply the first term in each bracket.

  Outer — multiply the two outer terms of each bracket.

  Inner — multiply the two inner terms of each bracket.

  Last — multiply the last term of each bracket.

FOIL – First Outer Inner Last

$(a + b)(c + d)$

$(a + b)(c + d) = ac + ad + bc + bd$

OR

'eyebrow, eyebrow, nose and mouth'

eyebrow — eyebrow

$(a + b)(c + d)$

nose — mouth

$(a + b)(c + d) = ac + bd + bc + ad$

Expanding binomial factors

Using FOIL, the expanded product of two binomial factors is given by:

$$(a + b)(c + d) = a \times c + a \times d + b \times c + b \times d = ac + ad + bc + bd$$

It is expected that the result will have 4 terms.

Note: It may be possible to simplify like terms after expanding.

Expand and simplify each of the following expressions.

a. $(x - 5)(x + 3)$ b. $(x + 2)(x + 3)$ c. $(2x + 2)(2x + 3)$

THINK	WRITE
a. 1. Expand the brackets using FOIL.	a. 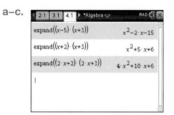
2. Simplify the expression by collecting like terms.	$= x^2 + 3x - 5x - 15$ $= x^2 - 2x - 15$
b. 1. Expand the brackets using FOIL.	b. $(x + 2)(x + 3)$
2. Simplify the expression by collecting like terms.	$= x^2 + 3x + 2x + 6$ $= x^2 + 5x + 6$
c. 1. Expand the brackets using FOIL.	c. $(2x + 2)(2x + 3)$
2. Simplify the expression by collecting like terms.	$= 4x^2 + 6x + 4x + 6$ $= 4x^2 + 10x + 6$

TI | THINK	WRITE	CASIO | THINK	WRITE
a–c. In a new problem, on a Calculator page, press: • MENU • 3: Algebra 3 • 3: Expand 3. Complete the entry lines as: expand $((x - 5) \times (x + 3))$ expand $((x + 2) \times (x + 3))$ expand $((2x + 2) \times (2x + 3))$ Press ENTER after each entry. *Note:* Remember to include the multiplication sign between the brackets.	a–c. 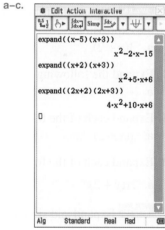 a. $(x - 5)(x + 3)$ $= x^2 - 2x - 15$ b. $(x + 2)(x + 3)$ $= x^2 + 5x + 6$ c. $(2x + 2)(2x + 3)$ $= 4x^2 + 10x + 6$	a–c. On the Main screen, press: • Action • Transformation • Expand Complete the entry lines as: expand $((x - 5)(x + 3))$ expand $((x + 2)(x + 3))$ expand $((2x + 2)(2x + 3))$ Press EXE after each entry line.	a–c. a. $(x - 5x + 3) = x^2 - 2x - 15$ b. $(x + 2x + 3) = x^2 + 5x + 6$ c. $(2x + 22x + 3)$ $= 4x^2 + 10x + 6$

DISCUSSION

Explain why, when expanded, $(x + y)(2x + y)$ gives the same result as $(2x + y)(x + y)$.

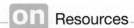
Exercise 2.5 Expanding brackets

learn on

Individual pathways

■ PRACTISE	■ CONSOLIDATE	■ MASTER
1, 2, 7, 9, 12, 15, 18, 22, 24, 26, 30, 33	3, 4, 6, 10, 13, 17, 20, 21, 25, 27, 31, 34	5, 8, 11, 14, 16, 19, 23, 28, 29, 32, 35, 36

To answer questions online and to receive **immediate corrective feedback** and **fully worked solutions** for all questions, go to your learnON title at www.jacplus.com.au.

Fluency

1. **WE10** Use 2 different methods to find the value of the following expressions.
   a. $8(10-2)$
   b. $11(99+1)$
   c. $-5(3+1)$
   d. $7(100-1)$

2. **WE11** Expand the following expressions. For the first two examples, draw a diagram to represent the expression.
   a. $3(x+2)$
   b. $4(x+3)$
   c. $4(x+1)$
   d. $7(x-1)$

3. Expand the following expressions.
   a. $-3(p-2)$
   b. $-(x-1)$
   c. $3(2b-4)$
   d. $8(3m-2)$

4. Expand each of the following. For the first two examples, draw a diagram to represent the expression.
   a. $x(x+2)$
   b. $a(a+5)$
   c. $x(4+x)$
   d. $m(7-m)$

5. Expand each of the following.
   a. $2x(y+2)$
   b. $-3y(x+4)$
   c. $-b(3-a)$
   d. $-6a(5-3a)$

6. **WE12** Expand and simplify the following expressions by collecting like terms.
   a. $2(p-3)+4$
   b. $5(x-5)+8$
   c. $-7(p+2)-3$
   d. $-4(3p-1)-1$
   e. $6x(x-3)-2x$

7. Expand and simplify the following expressions by collecting like terms.
   a. $2m(m+5)-3m$
   b. $3x(p+2)-5$
   c. $4y(y-1)+7$
   d. $-4p(p-2)+5p$
   e. $5(x-2y)-3y-x$

8. Expand and simplify the following expressions by collecting like terms.
   a. $2m(m-5)+2m-4$
   b. $-3p(p-2q)+4pq-1$
   c. $-7a(5-2b)+5a-4ab$
   d. $4c(2d-3c)-cd-5c$
   e. $6p+3-4(2p+5)$

9. **WE13** Expand and simplify the following expressions.
   a. $2(x+2y)+3(2x-y)$
   b. $4(2p+3q)+2(p-2q)$
   c. $7(2a+3b)+4(a+2b)$
   d. $5(3c+4d)+2(2c+d)$
   e. $-4(m+2n)+3(2m-n)$

10. Expand and simplify the following expressions.
    a. $-3(2x+y)+4(3x-2y)$
    b. $-2(3x+2y)+3(5x+3y)$
    c. $-5(4p+2q)+2(3p+q)$
    d. $6(a-2b)-5(2a-3b)$
    e. $5(2x-y)-2(3x-2y)$

11. Expand and simplify the following expressions.
    a. $4(2p-4q)-3(p-2q)$
    b. $2(c-3d)-5(2c-3d)$
    c. $7(2x-3y)-(x-2y)$
    d. $-5(p-2q)-(2p-q)$
    e. $-3(a-2b)-(2a+3b)$

12. Expand and simplify the following expressions.
    a. $a(b+2)+b(a-3)$
    b. $x(y+4)+y(x-2)$
    c. $c(d-2)+c(d+5)$
    d. $p(q-5)+p(q+3)$
    e. $3c(d-2)+c(2d-5)$

13. Expand and simplify the following expressions.
    a. $7a(b-3)-b(2a+3)$
    b. $2m(n+3)-m(2n+1)$
    c. $4c(d-5)+2c(d-8)$
    d. $3m(2m+4)-2(3m+5)$
    e. $5c(2d-1)-(3c+cd)$

14. Expand and simplify the following expressions.
    a. $-3a(5a+b)+2b(b-3a)$
    b. $-4c(2c-6d)+d(3d-2c)$
    c. $6m(2m-3)-(2m+4)$
    d. $7x(5-x)+6(x-1)$
    e. $-2y(5y-1)-4(2y+3)$

15. **MC** Choose the expression that is the equivalent of $3(a+2b)+2(2a-b)$.
    A. $5a+6b$
    B. $7a+4b$
    C. $5(3a+b)$
    D. $7a+8b$
    E. $5a+2b$

16. **MC** Choose the expression that is the equivalent of $-3(x-2y)-(x-5y)$.
    A. $-4x+11y$
    B. $-4x-11y$
    C. $4x+11y$
    D. $4x+7y$
    E. $-4x-7y$

17. **MC** Choose the expression that is the equivalent of $2m(n+4)+m(3n-2)$.
    A. $3m+4n-8$
    B. $5mn+4m$
    C. $5mn+10m$
    D. $5mn+6m$
    E. $3mn+4m$

18. **WE14** Expand and simplify each of the following expressions.
    a. $(a+2)(a+3)$
    b. $(x+4)(x+3)$
    c. $(y+3)(y+2)$
    d. $(m+4)(m+5)$
    e. $(b+2)(b+1)$

19. Expand and simplify each of the following expressions.
    a. $(p+1)(p+4)$
    b. $(a-2)(a+3)$
    c. $(x-4)(x+5)$
    d. $(m+3)(m-4)$
    e. $(y+5)(y-3)$

20. Expand and simplify each of the following expressions.
    a. $(y-6)(y+2)$
    b. $(x-3)(x+1)$
    c. $(x-3)(x-4)$
    d. $(p-2)(p-3)$
    e. $(x-3)(x-1)$

21. Expand and simplify each of the following expressions.
    a. $(2a+3)(a+2)$
    b. $(c-6)(4c-7)$
    c. $(7-2t)(5-t)$
    d. $(2+3t)(5-2t)$
    e. $(7-5x)(2-3x)$
    f. $(5x-2)(5x-2)$

22. Expand and simplify each of the following expressions.
    a. $(x+y)(z+1)$
    b. $(2x+y)(z+4)$
    c. $(3p+q)(r+1)$
    d. $(a+2b)(a+b)$
    e. $(2c+d)(c-3d)$
    f. $(x+y)(2x-3y)$

23. Expand and simplify each of the following expressions.
    a. $(4p-3q)(p+q)$
    b. $(a+2b)(b+c)$
    c. $(3p-2q)(1-3r)$
    d. $(4x-y)(3x-y)$
    e. $(p-q)(2p-r)$
    f. $(5-2j)(3k-1)$

24. Choose the expression that is the equivalent of $(4-y)(7+y)$.

    A. $28-y^2$ **B.** $28-3y+y^2$ **C.** $28-3y-y^2$ **D.** $11-2y$ **E.** $28-11y-y^2$

25. **MC** Choose the expression that is the equivalent of $(2p+1)(p-5)$.

    A. $2p^2-5$ **B.** $2p^2-11p-5$ **C.** $2p^2-9p-5$ **D.** $2p^2-6p-5$ **E.** $2p^2-4p-5$

Understanding

26. Expand the following expressions using FOIL, then simplify.

    a. $(x+3)(x-3)$ **b.** $(x+5)(x-5)$ **c.** $(x+7)(x-7)$
    d. $(x-1)(x+1)$ **e.** $(x-2)(x+2)$ **f.** $(2x-1)(2x+1)$

27. Expand the following expressions using FOIL, then simplify.

    a. $(x+1)(x+1)$ **b.** $(x+2)(x+2)$ **c.** $(x+8)(x+8)$
    d. $(x-3)(x-3)$ **e.** $(x-5)(x-5)$ **f.** $(x-9)(x-9)$

28. Simplify the following expressions.

    a. $2.1x(3x+4.7y)-3.1y(1.4x+y)$ **b.** $(2.1x-3.2y)(2.1x+3.2y)$
    c. $(3.4x+5.1y)^2$

29. For the box shown, calculate the following in expanded form:

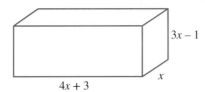

    a. the total surface area   **b.** the volume.

Reasoning

30. For each of the following shapes:

    i. write down the area in factor form
    ii. expand and simplify the expression
    iii. discuss any limitations on the value of x.

    a.

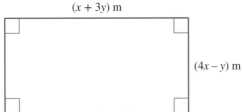

    b.

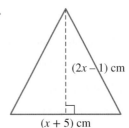

31. Show that the following is true.

$$(a-x)(a+x)-2(a-x)(a-x)-2x(a-x)$$
$$=-(a-x)^2$$

32. A series of incorrect expansions of $(x+8)(x-3)$ are shown below. For each of these incorrect expansions, explain the mistake that has been made.

    a. $x^2+11x+24$ **b.** $x^2+5x+24$ **c.** x^2-3x

Problem solving

33. Three students gave the following incorrect answers when expanding $-5(3x - 20)$.

    i. $-5x - 20$ ii. $-8x + 25$ iii. $15x - 100$

    a. Explain the errors made by each student.   b. Determine the correct answer.

34. In a test, a student expanded brackets and obtained the following answers. Identify and correct the student's errors and write the correct expansions.

    a. $-2(a - 5) = -2a - 10$ b. $2b(3b - 1) = 6b^2 - 1$ c. $-2(c - 4) = 2c + 8$

35. Three students' attempts at expanding $(3x + 4)(2x + 5)$ are shown.

Student A

$$(3x + 4)(2x + 5)$$
$$= 3x \times 2x + 3x \times 5 + 4 \times 2x + 4 \times 5$$
$$= 6x + 15x + 8x + 20$$
$$= 29x + 20$$
$$= 49x$$

Student B

$$(3x + 4)(2x + 5)$$
$$= 3x \times 2x + 4 \times 2x + 4 \times 5$$
$$= 6x^2 + 8x + 20$$

Student C

$$(3x + 4)(2x + 5)$$
$$= 3x \times 2x + 3x \times 5 + 4 \times 2x + 4 \times 5$$
$$= 6x^2 + 15x + 8x + 20$$
$$= 6x^2 + 23x + 20$$

    a. State which student's work is correct.

    b. Copy each of the incorrect answers into your workbook and correct the mistakes in each one as though you were their teacher.

36. Rectangular floor mats have an area of $(x^2 + 2x - 15)$ cm^2.

    a. The length of a mat is $(x + 5)$ cm. Determine an expression for the width of this mat.

    b. If the length of a mat is 70 cm, evaluate its width.

    c. If the width of a mat is 1 m, evaluate its length.

2.6 Difference of two squares and perfect squares

LEARNING INTENTION

At the end of this subtopic you should be able to:

- recognise and use the difference of two squares rule to quickly expand and simplify binomial products with the form $(a + b)(a - b)$
- recognise and use the perfect square rule to quickly expand and simplify binomial products with the form $(a \pm b)^2$.

2.6.1 Difference of two squares

eles-4598

- Consider the expansion of $(x + 4)(x - 4)$:

$$(x + 4)(x - 4) = x^2 - 4x + 4x - 16 = x^2 - 16$$

- Now consider the expansion of $(x-6)(x+6)$:

$$(x-6)(x+6)=x^2+6x-6x-36=x^2-36$$

- In both cases the middle two terms cancel each other out, leaving two terms, both of which are perfect squares.
- The two terms that are left are the first value squared minus the second value squared. This is where the phrase **difference of two squares** originates.
- In both cases the binomial terms can be written in the form $(x+a)$ and $(x-a)$. If we can recognise expressions that have this form, we can use the pattern above to quickly expand those expressions. For example: $(x+12)(x-12)=x^2-12^2=x^2-144$

> **Difference of two squares**
>
> The difference of two squares rule is used to quickly expand certain binomial products, as long as they are in the forms shown below:
> - $(a+b)(a-b)=a^2-b^2$
> - $(a-b)(a+b)=a^2-b^2$
>
> Because the two binomial brackets are being multiplied, the order of the brackets does not affect the final result.

WORKED EXAMPLE 15 Expanding using difference of two squares

Use the difference of two squares rule to expand and simplify each of the following expressions.
a. $(x+8)(x-8)$ **b.** $(6-3)(6+x)$ **c.** $(2x-3)(2x+3)$ **d.** $(3x+5)(5-3x)$

THINK	WRITE
a. 1. Write the expression.	**a.** $(x+8)(x-8)$
2. This expression is in the form $(a+b)(a-b)$, so the difference of two squares rule can be used. Expand using the formula.	$= x^2-8^2$ $= x^2-64$
b. 1. Write the expression.	**b.** $(6-x)(6+x)$
2. This expression is in the form $(a+b)(a-b)$, so the difference of two squares rule can be used. Expand using the formula. *Note:* $36-x^2$ is not the same as x^2-36.	$= 6^2-x^2$ $= 36-x^2$
c. 1. Write the expression.	**c.** $(2x-3)(2x+3)$
2. This expression is in the form $(a-b)(a+b)$, so the difference of two squares rule can be used. Expand using the formula. *Note:* $(2x)^2$ and $2x^2$ are not the same. In this case $a=2x$, so $a^2=(2x)^2$.	$= (2x)^2-3^2$ $= 4x^2-9$
d. 1. Write the expression.	**d.** $(3x+5)(5-3x)$
2. The difference of two squares rule can be used if we rearrange the terms, since $3x+5=5+3x$. Expand using the formula.	$(5+3x)(5-3x)$ $= 5^2-(3x)^2$ $= 25-9x^2$

▶ 2.6.2 Perfect squares

eles-4599

- A **perfect square** is the result of the square of a whole number. $1 \times 1 = 1, 2 \times 2 = 4$ and $3 \times 3 = 9$, showing that $1, 4$ and 9 are all perfect squares.
- Similarly, $(x + 3)(x + 3) = (x + 3)^2$ is a perfect square because it is the result of a binomial factor multiplied by itself.
- Consider the diagram illustrating $(x + 3)^2$. What shape is it?
- The area is given by $x^2 + 3x + 3x + 9 = x^2 + 6x + 9$.
- We can see from the diagram that there are two squares produced $(x^2$ and $3^2 = 9)$ and two rectangles that are identical to each other $(2 \times 3x)$.
- Compare this with the expansion of $(x + 6)^2$:

$$(x + 6)(x + 6) = x^2 + 6x + 6x + 36 = x^2 + 12x + 36$$

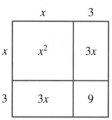

- A pattern begins to emerge after comparing these two expansions. The square of a binomial equals the square of the first term, plus double the product of the two terms plus the square of the second term. For example: $(x + 10)(x + 10) = x^2 + 2 \times 10 \times x + 10^2 = x^2 + 20x + 100$

Perfect squares

The rule for the expansion of the square of a binomial is given by:
- $(a + b)^2 = (a + b)(a + b) = a^2 + 2ab + b^2$
- $(a - b)^2 = (a - b)(a - b) = a^2 - 2ab + b^2$

We can use the above rules to quickly expand binomial products that are presented as squares.

WORKED EXAMPLE 16 Expanding perfect squares

Use the rules for expanding perfect binomial squares to expand and simplify the following.
a. $(x + 1)(x + 1)$
b. $(x - 2)^2$
c. $(2x + 5)^2$
d. $(4x - 5y)^2$

THINK	WRITE
a. 1. This expression is the square of a binomial.	a. $(x + 1)(x + 1)$
2. Apply the formula for perfect squares: $(a + b)^2 = a^2 + 2ab + b^2$	$= x^2 + 2 \times x \times 1 + 1^2$ $= x^2 + 2x + 1$
b. 1. This expression is the square of a binomial.	b. $(x - 2)^2$
2. Apply the formula for perfect squares: $(a - b)^2 = a^2 - 2ab + b^2$.	$= (x - 2)(x - 2)$ $= x^2 - 2 \times x \times 2 \times 2^2$ $= x^2 - 4x + 4$
c. 1. This expression is the square of a binomial.	c. $(2x + 5)^2$
2. Apply the formula for perfect squares: $(a + b)^2 = a^2 + 2ab + b^2$.	$= (2x)^2 + 2 \times 2x \times 5 + 5^2$ $= 4x^2 + 20x + 25$
d. 1. This expression is the square of a binomial.	d. $(4x - 5y)^2$
2. Apply the formula for perfect squares: $(a - b)^2 = a^2 - 2ab + b^2$.	$= (4x)^2 - 2 \times 4x \times 5y + (5y)^2$ $= 16x^2 - 40xy + 25y^2$

on Resources

eWorkbook   Topic 2 Workbook (worksheets, code puzzle and project) (ewbk-2002)

Digital document SkillSHEET Recognising expansion patterns (doc-10821)

Interactivities   Individual pathway interactivity: Difference of two squares and perfect squares (int-4484)
Difference of two squares (int-6036)

Exercise 2.6 Difference of two squares and perfect squares   learn on

Individual pathways

■ PRACTISE	■ CONSOLIDATE	■ MASTER
1, 3, 5, 8, 13, 16, 20, 21	2, 6, 9, 11, 14, 17, 22, 23	4, 7, 10, 12, 15, 18, 19, 24, 25

To answer questions online and to receive **immediate corrective feedback** and **fully worked solutions** for all questions, go to your learnON title at www.jacplus.com.au.

Fluency

1. **WE15** Use the difference of two squares rule to expand and simplify each of the following.
   a. $(x+2)(x-2)$ b. $(y+3)(y-3)$ c. $(m+5)(m-5)$ d. $(a+7)(a-7)$

2. Use the difference of two squares rule to expand and simplify each of the following.
   a. $(x+6)(x-6)$ b. $(p-12)(p+12)$ c. $(a+10)(a-10)$ d. $(m-11)(m+11)$

3. Use the difference of two squares rule to expand and simplify each of the following.
   a. $(2x+3)(2x-3)$ b. $(3y-1)(3y+1)$ c. $(5d-2)(5d+2)$ d. $(7c+3)(7c-3)$ e. $(2+3p)(2-3p)$

4. Use the difference of two squares rule to expand and simplify each of the following.
   a. $(d-9x)(d+9x)$ b. $(5-12a)(5+12a)$ c. $(3x+10y)(3x-10y)$
   d. $(2b-5c)(2b+5c)$ e. $(10-2x)(2x+10)$

5. **WE16** Use the rule for the expansion of the square of a binomial to expand and simplify each of the following.
   a. $(x+2)(x+2)$ b. $(a+3)(a+3)$ c. $(b+7)(b+7)$ d. $(c+9)(c+9)$

6. Use the rule for the expansion of the square of a binomial to expand and simplify each of the following.
   a. $(m+12)^2$ b. $(n+10)^2$ c. $(x-6)^2$ d. $(y-5)^2$

7. Use the rule for the expansion of the square of a binomial to expand and simplify each of the following.
   a. $(9-c)^2$ b. $(8+e)^2$ c. $2(x+y)^2$ d. $(u-v)^2$

8. Use the rule for expanding perfect binomial squares rule to expand and simplify each of the following.
   a. $(2a+3)^2$ b. $(3x+1)^2$ c. $(2m-5)^2$ d. $(4x-3)^2$

9. Use the rule for expanding perfect binomial squares to expand and simplify each of the following.
   a. $(5a-1)^2$ b. $(7p+4)^2$ c. $(9x+2)^2$ d. $(4c-6)^2$

10. Use the rule for expanding perfect binomial squares to expand and simplify each of the following.
    a. $(5+3p)^2$ b. $(2-5x)^2$ c. $(9x-4y)^2$ d. $(8x-3y)^2$

Understanding

11. Use the difference of two squares rule to expand and simplify each of the following.

    a. $(x+3)(x-3)$
    b. $(2x+3)(2x-3)$
    c. $(7x-4)(7x+4)$
    d. $(2x+7y)(2x-7y)$
    e. $(x^2+y^2)(x^2-y^2)$

12. Expand and simplify the following perfect squares.

    a. $(4x+5)^2$
    b. $(7x-3y)^2$
    c. $(5x^2-2y)^2$
    d. $2(x-y)^2$
    e. $\left(\dfrac{2}{x}+4x\right)^2$

13. A square has a perimeter of $4x+12$. Calculate its area.

14. Francis has fenced off a square in her paddock for spring lambs. The area of the paddock is $(9x^2+6x+1)\,\text{m}^2$.
    Using pattern recognition, determine the side length of the paddock in terms of x.

15. A square has an area of $x^2+18x+81$. Determine an expression for the perimeter of this square.

Reasoning

16. Show that $a^2-b^2=(a+b)(a-b)$ is true for each of the following.
    a. $a=5, b=4$
    b. $a=9, b=1$
    c. $a=2, b=7$
    d. $a=-10, b=-3$

17. Lin has a square bedroom. Her sister Tasneem has a room that is 1 m shorter in length than Lin's room, but 1 m wider.

    a. Show that Lin has the larger bedroom.
    b. Determine how much bigger Lin's bedroom is than Tasneem's bedroom.

18. Expand each of the following pairs of expressions.

    a. i. $(x-4)(x+4)$ and $(4-x)(4+x)$
        ii. $(x-11)(x+11)$ and $(11-x)(11+x)$
        iii. $(2x-9)(2x+9)$ and $(9-2x)(9+2x)$
    b. State what you notice about the answers to the pairs of expansions above.
    c. Explain how this is possible.

19. Answer the following questions.

    a. Expand $(10k+5)^2$.
    b. Show that $(10k+5)^2=100k(k+1)+25$.
    c. Using part b, evaluate 25^2 and 85^2.

Problem solving

20. A large square has been subdivided into 2 squares and 2 rectangles.

    a. Write formulas for the areas of these 4 pieces, using the dimensions a and b marked on the diagram.
    b. Write an equation that states that the area of the large square is equal to the combined area of its 4 pieces. Do you recognise this equation?

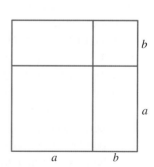

21. Expand each of the following pairs of expressions.

   a. i. $(x-3)^2$ and $(3-x)^2$
       ii. $(x-15)^2$ and $(15-x)^2$
       iii. $(3x-7)^2$ and $(7-3x)^2$
   b. State what you notice about the answers to the pairs of expansions above.
   c. Explain how this is possible.

22. Use the perfect squares rule to quickly evaluate the following.

   a. 27^2 b. 33^2 c. 39^2 d. 47^2

23. Allen is creating a square deck with a square pool installed in the middle of it. The side length of the deck is $(2x+3)$ m and the side length of the pool is $(x-2)$ m. Evaluate the area of the decking around the pool.

24. Ram wants to create a rectangular garden that is 10 m longer than it is wide. Write an expression for the area of the garden in terms of x, where x is the average length of the two sides.

25. The expansion of perfect squares $(a+b)^2 = a^2 + 2ab + b^2$ and $(a-b)^2 = a^2 - 2ab + b^2$ can be used to simplify some arithmetic calculations. For example:

$$97^2 = (100 - 3)^2$$
$$= 100^2 - 2 \times 100 \times 3 + 3^2$$
$$= 9409$$

   Use this method to evaluate the following.

   a. 103^2 b. 62^2 c. 997^2 d. 1012^2 e. 53^2 f. 98^2

2.7 Further expansions

LEARNING INTENTION

At the end of this subtopic you should be able to:
   • expand multiple sets of brackets and simplify the result.

▶ 2.7.1 Expanding multiple sets of brackets

eles-4600

   • When expanding expressions with more than two sets of brackets, expand all brackets first before simplifying the like terms.
   • It is important to be careful with signs, particularly when subtracting one expression from another. For example:
   $3x(x+6) - (x+5)(x-3)$
   $= 3x^2 + 18x - (x^2 + 5x - 3x - 15)$
   $= 3x^2 + 18x - (x^2 + 2x - 15)$
   $= 3x^2 + 18x - x^2 - 2x + 15$
   $= 2x^2 + 16x + 15$
   • When expanding this, since the entire expression $(x+5)(x-3)$ is being subtracted from $3x(x+6)$, its expanded form needs to be kept inside its own set of brackets. This helps us to remember that, as each term is being subtracted, the sign of each term will switch when opening up the brackets.
   • When expanding, it can help to treat the expression as though it has a -1 at the front, as shown.

$$-(x^2 + 2x - 15) = -1(x^2 + 2x - 15) = -x^2 - 2x + 15$$

WORKED EXAMPLE 17 Expanding and simplifying multiple sets of brackets

Expand and simplify each of the following expressions.

a. $(x+3)(x+4)+4(x-2)$

b. $(x-2)(x+3)-(x-1)(x+2)$

THINK

a. 1. Expand each set of brackets.

2. Simplify by collecting like terms.

b. 1. Expand and simplify each pair of brackets. Because the second expression is being subtracted, keep it in a separate set of brackets.

2. Subtract all of the second result from the first result. Remember that $-\left(x^2+x-2\right) = -1\left(x^2+x-2\right)$. Simplify by collecting like terms.

WRITE

a.

$(x+3)(x+4)+4(x-2)$

$= x^2+4x+3x+12+4x-8$

$= x^2+11x+4$

b.

$(x-2)(x+3)-(x-1)(x+2)$

$= x^2+3x-2x-6-\left(x^2+2x-x-2\right)$

$= x^2+x-6-\left(x^2+x-2\right)$

$= x^2+x-6-x^2-x+2$

$= -4$

WORKED EXAMPLE 18 Further expanding with multiple sets of brackets

Consider the expression $(4x+3)(4x+3)-(2x-5)(2x-5)$.

a. **Expand and simplify this expression.**

b. **Apply the difference of perfect squares rule to verify your answer.**

THINK

a. 1. Expand each set of brackets.

2. Simplify by collecting like terms.

b. 1. Rewrite as the difference of two squares.

2. Apply the difference of two squares rule, where $a=(4x+3)$ and $b=(2x-5)$.

3 Simplify, then expand the new brackets.

4. Simplify by collecting like terms.

WRITE

a. $(4x+3)(4x+3)-(2x-5)(2x-5)$

$= 16x^2+12x+12x+9-\left(4x^2-10x-10x+25\right)$

$= 16x^2+24x+9-\left(4x^2-20x+25\right)$

$= 16x^2+24x+9-4x^2+20x-25$

$= 12x^2+44x-16$

b. $(4x+3)(4x+3)-(2x-5)(2x-5)$

$= (4x+3)^2-(2x-5)^2$

$= ((4x+3)+(2x-5))((4x+3)-(2x-5))$

$= (6x-2)(2x+8)$

$= 12x^2+48x-4x-16$

$= 12x^2+44x-16$

DISCUSSION

On a diagram, how would you show $(m+2)(m+3)-(m+2)(m+4)$?

Exercise 2.7 Further expansions

learn on

Individual pathways

■ PRACTISE	■ CONSOLIDATE	■ MASTER
1, 4, 8, 11, 14	2, 5, 7, 9, 12, 15	3, 6, 10, 13, 16

To answer questions online and to receive **immediate corrective feedback** and **fully worked solutions** for all questions, go to your learnON title at www.jacplus.com.au.

Fluency

1. **WE17** Expand and simplify each of the following expressions.

   a. $(x+3)(x+5) + (x+2)(x+3)$
   b. $(x+4)(x+2) + (x+3)(x+4)$
   c. $(x+5)(x+4) + (x+3)(x+2)$
   d. $(x+1)(x+3) + (x+2)(x+4)$

2. Expand and simplify each of the following expressions.

   a. $(p-3)(p+5) + (p+1)(p-6)$
   b. $(a+4)(a-2) + (a-3)(a-4)$
   c. $(p-2)(p+2) + (p+4)(p-5)$
   d. $(x-4)(x+4) + (x-1)(x+20)$

3. Expand and simplify each of the following expressions.

   a. $(y-1)(y+3) + (y-2)(y+2)$
   b. $(d+7)(d+1) + (d+3)(d-3)$
   c. $(x+2)(x+3) + (x-4)(x-1)$
   d. $(y+6)(y-1) + (y-2)(y-3)$

4. Expand and simplify each of the following expressions.

   a. $(x+2)^2 + (x-5)(x-3)$
   b. $(y-1)^2 + (y+2)(y-4)$
   c. $(p+2)(p+7) + (p-3)^2$
   d. $(m-6)(m-1) + (m+5)^2$

5. Expand and simplify each of the following expressions.

   a. $(x+3)(x+5) - (x+2)(x+5)$
   b. $(x+5)(x+2) - (x+1)(x+2)$
   c. $(x+3)(x+2) - (x+4)(x+3)$
   d. $(m-2)(m+3) - (m+2)(m-4)$

6. Expand and simplify each of the following expressions.

   a. $(b+4)(b-6) - (b-1)(b+2)$
   b. $(y-2)(y-5) - (y+2)(y+6)$
   c. $(p-1)(p+4) - (p-2)(p-3)$
   d. $(x+7)(x+2) - (x-3)(x-4)$

7. Expand and simplify each of the following expressions.

   a. $(m+3)^2 - (m+4)(m-2)$
   b. $(a-6)^2 - (a-2)(a-3)$
   c. $(p-3)(p+1) - (p+2)^2$
   d. $(x+5)(x-4) - (x-1)^2$

Understanding

8. **WE18** Consider the expression $(x+3)(x+3) - (x-5)(x-5)$.

   a. Expand and simplify this expression.
   b. Apply the difference of perfect squares rule to verify your answer.

9. Consider the expression $(4x-5)(4x-5)-(x+2)(x+2)$.

   a. Expand and simplify this expression.
   b. Apply the difference of perfect squares rule to verify your answer.

10. Consider the expression $(3x-2y)(3x-2y)-(y-x)(y-x)$.

   a. Expand and simplify this expression.
   b. Apply the difference of perfect squares rule to verify your answer.

Reasoning

11. Determine the value of x for which $(x+3)+(x=4)^2=(x+5)^2$ is true.

12. Show that $(p-1)(p+2)+(p-3)(p+1)=2p^2-p-5$.

13. Show that $(x+2)(x-3)-(x+1)2=-3x-7$.

Problem solving

14. Answer the following questions.

   a. Show that $(a^2+b^2)(c^2+d^2)=(ac-bd)^2+(ad+bc)^2$.
   b. Using part a, write $(2^2+1^2)(3^2+4^2)$ as the sum of two squares and evaluate.

15. Answer the following questions.

   a. Expand $(x^2+x-1)^2$.
   b. Show that $(x^2+x-1)^2=(x-1)x(x+1)(x+2)+1$.
   c. i. Evaluate $4\times3\times2\times1+1$.
      ii. Determine the value of x if $4\times3\times2\times1+1=(x-1)x(x+1)(x+2)+1$.

16. Answer the following questions.

   a. Expand $(a+b)(d+e)$.
   b. Expand $(a+b+c)(d+e+f)$. Draw a diagram to illustrate your answer.

2.8 The highest common factor

> **LEARNING INTENTION**
>
> At the end of this subtopic you should be able to:
> - determine the highest common factor for a group of terms
> - factorise expressions by dividing out the highest common factor from each term.

⊙ 2.8.1 Factorising expressions

- **Factorising** a number or term means writing it as the product of a pair of its factors. For example, $18ab$ could be factorised as $3a\times6b$.
- Factorising an algebraic expression is the opposite process to expanding.
- $3(x+6)$ is the product of 3 and $(x+6)$ and is the factorised form of $3x+18$.
- In this section we will start with the expanded form of an expression, for example $6x+12$, and work to re-write it in factorised form, in this case $6(x+2)$.

TOPIC 2 Algebra **107**

WORKED EXAMPLE 19 Factorising terms

Complete the following factorisations.

a. $15x = 5x \times$ _____

b. $15x^2 = 3x \times$ _____

c. $-18ab^2 = 6a \times$ _____

THINK	WRITE
a. Divide $15x$ by $5x$.	a. $\dfrac{15x}{5x} = 3$ So, $15x = 5x \times 3$.
b. Divide $15x^2$ by $3x$.	b. $\dfrac{15x^2}{3x} = 5x$ So, $15x^2 = 3x \times 5x$.
c. Divide $-18ab^2$ by $6a$.	c. $\dfrac{-18ab^2}{6a} = -3b^2$ So, $-18ab^2 = 6a \times -3b^2$.

- The **highest common factor (HCF)** of two or more numbers is the largest factor that divides into all of them. The highest common factor of 18 and 27 is 9, since it is the biggest number that divides evenly into both 18 and 27.
- We can determine the HCF of two or more algebraic terms by determining the highest common factor of the coefficients, as well as any pronumerals that are common to all terms. For example, the HCF of $30xyz$, $25x^2z$ and $15xz^2$ is $5xz$. It can help to write each term in expanded form so that you can determine all of the common pronumerals.

WORKED EXAMPLE 20 Determining the highest common factor of two terms

Determine the highest common factor for each of the following pairs of terms.

a. $25a^2b$ and $10ab$

b. $3xy$ and $-3xz$

THINK	WRITE
a. 1. Write $25a^2b$ in expanded form.	a. $25a^2b = 5 \times 5 \times a \times a \times b$
2. Write $10ab$ in expanded form.	$10ab = 2 \times 5 \times a \times b$
3. Write the HCF.	$HCF = 5ab$
b. 1. Write $3xy$ in expanded form.	b. $3xy = 3 \times x \times y$
2. Write $-3xz$ in expanded form.	$-3xz = -1 \times 3 \times x \times z$
3. Write the HCF.	$HCF = 3x$

2.8.2 Factorising expressions by determining the highest common factor

eles-4602

- To factorise an expression such as $15x^2yz + 12xy^2$, determine the highest common factor of both terms. Writing each term in expanded form can help identify what is common to both terms.

$$15x^2yz + 12xy^2 = 3 \times 5 \times x \times x \times y \times z + 3 \times 4 \times x \times x \times y \times y$$
$$= 3 \times x \times x \times y \times 5 \times x \times x \times z + 3 \times x \times x \times y \times 4 \times x \times y$$

- We can see that $3xy$ is common to both terms. To complete the factorisation, place $3xy$ on the outside of a set of brackets and simplify what is left inside the brackets.

$$= 3 \times x \times y \,(5 \times x \times z + 4 \times y)$$
$$= 3xy\,(5xz + 4y)$$

- By removing the HCF from both terms, we have factorised the expression.
- The answer can always be checked by expanding out the brackets and making sure it produces the original expression.

WORKED EXAMPLE 21 Factorising by first determining the HCF

Factorise each expression by first determining the HCF.
a. $5x + 15y$
b. $-14xy - 7y$
c. $6x^2y + 9xy^2$

THINK	WRITE
a. 1. The HCF is 5.	a. $5x + 15y = 5(\quad)$
2. Divide each term by 5 to determine the binomial.	$= 5(x + 3y)$
3. Check the answer by expanding.	$5(x + 3y) = 5x + 15y$ (correct)
b. 1. The HCF is $7y$ or $-7y$, but $-7y$ makes things a little simpler.	b. $-14xy - 7y = -7y(\quad)$
2. Divide each term by $-7y$ to determine the binomial.	$= -7y(2x + 1)$
3. Check the answer by expanding.	$-7y(2x + 1) = -14xy - 7y$ (correct)
c. 1. The HCF is $3xy$.	c. $6x^2y + 9xy^2 = 3xy(\quad)$
2. Divide each term by $3xy$ to determine the binomial.	$= 3xy(2x + 3y)$
3. Check the answer by expanding.	$3xy(2x + 3y) = 6x^2y + 9xy^2$ (correct)

| TI | THINK | WRITE | CASIO | THINK | WRITE |

TI | THINK **WRITE**

a–c. In a new problem, on a Calculator page, press:
• MENU
• 3: Algebra 3
• 2: Factor 2.
Complete the entry lines as:
factor$(5x + 15y)$
factor $(-14x \times y - 7y)$
factor $(6x^2 \times y + 9x \times y^2)$

Press ENTER after each entry.

a–c.

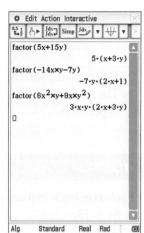

a. $5x + 15y = 5(x + 3)$
b. $-14xy - 7y = -7y(2x + 1)$
c. $6x^2y + 9xy^2 = 3xy(2x + 3y)$

CASIO | THINK **WRITE**

a–c. On the Main screen, press:
• Action
• Transformation
• Factor.
Complete the entry lines as:
factor$(5x + 15y)$
factor $(-14x \times y - 7y)$
factor $(6x^2 \times y + 9x \times y^2)$

Press EXE after each entry.

a–c.

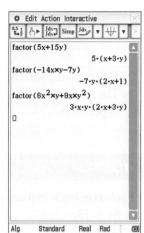

a. $(5x + 15y) = 5(x + 3y)$
b. $(-14xy - 7y) = -7y(2x + 1)$
c. $(6x^2y + 9xy^2) = 3xy(2x + 3y)$

DISCUSSION

How do you find the factors of terms within algebraic expressions?

Resources

eWorkbook	Topic 2 Workbook (worksheets, code puzzle and project) (ewbk-2002)	
Digital documents	SkillSHEET Finding the highest common factor (doc-10822)	
	SkillSHEET Factorising by finding the HCF (doc-10823)	
Video elesson	Factorisation (eles-1887)	
Interactivities	Individual pathway interactivity: The highest common factor (int-4486)	
	Highest common factor (int-6037)	

Exercise 2.8 The highest common factor   learn on

Individual pathways

■ PRACTISE	■ CONSOLIDATE	■ MASTER
1, 3, 7, 11, 13, 16, 21, 23, 26	2, 4, 6, 8, 10, 14, 17, 18, 22, 24, 27	5, 9, 12, 15, 19, 20, 25, 28

To answer questions online and to receive **immediate corrective feedback** and **fully worked solutions** for all questions, go to your learnON title at www.jacplus.com.au.

Fluency

1. **WE19** Complete each of the following factorisations by writing in the missing factor.

a. $8a = 4 \times$ _____
b. $8a = 2a \times$ _____
c. $12x^2 = 4x \times$ _____
d. $-12x^2 = 3x^2 \times$ _____
e. $3x^2 = x \times$ _____
f. $15a^2b = ab \times$ _____

2. Complete each of the following factorisations by writing in the missing factor.

   a. $12x = -4 \times$ _____

   b. $10\,mn = 10\,n \times$ _____

   c. $10\,mn = -10 \times$ _____

   d. $a^2b^2 = ab \times$ _____

   e. $30x^2 = 10x \times$ _____

   f. $-15\,mn^2 = -3\,m \times$ _____

3. **WE20** Determine the highest common factor (HCF) of each of the following.

   a. 4 and 12   b. 6 and 15   c. 10 and 25   d. 24 and 32   e. 12, 15 and 21

4. Determine the highest common factor (HCF) of each of the following.

   a. 25, 50 and 200

   b. 17 and 23

   c. $6a$ and $12ab$

   d. $14xy$ and $21xz$

   e. $60pq$ and $30q$

5. Determine the highest common factor (HCF) of each of the following.

   a. $50cde$ and $70fgh$

   b. $6x^2$ and $15x$

   c. $6a$ and $9c$

   d. $5ab$ and 25

   e. $3x^2y$ and $4x^2z$

6. **MC** Choose which of the following pairs has a highest common factor of $5m$.

   A. $2m$ and $5m$

   B. $5m$ and m

   C. $25mn$ and $15lm$

   D. $20m$ and $40m$

   E. $5m$ and lm

7. **WE21** Factorise each of the following expressions.

   a. $4x + 12y$ b. $5m + 15n$ c. $7a + 14b$ d. $7m - 21n$ e. $-8a - 24b$ f. $8x - 4y$

8. Factorise each of the following expressions.

   a. $-12p - 2q$

   b. $6p + 12pq + 18q$

   c. $32x + 8y + 16z$

   d. $16m - 4n + 24p$

   e. $72x - 8y + 64\,pq$

   f. $15x^2 - 3y$

9. Factorise each of the following expressions.

   a. $5p^2 - 20q$

   b. $5x + 5$

   c. $56q + 8p^2$

   d. $7p - 42x^2y$

   e. $16p^2 + 20q + 4$

   f. $12 + 36a^2b - 24b^2$

10. Factorise each of the following expressions.

    a. $9a + 21b$ b. $4c + 18d^2$ c. $12p^2 + 20q^2$ d. $35 - 14m^2n$ e. $25y^2 - 15x$

11. Factorise each of the following expressions.

    a. $16a^2 + 20b$

    b. $42m^2 + 12n$

    c. $63p^2 + 81 - 27y$

    d. $121a^2 - 55b + 110c$

    e. $10 - 22x^2y^3 + 14xy$

12. Factorise each of the following expressions.

    a. $18a^2bc - 27ab - 90c$

    b. $144p + 36q^2 - 84pq$

    c. $63a^2b^2 - 49 + 56ab^2$

    d. $22 + 99p^3q^2 - 44p^2r$

    e. $36 - 24ab^2 + 18b^2c$

13. Factorise each of the the following expressions.

    a. $-x + 5$ b. $-a + 7$ c. $-b + 9$ d. $-2m - 6$ e. $-6\,p - 12$ f. $-4a - 8$

14. Factorise each of the following expressions.

    a. $-3n^2 + 15m$

    b. $-7x^2y^2 + 21$

    c. $-7y^2 - 49z$

    d. $-12p^2 - 18q$

    e. $-63m + 56$

    f. $-12m^3 - 50x^3$

15. Factorise each of the following expressions.

    a. $-9a^2b + 30$

    b. $-15p - 12q$

    c. $-18x^2 + 4y^2$

    d. $-3ab + 18m - 21$

    e. $-10 - 25p^2 - 45q$

    f. $-90m^2 + 27n + 54p^3$

16. Factorise each of the following expressions.

    a. $a^2 + 5a$

    b. $14q - q^2$

    c. $18m + 5m^2$

    d. $6p + 7p^2$

    e. $7n^2 - 2n$

    f. $7p - p^2q + pq$

17. Factorise each of the following expressions.

a. $xy + 9y - 3y^2$
b. $5c + 3c^2d - cd$
c. $3ab + a^2b + 4ab^2$
d. $2x^2y + xy + 5xy^2$
e. $5p^2q^2 - 4pq + 3p^2q$
f. $6x^2y^2 - 5xy + x^2y$

18. Factorise each of the following expressions.

a. $5x^2 + 15x$
b. $24m^2 - 6m$
c. $32a^2 - 4a$
d. $-2m^2 + 8m$
e. $-5x^2 + 25x$
f. $-7y^2 + 14y$

19. Factorise each of the following expressions.

a. $-3a^2 + 9a$
b. $-12p^2 - 2p$
c. $-26y^2 - 13y$
d. $4m - 18m^2$
e. $-6t + 36t^2$
f. $-8p - 24p^2$

Understanding

20. A large billboard display is in the shape of a rectangle as shown. The billboard has 3 regions (A, B, C) with dimensions in terms of x, as shown.

a. Calculate the total area of the billboard. Give your answer in factorised form.
b. Determine an expression for the area of each region of the billboard. Write the expression in its simplest form.

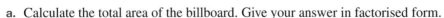

21. Consider the expression $3ab^2 + 18ab + 27a$.

a. Factorise this expression by taking out the highest common factor from each term.
b. Use the rule for perfect squares to fully factorise this expression.

22. Consider the expression $x^2(x+4) + 8x(x+4) + 16(x+4)$.

a. Factorise this expression by taking out the highest common factor from each term.
b. Use the rule for perfect squares to fully factorise this expression.

Reasoning

23. A question on Marcia's recent Algebra test was, 'Using factorisation, simplify the following expression: $a^2(a-b) - b^2(a-b)$'. Marcia's answer was $(a+b)(a-b)^2$. If Marcia used the difference of two squares rule to get her solution, explain the steps she took to get that answer.

24. Prove that, as long as $a \neq 0$, then $(x+a)^2 \neq x^2 + a^2$.

25. Using the fact that any positive number, n, can be written as $\left(\sqrt{n}\right)^2$, factorise the following expressions using the difference of two squares rule.

a. $x^2 - 13$
b. $4x^2 - 17$
c. $(x+3)^2 - 10$

Problem solving

26. Evaluate $\dfrac{x^2 + 2xy + y^2}{4x^2 - 4y^2} \div \dfrac{x^2 + xy}{8xy - 8x^2}$.

27. It has been said that, for any 2 numbers, the product of their LCM and HCF is the same as the product of the 2 numbers themselves. Show whether this is true.

28. a. Factorise $36x^2 - 100y^2$ by first taking out the common factor and then using the difference of two squares rule.
b. Factorise $36x^2 - 100y^2$ by first using the difference of two squares rule and then taking out the common factor.
c. Comment on whether you got the same answer for parts a and b.

2.9 The highest common binomial factor

> **LEARNING INTENTION**
>
> At the end of this subtopic you should be able to:
> - factorise expressions by taking out a common binomial factor
> - factorise expressions by grouping terms.

▶ 2.9.1 The common binomial factor

eles-4603

- When factorising an expression, we look for the highest common factor(s) first.
- It is possible for the HCF to be a binomial expression.
- Consider the expression $7(a - b) + 8x(a - b)$. The binomial expression $(a - b)$ is a common factor to both terms. Factorising this expression looks like this:

$$7(a - b) + 8x(a - b) = 7 \times (a-b) + 8x \times (a-b)$$
$$= (a-b)(7 + 8x)$$

WORKED EXAMPLE 22 Factorising by finding a binomial common factor

Factorise each of the following expressions.

a. $5(x + y) + 6b(x + y)$ **b.** $2b(a - 3b) - (a - 3b)$

Note: In both of these expressions the HCF is a binomial factor.

THINK

a. 1. The HCF is $(x + y)$.

   2. Divide each term by $(x + y)$ to determine the binomial.

b. 1. The HCF is $(a - 3b)$.

   2. Divide each term by $(a - 3b)$ to determine the binomial.

WRITE

a. $\dfrac{5(x + y)}{x + y} = 5, \dfrac{6b(x + y)}{x + y} = 6b$

Therefore,

$5(x + y) + 6b(x + y)$
$= (x + y)(5 + 6b)$

b. $\dfrac{2b(a - 3b)}{a - 3b} = 2b, \dfrac{-1(a - 3b)}{a - 3b} = -1$

Therefore,

$2b(a - 3b) - (a - 3b)$
$= 2b(a - 3b) - 1(a - 3b)$
$= (a - 3b)(2b - 1)$

▶ 2.9.2 Factorising by grouping in pairs

eles-4604

- If an algebraic expression has four terms and no common factors in any of its terms, it may be possible to group the terms in pairs and find a common factor in each pair.
- Consider the expression $10x + 15 - 6ax - 9a$.
- We can attempt to factorise by grouping the first two terms and the last two terms:

$$10x + 15 - 6ax - 9a = 5 \times 2x + 5 \times 3 - 3a \times 2x - 3a \times 3$$
$$= 5(2x + 3) - 3a(2x + 3)$$

- Once a common factor has been taken out from each pair of terms, a common binomial factor will appear. This common binomial factor can also be factorised out.

$$5(2x+3) - 3a(2x+3) = (2x+3)(5-3a)$$

- Thus the expression $10x + 15 - 6ax - 9a$ can be factorised to become $(2x+3)(5-3a)$.
- It is worth noting that it doesn't matter which terms are paired up first — the final result will still be the same.

$$
\begin{aligned}
10x + 15 - 6ax - 9a &= 10x - 6ax + 15 - 9a \\
&= 2x \times 5 + 2x \times -3a + 3 \times 5 + 3 \times -3a \\
&= 2x(5 - 3a) + 3(5 - 3a) \\
&= (5 - 3a)(2x + 3) \\
&= (2x + 3)(5 - 3a)
\end{aligned}
$$

WORKED EXAMPLE 23 Factorising by grouping in pairs

Factorise each of the following expressions by grouping the terms in pairs.

a. $5a + 10b + ac + 2bc$ **b.** $x - 3y + ax - 3ay$ **c.** $5p + 6q + 15pq + 2$

THINK

a. 1. Write the expression.

2. Take out the common factor $a + 2b$.

b. 1. Write the expression.

2. Take out the common factor $x - 3y$.

c. 1. Write the expression.

2. There are no simple common factors. Write the terms in a different order.

3. Take out the common factor $1 + 3q$.
Note: $1 + 3q = 3q + 1$.

WRITE

a. $5a + 10b + ac + 2bc$
$5a + 10b = 5(a + 2b)$
$ac + 2bc = c(a + 2b)$

$= 5(a + 2b) + c(a + 2b)$
$= (a + 2b)(5 + c)$

b. $x - 3y + ax - 3ay$
$x - 3y = 1(x - 3y)$
$ax - 3ay = a(x - 3y)$

$= 1(x - 3y) + a(x - 3y)$
$= (x - 3y)(1 + a)$

c. $5p + 6q + 15pq + 2$
$= 5p + 15pq + 6q + 2$
$5p + 15pq = 5p(1 + 3q)$
$6q + 2 = 2(3q + 1)$

$= 5p(1 + 3q) + 2(3q + 1)$
$= 5p(1 + 3q) + 2(1 + 3q)$

$= (1 + 3q)(5p + 2)$

DISCUSSION

How do you factorise expressions with four terms?

Exercise 2.9 The highest common binomial factor   **learn**on

Individual pathways

■ PRACTISE	■ CONSOLIDATE	■ MASTER
1, 3, 6, 9, 12	2, 4, 7, 10, 13	5, 8, 11, 14

To answer questions online and to receive **immediate corrective feedback** and **fully worked solutions** for all questions, go to your learnON title at www.jacplus.com.au.

Fluency

1. **WE22** Factorise each of the following expressions.

   a. $2(a+b) + 3c(a+b)$
   d. $4a(3b+2) - b(3b+2)$

   b. $4(m+n) + p(m+n)$
   e. $z(x+2y) - 3(x+2y)$

   c. $7x(2m+1) - y(2m+1)$

2. Factorise each of the following expressions.

   a. $12p(6-q) - 5(6-q)$
   d. $p^2(q+2p) - 5(q+2p)$

   b. $3p^2(x-y) + 2q(x-y)$
   e. $6(5m+1) + n^2(5m+1)$

   c. $4a^2(b-3) + 3b(b-3)$

3. **WE23** Factorise each of the following expressions by grouping the terms in pairs.

   a. $xy + 2x + 2y + 4$
   d. $2xy + x + 6y + 3$

   b. $ab + 3a + 3b + 9$
   e. $3ab + a + 12b + 4$

   c. $xy - 4y + 3x - 12$
   f. $ab - 2a + 5b - 10$

4. Factorise each of the following expressions by grouping the terms in pairs.

   a. $m - 2n + am - 2an$
   d. $10pq - q - 20p + 2$

   b. $5 + 3p + 15a + 9ap$
   e. $6x - 2 - 3xy + y$

   c. $15mn - 5n - 6m + 2$
   f. $16p - 4 - 12pq + 3q$

5. Factorise each of the following expressions by grouping the terms in pairs.

   a. $10xy + 5x - 4y - 2$
   d. $4x + 12y - xz - 3yz$

   b. $6ab + 9b - 4a - 6$
   e. $5pr + 10qr - 3p - 6q$

   c. $5ab - 10ac - 3b + 6c$
   f. $ac - 5bc - 2a + 10b$

Understanding

6. Simplify the following expressions using factorising.

   a. $\dfrac{ax + 2ay + 3az}{bx + 2by + 3bz}$

   b. $\dfrac{10(3x-4) + 2y(3x-4)}{7a(10+2y) - 5(10+2y)}$

7. Use factorising by grouping in pairs to simplify the following expressions.

   a. $\dfrac{3x + 6 + xy + 2y}{6 + 2y + 18x + 6xy}$

   b. $\dfrac{5xy + 10x + 3ay + 6a}{15bx - 10x + 9ab - 6a}$

8. Use factorising by grouping in pairs to simplify the following expressions.

   a. $\dfrac{6x^2 + 15xy - 4x - 10y}{6xy + 4x + 15y^2 + 10y}$

   b. $\dfrac{mp + 4mq - 4np - 16nq}{mp + 4mq + 4np + 16nq}$

Reasoning

9. Using the method of rectangles to expand, show how $a(m+n)+3(m+n)$ equals $(a+3)(m+n)$.

10. Fully factorise $6x+4x^2+6x+9$ by grouping in pairs. Discuss what you noticed about this factorisation.

11. **a.** Write out the product $5(x+2)(x+3)$ and show that it also corresponds to the diagram shown.

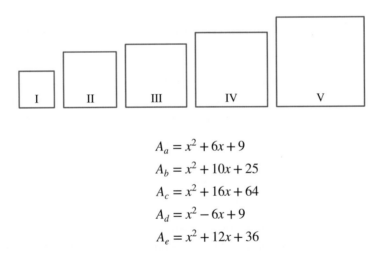

b. Explain why $5(x+2)(x+3)$ is equivalent to $(5x+10)(x+3)$. Use bracket expansion and a labelled diagram to support your answer.

c. Explain why $5(x+2)(x+3)$ is equivalent to $(x+2)(5x+15)$. Use bracket expansion and a labelled diagram to support your answer.

Problem solving

12. A series of five squares of increasing size and a list of five area formulas are shown.

a. Use factorisation to calculate the side length that correlates to each area formula.

b. Using the area given and the side lengths found, match the squares below with the appropriate algebraic expression of their area.

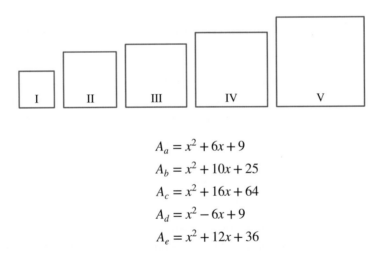

$$A_a = x^2 + 6x + 9$$
$$A_b = x^2 + 10x + 25$$
$$A_c = x^2 + 16x + 64$$
$$A_d = x^2 - 6x + 9$$
$$A_e = x^2 + 12x + 36$$

c. If $x = 5$ cm, use the formula given to calculate the area of each square.

13. Fully factorise both sets and brackets in the expression $(-12xy+27x+8y-18)-(-8xy+18x+12y-27)$ using grouping in pairs, then factorise the result.

14. The area formulas shown relate to either squares or rectangles.

    i. $9s^2 + 48s + 64$
    ii. $25s^2 - 4$
    iii. $s^2 + 4s + 3$
    iv. $4s^2 - 28s - 32$

a. Without completing any algebraic operations, examine these formulas and work out which ones belong to squares and which ones belong to rectangles. Explain your answer.

b. Factorise each formula and classify it as a square or rectangle. Check your classifications against your answer to part **a.**

2.10 Solving worded problems

2.10.1 Converting worded problems into algebraic expressions

eles-4605

- It is important to be able to convert worded problems (or 'real-world problems') into algebraic expressions. The key words that were identified in subtopic 2.3 can help with this process.
- Drawing diagrams is a useful strategy for understanding and solving worded problems.
- If a worded problem uses units in its questions, make sure that any answers also use those units.

WORKED EXAMPLE 24 Converting worded problems

A rectangular swimming pool measures 30 m by 20 m. A path around the edge of the pool is x m wide on each side.
  a. Determine the area of the pool.
  b. Write an expression for the area of the pool plus the area of the path.
  c. Write an expression for the area of the path.
  d. If the path is 1.5 m wide, calculate the area of the path.

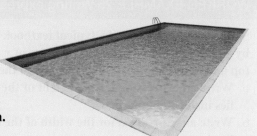

THINK

a. 1. Construct a drawing of the pool.

2. Calculate the area of the pool.

b. 1. Write an expression for the total length from one edge of the path to the other. Write another expression for the total width from one edge of the path to the other.

2. Area = length × width

WRITE

a.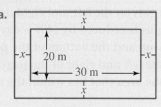

$$\text{Area} = \text{length} \times \text{width}$$
$$= 20 \times 30$$
$$= 600 \, \text{m}^2$$

b. $\text{Length} = 30 + x + x$
$$= 30 + 2x$$
$\text{Width} = 20 + x + x$
$$= 20 + 2x$$

$$\text{Area} = \text{length} \times \text{width}$$
$$= (30 + 2x)(20 + 2x)$$
$$= 600 + 60x + 40x + 4x^2$$
$$= (600 + 100x + 4x^2) \, \text{m}^2$$

c. Determine an expression for the area of the path by subtracting the area of the pool from the total area of the pool and the path combined.

c. Area of path = total area − area of pool

$$= 600 + 100x + 4x^2 - 600$$

$$= (100x + 4x^2)\text{m}^2$$

d. Substitute 1.5 for x in the expression you have found for the area of the path.

d. When $x = 1.5$,

Area of path $= 100\,(1.5) + 4(1.5)^2$

$$= 159\,\text{m}^2$$

- The algebraic expression found in part **c** of Worked example 24 allows us to calculate the area of the path for any given width.

WORKED EXAMPLE 25 Writing expressions for worded problems

Suppose that the page of a typical textbook is 24 cm high by 16 cm wide. Each page has margins of x cm at the top and bottom, and margins of y cm on the left and right.

a. Write an expression for the height of the section of the page that lies inside the margins.

b. Write an expression for the width of the section of the page that lies inside the margins.

c. Write an expression for the area of the section of the page that lies inside the margins.

d. Show that, if the margins on the left and right are doubled, the area within the margins is reduced by $(48y - 4xy)$ cm.

e. If the margins at the top and the bottom of the page are 1.5 cm and the margins on the left and right of the page are 1 cm, calculate the size of the area that lies within the margins.

THINK

a. 1. Construct a drawing that shows the key dimensions of the page and its margins.

WRITE

a.

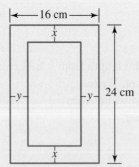

2. The total height of the page (24 cm) is effectively reduced by x cm at the top and x cm at the bottom.

Height $= 24 - x - x$

$$= (24 - 2x)\,\text{cm}$$

b. The total width of the page (16 cm) is reduced by y cm on the left and y cm on the right.

b. Width $= 16 - y - y$
$= (16 - 2y)$ cm

c. The area of the page that lies within the margins is the product of the width and height.

c. $\text{Area}_1 = (24 - 2x)(16 - 2y)$
$= (384 - 48y - 32x + 4xy)\text{cm}^2$

d. 1. If the left and right margins are doubled they both become $2y$ cm. Determine the new expression for the new width of the page.

d. Width $= 16 - 2y - 2y$
$= (16 - 4y)$ cm

2. Determine the new expression for the reduced area.

$\text{Area}_2 = (24 - 2x)(16 - 4y)$
$= 24 \times 16 + 24 \times -4y - 2x \times 16 - 2x \times -2y$
$= (384 - 96y - 32x + 8xy)\,\text{cm}^2$

3. Determine the difference in area by subtracting the reduced area obtained in part **d** from the original area obtained in part **c**.

Difference in area $= \text{Area}_1 - \text{Area}_2$
$= 384 - 48y - 32x + 4xy$
$\quad -(384 - 96y - 32x + 8xy)$
$= 384 - 48y - 32x + 4xy$
$\quad -384 + 96y + 32x - 8xy$
$= 48y - 4xy$

So the amount by which the area is reduced is $(48y - 4xy)\text{cm}^2$.

e. Using the area found in **c**, substitute 1.5 for x and 1 for y. Solve the expression.

e. Area $= (384 - 48y - 32x + 4xy)\,\text{cm}^2$
$= 384 - 48(1) - 32(1.5) + 4(1.5)(1)$
$294\,\text{cm}^2$

DISCUSSION

How can you use algebraic skills in real-life situations?

 Resources

 eWorkbook Topic 2 Workbook (worksheets, code puzzle and project) (ewbk-2002)

 Interactivity Individual pathway interactivity: Solving worded problems (int-4488)

Exercise 2.10 Solving worded problems

Individual pathways

■ PRACTISE	■ CONSOLIDATE	■ MASTER
1, 2, 6, 9, 12	3, 5, 7, 10, 13	4, 8, 11, 14, 15

To answer questions online and to receive **immediate corrective feedback** and **fully worked solutions** for all questions, go to your learnON title at www.jacplus.com.au.

Fluency

1. Answer the following for each shape shown.
   - i. Determine an expression for the perimeter.
   - ii. Determine the perimeter when $x = 5$.
   - iii. Determine an expression for the area. If necessary, simplify the expression by expanding.
   - iv. Determine the area when $x = 5$.

   a. $3x$

   b. $x + 2$

   c. 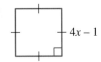 $4x - 1$

   d. $3x + 1$ $2x$

   e. $5x + 2$ $x - 3$

   f. 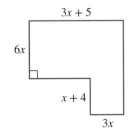 $3x + 5$ ... $6x$ ... $x + 4$ ... $3x$

Understanding

2. **WE24** A rectangular swimming pool measures 50 m by 25 m. A path around the edge of the pool is x m wide on each side.

   a. Determine the area of the pool.
   b. Write an expression for the area of the pool plus the area of the path.
   c. Write an expression for the area of the path.
   d. If the path is 2.3 m wide, calculate the area of the path.
   e. If the area of the path is 200 m^2, write an equation that can be solved to calculate the width of the path.

3. **WE25** The pages of a book are 20 cm high by 15 cm wide. The pages have margins of x cm at the top and bottom and margins of y cm on the left and right.

   a. Write an expression for the height of the section of the pages that lie inside the margins.
   b. Write an expression for the width of the section of the pages that lie inside the margins.
   c. Write an expression for the area of the section of the pages that lie inside the margins.
   d. Show that, if the margins on the left and right are doubled, the area between the margins is reduced by $(40y - 4xy)$ cm.

4. A rectangular book cover is 8 cm long and 5 cm wide.

   a. Calculate the area of the book cover.

   b. i. If the length of the book cover is increased by v cm, write an expression for its new length.

      ii. If the width of the book cover is increased by v cm, write an expression for its new width.

      iii. Write an expression for the new area of the book cover. Expand this expression.

      iv. Calculate the increased area of the book cover if $v = 2$ cm.

   c. i. If the length of the book cover is decreased by d cm, write an expression for its new length.

      ii. If the width of the book cover is decreased by d cm, write an expression for its new width.

      iii. Write an expression for the new area of the book cover. Expand this expression.

      iv. Calculate the new area of the book cover if $d = 2$ cm.

   d. i. If the length of the book cover is made x times longer, write an expression for its new length.

      ii. If the width of the book cover is increased by x cm, write an expression for its new width.

      iii. Write an expression for the new area of the book cover. Expand this expression.

      iv. Calculate the new area of the book cover if $x = 5$ cm.

5. A square has sides of length $5x$ m.

   a. Write an expression for its perimeter.

   b. Write an expression for its area.

   c. i. If the square's length is decreased by 2 m, write an expression for its new length.

      ii. If the square's width is decreased by 3 m, write an expression for its new width.

      iii. Write an expression for the square's new area. Expand this expression.

      iv. Calculate the square's area when $x = 6$ m.

6. A rectangular sign has a length of $2x$ cm and a width of x cm.

   a. Write an expression for the sign's perimeter.

   b. Write an expression for the sign's area.

   c. i. If the sign's length is increased by y cm, write an expression for its new length.

      ii. If the sign's width is decreased by y cm, write an expression for its new width.

      iii. Write an expression for the sign's new area and expand.

      iv. Calculate the sign's area when $x = 4$ cm and $y = 3$ cm using your expression.

7. A square has a side length of x cm.

   a. Write an expression for its perimeter.

   b. Write an expression for its area.

   c. i. If the square's side length is increased by y cm, write an expression for its new side length.

      ii. Write an expression for the square's new perimeter. Expand this expression.

      iii. Calculate the square's perimeter when $x = 5$ cm and $y = 9$ cm.

      iv. Write an expression for the square's new area and expand.

      v. Calculate the square's area when $x = 3.2$ cm and $y = 4.6$ cm.

8. A swimming pool with length $(4p + 2)$ m and width $3p$ m is surrounded by a path of width p m.
Write the following in expanded form:

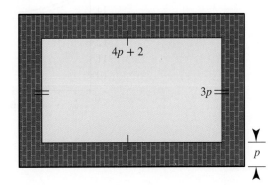

   a. An expression for the perimeter of the pool

   b. An expression for the area of the pool

   c. An expression for the length of the pool and path

   d. An expression for the width of the pool and path

   e. An expression for the perimeter of the pool and path

   f. An expression for the area of the pool and path

   g. An expression for the area of the path

   h. The area of the path when $p = 2$ m

Reasoning

9. The Body Mass Index (B) is used as an indicator of whether or not a person is in a healthy weight range for their height. It can be calculated using the formula $B = \dfrac{m}{h^2}$, where m is the person's mass in kilograms and h is the person's height in metres. Use this formula to answer the following questions, correct to 1 decimal place.

   a. Calculate Samir's Body Mass Index if they weigh 85 kg and are 1.75 m tall.

   b. A person is considered to be in a healthy weight range if their Body Mass Index is between 21 to 25 inclusive. Comment on Samir's weight for a person of their height.

   c. Calculate the Body Mass Index, correct to 1 decimal place, for each of the following people:

      i. Sammara, who is 1.65 m tall and has a mass of 52 kg

      ii. Nimco, who is 1.78 m tall and has a mass of 79 kg

      iii. Manuel, who is 1.72 m tall and has a mass of 65 kg.

   d. The healthy Body Mass Index range for children changes up until adulthood. These graphs show the Body Mass Index for boys and girls aged from 2 to 20 years of age.
   Study these graphs carefully and decide on possible age ranges (between 2 and 20) for Sammara, Nimco and Manuel if their BMI was in the satisfactory range. Give the age ranges in half-year intervals.

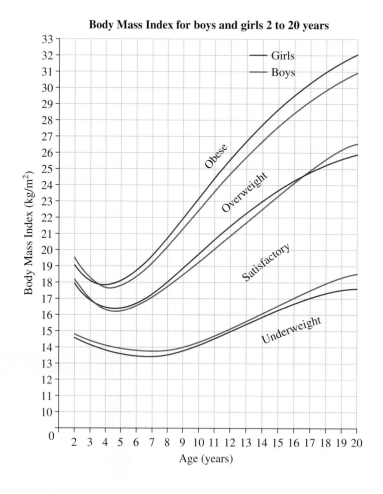

Body Mass Index for boys and girls 2 to 20 years

10. The figure shown is a rectangle. For the purpose of this activity we are going to call this rectangle an 'algebra rectangle'.

The length of each algebra rectangle is x cm and the width is 1 cm.

These algebra rectangles are put together to form larger rectangles in one of two ways.

Long algebra rectangle (3-long)　　　　**Tall algebra rectangle (3-tall)**

To calculate the perimeter of each of these algebra rectangles:
- perimeter of 3-long algebra rectangle $P = (x + x + x + x + x + x) + (1 + 1) = 6x + 2$
- perimeter of 3-tall algebra rectangle $P = (x + x) + (1 + 1 + 1 + 1 + 1 + 1) = 2x + 6$.

a. Using the expressions for the perimeters of a 3-long algebra rectangle and a 3-tall algebra rectangle as a basis, find the perimeter of each of the following algebra rectangles and add them to the table shown.

Type of algebra rectangle	Perimeter	Type of algebra rectangle	Perimeter
1-long		1-tall	
2-long		2-tall	
3-long	$6x + 2$	3-tall	$2x + 6$
4-long		4-tall	
5-long		5-tall	

b. Can you see the pattern? Based on this pattern, calculate the perimeter of a 20-long algebra rectangle and a 20-tall algebra rectangle.

c. Determine the values of x that will result in a tall algebra rectangle that has a larger perimeter than a long algebra rectangle.

11. Can you evaluate 997^2 without a calculator in less than 90 seconds? It is possible to work out the answer using long multiplication, but it would take a fair amount of time and effort. Mathematicians are always looking for quick and simple ways of solving problems.

What if we consider the expanding formula that produces the difference of two squares?

$(a + b)(a - b) = a^2 - b^2$

Adding b^2 to both sides gives $(a + b)(a - b) + b^2 = a^2 - b^2 + b^2$.

Simplifying and swapping sides gives $a^2 = (a + b)(a - b) + b^2$.

We can use this formula, combined with the fact that multiplying by 1000 is an easy operation, to evaluate 997^2.

a. If $a = 997^2$, determine what the value of b should be so that $(a + b)$ equals 1000.

b. Substitute these a and b values into the formula to evaluate 997^2.

c. Try the above method to evaluate the following.

　i. 995^2　　　　　　　　　　　　　ii. 990^2

Problem solving

12. In the picture of the phone shown, the dimensions are in cm.

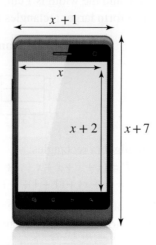

a. Use the information in the picture to write an expression for the area of:

   i. the viewing screen

   ii. the entire front face of the phone.

b. Your friend has a phone that is 4 cm longer than the one shown, but also 1 cm narrower. Write expressions in expanded form for:

   i. the length and width of your friend's phone

   ii. the area of your friend's phone.

c. If the phone in the picture is 5 cm wide, use your answer to part b to calculate the area of the front face of your friend's phone.

13. A proposed new flag for Australian schools will have the Australian flag in the top left-hand corner. The dimensions of this new flag are given in metres.

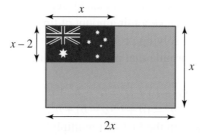

a. Write an expression in factorised form for the area of:

   i. the Australian flag section of the proposed flag

   ii. the whole area of the proposed flag.

b. Use the answers from part a to write the area of the Australian flag as a fraction of the school flag. Simplify this fraction.

c. Use the fraction from part b to express the area of the Australian flag as a percentage of the proposed school flag.

d. Use the formula for the percentage of the area taken up by the Australian flag to find the percentages for the following suggested widths for the proposed school flag.

   i. 4 m

   ii. 4.5 m

   iii. 4.8 m

e. If the percentage of the school flag taken up by the Australian flag measures the importance a school places on Australia, determine what can be said about the three suggested flag widths.

14. A new game has been created by students for the school fair. To win the game you need to hit the shaded region of the target shown with 5 darts.

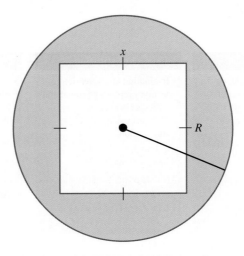

a. Write an expression to calculate the area of the shaded region of the game board.
b. Calculate the area of the shaded region of the game board if $R = 7.5$ cm and $x = 4$ cm.

c. Show that $R = \sqrt{\dfrac{A + x^2}{\pi}}$ by transposing the formula found in part a.

d. If $A = 80$ cm^2 and $x = 3$ cm, calculate the value of R.
e. The students found that the best size for the game board is when $R = 10$ and $x = 5$. For a board of this size, determine the percentage of the total board that is shaded.

15. Cubic expressions are the expanded form of expressions with three linear factors. The expansion process for three linear factors is called trinomial expansion.

a. Explain how the area model can be altered to show that binomial expansion can also be altered to become a model for trinomial expansion.
b. Investigate the powers in cubic expressions by expanding the following expressions.

   i. $(3x + 2)(-x + 1)(x + 1)$
   ii. $(x + 5)(2x - 2)(4x - 8)$
   iii. $(3 - x)(x + 8)(5 - x)$

c. Describe the patterns in the powers in a cubic expression.

2.11 Review

2.11.1 Topic summary

Expressions
- An expression is a group of terms in which each term is separated by a + or a − sign.
- A pronumeral is a letter that can take the place of a number. Pronumerals are also referred to as variables.
- Consider the expression $3x + 5y^2 - 8 - 9x^2z$.
  - This expression has 4 terms.
  - The constant term is −8.
  - The smallest coefficient is −9.
  - The coefficient of x is 3.
  - The pronumerals used are x, y and z.

Substitution
- Evaluating an expression involves replacing pronumerals with specific values.
- This process is called substitution.

e.g. Evaluate $3x^2 - 4y$ when $x = 2$ and $y = 5$:

$$3x^2 - 4y = 3(2^2) - 4(5)$$
$$= 12 - 20$$
$$= -8$$

Formula
- A formula is a mathematical rule that represents a real-world relationship.
- The area of a triangle is given by the rule:

$$\text{Area} = \frac{1}{2} \times b \times h$$

e.g. The area of a triangle when $b = 10$ cm and $h = 15$ cm:

$$\text{Area} = \frac{1}{2} \times 10 \times 15 = 75 \text{ cm}^2$$

ALGEBRA

Simplifying algebraic expressions
- An algebraic expression can be simplified by adding or subtracting like terms.
- Two terms are considered like terms if they have exactly the same pronumeral component.
  - $3x$ and $-10x$ are **like** terms.
  - $5x^2$ and $11x$ are **not like** terms.
  - $4abc$ and $-3cab$ are **like** terms.

e.g. The expression $3xy + 10xy - 5xy$ has 3 like terms and can be simplified to $8xy$.

The expression $4a \times 2b \times a$ can be simplified to $8a^2b$.

The expression $12xy \div 4xz$ can be simplified to $\frac{3y}{z}$.

Setting up worded problems
- In order to turn a worded problem into an algebraic expression it is important to look out for the following words.
  - **Words that mean addition:** sum, altogether, add, more than, and, in total
  - **Words that mean subtraction:** difference, less than, take away, take off, fewer than,
  - **Words that mean multiplication:** product, groups of, times, of, for each, double, triple
  - **Words that mean division:** quotient, split into, halve, thirds.

Expanding brackets
- We can expand a single set of brackets using the **Distributive Law:**

$$a(b + c) = ab + ac$$

- The product of binomial factors can be expanded using **FOIL:**

$$(a + b)(c + d) = ac + ad + bc + bd$$

e.g. $(x + 3)(y - 4) = xy - 4x + 3y - 12$

Factorising
- Factorising is the opposite process to expanding.
- An expression is factorised by determining the highest common factor of each term.

e.g. $9xy + 15xz - 21x^2$
$$= 3x(3y + 5z - 7x)$$

The highest common factor of each term is $3x$.

Solving application questions
- Drawing diagrams will help set up expressions to solve worded problems.
- Make sure you include units in your answers.

Special cases
- We can also expand if we recognise the following special cases.
  - Difference of two squares:

$$(a + b)(a - b) = a^2 - b^2$$

  - Perfect squares:

$$(a + b)^2 = a^2 + 2ab + b^2$$
$$(a - b)^2 = a^2 - 2ab + b^2$$

e.g. $(x - 6)(x + 6) = x^2 - 36$
$$(2x + 5)^2 = 4x^2 + 20x + 25$$
$$(x - 2)^2 = x^2 - 4x + 4$$

Grouping in pairs
- When presented with an expression that has 4 terms with no common factor, we can factorise by grouping the terms in pairs.

e.g. $6xy + 8y - 12xz - 16z$
$$= 2y(3x + 4) - 4z(3x + 4)$$
$$= (3x + 4)(2y - 4z)$$

2.11.2 Success criteria

Tick the column to indicate that you have completed the subtopic and how well you have understood it using the traffic light system.

(**Green:** I understand; **Yellow:** I can do it with help; **Red:** I do not understand)

Subtopic	Success criteria			
2.2	I can identify the number of terms in an algebraic expression.			
	I can identify the variables, constant terms and coefficients of the terms in an expression.			
	I can evaluate an expression or formula by substituting values for each variable.			
2.3	I can assign pronumerals to the unknown quantities in a worded problem.			
	I can convert the worded statements presented in a question into algebraic expressions.			
2.4	I can identify like terms.			
	I can simplify algebraic expressions by adding and subtracting like terms.			
	I can simplify algebraic expressions that involve multiplication and division of multiple algebraic terms.			
2.5	I can use the distributive law to expand and simplify an expression with a single set of brackets.			
	I can use the distributive law to expand and simplify expressions containing two or more sets of brackets.			
2.6	I can quickly expand and simplify binomial products using the rule for difference of two squares.			
	I can quickly expand and simplify binomial products using the rule for perfect squares.			
2.7	I can expand multiple sets of brackets and simplify the result.			
2.8	I can determine the highest common factor for a group of terms.			
	I can factorise expressions by dividing out the highest common factor from each term.			
2.9	I can factorise an expression by taking out a common binomial factor			
	I can factorise expressions by grouping terms in pairs.			
2.10	I can write an algebraic expression for a worded problem and use it to solve that problem.			

2.11.3 Project

Quilt squares

People all over the world are interested in quilt making, which involves stitching pieces of fabric together and inserting stuffing between layers of stitched-together fabric. When making a quilt, the fabric can be arranged and sewn in a variety of ways to create attractive geometric designs. Because of the potential for interesting designs in quilt-making, quilts are often used as decorative objects.

Medini is designing a quilt. She is sewing pieces of differently coloured fabric together to make a block, then copying the block and sewing the blocks together in a repeated pattern.

Making your own quilt

A scaled diagram of the basic block that Medini is using to make her quilt is shown. The letters indicate the colours of the fabric that make up the block: yellow, black and white. The yellow and white pieces are square, while the black pieces are rectangular. The finished blocks are sewn together in rows and columns.

y	b	y
b	w	b
y	b	y

Trace or copy the basic block shown onto a sheet of paper. Repeat this process until you have 9 blocks. Colour in each section and cut out all 9 blocks.

1. Place all 9 blocks together to form a 3 × 3 square. Draw and colour a scaled diagram of your result.
2. Describe the feature created by arranging the blocks in the manner described in question 1. Observe the shapes created by the different colours.

Medini sold her design to a company that now manufactures quilts made from 100 of these blocks. Each quilt covers an area of 1.44 m^2 Each row and column has the same number of blocks. Answer the following, ignoring any seam allowances.

3. Calculate the side length of each square block.
4. If the entire quilt has an area of 1.44 m^2, what is the area of each block?
5. Determine the dimensions of the yellow, black and white pieces of fabric of each block.
6. Calculate the area of the yellow, black and white pieces of fabric of each block.
7. Determine the total area of each of the three different colours required to construct this quilt.
8. Due to popular demand, the company that manufactures these quilts now makes them in different sizes. A customer can specify the approximate quilt area, the three colours they want and the number of blocks in the quilt, but the quilt must be either square or rectangular. Come up with a general formula that would let the company quickly work out the areas of the three coloured fabrics in each block. Give an example of this formula. Draw a diagram on a separate sheet of paper to illustrate your formula.

Resources

eWorkbook   Topic 2 Workbook (worksheets, code puzzle and project) (ewbk-2002)

Interactivities Crossword (int-0699)
              Sudoku puzzle (int-3203)

Exercise 2.11 Review questions

To answer questions online and to receive **immediate corrective feedback** and **fully worked solutions** for all questions, go to your learnON title at www.jacplus.com.au.

Fluency

1. **MC** Select the coefficient of the second term in the expression $-5xy^2 + 2y + 8y^2 + 6$.
   A. -5 B. 2 C. 8 D. 6 E. 5

2. **MC** Jodie had $55 in her purse. She spent $x. Select the expression for the amount of money she has left.
   A. $\$(x - 55)$ B. $\$(x + 55)$ C. $\$55x$ D. $\$(55 - x)$ E. $\$\left(\dfrac{55}{x}\right)$

3. **MC** $\dfrac{3x}{5} - \dfrac{x}{4}$ can be simplified to:
   A. $\dfrac{7x}{20}$ B. $\dfrac{2x}{20}$ C. $\dfrac{2x}{1}$ D. $\dfrac{4x}{20}$ E. $\dfrac{x}{20}$

4. **MC** $\dfrac{3}{p} \div \dfrac{6}{p}$ is equal to:
   A. 2 B. $\dfrac{1}{2}$ C. $12p$ D. $12p^2$ E. $\dfrac{1}{2p^2}$

5. **MC** The volume of a sphere, V, is given by the formula $V = \dfrac{4\pi r^3}{3}$, where r is the radius of the sphere. If $r = 3$, select the volume of the sphere from the following options.
   A. 25.12 B. 38.37 C. 64.05 D. 113.10 E. 4.19

6. **MC** Choose the equivalent of $6 - 4x(x + 2) + 3x$.
   A. $6 - 4x^2 - 5x$ B. $6 + 4x^2 + 5x$ C. $6 - 4x^2 + 5x$ D. $6 + 4x^2 + 5x$ E. $6 + 4x^2 - 5x$

7. **MC** Choose the equivalent of $(3 - a)(3 + a)$.
   A. $9 + a^2$ B. $9 - a^2$ C. $3 + a^2$ D. $3 - a^2$ E. $9 - a$

8. **MC** Choose the equivalent of $(2y + 5)^2$.
   A. $4y^2 + 20y + 25$ B. $4y^2 + 10y + 25$ C. $2y^2 + 20y + 25$ D. $2y^2 + 20y + 5$ E. $2y^2 + 10y + 25$

9. **MC** Select what $6(a + 2b) - x(a + 2b)$ equals when it is factorised.
   A. $6 - x(a + 2b)$ B. $(6 - x)(a + 2b)$ C. $6(a + 2b - x)$ D. $(6 + x)(a - 2b)$ E. $(6 + x)(a + 2b)$

10. **a.** For the expression $-8xy^2 + 2x + 8y^2 - 5$:
    i. state the number of terms
    ii. state the coefficient of the first term
    iii. state the constant term
    iv. state the term with the lowest coefficient.
    b. Write expressions for the following, where x and y represent the following numbers.
    i. A number 8 more than y
    ii. The difference between x and y
    iii. The sum of x and y
    iv. A number that is 7 times the product of x and y
    v. The number given when 2 times x is subtracted from 5 times y

11. **a.** Leo receives x dollars for each car he washes. If he washes y cars, calculate how much he earn.
    b. A piece of rope is 24 m long.
      i. If Rawiri cuts k m off, calculate how much is left.
      ii. After Rawiri has cut k m off, he divides the rest of the rope into three pieces of equal length. Calculate the length of each piece.

12. Simplify the following expressions by collecting like terms.
    a. $8p + 9p$
    b. $5y^2 + 2y - 4y^2$
    c. $9s^2t - 12s^2t$
    d. $11c^2d - 2cd + 5dc^2$
    e. $n^2 - p^2q - 3p^2q + 6$
    f. $8ab + 2a^2b^2 - 5a^2b^2 + 7ab$

13. Simplify the following expressions.
    a. $6a \times 2b$
    b. $2ab \times b$
    c. $2xy \times 4yx$
    d. $\dfrac{4x}{12}$
    e. $18 \div 4b$

14. Expand the following expressions.
    a. $5(x + 3)$
    b. $-(y + 5)$
    c. $-x(3 - 2x)$
    d. $-4x(2m + 1)$

15. Expand and simplify the following expressions by collecting like terms.
    a. $3(x - 2) + 9$
    b. $-2(5m - 1) - 3$
    c. $4m(m - 3) + 3m - 5$
    d. $7p - 2 - (3p + 4)$

16. Expand and simplify the following expressions.
    a. $3(a + 2b) + 2(3a + b)$
    b. $-4(2x + 3y) + 3(x - 2y)$
    c. $2m(n + 6) - m(3n + 1)$
    d. $-2x(3 - 2x) - (4x - 3)$

17. Expand and simplify the following expressions.
    a. $(x + 4)(x + 5)$
    b. $(m - 2)(m + 1)$
    c. $(3m - 2)(m - 5)$
    d. $(2a + b)(a - 3b)$

18. Expand and simplify the following expressions.
    a. $(x + 4)(x - 4)$
    b. $(9 - m)(9 + m)$
    c. $(x + y)(x - y)$
    d. $(1 - 2a)(1 + 2a)$

19. Expand and simplify the following expressions.
    a. $(x + 5)^2$
    b. $(m - 3)^2$
    c. $(4x + 1)^2$
    d. $(2 - 3y)^2$

20. Expand and simplify the following expressions.
    a. $(x + 2)(x + 1) + (x + 3)(x + 2)$
    b. $(m + 7)(m - 2) + (m + 3)^2$
    c. $(x + 6)(x + 2) - (x + 3)(x - 1)$
    d. $(b - 7)^2 - (b - 3)(b - 4)$

21. Expand and simplify the following expressions.
    a. $(x + 2)^2 + (x + 3)^2$
    b. $(x - 2)^2 - (x - 3)^2$
    c. $(x + 4)^2 - (x - 4)^2$

22. If $y = 5x^2 + 2x - 1$, determine the value of y when:
    a. $x = 2$
    b. $x = 5$.

23. The volume (V) of each of the following paint tins is given by the formula $V = \pi r^2 h$, where r is the radius and h is the height of the cylinder. Determine the value of V (to 2 decimal places) when:
    a. $r = 7$ cm and $h = 2$ cm.
    b. $r = 9.3$ cm and $h = 19.8$ cm.

24. Factorise each of the following expressions by determining common factors.

 a. $6x + 12$ **b.** $6x^2 + 12x^2y$ **c.** $8a^2 - 4b$

 d. $16x^2 - 24xy$ **e.** $-2x - 4$ **f.** $b^2 - 3b + 4bc$

25. Simplify each of the following expressions using common factor techniques.

 a. $5(x + y) - 4a(x + y)$ **b.** $7a(b + 5c) - 6c(b + 5c)$ **c.** $15x(d + 2e) + 25xy(d + 2e)$

 d. $2x + 2y + ax + ay$ **e.** $6xy + 4x - 6y - 4$ **f.** $pq - r + p - rq$

Problem solving

26. A rectangular rug has a length of $3x$ cm and a width of x cm.

 a. Write an expression for the rug's perimeter.

 b. Write an expression for the rug's area.

 c. The rug's side length is increased by y cm.

  i. Write an expression for its new side length.

  ii. Write an expression for the rug's new perimeter. Expand this expression.

  iii. Calculate the rug's perimeter when $x = 90$ cm and $y = 30$ cm.

  iv. Write an expression for the rug's new area. Expand this expression.

  v. Calculate the rug's area when $x = 90$ cm and $y = 30$ cm.

27. A rectangular garden bed has a length of 15 m and a width of 8 m. It is surrounded by a path with a width of p m

 a. Write down the total area of the garden bed and path in factorised form.

 b. Expand the expression you wrote down in part **a.**

 c. Calculate the area of the path in terms of p.

 d. Write an equation that can be solved to find the width of the path if the area of the path is 200 m^2.

28. Calculate the circumference of a circle whose area is $\pi \left(\dfrac{x}{y} \right)$ cm^2. Give your answer in terms of π.

29. The large sign shown appears in a parking lot at the entrance to car park 5. It has a uniform width, with dimensions shown.

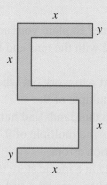

Write an algebraic expression for the area of the front of the sign.

30. At the end of a lesson involving boxes of Smarties, the Smarties left over are shared equally among the class. There are 6 boxes full of Smarties left over and 42 Smarties that are no longer in boxes.

  a. Write an expression for the total number of Smarties left over.
  (*Hint:* Let $n =$ the number of Smarties in a box.)
  b. Write the expression from part **a** in factorised form.
  c. If there are only three people in the class, write an expression for the number of Smarties each person receives.
  d. If there are x people in the class, write an expression for the number of Smarties each person receives.

31. When entering numbers into an electronic device, or even writing numbers down, we can often make mistakes. A common type of mistake is called a transposition error. This is when two digits are written in the reverse order. For example, the number 2869 could be written as 8269, 2689 or 2896.

A common rule for checking if these mistakes have been made is to see if the difference between the correct number and the recorded number is a multiple of 9. If it is, a transposition error has occurred.

We can use algebraic expressions to check this rule. Let the digit in the thousands position be represented by a, the digit in the hundreds position by b, the digit in the tens position by c, and the digit in the ones position by d.

In this way the real number can be represented as $1000a + 100b + 10c + d$.

  a. If the digits in the ones position and the tens position were written in the reverse order, the number would be $1000a + 100b + 10d + c$. The difference between the correct number and the incorrect one would then be $1000a + 100b + 10c + d - (1000a + 100b + 10d + c)$.
  i. Simplify this expression.
  ii. Is the expression a multiple of nine? Explain your answer.
  b. If a transposition error had occurred with the tens and hundreds positions, the incorrect number would be $1000a + 100c + 10b + d$.
  Perform the procedure shown in part **a** to work out whether the difference between the correct number and the incorrect one is a multiple of 9.
  c. Consider a transposition error in the thousands and hundreds positions. Explain whether the difference between the two numbers is a multiple of 9.
  d. Comment on this rule for transposition errors.
  e. The price marked on the tag of a CD in a music store is x dollars and y cents. Its value is less than $100. The person on the counter makes a mistake and enters a value of y dollars and x cents into the cash register.
  Show whether the multiple of 9 rule for checking transposition errors applies in this case.

on To test your understanding and knowledge of this topic, go to your learnON title at www.jacplus.com.au and complete the **post-test**.

Online Resources

 Resources

Below is a full list of **rich resources** available online for this topic. These resources are designed to bring ideas to life, to promote deep and lasting learning and to support the different learning needs of each individual.

eWorkbook

Download the workbook for this topic, which includes worksheets, a code puzzle and a project (ewbk-2002) ☐

Solutions

Download a copy of the fully worked solutions to every question in this topic (sol-0722) ☐

Digital documents

2.2 SkillSHEET Alternative expressions used for the four operations (doc-6122) ☐
2.3 SkillSHEET Algebraic expressions (doc-6123)
SkillSHEET Substitution into algebraic expressions (doc-6124) ☐
2.4 SkillSHEET Like terms (doc-6125) ☐
SkillSHEET Collecting like terms (doc-6126) ☐
SkillSHEET Multiplying algebraic terms (doc-6127) ☐
SkillSHEET Dividing algebraic terms (doc-6128) ☐
SkillSHEET Adding and subtracting integers (doc-10817) ☐
SkillSHEET Multiplying and dividing integers (doc-10818) ☐
2.5 SkillSHEET Expanding brackets (doc-10819) ☐
SkillSHEET Expanding a pair of brackets (doc-10820) ☐
2.6 SkillSHEET Recognising expansion patterns (doc-10821) ☐
2.8 SkillSHEET Finding the highest common factor (doc-10822) ☐
SkillSHEET Factorising by finding the HCF (doc-10823) ☐

Video eLessons

2.2 Expressions (eles-4588) ☐
Substitution 1 (eles-4589) ☐
Substitution into formulas (eles-4590) ☐
Substitution 2 (eles-1892) ☐
2.3 Algebra in worded problems (eles-4591) ☐
2.4 Addition and subtraction of like terms (eles-4592) ☐
Multiplication of algebraic terms (eles-4593) ☐
Division of algebraic terms (eles-4594) ☐
Simplification of expressions (eles-1884) ☐
2.5 The Distributive Law (eles-4595) ☐
Expanding and simplifying (eles-4596) ☐
Expanding binomial factors (eles-4597) ☐
Expanding brackets (eles-1888) ☐
Expansion of binomial expressions (eles-1908) ☐
2.6 Difference of two squares (eles-4598) ☐
Perfect squares (eles-4599) ☐
2.7 Expanding multiple sets of brackets (eles-4600) ☐
2.8 Factorising expressions (eles-4601) ☐
Factorising expressions by finding the highest common factor (eles-4602) ☐
Factorisation (eles-1887) ☐
2.9 The common binomial factor (eles-4603) ☐
Factorising by grouping in pairs (eles-4604) ☐
2.10 Converting worded problems into algebra (eles-4605) ☐

Interactivities

2.2 Individual pathway interactivity: Using pronumerals (int-4480) ☐
Substituting positive and negative numbers (int-3765) ☐
2.3 Individual pathway interactivity: Algebra in worded problems (int-4481) ☐
2.4 Individual pathway interactivity: Simplifying algebraic expressions (int-4482) ☐
Multiplying variables (int-3772) ☐
Dividing expressions with variables (int-3773) ☐
2.5 Individual pathway interactivity: Expanding brackets (int-4483) ☐
Expanding brackets (int-6034) ☐
Expanding binomial factors (int-6033) ☐
Like terms (int-6035) ☐
2.6 Individual pathway interactivity: Difference of two squares and perfect squares (int-4484) ☐
Difference of two squares (int-6036) ☐
2.7 Individual pathway interactivity: Further expansions (int-4485) ☐
2.8 Individual pathway interactivity: The highest common factor (int-4486) ☐
Highest common factor (int-6037) ☐
2.9 Individual pathway interactivity: The highest common binomial factor (int-4487) ☐
Common binomial factor (int-6038) ☐
2.10 Individual pathway interactivity: Solving worded problems (int-4488) ☐
2.11 Crossword (int-0699) ☐
Sudoku puzzle (int-3203) ☐

Teacher resources

There are many resources available exclusively for teachers online.

To access these online resources, log on to **www.jacplus.com.au**.

Answers

Topic 2 Algebra

Exercise 2.1 Pre-test

1. D
2. $2x^2 - y^2 - xyz = 11$
3. $m = 2$
4. E
5. A
6. a. $2m + 13n$ b. $-2xy - 3x + 5y$
7. $6q^2 + qr - r^2$
8. D
9. Yes
10. False
11. $3b^2 - 21b + 37$
12. a. $4p(2p + 5)$ b. $5(5x^2 - 2y)$ c. $xy(6 + 8x - 3y)$
13. a. $(7a - 1)(b + 3)$ b. $(5x - 2)(2y + 1)$
14. $2x(2x - 1) + 2x(3x + 2) + 2(3x + 2)(2x - 1)$
15. $(x + 4)(x - 1)$

Exercise 2.2 Using pronumerals

1. a. 3   b. -2 c. $\dfrac{1}{3}$ d. $-\dfrac{1}{4}$
2. a. 5   b. -2 c. $\dfrac{1}{6}$ d. $-\dfrac{2}{7}$
3. a. 15   b. $-\dfrac{2}{5}$ c. $-\dfrac{7}{2}$ d. $-\dfrac{1}{7}$
4. a.i. 3   ii. 5   iii. 8   iv. $5x^2$
   b.i. 3   ii. -9 iii. -6 iv. $-9m^2$
   c.i. 4   ii. 5   iii. 5   iv. $-7x^2$
   d.i. 4   ii. 9   iii. 4   iv. $-9b^2$
5. a.i. 4   ii. 4   iii. -2 iv. $-3ac$
   b.i. 5   ii. 5   iii. 9   iv. $-3u$
   c.i. 5   ii. -1 iii. 8   iv. $-m$
   d.i. 5   ii. 7   iii. 14   iv. $-3cd^2$
6. a.i. 3   ii. 3   iii. -7 iv. $-10x^2$
   b.i. 5   ii. -3 iii. 7   iv. $-6x^2$
   c.i. 5   ii. -5 iii. 8   iv. $-5x$
   d.i. 6   ii. 6   iii. 6   iv. $-15xy$
7. a. 2, 12, -8 b. 0, 2, 6
   c. -1, 9, -3 d. 0, 1, -1
8. a. 4, -11, 19   b. 1, 4, 16
   c. 0, 21, -3 d. 8, 2, -4
9. a. $\sqrt{3}$, 2, 1   b. 0, 3, -2
   c. -12, -12, -6 d. $\dfrac{3}{5}$, 1, $-\dfrac{1}{5}$
10. a. B   b. D   c. A   d. B
11. a. 4   b. -6 c. 6
    d. 7   e. -12 f. -3
12. a. 7   b. -40 c. 7
13. a. i. 32   ii. 250
    b. i. 4   ii. 3

14. a. 15   b. 10   c. 18
    d. 45   e. 30
15. a. 10.72   b. 21   c. 22.4
    d. 50   e. 171
16. a. 4   b. -30 c. 32.75
    d. 38.33   e. 34.56
17. a. Sample responses can be found in the worked solutions in the online resources.
    b. 12 cm
18. a. 10   b. 10
19. a. 15
    b. Sample responses can be found in the worked solutions in the online resources.
20. a. 5
    b. Sample responses can be found in the worked solutions in the online resources.
21. a. 19.44
    b. Sample responses can be found in the worked solutions in the online resources.
22. a. 63 cm^3 b. 4 cm
23. a. 461.81   b. 9.55
24. a. 1671.33   b. 2.64
25. a.

d	$2d + 2g$	g
$2g$	$d + g$	$2d$
$2d + g$	0	$d + 2g$

Magic number $= 3d + 3g$

b., c Sample responses can be found in the worked solutions in the online resources.
26. a. mn
    b. $(m - 2)(n - 2)$
    c. $(2m + 2n)$ units
27. $y - (z - x)$

Exercise 2.3 Algebra in worded problems

1. a. 11   b. $x + 5$ c. $y + 5$
2. a. 3.5 km   b. $(x + 2)$ km
3. a. 4   b. $2.5 + t$ c. $2.5 - y$ hours
4. a. $d - x$ km   b. $d + y$ km   c. $2d + 4$ km
5. a. $10n$, where n = number of shirts
   b. $N + 30$, where N = the number of Nick's dollars
   c. i. $g - 4$ ii. $7g - 4$
6. a. i. 94   ii. $6x + y$
   b. i. 66   ii. $6p + q$
   c. i. 28   ii. $6x + y - 6p - q$
7. xy dollars
8. a. $\dfrac{x}{n}$ b. $\dfrac{x}{2n}$
9. a. $(30 - d)$ cm   b. $\dfrac{30 - d}{4}$ cm   c. $\dfrac{3(30 - d)}{4}$ cm
10. a. $\dfrac{x}{4}$ b. $\dfrac{x}{6}$ c. $\dfrac{x}{4} - \dfrac{x}{6} = \dfrac{x}{12}$

11. a. $\dfrac{24-c}{7}$ hours   **b.** $\left(c + \dfrac{24-c}{7}\right) = \dfrac{6c+24}{7}$

12. a. 1   **b.** 7:30 pm and 8:00 pm
   c. 5   **d.** 24
   e. 8:30 pm and 9:00 pm

13. a. $\$(y-30)$

   b. $\$\left(\dfrac{1}{3}(y-30)\right)$

   c. $\$930$

14. a. $b = 2h - 1$ **b.** 19 blocks

15. 15 red vases and 10 clear vases

16. The father is 32 years old.

17. 4 dots $= 4$ triangles
5 dots $= 10$ triangles
6 dots $= 20$ triangles
7 dots $= 35$ triangles
$t = \dfrac{d!}{6(d-3)!}$ where $d!$ is the product of all positive integers less than or equal to d.

18.   **a.** $x = \dfrac{T-3}{T-1}$
   b., c. Answer is undefined.

Exercise 2.4 Simplification of algebraic expressions

1. a. $9ab,\ -ab$ **b.** $4x$
   c. $-3za,\ -az$ **d.** $2x^2,\ -x^2$

2. a. $-2x^2y$ **b.** $-x^2y^5,\ -3x^2y^5$
   c. $p^3x^2w^5,\ -5x^2p^3w^5$ **d.** $4y^5z^4x^2,\ -2x^2z^4y^5$

3. a. $7x$ **b.** $11y$ **c.** $19m$ **d.** $11q$ **e.** $8r$ **f.** $3x$

4. a. $8a$ **b.** $8y$ **c.** $13x$ **d.** $16p$ **e.** $9q^2$ **f.** $3x^2$

5. a. $8x^2 - 3y$ **b.** $2m^2 + 2n$
   c. $-2g^2 + g - 12$ **d.** $-5m^2 + m + 15$
   e. $12a^2 + b + 4b^2$ **f.** $3m + 7n^2$

6. a. $12xy + 2y^2$ **b.** $2ab + 5a^2b$
   c. $16x^2y - 3xy$ **d.** $m^2n + 11n$
   e. $-7x^2 + 2x^2y$ **f.** $-3a^2b - 9a^2 + 5b + 4$

7. B

8. D

9. A

10. D

11. a. $6mn$ **b.** $20xy$ **c.** $8pq$
   d. $-10xy$ **e.** $-12xy$

12. a. $15mn$ **b.** $10a^2$ **c.** $6mnp$
   d. $-6ab^2$ **e.** $10m^2n$

13. a. $-18a^2b$ **b.** $30x^3y^2$ **c.** $-12p^2q^3$
   d. $-56c^3d$ **e.** $30a^4b^7$

14. a. $3x$ **b.** $3m$ **c.** $2y$ **d.** $4m$ **e.** $4m$

15. a. $2x$ **b.** $-7x$ **c.** $-4m$ **d.** $\dfrac{m}{2}$ **e.** $\dfrac{x}{3}$

16. a. $\dfrac{4m}{9}$ **b.** $\dfrac{1}{2a}$ **c.** $2yz$
   d. $\dfrac{xy}{4}$ **e.** $\dfrac{-7yz}{11}$

17. a. $40x^2y^2$ **b.** $56ax^2y^2$
   c. $\dfrac{x^2}{2y}$ **d.** $\dfrac{-5a}{4b}$

18. a. 2 **b.** $40a^3b^2$
   c. $-8a^3b^3$ **d.** $16a^4$

19. a.   **i.** $(0.01\,mp + nq)$ dollars   **ii.** $(mp + 100\,nq)$ cents
   b. $20 - (0.01mp + nq)$ dollars

20. Shirt $= \$12$ each
Shorts $= \$30$ each

21. $\$45$

22. The correct answer is $7a^2c^2$. Sample responses can be found in the worked solutions in the online resources.

23. $\dfrac{3b^9}{2c^6}$

24. a. $\dfrac{n}{(n+2)}$ **b.** $\dfrac{n}{(n+3)}$ **c.** $\dfrac{1}{100}$

25. a.   **i.**

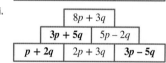

   ii.

   b.   **i.**

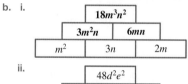

   ii.

26. a. $P = \dfrac{x}{3} + x + \dfrac{5x}{7}$
   b. Sample responses can be found in the worked solutions in the online resources.
   c. $\$18$

27. a. $P = 2w + \dfrac{32w}{7}$
   b. Sample responses can be found in the worked solutions in the online resources.
   c. $\$72.22$

28. a. $P = 2x + 2(4x - 7)$
   b. $P = 10x - 14$
   c. 16 units
   d. If $x = 1$ the rectangle will have a negative value perimeter, which is not possible.

29. a. $l = 4(x + 40) + 8(x + 20)$
   b. $l = 12x + 320$
   c. $\$19.20$

30. a. $P_1 = (10x + 12)$ cm   **b.** $x = 2$
   $P_2 = 16x$ cm

Exercise 2.5 Expanding brackets

1. a. 64   **b.** 1100
   c. -20 **d.** 693

2. a. $3x + 6$

3	$3x$	6

      x 2

   b. $4x + 12$

4	$4x$	12

      x 3

   c. $4x + 4$ d. $7x - 7$

3. a. $-3p + 6$ b. $-x + 1$
   c. $6b - 12$ d. $24m - 16$

4. a. $x^2 + 2x$

x	x^2	$2x$

      x 2

   b. $a^2 + 5a$

a	a^2	$5a$

      a 5

   c. $4x + x^2$ d. $7m - m^2$

5. a. $2xy + 4x$ b. $-3xy - 12y$
   c. $-3b + ab$ d. $-30a + 18a^2$

6. a. $2p - 2$ b. $5x - 17$ c. $-7p - 17$
   d. $-12p + 3$ e. $6x^2 - 20x$

7. a. $2m^2 + 7m$ b. $3px + 6x - 5$
   c. $4y^2 - 4y + 7$ d. $-4p^2 + 13p$
   e. $4x - 13y$

8. a. $2m^2 - 8m - 4$ b. $-3p^2 + 10pq - 1$
   c. $-30a + 10ab$ d. $7cd - 12c^2 - 5c$
   e. $-2p - 17$

9. a. $8x + y$ b. $10p + 8q$ c. $18a + 29b$
   d. $19c + 22d$ e. $2m - 11n$

10. a. $6x - 11y$ b. $9x + 5y$ c. $-14p - 8q$
    d. $-4a + 3b$ e. $4x - y$

11. a. $5p - 10q$ b. $-8c + 9d$ c. $13x - 19y$
    d. $-7p + 11q$ e. $-5a + 3b$

12. a. $2ab + 2a - 3b$ b. $2xy + 4x - 2y$
    c. $2cd + 3c$ d. $2pq - 2p$
    e. $5cd - 11c$

13. a. $5ab - 21a - 3b$ b. $5m$
    c. $6cd - 36c$ d. $6m^2 + 6m - 10$
    e. $9cd - 8c$

14. a. $-15a^2 + 2b^2 - 9ab$
    b. $-8c^2 + 3d^2 + 22cd$
    c. $12m^2 - 20m - 4$
    d. $-7x^2 + 41x - 6$
    e. $-10y^2 - 6y - 12$

15. B

16. A

17. D

18. a. $a^2 + 5a + 6$ b. $x^2 + 7x + 12$
    c. $y^2 + 5y + 6$ d. $m^2 + 9m + 20$
    e. $b^2 + 3b + 2$

19. a. $p^2 + 5p + 4$ b. $a^2 + a - 6$
    c. $x^2 + x - 20$ d. $m^2 - m - 12$
    e. $y^2 + 2y - 15$

20. a. $y^2 - 4y - 12$ b. $x^2 - 2x - 3$
    c. $x^2 - 7x + 12$ d. $p^2 - 5p + 6$
    e. $x^2 - 4x + 3$

21. a. $2a^2 + 7a + 6$ b. $4c^2 - 31c + 42$
    c. $2t^2 - 17t + 35$ d. $-6t^2 + 11t + 10$
    e. $15x^2 - 31x + 14$ f. $25x^2 - 20x + 4$

22. a. $xz + x + yz + y$ b. $2xz + 8x + yz + 4y$
    c. $3pr + 3p + qr + q$ d. $a^2 + 3ab + 2b^2$
    e. $2c^2 - 5cd - 3d^2$ f. $2x^2 - xy - 3y^2$

23. a. $4p^2 + pq - 3q^2$ b. $ab + ac + 2b^2 + 2bc$
    c. $3p - 9pr - 2q + 6qr$ d. $12x^2 - 7xy + y^2$
    e. $2p^2 - pr - 2pq + qr$ f. $15k - 5 - 6jk + 2j$

24. C

25. C

26. a. $x^2 - 9$ b. $x^2 - 25$ c. $x^2 - 49$
    d. $x^2 - 1$ e. $x^2 - 4$ f. $4x^2 - 1$

27. a. $x^2 + 2x + 1$ b. $x^2 + 4x + 4$
    c. $x^2 + 16x + 64$ d. $x^2 - 6x + 9$
    e. $x^2 - 10x + 25$ f. $x^2 - 18x + 81$

28. a. $6.3x^2 + 5.53xy - 3.1y^2$
    b. $4.41x^2 - 10.24y^2$
    c. $11.56x^2 + 34.68xy + 26.01y^2$

29. a. Surface area $= 38x^2 + 14x - 6$
    b. Volume $= 12x^3 + 5x^2 - 3x$

30. a. i. $((x + 3y)(4x - y)) \, \text{m}^2$
      ii. $(4x^2 + 11xy - 3y^2) \, \text{cm}^2$
      iii. Both brackets must be positive; therefore, $x > -3y$, $x > \dfrac{y}{4}$
    b. i. $\dfrac{(2x - 1)(x + 5)}{2}$
      ii. $\dfrac{2x^2 + 9x - 5}{2}$
      iii. Sample responses can be found in the worked solutions in the online resources.

31. Sample responses can be found in the worked solutions in the online resources.

32. a. Negative sign ignored
    b. Negative sign ignored
    c. Distributive law not used

33. a. i. The student did not multiply both terms.
      ii. The student used addition instead of multiplication.
      iii. The student did not change negative and positive signs.
    b. $100 - 15x$

34. a. $-2(a - 5) = 2a - 10$ is incorrect because the student did not change the multiplied negative signs for -2×-5 to a positive sign. The correct answer is $-2a + 10$.
    b. $2b(3b - 1) = 6b^2 - 1$ is incorrect because the student did not multiply -1 and $2b$ together. The correct answer is $6b^2 - 2b$.
    c. $-2(c - 4) = 2c + 8$ is incorrect because the student left out the negative sign when multiplying -2 and c. The correct answer is $-2c + 8$.

35. a. Student C

   b. Corrections to the students' answers are shown in bold.
   Student A:
   $(3x + 4)(2x + 5)$
   $= 3x \times 2x + 3x \times 5 + 4 \times 2x + 4 \times 5$
   $= \mathbf{6x^2} + \mathbf{23x} + 20$
   (Also, $29x + 20$ does not equal $49x$.)
   Student B:
   $(3x + 4)(2x + 5)$
   $= 3x \times 2x + \mathbf{3x \times 5} + 4 \times 2x + 4 \times 5$
   $= 6x^2 + \mathbf{15x} + 8x + 20$
   $= 6x^2 + \mathbf{23x} + 20$

36. a. $w = (x - 3)\,\text{cm}$
   b. $w = 62\,\text{cm}$
   c. $l = 108\,\text{cm}$

Exercise 2.6 Difference of two squares and perfect squares

1. a. $x^2 - 4$ b. $y^2 - 9$ c. $m^2 - 25$ d. $a^2 - 49$

2. a. $x^2 - 36$ b. $p^2 - 144$ c. $a^2 - 100$ d. $m^2 - 121$

3. a. $4x^2 - 9$ b. $9y^2 - 1$ c. $25d^2 - 4$
   d. $49c^2 - 9$ e. $4 - 9p^2$

4. a. $d^2 - 81x^2$ b. $25 - 144a^2$ c. $9x^2 - 100y^2$
   d. $4b^2 - 25c^2$ e. $100 - 4x^2$

5. a. $x^2 + 4x + 4$ b. $a^2 + 6a + 9$
   c. $b^2 + 14b + 49$ d. $c^2 + 18c + 81$

6. a. $m^2 + 24m + 144$ b. $n^2 + 20n + 100$
   c. $x^2 - 12x + 36$ d. $y^2 - 10y + 25$

7. a. $81 - 18c + c^2$ b. $64 + 16e + e^2$
   c. $2x^2 + 4xy + 2y^2$ d. $u^2 - 2uv + v^2$

8. a. $4a^2 + 12a + 9$ b. $9x^2 + 6x + 1$
   c. $4m^2 - 20m + 25$ d. $16x^2 - 24x + 9$

9. a. $25a^2 - 10a + 1$ b. $49p^2 + 56p + 16$
   c. $81x^2 + 36x + 4$ d. $16c^2 - 48c + 36$

10. a. $25 + 30p + 9p^2$ b. $4 - 20x + 25x^2$
    c. $81x^2 - 72xy + 16y^2$ d. $64x^2 - 48xy + 9y^2$

11. a. $x^2 - 9$ b. $4x^2 - 9$ c. $49x^2 - 16$
    d. $4x^2 - 49y^2$ e. $x^4 - y^4$

12. a. $16x^2 + 40x + 25$ b. $49x^2 - 42xy + 9y^2$
    c. $25x^4 - 20x^2y + 4y^2$ d. $2x^2 - 4xy + 2y^2$
    e. $\dfrac{4}{x^2} + 16 + 16x^2$

13. $(x^2 + 6x + 9)\,\text{units}^2$

14. $(3x + 1)\,\text{m}$

15. Perimeter $= 4x + 36$

16. Sample responses can be found in the worked solutions in the online resources.

17. a. Sample responses can be found in the worked solutions in the online resources.
    b. Lin's bedroom is larger by $1\,\text{m}^2$.

18. a. i. $x^2 - 16$ and $16 - x^2$ iii. $4x^2 - 81$ and $81 - 4x^2$
       ii. $x^2 - 121$ and $121 - x^2$

    b. The answers to the pairs of expansions are the same, except that the negative and positive signs are reversed.

c. This is possible because when a negative number is multiplied by a positive number, it becomes negative. When expanding a DOTS in which the expressions have different signs, the signs will be reversed.

19. a. $100k^2 + 100k + 25$
    b. $(10k + 5)^2 = 100 \times k \times k + 100 \times k + 25$
       $\qquad\qquad = 100k(k + 1) + 25$

    c. $25^2 = (10 \times 2 + 5)^2$
       Let $k = 2$.
       $25^2 = 100k(k + 1) + 25$
       $\quad = 100 \times 2 \times (2 + 1) + 25$
       $\quad = 625$
       $85^2 = (10 \times 8 + 5)^2$
       Let $k = 8$.
       $85^2 = 100k(k + 1) + 25$
       $\quad = 100 \times 8 \times (8 + 1) + 25$
       $\quad = 7225$

20. a. $A_1 = a^2\ \text{units}^2$
       $A_2 = ab\ \text{units}^2$
       $A_3 = ab\ \text{units}^2$
       $A_4 = b^2\ \text{units}^2$

    b. $A = a^2 + ab + ab + b^2$
       $\quad = a^2 + 2ab + b^2$
       This is the equation for perfect squares.

21. a. i. $x^2 - 6x + 9$ and $9 - 6x + x^2$
       ii. $x^2 - 30x + 225$ and $225 - 30x + x^2$
       iii. $9x^2 - 42x + 49$ and $49 - 42x + 9x^2$

    b. The answers to the pairs of expansions are the same.

    c. This is possible because when a negative number is multiplied by itself, it becomes positive. When expanding a perfect square in which the two expressions are the same, the negative signs cancel out and result in the same answer.

22. a. 729   b. 1089   c. 1521   d. 2209

23. $(3x^2 + 16x + 5)\,\text{m}^2$

24. $(x^2 - 25)\,\text{m}^2$

25. a. 10609   b. 3844   c. 994 009
    d. 1 024 144   e. 2809   f. 9604

Exercise 2.7 Further expansions

1. a. $2x^2 + 13x + 21$ b. $2x^2 + 13x + 20$
   c. $2x^2 + 14x + 26$ d. $2x^2 + 10x + 11$

2. a. $2p^2 - 3p - 21$ b. $2a^2 - 5a + 4$
   c. $2p^2 - p - 24$ d. $2x^2 + 19x - 36$

3. a. $2y^2 + 2y - 7$ b. $2d^2 + 8d - 2$
   c. $2x^2 + 10$ d. $2y^2$

4. a. $2x^2 - 4x + 19$ b. $2y^2 - 4y - 7$
   c. $2p^2 + 3p + 23$ d. $2m^2 + 3m + 31$

5. a. $x + 5$ b. $4x + 8$
   c. $-2x - 6$ d. $3m + 2$

6. a. $-3b - 22$ b. $-15y - 2$
   c. $8p - 10$ d. $16x + 2$

7. a. $4m + 17$ b. $-7a + 30$
   c. $-6p - 7$ d. $3x - 21$

8. a. $16x - 16$

b. Sample responses can be found in the worked solutions in the online resources.

9. a. $15x^2 - 44x + 21$

b. Sample responses can be found in the worked solutions in the online resources.

10. a. $8x^2 - 10xy + 3y^2$

b. Sample responses can be found in the worked solutions in the online resources.

11. $x = -4$

12. $(p^2 + p - 2) + (p^2 - 2p - 3)$
$$= p^2 + p^2 + p - 2p - 2 - 3$$
$$= 2p^2 - p - 5$$

13. $(x + 2)(x - 3) - (x + 1)^2 = (x^2 - x - 6) - (x^2 + 2x + 1)$
$$= -3x - 7$$

14. a. Sample responses can be found in the worked solutions in the online resources.

b. $a = 2, b = 1, c = 3, d = 4$
$$(2^2 + 1^2)(3^2 + 4^2) = (2 \times 3 - 1 \times 4)^2 + (2 \times 4 - 1 \times 3)^2$$
$$= 4 + 121$$
$$= 125$$

15. a. $x^4 + 2x^3 - x^2 - 2x + 1$

b. Sample responses can be found in the worked solutions in the online resources.

c. i. 25

ii. $x = 2$

16. a. $ad + ae + bd + be$

b. $(a + b + c)(d + e + f) = ad + ae + bd + be + cd + ce + af + bf + cf$

	a	b	c
d	ad	bd	cd
e	ae	be	ce
f	af	bf	cf

Exercise 2.8 The highest common factor

1. a. $2a$ **b.** 4 **c.** $3x$
   d. -4 **e.** $3x$ **f.** $15a$

2. a. $-3x$ **b.** m **c.** $-mn$
   d. ab **e.** $3x$ **f.** $5n^2$

3. a. 4 **b.** 3 **c.** 5 **d.** 8 **e.** 3

4. a. 25 **b.** 1 **c.** $6a$ **d.** $7x$ **e.** $30q$

5. a. 10 **b.** $3x$ **c.** 3 **d.** 5 **e.** x^2

6. C

7. a. $4(x + 3y)$ **b.** $5(m + 3n)$
   c. $7(a + 2b)$ **d.** $7(m - 3n)$
   e. $-8(a + 3b)$ **f.** $4(2x - y)$

8. a. $-2(6p + q)$ **b.** $6(p + 2pq + 3q)$
   c. $8(4x + y + 2z)$ **d.** $4(4m - n + 6p)$
   e. $8(9x - y + 8pq)$ **f.** $3(5x^2 - y)$

9. a. $5(p^2 - 4q)$ **b.** $5(x + 1)$
   c. $8(7q + p^2)$ **d.** $7(p - 6x^2y)$
   e. $4(4p^2 + 5q + 1)$ **f.** $12(1 + 3a^2b - 2b^2)$

10. a. $3(3a + 7b)$ **b.** $2(2c + 9d^2)$
   c. $4(3p^2 + 5q^2)$ **d.** $7(5 - 2m^2n)$
   e. $5(5y^2 - 3x)$

11. a. $4(4a^2 + 5b)$ **b.** $6(7m^2 + 2n)$
   c. $9(7p^2 + 9 - 3y)$ **d.** $11(11a^2 - 5b + 10c)$
   e. $2(5 - 11x^2y^3 + 7xy)$

12. a. $9(2a^2bc - 3ab - 10c)$ **b.** $12(12p + 3q^2 - 7pq)$
   c. $7(9a^2b^2 - 7 + 8ab^2)$ **d.** $11(2 + 9p^3q^2 - 4p^2r)$
   e. $6(6 - 4ab^2 + 3b^2c)$

13. a. $-(x - 5)$ **b.** $-(a - 7)$ **c.** $-(b - 9)$
   d. $-2(m + 3)$ **e.** $-6(p + 2)$ **f.** $-4(a + 2)$

14. a. $-3(n^2 - 5m)$ **b.** $-7(x^2y^2 - 3)$ **c.** $-7(y^2 + 7z)$
   d. $-6(2p^2 + 3q)$ **e.** $-7(9m - 8)$ **f.** $-2(6m^3 + 25x^3)$

15. a. $-3(3a^2b - 10)$ **b.** $-3(5p + 4q)$
   c. $-2(9x^2 - 2y^2)$ **d.** $-3(ab - 6m + 7)$
   e. $-5(2 + 5p^2 + 9q)$ **f.** $-9(10m^2 - 3n - 6p^3)$

16. a. $a(a + 5)$ **b.** $q(14 - q)$ **c.** $m(18 + 5m)$
   d. $p(6 + 7p)$ **e.** $n(7n - 2)$ **f.** $p(7 - pq + q)$

17. a. $y(x + 9 - 3y)$ **b.** $c(5 + 3cd - d)$
   c. $ab(3 + a + 4b)$ **d.** $xy(2x + 1 + 5y)$
   e. $pq(5pq - 4 + 3p)$ **f.** $xy(6xy - 5 + x)$

18. a. $5x(x + 3)$ **b.** $6m(4m - 1)$ **c.** $4a(8a - 1)$
   d. $-2m(m - 4)$ **e.** $-5x(x - 5)$ **f.** $-7y(y - 2)$

19. a. $-3a(a - 3)$ **b.** $-2p(6p + 1)$ **c.** $-13y(2y + 1)$
   d. $2m(2 - 9m)$ **e.** $-6t(1 - 6t)$ **f.** $-8p(1 + 3p)$

20. a. $2(x + 3)(4x + 1)$

b. $A = x(x + 3)$
$B = (5x + 2)(x + 3)$
$C = 2x(x + 3)$

21. a. $3a(b^2 + 6b + 9)$
b. $3a(b + 3)^2$

22. a. $(x + 4)(x^2 + 8x + 16)$
b. $(x + 4)^3$

23. Sample responses can be found in the worked solutions in the online resources.

24. Sample responses can be found in the worked solutions in the online resources.

25. a. $\left(x + \sqrt{13}\right)\left(x - \sqrt{13}\right)$

b. $\left(2x - \sqrt{17}\right)\left(2x + \sqrt{17}\right)$

c. $\left(x + 3 + \sqrt{10}\right)\left(x + 3 - \sqrt{10}\right)$

26. -2

27. True. Sample responses can be found in the worked solutions in the online resources.

28. a. $4(3x - 5y)(3x + 5y)$
b. $4(3x - 5y)(3x + 5y)$
c. Yes, the answers are the same.

Exercise 2.9 The highest common binomial factor

1. a. $(a + b)(2 + 3c)$ **b.** $(m + n)(4 + p)$
   c. $(2m + 1)(7x - y)$ **d.** $(3b + 2)(4a - b)$
   e. $(x + 2y)(z - 3)$

2. a. $(6-q)(12p-5)$ **b.** $(x-y)(3p^2+2q)$
   c. $(b-3)(4a^2+3b)$ **d.** $(q+2p)(p^2-5)$
   e. $(5m+1)(6+n^2)$

3. a. $(y+2)(x+2)$ **b.** $(b+3)(a+3)$
   c. $(x-4)(y+3)$ **d.** $(2y+1)(x+3)$
   e. $(3b+1)(a+4)$ **f.** $(b-2)(a+5)$

4. a. $(m-2n)(1+a)$ **b.** $(5+3p)(1+3a)$
   c. $(3m-1)(5n-2)$ **d.** $(10p-1)(q-2)$
   e. $(3x-1)(2-y)$ **f.** $(4p-1)(4-3q)$

5. a. $(2y+1)(5x-2)$ **b.** $(2a+3)(3b-2)$
   c. $(b-2c)(5a-3)$ **d.** $(x+3y)(4-z)$
   e. $(p+2q)(5r-3)$ **f.** $(a+5b)(c-2)$

6. a. $\dfrac{a}{b}$ **b.** $\dfrac{3x-4}{7a-5}$

7. a. $\dfrac{x+2}{2+6x}$ **b.** $\dfrac{y+2}{3b-2}$

8. a. $\dfrac{3x-2}{3y+2}$ **b.** $\dfrac{m-4n}{m+4n}$

9. Sample responses can be found in the worked solutions in the online resources.

10. $(2x+3)^2$ This is a perfect square.

11. a. $5(x+2)(x+3)=5(x^2+5x+6)=5x^2+25x+30$

	x	3	x	3	x	3	x	3	x	3
x	x^2	$3x$	x^2	$3x$	x^2	$3x$	x^2	$3x$	x^2	$3x$
2	$2x$	6	$2x$	6	$2x$	6	$2x$	6	$2x$	6

b. $5(x+2)(x+3)=(5\times x+2\times5)(x+3)$
   $=(5x+10)(x+3)$

	x	2	x	2	x	2	x	2	x	2
x	x^2	$2x$	x^2	$2x$	x^2	$2x$	x^2	$2x$	x^2	$2x$
3	$3x$	6	$3x$	6	$3x$	6	$3x$	6	$3x$	6

c. $5(x+2)(x+3)=(5\times x+3\times5)(x+2)$
   $=(5x+15)(x+2)$

	$5x$	15
x	$5x^2$	$15x$
2	$10x$	30

12. a. Side length$_a=x+3$
   Side length$_b=x+5$
   Side length$_c=x+8$
   Side length$_d=x-3$
   Side length$_e=x+6$

b. $A_a=\text{II}=(x+3)^2$
   $A_b=\text{III}=(x+5)^2$
   $A_c=\text{V}=(x+8)^2$
   $A_d=\text{I}=(x-3)^2$
   $A_e=\text{IV}=(x+6)^2$

c. $A_a=\text{II}=64\,\text{cm}^2$
   $A_b=\text{III}=100\,\text{cm}^2$
   $A_c=\text{V}=169\,\text{cm}^2$
   $A_d=\text{I}=4\,\text{cm}^2$
   $A_e=\text{IV}=121\,\text{cm}^2$

13. $-(x+1)(4y-9)$

14. a. i. Square, because it is a perfect square.
   ii. Rectangle, because it is a DOTS.
   iii. Rectangle, because it is a trinomial.
   iv. Rectangle, because it is a trinomial.
   b. i. $(3s+8)^2$ ii. $(5s+2)(5s-2)$
   iii. $(s+1)(s+3)$ iv. $4(s-8)(s+1)$

Exercise 2.10 Solving worded problems

1. a. i. $12x$ ii. 60   iii. $9x^2$ iv. 225
   b. i. $4x+8$ ii. 28
   iii. x^2+4x+4 iv. 49
   c. i. $16x-4$ ii. 76
   iii. $16x^2-8x+1$ iv. 361
   d. i. $10x+2$ ii. 52
   iii. $6x^2+2x$ iv. 160
   e. i. $12x-2$ ii. 58
   iii. $5x^2-13x-6$ iv. 54
   f. i. $20x+18$ ii. 118
   iii. $21x^2+42x$ iv. 735

2. a. $1250\,\text{m}^2$
   b. $(1250+150x+4x^2)\,\text{m}^2$
   c. $(150x+4x^2)\,\text{m}^2$
   d. $366.16\,\text{m}^2$
   e. $150x+4x^2=200$

3. a. $(20-2x)\,\text{cm}$
   b. $(15-2y)\,\text{cm}$
   c. $(300-40y-30x+4xy)\,\text{cm}^2$
   d. Sample responses can be found in the worked solutions in the online resources.

4. a. $40\,\text{cm}^2$
   b. i. $(8+v)\,\text{cm}$
   ii. $(5+v)\,\text{cm}$
   iii. $(v^2+13v+40)\,\text{cm}^2$
   iv. $70\,\text{cm}^2$
   c. i. $(8-d)\,\text{cm}$
   ii. $(5-d)\,\text{cm}$
   iii. $(40-13d+d^2)\,\text{cm}^2$
   iv. $18\,\text{cm}^2$
   d. i. $8x\,\text{cm}^2$ ii. $(5+x)\,\text{cm}$
   iii. $(8x^2+40x)\,\text{cm}^2$ iv. $400\,\text{cm}^2$

5. a. $20x\,\text{m}$ **b.** $25x^2\text{m}^2$
   c.
   i. $(5x-2)\,\text{m}$
   ii. $(5x-3)\,\text{m}$
   iii. $(25x^2-25x+6)\,\text{m}^2$
   iv. 756m^2

6. a. $6x\,\text{cm}$
   b. $2x^2\text{cm}^2$
   c. i. $(2x+y)\,\text{cm}$
   ii. $(x-y)\,\text{cm}$
   iii. $(2x^2-xy-y^2)\,\text{cm}^2$
   iv. $11\,\text{cm}^2$

7. a. $4x$ cm

  b. x^2 cm^2

  c. i. $(x + y)$ cm

   ii. $(4x + 4y)$ cm

   iii. 56 cm

   iv. $(x^2 + 2xy + y^2)$ cm^2

   v. 60.84 cm^2

8. a. $(14p + 4)$ m **b.** $(12p^2 + 6p)$ m^2

  c. $(6p + 2)$ m **d.** $5p$ m

  e. $(22p + 4)$ m **f.** $(30p^2 + 10p)$ m^2

  g. $(18p^2 + 4p)$ m^2 **h.** 80 m^2

9. a. 27.8

  b. Samir is not within the healthy weight range.

  c. i. 19.1 **ii.** 24.9 **iii.** 22.0

  d. Sammara: 9 to 20 years of age; Nimco: 17.5 to 20 years of age; Manuel: 13.5 to 20 years of age

10. a.

Type of algebra	Perimeter	Type of algebra rectangle	Perimeter
1-long	$2x + 2$	1-tall	$2x + 2$
2-long	$4x + 2$	2-tall	$2x + 4$
3-long	$6x + 2$	3-tall	$2x + 6$
4-long	$8x + 2$	4-tall	$2x + 8$
5-long	$10x + 2$	5-tall	$2x + 10$

  b. Perimeter: long $40x + 2$, tall $2x + 40$

  c. $0 < x < 1$

11. a. 3

  b. 994009

  c. i. 990025 **ii.** 980100

12. a. i. $A_{\text{screen}} = x(x + 2) = x^2 + 2x$

   ii. $A_{\text{phone 1}} = (x + 1)(x + 7) = x^2 + 8x + 7$

  b. i. $l = x + 11,\ w = x$

   ii. $A_{\text{phone 2}} = x(x + 11) = x^2 + 11x$

  c. $x = 4$cm, $A_{\text{phone 2}} = 60$ cm^2

13. a. i. $x(x - 2)$ m^2 **ii.** $x(2x)$ m^2

  b. $\dfrac{x - 2}{2x}$

  c. $\dfrac{50(x - 2)}{x}$%

  d. i. 25% **ii.** 27.8% **iii.** 29.17%

  e. The third flag places the most importance on Australia.

14. a. $A = \pi R^2 - x^2$

  b. $A = 160.71$ cm^2

  c. Sample responses can be found in the worked solutions in the online resources.

  d. $R = 5.32$ cm

  e. 92.0%

15. a. Sample responses can be found in the worked solutions in the online resources.

  b. i. $-3x^3 - 2x^2 + 3x + 2$

   ii. $8x^3 + 16x^2 - 104x + 80$

   iii. $x^3 - 49x + 120$

  c. In a cubic expression there are four terms with descending powers of x and ascending values of the pronumeral.
  For example, $(x + a)^3 = x^3 + 3ax^2 + 3a^2x + a^3$, where the powers of x are descending and the values of a are ascending.

Project

1. Solution not required.

2. Students need to describe the feature created by arranging the blocks.

3. 0.12 m

4. 0.0144 m^2

5. Yellow: 0.03 m $\times$ 0.03 m
  Black: 0.03 m $\times$ 0.06 m
  White: 0.06 m $\times$ 0.06 m

6. Yellow: 0.0009 m^2
  Black: 0.0018 m^2
  White: 0.0036 m^2

7. Yellow: 0.0036 m^2
  Black: 0.0072 m^2
  White: 0.0036 m^2

8. Students are required to derive a general formula that would allow the company to determine the area of the three coloured fabrics in each block. They should give an example and illustrate it with the help of a diagram.

Exercise 2.11 Review questions

1. B

2. D

3. A

4. B

5. D

6. A

7. B

8. A

9. B

10. a. i. 4

   ii. -8

   iii. -5.

   iv. $-8xy^2$

  b. i. $y + 8$

   ii. $x - y$

   iii. $x + y$

   iv. $7xy$

   v. $5y - 2x$

11. a. $\$xy$

  b. i. $(24 - k)$m

   ii. $\dfrac{24 - k}{3}$m

12. a. $17p$ **b.** $y^2 + 2y$

  c. $-3s^2t$ **d.** $16c^2d - 2cd$

  e. $n^2 - 4p^2q + 6$ **f.** $15ab - 3a^2b^2$

13. a. $12ab$ **b.** $2ab^2$ **c.** $8x^2y^2$

  d. $\dfrac{x}{3}$ **e.** $\dfrac{9}{2b}$

14. a. $5x + 15$ b. $-y - 5$
    c. $-3x + 2x^2$ d. $-8mx - 4x$

15. a. $3x + 3$ b. $-10m - 1$
    c. $4m^2 - 9m - 5$ d. $4p - 6$

16. a. $9a + 8b$ b. $-5x - 18y$
    c. $-mn + 11m$ d. $4x^2 - 10x + 3$

17. a. $x^2 + 9x + 20$ b. $m^2 - m - 2$
    c. $3m^2 - 17m + 10$ d. $2a^2 - 5ab - 3b^2$

18. a. $x^2 - 16$ b. $81 - m^2$
    c. $x^2 - y^2$ d. $1 - 4a^2$

19. a. $x^2 + 10x + 25$ b. $m^2 - 6m + 9$
    c. $16x^2 + 8x + 1$ d. $4 - 12y + 9y^2$

20. a. $2x^2 + 8x + 8$ b. $2m^2 + 11m - 5$
    c. $6x + 15$ d. $-7b + 37$

21. a. $2x^2 + 10x + 13$ b. $2x - 5$
    c. $16x$

22. a. 23 b. 134

23. a. $307.88\,\text{cm}^3$ b. $5379.98\,\text{cm}^3$

24. a. $6(x + 2)$ b. $6x^2(1 + 2xy)$
    c. $4(2a^2 - b)$ d. $8x(2x - 3y)$
    e. $-2(x + 2)$ f. $b(b - 3 + 4c)$

25. a. $(5 - 4a)(x + y)$ b. $(7a - 6c)(b + 5c)$
    c. $5x(3 + 5y)(d + 2e)$ d. $(2 + a)(x + y)$
    e. $2(x - 1)(3y + 2)$ f. $(p - r)(q + 1)$

26. a. $8x\,\text{cm}$

    b. $3x\,\text{cm}^2$

    c. i. $(3x + y)\,\text{cm}$

       ii. $(8x + 2y)\,\text{cm}$

       iii. $780\,\text{cm}$

       iv. $(3x^2 + xy)\,\text{cm}^2$

       v. $27\,000\,\text{cm}^2$

27. a. $(8 + 2p)(15 + 2p)\,\text{m}^2$ b. $(4p^2 + 46p + 120)\,\text{m}^2$
    c. $(4p^2 + 46p)\,\text{m}^2$ d. $4p^2 + 46p = 200$

28. $\dfrac{2\pi x}{y}\,\text{cm}$

29. $5xy - 4y^2$

30. a. $6n + 42$ b. $6(n + 7)$

    c. $2(n + 7)$ d. $\dfrac{6(n + 7)}{x}$

31. a. i. $9(c - d)$

       ii. Yes, this is a multiple of 9 because the number that multiplies the brackets is 9.

    b. $90(b - c)$ –90 is a multiple of 9, so the difference between the correct and incorrect numbers is a multiple of 9.

    c. $900\,(a - b)$ is a multiple of 9, so the difference between the correct and incorrect numbers is a multiple of 9.

d. If two adjacent digits are transposed, the difference between the correct number and the transposed number is a multiple of 9.

e. Let the correct price of the CD be $ab.cd$. This can be written as $10a + b + 0.1c + 0.01d$.
   The cashier enters $cd.ab$ as the price. This can be written as $10c + d + 0.1$.
   The difference is $10a + b + 0.1c + 0.01d - (10c + d + 0.1a + 0.01b)$.
   This is equal to $9.9a + 0.99b - 9.9c - 0.99d$, which is a multiple of 9.
   This means that the rule for checking transposition errors applies in this case.

3 Linear equations

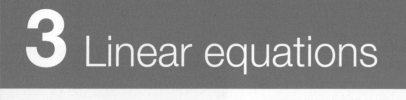

3.1 Overview

Why learn this?

Mathematical equations are all around us. Learning how to solve equations lets us work out unknown values. Being able to manipulate and solve equations has many uses in many fields, including chemistry, medicine, economics and commerce, to name just a few.

Being able to solve linear equations is a useful skill for many everyday tasks. Linear equations can be used when converting temperature from Celsius to Fahrenheit, when working out how to balance chemical equations, or when converting between different currencies. Linear equations can also be used when working out fuel consumption for a road trip, deciding on the right dosage of medicine, or budgeting for a holiday.

Any calculation of rates will use some kind of linear equation. Linear equations can also be used to help predict future trends involving growth or decay. An understanding of the underlying principles for solving linear equations can also be applied to the solution of other mathematical models.

Where to get help

Go to your learnON title at **www.jacplus.com.au** to access the following digital resources. The Online Resources Summary at the end of this topic provides a full list of what's available to help you learn the concepts covered in this topic.

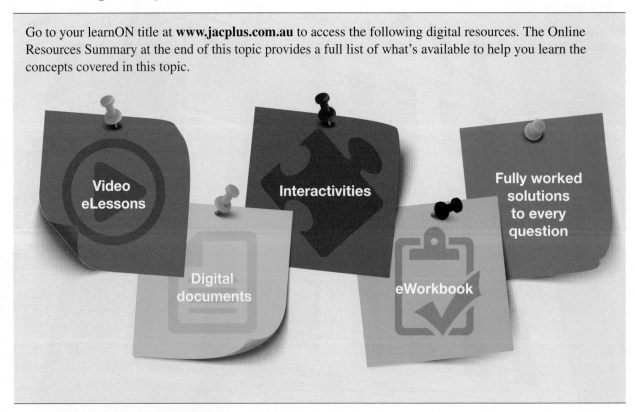

Complete this pre-test in your learnON title at www.jacplus.com.au and receive **automatic marks**, **immediate corrective feedback** and **fully worked solutions**.

1. Solve the linear equation $\dfrac{x}{3} - 2 = -1$.

2. Solve the linear equation $0.4x - 2.6 = 6.2$.

3. **MC** Select which of the following equations has the solution $x = 2$.

   A. $x + 3 = -5$ B. $3 - x = 5$ C. $3(x + 1) = 2x + 5$

   D. $\dfrac{3}{x} = 6$ E. $x^2 - x = 1$

4. Solve the following linear equations.

   a. $\dfrac{3x + 1}{4} = -5$ b. $\dfrac{3 - z}{2} = -4$

5. Solve the equation $-3(2a - 4) = -12$.

6. Dylan is solving the equation $4(y + 3) = -10$. State whether it is true or false that Dylan will calculate the correct value of y if he subtracts 3 from both sides of the equation then divides both sides by 4.

7. **MC** Select the correct value for m in the equation $0.4(m - 8) = 0.6$.

   A. $m = -6.5$ B. $m = 6.5$ C. $m = 7.8$ D. $m = 8.2$ E. $m = 9.5$

8. **MC** Select the first step to solve the equation $2x + 1 = 11 - 3x$.

   A. Add $2x$ to both sides.   B. Add 1 to both sides.

   C. Add $3x$ to both sides.   D. Add 11 to both sides.

   E. Subtract $3x$ from both sides.

9. Solve the following equations.

   a. $5(a + 2) = 4a + 12$ b. $0.2(b - 6) = 5.2 + b$

10. **MC** Choose the equation that matches the statement 'subtracting 6 from 5 multiplied by a certain number gives a result of -8'.

    A. $6 - 5x = -8$ B. $5x - 6 = -8$ C. $5 - 6x = -8$

    D. $6x - 5 = -8$ E. $-6 - 5x = -8$

11. The cost of renting a car is given by $c = 75d + 0.4k$, where $d =$ number of days rented and $k =$ number of kilometres driven. Toni has \$340 to spend on a car for three days. Calculate the total distance she can travel.

12. Solve the value of x in the statement 'x is multiplied by 5 and then 4 is subtracted. The result is the same as three times x minus 10'.

13. **MC** Select the correct equation in which $\dfrac{y + b}{x + b} = m$ is rearranged to make x the subject.

    A. $y = \dfrac{4x + 3}{x}$ B. $x = \dfrac{m - b}{y + b}$ C. $x = \dfrac{y + b}{m} + b$

    D. $x = \dfrac{y + b - bm}{m}$ E. $x = \dfrac{y}{m}$

14. **MC** Select the correct equation in which $x(y-4) = y+3$ is rearranged to make y the subject.

A. $y = \dfrac{4x+3}{x}$

B. $y = \dfrac{4x+3}{x-1}$

C. $y = 3x+3$

D. $y = \dfrac{y+3}{x} + 4$

E. $y = y+3-4x$

15. Rearrange $\dfrac{1}{a} = \dfrac{1}{b} + \dfrac{1}{c}$ to make:

a. a the subject

b. b the subject.

3.2 Solving linear equations

LEARNING INTENTION

At the end of this subtopic you should be able to:
- recognise inverse operations
- solve multistep equations with pronumerals on one side of the equation
- solve equations with algebraic fractions.

▶ 3.2.1 What is a linear equation?

- An **equation** is a mathematical statement that contains an equals sign (=).
- The **expression** on the left-hand side of the equals sign has the same value as the expression on the right-hand side of the equals sign.
- Solving a **linear equation** means determining a value for the pronumeral that makes the statement true.
- 'Doing the same thing' to both sides of an equation (also known as applying the **inverse operation**) ensures that the expressions on either side of the equals sign remain equal, or **balanced**.

WORKED EXAMPLE 1 Using substitution to check equations

For each of the following equations, determine whether $x = 10$ is a solution.

a. $\dfrac{x+2}{3} = 6$

b. $2x+3 = 3x-7$

c. $x^2 - 2x = 9x - 10$

THINK

WRITE

a. 1. Substitute 10 for x in the left-hand side of the equation.

a. LHS $= \dfrac{x+2}{3}$

$= \dfrac{10+2}{3}$

$= \dfrac{12}{3}$

$= 4$

2. Write the right-hand side.

RHS $= 6$

3. Is the equation true? Does the left-hand side equal the right-hand side?

LHS $\neq$ RHS

4. State whether $x = 10$ is a solution.

$x = 10$ is not a solution.

146 Jacaranda Maths Quest 9

b. **1.** Substitute 10 for x in the left-hand side of the equation.

b. LHS $= 2x + 3$
$= 2(10) + 3$
$= 23$

2. Substitute 10 for x in the right-hand side of the equation.

RHS $= 3x - 7$
$= 3(10) - 7$
$= 23$

3. Is the equation true? Does the left-hand side equal the right-hand side?

LHS $=$ RHS

4. State whether $x = 10$ is a solution.

$x = 10$ is a solution.

c. **1.** Substitute 10 for x in the left-hand side of the equation.

c. LHS $= x^2 - 2x$
$= 10^2 - 2(10)$
$= 100 - 20$
$= 80$

2. Substitute 10 for x in the right-hand side of the equation.

RHS $= 9x - 10$
$= 9(10) - 10$
$= 90 - 10$
$= 80$

3. Is the equation true? Does the left-hand side equal the right-hand side?

LHS $=$ RHS

4. State whether $x = 10$ is a solution.

$x = 10$ is a solution.

Inverse operations

The inverse operation of an operation has the effect of undoing the original operation.

Operation	Inverse operation
$+$	$-$
$-$	$+$
$\times$	$\div$
$\div$	$\times$

▶ 3.2.2 Solving multistep equations

eles-4690

- If an equation performs two operations on a pronumeral, it is known as a **two-step equation**.
- If an equation performs more than two operations on a pronumeral, it is known as a **multistep equation**.
- To solve multistep equations, first determine the order in which the operations were performed.
- Once the order of operations is determined, perform the inverse of those operations, in the reverse order, to both sides of the equation. This will keep the equation balanced.
- Each inverse operation must be performed one step at a time.
- These steps can be applied to any equation with two or more steps, as shown in the worked examples that follow.

Solve the following linear equations.

a. $3x + 1 = 11$

b. $1 - \dfrac{x}{3} = 11$

c. $\dfrac{2x}{3} + \dfrac{x}{4} = -11$

d. $\dfrac{-x - 0.3}{2} = -1.1$

THINK	WRITE
a. 1. Subtract 1 from both sides of the equation.	a. $3x + 1 = 11$ $3x + 1 - 1 = 11 - 1$
2. Divide both sides by 3.	$\dfrac{3x}{3} = \dfrac{10}{3}$
3. Write the value for x. *Note:* It is preferable to write fractions in improper form rather than as mixed numbers.	$x = \dfrac{10}{3}$
b. 1. Subtract 1 from both sides of the equation.	b. $1 - \dfrac{x}{3} = 11$ $1 - \dfrac{x}{3} - 1 = 11 - 1$
2. Multiply both sides by -3.	$\dfrac{-x}{3} \times (-3) = 10 \times (-3)$
3. Write the value for x.	$x = -30$

Alternative method:

b. 1. Subtract 1 from both sides.	b. $1 - \dfrac{x}{3} = 11$ $1 - \dfrac{x}{3} - 1 = 11 - 1$
2. Multiply both sides by 3.	$\dfrac{-x}{3} \times 3 = 10 \times 3$
3. Divide both sides by -1.	$\dfrac{-x}{-1} = \dfrac{30}{-1}$
4. Write the value for x.	$x = -30$
c. 1. Add like terms to simplify the left-hand side of the equation. To do this, you will need to find the lowest common denominator.	c. $\dfrac{2x}{3} + \dfrac{x}{4} = -11$ $\dfrac{2x}{3} \times \dfrac{4}{4} + \dfrac{x}{4} \times \dfrac{3}{3} = -11$ $\dfrac{8x + 3x}{12} = -11$
2. Multiply both sides by 12.	$\dfrac{11x}{12} \times 12 = -11 \times 12$
3. Divide both sides by 11.	$\dfrac{11x}{11} = \dfrac{-132}{11}$
4. Write the value for x.	$x = -12$

d. 1. Multiply both sides of the equation by 2.

d. $$\frac{-x-0.3}{2}=-1.1$$

$$\frac{-x-0.3}{2}\times 2=-1.1\times 2$$

2. Add 0.3 to both sides.

$$-x-0.3+0.3=-2.2+0.3$$

3. Divide both sides by -1.

$$\frac{-x}{-1}=\frac{-1.9}{-1}$$

4. Write the value for x.

$$x=1.9$$

TI | THINK	**DISPLAY/WRITE**	**CASIO | THINK**	**DISPLAY/WRITE**
a–d. On a Calculator page, press: • MENU • 3: Algebra • 1: Solve Complete the entry lines as: solve $(3x+1=11,x)$ solve $\left(1-\dfrac{x}{3}=11,x\right)$ solve $\left(\dfrac{2x}{3}+\dfrac{x}{4}=-11,x\right)$ solve $\left(\dfrac{-x-0.3}{2}=-1.1,x\right)$ Press ENTER after each entry.	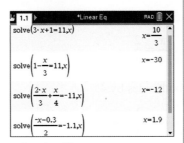 **a.** $3x+1=11$ $\quad x=\dfrac{10}{3}$ **b.** $1-\dfrac{x}{3}=11$ $\quad x=-30$ **c.** $\dfrac{2x}{3}+\dfrac{x}{4}=-11$ $\quad x=-12$ **d.** $\dfrac{-x-0.3}{2}=-1.1$ $\quad x=1.9$	**a–d.** On a standard Main screen, press: • Action • Advanced • Solve Complete the entry lines as: solve $(3x+1=11)$ solve $\left(1-\dfrac{x}{3}=11\right)$ solve $\left(\dfrac{2x}{3}+\dfrac{x}{4}=-11\right)$ solve $\left(\dfrac{-x-0.3}{2}=-1.1\right)$ Press EXE after each entry. *Note:* As x is the only variable in each equation there is no need to specify 'solve for x', so the 'x' at the end of each entry line may be left out. This only applies if x is the variable.	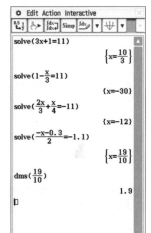

▶ 3.2.3 Algebraic fractions with the pronumeral in the denominator

eles-4691

- If an equation uses fractions and the pronumeral is in the denominator, there is an extra step involved in determining the solution.
- Consider the following equation.

$$\frac{4}{x}=\frac{3}{2}$$

- In order to solve this equation, first multiply both sides by x.

$$\frac{4}{x}\times x=\frac{3}{2}\times x$$

$$4=\frac{3x}{2}$$

$$\text{or } \frac{3x}{2} = 4$$

- The pronumeral is now in the numerator, and the equation is easy to solve.

$$\frac{3x}{2} = 4$$
$$3x = 8$$
$$x = \frac{8}{3}$$

- Alternatively, the equation can be solved using the equivalent ratios method.

$$\frac{4}{x} = \frac{3}{2}$$

- This equation can be written as the equivalent ratio.

$$\frac{x}{4} = \frac{2}{3}$$

- Now the equation can be solved by multiplying both sides by 4.

$$\frac{x}{4} \times 4 = \frac{2}{3} \times 4$$
$$x = \frac{8}{3}$$

WORKED EXAMPLE 3 Solving equations with pronumerals in the denominator

Solve each of the following linear equations.

a. $\dfrac{3}{a} = \dfrac{4}{5}$
b. $\dfrac{5}{b} = -2$

THINK	WRITE
a. 1. Write the equation	a. $\dfrac{3}{a} = \dfrac{4}{5}$
2. Multiply both sides by a.	$3 = \dfrac{4a}{5}$
3. Multiply both sides by 5.	$15 = 4a$
4. Divide both sides by 4.	$a = \dfrac{15}{4}$
b. 1. Write the equation.	b. $\dfrac{5}{b} = -2$
2. Multiply both sides by b.	$5 = -2b$
3. Divide both sides by -2.	$\dfrac{5}{-2} = b$ or $\dfrac{-5}{2} = b$

Exercise 3.2 Solving linear equations

learnon

Individual pathways

■ PRACTISE	■ CONSOLIDATE	■ MASTER
1, 3, 6, 8, 12, 16, 21, 23, 26, 29	2, 4, 9, 10, 13, 14, 17, 19, 20, 24, 27, 30	5, 7, 11, 15, 18, 22, 25, 28, 31, 32

To answer questions online and to receive **immediate corrective feedback** and **fully worked solutions** for all questions, go to your learnON title at www.jacplus.com.au.

Fluency

1. **WE1** For each of the following equations, determine whether $x = 6$ is a solution.

   a. $x + 3 = 7$
   b. $2x - 5 = 7$
   c. $x^2 - 2 = 38$

   d. $\dfrac{6}{x} + x = 7$
   e. $\dfrac{2(x + 1)}{7} = 2$
   f. $3 - x = 9$

2. For each of the following equations, determine whether $x - 6$ is a solution.

   a. $x^2 + 3x = 39$
   b. $3(x + 2) = 5(x - 4)$
   c. $x^2 + 2x = 9x - 6$
   d. $x^2 = (x + 1)^2 - 14$
   e. $(x - 1)^2 = 4x + 1$
   f. $5x + 2 = x^2 + 4$

3. **WE2a** Solve each of the following linear equations. Check your answers by substitution.

   a. $x - 43 = 167$
   b. $x + 286 = 516$
   c. $58 + x = 81$
   d. $209 - x = 305$
   e. $5x = 185$
   f. $60x = 1200$

4. **WE2b** Solve each of the following linear equations. Check your answers by substitution.

   a. $5x = 250$
   b. $\dfrac{x}{23} = 6$
   c. $\dfrac{x}{17} = 26$
   d. $\dfrac{x}{9} = 27$
   e. $y - 16 = -31$
   f. $5.5 + y = 7.3$

5. **WE2c** Solve each of the following linear equations. Check your answers by substitution.

   a. $y - 7.3 = 5.5$
   b. $6y = 14$
   c. $0.9y = -0.05$
   d. $\dfrac{y}{5} = 4.3$
   e. $\dfrac{y}{7.5} = 23$
   f. $\dfrac{y}{8} = -1.04$

6. Solve each of the following linear equations.

   a. $2y - 3 = 7$
   b. $2y + 7 = 3$
   c. $5y - 1 = 0$
   d. $6y + 2 = 8$
   e. $7 + 3y = 10$
   f. $8 + 2y = 12$

7. Solve each of the following linear equations.

   a. $15 = 3y - 1$
   b. $-6 = 3y - 1$
   c. $6y - 7 = 140$
   d. $4.5y + 2.3 = 7.7$
   e. $0.4y - 2.7 = 6.2$
   f. $600y - 240 = 143$

8. Solve each of the following linear equations.
   a. $3 - 2x = 1$
   b. $-3x - 1 = 5$
   c. $-4x - 7 = -19$
   d. $1 - 3x = 19$
   e. $-5 - 7x = 2$
   f. $-8 - 2x = -9$

9. Solve each of the following linear equations.
   a. $9 - 6x = -1$
   b. $-5x - 4.2 = 7.4$
   c. $2 = 11 - 3x$
   d. $-3 = -6x - 8$
   e. $-1 = 4 - 4x$
   f. $35 - 13x = -5$

10. Solve each of the following linear equations.
    a. $7 - x = 8$
    b. $8 - x = 7$
    c. $5 - x = 5$
    d. $5 - x = 0$
    e. $15.3 = 6.7 - x$
    f. $5.1 = 4.2 - x$

11. Solve each of the following linear equations.
    a. $9 - x = 0.1$
    b. $140 - x = 121$
    c. $-30 - x = -4$
    d. $-5 = -6 - x$
    e. $-x + 1 = 2$
    f. $-2x - 1 = 0$

12. Solve each of the following linear equations.
    a. $\dfrac{x}{4} + 1 = 3$
    b. $\dfrac{x}{3} - 2 = -1$
    c. $\dfrac{x}{8} = \dfrac{1}{2}$
    d. $-\dfrac{x}{3} = 5$
    e. $5 - \dfrac{x}{2} = -8$
    f. $4 - \dfrac{x}{6} = 11$

13. Solve each of the following linear equations.
    a. $\dfrac{2x}{3} = 6$
    b. $\dfrac{5x}{2} = -3$
    c. $-\dfrac{3x}{4} = -7$
    d. $-\dfrac{8x}{3} = 6$
    e. $\dfrac{2x}{7} = -2$
    f. $-\dfrac{3x}{10} = -\dfrac{1}{5}$

14. Solve each of the following linear equations.
    a. $\dfrac{z - 1}{3} = 5$
    b. $\dfrac{z + 1}{4} = 8$
    c. $\dfrac{z - 4}{2} = -4$
    d. $\dfrac{6 - z}{7} = 0$
    e. $\dfrac{3 - z}{2} = 6$
    f. $\dfrac{-z - 50}{22} = -2$

15. Solve each of the following linear equations.
    a. $\dfrac{z - 4.4}{2.1} = -3$
    b. $\dfrac{z + 2}{7.4} = 1.2$
    c. $\dfrac{140 - z}{150} = 0$
    d. $\dfrac{-z - 0.4}{2} = -0.5$
    e. $\dfrac{z - 6}{9} = -4.6$
    f. $\dfrac{z + 65}{73} = 1$

16. Solve each of the following linear equations.
    a. $\dfrac{5x + 1}{3} = 2$
    b. $\dfrac{2x - 5}{7} = 3$
    c. $\dfrac{3x + 4}{2} = -1$
    d. $\dfrac{4x - 13}{9} = -5$
    e. $\dfrac{4 - 3x}{2} = 8$
    f. $\dfrac{1 - 2x}{6} = -10$

17. Solve each of the following linear equations.
    a. $\dfrac{-5x - 3}{9} = 3$
    b. $\dfrac{-10x - 4}{3} = 1$
    c. $\dfrac{4x + 2.6}{5} = 8.8$
    d. $\dfrac{5x - 0.7}{-0.3} = -3.1$
    e. $\dfrac{1 - 0.5x}{4} = -2.5$
    f. $\dfrac{-3x - 8}{14} = \dfrac{1}{2}$

18. WE3 Solve each of the following linear equations.
    a. $\dfrac{2}{x} = \dfrac{1}{2}$
    b. $\dfrac{3}{x} = 7$
    c. $\dfrac{-4}{x} = \dfrac{7}{2}$
    d. $\dfrac{5}{x} = \dfrac{-3}{4}$
    e. $\dfrac{0.4}{x} = \dfrac{9}{2}$
    f. $\dfrac{8}{x} = 1$

19. Solve each of the following linear equations.

a. $\dfrac{-4}{x} = \dfrac{2}{3}$

b. $\dfrac{-6}{x} = \dfrac{-4}{5}$

c. $\dfrac{1.7}{x} = \dfrac{1}{3}$

d. $\dfrac{6}{x} = -1$

e. $\dfrac{4}{x} = \dfrac{-15}{22}$

f. $\dfrac{50}{x} = \dfrac{-35}{43}$

20. **MC** Answer the following questions.

a. Determine the solution to the equation $82 - x = 44$.

A. $x = 126$ B. $x = -126$ C. $x = 122$ D. $x = 38$ E. $x = 22$

b. Determine the solution to the equation $5x - 12 = -62$.

A. $x = -14.8$ B. $x = 14.8$ C. $x = 10$ D. $x = -10$ E. $x = 62$

c. Determine the solution to the equation $\dfrac{x-1}{2} = 5.3$.

A. $x = 9.6$ B. $x = 10.6$ C. $x = 11.6$ D. $x = 2$ E. $x = 5$

21. Solve each of the following linear equations.

a. $3a + 7 = 4$

b. $5 - b = -5$

c. $4c - 4.4 = 44$

d. $\dfrac{d-4}{67} = 0$

e. $5 - 3e = -10$

f. $\dfrac{2f}{3} = 8$

22. Solve each of the following linear equations.

a. $100 = 6g + 4.2$

b. $\dfrac{h+2}{6} = 5.5$

c. $452i - 124 = -98$

d. $\dfrac{6j-1}{17} = 0$

e. $\dfrac{12-k}{5} = 4$

f. $\dfrac{l-5.2}{3.4} = 1.5$

Understanding

23. Write each of the following worded statements as a mathematical sentence and then solve for the unknown value.

a. Seven is added to the product of x and 3, which gives the result of 4.

b. Four is divided by x and this result is equivalent to $\dfrac{2}{3}$.

c. Three is subtracted from x and this result is divided by 12 to give 25.

24. Driving lessons are usually quite expensive, but a discount of $15 per lesson is given if a family member belongs to the automobile club.
If 10 lessons cost $760 after the discount, calculate the cost of each lesson before the discount.

25. Anton lives in Australia. His friend Utan lives in the USA. Anton's home town of Horsham experienced one of the hottest days on record with a temperature of 46.7 °C. Utan said that his home town had experienced a day hotter than that, with the temperature reaching 113 °F.
The formula for converting Celsius to Fahrenheit is $F = \dfrac{9}{5}C + 32$. State whether Utan was correct. Show full working.

Reasoning

26. If the expression $\dfrac{12}{x-4}$ always results in a positive integer value, explain how you would determine the possible values for x.

27. Santo solved the linear equation $9 = 5 - x$. His second step was to divide both sides by -1. Trudy, his mathematics buddy, said she multiplied both sides by -1. Explain why they are both correct.

28. Determine the mistake in the following working and explain what is wrong.

$$\frac{x}{5} - 1 = 2$$
$$x - 1 = 10$$
$$x = 11$$

Problem solving

29. Sweet-tooth Sami goes to the corner store and buys an equal number of 25-cent and 30-cent lollies for a total of $16.50. Determine the amount of lollies he bought.

30. In a cannery, cans are filled by two machines that together produce 16 000 cans during an 8-hour shift. If the newer machine of the two produces 340 more cans per hour than the older machine, evaluate the number of cans produced by each machine in an eight-hour shift.

31. General admission to a music festival is $55 for an adult ticket, $27 for a child and $130 for a family of two adults and two children.

   a. Evaluate how much you would save by buying a family ticket instead of two adult and two child tickets.

   b. Determine if it is worthwhile buying a family ticket if a family has only one child.

32. A teacher asks her students to determine the value of n in the diagram shown. Use your knowledge of linear equations to solve this problem.

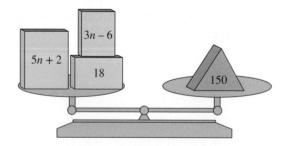

3.3 Solving linear equations with brackets

▶ 3.3.1 Linear equations with brackets

eles-4692

- Consider the equation $3(x + 5) = 18$. There are two methods for solving this equation.

Method 1:

Start by dividing both sides by 3.
$$\frac{3(x+5)}{3} = \frac{18}{3}$$
$$x + 5 = 6$$
$$x = 1$$

Method 2:

Start by expanding the brackets.
$$3(x + 5) = 18$$
$$3x + 15 = 18$$
$$3x = 3$$
$$x = 1$$

- In this case, method 1 works well because 3 divides exactly into 18.
- Now try the equation $7(x + 2) = 10$.

Method 1:

Start by dividing both sides by 7.
$$\frac{7(x+2)}{7} = \frac{10}{7}$$
$$x + 2 = \frac{10}{7}$$
$$x = -\frac{4}{7}$$

Method 2:

Start by expanding the brackets.
$$7(x + 2) = 10$$
$$7x + 14 = 10$$
$$7x = -4$$
$$x = \frac{-4}{7}$$

- In this case, method 2 works well because it avoids the addition or subtraction of fractions.
- For each equation you need to solve, try both methods and choose the method that works best.

WORKED EXAMPLE 4 Solving equations with brackets

Solve each of the following linear equations.
a. $7(x - 5) = 28$
b. $6(x + 3) = 7$

THINK	WRITE
a. 1. 7 is a factor of 28, so it will be easier to divide both sides by 7.	a. $7(x - 5) = 28$ $\frac{7(x-5)}{7} = \frac{28}{7}$
2. Add 5 to both sides.	$x - 5 = 4$
3. Write the value of x.	$x = 9$
b. 1. 6 is not a factor of 7, so it will be easier to expand the brackets first.	b. $6(x + 3) = 7$ $6x + 18 = 7$

| 2. Subtract 18 from both sides. | $6x + 18 = 7 - 18$ |
| | $6x = -11$ |

| 3. Divide both sides by 6. | $x = \dfrac{11}{6}$ |

TI | THINK	DISPLAY/WRITE	CASIO | THINK	DISPLAY/WRITE
a–b. In a new problem, on a Calculator page, press: • MENU • 3: Algebra • 1: Solve Complete the entry lines as: solve $(7(x-5)=28, \ x)$ solve $(6(x+3)=7, \ x)$ Press ENTER after each entry.	**a–b.**  **a.** $7(x-5)=28$ $x=9$ **b.** $6(x+3)=7$ $x=-\dfrac{11}{6}$	**a–b.** On the Main screen, press: • Action • Advanced • Solve Complete the entry lines as: solve $(7(x-5)=28)$ solve $(6(x+3)=7)$ Press EXE after each entry.	**a–b.** 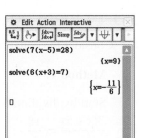 **a.** $7(x-5)=28$ $x=9$ **b.** $6(x+3)=7$ $x=-\dfrac{11}{6}$

 Resources

 eWorkbook   Topic 3 Workbook (worksheets, code puzzle and project) (ewbk-2003)

 Digital document SkillSHEET Expanding brackets (doc-10827)

Interactivities   Individual pathway interactivity: Solving linear equations with brackets (int-4490)

                  Linear equations with brackets (int-6039)

Exercise 3.3 Solving linear equations with brackets   learn

Individual pathways

■ PRACTISE	■ CONSOLIDATE	■ MASTER
1, 3, 5, 11, 13, 16	2, 7, 9, 12, 14, 17	4, 6, 8, 10, 15, 18

To answer questions online and to receive **immediate corrective feedback** and **fully worked solutions** for all questions, go to your learnON title at www.jacplus.com.au.

Fluency

1. **WE4** Solve each of the following linear equations.

    a. $5(x-2)=20$ **b.** $4(x+5)=8$ **c.** $6(x+3)=18$

    d. $5(x-41)=75$ **e.** $8(x+2)=24$ **f.** $3(x+5)=15$

2. Solve each of the following linear equations.

a. $5(x+4)=15$
b. $3(x-2)=-12$
c. $7(x-6)=0$
d. $-6(x-2)=12$
e. $4(x+2)=4.8$
f. $16(x-3)=48$

3. Solve each of the following linear equations.

a. $6(b-1)=1$
b. $2(m-3)=3$
c. $2(a+5)=7$
d. $3(m+2)=2$
e. $5(p-2)=-7$
f. $6(m-4)=-8$

4. Solve each of the following linear equations.

a. $-10(a+1)=5$
b. $-12(p-2)=6$
c. $-9(a-3)=-3$
d. $-2(m+3)=-1$
e. $3(2a+1)=2$
f. $4(3m+2)=5$

5. Solve each of the following linear equations.

a. $9(x-7)=82$
b. $2(x+5)=14$
c. $7(a-1)=28$
d. $4(b-6)=4$
e. $3(y-7)=0$
f. $-3(x+1)=7$

6. Solve each of the following linear equations.

a. $-6(m+1)=-30$
b. $-4(y+2)=-12$
c. $-3(a-6)=3$
d. $-2(p+9)=-14$
e. $3(2m-7)=-3$
f. $2(4p+5)=18$

7. Solve each of the following linear equations. Round the answers to 3 decimal places where appropriate.

a. $2(y+4)=-7$
b. $0.3(y+8)=1$
c. $4(y+19)=-29$
d. $7(y-5)=25$
e. $6(y+3.4)=3$
f. $7(y-2)=8.7$

8. Solve each of the following linear equations. Round the answers to 3 decimal places where appropriate.

a. $1.5(y+3)=10$
b. $2.4(y-2)=1.8$
c. $1.7(y+2.2)=7.1$
d. $-7(y+2)=0$
e. $-6(y+5)=-11$
f. $-5(y-2.3)=1.6$

9. **MC** Select the best first step for solving the equation $7(x-6)=23$.

A. Add 6 to both sides.
B. Subtract 7 from both sides.
C. Divide both sides by 23.
D. Expand the brackets.
E. Multiply both sides by 7.

10. **MC** Select which one of the following is closest to the solution for the equation $84(x-21)=782$.

A. $x=9.31$
B. $x=9.56$
C. $x=30.31$
D. $x=-11.69$
E. $x=30.32$

Understanding

11. In 1974 a mother was 6 times as old as her daughter. If the mother turned 50 in the year 2000, calculate the year in which the mother's age was double her daughter's age.

12. New edging is to be placed around a rectangular children's playground. The width of the playground is x metres and the length is 7 metres longer than the width.

a. Write down an expression for the perimeter of the playground. Write your answer in factorised form.
b. If the amount of edging required is 54 m, determine the dimensions of the playground.

Reasoning

13. Explain the two possible methods for solving equations in factorised form.

14. Juanita is solving the equation $2(x-8)=10$. She performs the following operations to both sides of the equation in order: $+8$, $÷2$. Explain why Juanita will not find the correct value of x using her order of inverse operations, then solve the equation correctly.

15. As your first step to solve the equation $3(2x - 7) = 18$, you are given three options:
    - Expand the brackets on the left-hand side.
    - Add 7 to both sides.
    - Divide both sides by 3.

    Explain which of these options is your least preferred.

Problem solving

16. Five times the sum of four and an unknown number is equal to 35. Evaluate the value of the unknown number.

17. Oscar earns $55 more than Josue each week, but Hector earns three times as much as Oscar. If Hector earns $270 a week, determine how much Oscar and Josue earn each week.

18. A school wants to hire a bus to travel to a football game. The bus will take 28 passengers, and the school will contribute $48 towards the cost of the trip. The price of each ticket is $10. If the hiring of the bus is $300 + 10% of the total cost of all the tickets, evaluate the cost per person.

3.4 Solving linear equations with pronumerals on both sides

LEARNING INTENTION

At the end of this subtopic you should be able to:
- solve equations by collecting pronumerals on one side of the equation before solving.

3.4.1 Linear equations with pronumerals on both sides

eles-4693

- If pronumerals occur on both sides of an equation, the first step in solving the equation is to move all of the pronumerals so that they appear on only one side of the equation.
- When solving equations, it is important to remember that whatever you do to one side of an equation you must also do to the other side.

WORKED EXAMPLE 5 Solving equations with pronumerals on both sides

Solve each of the following linear equations.
a. $5y = 3y + 3$
b. $7x + 5 = 2 - 4x$
c. $3(x + 1) = 14 - 2x$
d. $2(x + 3) = 3(x + 7)$

THINK		WRITE
a. 1. $3y$ is smaller than $5y$. This means it is easiest to subtract $3y$ from both sides of the equation.	a.	$5y = 3y + 3$
		$5y - 3y = 3y + 3 - 3y$
		$2y = 3$
2. Divide both sides by 2.		$y = \dfrac{3}{2}$

b. 1. $-4x$ is smaller than $7x$. This means it is easiest to add $4x$ to both sides of the equation.

2. Subtract 5 from both sides.

3. Divide both sides by 11.

c. 1. Expand the brackets.

2. $-2x$ is smaller than $3x$. This means it is easiest to add $2x$ to both sides of the equation.

3. Subtract 3 from both sides.

4. Divide both sides by 5.

d. 1. Expand the brackets.

2. $2x$ is smaller than $3x$. This means it is easiest to subtract $2x$ from both sides of the equation.

3. Subtract 21 from both sides.

4. Write the answer with the pronumeral written on the left-hand side.

b.
$$7x + 5 = 2 - 4x$$
$$7x + 5 + 4x = 2 - 4x + 4x$$
$$11x + 5 = 2$$

$$11x + 5 - 5 = 2 - 5$$
$$11x = -3$$

$$x = \frac{-3}{11}$$

c.
$$3(x + 1) = 14 - 2x$$
$$3x + 3 = 14 - 2x$$

$$3x + 3 + 2x = 14 - 2x + 2x$$
$$5x + 3 = 14$$

$$5x + 3 - 3 = 14 - 3$$
$$5x = 11$$

$$x = \frac{11}{5}$$

d.
$$2(x + 3) = 3(x + 7)$$
$$2x + 6 = 3x + 21$$

$$2x + 6 - 2x = 3x + 21 - 2x$$
$$6 = x + 21$$

$$6 - 21 = x + 21 - 21$$
$$-15 = x$$

$$x = -15$$

TI | THINK

a–d. In a new problem, on a Calculator page, press:
- MENU
- 3: Algebra
- 1: Solve

Complete the entry lines as:
solve $(5y = 3y + 3, y)$
solve $(7x + 5 = 2 - 4x, x)$
solve $(3(x + 1) = 14 - 2x, x)$
solve $(2(x + 3) = 3(x + 7), x)$
Press ENTER after each entry.

DISPLAY/WRITE

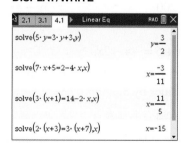

a. $y = \dfrac{3}{2}$

b. $x = \dfrac{-3}{11}$

c. $x = \dfrac{11}{5}$

d. $x = -15$

CASIO | THINK

a–d. On the Main screen, press:
- Action
- Advanced
- Solve

Complete the entry lines as:
solve $(5y = 3y + 3, y)$
solve $(7x + 5 = 2 - 4x, x)$
solve $(3(x + 1) = 14 - 2x, x)$
solve $(2(x + 3) = 3(x + 7), x)$
Press EXE after each entry. *Note:* If the variable is *not* x, it must be specified.

DISPLAY/WRITE

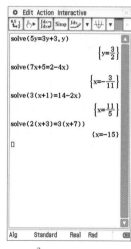

a. $y = \dfrac{3}{2}$

b. $x = \dfrac{-3}{11}$

c. $x = \dfrac{11}{5}$

d. $x = -15$

on Resources

eWorkbook	Topic 3 Workbook (worksheets, code puzzle and project) (ewbk-2003)
Digital document	SkillSHEET Simplifying like terms (doc-10828)
Video eLesson	Solving linear equations with pronumerals on both sides (eles-1901)
Interactivity	Individual pathway interactivity: Solving linear equations with pronumerals on both sides (int-4491)

Exercise 3.4 Solving linear equations with pronumerals on both sides

learn**on**

Individual pathways

■ PRACTISE	■ CONSOLIDATE	■ MASTER
1, 3, 5, 8, 12, 14, 17	2, 6, 10, 13, 15, 18	4, 7, 9, 11, 16, 19, 20

To answer questions online and to receive **immediate corrective feedback** and **fully worked solutions** for all questions, go to your learnON title at www.jacplus.com.au.

Fluency

1. **WE5a** Solve each of the following linear equations.

   a. $5y = 3y - 2$
   b. $6y = -y + 7$
   c. $10y = 5y - 15$
   d. $25 + 2y = -3y$
   e. $8y = 7y - 45$
   f. $15y - 8 = -12y$

2. Solve each of the following linear equations.

   a. $7y = -3y - 20$
   b. $23y = 13y + 200$
   c. $5y - 3 = 2y$
   d. $6 - 2y = -7y$
   e. $24 - y = 5y$
   f. $6y = 5y - 2$

3. **MC** Select the first step for solving the equation $3x + 5 = -4 - 2x$.

   A. Add $3x$ to both sides.
   B. Add 5 to both sides.
   C. Add $2x$ to both sides.
   D. Subtract $2x$ from both sides.
   E. Divide both sides by 3.

4. **MC** Select the first step for solving the equation $6x - 4 = 4x + 5$.

   A. Subtract $4x$ from both sides.
   B. Add $4x$ to both sides.
   C. Subtract 4 from both sides.
   D. Add 5 to both sides.
   E. Divide both sides by 6.

5. **WE5b** Solve each of the following linear equations.

   a. $2x + 3 = 8 - 3x$
   b. $4x + 11 = 1 - x$
   c. $x - 3 = 6 - 2x$
   d. $4x - 5 = 2x + 3$
   e. $3x - 2 = 2x + 7$
   f. $7x + 1 = 4x + 10$

6. Solve each of the following linear equations.

   a. $5x + 3 = x - 5$
   b. $6x + 2 = 3x + 14$
   c. $2x - 5 = x - 9$
   d. $10x - 1 = -2x + 5$
   e. $7x + 2 = -5x + 2$
   f. $15x + 3 = 7x - 3$

7. Solve each of the following linear equations.

   a. $x - 4 = 3x + 8$
   b. $2x + 9 = 7x - 1$
   c. $-2x + 7 = 4x + 19$
   d. $-3x + 2 = -2x - 11$
   e. $11 - 6x = 18 - 5x$
   f. $6 - 9x = 4 + 3x$

8. **MC** Determine the solution to $5x + 2 = 2x + 23$.

   A. $x = 3$ **B.** $x = -3$ **C.** $x = 5$ **D.** $x = 7$ **E.** $x = -5$

9. **MC** Determine the solution to $3x - 4 = 11 - 2x$.

   A. $x = 15$ **B.** $x = 7$ **C.** $x = 3$ **D.** $x = 5$ **E.** $x = -3$

10. **WE5c,d** Solve each of the following.

    a. $5(x - 2) = 2x + 5$ **b.** $7(x + 1) = x - 11$ **c.** $2(x - 8) = 4x$
    d. $3(x + 5) = x$ **e.** $6(x - 3) = 14 - 2x$ **f.** $9x - 4 = 2(3 - x)$

11. Solve each of the following.

    a. $4(x + 3) = 3(x - 2)$ **b.** $5(x - 1) = 2(x + 3)$ **c.** $8(x - 4) = 5(x - 6)$
    d. $3(x + 6) = 4(2 - x)$ **e.** $2(x - 12) = 3(x - 8)$ **f.** $4(x + 11) = 2(x + 7)$

Understanding

12. Aamir's teacher gave him the following algebra problem and told him to solve it.

    $$3x + 7 = x^2 + k = 7x + 15$$

    Suggest how you can you help Aamir calculate the value of k.

13. A classroom contained an equal number of boys and girls. Six girls left to play hockey, leaving twice as many boys as girls in the classroom. Determine the original number of students present.

Reasoning

14. Express the information in the diagram shown as an equation, then show that $n = 29$ is the solution.

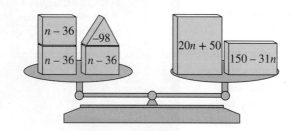

15. The block shown has a width of 1 unit and length of $(x + 1)$ units.

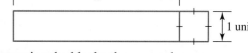

    a. Draw two rectangles with different sizes, each using 3 of the blocks shown.
    b. Show that the areas of both of the rectangles you have drawn using the blocks shown are the same. Explain how the areas of these rectangles relate to expanding brackets.

16. Explain what the difficulty is when trying to solve the equation $4(3x - 5) = 6(4x + 2)$ without expanding the brackets first.

Problem solving

17. This year Tom is 4 times as old as his daughter. In 5 years' time he will be only 3 times as old as his daughter. Determine the age of Tom and his daughter now.

18. If you multiply an unknown number by 6 and then add 5, the result is 7 less than the unknown number plus 1, multiplied by 3. Evaluate the unknown number.

19. You are looking into getting a business card printed for your new game store. A local printing company charges $250 for the materials and an hourly rate for labour of $40.

a. If h is the number of hours of labour required to print the cards, construct an equation for the cost of the cards, C.

b. You have budgeted $1000 for this printing job. Determine the number of hours of labour that you can afford. Give your answer to the nearest minute.

c. The company estimates that it can print 1000 cards an hour. Determine the number of cards that you can afford to get printed with your current budget.

d. An alternative to printing is photocopying. The company charges 15 cents per side for the first 10 000 cards, then 10 cents per side for additional cards. Determine the cheaper option to get 18 750 single-sided cards made. Determine how much cheaper this is than the other option.

20. A local games arcade offers its regular customers the following deal. For a monthly fee of $40, players can play 25 $2 games of their choice. Extra games cost $2 each. After a player has played 50 games in a month, all further games are $1.

a. If Iman has $105 to spend in a month, determine the number of games she can play if she takes up the special deal.

b. Determine the amount of money Iman will save by taking up the special deal compared to playing the same number of games at $2 a game.

3.5 Solving problems with linear equations

LEARNING INTENTION

At the end of this subtopic you should be able to:
- define a pronumeral and write an equation to represent a problem
- solve a problem and answer the question (or questions) posed in the problem.

▶ 3.5.1 Solving problems with linear equations

eles-4694

- When pronumerals are not provided in a problem, you need to introduce a pronumeral to the problem.
- When translating a worded problem into an equation, make sure that the order of the operations is presented correctly.

WORKED EXAMPLE 6 Translating worded expressions into linear equations

Write linear equations for each of the following statements, using x to represent the unknown value.
Solve the equations.
a. Three more than seven times a certain number is zero. Determine the number.
b. The sum of three consecutive integers is 102. Determine the three numbers.
c. The difference between two numbers is 11. Determine the numbers if, when eight is added to each
of the two numbers, the larger result will be double the value of the smaller result.

THINK

a. 1. Let x be the number.

2. 7 times the number is $7x$. Increasing the
value of $7x$ by 3 gives $7x + 3$. This expression
equals 0. Write the equation.

3. Solve the equation using inverse operations.

4. Write the value for x.

b. 1. Let x be the first number.

2. The next two consecutive integers are $(x + 1)$
and $(x + 2)$. The sum of all three integers
equals 102. Write the equation.

3. Solve the equation using inverse operations.

4. Write the value for x and the next 2 integers.

c. 1. Let x be the larger number.

2. The difference between the 2 numbers is 11.
This means the smaller number is 11 less
than x. Write this as $(x - 11)$.

3. Increasing both numbers by 8 makes the
larger number $(x + 8)$ and the smaller number
$(x - 11 + 8)$.

The larger number becomes double the
smaller number.

4. Solve the equation using inverse operations.

5. Write the value of x and the other number.

WRITE

a. $x =$ unknown number

$$7x + 3 = 0$$

$$7x + 3 - 3 = 0 - 3$$
$$\frac{7x}{7} = \frac{-3}{7}$$

$$x = \frac{-3}{7}$$

b. $x =$ smallest number

$$x + (x + 1) + (x + 2) = 102$$

$$3x + 3 - 3 = 102 - 3$$
$$\frac{3x}{3} = \frac{99}{3}$$

$$x = 33$$

The three numbers are 33, 34 and 35.

c. $x =$ larger number

$(x - 11) =$ smaller number

$$(x + 8) = 2(x - 11 + 8)$$
$$x + 8 = 2(x - 3)$$

$$x + 8 - 8 = 2x - 6 - 8$$
$$x - x = 2x - 14 - x$$
$$0 + 14 = x - 14 + 14$$
$$x = 14$$

The two numbers are 14 and 3.

WORKED EXAMPLE 7 Solving problems using linear equations

A taxi charges $3.60 plus $1.38 kilometre for any trip in Melbourne. If Elena's taxi ride cost $38.10, calculate the distance she travelled.

THINK	WRITE
1. The distance travelled by Elena has to be found. Define the pronumeral.	Let $x =$ the distance travelled (in kilometres).
2. It costs 1.38 to travel 1 kilometre, so the cost to travel x kilometres $= 1.38x$. The fixed cost is $3.60. Write an expression for the total cost.	Total cost $= 3.60 + 1.38x$
3. Let the total cost $= 38.10$.	$3.60 + 1.38x = 38.10$
4. Solve the equation.	$3.60 + 1.38x = 38.10$ $1.38x = 34.50$ $x = \dfrac{34.50}{1.38}$ $= 25$
5. Write the solution in words.	Elena travelled 25 kilometres.

WORKED EXAMPLE 8 Solving problems using linear equations

In a basketball game, Hao scored 5 more points than Seve. If they scored a total of 27 points between them, calculate how many points each of them scored.

THINK	WRITE
1. Define the pronumeral.	Let Seve's score be x.
2. Hao scored 5 more than Seve. Write an expression for Hao's score.	Hao's score is $x + 5$.
3. Hao and Seve scored a total of 27 points. Write this as an equation.	$x + (x + 5) = 27$
4. Solve the equation.	$2x + 5 = 27$ $2x = 22$ $x = 11$

5. Since $x = 11$, Seve's score is 11. Substitute this into the expression to work out Hao's score.	Hao's score $= x + 5$ $= 11 + 5$ $= 16$
6. Write the answer in words.	Seve scored 11 points and Hao scored 16 points.

WORKED EXAMPLE 9 Solving problems using linear equations

A collection of 182 marbles is owned by four friends. Pat has twice the number of marbles that Quentin has, and Rachel has 20 fewer marbles than Pat. If Sam has two-thirds the number of marbles that Rachel has, determine who has the second-largest number of marbles.

THINK	WRITE
a. 1. Let p represent the number of marbles Pat has.	$p =$ number of Pat's marbles
Pat has twice the number of marbles that Quentin has.	$\left(\dfrac{p}{2}\right) =$ number of Quentin's marbles
Rachel has 20 fewer marbles than Pat.	$p - 20 =$ number of Rachael's marbles
Sam has $\dfrac{2}{3}$ of the number of marbles that Rachel has.	$\dfrac{2}{3}(p - 20) =$ number of Sam's marbles
2. Let the sum of all marbles equal 182.	$p + \left(\dfrac{p}{2}\right) + (p - 20) + \dfrac{2}{3}(p - 20) = 182$
3. Simplify the left-hand side of the equation by adding like terms.	$p + \dfrac{p}{2} + p + \dfrac{2p}{3} - 20 - \dfrac{40}{3} = 182$ $\dfrac{6p}{6} + \dfrac{3p}{6} + \dfrac{6p}{6} + \dfrac{4p}{6} - \dfrac{120}{6} - \dfrac{80}{6} = 182$
4. Solve using inverse operations.	$\dfrac{19p - 200}{6} \times 6 = 182 \times 6$ $19p - 200 + 200 = 1092 + 200$ $\dfrac{19p}{19} = \dfrac{1292}{19}$ $p = 68$
5. Calculate the number of marbles that each friend has.	Pat has 68 marbles. Quentin has $\dfrac{68}{2} = 34$ marbles. Rachel has $68 - 20 = 48$ marbles. Sam has $\dfrac{2}{3} \times 48 = 32$ marbles.
6. Write down the second-highest number of marbles and the name of the person with that number of marbles.	Rachel has the second-highest number of marbles. She has 48 marbles.

Exercise 3.5 Solving problems with linear equations   **learn**on

Individual pathways

■ PRACTISE	■ CONSOLIDATE	■ MASTER
1, 3, 7, 11, 12, 16	2, 4, 8, 9, 13, 17	5, 6, 10, 14, 15, 18

To answer questions online and to receive **immediate corrective feedback** and **fully worked solutions** for all questions, go to your learnON title at www.jacplus.com.au.

Fluency

1. **WE6** Write linear equations for each of the following statements, using x to represent the unknown, without solving the equations.

   a. When 3 is added to a certain number, the answer is 5.
   b. Subtracting 9 from a certain number gives a result of 7.
   c. Multiplying a certain number by 7 gives 24.
   d. A certain number divided by 5 gives a result of 11.
   e. Dividing a certain number by 2 equals -9.

2. Write linear equations for each of the following statements, using x to represent the unknown, without solving the equations.

   a. Subtracting 3 from 5 times a certain number gives a result of -7.
   b. When a certain number is subtracted from 14 and the result is then multiplied by 2, the final result is -3.
   c. When 5 is added to 3 times a certain number, the answer is 8.
   d. When 12 is subtracted from 2 times a certain number, the result is 15.
   e. The sum of 3 times a certain number and 4 is divided by 2. This gives a result of 5.

3. **MC** Select the equation that matches the following statement.
   A certain number, when divided by 2, gives a result of -12.

   A. $x = \dfrac{-12}{2}$ **B.** $2x = -12$ **C.** $\dfrac{x}{2} = -12$ **D.** $\dfrac{x}{12} = -2$ **E.** $\dfrac{2}{x} = 12$

4. **MC** Select the equation that matches the following statement.
   Dividing 7 times a certain number by -4 equals 9.

   A. $\dfrac{x}{-4} = 9$ **B.** $\dfrac{-4x}{7} = 9$ **C.** $\dfrac{7+x}{-4} = 9$ **D.** $\dfrac{7x}{-4} = 9$ **E.** $\dfrac{7x}{-4} = -9$

5. **MC** Select the equation that matches the following statement.
   Subtracting twice a certain number from 8 gives 12.

   A. $2x - 8 = 12$ **B.** $8 - 2x = 12$ **C.** $2 - 8x = 12$ **D.** $8 - (x+2) = 12$ **E.** $2(8-x) = 12$

6. **MC** Select the equation that matches the following statement.
   When 15 is added to a quarter of a number, the answer is 10.

   A. $15 + 4x = 10$ **B.** $10 = \dfrac{x}{4} + 15$ **C.** $\dfrac{x + 15}{4} = 10$ **D.** $15 + \dfrac{4}{x} = 10$ **E.** $15 = \dfrac{x}{4} + 10$

Understanding

7. When a certain number is added to 3 and the result is multiplied by 4, the answer is the same as when that same number is added to 4 and the result is multiplied by 3. Determine the number.

8. **WE7** John is three times as old as his son Jack. The sum of their ages is 48. Calculate how old John is.

9. In one afternoon's shopping Seedevi spent half as much money as Georgia, but $6 more than Amy. If the three of them spent a total of $258, calculate how much Seedevi spent.

10. The rectangular blocks of land shown have the same area. Determine the dimensions of each block and use these dimensions to calculate the area.

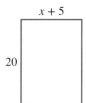
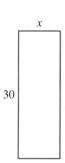

Reasoning

11. **WE 8** A square pool is surrounded by a paved area that is 2 metres wide. If the area of the paving is $72\,\text{m}^2$, determine the length of the pool. Show all working.

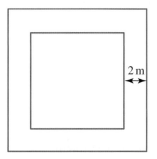

12. Maria is paid $11.50 per hour plus $7 for each jacket that she sews. If she earned $176 for a particular eight-hour shift, determine the number of jackets she sewed during that shift.

13. Mai hired a car for a fee of $120 plus $30 per day. Casey's rate for his car hire was $180 plus $26 per day. If their final cost and rental period is the same for both Mai and Casey, determine the rental period.

14. **WE9** The cost of producing an album on CD is quoted as $1200 plus $0.95 per CD. If Maya has a budget of $2100 for her debut album, determine the number of CDs she can make.

15. Joseph wants to have some flyers for his grocery business delivered. Post Quick quotes a price of $200 plus 50 cents per flyer, and Fast Box quotes $100 plus 80 cents per flyer.

    a. If Joseph needs to order 1000 flyers, determine the distributor that would be cheaper to use.

    b. Determine the number of flyers that Joseph need to get delivered for the cost to be the same for either distributor.

Problem solving

16. A certain number is multiplied by 8 and then 16 is subtracted. The result is the same as 4 times the original number minus 8. Evaluate the number.

17. Carmel sells three different types of healthy drinks: herbal, vegetable and citrus fizz. In an hour she sells 4 herbal, 3 vegetable and 6 citrus fizz drinks for $60.50.

    In the following hour she sells 2 herbal, 4 vegetable and 3 citrus fizz drinks. In the third hour she sells 1 herbal, 2 vegetable and 4 citrus. The total amount in cash sales for the three hours is $136.50. Carmel made $7 less in the third hour than she did in the second hour of sales.

    Evaluate Carmel's sales in the fourth hour if she sells 2 herbal, 3 vegetable and 4 citrus fizz drinks.

18. A rectangular swimming pool is surrounded by a path that is enclosed by a pool fence. All measurements given are in metres and are not to scale in the diagram shown.

    a. Write an expression for the area of the entire fenced-off section.

    b. Write an expression for the area of the path surrounding the pool.

    c. If the area of the path surrounding the pool is 34 m², evaluate the dimensions of the swimming pool.

    d. Determine the fraction of the fenced-off area taken up by the pool.

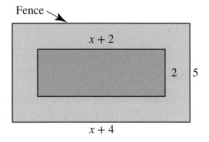

3.6 Rearranging formulas

LEARNING INTENTION

At the end of this subtopic you should be able to:
- rearrange formulas with two or more pronumerals to make only one of the pronumerals the subject.

▶ 3.6.1 Rearranging formulas

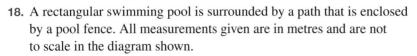
eles-4695

- Formulas are equations that are used for specific purposes. They are generally written in terms of two or more pronumerals (also known as 'variables').
- One pronumeral is usually written on the left of the equals sign. This pronumeral is called the **subject**.
- Solving a linear equation requires finding a value for the pronumeral (or pronumerals), but rearranging formulas does not. For this reason formulas are defined as **literal equations**.

Inverse operations

When rearranging formulas, use inverse operations in reverse order, just as you do when you are solving linear equations.

For example, $C = 2\pi r$ is a formula describing the relationship of the circumference of a circle (C) to its radius (r). To make r the subject of this formula, the following steps are required:

Divide both sides by 2π.	$C = 2\pi r$ $$\frac{C}{2\pi} = \frac{2\pi r}{2\pi}$$ $$\frac{C}{2\pi} = r$$
Now r is the subject.	$$r = \frac{C}{2\pi}$$

DISCUSSION

In pairs, share some formulas that you know. Discuss each of the formulas by thinking about why we need them, how and when they are used in the real world, and why rearranging the pronumerals in these formulas can be helpful.

WORKED EXAMPLE 10 Rearranging formulas

Rearrange each formula to make x the subject.
a. $y = kx + m$
b. $6(y + 1) = 7(x - 2)$

THINK	WRITE
a. 1. Subtract m from both sides.	a. $\quad y = kx + m$ $\quad y - m = kx$
2. Divide both sides by k.	$$\frac{y - m}{k} = \frac{kx}{k}$$ $$\frac{y - m}{k} = x$$
3. Rewrite the equation so that x is on the left-hand side.	$$x = \frac{y - m}{k}$$
b. 1. Expand the brackets.	b. $6(y + 1) = 7(x - 2)$ $\quad 6y + 6 = 7x - 14$
2. Add 14 to both sides.	$6y + 20 = 7x$
3. Divide both sides by 7.	$$\frac{6y + 20}{7} = x$$
4. Rewrite the equation so that x is on the left-hand side.	$$x = \frac{6y + 20}{7}$$

TI	THINK	DISPLAY/WRITE	CASIO	THINK	DISPLAY/WRITE

a–b. In a new problem, on a Calculator page, press:
- MENU
- 3: Algebra
- 1: Solve

Complete the entry lines as:
solve
$(y = k \times x + m, x)$
solve $(6(y + 1) = 7(x - 2), x)$

Press ENTER after each entry.

a–b.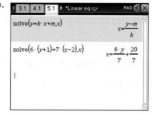

$$x = \frac{y - m}{k}$$

$$x = \frac{6y}{7} + \frac{20}{7}$$

a–b. On the Main screen, press:
- Action
- Advanced
- Solve

Complete the entry lines as:
solve
$(y = k \times x + m, x)$
solve $(6(y + 1) = 7(x - 2), x)$

Press EXE after each entry.

a–b.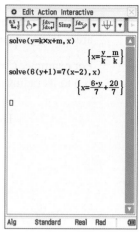

$$x = \frac{y - m}{k}$$

$$x = \frac{6y}{7} + \frac{20}{7}$$

WORKED EXAMPLE 11 Solving literal equations

a. If $g = 6d - 3$, solve for d.
b. Solve for v, given that $a = \dfrac{v - u}{t}$.

THINK	WRITE
a. 1. Add 3 to both sides.	**a.** $g = 6d - 3$ $g + 3 = 6d$
2. Divide both sides by 6.	$\dfrac{g + 3}{6} = d$
3. Rewrite the equation so that d is on the left-hand side.	$d = \dfrac{g + 3}{6}$
b. 1. Multiply both sides by t.	**b.** $a = \dfrac{v - u}{t}$ $t = v - u$
2. Add u to both sides.	$at + u = v$
3. Rewrite the equation so that v is on the left-hand side.	$v = at + u$

---

  on Resources

eWorkbook   Topic 3 Workbook (worksheets, code puzzle and project) (ewbk-2003)

Digital document SkillSHEET Transposing and substituting into a formula (doc-10829)

Interactivities   Individual pathway interactivity: Rearranging formulas (int-4493)

                   Rearranging formulas (int-6040)

Exercise 3.6 Rearranging formulas

Individual pathways

■ PRACTISE	■ CONSOLIDATE	■ MASTER
1, 3, 6, 8, 11	2, 4, 7, 9, 12	5, 10, 13

To answer questions online and to receive **immediate corrective feedback** and **fully worked solutions** for all questions, go to your learnON title at www.jacplus.com.au.

Fluency

1. **WE10** Rearrange each formula to make x the subject.

   a. $y = ax$
   b. $y = ax + b$
   c. $y = 2ax - b$
   d. $y + 4 = 2x - 3$
   e. $6(y + 2) = 5(4 - x)$
   f. $x(y - 2) = 1$

2. Rearrange each formula to make x the subject.

   a. $x(y - 2) = y + 1$
   b. $5x - 4y = 1$
   c. $6(x + 2) = 5(x - y)$
   d. $7(x - a) = 6x + 5a$
   e. $5(a - 2x) = 9(x + 1)$
   f. $8(9x - 2) + 3 = 7(2a - 3x)$

3. **WE11** For each of the following, make the variable shown in brackets the subject of the formula.

   a. $g = 4P - 3$ $\quad (P)$
   b. $f = \dfrac{9c}{5}$ $\quad (c)$
   c. $f = \dfrac{9c}{5} + 32$ $\quad (c)$
   d. $v = u + at$ $\quad (t)$

4. For each of the following, make the variable shown in brackets the subject of the formula.

   a. $d = b^2 - 4ac$ $\quad (c)$
   b. $m = \dfrac{y - k}{h}$ $\quad (y)$
   c. $m = \dfrac{y - a}{x - b}$ $\quad (a)$
   d. $m = \dfrac{y - a}{x - b}$ $\quad (x)$

5. For each of the following, make the variable shown in brackets the subject of the formula.

   a. $C = \dfrac{2\pi}{r}$ $\quad (r)$
   b. $f = ax + by$ $\quad (x)$
   c. $s = ut + \dfrac{1}{2}at^2$ $\quad (a)$
   d. $F = \dfrac{GMm}{r^2}$ $\quad (G)$

Understanding

6. The cost to rent a car is given by the formula $C = 50d + 0.2k$, where $d =$ the number of days rented and $k =$ the number of kilometres driven. Lin has \$300 to spend on car rental for her four-day holiday.
   Calculate the distance that she can drive on her holiday.

7. A cyclist pumps up a bike tyre that has a slow leak. The volume of air (in cm³) after t minutes is given by the formula $V = 24\,000 - 300t$.

   a. Determine the volume of air in the tyre when it is first filled.
   b. Write out the equation for the tyre's volume and solve it to work out how long it would take the tyre to go completely flat.

Reasoning

8. The total surface area of a cylinder is given by the formula $T = 2\pi r^2 + 2\pi rh$, where r = radius and h = height. A car manufacturer wants its engines' cylinders to have a radius of 4 cm and a total surface area of 400 cm². Show that the height of the cylinder is approximately 11.92 cm, correct to 2 decimal places. (*Hint:* Express h in terms of T and r.)

9. If $B = 3x - 6xy$, write x as the subject. Explain the process you followed by showing your working.

10. The volume, V, of a sphere can be calculated using the formula $V = \dfrac{4}{3}\pi r^3$, where r is the radius of the sphere. Explain how you would work out the radius of a spherical ball that has the capacity to hold 5 litres of air.

Problem solving

11. Use algebra to show that $\dfrac{1}{v} = \dfrac{1}{u} - \dfrac{1}{f}$ can also be written as $u = \dfrac{fv}{v+f}$.

12. Consider the formula $d = \sqrt{b^2 - 4ac}$.
    Rearrange the formula to make a the subject.

13. Evaluate the values for a and b such that:
$$\frac{4}{x+1} - \frac{3}{x+2} = \frac{ax+b}{(x+1)(x+2)}.$$

3.7 Review

3.7.1 Topic summary

LINEAR EQUATIONS

Inverse operations

- Solving a linear equation means determining a value for the pronumeral that makes the statement true.
- Solving equations requires inverse operations.
- The inverse operation has the effect of undoing the original operation.

Operation	Inverse operation
+	−
−	+
×	÷
÷	×

- To keep an equation balanced, inverse operations are done on both sides of the equation.

  e.g.
  $$3x + 1 = 11$$
  $$3x + 1 - 1 = 11 - 1$$
  $$\frac{3x}{3} = \frac{10}{3}$$
  $$x = \frac{10}{3}$$

Equations with brackets

- The number in front of the brackets indicates the multiple of the expression inside the brackets.
- Brackets can be removed by dividing both sides by the multiple.

  e.g.
  $$3(x + 5) = 18$$
  $$\frac{3(x + 5)}{3} = \frac{18}{3}$$
  $$(x + 5) = 6$$

- Alternatively, brackets can be removed by expanding (multiplying) out the expression.

  e.g.
  $$3(x + 5) = 18$$
  $$3x + 15 = 18$$

Equations with pronumerals on both sides

- Pronumerals must be moved so that they appear only on one side using inverse operations.
- Once this is done, you can solve the equation.

  e.g.
  $$5y = 3y + 3$$
  $$5y - 3y = 3y + 3 - 3y$$
  $$2y = 3$$
  $$y = \frac{3}{2}$$

Pronumeral in the denominator

- The pronumeral needs to be expressed as a numerator.
- This can be done by multiplying both sides of the equation by the pronumeral.

  e.g.
  $$\frac{4}{x} \times x = \frac{3}{2} \times x$$

- Alternatively, you can flip both sides of the equation.

  e.g.
  $$\frac{4}{x} = \frac{3}{2}$$

  becomes $\frac{x}{4} = \frac{2}{3}$.

Worded problems

- Worded problems should be translated into equations.
- Pronumerals must be defined in the answer if this is not already done in the question.

  e.g. The equation for the worded problem 'When 6 is subtracted from a certain number, the result is 15' is '$x - 6 = 15$'.

Rearranging formulas

- The subject of the formula is the pronumeral that is written by itself.
- The difference between rearranging formulas and solving linear equations is that rearranging formulas does not require a value for the pronumeral to be found.

  e.g. Rearrange the following to make x the subject:
  $$y = kx + m$$
  $$y - m = kx$$
  $$\frac{y - m}{k} = \frac{kx}{k}$$
  $$y - \frac{m}{k} = x$$

3.7.2 Success criteria

Tick the column to indicate that you have completed the subtopic and how well you have understood it using the traffic light system.

(**Green:** I understand; **Yellow:** I can do it with help; **Red:** I do not understand)

Subtopic	Success criteria	⬤	◯	⬤
3.2	I can recognise inverse operations.			
	I can solve multistep equations with pronumerals on one side of the equation.			
	I can solve equations with algebraic fractions.			
3.3	I can solve equations involving brackets.			
3.4	I can solve equations by collecting pronumerals on one side of an equation before solving.			
3.5	I can define a pronumeral and write an equation to represent the problem.			
	I can solve a problem and answer the question (or questions) posed in the problem.			
3.6	I can rearrange formulas with two or more pronumerals to make only one of the pronumerals the subject.			

3.7.3 Project

Forensic science

Scientific studies have been conducted on the relationship between a person's height and the measurements of a variety of body parts. One study has suggested that there is a general relationship between a person's height and the humerus bone in their upper arm, and that this relationship is slightly different for men and women.

According to this study, there is a general trend indicating that $h = 3.08l + 70.45$ for men, and $h = 3.36l + 57.97$ for women, where h represents body height in centimetres and l the length of the humerus in centimetres.

Imagine the following situation.

A decomposed body has been in the bushland outside your town. A team of forensic scientists suspects that the body could be the remains of either Alice Brown or James King, both of whom have been missing for several years. From the descriptions provided by their Missing Persons files, Alice is 162 cm tall and James is 172 cm tall. The forensic scientists hope to identify the body based on the length of the body's humerus.

1. Based on the relationship suggested by the study previously mentioned, complete both of the following tables using the equations provided. Calculate body height to the nearest centimetre.

Body height (men)

Length of humerus, l (cm)	20	25	30	35	40
Body height, h (cm)					

Body height (women)

Length of humerus, l (cm)	20	25	30	35	40
Body height, h (cm)					

2. On a piece of graph paper, draw the first quadrant of a Cartesian plane. Since the length of the humerus is the independent variable, place it on the x-axis. Place the dependent variable, body height, on the y-axis.
3. Plot the points from the two tables onto the set of axes drawn in question 2. Join the points with straight lines, using different colours to represent men and women.
4. Describe the shape of the two graphs.
5. Measure the length of your humerus. Use your graph to predict your height. How accurate is the measurement?
6. The two lines of your graph will intersect if extended. At what point does this occur? Comment on this value.

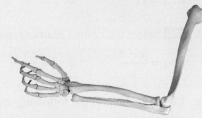

The forensic scientists measured the length of the humerus of the decomposed body and found it to be 33 cm.

7. Using methods covered in this activity, is it more probable that the body is that of Alice or of James? Justify your decision with mathematical evidence.

---

  on Resources

eWorkbook　　Topic 3 Workbook (worksheets, code puzzle and project) (ewbk-2003)

Interactivities Crossword (int-0700)

　　　　　　　　Sudoku puzzle (int-3204)

Exercise 3.7 Review questions

 learn on

To answer questions online and to receive **immediate corrective feedback** and **fully worked solutions** for all questions, go to your learnON title at www.jacplus.com.au.

Fluency

1. **MC** Select the linear equation represented by the sentence 'When a certain number is multiplied by 3, the result is 5 times that certain number plus 7'.
   A. $3x + 7 = 5x$　　　　　　B. $5(x + 7) = 3x$　　　　　　C. $5x + 7 = 3x$
   D. $5x = 3x + 7$　　　　　　E. $3x = 7x + 5$

2. **MC** Calculate the solution to the equation $\dfrac{x}{3} = 5$.
   A. $x = -15$　　　B. $x = 15$　　　C. $x = 1\dfrac{2}{3}$　　　D. $x = 3$　　　E. $x = -3$

3. **MC** Determine the solution to the equation $7 = 21 + x$.
   A. $x = 28$　　　B. $x = -28$　　　C. $x = -14$　　　D. $x = 14$　　　E. $x = 7$

4. **MC** Calculate the solution to the equation $5x + 3 = 37$.
   A. $x = 8$　　　B. $x = -8$　　　C. $x = 6.8$　　　D. $x = 106$　　　E. $x = -6.8$

5. **MC** Determine the solution to the equation $8 - 2x = 22$.
   A. $x = 11$　　　B. $x = 15$　　　C. $x = -15$　　　D. $x = -7$　　　E. $x = 7$

6. **MC** Determine the solution to the equation $4x + 3 = 7x - 33$.
   A. $x = -12$　　　B. $x = 12$　　　C. $x = \dfrac{36}{11}$　　　D. $x = \dfrac{30}{11}$　　　E. $x = \dfrac{11}{30}$

7. **MC** Calculate the solution to the equation $7(x - 15) = 28$.
   A. $x = 11$　　　B. $x = 19$　　　C. $x = 20$　　　D. $x = 6.14$　　　E. $x = 6$

8. **MC** Select the correct rearrangement of $y = ax + b$ in terms of x.
   A. $x = \dfrac{y - a}{b}$　　　B. $x = \dfrac{y - b}{a}$　　　C. $x = \dfrac{b - y}{a}$　　　D. $x = \dfrac{y + b}{a}$　　　E. $x = \dfrac{x - b}{a}$

9. Identify which of the following are linear equations.
   a. $5x + y^2 = 0$　　　　　　　　　　b. $2x + 3 = x - 2$
   c. $\dfrac{x}{2} = 3$　　　　　　　　　　d. $x^2 = 1$
   e. $\dfrac{1}{x} + 1 = 3x$　　　　　　　　f. $8 = 5x - 2$
   g. $5(x + 2) = 0$　　　　　　　　　h. $x^2 + y = -9$
   i. $r = 7 - 5(4 - r)$

10. Solve each of the following linear equations.
    a. $3a = 8.4$　　　　　　　　　　b. $a + 2.3 = 1.7$
    c. $\dfrac{b}{21} = -0.12$　　　　　　　　d. $b - 1.45 = 1.65$
    e. $b + 3.45 = 0$　　　　　　　　f. $7.53b = 5.64$

11. Solve each of the following linear equations.
    a. $\dfrac{2x - 3}{7} = 5$　　　b. $\dfrac{5 - x}{2} = -4$　　　c. $\dfrac{-3x - 4}{5} = 3$
    d. $\dfrac{6}{x} = 5$　　　e. $\dfrac{4}{x} = \dfrac{3}{5}$　　　f. $\dfrac{x + 1.7}{2.3} = -4.1$

12. Solve each of the following linear equations.
    a. $5(x - 2) = 6$　　　b. $7(x + 3) = 40$　　　c. $4(5 - x) = 15$
    d. $6(2x + 3) = 1$　　　e. $4(x + 5) = 2x - 5$　　　f. $3(x - 2) = 7(x + 4)$

13. Liz has a packet of 45 lolly snakes. She saves 21 to eat tomorrow, but she rations the remainder so that she can eat 8 snakes every hour until today's share of snakes is gone.
    a. Write a linear equation in terms of the number of hours, h, to represent this situation.
    b. Determine the number of hours it will take Liz to eat today's share.

14. Solve each of the following linear equations.
    a. $11x = 15x - 2$
    b. $3x + 4 = 16 - x$
    c. $5x + 2 = 3x + 8$
    d. $8x - 9 = 7x - 4$
    e. $2x + 5 = 8x - 7$
    f. $3 - 4x = 6 - x$

15. Translate the following sentences into algebraic equations. Use x to represent the number in question.
    a. Twice a certain number is equal to 3 minus that certain number.
    b. When 8 is added to 3 times a certain number, the result is 19.
    c. Multiplying a certain number by 6 equals 4.
    d. Dividing 10 by a certain number equals one more than dividing that number by 6.
    e. Multiply a certain number by 2, then add 5. Multiply this result by 7. This expression equals 0
    f. Twice a certain distance travelled is 100 metres more than that certain distance travelled plus 50 metres.

16. Takanori decides to go on a holiday. He travels a certain distance on the first day, twice that distance on the second day, three times that distance on the third day and four times that distance on the fourth day. If Takanori's total journey is 2000 km, calculate the distance he travelled on the third day.

17. For each of the following, make the variable shown in brackets the subject of the formula.
    a. $y = 6x - 4$ (x)
    b. $y = mx + c$ (x)
    c. $q = 2(P - 1) + 2r$ (P)
    d. $P = 2l + 2w$ (w)

18. For each of the following, make the variable shown in brackets the subject of the formula.
    a. $v = u + at$ (a)
    b. $s = \left(\dfrac{u + v}{2}\right)t$ (t)
    c. $v^2 = u^2 + 2as$ (a)
    d. $2A = h(a + b)$ (b)

Problem solving

19. Saeed is comparing two car rental companies, Golden Ace Rental Company and Silver Diamond Rental Company. Golden Ace Rental Company charges a flat rate of $38 plus $0.20 per kilometre. The Silver Diamond Rental Company charges a flat rate of $30 plus $0.32 per kilometre. Saeed plans to rent a car for three days.

    a. Write an equation for the cost of renting a car for three days from the Golden Ace Rental Company, in terms of the number of kilometres travelled, k.
    b. Write an algebraic equation for the cost of renting a car for three days from the Silver Diamond Rental Company, in terms of the number of kilometres travelled, k.
    c. Evaluate the number of kilometres Saeed would have to travel so that the cost of hiring from each company is the same.

20. Frederika has $24 000 saved to pay for a holiday. Her travel expenses are $5400 and her daily expenses are $260.
    a. Write down an equation for the cost of her holiday if she stays for d days.
    b. Determine the number of days Frederika can spend on holiday if she wants to be able to buy a new laptop for $ 2500 when she gets back from her holidays.

21. A company that makes bottled orange juice buys their raw materials from two sources. The first source provides liquid with 6% orange juice, whereas the second source provides liquid with 3% orange juice. The company wants to make 1-litre bottles that have 5% orange juice. Let x = the amount of liquid (in litres) that the company buys from the first source.

   a. Write an expression for the amount of orange juice from the first supplier, given that x is the amount of liquid.

   b. Write an expression for the amount of liquid from the second supplier, given that x is the amount of liquid used from the first supplier.

   c. Write an expression for the amount of orange juice from the second supplier.

   d. Write an equation for the total amount of orange juice in a mixture of liquids from the two suppliers, given that 1 litre of bottled orange juice will be mixed to contain 5% orange juice.

   e. Determine the quantity of the first supplier's liquid that the company uses.

22. Jayani goes on a four-day bushwalk. She travels a certain distance on the first day, half of that distance on the second day, a third of that distance on the third day and a quarter of that distance on the fourth day. If Jayani's total journey is 50 km, evaluate the distance that she walked on the first day.

23. Svetlana goes on a five-day bushwalk, travelling the same relative distances as Jayani travelled in question **22** (a certain amount, then half that amount, then one third of that, then one quarter of that, and finally one fifth of that). If Svetlana's journey is also 50 km, determine the distance that she travelled on the first day.

24. An online bookstore advertises its shipping cost to Australia as a flat rate of $20 for up to 10 books. Their major competitor offers a flat rate of $12 plus $1.60 per book. Determine the number of books you would have to buy (6, 7, 8, 9 or 10) for the first bookstore's shipping cost to be a better deal.

on To test your understanding and knowledge of this topic, go to your learnON title at www.jacplus.com.au and complete the **post-test**.

Online Resources

Below is a full list of **rich resources** available online for this topic. These resources are designed to bring ideas to life, to promote deep and lasting learning and to support the different learning needs of each individual.

eWorkbook

Download the workbook for this topic, which includes worksheets, a code puzzle and a project (ewbk-2003) ☐

Solutions

Download a copy of the fully worked solutions to every question in this topic (sol-0723) ☐

Digital documents

3.2 SkillSHEET Solving one-step equations (doc-6150) ☐
SkillSHEET Checking solutions to equations (doc-6151) ☐
SkillSHEET Solving equations (doc-6152) ☐
3.3 SkillSHEET Expanding brackets (doc-10827) ☐
3.4 SkillSHEET Simplifying like terms (doc-10828) ☐
3.5 SkillSHEET Writing equations from worded statements (doc-10826) ☐
3.6 SkillSHEET Transposing and substituting into a formula (doc-10829) ☐

Video eLessons

3.1 What is a linear equation? (eles-4689) ☐
Solving multistep equations (eles-4690) ☐
Algebraic fractions with the pronumeral in the denominator (eles-4691) ☐
Solving linear equations (eles-1895) ☐
3.2 Linear equations with brackets (eles-4692) ☐
3.3 Linear equations with pronumerals on both sides (eles-4693) ☐
Solving linear equations with pronumerals on both sides (eles-1901) ☐
3.4 Solving problems with linear equations (eles-4694) ☐
3.5 Rearranging formulas (eles-4695) ☐

Interactivities

3.2 Individual pathway interactivity: Solving linear equations (int-4489) ☐
Using algebra to solve problems (int-3805) ☐
3.3 Individual pathway interactivity: Solving linear equations with brackets (int-4490) ☐
Linear equations with brackets (int-6039) ☐
3.4 Individual pathway interactivity: Solving linear equations with pronumerals on both sides (int-4491) ☐
3.5 Individual pathway interactivity: Solving problems with linear equations (int-4492) ☐
3.6 Individual pathway interactivity: Rearranging formulas (int-4493) ☐
Rearranging formulas (int-6040) ☐
3.7 Crossword (int-0700) ☐
Sudoku puzzle (int-3204) ☐

Teacher resources

There are many resources available exclusively for teachers online.

To access these online resources, log on to **www.jacplus.com.au**.

Answers

Topic 3 Linear equations

Exercise 3.1 Pre-test

1. 3
2. 22
3. C
4. a. $x = -7$ b. $z = 11$
5. 4
6. False
7. E
8. C
9. a. $a = 2$ b. $b = -8$
10. B
11. 287.5 km
12. -3
13. D
14. B
15. a. $a = \dfrac{bc}{(b+c)}$ b. $b = \dfrac{ac}{(c-a)}$

Exercise 3.2 Solving linear equations

1. a. No b. Yes c. No
   d. Yes e. Yes f. No

2. a. No b. No c. Yes
   d. No e. Yes f. No

3. a. $x = 210$ b. $x = 230$ c. $x = 23$
   d. $x = -96$ e. $x = 37$ f. $x = 20$

4. a. $x = 50$ b. $x = 138$ c. $x = 442$
   d. $x = 243$ e. $y = -15$ f. $y = 1.8$

5. a. $y = 12.8$ b. $y = 2\dfrac{1}{3}$ c. $y = -\dfrac{1}{18}$
   d. $y = 21.5$ e. $y = 172.5$ f. $y = -8.32$

6. a. $y = 5$ b. $y = -2$ c. $y = 0.2$
   d. $y = 1$ e. $y = 1$ f. $y = 2$

7. a. $y = 5\dfrac{1}{3}$ b. $y = -1\dfrac{2}{3}$ c. $y = 24.5$
   d. $y = 1.2$ e. $y = 22.25$ f. $y = \dfrac{383}{600}$

8. a. $x = 1$ b. $x = -2$ c. $x = 3$
   d. $x = -6$ e. $x = -1$ f. $x = \dfrac{1}{2}$

9. a. $x = 1\dfrac{2}{3}$ b. $x = -2.32$ c. $x = 3$
   d. $x = -\dfrac{5}{6}$ e. $x = 1\dfrac{1}{4}$ f. $x = 3\dfrac{1}{13}$

10. a. $x = -1$ b. $x = 1$ c. $x = 0$
    d. $x = 5$ e. $x = -8.6$ f. $x = -0.9$

11. a. $x = 8.9$ b. $x = 19$ c. $x = -26$
    d. $x = -1$ e. $x = -1$ f. $x = -\dfrac{1}{2}$

12. a. $x = 8$ b. $x = 3$ c. $x = 4$
    d. $x = -15$ e. $x = 26$ f. $x = -42$

13. a. $x = 9$ b. $x = -1\dfrac{1}{5}$ c. $x = 9\dfrac{1}{3}$
    d. $x = -2\dfrac{1}{4}$ e. $x = -7$ f. $x = \dfrac{2}{3}$

14. a. $z = 16$ b. $z = 31$ c. $z = -4$
    d. $z = 6$ e. $z = -9$ f. $z = -6$

15. a. $z = -1.9$ b. $z = 6.88$ c. $z = 140$
    d. $z = 0.6$ e. $z = -35.4$ f. $z = 8$

16. a. $x = 1$ b. $x = 13$ c. $x = -2$
    d. $x = -8$ e. $x = -4$ f. $x = 30\dfrac{1}{2}$

17. a. $x = -6$ b. $x = -\dfrac{7}{10}$ c. $x = 10.35$
    d. $x = 0.326$ e. $x = 22$ f. $x = -5$

18. a. $x = 4$ b. $x = \dfrac{3}{7}$ c. $x = -1\dfrac{1}{7}$
    d. $x = -6\dfrac{2}{3}$ e. $x = \dfrac{4}{45}$ f. $x = 8$

19. a. $x = -6$ b. $x = 7.5$ c. $x = 5.1$
    d. $x = -6$ e. $x = -5\dfrac{13}{15}$ f. $x = -61\dfrac{3}{7}$

20. a. D b. D c. C

21. a. $a = -1$ b. $b = 10$ c. $c = 12.1$
    d. $d = 4$ e. $e = 5$ f. $f = 12$

22. a. $g = 15\dfrac{29}{30}$ b. $h = 31$ c. $i = \dfrac{13}{226}$
    d. $j = \dfrac{1}{6}$ e. $k = -8$ f. $l = 10.3$

23. a. -1 b. 6 c. 303

24. $91

25. No. 46.7 °C $\approx$ 116.1 °F

26. For a positive integer, $(x-4)$ must be a factor of 12.
    $x = 5, 6, 7, 8, 10, 16$

27. Sample responses can be found in the worked solutions in the online resources.

28. The mistake is in the second line: the -1 should have been multiplied by 5.

29. 60 lollies

30. Old machine: 6640 cans; new machine: 9360 cans

31. a. $34
    b. Yes, a saving of $7

32. 17

Exercise 3.3 Solving linear equations with brackets

1. a. $x = 6$ b. $x = -3$ c. $x = 0$
   d. $x = 56$ e. $x = 1$ f. $x = 0$

2. a. $x = -1$ b. $x = -2$ c. $x = 6$
   d. $x = 0$ e. $x = -0.8$ f. $x = 6$

3. a. $b = 1\dfrac{1}{6}$ b. $m = 4\dfrac{1}{2}$ c. $a = -1\dfrac{1}{2}$
   d. $m = -1\dfrac{1}{3}$ e. $p = \dfrac{3}{5}$ f. $m = 2\dfrac{2}{3}$

4. a. $a = -1\dfrac{1}{2}$ b. $p = 1\dfrac{1}{2}$ c. $a = 3\dfrac{1}{3}$

d. $m = -2\dfrac{1}{2}$ e. $a = -\dfrac{1}{6}$ f. $m = -\dfrac{1}{4}$

5. a. $x = 16\dfrac{1}{9}$ b. $x = 2$ c. $a = 5$

   d. $b = 7$ e. $y = 7$ f. $x = -3\dfrac{1}{3}$

6. a. $m = 4$ b. $y = 1$ c. $a = 5$
   d. $p = -2$ e. $m = 3$ f. $p = 1$

7. a. $y = -7.5$ b. $y = -4.667$ c. $y = -26.25$
   d. $y = 8.571$ e. $y = -2.9$ f. $y = 3.243$

8. a. $y = 3.667$ b. $y = 2.75$ c. $y = 1.976$
   d. $y = -2$ e. $y = -3.167$ f. $y = 1.98$

9. D

10. C

11. 1990

12. a. $[2(2x + 7)]$ m

    b. Width 10 m, length 17 m

13. Expand out the brackets or divide by the factor.

14. $x = 13$. Sample responses can be found in the worked solutions in the online resources.

15. Adding 7 to both sides is the least preferred option, as it does not resolve the subtraction of 7 within the brackets.

16. 3

17. Oscar: \$90, Josue: \$35

18. \$20

Exercise 3.4 Solving linear equations with pronumerals on both sides

1. a. $y = -1$ b. $y = 1$ c. $y = -3$
   d. $y = -5$ e. $y = -45$ f. $y = \dfrac{8}{27}$

2. a. $y = -2$ b. $y = 20$ c. $y = 1$
   d. $y = -1\dfrac{1}{5}$ e. $y = 4$ f. $y = -2$

3. C

4. A

5. a. $x = 1$ b. $x = -2$ c. $x = 3$
   d. $x = 4$ e. $x = 9$ f. $x = 3$

6. a. $x = -2$ b. $x = 4$ c. $x = -4$
   d. $x = \dfrac{1}{2}$ e. $x = 0$ f. $x = -\dfrac{3}{4}$

7. a. $x = -6$ b. $x = 2$ c. $x = -2$
   d. $x = 13$ e. $x = -7$ f. $x = \dfrac{1}{6}$

8. D

9. C

10. a. $x = 5$ b. $x = -3$ c. $x = -8$
    d. $x = -7\dfrac{1}{2}$ e. $x = 4$ f. $x = \dfrac{10}{11}$

11. a. $x = -18$ b. $x = 3\dfrac{2}{3}$ c. $x = \dfrac{2}{3}$
    d. $x = -1\dfrac{3}{7}$ e. $x = 0$ f. $x = -15$

12. $k = -3$

13. 24

14. $3(n - 36) - 98 = -11n + 200$

15. a.

b. Area of first rectangle $= (x + 1) \times 3$
$$= 3x + 3$$
Area of second rectangle $= (x + 1 + x + 1 + x + 1) \times 1$
$$= 3x + 3$$
Both rectangles have an area of $3x + 3$.

16. You cannot easily divide the left-hand side by 6 or the right-hand side by 4.

17. Daughter = 10 years, Tom = 40 years

18. The unknown number is -3.

19. a. $C = 40h + 250$

    b. 18 hours, 45 minutes

    c. 18 750

    d. The printing is cheaper by \$1375.

20. a. 65 games   b. \$25

Exercise 3.5 Solving problems with linear equations

1. a. $x + 3 = 5$ b. $x - 9 = 7$ c. $7x = 24$
   d. $\dfrac{x}{5} = 11$ e. $\dfrac{x}{2} = -9$

2. a. $5x - 3 = -7$ b. $2(14 - x) = -3$
   c. $3x + 5 = 8$ d. $2x - 12 = 15$
   e. $\dfrac{3x + 4}{2} = 5$

3. C

4. D

5. B

6. B

7. 0

8. 36 years old

9. \$66

10. 20×15; 30×10; area $= 300$ square units

11. 7 m

12. 12 jackets

13. 15 days

14. 947 CDs

15. a. Post Quick (cost $=$ \$700)

    b. The cost is nearly the same for 333 flyers (\$366.50 and \$366.40).

16. 2

17. \$42.50

18. a. $A_{\text{fenced}} = (5x + 20)$ m^2 b. $A_{\text{path}} = (3x + 16)$ m^2

    c. $l = 8$ m, $w = 2$ m   d. $\dfrac{8}{25}$ m

Exercise 3.6 Rearranging formulas

1. a. $x = \dfrac{y}{a}$ b. $x = \dfrac{y-b}{a}$

 c. $x = \dfrac{y+b}{2a}$ d. $x = \dfrac{y+7}{2}$

 e. $x = \dfrac{8-6y}{5}$ f. $x = \dfrac{1}{y-2}$

2. a. $x = \dfrac{y+1}{y-2}$ b. $x = \dfrac{4y+1}{5}$ c. $x = -5y - 12$

 d. $x = 12a$ e. $x = \dfrac{5a-9}{19}$ f. $x = \dfrac{14a+13}{93}$

3. a. $P = \dfrac{g+3}{4}$ b. $c = \dfrac{5f}{9}$

 c. $c = \dfrac{5(f-32)}{9}$ d. $t = \dfrac{v-u}{a}$

4. a. $c = \dfrac{b^2 - d}{4a}$ b. $y = hm + k$

 c. $a = y - m(x - b)$ d. $x = \dfrac{y-a+mb}{m}$

5. a. $r = \dfrac{2\pi}{C}$ b. $x = \dfrac{f-by}{a}$

 c. $a = \dfrac{2(s-ut)}{t^2}$ d. $G = \dfrac{Fr^2}{Mm}$

6. $500\,\text{km}$

7. a. $24\,000\,\text{cm}^3$

 b. $t = 80\ \text{min} = 1\,\text{h}\,20\,\text{min}$

8. Sample responses can be found in the worked solutions in the online resources.

9. $\dfrac{B}{3(1-2y)} = x$

10. radius $= 10.6\,\text{cm}$

11. Sample responses can be found in the worked solutions in the online resources.

12. $a = \dfrac{b^2 - d^2}{4c}$

13. $a = 1;\ b = 5$

Project

1. Body height (men)

Length of humerus, l (cm)	20	25	30	35	40
Body height, h (cm)	132	147	163	178	194

Body height (women)

Length of humerus, l (cm)	20	25	30	35	40
Body height, h (cm)	125	142	159	176	192

2. and 3.

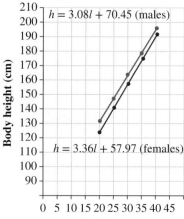

3. Linear

4. Results will vary for each student.

5. $(44.6,\ 207.8)$

6. The height of the male body with $l = 33\,\text{cm}$:
$h = 3.08 \times 33 + 70.45 = 172.09\,\text{cm}$
The height of the female body with $l = 33\,\text{cm}$:
$h = 3.36 \times 33 + 57.97 = 168.85\,\text{cm}$
The estimated height for a male body is very close to the known height of James King; the estimated height for a female body is more different from the known height of Alice Brown. Therefore, the body is more likely to be that of James King.

Exercise 3.7 Review questions

1. C
2. B
3. C
4. C
5. D
6. B
7. B
8. B
9. b, c, f, g, i

10. a. $a = 2.8$ b. $a = -0.6$
 c. $b = -2.52$ d. $b = 3.1$
 e. $b = -3.45$ f. $b = 0.749$

11. a. $x = 19$ b. $x = 13$ c. $x = -6\dfrac{1}{3}$
 d. $x = 1\dfrac{1}{5}$ e. $x = 6\dfrac{2}{3}$ f. $x = -11.13$

12. a. $x = 3\dfrac{1}{5}$ b. $x = 2\dfrac{5}{7}$
 c. $x = 1\dfrac{1}{4}$ d. $x = -1\dfrac{5}{12}$
 e. $x = -12\dfrac{1}{2}$ f. $x = -8\dfrac{1}{2}$

13. a. $8h + 21 = 45$ b. 3 hours

14. a. $x = \dfrac{1}{2}$ b. $x = 3$ c. $x = 3$
 d. $x = 5$ e. $x = 2$ f. $x = -1$

15. a. $2x = 3 - x$ b. $3x + 8 = 19$

   c. $6x = 4$ d. $\dfrac{10}{x} - 1 = \dfrac{x}{6}$

   e. $7(2x + 5) = 0$ f. $2x - 100 = x + 50$

16. $600\,\text{km}$

17. a. $x = \dfrac{y + 4}{6}$ b. $x = \dfrac{y - c}{m}$

   c. $P = \dfrac{q - 2r}{2} + 1$ d. $w = \dfrac{P - 2l}{2}$

18. a. $a = \dfrac{v - u}{t}$ b. $t = \dfrac{2s}{u + v}$

   c. $a = \dfrac{v^2 - u^2}{2s}$ d. $b = \dfrac{2A - ah}{h}$

19. a. $C_G = 114 + 0.20k$

   b. $C_S = 90 + 0.32k$

   c. $200\,\text{km}$

20. a. $5400 + 260d = C_H$ b. 61 days

21. a. $0.06x$

   b. $(1 - x)$

   c. $0.03(1 - x)$

   d. $0.06x + 0.03(1 - x) = 0.05$

   e. 0.667 or 66.7%

22. $24\,\text{km}$

23. $21\dfrac{123}{137} \approx 21.9\,\text{km}$

24. The online bookstore's fee is a better price than their major competitor's fee for all of the options given: 6, 7, 8, 9 and 10 books.

4 Congruence and similarity

LEARNING SEQUENCE

4.1 Overview

Why learn this?

Geometry allows us to explore our world in very precise ways. It is also one of the oldest areas of mathematics. It involves the study of points, lines and angles and how they can be combined to make different shapes. Similarity and congruence are two important concepts in geometry. When trying to determine whether two shapes are exactly the same, or if they are enlargements of each other, the answers can be found by considering the sides and angles of those shapes.

When you take a photo on your phone it is very small. If you save it to your computer you can make it larger. The larger photo is an enlargement of the original photo — this is an example of how similar figures work in the everyday world.

The principle of similar triangles can be used to work out the height of tall objects by calculating the length of their shadows. This technique was extremely important in early engineering and architecture. Today, architects and designers still prepare scale diagrams before starting the building process.

In manufacturing, the products that come off an assembly line all have exactly the same shape and size. These products can be described as congruent. Designers, engineers and surveyors all use the concepts of congruence and similarity in their daily work.

Where to get help

Go to your learnON title at **www.jacplus.com.au** to access the following digital resources. The Online Resources Summary at the end of this topic provides a full list of what's available to help you learn the concepts covered in this topic.

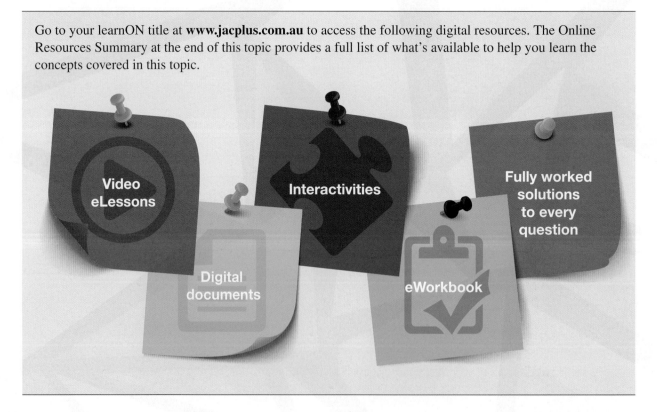

Video eLessons

Digital documents

Interactivities

eWorkbook

Fully worked solutions to every question

Complete this pre-test in your learnON title at www.jacplus.com.au and receive **automatic marks**, **immediate corrective feedback** and **fully worked solutions**.

1. Express the ratio $1\frac{2}{3} : 2\frac{3}{5}$ in simplest form.

$$1\frac{2}{3} : 2\frac{3}{5} = \boxed{} : \boxed{}$$

2. Determine the value of x in the proportion $2 : 3 = 9 : x$

3. **MC** Select the correct symbol for congruence.
   A. $=$
   B. $\approx$
   C. $\sim$
   D. $\simeq$
   E. $\cong$

4. **MC** If OB $=$ OC, choose which of the following congruency tests can be used to prove $\triangle AOB \cong \triangle DOC$.
   A. AAA
   B. RHS
   C. SSS
   D. ASA
   E. SAS

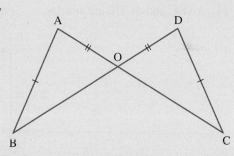

5. Determine the simplest ratio of $y : z$ if $3x = 2y$ and $4x = 3z$.

6. The ratio $1.4 : 0.2$ in its simplest form is $14 : 2$. State whether this statement is True or False.

7. **MC** Select the only pair of congruent triangles from the following.
   A. $\triangle ABC \cong \triangle BCD$
   B. $\triangle ABD \cong \triangle BCD$
   C. $\triangle BCD \cong \triangle DEF$
   D. $\triangle ABC \cong \triangle DEF$
   E. $\triangle ABD \cong \triangle DEF$

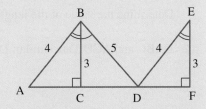

8. For the diagram given:
   a. name the angle that is equal to angle ABC
   b. state the congruency test that can be used to prove
      $\triangle ABC \cong \triangle CED$.

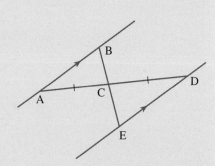

9. **MC** Select which similarity test can be used to prove $\triangle ABC = \triangle ADE$.
   A. AAA
   B. RHS
   C. SSS
   D. ASA
   E. SAS

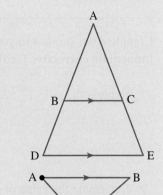

10. **MC** $\triangle ABC$ and $\triangle CDE$ are similar.
    Choose which of the following statements is true.

    A. $\dfrac{AB}{ED} = \dfrac{BC}{CD}$

    B. $\dfrac{AB}{ED} = \dfrac{AC}{CD}$

    C. $\dfrac{AC}{EC} = \dfrac{BC}{CD}$

    D. $\dfrac{AB}{ED} = \dfrac{BC}{CE}$

    E. $\dfrac{AB}{ED} = \dfrac{CD}{AC}$

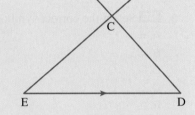

11. $\triangle ABC$ and $\triangle CDE$ are similar.

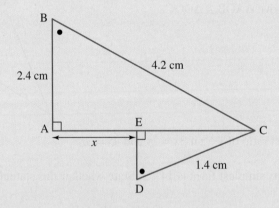

Determine the value of the length x correct to 1 decimal place.

12. $\triangle ABC$ and $\triangle CDE$ are similar. Determine the value of b.

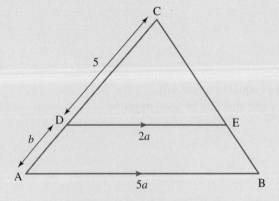

13. A pair of cones are similar. The ratio of their volumes is 27 : 125.
    a. The perpendicular height ratio of the two cones is ☐ : ☐.
    b. The areas of the base ratio of the two cones is ☐ : ☐.

14. A pair of rectangles are similar. If the width of the first rectangle is 3 times the width of the other, the ratio of their areas is:
    larger area : smaller area = ☐ : ☐.

15. **MC** A rectangular box has a surface area of 94 cm² and volume of 60 cm³. Select the volume and surface area of a similar box that has side lengths that are twice the size of the original.
    A. Volume = 480 cm³ and surface area = 752 cm²
    B. Volume = 480 cm³ and surface area = 376 cm²
    C. Volume = 120 cm³ and surface area = 188 cm²
    D. Volume = 240 cm³ and surface area = 376 cm²
    E. Volume = 240 cm³ and surface area = 752 cm²

4.2 Ratio and scale

LEARNING INTENTION

At the end of this subtopic you should be able to:
- compare two quantities of the same type using ratios
- simplify ratios
- enlarge a figure using a scale factor
- calculate and use a scale factor.

▶ 4.2.1 Ratio

eles-4727

- **Ratios** are used to compare quantities of the same kind, measured in the same unit.
- The ratio '1 is to 4' can be written in two ways: as 1 : 4 or as $\frac{1}{4}$.
- The order of the numbers in a ratio is important.
- In simplest form, a ratio is written using the smallest whole numbers possible.

WORKED EXAMPLE 1 Expressing ratios in simplest form

A lighthouse is positioned on a cliff that is 80 m high. A ship at sea is 3600 m from the base of the cliff.
a. **Write the following ratios in simplest form.**
    i. **The height of the cliff to the distance of the ship from shore**
    ii. **The distance of the ship from shore to the height of the cliff**
b. **Compare the distance of the ship from shore with the height of the cliff.**

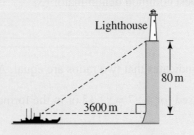

THINK	WRITE
a. i. 1. The height and distance are in the same units (m). Write the height first and the distance second.	**a. i.** Height of cliff : distance of ship from shore $= 80 : 3600$
2. Simplify the ratio by dividing both terms by the highest common factor (80).	$= 1 : 45$
ii. 1. Write the distance first and the height second.	**ii.** Distance of ship from shore : height of cliff $= \dfrac{3600}{80}$
2. Simplify by dividing both terms by the highest common factor (80). *Note:* Do not write $\dfrac{45}{1}$ as 45, because a ratio is a comparison between two numbers.	$= \dfrac{45}{1}$
b. 1. Write the ratio 'distance of the ship from shore to height of the cliff'.	**b.** $45 : 1$
2. Write the answer.	The distance of the ship from shore is 45 times the height of the cliff.

WORKED EXAMPLE 2 Simplifying ratios into simplest form

Express each of the following ratios in simplest form.

a. $24 : 8$ b. $3.6 : 8.4$ c. $1\dfrac{4}{9} : 1\dfrac{2}{3}$

THINK	WRITE
a. Divide both terms by the highest common factor (8).	**a.** $24 : 8$ $= 3 : 1$
b. 1. Multiply both terms by 10 to obtain whole numbers.	**b.** $3.6 : 8.4$ $= 36 : 84$
2. Divide both terms by the highest common factor (12).	$= 3 : 7$
c. 1. Change both mixed numbers into improper fractions.	**c.** $1\dfrac{4}{9} : 1\dfrac{2}{3}$ $= \dfrac{13}{9} : \dfrac{5}{3}$
2. Multiply both terms by the lowest common denominator (3) to obtain whole numbers.	$= 13 : 15$

- A **proportion** is a statement that indicates that two ratios are equal. A proportion can be written in two ways, for example in the format used for $4 : 7 = x : 15$ or in the format used for $\dfrac{2}{3} = \dfrac{11.5}{x}$.

WORKED EXAMPLE 3 Calculating a value in a proportion

Find the value of x in the proportion $4:9 = 7:x$.

THINK	WRITE
1. Write the ratios as equal fractions.	$\dfrac{4}{9} = \dfrac{7}{x}$
2. Multiply both sides by x.	$\dfrac{4x}{9} = 7$
3. Solve the equation to obtain the value of x.	$4x = 63$
4. Write the answer.	$x = 15.75$

▶ 4.2.2 Scale

eles-4728

- Ratios are used when creating scale drawings or maps.
- Consider the situation in which we want to enlarge a triangle ABC (the **object**) by a **scale factor** of 2 (this means we want to make it twice its size). The following is one method that we can use.

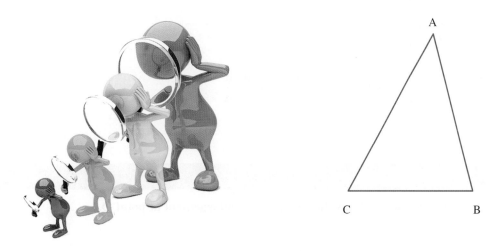

1. Mark a point O somewhere outside the triangle and draw the lines OA, OB and OC as shown.

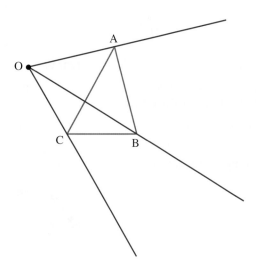

2. Measure the length of OA and mark in the point A′ (this is called the **image** of A) so that the distance OA′ is twice the distance of OA.

3. In the same way, mark in points B′ and C′. $(OB' = 2 \times OB,\ and\ OC' = 2 \times OC.)$

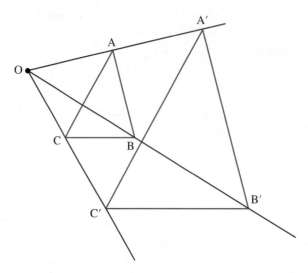

4. Joining A′B′C′ gives a triangle that has side lengths double those of △ABC. △A′B′C′ is called the image of △ABC.

Scale factor

$$\textbf{scale factor} = \frac{\textbf{image length}}{\textbf{object length}}$$

WORKED EXAMPLE 4 Enlarging a figure

Enlarge triangle ABC by a scale factor of 3, with the centre of enlargement at point O.

THINK

1. Join each vertex of the triangle to the centre of enlargement (O) with straight lines, then extend them.

2. Locate points A′, B′ and C′ along the lines so that OA′ = 3OA, OB′ = 3OB and OC′ = 3OC.

3. Join points A′, B′ and C′ to complete the image.

DRAW

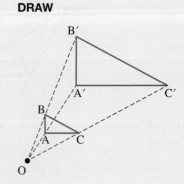

- Enlargements have the following properties.
    - The corresponding side lengths of the enlarged figure are changed in a fixed ratio (they have same ratio).
    - The corresponding angles are the same.
    - A scale factor greater than 1 produces an enlarged figure.
    - If the scale factor is a positive number less than 1, the image is smaller than the object (this means that a reduction has taken place).

WORKED EXAMPLE 5 Calculating the scale factor

A triangle PQR has been enlarged to create triangle P′Q′R′. PQ = 4 cm, PR = 6 cm, P′Q′ = 10 cm and Q′R′ = 20 cm. Calculate:
a. the scale factor for the enlargement
b. the length of P′R′
c. the length of QR.

THINK	WRITE/DRAW
a. 1. Draw a diagram.	a. 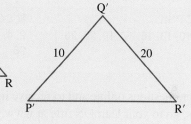
2. Find two corresponding sides. P′Q′ corresponds to PQ.	$\text{Scale factor} = \dfrac{\text{image length}}{\text{object length}}$ $= \dfrac{P'Q'}{PQ}$ $= \dfrac{10}{4}$ $= 2.5$
b. 1. Apply the scale factor. P′R′ = 2.5 × PR	b. $P'R' = 2.5 \times PR$ $= 2.5 \times 6$ $= 15$
2. Write the answer.	P′R′ is 15 cm long.
c. 1. Apply the scale factor. Q′R′ = 2.5 × QR	c. $\dfrac{Q'R'}{QR} = \dfrac{20\,\text{cm}}{x\,\text{cm}} = 2.5$ $Q'R' = 2.5 \times QR$ $20 = 2.5 \times QR$ $QR = \dfrac{20}{2.5}$ $= 8$
2. Write the answer.	QR is 8 cm long.

Exercise 4.2 Ratio and scale

learn on

Individual pathways

■ PRACTISE	■ CONSOLIDATE	■ MASTER
1, 3, 5, 10, 13, 16, 17, 19, 22, 26	2, 4, 6, 8, 11, 15, 18, 21, 23, 27, 28	7, 9, 12, 14, 20, 24, 25, 29, 30, 31

To answer questions online and to receive **immediate corrective feedback** and **fully worked solutions** for all questions, go to your learnON title at www.jacplus.com.au.

Fluency

1. **WE1** The horse track shown is 1200 m long and 35 m wide.

   a. Write the following ratios in simplest form.

      i. Track length to track width
      ii. Track width to track length

   b. Compare the distance of the length of the track with the width of the track.

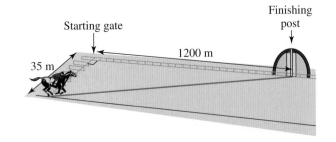

Starting gate — 1200 m — Finishing post
35 m

2. A dingo perched on top of a cliff spots an emu on the ground below.

   a. Write the following ratios in simplest form.

      i. Cliff height to distance between cliff base and emu
      ii. Distance between emu and cliff base to cliff height

   b. Compare the height of the cliff with the horizontal distance between the base of the cliff and the emu.

3. **WE2a** Express each of the following ratios in simplest form.

   a. $12 : 18$
   b. $8 : 56$
   c. $9 : 27$
   d. $14 : 35$

4. Express each of the following ratios in simplest form.

   a. $16 : 60$
   b. $200 : 155$
   c. $32 : 100$
   d. $800 : 264$

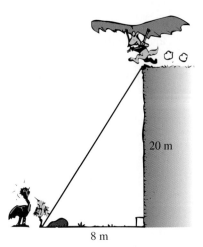

20 m

8 m

5. **WE2b** Express each of the following ratios in simplest form.
    a. $1.2 : 0.2$
    b. $3.9 : 4.5$
    c. $9.6 : 2.4$
    d. $18 : 3.6$

6. Express each of the following ratios in simplest form.
    a. $1.8 : 3.6$
    b. $4.4 : 0.66$
    c. $0.9 : 5.4$
    d. $0.35 : 0.21$

7. Express each of the following ratios in simplest form.
    a. $6 : 1.2$
    b. $12.1 : 5.5$
    c. $8.6 : 4$
    d. $0.07 : 14$

8. **WE2c** Write each of the following ratios in the simplest form.
    a. $1\frac{1}{2} : 2$
    b. $2 : 1\frac{3}{4}$
    c. $1\frac{1}{3} : 2$
    d. $1\frac{2}{5} : 1\frac{1}{4}$

9. Write each of the following ratios in the simplest form.
    a. $\frac{4}{7} : 2$
    b. $5 : 1\frac{1}{2}$
    c. $2\frac{3}{4} : 1\frac{1}{3}$
    d. $3\frac{5}{6} : 2\frac{1}{2}$

10. **WE3** Determine the value of the pronumeral in each of the following proportions.
    a. $a : 15 = 3 : 5$
    b. $b : 18 = 4 : 3$
    c. $24 : c = 3 : 4$

11. Determine the value of the pronumeral in each of the following proportions.
    a. $e : 33 = 5 : 44$
    b. $6 : f = 5 : 12$
    c. $3 : 4 = g : 5$

12. Determine the value of the pronumeral in each of the following proportions.
    a. $11 : 3 = i : 8$
    b. $7 : 20 = 3 : j$
    c. $15 : 13 = 12 : k$

For questions **13** to **15**, enlarge the figures shown by the given scale factor and the centre of enlargement marked with O. Show the image of each of these figures.

13. **WE4** Scale factor = 2

14. Scale factor = 1.5

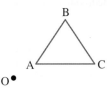

15. Scale factor = $\frac{1}{2}$

16. **WE5** A quadrilateral ABCD is enlarged to A′B′C′D′. AB = 7 cm, AD = 4 cm, A′B′ = 21 cm, B′C′ = 10.5 cm. Determine:

    a. the scale factor for enlargement
    b. A′D′
    c. BC.

Understanding

17. The estimated volume of Earth's salt water is 1 285 600 000 cubic kilometres. The estimated volume of fresh water is about 35 000 000 cubic kilometres.

    a. Determine the ratio of fresh water to salt water (in simplest form).
    b. Determine the value of x, to the nearest whole number, when the ratio found in part a. is expressed in the form $1 : x$.

18. Super strength glue comes in two tubes that contain 'Part A' and 'Part B' pastes. These pastes have to be mixed in the ratio 1 : 4 for maximum strength. Determine how many mL of Part A should be mixed with 10 mL of Part B.

19. A recipe for a tasty cake says the butter and flour needs to be combined in the ratio 2 : 7. Determine the amount of butter that should be mixed with 3.5 kg of flour.

20. The diagram shown lays out the floor plan of a house. The actual size of bedroom 1 is 8 m × 4 m.

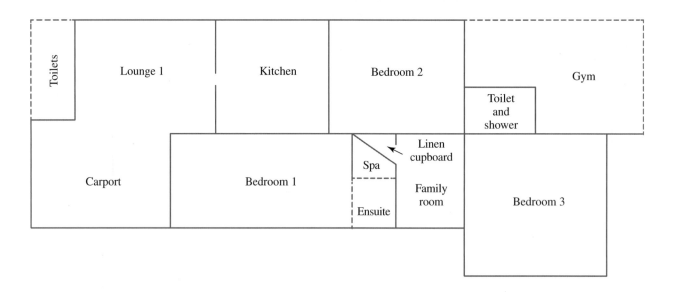

    a. If the dimensions of bedroom 1 as it appears on the ground plan are 4 cm by 2 cm, calculate the scale factor when the actual house (object) is built from the plan (image).
    b. Determine the real-life dimensions of bedroom 3 if the dimensions as shown on the ground plan are 3 cm × 3 cm.
    c. Determine the real-life dimensions of the kitchen if the dimensions as shown on the ground plan are 2.5 cm × 2.5 cm.

Reasoning

21. Pure gold is classed as 24-carat gold. This kind of gold is too soft to use for making jewellery, so it gets combined with other metals to form an alloy. The ratio of gold to other metals in 18-carat gold is 18 : 6.
The composition of 18-carat rose gold is 75% gold, 22.25% copper and 2.75% silver.

    a. Show that the mass of silver in a 2.5-gram rose gold bracelet is 0.07 g.
    b. Determine the composition of metals in a rose gold bracelet that contains 0.5 g of copper.

22. The angles of a triangle have the ratio 3 : 4 : 5. Show that the sizes of the three angles are 45°, 60° and 75°.

23. The dimensions of a rectangular box have the ratio 2 : 3 : 5. The box's volume is 21 870 cm³. Show that the dimensions of the box are 18, 27 and 45 cm.

24. Tyler, Dylan and Ari invested money in the ratio 11 : 9 : 4. If their profits are shared in the ratio 17 : 13 : 6, is this fair to each person? Explain your answer.

25. It costs the same amount to buy either 5 pens or 2 pens and 6 pencils. This is also the cost of 6 sharpeners and a pencil. Show a relationship between the cost of each kind of item.

Problem solving

26. Sharnee, a tourist at Kakadu National Park, takes a picture of a 2-metre-long crocodile beside a cliff. When they develops the pictures, they can see that on the photo the crocodile is 2.5 cm long and the cliff is 8.5 cm high.
Determine the actual height of the cliff in cm.

27. Evaluate the ratio of $y : z$ if $2x = 3y$ and $3x = 4z$.

28. The quantities P and Q are in the ratio 2 : 3. If P is reduced by 1, the ratio becomes $\frac{1}{2}$. Determine the values of P and Q.

29. The ratio of boys to girls among the students who signed up for a basketball competition is 4 : 3.
If 3 boys drop out of the competition and 4 girls join, there will be the same number of boys and girls.
Evaluate the number of students who have signed up for the basketball competition.

30. In a group of students who voted in a Year 9 class president election, the ratio of girls to boys is 2 : 3.
If 10 more girls and 5 more boys had voted, the ratio would have been 3 : 4.
Evaluate the number of students who voted altogether.

31. The ratio of the base radii of two cylinders is 2 : 1. The the ratio of their heights is 3 : 1. Determine the ratio of the volumes of these two cylinders.

4.3 Congruent figures

▶ 4.3.1 Congruent figures

eles-4729

- **Congruent figures** are identical figures. They have exactly the same shape and size.
- Congruent figures can be superimposed exactly on top of each other using reflection, rotation or translation (or a combination of some or all of these actions).

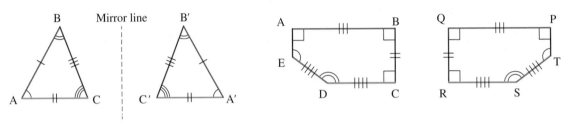

- The symbol for congruence is ≅. When reading this symbol out loud we say, 'is congruent to'.
- For the diagrams shown, ABC ≅ A′B′C′ and ABCDE ≅ PQRST.
- When writing congruence statements, the vertices of each figure are named in corresponding order.

WORKED EXAMPLE 6 Identifying congruent shapes

Identify a pair of congruent shapes from the following set.

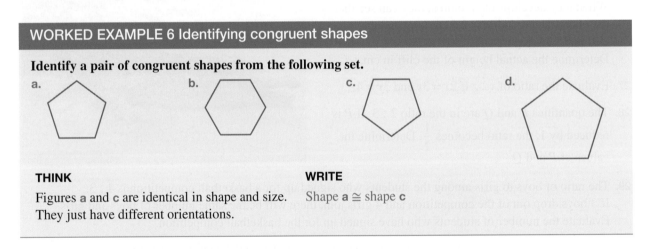

a.　　　　　　b.　　　　　　c.　　　　　　d.

THINK

Figures **a** and **c** are identical in shape and size. They just have different orientations.

WRITE

Shape **a** ≅ shape **c**

▶ 4.3.2 Testing triangles for congruence

eles-4730

- To determine whether two triangles are congruent, it is not necessary to know that all three sides and all three angles of a each triangle are equal to the corresponding sides and angles of the other triangle.
- Certain minimum conditions can guarantee that two triangles are congruent.

Side-side-side condition for congruence (SSS)

- If all of the corresponding sides of two triangles are equal, the angles opposite those corresponding sides will also be equal. This means that the two triangles are congruent.
- This is known as the *side-side-side* (*SSS*) condition for congruence.

Side-angle-side condition for congruence (SAS)

- If two triangles have two corresponding sides that are equal, and the angles between those corresponding sides are equal, then the two triangles are congruent.
- This is known as the *side–angle–side* (*SAS*) condition for congruence.

Angle-side-angle condition for congruence (ASA)

- If two triangles have two corresponding angles that are equal, and 1 side that is also equal, then the two triangles are congruent. (*Note:* The third corresponding angle will also be equal.)
- This is known as the *angle–side–angle* (*ASA*) condition for congruence.

Right angle-hypotenuse-side condition for congruence (RHS)

- If the hypotenuse and one other side of two right-angled triangles are equal, then the two triangles are congruent.
- This is known as the *right angle–hypotenuse–side* (*RHS*) condition for congruence.

Summary of congruence tests

Congruence test	Description	Abbreviation
	• All corresponding sides are equal in length.	SSS (side–side–side)
	• Two corresponding sides are equal in length • The corresponding angles between them are equal in size.	SAS (side–angle–side)
	• Two corresponding angles are equal in size • One corresponding side is equal in length.	ASA (angle–side–angle)
	• The hypotenuse and one other corresponding side of two right-angled triangles are equal in length.	RHS (right angle–hypotenuse–side)

Identify which of the following triangles are congruent. Give reasons for your answer.

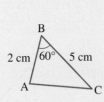
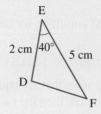
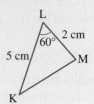

THINK

In all three triangles two given sides are of equal length (2 cm and 5 cm). The included angles in triangles ABC and KLM are also equal (60°). B corresponds to L, and A corresponds to M.

WRITE

$\triangle ABC \cong \triangle MLK$ (SAS)

Given that $\triangle ABD \cong \triangle CBD$, determine the values of the pronumerals in the figure shown.

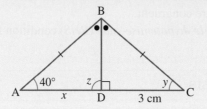

THINK

1. Congruent triangles have corresponding sides that are equal in length. Side AD (marked x) corresponds to side CD.

2. Since these triangles are congruent, the corresponding angles are equal.

WRITE

$\triangle ABD \cong \triangle CBD$
$AD = CD$
$x = 3 \text{ cm}$

$\angle A = \angle C$
$y = 40°$
$\angle BDA = \angle BDC$
$z = 90°$

Prove that $\triangle PQS$ is congruent to $\triangle RQS$.

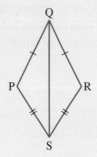

THINK	WRITE
1. Study the diagram and state which sides and/or angles are equal.	QP = QR (given) PS = RS (given) QS is common.
2. This fits the SSS condition, which proves congruence.	$\triangle$PQS $\cong$ $\triangle$RQS (SSS)

COLLABORATIVE TASK

On a piece of paper, draw the trapezium shown as per the dimensions given in the diagram. Working in pairs or small teams, try to divide the trapezium into four congruent trapeziums that are similar in shape to the original trapezium.

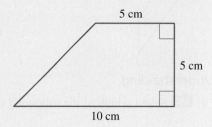

5 cm

5 cm

10 cm

On Resources

eWorkbook	Topic 4 Workbook (worksheets, code puzzle and project) (ewbk-2004)
Digital documents	SkillSHEET Naming angles (doc-6195)
	SkillSHEET Complementary and supplementary angles (doc-6196)
	SkillSHEET Angles in a triangle (doc-6197)
Interactivities	Individual pathway interactivity: Congruent figures (int-4495)
	Congruency tests (int-3755)
	Congruent figures (int-3749)
	Congruent triangles (int-3754)

Exercise 4.3 Congruent figures

learn on

Individual pathways

■ PRACTISE	■ CONSOLIDATE	■ MASTER
1, 4, 5, 8, 11, 12, 16, 17, 22	2, 6, 9, 13, 14, 18, 19, 23	3, 7, 10, 15, 20, 21, 24, 25

To answer questions online and to receive **immediate corrective feedback** and **fully worked solutions** for all questions, go to your learnON title at www.jacplus.com.au.

Fluency

1. **WE6** Select a pair of congruent shapes from the figures shown.

a.

b.

c.

d.

2. Select a pair of congruent shapes from the figures shown.

a.
2 cm
5 cm

b.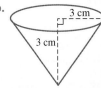
3 cm
3 cm

c.
2 cm
5 cm

d.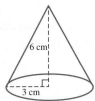
6 cm
3 cm

3. Select a pair of congruent shapes from the figures shown.

a.

b.

c.

d.

Understanding

4. **MC** Select which of the following is congruent to the triangle shown.

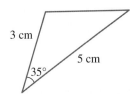
3 cm
5 cm
35°

A.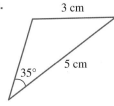
3 cm
5 cm
35°

B.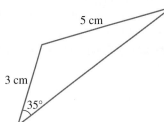
5 cm
3 cm
35°

C.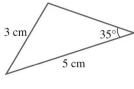
3 cm
35°
5 cm

D.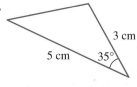
3 cm
5 cm
35°

E.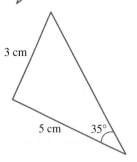
3 cm
5 cm
35°

5. **WE7** Identify which of the following triangles are congruent. Give a reason for your answer.

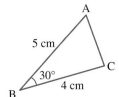
A
5 cm
30°
C
B
4 cm

N
5 cm
4 cm
L
30°
M

P
5 cm
R
30°
4 cm
Q
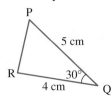

6. Identify which of the following triangles are congruent. Give a reason for your answer.

7. Identify which of the triangles are congruent. Give a reason for your answer.

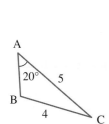 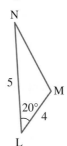 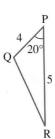

8. Identify which of the following triangles are congruent. Give a reason for your answer.

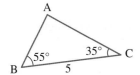 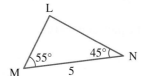 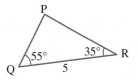

9. Identify which of the following triangles are congruent. Give a reason for your answer.

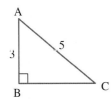 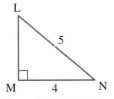 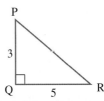

10. Identify which of the following triangles are congruent. Give a reason for your answer.

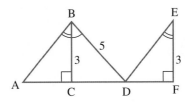

11. **WE8** Determine the value of the pronumeral in the following pair of congruent triangles. All side lengths are in centimetres.

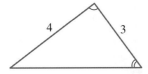 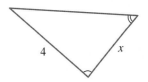

12. Determine the value of the pronumeral in the pair of congruent triangles shown. All side lengths are in centimetres.

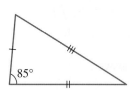 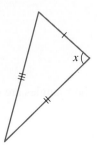

13. Determine the value of the pronumerals in the pair of congruent triangles shown. All side lengths are in centimetres.

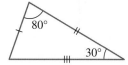 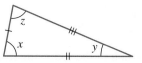

14. Determine the value of the pronumerals in the pair of congruent triangles shown. All side lengths are in centimetres.

a.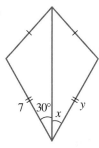

b.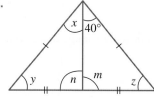

15. Calculate the length of the side marked with the pronumeral using the following congruent triangles.

a.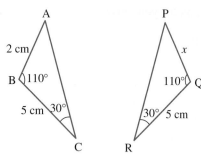

b.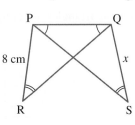

c.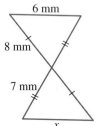

d.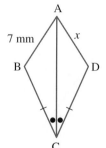

Reasoning

16. Do congruent figures have the same area? Explain your answer.

17. If two congruent triangles have a right angle, is the reason for their congruence always due to the RHS condition of congruence? Justify your answer.

18. Give an example to show that triangles with two angles of equal size and a pair of non-corresponding sides of equal length may not be congruent.

19. **WE9** For each of the following, prove that the statement accompanying the diagram is true.

a.
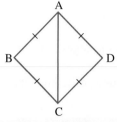
$\triangle ABC \cong \triangle ADC$

b.
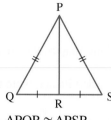
$\triangle PQR \cong \triangle PSR$

c.
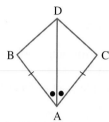
$\triangle DBA \cong \triangle DCA$

20. ABCD is a trapezium with both AD and BC perpendicular to AB. If a right-angled triangle DEC is constructed with an angle $\angle ECD$ equal to 45°, prove that $\triangle EDA \cong \triangle ECB$.

21. A teacher asked their class to each draw a triangle with side lengths of 5 cm and 4 cm, and an angle of 45° that is not formed at the point joining the 5 cm and 4 cm side. Explain why the triangles drawn by every member of the class would be congruent.

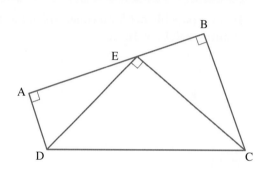

Problem solving

22. Construct five congruent triangles from nine matchsticks.

23. Construct seven congruent triangles from nine matchsticks.

24. Demonstrate how the figure shown can be cut into four congruent pieces.

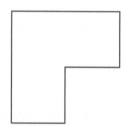

25. Determine the ratio of the outer (unshaded) area to the inner (shaded) area of the six-pointed star shown.

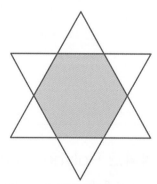

4.4 Nets, polyhedra and Euler's formula

4.4.1 Nets

eles-4731

- The faces forming a solid can be drawn as plane shapes, which are joined across the edges to form the solid.
- The complete set of faces forming a solid is called its net. Note that for some figures, different nets can be drawn.

WORKED EXAMPLE 10 Drawing a possible net for a cone

Draw a possible net for a cone, which has a radius of 7 cm and a slanting height of 10 cm.

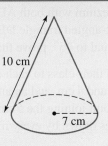

THINK	DRAW
1. The base of a cone is a circle of radius 7 cm.	
2. The other (slanted) part of the cone when split open will form a sector of a circle of radius 10 cm.	
3. If the two parts (from steps 1 and 2) are put together, the complete net of a cone is obtained.	

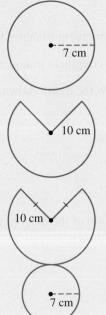

4.4.2 Polyhedra construction

eles-4839

- A 3-dimensional solid where each of the faces is a polygon is called a **polyhedron**. If all the faces are congruent, the solid is called a regular polyhedron or a Platonic solid.

- Platonic solids are shown below.
- A **cube**, with 6 faces, each of which is a square.

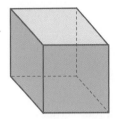

- A **tetrahedron**, with 4 faces, each of which is an equilateral triangle.

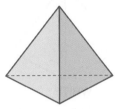

- An **octahedron**, with 8 faces, each of which is an equilateral triangle.

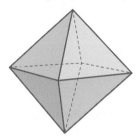

- A **dodecahedron**, with 12 faces, each of which is a regular pentagon.

- An **icosahedron**, with 20 faces, each of which is an equilateral triangle.

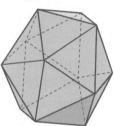

- Construction of polyhedra is conveniently done from nets.

WORKED EXAMPLE 11 Drawing a net of an octahedron

Draw a net of an octahedron.

THINK

An octahedron is a polyhedron with 8 faces, each of which is an equilateral triangle. So its net will consist of 8 equilateral triangles. Draw a possible net of an octahedron.

DRAW

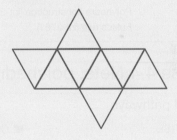

4.4.3 Euler's formula

eles-4840

- **Euler's formula** shows the relationship between the number of **edges**, the number of **faces** and the number of vertices in any polyhedron. Note that a **vertex** (the singular of vertices) is a point or a corner of a shape where the straight edges meet.

Euler's formula

Euler's formula states that for any polyhedron:

$$\textbf{number of faces} \, (F) + \textbf{number of vertices} \, (V) - 2 = \textbf{number of edges} \, (E)$$
$$F + V - 2 = E$$

Verify Euler's formula for a tetrahedron.

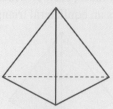

THINK	WRITE
1. Look at the shape of a tetrahedron and state the number of faces, vertices and edges.	In a tetrahedron there are four faces, four vertices and six edges.
2. Write Euler's formula.	number of faces (F) + number of vertices (V) − 2 = number of edges (E)
3. To verify Euler's formula, show that the left-hand side (LHS) is equal to right-hand side (RHS). Substitute the values of F, V and E in the formula.	$F = 4$, $V = 4$ and $E = 6$ $\begin{aligned} \text{LHS} &= F + V - 2 \\ &= 4 + 4 - 2 \\ &= 6 \end{aligned}$ $\text{RHS} = 6$ $\text{LHS} = \text{RHS}$
4. Write the conclusion.	Euler's formula holds for a tetrahedron.

---

 on Resources

 eWorkbook   Topic 4 Workbook (worksheets, code puzzle and project) (ewbk-2004)

Interactivities Individual pathway interactivity: Nets, polyhedra and Euler's rule (int-4428)

Nets (int-3759)

Polyhedra construction (int-3760)

Euler's rule (int-3761)

Exercise 4.4 Nets, polyhedra and Euler's formula

learn on

Individual pathways

■ PRACTISE	■ CONSOLIDATE	■ MASTER
1, 3, 7, 11, 14	2, 5, 8, 10, 12, 15	4, 6, 9, 13, 16

To answer questions online and to receive **immediate corrective feedback** and **fully worked solutions** for all questions, go to your learnON title at www.jacplus.com.au.

Fluency

1. **WE10** Draw a possible net for the cylinder shown.

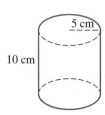

2. Draw a possible net for each of the following solids.

a.
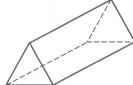

b.
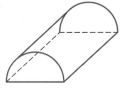

c.

3. **WE11** Draw 2 different nets to form a cube.

4. Draw a net for the octahedron that is different from the one shown. Cut out your net and fold it to form the octahedron.

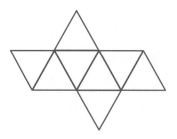

5. Using nets generated from geometry software or elsewhere, construct:

a. a cube
b. a tetrahedron
c. a dodecahedron.

6. Construct the pyramids shown in the figures.

a.
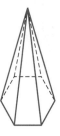

b.
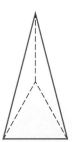

Understanding

7. Draw some 3-dimensional shapes of your choice. State how many of these shapes are polyhedra, and name them.

8. **WE12** Verify Euler's formula for the following Platonic solids.

a. A cube
b. An octahedron

9. Show that Euler's formula holds true for these solids.

a.
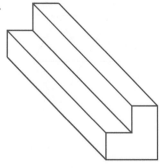

b.
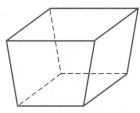

c.

10. Make the prisms shown. Verify Euler's formula for each shape.

a.

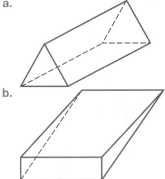

b.

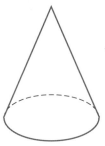

Reasoning

11. Discuss why a knowledge of 3D shapes is important.

12. Explain whether you can verify Euler's formula for the following shapes. Give reasons.

a.    b. 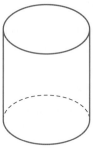   c.

13. A spherical scoop of ice-cream sits on a cone. Draw this shape. Explain whether you can verify Euler's formula for this shape.

Problem solving

14. Renee knows that a polyhedron has 12 faces and 8 vertices. Show how she can determine the number of edges.

15. Determine which of the following compound shapes are nets that can be folded into a 3D solid. You may wish to enlarge them onto paper, cut them out and fold them.

a. 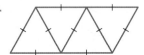   b. 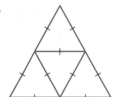   c.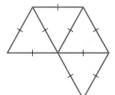

16. The net forms an icosahedron as shown. Verify Euler's formula for this shape, showing your working.

 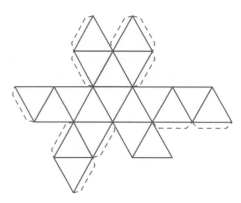

4.5 Similar figures

LEARNING INTENTION

At the end of this subtopic you should be able to:
- identify similar figures
- determine the scale factor between similar figures
- show that two triangles are similar using the appropriate similarity test.

▶ 4.5.1 Similar figures and similarity condition

eles-4732

- **Similar figures** are identical in shape, but different in size.
- The corresponding angles in similar figures are equal in size and the corresponding sides are in the same ratio.
- The symbol used to denote similarity is ~. When reading this symbol out loud we say, 'is similar to'.
- In the triangles shown ΔABC is similar to ΔUVW. That is, ΔABC~ΔUVW.
- The ratio of side lengths is known as the scale factor.

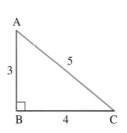 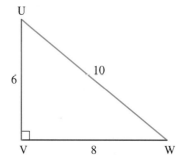

Scale factor

$$\text{scale factor} = \frac{\text{length of image}}{\text{length of object}}$$

- If the scale factor is > 1, an enlargement has occurred.
- If the scale factor is < 1, a reduction has occurred.

- The scale factor for the triangles shown is 2. Each side in UVW is twice the length of the corresponding side in ABC.
- Enlargements and reductions are transformations that create similar figures.
- The method for creating enlarged figures that is explained in subtopic 4.2 can also be used to create similar figures.

Testing triangles for similarity

- As with congruent triangles, to determine whether two triangles are similar, it is not necessary to know that all pairs of corresponding sides are in the same ratio and that all corresponding angles are equal.
- Certain minimum conditions can guarantee that two triangles are similar.

Angle–angle–angle condition of similarity (AAA)

- If all of the angles of two triangles are equal, then the triangles are similar.
- This is known as the angle–angle–angle (AAA) condition for similarity.
- In the diagram shown, $\triangle ABC \sim \triangle RST$ (AAA).

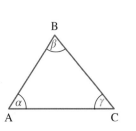 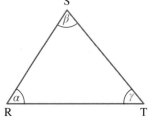

Side–side–side condition for similarity (SSS)

- If two triangles have a constant ratio for all corresponding side lengths, then the two triangles are similar.
- This is known as the side–side–side (SSS) condition for similarity.
- In the diagram shown, the ratios of all corresponding side lengths are equal $\left(\dfrac{9}{6} = \dfrac{15}{10} = \dfrac{10.5}{7} = 1.5 \right)$, therefore $\triangle ABC \sim \triangle RST$ (SSS).

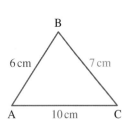 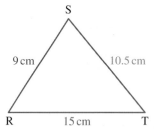

Side–angle–side condition for similarity (SAS)

- If two triangles have two corresponding sides in the same ratio and the included angles of those sides are equal, then the two triangles are similar.
- This is known as the side–angle–side (SAS) condition for similarity.
- In the diagram shown, the ratio of the triangles' two corresponding side lengths are equal $\left(\dfrac{9}{6} = \dfrac{15}{10} = 1.5 \right)$ and the included angles are also the same, therefore $\triangle ABC \sim \triangle RST$ (SAS).

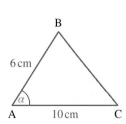 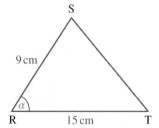

Right angle–hypotenuse–side condition for similarity (RHS)

- If the hypotenuse and one other corresponding side of two right-angled triangles are in the same ratio, then the two triangles are similar.
- This is known as the the right angle–hypotenuse–side (RHS) condition for similarity.
- In the diagram shown, the ratio of the hypotenuses and one other pair of corresponding sides are equal $\left(\frac{12}{6} = \frac{10}{5} = 2 \right)$, therefore $\triangle ABC \sim \triangle RST$ (RHS).

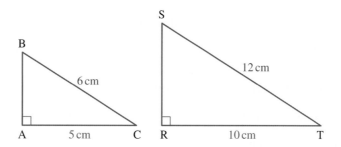

Summary of similarity tests

- Triangles can be checked for similarity using one of the tests described in the table shown.

Test description	Abbreviation
All corresponding angles are equal in size	AAA (angle–angle–angle)
All corresponding sides are in the same ratio	SSS (side–side–side)
Two corresponding sides are in the same ratio and the included angles are equal in size	SAS (side–angle–side)
The hypotenuse and one other corresponding side of two right-angled triangles are in the same ratio	RHS (right angle–hypotenuse–side)

Note: When using the AAA test, it is sufficient to show that two corresponding angles are equal. Since the sum of the interior angles in any triangle is 180°, the third corresponding angle will automatically be equal.

WORKED EXAMPLE 13 Identifying similar triangles

Identify a pair of similar triangles from the triangles shown. Give a reason for your answer.

a.

b.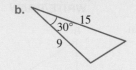

c.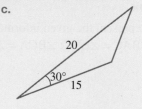

THINK	WRITE
1. In each triangle we know the size of two sides and the included angle, so the SAS test can be applied. Since all included angles are equal (30°), we need to determine the ratios of the corresponding sides, looking at two triangles at a time.	For triangles **a** and **b**: $$\frac{15}{10} = \frac{9}{6} = 1.5$$ For triangles **a** and **c**: $$\frac{20}{10} = 2, \frac{15}{6} = 2.5$$
2. Write the answer.	Triangle **a** ~ triangle **b** (SAS)

WORKED EXAMPLE 14 Proving that two triangles are similar

Prove that ΔABC is similar to ΔEDC.

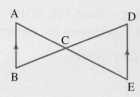

THINK

1. AB is parallel to DE. ∠ABC and ∠EDC are alternate angles.

2. ∠BAC and ∠DEC are alternate angles.

3. The third pair of angles must be equal.

4. This proves that the triangles are similar.

WRITE

∠ABC = ∠EDC (alternate angles)

∠BAC = ∠DEC (alternate angles)

∠BCA = ∠DCE (vertically opposite angles)

ΔABC ~ ΔEDC (AAA)

- The ratio of the corresponding sides in similar figures can be used to calculate missing side lengths or angles in those figures.

WORKED EXAMPLE 15 Solving worded problems using similar triangles

A pole 1.5 metres high casts a shadow 3 metres long, as shown. Calculate the height of a building that casts a shadow 15 metres long at the same time of the day.

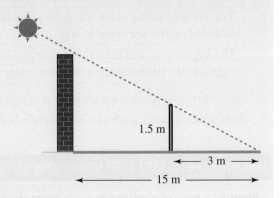

THINK

1. Represent the given information on a diagram.
∠BAC = ∠EDC; ∠BCA = ∠ECD

WRITE/DRAW

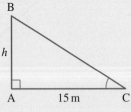

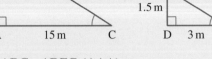

2. Triangles ABC and DEC are similar. This means the ratios of corresponding sides are the same. Write the ratios.

$$\Delta ABC \sim \Delta DEC \ (AAA)$$
$$\frac{h}{1.5} = \frac{15}{3}$$

3. Solve the equation for h.

$$h = \frac{15 \times 1.5}{3}$$
$$= 7.5$$

4. Write the answer in words, including units.

The building is 7.5 metres high.

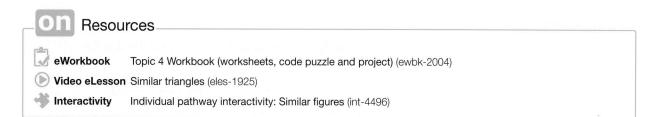

Resources

eWorkbook   Topic 4 Workbook (worksheets, code puzzle and project) (ewbk-2004)

Video eLesson Similar triangles (eles-1925)

Interactivity   Individual pathway interactivity: Similar figures (int-4496)

Exercise 4.5 Similar figures

learn on

Individual pathways

■ PRACTISE	■ CONSOLIDATE	■ MASTER
1, 2, 6, 7, 11, 14, 15, 19	3, 4, 8, 12, 16, 17, 20, 21	5, 9, 10, 13, 18, 22, 23, 24

To answer questions online and to receive **immediate corrective feedback** and **fully worked solutions** for all questions, go to your learnON title at www.jacplus.com.au.

Fluency

WE13 For questions 1 to 5, Identify the pair of similar triangles among those shown. Give reasons for your answers.

1. a.   b.   c.

2. a. 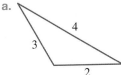   b. 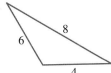   c.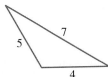

3. a. 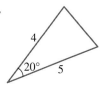   b.   c.

4. a. 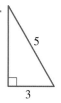   b.   c.

5. a.   b.   c.

Understanding

6. Name two similar triangles in each of the following figures, ensuring that vertices are listed in the correct order.

a.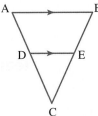

b.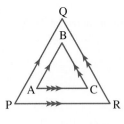

c.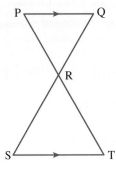

d.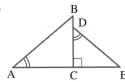

e.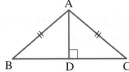

7. In the diagram shown, C is the centre of the circle. Complete this statement: △ABC is similar to …

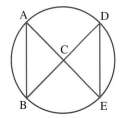

8. For the diagram shown:

a. complete the statement: $\dfrac{AB}{AD} = \dfrac{BC}{\Box} = \dfrac{\Box}{AE}$

b. determine the value of the pronumerals.

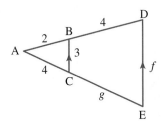

9. Using the diagram shown:

a. determine the values of h and i

b. determine the values of j and k.

10. Determine the value of the pronumeral in the diagram shown.

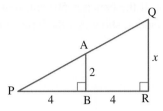

11. If the two triangles shown are similar, determine the values of the pronumerals x and y.

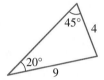

12. Determine the values of the pronumerals x and y in the diagram shown.

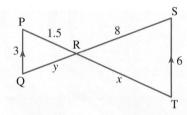

13. Determine the value of each pronumeral in the following triangles. Show how you arrived at your answers.

a.

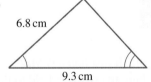

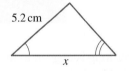

b.

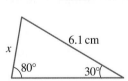

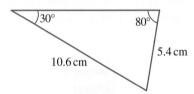

c.

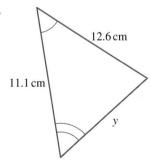

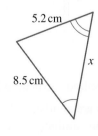

Reasoning

14. **WE15** A ladder just touches a bench and also leans on a wall that is 4 metres high, as shown in the diagram. If the bench is 50 centimetres high and 1 metre from the base of the ladder, show that the base of the ladder is 8 metres from the wall.

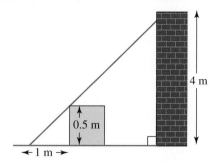

15. **WE14** Prove that ΔABC is similar to ΔEDC in each of the following.

a.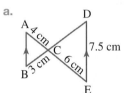

b.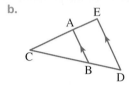

c.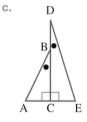

d.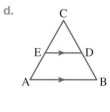

16. Natalie is 1.5 metres tall. They cast a shadow 2 metres long at a certain time of the day. If Alex is 1.8 metres tall, show that their shadow would be 2.4 m long at the same time of day.

17. A string 50 metres long is pegged to the ground and tied to the top of a flagpole. It just touches Maz on the top of her head. If Maz is 1.5 metres tall and 5 metres away from the point where the string is held to the ground, show that the height, h, of the flagpole is 14.37 m.

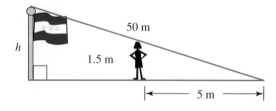

18. Using diagrams or other methods, explain whether the following statements are True or False.

a. All equilateral triangles are similar.
b. All isosceles triangles are similar.
c. All right-angled triangles are similar.
d. All right-angled isosceles triangles are similar.

Problem solving

19. Paw and Thuy play tennis at night under floodlights. When Paw stands 2.5 m from the base of the floodlight, her shadow is 60 cm long.

    a. If Paw is 1.3 m tall, evaluate the height of the floodlight in metres, correct to 2 decimal places.
    b. If Thuy, who is 1.6 m tall, stands in the same place, calculate her shadow length in cm.

20. To determine the height of a flagpole, Jenna and Mia decide to measure the shadow cast by the flagpole. They place a 1 m ruler at a distance of 3 m from the base of the flagpole and measure the shadows that both the ruler and flagpole cast. Both shadows finish at the same point. After measuring the shadow of the flagpole, Jenna and Mia calculate that the height of the flagpole is 5 m.
    Determine the length of the shadow cast by the flagpole, as measured by Jenna and Mia. Give your answer in metres.

21. Use the diagram shown to determine the value of a if $XZ = 8$ cm, $X'Z' = 12$ cm, $X'X = a$ cm and $XY = (a + 1)$ cm.

    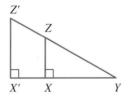

22. AB and CD are parallel lines in the figure shown.

    a. State the similar triangles.
    b. Determine the values of x and y.

    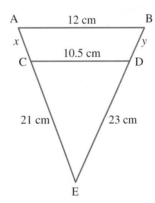

23. PQ is the diameter of the circle shown. The circle's centre is located at S. R is any point on the circumference. T is the midpoint of PR.

    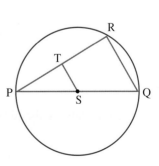

    a. Write down everything you know about this figure.
    b. Explain why ΔPTS is similar to ΔPRQ.
    c. Determine the length of TS if RQ is 8 cm.
    d. Determine the length of every other side, given that PT is 3 cm and the angle PRQ is a right angle.

24. For the diagram shown, show that, if the base of the triangle is raised to half of the height of the triangle, the length of the base of the newly formed triangle will be half of its original length.

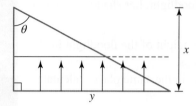

4.6 Area and volume of similar figures

LEARNING INTENTION

At the end of this subtopic you should be able to:
- convert between different units of length, area and volume
- determine the area of similar figures, given a scale factor of n^2
- determine the volume of similar figures, given a scale factor of n^3.

⏵ 4.6.1 Converting between units of length, area and volume

eles-4733

Units of length

- Metric units of length include millimetres (mm), centimetres (cm), metres (m) and kilometres (km).
- Length units can be converted using the chart shown.

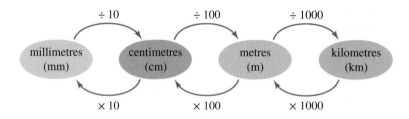

Units of area

- **Area** is measured in square units, such as square millimetres (mm^2), square centimetres (cm^2), square metres (m^2) and square kilometres (km^2).
- Area units can be converted using the chart shown.

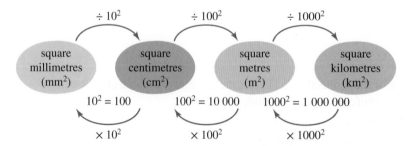

- Area units are the squares of the corresponding length units.

Units of volume

- **Volume** is measured in cubic units, such as cubic millimetres (mm^3), cubic centimetres (cm^3) and cubic metres (m^3).
- Volume units can be converted using the chart shown.

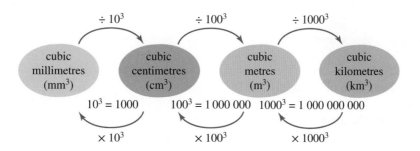

- Volume units are the cubes of the corresponding length units.

WORKED EXAMPLE 16 Converting between units of measurements

a. **Convert 9 m into mm.**
b. **Convert 150 cm² into mm².**
c. **A cube has a side length of 8 cm. Calculate the volume of the cube in m³.**

THINK	WRITE
a. 1. To convert m to mm:	a. $1\,\text{m} = 1000\,\text{mm}$

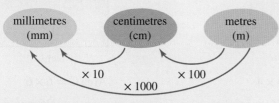

THINK	WRITE
2. To convert 9 m to mm, multiply 9 with 1000.	$9\,\text{m} = 9 \times 1000\,\text{mm}$
3. Simplify and write the answer.	$9000\,\text{mm}$
b. 1. To convert cm² into mm²:	b. $1\,\text{cm}^2 = \text{cm}^2\,\text{mm}^2$

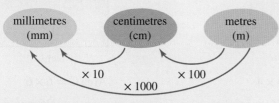

THINK	WRITE
2. To convert 150 cm² into mm², multiply 150 by 10^2 (100).	$150\,\text{cm}^2 = 150 \times 100\,\text{mm}^2$
3. Simplify and write the answer.	$15\,000\,\text{mm}^2$
c. 1. Write the formula for the volume of a cube.	c. Volume of a cube $(V) = l \times l \times l$, where l is the side length.
2. Substitute the value of the side length (l) in the volume formula.	$V = 8 \times 8 \times 8$ $= 512\,\text{cm}^3$

3. The question states that the answer should be given in m^3. $1\,cm^3 = \dfrac{1}{100^3}\,m^3$
   To convert cm^3 into m^3:

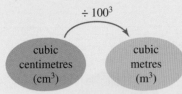

$\div 100^3$

4. To convert $512\,cm^3$ into m^3, multiply 512 by $\dfrac{1}{100^3}\,m^3$. $\qquad 512\,cm^3 = \dfrac{512}{100^3}\,m^3$

5. Simplify and write the answer. $\qquad\qquad\qquad\qquad\qquad\qquad\quad 0.000\,512\,m^3$

▶ 4.6.2 Area and volume of similar figures

eles-4734

Area and surface area of similar figures

- If the side lengths of any figure are increased by a scale factor of n, then the area of similar figures increases by a scale factor of n^2.
  For example, consider the squares shown.

A

2 cm

B

4 cm

C

6 cm

$$\text{Area A} = 2 \times 2$$
$$= 4\,cm^2$$

$$\text{Area B} = 4 \times 4$$
$$= 16\,cm^2$$

$$\text{Area C} = 6 \times 6$$
$$= 36\,cm^2$$

- The scale factors for the side lengths and the scale factors for the areas are calculated in the table shown.

Squares	Scale factor for side length	Scale factor for area
A and B	$\dfrac{4}{2} = 2$	$\dfrac{16}{4} = 4 = 2^2$
A and C	$\dfrac{6}{2} = 3$	$\dfrac{36}{4} = 9 = 3^2$
B and C	$\dfrac{6}{4} = \dfrac{3}{2}$	$\dfrac{36}{16} = \dfrac{9}{4} = \left(\dfrac{3}{2}\right)^2$

- The surface area of a 3D object also increases by the square of the length scale factor. For example, consider the cubes shown.

A

2 cm

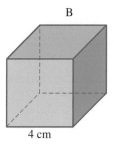

B

4 cm

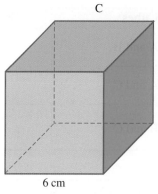

C

6 cm

Surface area A $= 6 \times 4$
$$= 24 \, \text{cm}^2$$

Surface area B $= 6 \times 16$
$$= 96 \, \text{cm}^2$$

Surface area C $= 6 \times 36$
$$= 216 \, \text{cm}^2$$

- The scale factors for the side lengths and the scale factors for the surface areas are calculated in the table shown.

Cubes	Scale factor for side length	Scale factor for surface area
A and B	$\dfrac{4}{2} = 2$	$\dfrac{96}{24} = 4 = 2^2$
A and C	$\dfrac{6}{2} = 3$	$\dfrac{216}{24} = 9 = 3^2$
B and C	$\dfrac{6}{4} = \dfrac{3}{2}$	$\dfrac{216}{96} = \dfrac{9}{4} = \left(\dfrac{3}{2}\right)^2$

Areas of similar figures

When side lengths are increased by a factor of n, the area increases by a factor of n^2.

Volume of similar figures

- If the side lengths of any solid are increased by a scale factor of n, then the volume of similar solids increases by a scale factor of n^3.
For example, consider the cubes shown.

A

2 cm

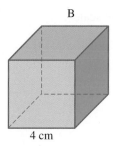

B

4 cm

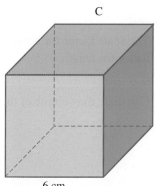

C

6 cm

Volume A $= 2 \times 2 \times 2$
$$= 8 \, \text{cm}^3$$

Volume B $= 4 \times 4 \times 4$
$$= 64 \, \text{cm}^3$$

Volume C $= 6 \times 6 \times 6$
$$= 216 \, \text{cm}^3$$

- The scale factors for the side lengths and the scale factors for the volumes are calculated in the table shown.

Cubes	Scale factor for side length	Scale factor for volume
A and B	$\dfrac{4}{2} = 2$	$\dfrac{64}{8} = 8 = 2^3$
A and C	$\dfrac{6}{2} = 3$	$\dfrac{216}{8} = 27 = 3^3$
B and C	$\dfrac{6}{4} = \dfrac{3}{2}$	$\dfrac{216}{64} = \dfrac{27}{8} = \left(\dfrac{3}{2}\right)^3$

Volumes of similar figures

When side lengths are increased by a factor of n, the volume increases by a factor of n^3.

WORKED EXAMPLE 17 Calculating area and volume of similar figures

The side lengths of a box have been increased by a factor of 3.
  a. Calculate the surface area of the new box if the original surface area was $94\,\text{cm}^2$.
  b. Determine the volume of the new box if the original volume was $60\,\text{cm}^3$.

THINK	WRITE
a. 1. State the scale factor for side length used to increase the size of the original box.	a. Scale factor for side length $= 3$
2. The scale factor for surface area is the square of the scale factor for length.	Scale factor for surface area $= 3^2$ $= 9$
3. Calculate the surface area of the new box.	Surface area of new box $= 94 \times 9$ $= 846\,\text{cm}^2$
b. 1. The scale factor for volume is the cube of the scale factor for length.	b. Scale factor for volume $= 3^3$ $= 27$
2. Calculate the volume of the new box.	Volume of new box $= 60 \times 27$ $= 1620\,\text{cm}^3$

A small cone has a radius of 5 cm and a height of 15 cm.
a. If the scale factor is 4, determine the dimensions of a larger similar cone.
b. Giving your answers both in exact form and correct to 3 decimal places, calculate the volume of:
    i. the smaller cone   ii. the larger cone.
c. Show that the volumes of the two cones are in the ratio 1 : 64.

THINK	WRITE
a. The scale factor is 4, so multiply the radius and height by 4. Write the answers.	For the larger cone: radius $= 5\,\text{cm} \times 4 = 20\,\text{cm}$ height $= 15\,\text{cm} \times 4 = 60\,\text{cm}$
b. i. Use your calculator to work out the volume of the smaller cone by substituting the following into the formula: $r = 5$ $h = 15$.	$\begin{aligned} V_s &= \frac{1}{3}\pi r^2 h \\ &= \frac{1}{3}\pi \times 5^2 \times 15 \\ &= 125\pi \ \text{cm}^3 \\ &\approx 392.699 \ \text{cm}^3 \ (\text{to 3 d.p}) \end{aligned}$
ii. Use your calculator to work out the volume of the larger cone (V_L) by substituting the following into the formula: $r = 20$ $h = 60$.	$\begin{aligned} V_L &= \frac{1}{3}\pi r^2 h \\ &= \frac{1}{3}\pi \times 20^2 \times 60 \\ &= 8000\pi \\ &\approx 25\,132.741 \ \text{cm}^3 \ (\text{to 3 d.p}) \end{aligned}$
c. Use your calculator to evaluate the ratio of the volumes of the smaller cone and the larger cone. $\dfrac{V_s}{V_L} = \dfrac{125\pi}{8000\pi}$	$\dfrac{V_s}{V_L} = \dfrac{1}{64}$ The volumes are in the ratio 1 : 64.

TI | THINK	DISPLAY/WRITE	CASIO | THINK	DISPLAY/WRITE
a-b. In a new problem, on a Calculator page, complete the entry line as: $\frac{1}{3} \times \pi \times 5^2 \times 15$ Then press ENTER. $\frac{1}{3} \times \pi \times 20^2 \times 60$ Then press ENTER. To convert the answer to decimal: • Menu • 2: Number • 1: Convert to Decimal Then press ENTER.	**a-b.** 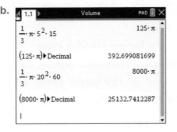 $V_s = 125\pi \text{ cm}^3$ $V_s \approx 392.699 \text{ cm}^3$ (to 3 d.p.) $V_L = 8000\pi \text{ cm}^3$ $V_L \approx 25\,132.741 \text{ cm}^3$ (to 3 d.p.)	**a-b.** On the Main screen, complete the entry line as: $\frac{1}{3} \times \pi \times 5^2 \times 15$ Then press EXE. $\frac{1}{3} \times \pi \times 20^2 \times 60$ Then press EXE. *Note*: Change Standard to Decimal	**a-b.**  $V_s = 125\pi \text{ cm}^3$ $V_s \approx 392.699 \text{ cm}^3$ (to 3 d.p.) $V_L = 8000\pi \text{ cm}^3$ $V_L \approx 25\,132.741 \text{ cm}^3$ (to 3 d.p.)

on Resources

 eWorkbook   Topic 4 Workbook (worksheets, code puzzle and project) (ewbk-2004)

 Interactivities Individual pathway interactivity: Area and volume of similar figures (int-4497)

Units of length (int-3779)

Area of similar figures (int-6043)

Volume and surface area of similar figures (int-6044)

Exercise 4.6 Area and volume of similar figures

learn on

Individual pathways

■ PRACTISE	■ CONSOLIDATE	■ MASTER
1, 3, 4, 8, 9, 13	2, 5, 7, 10, 14, 15	6, 11, 12, 16, 17, 18

To answer questions online and to receive **immediate corrective feedback** and **fully worked solutions** for all questions, go to your learnON title at www.jacplus.com.au.

Fluency

1. **WE17a** The side lengths of the following shapes have all been increased by a factor of **3**. Copy and complete the table shown.

Original surface area	Enlarged surface area
100 cm^2	a.
7.5 cm^2	b.
95 mm^2	c.
d.	918 cm^2
e.	45 m^2
f.	225 mm^2

2. A rectangular box has a surface area of 96 cm^2 and volume of 36 cm^3. Calculate the volume and surface area of a similar box that has side lengths that are double the size of the original box.

3. **WE17b** The side lengths of the following shapes have all been increased by a factor of 3. Copy and complete the table shown.

Original volume	Enlarged volume
200 cm^3	a.
12.5 cm^3	b.
67 mm^3	c.
d.	2700 cm^3
e.	67.5 m^3
f.	27 mm^3

Understanding

4. The area of a bathroom as drawn on a house plan is 5 cm^2. Calculate the area of the actual bathroom if the map has a scale of 1 : 100.

5. The area of a kitchen is 25 m^2.
   a. **WE16** Convert 25 m^2 to cm^2.
   b. Calculate the area of the kitchen as drawn on a plan if the scale of the plan is 1 : 120. (Give your answer correct to 1 decimal place.)

6. The volume of a swimming pool as it appears on its construction plan is 20 cm^3. Determine the actual volume of the pool if the plan has a scale of 1 : 75.

7. The total surface area of an aeroplane's wings is 120 m^3.
   a. Convert 120 m^2 to cm^2.
   b. Calculate the total surface area of the wings of a scale model of the aeroplane if the model is built using the scale 1 : 80.

Reasoning

8. A triangle ABC maps to triangle A$'$B$'$C$'$ under an enlargement.
   AB = 7 cm, AC = 5 cm, A$'$B$'$ = 21 cm, B$'$C$'$ = 30 cm.
   a. Show that the scale factor for enlargement is 3.
   b. Determine BC.
   c. Determine A$'$C$'$.
   d. If the area of $\triangle$ABC is 9 cm^2, show that the area of $\triangle$A$'$B$'$C$'$ is 81 cm^2.

9. A pentagon has an area of 20 cm^2. If all the side lengths are doubled, show that the area of the enlarged pentagon is 80 cm^2.

10. Two rectangles are similar. If the width of the first rectangle is twice of width of the other, prove that the ratio of their areas is 4 : 1.

11. A cube has a surface area of 253.5 cm^2. (Give answers correct to 1 decimal place where appropriate.)
    a. Show that the side length of the cube is 6.5 cm.
    b. Show that the volume of the cube is 274.625 cm^3.
    c. Determine the volume of a similar cube that has side lengths twice as long.
    d. Determine the volume of a similar cube that has side lengths half as long.
    e. Determine the surface area of a similar cube that has side lengths one third as long.

12. In the diagram shown, a light is shining through a hole, resulting in a circular bright spot with a radius of 5 cm on the screen. The hole is 10 mm wide. If the light is 1 m behind the hole, show that the light is 10 m from the screen.

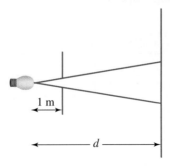

Problem solving

13. The areas of two similar trapeziums are 9 and 25. Determine the ratio of one pair of these trapeziums' corresponding side lengths.

14. Answer the following questions.
    a. Calculate the areas of squares with sides 2 cm, 5 cm, 10 cm and 20 cm.
    b. State in words how the ratio of the areas of these squares is related to the ratio of their side lengths.

15. Two cones are similar. The ratio of these cones' volumes is 27 : 64. Determine the ratio of:
    a. the perpendicular heights of the cones
    b. the areas of the bases of the cones.

16. Rectangle A has dimensions 5 by 4 units, rectangle B has the dimensions 4 by 3 units, and rectangle C has the dimensions 3 by 2.4 units.
    a. Determine which of these rectangles are similar. Explain your answer.
    b. Evaluate the area scale factor for the similar rectangles that you have identified.

17. A balloon in the shape of a sphere has an initial volume of 840 cm^3. Its volume is then increased to $430\,080 \text{ cm}^3$. Determine the increase in the radius of the balloon.

18. **WE18** The bottom half of an egg timer, which is shaped like two cones connected at their apexes, has sand poured into it as shown by the blue section of this diagram. Using the measurements given, evaluate the ratio of the volume of sand in the bottom half of the egg timer to the volume of empty space that is left in the bottom half of the egg timer. You could also use technology to evaluate the answer.

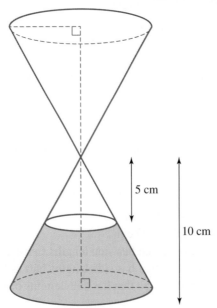

4.7 Review

4.7.1 Topic summary

Units of measurement

When converting units of **length**:

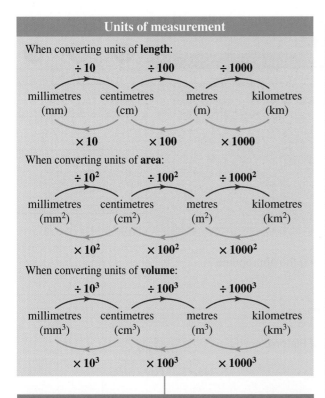

$\div 10$ $\div 100$ $\div 1000$

millimetres   centimetres   metres   kilometres
(mm)   (cm)   (m)   (km)

$\times 10$ $\times 100$ $\times 1000$

When converting units of **area**:

$\div 10^2$ $\div 100^2$ $\div 1000^2$

millimetres   centimetres   metres   kilometres
(mm^2) (cm^2) (m^2) (km^2)

$\times 10^2$ $\times 100^2$ $\times 1000^2$

When converting units of **volume**:

$\div 10^3$ $\div 100^3$ $\div 1000^3$

millimetres   centimetres   metres   kilometres
(mm^3) (cm^3) (m^3) (km^3)

$\times 10^3$ $\times 100^3$ $\times 1000^3$

CONGRUENCE AND SIMILARITY

Congruent figures

- Congruent figures have exactly the same shape and size.
- The symbol for congruence is $\cong$.

Tests for congruent triangles

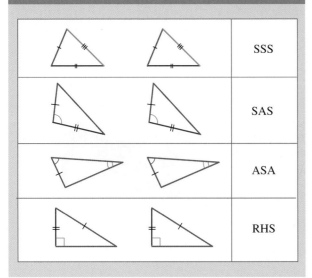

	SSS
	SAS
	ASA
	RHS

Ratio and scale factors

- Ratios compare quantities of the same type.
- Always simplify ratios. For example: $8 : 24 = 1 : 3$
- Scale factor $= \dfrac{\text{image length}}{\text{object length}}$.
- For a scale factor of n:
  - if $n > 1$ the image is larger than the object
  - if $n < 1$ the image is smaller than the object.

Similar figures

- Similar figures have the same shape, but different sizes.
- The symbol for similarity is ~.

If the scale factor of similar figures' sides is n, then:
- the scale factor of their areas is n^2
- the scale factor of their volumes is n^3.

Tests for similar triangles

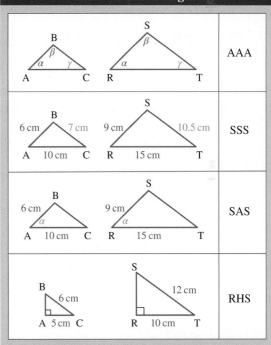

	AAA
	SSS
	SAS
	RHS

- *Note:* Sides of similar triangles are not equal. They are proportional, or have the same scale factor.

Nets, polyhedra and Euler's formula

- The complete set of faces forming a solid is called its **net**.
- A 3-dimensional solid where each of the faces is a polygon is called a **polyhedron**.
- **Euler's formula** shows the relationship between the number of **edges (E)**, **faces (F)** and **vertices (V)** in any polyhedron: $F + V - 2 = E$

4.7.2 Success criteria

Tick the column to indicate that you have completed the subtopic and how well you have understood it using the traffic light system.

(**Green:** I understand; **Yellow:** I can do it with help; **Red:** I do not understand)

Subtopic	Success criteria	⬤	⬤	⬤
4.2	I can compare two quantities of the same type using ratios.			
	I can simplify ratios.			
	I can enlarge a figure using a scale factor.			
	I can calculate and use a scale factor.			
4.3	I can identify congruent figures.			
	I can show that two triangles are congruent using the appropriate congruency test.			
4.4	I can identify and produce nets of various solids.			
	I can define polyhedra and identify examples of Platonic solids.			
	I can verify Euler's formula for various polyhedra.			
4.5	I can identify similar figures.			
	I can determine the scale factor between similar figures.			
	I can show that two triangles are similar using the appropriate similarity test.			
4.6	I can convert between different units of length, area and volume.			
	I can, for a given scale factor of n^2, determine the area of similar figures.			
	I can, for a given scale factor of n^3, determine the volume of similar figures.			

4.7.3 Project

What's this object?

When using geometrical tools to construct shapes, we have to make sure that our measurements are precise. Even a small error in one single step of the measuring process can result in an incorrect shape. The object you will be making in this task is a very interesting one. This object is made by combining three congruent shapes.

The instructions for making the congruent shape that you will use to then make the final object are given below.

Part 1: Making the congruent shape

1. Using a ruler, a protractor, a pencil and two compasses, draw a shape in your workbook by following the instructions below.
   • Measure a horizontal line (AB) that is 9.5 cm long. To make sure there is enough space for the whole of the final object, draw this first line close to the bottom of the page you are working on.
   • At point B, measure an angle of 120° above the line AB. Draw a line that follows this angle for 2 cm. Mark the end of this line as point C, creating the line BC.
   • At point C, measure an angle of 60° on the same side of line BC as point A. Draw a line that follows this angle for 7.5 cm. Mark the end of this line as point D, creating the line CD.
   • At point D, measure an angle of 60° above the line CD. Draw a line that follows this angle for 3.5 cm. Mark the end of this line as point E, creating the line DE.
   • At point E, measure an angle of 120° above the line DE. Draw a line that follows this angle for 2 cm. Mark the end of this line as point F, creating the line EF.
   • Join point F to point A.
     Colour or shade this shape using any colour you wish.
2. Determine the length of the line joining point A to point F.
3. Describe what you notice about the size of the angles FAB and AFE.

Part 2: Making the final object

You have now constructed the shape that will be used three times to make the final object. To make this object, follow the instructions below.

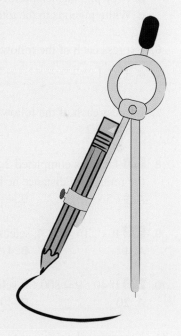

   • Trace the original shape onto a piece of tracing paper twice. Cut around the edges of both of the shapes you have just drawn. Label the shapes with the same letters you used to mark the points A–F when making the original shape.
   • Place line AF of the first traced copy of the original shape so that it is covering up line CD of the shape you first drew. Use the tracing paper to transfer this shape onto the original shape. Colour or shade the traced shape using a colour that is different to the colour you used for the first shape.
   • Place the line DC of your second traced copy so that it is covering up line FA of the shape you first drew. Use the tracing paper to transfer this shape onto the original shape. Colour or shade this third section with a third colour.
4. Describe the object you have drawn.
5. Using the internet, your school library or other references, investigate other 'impossible objects' that can be drawn as 2-dimensional shapes. Recreate these shapes on a separate sheet of paper. Beside these shapes, briefly write some reasons for these shapes being referred to as 'impossible'.

To answer questions online and to receive **immediate corrective feedback** and **fully worked solutions** for all questions, go to your learnON title at www.jacplus.com.au.

Fluency

1. Express each of the following ratios in simplest form.
   a. 8 : 16
   b. 24 : 16
   c. 27 : 18

2. Express each of the following ratios in simplest form.
   a. 56 : 80
   b. 8 : 20
   c. 49 : 35

3. **MC** There are 9 girls and 17 boys in a Year 9 maths class. Select the ratio of boys to girls.
   A. 9 : 17
   B. 17 : 9
   C. 17 : 26
   D. 9 : 26
   E. 26 : 9

4. Jan raised $15 for a charity fundraising event, while her friend Lara raised $25. Calculate the ratio of:
   a. the amount Jan raised to the amount Lara raised
   b. the amount Jan raised to the total amount raised by the pair
   c. the amount the pair raised to the amount Lara raised.

5. A pigeon breeder has 45 pigeons. These include 15 white, 21 speckled pigeons, and the rest grey. Calculate the following ratios.
   a. White pigeons to grey pigeons
   b. Grey pigeons to speckled pigeons
   c. White pigeons to the total number of pigeons

6. Express each of the following ratios in simplest form.
   a. $4\frac{1}{2} : 1$
   b. $6\frac{1}{8} : 9$
   c. $7\frac{1}{4} : 3\frac{1}{2}$

7. Express each of the following ratios in simplest form.
   a. 8.4 : 7.2
   b. 0.2 : 2.48
   c. 6.6 : 0.22

8. **MC** Jack has completed 2.5 km of a 4.5 km race. Choose the ratio of the distance Jack has completed to the remaining distance he has left to run.
   A. 9 : 5
   B. 4 : 5
   C. 4 : 9
   D. 5 : 4
   E. 5 : 9

9. **MC** If $x : 16 = 5 : 4$, select the value of the pronumeral x.
   A. 1
   B. 4
   C. 16
   D. 20
   E. 25

10. **MC** If $40 : 9 = 800 : y$, select the value of the pronumeral y.
    A. 20
    B. 40
    C. 180
    D. 450
    E. 761

11. Determine the value of the pronumeral in each of the following.
    a. $a : 15 = 2 : 5$
    b. $b : 20 = 5 : 8$
    c. $9 : 10 = 12 : c$

12. Determine the value of the pronumeral in each of the following.
    a. $11 : 9 = d : 5$
    b. $7 : e = 4 : 5$
    c. $3 : 4 = 8 : f$

13. $\triangle PQR \sim \triangle DEF$. Determine the length of the missing side in each of the following combinations.
    a. Calculate DF if PQ = 10 cm, DE = 5 cm and PR = 6 cm
    b. Calculate EF if PQ = 4 cm, DE = 12 cm and QR = 5 cm
    c. Calculate QR if DE = 4 cm, PQ = 6 cm and EF = 8 cm
    d. Calculate PQ if DF = 5 cm, PR = 8 cm and DE = 6 cm
    e. Calculate DE if QR = 16 cm, EF = 6 cm and PQ = 12 cm

14. Identify which of the following pairs of shapes are congruent.
    a.
    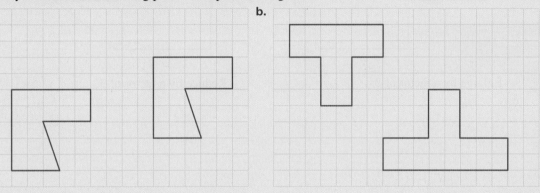
    b.

    c.
    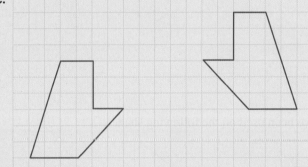

15. Name the congruent triangles in these figures and determine the value of the pronumerals in each case.

    a.
    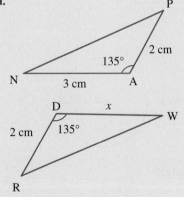

    b.
    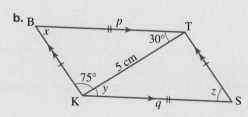

16. Copy each of the following shapes and enlarge (or reduce) them by the given factor.

a.

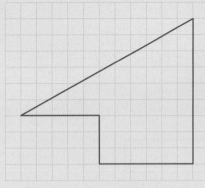

Enlarge by a factor of 2.

b.

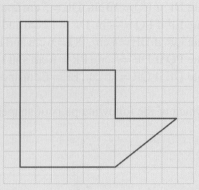

Reduce by a factor of 3.

c.

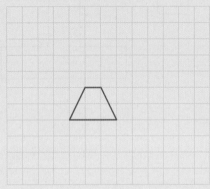

Enlarge by a factor of 4.

17. Determine the enlargement factors that have been used on the following shapes.

a.

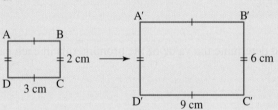

b.

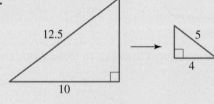

c.

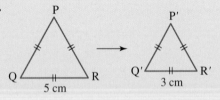

d.

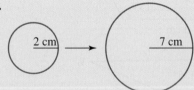

18. Evaluate the value of the pronumerals in the pair of congruent triangles shown.

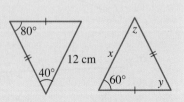

19. The area of a family room is $16\,\text{m}^2$ and the length of the room is 6.4 m. Calculate the area of the room if it is drawn on a plan that uses a scale of 1 : 20.

20. Each of the diagrams shown shows a pair of similar triangles. Calculate the value of x in each case.

a.

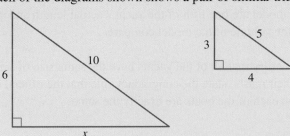

b.

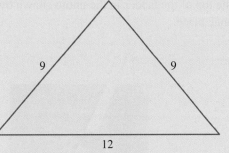

c.

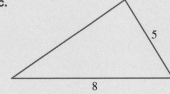

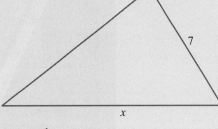

d.

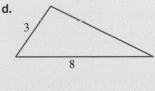

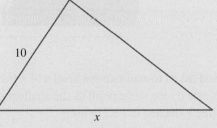

Problem solving

21. The diagram shows a ramp made by Jinghua for her automotive class. The first post has a height of 0.25 m and is placed 2 m from the end of the ramp.
If the second post is 1.5 m high, determine the distance it should be placed from the first post.

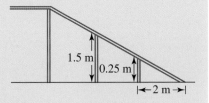

22. Consider the figure shown.
  a. Write a formal proof for a similarity relationship between two triangles in this figure. Give a reason for the similarity.
  b. Determine the values of the pronumerals.

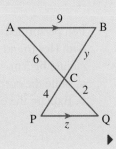

23. *Slocum* is a yacht with a length of 12 m and a beam (width) of 2.5 m. A model of *Slocum* is constructed for a museum. If the length and width of the model are one fifth of the yacht's actual length and width, determine how the volume of the yacht and the volume of the model compare.

24. Before the start of a yacht race, judges must certify that all of the yachts have the same size of sail. Without removing the triangular sails from their masts, state the congruency rule that the official could most efficiently use to work out if the sails on each of the boats are exactly the same.
Explain why this is the most appropriate rule.

25. Calculate the height of the top of the ladder in the photo shown by using similar triangles. Give your answer correct to 1 decimal place.

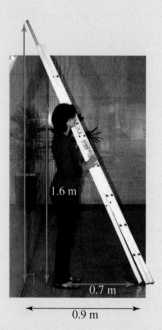

1.6 m

0.7 m

0.9 m

26. Poh is given a 1 m ruler and asked to estimate the height of a palm tree. She places the ruler vertically so that its shadow ends at exactly the same point as the shadow of the palm tree. The ruler's shadow is 2.5 m long and the palm tree's shadow is 12.5 m long.

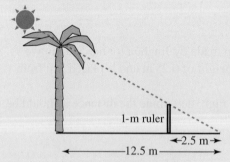

1-m ruler

2.5 m

12.5 m

Poh performed some calculations using similar triangles and calculated the height of the palm tree to be 4 m. Her friend Mikalya said that she thought Poh's calculations were incorrect, and that the answer should be 5 m.
a. State the correct answer.
b. Explain the error that was made by the person with the incorrect answer.

27. A flagpole casts a shadow 2 m long. If a 50 cm ruler is placed upright at the base of the flagpole it casts a shadow 20 cm long. Evaluate the height of the flagpole.

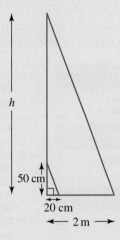

28. A student (S) uses a tape measure, backpack (B), water bottle (W), tree (T) and jacket (J) to help calculate the width of a river.

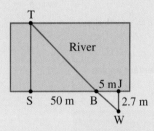

a. Copy and complete the following:
   i. ∠____ and ∠ ____ are both ____.
   ii. ∠SBT and ∠ ____ are equal because ____.
   iii. ∠STB and ∠ ____ are equal because ____.
b. Rewrite the following statement, selecting the correct alternatives from within each set of brackets.
   ∠STB and ∠JWB are (similar / congruent) (SSS / SAS / AAA / ASA / RHS/)
c. Copy and complete the following equation.

$$\frac{\overline{ST}}{\overline{JW}} = \frac{\square}{\square}$$

d. Use your answer to part **c** to evaluate the width of the river.

29. A polyhedron has a faces, $3a$ vertices and $(2a + 8)$ edges.
Evaluate a.

on To test your understanding and knowledge of this topic go to your learnON title at www.jacplus.com.au and complete the **post-test**.

Online Resources

 **Resources**

Below is a full list of **rich resources** available online for this topic. These resources are designed to bring ideas to life, to promote deep and lasting learning and to support the different learning needs of each individual.

📋 eWorkbook

Download the workbook for this topic, which includes worksheets, a code puzzle and a project (ewbk-2004) ☐

📋 Solutions

Download a copy of the fully worked solutions to every question in this topic (sol-0724) ☐

📄 Digital documents

4.2 SkillSHEET Simplifying fractions (doc-6190) ☐
SkillSHEET Simplifying ratios (doc-6191) ☐
SkillSHEET Finding and converting to the lowest common denominator (doc-6192) ☐
SkillSHEET Solving equations of the type $a = \dfrac{x}{b}$ to find x
(doc-6193) ☐
SkillSHEET Solving equations of the type $a = \dfrac{b}{x}$ to find x
(doc-6194) ☐

4.3 SkillSHEET Naming angles (doc-6195) ☐
SkillSHEET Complementary and supplementary angles (doc-6196) ☐
SkillSHEET Angles in a triangle (doc-6197) ☐

▶ Video eLessons

4.2 Ratio (eles-4727) ☐
Scale (eles-4728) ☐
4.3 Congruent figures (eles-4729) ☐
Testing triangles for congruence (eles-4730) ☐
4.4 Nets (eles-4731) ☐
Polyhedra construction (eles-4839) ☐
Euler's formula (eles-4840) ☐
4.5 Similar figures (eles-4732) ☐
Similar triangles (eles-1925) ☐
4.6 Converting between units of length, area and volume (eles-4733) ☐
Area and volume of similar figures (eles-4734) ☐

🧩 Interactivities

4.2 Individual pathway interactivity: Ratio and scale (int-4494) ☐
Proportion (int-3735) ☐
Introduction to ratios (int-3733) ☐
Scale factors (int-6041) ☐
4.3 Individual pathway interactivity: Congruent figures (int-4495) ☐
Congruency tests (int-3755) ☐
Congruent figures (int-3749) ☐
Congruent triangles (int-3754) ☐

4.4 Individual pathway interactivity: Nets, polyhedra and Euler's rule (int-4428) ☐
Nets (int-3759) ☐
Polyhedra construction (int-3760) ☐
Euler's rule (int-3761) ☐
4.5 Individual pathway interactivity: Similar figures (int-4496) ☐
4.6 Individual pathway interactivity: Area and volume of similar figures (int-4497) ☐
Units of length (int-3779) ☐
Area of similar figures (int-6043) ☐
Volume and surface area of similar figures (int-6044) ☐
4.7 Crossword (int-2693) ☐
Sudoku puzzle (int-3205) ☐

Teacher resources

There are many resources available exclusively for teachers online.

To access these online resources, log on to **www.jacplus.com.au**.

Answers

Topic 4 Congruence and similarity

Exercise 4.1 Pre-test

1. $25 : 39$
2. $x = 13.5$
3. E
4. C
5. $y : z = 9 : 8$
6. False
7. D
8. a. $\angle CED$ b. ASA
9. A
10. B, D
11. $x = 2.3\,\text{cm}$
12. $b = 7.5$
13. a. $3 : 5$ b. $9 : 25$
14. $9 : 1$
15. B

Exercise 4.2 Ratio and scale

1. a. i. $240 : 7$ ii. $7 : 240$
   b. The track is $34\frac{2}{7}$ times as long as it is wide.
2. a. i. $5 : 2$ ii. $2 : 5$
   b. The cliff is 2.5 times as high as the distance from the base of the cliff to the emu.
3. a. $2 : 3$ b. $1 : 7$ c. $1 : 3$ d. $2 : 5$
4. a. $4 : 15$ b. $40 : 31$ c. $8 : 25$ d. $100 : 33$
5. a. $6 : 1$ b. $13 : 15$ c. $4 : 1$ d. $5 : 1$
6. a. $1 : 2$ b. $20 : 3$ c. $1 : 6$ d. $5 : 3$
7. a. $5 : 1$ b. $11 : 5$ c. $43 : 20$ d. $1 : 200$
8. a. $3 : 4$ b. $8 : 7$ c. $2 : 3$ d. $28 : 25$
9. a. $2 : 7$ b. $10 : 3$ c. $33 : 16$ d. $23 : 15$
10. a. $a = 9$ b. $b = 24$ c. $c = 32$
11. a. $e = 3\frac{3}{4}$ b. $f = 14\frac{2}{5}$ c. $g = 3\frac{3}{4}$
12. a. $i = 29\frac{1}{3}$ b. $j = 8\frac{4}{7}$ c. $k = 10\frac{2}{5}$
13.
14.
15.

16. a. 3   b. $12\,\text{cm}$ c. $3.5\,\text{cm}$
17. a. $175 : 6428$ b. 37
18. $2.5\,\text{mL}$
19. $1000\,\text{g}$
20. a. 200   b. $6\,\text{m} \times 6\,\text{m}$ c. $5\,\text{m} \times 5\,\text{m}$
21. a. Sample responses can be found in the worked solutions in the online resources.
    b. 1.69 g gold, 0.5 g copper, 0.06 g silver
22. $3 + 4 + 5 = 12$
    $180 \div 12 = 15$
    $3 \times 15 = 45; \, 4 \times 15 = 60; \, 5 \times 15 = 75$
    The 3 angles are $45°$, $60°$ and $75°$.
23. $2k \times 3k \times 5k = 30k^3$
    $$30k^3 = 21870$$
    $$k^3 = 729$$
    $$k = 9$$
    Substituting k into the ratio $2k : 3k : 5k$, the dimensions are 18 cm, 27 cm and 45 cm.
24. The profits aren't shared in a fair ratio. Tyler gets more profit than his share and Dylan gets less profit than his share. Only Ari gets the correct share of the profit.
25. The cost of a pen is twice as much as a pencil.
    The cost of 2 sharpeners is the same as 3 pencils.
    The cost of 4 sharpeners is the same as 3 pens.
    Pen : pencil : sharpener $= 1 : 2 : 3$
26. $680\,\text{cm}$
27. $y : z = 9 : 8$
28. $P = 4, Q = 6$
29. 49 students
30. 125 students
31. $12 : 1$

Exercise 4.3 Congruent figures

1. **b** and **c**
2. **a** and **c**
3. **a** and **b**
4. D
5. $\triangle ABC$ and $\triangle PQR$, SAS
6. $\triangle ABC$ and $\triangle LNM$, SSS
7. $\triangle LMN$ and $\triangle PQR$, SAS
8. $\triangle ABC$ and $\triangle PQR$, ASA
9. $\triangle ABC$ and $\triangle LMN$, RHS
10. $\triangle ABC$ and $\triangle DEF$, ASA
11. $x = 3\,\text{cm}$
12. $x = 85°$
13. $x = 80°, y = 30°, z = 70°$
14. a. $x = 30°, y = 7$
    b. $x = 40°, y = 50°, z = 50°, n = 90°, m = 90°$
15. a. $2\,\text{cm}$ b. $8\,\text{cm}$ c. $6\,\text{mm}$ d. $7\,\text{mm}$
16. Yes, because they are identical.
17. No. The reason for congruence in this situation could alternatively be ASA.

18.

19. a. SSS   **b.** SSS   **c.** SAS

20. Sample responses can be found in the worked solutions in the online resources.

21. Because the angle is not between the two given sides, the general shape of the triangle is not set. This means that it is possible to draw many different shapes that fit these conditions.

22. This can be done by making an equilateral triangle and a regular tetrahedron.

23. This can be done by making a a double regular tetrahedron.

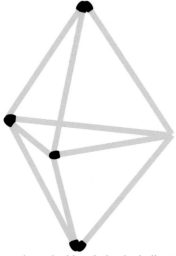

24. Each piece shown in this solution is similar to the original shape.

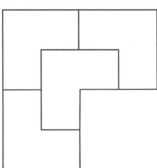

25. 1 : 1

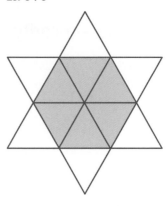

Exercise 4.4 Nets, polyhedra and Euler's formula

1.

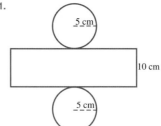

2. a.

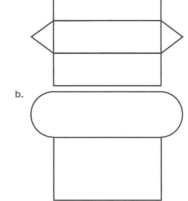

b.

c.

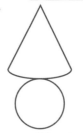

3. Two possible nets are:

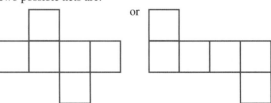

or

4. A possible net is:

5 to 7 Please see the worked solutions.

8. a. $E = 12, V = 8, F = 6, 8 + 6 - 2 = 12$

   b. $E = 12, V = 6, F = 8, 6 + 8 - 2 = 12$

9. a. $E = 18, V = 12, F = 8, 8 + 12 - 2 = 18$

   b. and c $E = 12, V = 8, F = 6, 6 + 8 - 2 = 12$

10. For each shape, $E = 9, V = 6$ and $F = 5$ where $5 + 6 - 2 = 9$.

11. Answers will vary. Sample response: A knowledge of 3D shapes reinforces key maths concepts such as volume relationships, geometrical models and shape recognition. Identify and orgainse visual information. Interpretations of data displays require an understanding of the representation of space.

12. No. Edges, vertices and faces can not be identified on a solid with curves.

13.

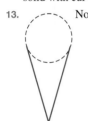

No

14. According to Euler's formula:
    $F + V - E = 2$
    $12 + 8 - E = 2$
    $20 - E = 2$
    $E = 18$
    Number of edges = 18

15. a. Yes   b. Yes   c. No

16. 20 triangular faces, 30 edges and 12 vertices
    $20 + 12 - 2 = 30$

Exercise 4.5 Similar figures

1. a and c, AAA

2. a and b, SSS

3. a and b, SAS

4. a and c, RHS

5. a and c, SSS

6. a. ΔABC and ΔDEC

   b. ΔPQR and ΔABC

   c. ΔPQR and ΔTSR

   d. ΔABC and ΔDEC

   e. ΔADB and ΔADC

7. a. ΔEDC

8. a. $\dfrac{AB}{AD} = \dfrac{BC}{DE} = \dfrac{AC}{AE}$

   b. $f = 9, g = 8$

9. a. $h = 3.75, i = 7.5$

   b. $j = 2.4, k = 11.1$

10. $x = 4$

11. $x = 20°, y = 2\dfrac{1}{4}$

12. $x = 3, y = 4$

13. a. $x = 7.1$ b. $x = 3.1$ c. $x = 7.5, y = 7.7$

14. Sample responses can be found in the worked solutions in the online resources.

15. Sample responses can be found in the worked solutions in the online resources.

16. Sample responses can be found in the worked solutions in the online resources.

17. Sample responses can be found in the worked solutions in the online resources.

18. a. True   b. False   c. False   d. True

19. a. 6.72 m   b. 78 cm

20. 3.75 m

21. $a = 1$ cm

22. a. ΔEDC and ΔEBA

    b. $x = 3$ cm, $y \approx 3.29$ cm

23. a. Sample response can be found in the worked solutions in the online resources.

    b. SAS

    c. 4 cm

    d. PR 6 cm, PS 5 cm and PQ 10 cm

24. The triangles are similar (AAA). $l = y/2$.

Exercise 4.6 Area and volume of similar figures

1. a. 900 cm^2 b. 67.5 cm^2 c. 855 mm^2
   d. 102 cm^2 e. 5 m^2 f. 25 mm^2

2. SA $= 384$ cm^2, V $= 288$ cm^3

3. a. 5400 cm^3 b. 337.5 cm^3 c. 1809 mm^3
   d. 100 cm^3 e. 2.5 m^3 f. 1 mm^3

4. $50\,000$ cm^2

5. a. $250\,000$ cm^2 b. 17.4 cm^2

6. $8\,437\,500$ cm^3

7. a. $1\,200\,000$ cm^2 b. 187.5 cm^2

8. a. Sample responses can be found in the worked solutions in the online resources.

   b. 10 cm

   c. 15 cm

   d. Sample responses can be found in the worked solutions in the online resources.

9. Sample responses can be found in the worked solutions in the online resources.

10. Sample responses can be found in the worked solutions in the online resources.

11. a. Sample responses can be found in the worked solutions in the online resources.

    b. Sample responses can be found in the worked solutions in the online resources.

    c. 2197 cm^3

    d. 34.3 cm^3

    e. 28.2 cm^2

12. Sample responses can be found in the worked solutions in the online resources.

13. 3 : 5

14. a. $4\,\text{cm}^2$, $25\,\text{cm}^2$, $100\,\text{cm}^2$, $400\,\text{cm}^2$

　　b. The ratio of the areas is equal to the square of the ratio of the side lengths.

15. a. 3 : 4　　　　b. 9 : 16

16. a. A and C are similar rectangles by the ratio 5 : 3 or scale factor $\dfrac{5}{3}$.

　　b. $\dfrac{25}{9}$

17. The new radius is 8 times the old radius.

18. The ratio is 7 : 1.

Project

1.

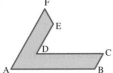

2. 7.5 cm

3. $\angle FAB = 60°$, $\angle AFE = 60°$

4. The impossible triangle

5. Students are required to research and investigate other 'impossible' objects drawn as two-dimensional shapes. These need to be created on a separate sheet of paper with reasons as to why they are termed 'impossible'.

Exercise 4.7 Review questions

1. a. 1 : 2　　　b. 3 : 2　　　c. 3 : 2

2. a. 7 : 10　　　b. 2 : 5　　　c. 7 : 5

3. B

4. a. 3 : 5　　　b. 3 : 8　　　c. 8 : 5

5. a. 5 : 3　　　b. 3 : 7　　　c. 1 : 3

6. a. 9 : 2　　　b. 49 : 72　　　c. 29 : 14

7. a. 7 : 6　　　b. 5 : 62　　　c. 30 : 1

8. D

9. D

10. C

11. a. $a = 6$　　　b. $b = 12.5$　　　c. $c = 13\dfrac{1}{3}$

12. a. $d = 6\dfrac{1}{9}$　　　b. $e = 8\dfrac{3}{4}$　　　c. $f = 10\dfrac{2}{3}$

13. a. 3 cm　　　b. 15 cm　　　c. 12 cm
　　d. 9.6 cm　　　e. 4.5 cm

14. a. Congruent　　　b. Not congruent　　　c. Congruent

15. a. $\triangle ANP \cong \triangle DWR$, $x = 3\,\text{cm}$

　　b. $\triangle TKB \cong \triangle KTS$ or $\triangle TKB \cong \triangle KST$,
　　　$x = 75°$, $y = 30°$, $z = 75°$, $p = 5\,\text{cm}$, $q = 5\,\text{cm}$

16. a.

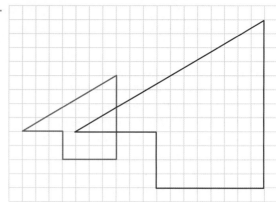

　　b.

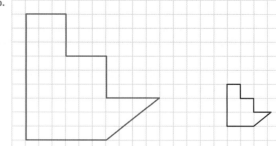

　　c.

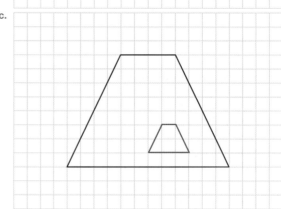

17. a. 3　　　b. 0.4　　　c. $\dfrac{3}{5}$　　　d. 3.5

18. $x = 12\,\text{cm}$, $y = 80°$, $z = 40°$

19. $400\,\text{cm}^2$

20. a. $x = 8$　　　b. $x = 4$　　　c. $x = 11\dfrac{1}{5}$　　　d. $x = 26\dfrac{2}{3}$

21. 10 m

22. a. $\triangle ABC \sim \triangle QPC$, AAA

　　b. $y = 12$, $z = 3$

23. The model is $\dfrac{1}{125}$ the volume of the yacht.

24. The easiest way to check for congruency is to measure the parts of each sail that are easiest to reach. This would be the bottom of each sail and the angles at the bottom of each of those sails. In this situation ASA would be the most appropriate rule to use.

25. 2.1 m

26. a. 5 m
    b. Mikalya had the correct answer. Poh used the distance of 10 m (from the ruler to the tree) in her calculations instead of 12.5 m (the whole length of the tree's shadow).

27. 5 m

28. a. i. ∠TSB and ∠WJB are both right angles.
        ii. ∠SBT and ∠JBW are equal because they are vertically opposite.
        iii. ∠STB and ∠JWB are equal because they are alternate.
    b. ∠SBT and ∠BJW are similar (AAA).
    c. $\dfrac{\overline{ST}}{\overline{JW}} = \dfrac{50}{5}$
    d. 27 m

29. $a = 5$

5 Pythagoras and trigonometry

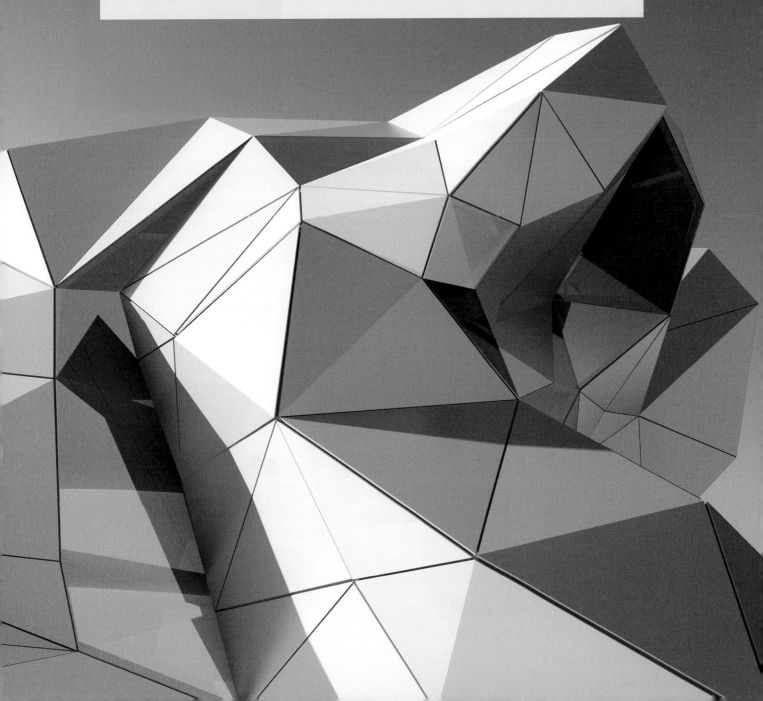

5.1 Overview

Why learn this?

Trigonometry is the branch of geometry that is concerned with triangles. The word 'trigonometry' was created in the sixteenth century from the Greek words *trigōnon* ('triangle') and *metron* ('measure'), but the study of the geometry of triangles goes back to at least the sixth century BCE, when the ancient Greek philosopher Pythagoras of Samos developed his famous theorem.

Pythagoras was particularly interested in right-angled triangles and the relationships between their sides. Later Greek mathematicians used Pythagoras' theorem and the trigonometric ratios to calculate all kinds of distances, including Earth's circumference. Trigonometry is still the primary tool used by surveyors and geographers today when working out distances between points on Earth's surface.

None of the structures we build would be possible without our understanding of geometry and trigonometry. Engineers apply the principles of geometry and trigonometry regularly to make sure that buildings are strong, stable and capable of withstanding extreme conditions. Triangles are particularly useful to engineers and architects because they are the strongest shape. Any forces applied to a triangular frame will be distributed equally to all of its sides and joins. This fact has been known for thousands of years — triangular building frames were used as far back as the sixth century BCE.

A truss is an example of a structure that relies on the strength of triangles. Trusses are often used to hold up the roofs of houses and to keep bridges from falling down. Triangular frames can even be applied to curved shapes. Geodesic domes, like the one shown here, are rounded structures that are made up of many small triangular frames connected together. This use of triangular frames makes geodesic domes very strong, but also very light and easy to build.

Where to get help

Go to your learnON title at **www.jacplus.com.au** to access the following digital resources. The Online Resources Summary at the end of this topic provides a full list of what's available to help you learn the concepts covered in this topic.

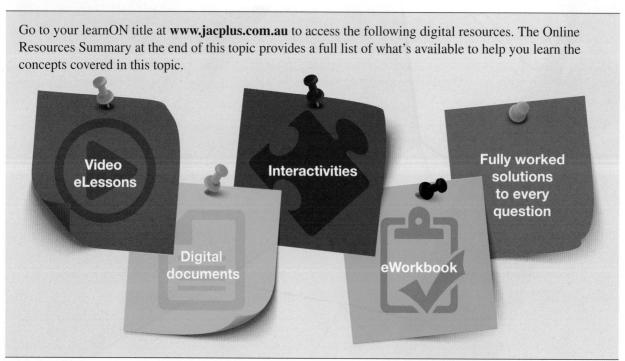

Video eLessons

Interactivities

Fully worked solutions to every question

Digital documents

eWorkbook

Complete this pre-test in your learnON title at www.jacplus.com.au, and receive **automatic marks**, **immediate corrective feedback** and **fully worked solutions**.

1. Calculate the value of x in the diagram shown.

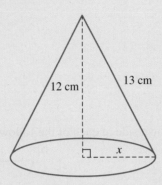

2. Calculate the value of a in the diagram shown, correct to 2 decimal places.

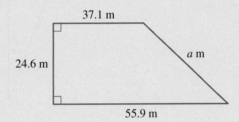

3. **MC** Identify which set of 3 numbers could represent the sides of a right-angled triangle.
   A. 2, 3, 4   **B.** 5, 12, 13   **C.** 1, 2, 3   **D.** 5, 6, 7   **E.** 3, 5, 6

4. If a right-angled triangle has side lengths a, $a + b$ and $a - b$, where both a and b are greater than 0, state which one of the lengths is the hypotenuse.

5. **MC** Identify the correct rule linking the sides of the right-angled triangle shown.
   A. $c^2 = (a + b)^2$
   B. $c = a + b$
   C. $c^2 = a^2 + b^2$
   D. $c = \dfrac{b}{a}$
   E. $c^2 = a^2 - b^2$

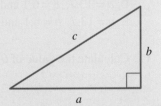

6. **MC** Consider the right-angled triangle shown. Identify the correct option for angle θ.
   A. a is the adjacent side.
   B. b is the opposite side.
   C. a is the hypotenuse
   D. b is the adjacent side.
   E. c is the opposite side.

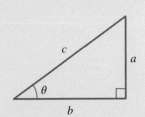

7. **MC** Identify the correct trigonometric ratio for the triangle shown.

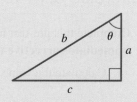

A. $\tan\theta = \dfrac{a}{c}$

B. $\sin\theta = \dfrac{b}{a}$

C. $\cos\theta = \dfrac{c}{b}$

D. $\sin\theta = \dfrac{c}{b}$

E. $\cos\theta = \dfrac{b}{c}$

8. Calculate the length of the unknown side of each of the following triangles, correct to 2 decimal places.

a.

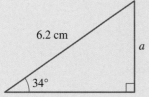

b.

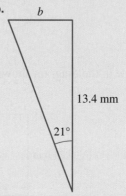

9. Evaluate, correct to 4 decimal places.

   a. $\sin(62.5°)$ b. $\cos(12.1°)$ c. $\tan(74.9°)$

10. **MC** Identify the lengths of the unknown sides in the triangle shown.

   A. $a = 6.1$, $b = 13.7$ and $c = 7.3$

   B. $a = 6.1$, $b = 13.7$ and $c = 5.1$

   C. $a = 5.1$, $b = 6.1$ and $c = 13.7$

   D. $a = 13.7$, $b = 6.1$ and $c = 5.1$

   E. $a = 13.7$, $b = 6.1$ and $c = 7.3$

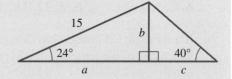

11. Calculate the value of θ in the triangle shown, correct to 1 decimal place.

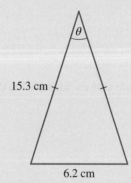

12. **MC** Identify the value of θ in the triangle shown, correct to 2 decimal places.
    A. 25.27°
    B. 28.16°
    C. 28.17°
    D. 61.83°
    E. 64.73°

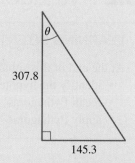

13. Determine the correct value of the side length *a*.
    A. 7.62
    B. 10.04
    C. 14.34
    D. 20.48
    E. 29.25

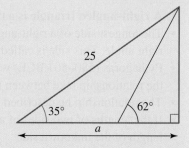

14. A person who is 1.54 m tall stands 10 m from the foot of a tree and records the angle of elevation (using an inclinometer) to the top of the tree as 30°.
    Evaluate the height of the tree, correct to 2 decimal places.

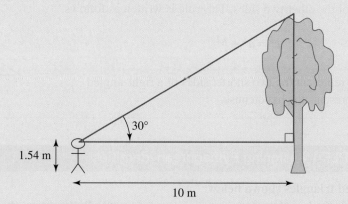

15. The angle of depression from a scuba diver who is floating on the water's surface to a shark swimming below them on the sea floor is 35.8°. The depth of the water is 35 m.
    Evaluate the horizontal distance from the scuba diver to the shark, correct to 2 decimal places.

5.2 Pythagoras' theorem

▶ 5.2.1 Introducing Pythagoras' theorem

eles-4750

- A **right-angled triangle** is a triangle that contains a right angle (90°).
- The longest side of a right-angled triangle is always found opposite the right angle. This side is called the **hypotenuse**.
- Pythagoras (580–501 BCE) was a Greek mathematician who explored the relationship take between the lengths of the sides of right-angled triangles.
- The relationship he described is known as **Pythagoras' theorem**.
- If the lengths of two sides of a right-angled triangle are given, then Pythagoras' theorem lets us find the length of the unknown side.

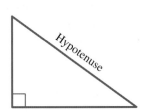

Pythagoras' theorem

In any right-angled triangle, the square of the hypotenuse is equal to the sum of the squares of the other two sides. This rule is written as follows.

$$c^2 = a^2 + b^2$$

In this rule, a and b represent the two shorter sides of a right-angled triangle, while c represents the hypotenuse.

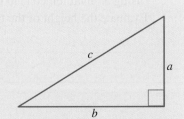

WORKED EXAMPLE 1 Assigning Pythagorean identities

For the right-angled triangles shown below:

i. state which side is the hypotenuse

ii. write Pythagoras' theorem.

a.

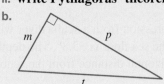

b.

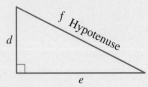

THINK

a. i. The hypotenuse is always opposite the right angle.

WRITE

Side f is opposite the right angle. Therefore, side f is the hypotenuse.

ii. If the triangle is labelled as usual with a, b and c, as shown in blue, Pythagoras' theorem can be written and then the letters can be replaced with their new values (names).

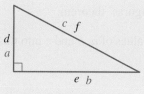

$$c = f; a = d; \ b = e$$
$$c^2 = a^2 + b^2$$
$$f^2 = d^2 + e^2$$

b. i. The hypotenuse is opposite the right angle.

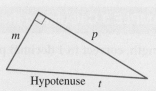

Side t is opposite the right angle.
Therefore, side t is the hypotenuse.

ii. If the labels on the triangle are replaced with a, b and c, as shown in blue, Pythagoras' theorem can be written out, and then the letters a, b and c can be changed back to the original labels.

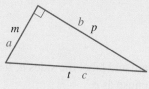

$$c = t; \ b = p; \ a = m$$
$$c^2 = a^2 + b^2$$
$$t^2 = m^2 + p^2$$

▶ 5.2.2 Calculating unknown side lengths

eles-4751

- Pythagoras' theorem can be used to calculate the unknown side length of a right-angled triangle if the other two side lengths are known.
- To calculate an unknown side length, substitute the known values into the equation $c^2 = a^2 + b^2$ and solve for the remaining pronumeral.
- Remember that the hypotenuse always corresponds to the pronumeral c.

WORKED EXAMPLE 2 Calculating the length of the hypotenuse

For the triangle shown, calculate the length of the hypotenuse, x, correct to 1 decimal place.

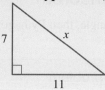

THINK

1. Copy the diagram and apply the labels a, b and c to the triangle, pairing them with the known values and the pronumeral. Remember to label the hypotenuse as c.

WRITE

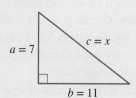

2. Write out Pythagoras' theorem.

$$c^2 = a^2 + b^2$$

3. Substitute the values of a, b and c into this rule and simplify.

$$x^2 = 7^2 + 11^2$$
$$= 49 + 121$$
$$= 170$$

4. Calculate x by taking the square root of 170. Round the answer to 1 decimal place.

$$x = \sqrt{170}$$
$$x = 13.0$$

WORKED EXAMPLE 3 Calculating the length of a shorter side

Calculate the length, correct to 1 decimal place, of the unmarked side of the triangle shown.

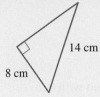

14 cm

8 cm

THINK

1. Copy the diagram and label the sides a, b and c. Remember to label the hypotenuse as c.

2. Write out Pythagoras' theorem for a shorter side.

3. Substitute the values of b and c into the equation and simplify.

4. Calculate the value of a by taking the square root of 132. Round the answer to 1 decimal place.

WRITE

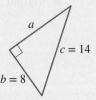

a

$c = 14$

$b = 8$

$$a^2 = c^2 - b^2$$

$$a^2 = 14^2 - 8^2$$
$$= 196 - 64$$
$$= 132$$

$$a = \sqrt{132}$$
$$= 11.5 \text{ cm}$$

---

eles-4752

⊙ 5.2.3 Applying Pythagoras' theorem

- If a problem involves a right-angled triangle, then Pythagoras' theorem can be applied to determine a solution.

Applying Pythagoras' theorem to a problem

1. Read the question carefully and draw a right-angled triangle to represent the problem.
2. Identify the known values and place them on the diagram in their correct positions.
3. Identify the variable that you will need to find.
4. Substitute the known values into Pythagoras' theorem. Make sure that they are all using the same units of length. You may need to convert some units.
5. Solve for the unknown value.
6. Use the result to write the answer as a complete sentence.

A wedge of cake has been cut so that it forms a right-angled triangle.
The longest edge of the wedge is 12 cm long and its shortest edge
is 84 mm long.
Calculate the length of the third side, correct to 1 decimal place.

THINK	WRITE
1. Draw a right-angled triangle. Identify the longest side (this will always be the hypotenuse) and place its value on the diagram. Identify the shortest side and place its value on the diagram. Label the side with unknown length as x.	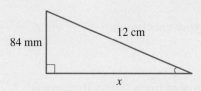
2. Identify the length of the hypotenuse and the two shorter sides. Check that the units for all measurements are the same. Convert 84 mm to centimetres by dividing by 10.	$c = 12 \text{ cm}$ $a = 84 \text{ mm} = 8.4 \text{ cm}$ $b = x$
3. Substitute the values into the equation and simplify.	$c^2 = a^2 + b^2$ $12^2 = (8.4)^2 + x^2$ $144 = 70.56 + x^2$
4. Solve the equation for x by: subtracting 70.56 from both sides and taking the square root of both sides.	$144 - 70.56 = 70.56 - 70.56 + x^2$ $73.44 = x^2$ $\sqrt{73.44} = \sqrt{x^2}$ $\pm 8.5697 \approx x$
5. It is not possible for a triangle to have a side with a negative length, therefore the solution must be the positive value. Round the answer to 1 decimal place and include the appropriate units.	$x = 8.6 \text{ cm}$
6. Write the answer as a sentence.	The third side of the slice of cake is 8.6 cm long.

▶ 5.2.4 Pythagorean triads

eles-4753

- A **Pythagorean triad** is a group of any three whole numbers that satisfy Pythagoras' theorem.
  For example, $\{3, 4, 5\}$ is a Pythagorean triad because $3^2 = 9, 4^2 = 16$ and $5^2 = 25$, and $9 + 16 = 25$.
- Pythagorean triads are useful when solving problems using Pythagoras' theorem. If two known side lengths in a triangle belong to a triad, the length of the third side can be stated without performing any calculations.
- Some well known Pythagorean triads are $\{3, 4, 5\}, \{5, 12, 13\}, \{8, 15, 17\}$ and $\{7, 24, 25\}$.

WORKED EXAMPLE 5 Identifying Pythagorean triads

Determine whether the following sets of numbers are Pythagorean triads.
a. $\{9, 10, 14\}$ **b.** $\{33, 65, 56\}$

THINK	WRITE
a. 1. Pythagorean triads satisfy Pythagoras' theorem. Substitute the values into the equation $c^2 = a^2 + b^2$ and determine whether the equation is true. Remember that c is always the longest side.	**a.** $c^2 = a^2 + b^2$ $\text{LHS} = c^2 \qquad \text{RHS} = a^2 + b^2$ $\qquad = 14^2 \qquad \qquad = 9^2 + 10^2$ $\qquad = 196 \qquad \qquad = 81 + 100$ $\qquad \qquad \qquad \qquad \quad = 181$
2. Write your conclusion.	Since $\text{LHS} \neq \text{RHS}$, the set $\{9, 10, 14\}$ is not a Pythagorean triad.
b. 1. Pythagorean triads satisfy Pythagoras' theorem. Substitute the values into the equation $c^2 = a^2 + b^2$ and determine whether the equation is true. Remember that c is always the longest side.	**b.** $c^2 = a^2 + b^2$ $\text{LHS} = 65^2 \qquad \text{RHS} = 333^2 + 56^2$ $\qquad = 4225 \qquad \qquad = 1089 + 3136$ $\qquad \qquad \qquad \qquad \qquad = 4225$
2. Write your conclusion.	Since $\text{LHS} = \text{RHS}$, the set $\{33, 65, 56\}$ is a Pythagorean triad.

- If each term in a Pythagorean triad is multiplied by the same number, the result is also a triad. For example, if we multiply each number in $\{5, 12, 13\}$ by 2, the result $\{10, 24, 26\}$ is also a triad.
- Builders and gardeners use multiples of the Pythagorean triad $\{3, 4, 5\}$ to make sure that walls and floors are at right angles to each other.

 Resources

Exercise 5.2 Pythagoras' theorem

learn on

Individual pathways

■ PRACTISE	■ CONSOLIDATE	■ MASTER
1, 5, 6, 8, 12, 15	2, 7, 9, 13, 16	3, 4, 10, 11, 14, 17

To answer questions online and to receive **immediate corrective feedback** and **fully worked solutions** for all questions, go to your learnON title at www.jacplus.com.au.

Fluency

1. **WE1** For the right-angled triangles shown below:
   - **i.** state which side is the hypotenuse
   - **ii.** write Pythagoras' theorem.

   a. 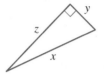   **b.** 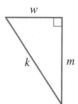   **c.**   **d.**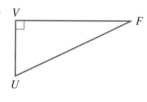

2. **WE2** For each of the following triangles, calculate the length of the hypotenuse, giving answers correct to 2 decimal places.

   a. 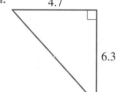   **b.**    **c.**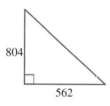

3. **WE3** Calculate the value of the pronumeral in each of the following triangles. Give your answers correct to 2 decimal places.

   a. 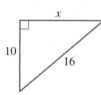   **b.** 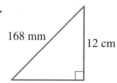   **c.** 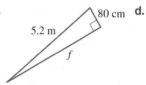   **d.**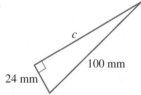

4. **WE4** Calculate the value of the missing side length. Give your answer correct to 1 decimal place.

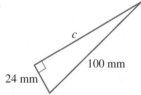

5. Determine whether the following sets of numbers are Pythagorean triads.

   a. {2, 5, 6}   b. {7, 10, 12}   c. {18, 24, 30}

   d. {72, 78, 30}   e. {13, 8, 15}   f. {50, 40, 30}

Understanding

6. In a right-angled triangle, the two shortest sides are 4.2 cm and 3.8 cm.

   a. Draw a sketch of the triangle.

   b. Calculate the length of the hypotenuse correct to 2 decimal places.

7. A right-angled triangle has a hypotenuse of 124 mm and another side of 8.5 cm.

   a. Draw a diagram of the triangle.

   b. Calculate the length of the third side of this triangle. Give your answer in millimetres, correct to 2 decimal places.

8. **MC** Identify which of the following sets is formed from the Pythagorean triad {21, 20, 29}.

   A. {95, 100, 125}   **B.** {105, 145, 100}   **C.** {84, 80, 87}

   D. {105, 80, 87}   **E.** {215, 205, 295}

9. **MC** Identify the length of the hypotenuse in the triangle shown.

   A. 25 cm

   B. 50 cm

   C. 50 mm

   D. 500 mm

   E. 2500 mm

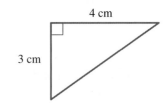

10. **MC** Identify the length of the third side in the triangle shown.

   A. 48.75 cm

   B. 0.698 m

   C. 0.926 m

   D. 92.6 cm

   E. 69.8 mm

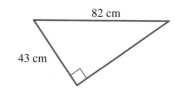

11. The lengths of two sides of a right-angled triangle are given in the following. Calculate the value of the third side in each of the following. Give your answer in the units specified (correct to 2 decimal places). The diagram shown illustrates how each triangle is to be labelled.

   Remember that c is always the hypotenuse.

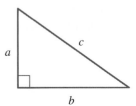

   a. $a = 37$ cm, $c = 180$ cm; calculate b in cm.

   b. $a = 856$ mm, $b = 1200$ mm; calculate c in cm.

   c. $b = 4950$ m, $c = 5.6$ km; calculate a in km.

   d. $a = 125\,600$ mm, $c = 450$ m; calculate b in m.

Reasoning

12. A ladder that is 7 m long leans up against a vertical wall. The top of the ladder reaches 6.5 m up the wall. Determine how far the foot of the ladder is from the wall. Give your answer correct to 2 decimal places.

13. A rectangular park measures 260 m by 480 m. Danny usually trains by running five circuits around the edge of the park. After heavy rain, two of the adjacent sides are too muddy to run along, so Danny runs along a triangular path that traces the other two sides and the diagonal. Danny does five circuits of this path for training. Show that Danny has run about 970 metres less than his usual training session.

14. A water park has hired Sharnee to build part of a ramp for a new water slide. She builds a 12-metre-long ramp that rises to a height of 250 cm. To meet safety regulations, the ramp can only have a gradient between 0.1 and 0.25. Show that the ramp Sharnee has built is within the regulations.

Problem solving

15. Spiridoula is trying to hang her newest painting on a hook in her hallway. She leans a ladder against the wall so that the distance between the foot of the ladder and the wall is 80 cm. The ladder is 1.2 m long.

    a. Draw a diagram showing the ladder leaning on the wall.
    b. Calculate how far up the wall the ladder reaches. Give your answer correct to 2 decimal places.
    c. Spiridoula climbs the ladder to check whether she can reach the hook from the step at the very top of the ladder. When she extends her arm, the distance from her feet to her fingertips is 1.7 m. If the hook is 2.5 m above the floor, determine whether she can reach it from the top step of the ladder.

16. a. The smallest numbers of four Pythagorean triads are given below. Determine the middle number of each triad, and use this answer to find the third number of each triad.

    i. 9 ii. 11 iii. 13 iv. 29

    b. Comment about the triads formed in part a.

17. We know that it is possible to find the exact square root of some numbers, but not others. For example, we can determine $\sqrt{4}$ exactly, but not $\sqrt{3}$ or $\sqrt{5}$. Our calculators can find decimal approximations of these square roots, but because they cannot be found exactly they are called 'irrational numbers'. However, there is a method of showing the exact location of irrational numbers on a number line.
    Using graph paper, draw a right-angled triangle with 2 equal sides of length 1 cm as shown below.

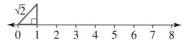

Using Pythagoras' theorem, we know that the longest side of this triangle is $\sqrt{2}$ units. Place the compass point at zero and make an arc that will show the location of $\sqrt{2}$ on the number line.

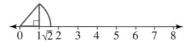

Draw another right-angled triangle using the longest side of the first triangle as one side, and make the other side 1 cm in length.

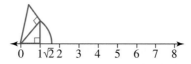

The longest side of this triangle will have a length of $\sqrt{3}$ units. Draw an arc to find the location of $\sqrt{3}$ on the number line.
Repeat these steps to draw triangles that will have sides of length $\sqrt{4}$, $\sqrt{5}$, $\sqrt{6}$ units, and so on.

5.3 Applications of Pythagoras' theorem

LEARNING INTENTIONS

At the end of this subtopic you should be able to:
- divide composite shapes into simple shapes
- calculate unknown side lengths in composite shapes and irregular triangles
- use Pythagoras' theorem to solve problems in 3D contexts.

5.3.1 Using composite shapes to solve problems

eles-4754

- Complex or unusual shapes can be thought of as composites of simpler shapes with known properties.
- Dividing a shape into a series of these simpler shapes lets us use the known properties of those shapes to determine unknown values.
- For example, to calculate the value of x in the trapezium shown, a vertical line can be added to the trapezium to create two shapes: a right-angled triangle and a rectangle.
- Using Pythagoras' theorem, we can now find the length of x, which has become the hypotenuse of triangle ABE.

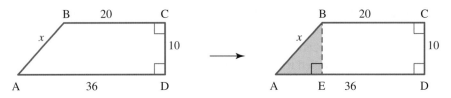

WORKED EXAMPLE 6 Calculating unknown side lengths in composite shapes

Calculate the value of x in the diagram shown, correct to 1 decimal place.

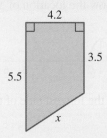

THINK

1. Add a horizontal line to divide the shape into 2 smaller shapes, one that is a right-angled triangle, as shown in orange. x is the hypotenuse of a right-angled triangle. To use Pythagoras' theorem, the length of the side shown in green must be known. This length can be calculated as the difference between the long and short vertical edges of the trapezium: $5.5 - 3.5 = 2$.

WRITE

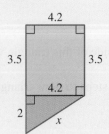

2. Write out Pythagoras' theorem and substitute the values of a, b and c.
$a = 2, b = 4.2, c = x$

$$c^2 = a^2 + b^2$$
$$x^2 = 2^2 + (4.2)^2$$
$$x^2 = 4 + 17.64$$
$$x^2 = 21.64$$
$$\sqrt{x^2} = \sqrt{21.64}$$
$$x \approx \pm 4.651\,88$$

3. It is not possible to have a side with a negative length. Therefore the solution is the positive value. Round the answer to 1 decimal place

$$x = 4.7$$

- To determine the value of y in the irregular triangle shown, the triangle can be split into two right-angled triangles: $\triangle ABD$ (pink) and $\triangle BDC$ (blue). There is now enough information to calculate the missing side length from $\triangle ABD$. This newly calculated length can then be used to determine the value of y.

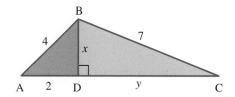

WORKED EXAMPLE 7 Applying Pythagoras' theorem to non-right-angled triangles

a. **Calculate the perpendicular height of an isosceles triangle with equal sides that are both 15 mm long and a third side that is 18 mm long.**

b. **Calculate the area of this triangle.**

THINK

a. 1. Draw the triangle and label all side lengths as described.

2. Draw an additional line to represent the height of the triangle (shown in pink) and label it appropriately. Because the triangle is an isosceles triangle, h bisects the base of 18 mm (shown in purple) to create two right-angled triangles.

3. Consider one of these right-angled triangles, which both have a height of h. The base length of each of these triangles is $\dfrac{18}{2} = 9$ mm.

WRITE

a.

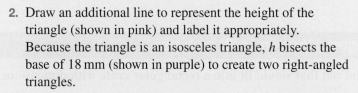

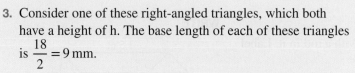

4. Substitute the values from this right-angled triangle into Pythagoras' theorem to calculate the value of h.

$$c^2 = a^2 + b^2$$
$$15^2 = h^2 + 9^2$$
$$225 = h^2 + 81$$
$$225 - 81 = h^2 + 81 - 81$$
$$144 = h^2$$
$$\sqrt{144} = \sqrt{h^2}$$
$$\pm 12 = h$$

5. It is not possible to have a side with a negative length. Therefore the solution is the positive value. Include the appropriate units in your answer.

$h = 12\,\text{mm}$
The perpendicular height of the triangle is 12 mm.

b. 1. Write the formula for the area of a triangle. Using the answers from part **a**, the base is 18 mm and the height is 12 mm.

b. $A_\triangle = \dfrac{1}{2}bh$

$\quad = \dfrac{1}{2} \times 18 \times 12$

$\quad = 108$

2. Answer the question with appropriate units.

The area of the triangle is 108 mm^2.

▶ 5.3.2 Pythagoras' theorem in 3D

eles-4755

- Pythagoras' theorem can be used to solve problems in three dimensions.
- 3D problems often involve multiple right-angled triangles.
- It is often helpful to redraw sections of a 3D diagram in two dimensions. This makes it easier to see any right angles.

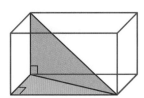

WORKED EXAMPLE 8 Using Pythagoras' theorem in 3D problems

Calculate the maximum length of a metal rod that would fit into a rectangular crate with dimensions $1\,\text{m} \times 1.5\,\text{m} \times 0.5\,\text{m}$.

THINK

1. Draw a diagram of a rectangular box with a rod in it. Label the dimensions of this diagram.

2. Draw a right-angled triangle that has the metal rod as one of its sides (x, shown in pink). The lengths of x and y in this right-angled triangle are not known.
Draw a second right-angled triangle (shown in green) to allow you to calculate the length of y.

WRITE

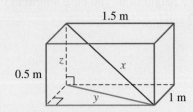

3. Calculate the length of y using Pythagoras' theorem.

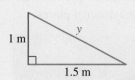

$$c^2 = a^2 + b^2$$
$$y^2 = 1.5^2 + 1^2$$
$$y^2 = 3.25$$
$$y = \sqrt{3.25}$$

4. Draw the right-angled triangle that has the rod as its hypotenuse. Use Pythagoras' theorem to calculate the length of x.
   $z = $ height of the crate $= 0.5\,\text{m}$

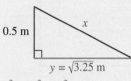

$$c^2 = a^2 + b^2$$
$$x^2 = \left(\sqrt{3.25}\right)^2 + 0.5^2$$
$$x^2 = 3.25 + 0.25$$
$$x^2 = 3.5$$
$$x = \sqrt{3.5}$$
$$\approx 1.87$$

5. Answer the question with appropriate units.

The maximum length of the metal rod is 1.87 m (correct to 2 decimal places).

 Resources

eWorkbook   Topic 5 Workbook (worksheets, code puzzle and project) (ewbk-2005)

Video eLesson Pythagoras' theorem in 3 dimensions (eles-1913)

Interactivities Individual pathway interactivity: Applications of Pythagoras' theorem (int-4475)
                    Composite shapes (int-3847)

Exercise 5.3 Applications of Pythagoras' theorem

learn on

Individual pathways

■ PRACTISE	■ CONSOLIDATE	■ MASTER
1, 2, 5, 8, 14, 18	3, 6, 7, 9, 13, 15, 16, 19	4, 10, 11, 12, 17, 20

To answer questions online and to receive **immediate corrective feedback** and **fully worked solutions** for all questions, go to your learnON title at www.jacplus.com.au.

Fluency

1. **WE6** Calculate the length of the side x in the figure shown, correct to 2 decimal places.

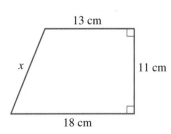

2. Calculate the values of the pronumerals in each of the following photographs and diagrams. Give your answers correct to 2 decimal places.

a.

b.

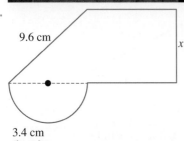

c.

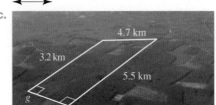

3. For each of the following diagrams, calculate the lengths of the sides marked x and y. Give your answer correct to 2 decimal places (where necessary).

a.

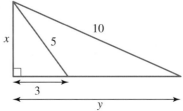

b.

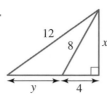

c.

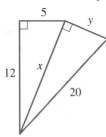

d.

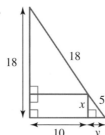

4. **WE8** Calculate the length of the longest metal rod that could fit diagonally into each of the boxes shown below. Give your answers correct to 2 decimal places.

a.

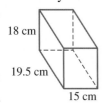

b.

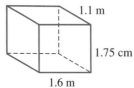

c.

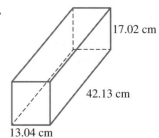

Understanding

5. **WE7** The height of an isosceles triangle is 3.4 mm and its equal sides are twice as long.

   a. Sketch the triangle, showing all the information given.
   b. Calculate the length of the third side, correct to 2 decimal places.
   c. Calculate the area of the triangle, correct to 2 decimal places.

6. The side length of an equilateral triangle is 1 m. Calculate:

   a. the height of the triangle in metres, correct to 2 decimal places
   b. the area of the triangle in m^2, correct to 3 decimal places.

7. **MC** Identify the longest length of an object that fits into a box with dimensions of $42\,cm \times 60\,cm \times 13\,cm$.
   A. 74.5 cm   **B.** 60 cm   **C.** 73.2 cm   **D.** 5533 cm   **E.** 74 cm

8. Priya wants to pack an umbrella into her suitcase. The umbrella is 1 metre long. If Priya's suitcase measures $89\,cm \times 21\,cm \times 44\,cm$, decide if their umbrella will fit. Give the length (correct to 3 decimal places) of the longest umbrella that Priya could fit in their suitcase.

9. Giang is packing her lunch into her lunchbox. She wants to pack a 22-cm-long straw. If her lunchbox is $15\,cm \times 5\,cm \times 8\,cm$, will it fit? Calculate the length of the longest object that will fit into Giang's lunchbox. Give your answer correct to 2 decimal places.

10. A cylindrical pipe is 2.4 m long. It has an internal diameter of 30 cm. Calculate the size of largest object that could fit inside the pipe. Give your answer correct to the nearest cm.

11. A classroom contains 20 identical desks shaped like the one shown.

    a. Calculate the width of each desktop (labelled w). Give your answer correct to 2 decimal places.
    b. Calculate the area of each desktop, correct to 2 decimal places.
    c. If the top surface of each desk needs to be painted, determine the total area of the desktops that need to be painted. Give your answer correct to 2 decimal places.
    d. Each desktop needs to be given 2 coats of paint. The paint being used is sold in 1-litre tins that cost $29.95 each.

    Each of these tins holds enough paint to cover an area of $12\,m^2$.
    Using your answer to part **c**, calculate how much it will cost to paint all 20 desks.

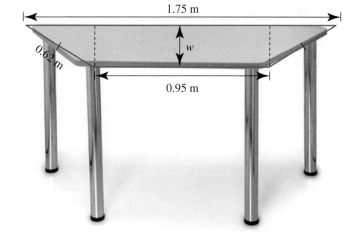

12. A utility knife has a blade shaped like a right-angled trapezium. Its sloping edge is 2 cm long and its parallel sides are 32 mm and 48 mm long.
    Calculate the width of the blade, then use your answer to calculate the blade's area.

13. A cylindrical pencil holder has an internal height of 16 cm. If its diameter is 8.5 cm and its width is 2 mm, calculate the length of the longest pen that could fit inside the holder. Give your answer correct to the nearest mm.

Reasoning

14. Consider the figure shown.

    a. Determine the order in which you are able to calculate the lengths of the sides AD, AC and DC.
    b. Calculate the lengths of the sides in part a. Give your answer correct to 2 decimal places.
    c. Explain whether the triangle ABC is right-angled.

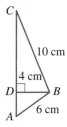

15. Explain in your own words how a 2D right-angled triangle can be seen in a 3D figure.

16. Katie goes on a hike. She walks 2.5 km north, then 3.1 km east. She then walks 1 km north and 2 km west. Finally, she walks in a straight line back to her starting point. Show that Katie has walked a total distance of 12.27 km.

17. Consider the following diagram:

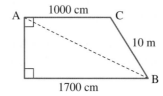

    a. Show that the distance AB is 18.44 metres.
    b. Show that the angle ∠ACB is not a right angle.

Problem solving

18. The diagram shown illustrates a cross-section of the roof of Gabriel's house.

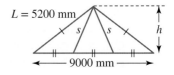

    a. Calculate the height of the roof, h, to the nearest millimetre.
    b. The roof's longer supports, labelled L, are 5200 mm long. Show that the shorter supports, labelled s, are 2193 mm shorter than the longer supports.

19. A flagpole is attached to the ground by two wires, as illustrated in the diagram shown. Use the information in the diagram to determine the length of the lower wire, x. Give your answer correct to 1 decimal place.

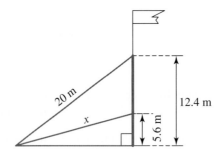

20. Evaluate the values of the pronumerals w, x, y and z in the diagram shown.

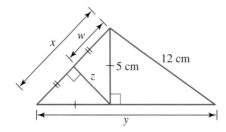

5.4 Trigonometric ratios

LEARNING INTENTION

At the end of this subtopic you should be able to:
- identify the hypotenuse, opposite and adjacent sides of a right-angled triangle with respect to a reference angle
- use the ratios of a triangle's sides to calculate the sine, cosine and tangent of an angle.

▶ 5.4.1 What is trigonometry?

eles-4756

- The word **trigonometry** is derived from the Greek words *trigonon* ('triangle') and *metron* ('measurement'). The literal translation of the word is 'to measure a triangle'.
- Trigonometry deals with the relationship between the sides and the angles of a triangle.
- As we have already seen, the longest side of a right-angled triangle is called the hypotenuse. It is always located opposite the right angle.
- In order to name the remaining two sides of a triangle, another angle, called the 'reference angle', θ, must be labelled on the triangle. θ is pronounced 'theta'; it is the eighth letter of the Greek alphabet.
- The side that is across from the reference angle is called the **opposite side**. You can see that the opposite side does not touch the reference angle at all.
- The remaining side is called the **adjacent side**.
- Note that the reference angle always sits between the hypotenuse and the adjacent side.

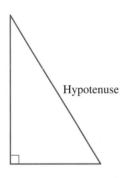
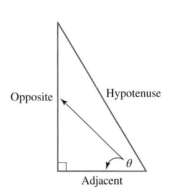

WORKED EXAMPLE 9 Labelling the sides of a right-angled triangle

Label the sides of the right-angled triangle shown using the words 'hypotenuse', 'adjacent' and 'opposite'.

THINK	WRITE
1. Identify the hypotenuse. Remember that the hypotenuse always lies opposite the right angle.	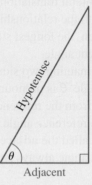
2. Look at the position of the reference angle. The reference angle always sits between the hypotenuse and the adjacent side; therefore, the adjacent side is the bottom side of this triangle.	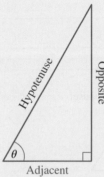
3. The opposite side does not touch the reference angle; therefore, it is the vertical side on the right of this triangle.	

▶ 5.4.2 Trigonometric ratios

eles-4757

- Trigonometry is based upon the ratios between pairs of side lengths. Each of these ratios is given a special name.
- In any right-angled triangle:

$$\textbf{sine}(\theta) = \frac{\text{opposite}}{\text{hypotenuse}}$$

$$\textbf{cosine}(\theta) = \frac{\text{adjacent}}{\text{hypotenuse}}$$

$$\textbf{tangent}(\theta) = \frac{\text{opposite}}{\text{adjacent}}$$

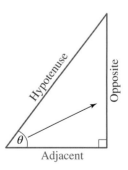

- These rules are often abbreviated as shown in the following.

> **Trigonometric ratios**
>
> $$\sin(\theta) = \frac{O}{H} \qquad \cos(\theta) = \frac{A}{H} \qquad \tan(\theta) = \frac{O}{A}$$

- The following mnemonic can be used to help remember the trigonometric ratios.

$$\text{SOH} - \text{CAH} - \text{TOA}$$

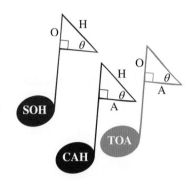

- In this mnemonic:
  - **SOH** refers to $\sin(\theta) = $ **O**pposite/**H**ypotenuse
  - **CAH** refers to $\cos(\theta) = $ **A**djacent/**H**ypotenuse
  - **TOA** refers to $\tan(\theta) = $ **O**pposite/**A**djacent.

WORKED EXAMPLE 10 Identifying trigonometric ratios

For the triangle shown, write the equations for the sine, cosine and tangent ratios of the given angle (θ).

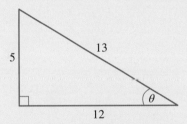

THINK	WRITE
1. Label the sides of the triangle.	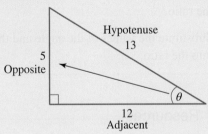
2. Write the trigonometric ratios.	$\sin(\theta) = \dfrac{O}{H}$, $\cos(\theta) = \dfrac{A}{H}$, $\tan(\theta) = \dfrac{O}{A}$.
3. Substitute the values of A, O and H into each formula.	$\sin(\theta) = \dfrac{5}{13}$, $\cos(\theta) = \dfrac{12}{13}$, $\tan(\theta) = \dfrac{5}{12}$

For each of the following triangles, write the trigonometric ratio that relates the two given sides and the reference angle.

a.

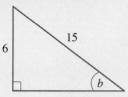

b.

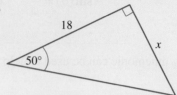

THINK

a. 1. Label the given sides.

2. We are given O and H. These are used in SOH. Write the ratio.

3. Substitute the values of the pronumerals into the ratio.

4. Simplify the fraction.

b. 1. Label the given sides.

2. We are given A and O. These are used in TOA. Write the ratio.

3. Substitute the values of the angle and the pronumerals into the ratio.

WRITE

a.

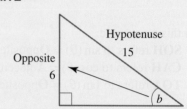

$$\sin(\theta) = \frac{O}{H}$$

$$\sin(b) = \frac{6}{15}$$

$$\sin(b) = \frac{2}{5}$$

b.

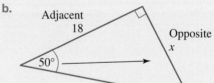

$$\tan(\theta) = \frac{O}{A}$$

$$\tan(50°) = \frac{x}{18}$$

 Resources

eWorkbook   Topic 5 Workbook (worksheets, code puzzle and project) (ewbk-2005)

Digital documents SkillSHEET Measuring angles with a protractor (doc-10831)

Interactivities   Individual pathway interactivity: Trigonometric ratios (int-4498)
                      Triangles (int-3843)
                      Trigonometric ratios (int-2577)

Exercise 5.4 Trigonometric ratios

learnon

Individual pathways

■ PRACTISE	■ CONSOLIDATE	■ MASTER
1, 3, 5, 8, 11, 13, 16, 19	2, 6, 9, 14, 17, 20	4, 7, 10, 12, 15, 18, 21

To answer questions online and to receive **immediate corrective feedback** and **fully worked solutions** for all questions, go to your learnON title at www.jacplus.com.au.

Fluency

WE9 For questions 1 and 2, label the sides of the following right-angled triangles using the words 'hypotenuse', 'adjacent' and 'opposite'.

1. a. 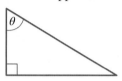 b. 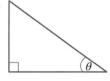 c.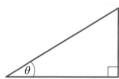

2. a.  b. 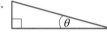 c.

3. For each of the following right-angled triangles, label the reference angle and, where appropriate, the hypotenuse, the adjacent side and the opposite side.

   a. 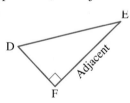 b. 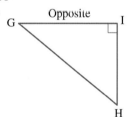 c.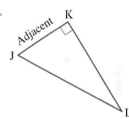

4. **MC** Identify which option correctly names the sides and angle of the triangle shown.

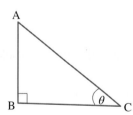

   A. $\angle B = \theta$, AB = adjacent side, AC = hypotenuse, BC = opposite side
   B. $\angle C = \theta$, AC = opposite side, BC = hypotenuse, AC = adjacent side
   C. $\angle A = \theta$, AB = opposite side, BC = hypotenuse, AC = adjacent side
   D. $\angle C = \theta$, AB = opposite side, AC = hypotenuse, BC = adjacent side
   E. $\angle C = \theta$, BC = adjacent side, AB = hypotenuse, AC = opposite side

WE10 For questions 5–10, based on the given angle in the triangles shown, write the expressions for the ratios of sine, cosine and tangent.

5.

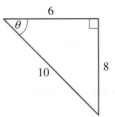

6.

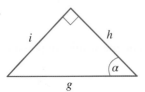

7.

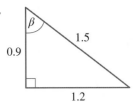

8.

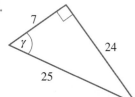

9.

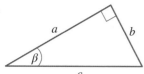

10.

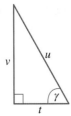

11. **WE11** Write the trigonometric ratio that relates the two given sides and the reference angle in each of the following triangles.

a.

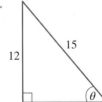

b.

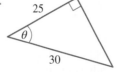

c.

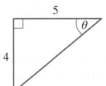

d.

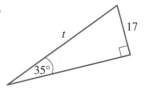

12. Write the trigonometric ratio that relates the two given sides and the reference angle in each of the following triangles.

a.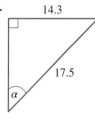
14.3
17.5
α

b.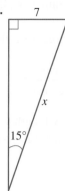
7
x
15°

c.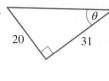
θ
20
31

d.
9.8
α
3.1

Understanding

13. **MC** Identify the correct trigonometric ratio for the triangle shown.

A. $\tan(\gamma) = \dfrac{a}{c}$ **B.** $\sin(\gamma) = \dfrac{c}{a}$ **C.** $\cos(\gamma) = \dfrac{c}{b}$

D. $\sin(\gamma) = \dfrac{c}{b}$ **E.** $\tan(\gamma) = \dfrac{a}{b}$

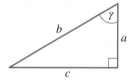

14. **MC** Identify the incorrect trigonometric ratio for the triangle shown.

A. $\sin(\alpha) = \dfrac{b}{c}$ **B.** $\sin(\alpha) = \dfrac{a}{c}$ **C.** $\cos(\alpha) = \dfrac{a}{c}$

D. $\tan(\alpha) = \dfrac{b}{a}$ **E.** $\cos(\alpha) = \dfrac{b}{c}$

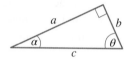

15. **MC** A right-angled triangle contains angles α and β as well as its right angle. Identify which of the following statements is always correct.

A. $\sin(\alpha) = \sin(\beta)$ **B.** $\tan(\alpha) = \tan(\beta)$ **C.** $\cos(\alpha) = 1 - \sin(\beta)$

D. $\tan(\alpha) = \dfrac{1}{\tan(\beta)}$ **E.** $\cos(\alpha) = 1 + \sin(\beta)$

Reasoning

16. If a right-angled triangle has side lengths m, $(m+n)$ and $(m-n)$, explain which of the lengths is the hypotenuse.

17. Given the triangle shown:

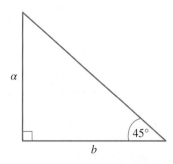
α
45°
b

a. explain why $a = b$

b. determine the value of $\tan(45°)$.

18. Using a protractor and ruler, carefully measure and draw a right-angled triangle with a base 10 cm long and an angle of 60°, as shown in the diagram. Measure the length of the other two sides of this triangle and mark these lengths on the diagram as well. Give the lengths of these sides to the nearest millimetre.

Use your measurements to calculate the following three ratios, correct to 2 decimal places.

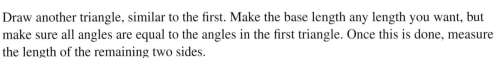

$$\frac{\text{opposite}}{\text{adjacent}} = \qquad , \qquad \frac{\text{opposite}}{\text{hypotenuse}} = \qquad , \qquad \frac{\text{adjacent}}{\text{hypotenuse}} = \qquad$$

Draw another triangle, similar to the first. Make the base length any length you want, but make sure all angles are equal to the angles in the first triangle. Once this is done, measure the length of the remaining two sides.

Calculate the following three ratios correct to 2 decimal places.

$$\frac{\text{opposite}}{\text{adjacent}} = \qquad , \qquad \frac{\text{opposite}}{\text{hypotenuse}} = \qquad , \qquad \frac{\text{adjacent}}{\text{hypotenuse}} = \qquad$$

Comment on the conclusions you can draw from these two sets of ratios.

Problem solving

19. A ladder leans on a wall as shown. In relation to the angle given, identify the part of the image that represents:

    a. the adjacent side
    b. the hypotenuse
    c. the opposite side.

20. Consider the right-angled triangle shown.

    a. Label each of the sides using the letters O, A and H with respect to the 41° angle.
    b. Determine the value of each of the following trigonometric ratios, correct to 2 decimal places.
       i. $\sin(41°)$ ii. $\cos(41°)$ iii. $\tan(41°)$
    c. Calculate the value of the unknown angle, α.
    d. Determine the value of each of the following trigonometric ratios, correct to 2 decimal places. (*Hint:* Start by re-labelling the sides of the triangle with respect to angle α.)
       i. $\sin(\alpha)$ ii. $\cos(\alpha)$ iii. $\tan(\alpha)$
    e. Comment about the relationship between $\sin(41°)$ and $\cos(\alpha)$.
    f. Comment about the relationship between $\sin(\alpha)$ and $\cos(41°)$.
    g. Make a general statement about the two angles.

21. In relation to right-angled triangles, investigate the following.

    a. As the acute angle increases in size, determine what happens to the ratio of the length of the opposite side to the length of the hypotenuse in any right-angled triangle.
    b. As the acute angle increases in size, determine what happens to the ratio of the length of the adjacent side to the length of the hypotenuse. Determine what happens to the ratio of the length of the opposite side to the length of the adjacent side.
    c. Evaluate the largest possible value for the following.

       i. $\sin(\theta)$ ii. $\cos(\theta)$ iii. $\tan(\theta)$

5.5 Calculating unknown side lengths

LEARNING INTENTION

At the end of this subtopic, you should be able to:
- use a calculator to calculate the trigonometric ratios for a given reference angle
- determine the lengths of unknown sides in a right-angled triangle given a reference angle and one known side length.

▶ 5.5.1 Values of trigonometric ratios

eles-4758

- The values of the trigonometric ratios are always the same for a given angle.
  For example, if a right-angled triangle has a reference angle of 30°, then the ratio of the opposite side to the hypotenuse will always be equal to $\sin(30°) = 0.5$.
- The ratio of the adjacent side to the hypotenuse will always be equal to $\cos(30°) \approx 0.87$. This relationship is demonstrated in the table.

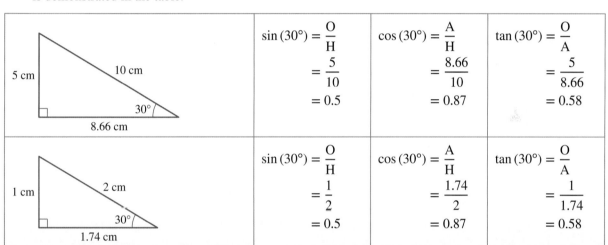

	$\sin(30°) = \dfrac{O}{H}$	$\cos(30°) = \dfrac{A}{H}$	$\tan(30°) = \dfrac{O}{A}$
5 cm, 10 cm, 8.66 cm, 30°	$= \dfrac{5}{10}$ $= 0.5$	$= \dfrac{8.66}{10}$ $= 0.87$	$= \dfrac{5}{8.66}$ $= 0.58$
1 cm, 2 cm, 1.74 cm, 30°	$\sin(30°) = \dfrac{O}{H}$ $= \dfrac{1}{2}$ $= 0.5$	$\cos(30°) = \dfrac{A}{H}$ $= \dfrac{1.74}{2}$ $= 0.87$	$\tan(30°) = \dfrac{O}{A}$ $= \dfrac{1}{1.74}$ $= 0.58$

WORKED EXAMPLE 12 Evaluating trigonometric ratios using a calculator

Evaluate each of the following. Give your answers correct to 4 decimal places.
a. $\sin(53°)$ **b.** $\cos(31°)$ **c.** $\tan(79°)$

THINK	WRITE
a. 1. Set the calculator to degree mode. Make the calculation and then write out the first 5 decimal places.	**a.** $\sin(53°) = 0.79863$
2. Round the answer to 4 decimal places.	≈ 0.7986
b. 1. Set the calculator to degree mode. Make the calculation and then write out the first 5 decimal places.	**b.** $\cos(31°) = 0.85716$
2. Round the answer to 4 decimal places.	≈ 0.8572
c. 1. Set the calculator to degree mode. Make the calculation and then write out the first 5 decimal places.	**c.** $\tan(79°) = 5.14455$
2. Round the answer to 4 places.	≈ 5.1446

TI	THINK	DISPLAY/WRITE	CASIO	THINK	DISPLAY/WRITE

a-c. 1. Ensure your calculator is set to Degree and Approximate modes. To do this, press:

• HOME
• 5: settings
• 2: Document settings

In the Display Digits, select Fix 4, tab to Angle and select Degree, then tab to Calculation Mode and select Approximate.

Tab to OK and press ENTER.

a-c.

a-c. Ensure the calculator is in Degree and Decimal modes. On the Main screen, complete the entry lines as:

sin(53)
cos(31)
tan(79)

Press EXE after each entry.

a. $\sin(53°) \approx 0.7986$
b. $\cos(31°) \approx 0.8572$
c. $\tan(79°) \approx 5.1446$

2. On a Calculator page, complete the entry lines as:

sin(53)
cos(31)
tan(79)

Press CTRL ENTER after each entry to get a decimal approximation.

a-c.

1.1 ▶	Trigonometry	DEG ☐ ✕
sin(53)		0.7986
cos(31)		0.8572
tan(79)		5.1446
I		

a. $\sin(53°) \approx 0.7986$
b. $\cos(31°) \approx 0.8572$
c. $\tan(79°) \approx 5.1446$

▶ 5.5.2 Determining side lengths using trigonometric ratios

eles-4759

- If a reference angle and any side length of a right-angled triangle are known, it is possible to calculate the lengths of the other sides using trigonometry.

WORKED EXAMPLE 13 Calculating an unknown side length using the tangent ratio

Use the appropriate trigonometric ratio to calculate the length of the unknown side in the triangle shown. Give your answer correct to 2 decimal places.

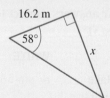

THINK

1. Label the given sides.

2. These sides can be used in TOA. Write the ratio.

3. Substitute the values of θ, O and A into the ratio.

WRITE

Adjacent
16.2 m
58°
Opposite
x

$\tan(\theta) = \dfrac{O}{A}$

$\tan(58°) = \dfrac{x}{16.2}$

4. Solve the equation for x.

$$16.2 \times \tan(58°) = x$$

5. Calculate the value of x to 3 decimal places, then round the answer to 2 decimal places.

$$x = 16.2 \ \tan(58°)$$
$$x = 25.925$$
$$x \approx 25.93 \ \text{m}$$

TI	THINK	DISPLAY/WRITE

TI | THINK

Ensure your calculator is set to Degree mode. In a new problem, on a Calculator page, complete the entry line as:

$$\text{solve}\left(\tan(58) = \frac{x}{16.2}, x\right)$$

Then press ENTER.

DISPLAY/WRITE

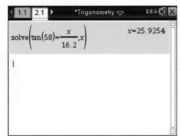

$x = 25.93$ m (correct to 2 decimal places)

CASIO | THINK

Ensure your calculator is set to Degree mode. On the Main screen, complete the entry line as:

$$\text{solve}\left(\tan(58) = \frac{x}{16.2}, x\right)$$

Then press EXE.

DISPLAY/WRITE

$x = 25.93$ m (correct to 2 decimal places)

WORKED EXAMPLE 14 Calculating an unknown side length using the cosine ratio

Use the appropriate trigonometric ratios to calculate the length of the side marked m in the triangle shown. Give your answer correct to 2 decimal places.

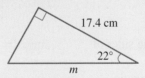

17.4 cm

22°

m

THINK

1. Label the given sides.

2. These sides can be used in CAH. Write the ratio.

3. Substitute the values of θ, A and H into the ratio.

4. Solve the equation for m.

WRITE

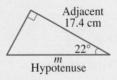

Adjacent
17.4 cm

22°

m
Hypotenuse

$$\cos(\theta) = \frac{A}{H}$$

$$\cos(22°) = \frac{17.4}{m}$$

$$m \cos(22°) = 17.4$$
$$m = \frac{17.4}{\cos(22°)}$$

5. Calculate the value of m to 3 decimal places, then round the answer to 2 decimal places.

$m = 18.766$

$m \approx 18.77\,\text{cm}$

WORKED EXAMPLE 15 Solving worded problems using trigonometric ratios

Bachar set out on a bushwalking expedition. Using a compass, he set off on a course N 70°E (or 070°T) and travelled a distance of 5 km from his base camp.

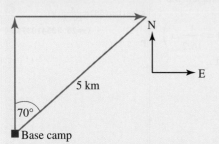
5 km
70°
Base camp
N
E

a. **Calculate how far east Bachar travelled.**
b. **Calculate how far north Bachar travelled from the base camp.**
   Give answers correct to 2 decimal places.

THINK	WRITE
a. 1. Label the eastern distance x. Label the northern distance y. Label the sides of the triangle as hypotenuse, opposite and adjacent.	a. 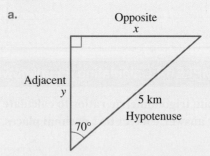 Opposite x Adjacent y 5 km Hypotenuse 70°
2. To calculate the value of x, use the sides of the triangle: $x = O$, $H = 5$. These are used in SOH. Write the ratio.	$\sin(\theta) = \dfrac{O}{H}$
3. Substitute the values of the angle and the pronumerals into the sine ratio.	$\sin(70°) = \dfrac{x}{5}$
4. Make x the subject of the equation.	$x = 5\sin(70°)$
5. Solve for x using a calculator. Write out the answer to 3 decimal places.	$= 4.698$
6. Round the answer to 2 decimal places.	$\approx 4.70\,\text{km}$
7. Write the answer in sentence form.	Bachar has travelled 4.70 km east of the base camp.
b. 1. To calculate the value of y, use the hypotenuse and the adjacent side: $y = A$, $H = 5$. These are used in CAH. Write the ratio.	b. $\cos(\theta) = \dfrac{A}{H}$

2. Substitute the values of the angle and the pronumerals into the cosine ratio.	$\cos(70°) = \dfrac{y}{5}$
3. Make y the subject of the equation.	$y = 5\cos(70°)$
4. Solve for y using a calculator. Write out the answer to 3 decimal places.	$= 1.710$
5. Round the answer to 2 decimal places.	$\approx 1.71\text{ km}$
6. Write the answer in sentence form.	Bachar has travelled 1.71 km north of the base camp.

 Resources

---

 eWorkbook   Topic 5 Workbook (worksheets, code puzzle and project) (ewbk-2005)

Digital documents SkillSHEET Solving equations of the type $a = \dfrac{x}{b}$ to find x (doc-10832)

SkillSHEET Solving equations of the type $a = \dfrac{b}{x}$ to find x (doc-10833)

Interactivity   Individual pathway interactivity: Calculating unknown side lengths (int-4499)

Exercise 5.5 Calculating unknown side lengths

Individual pathways

■ PRACTISE	■ CONSOLIDATE	■ MASTER
1, 4, 6, 9, 12, 13, 16, 19, 22	2, 5, 7, 10, 14, 17, 20, 23	3, 8, 11, 15, 18, 21, 24, 25

To answer questions online and to receive **immediate corrective feedback** and **fully worked solutions** for all questions, go to your learnON title at www.jacplus.com.au.

Fluency

1. a. **WE12** Calculate the following. Give your answers correct to 4 decimal places.
   i. $\sin(55°)$ ii. $\sin(11.6°)$

   b. Complete the table shown. Use your calculator to find each value of $\sin(\theta)$. Give your answers correct to 2 decimal places.

θ	0°	15°	30°	45°	60°	75°	90°
$\sin(\theta)$							

   c. Summarise the trend in these values.

2. a. Calculate the following. Give your answers correct to 4 decimal places.
   i. $\cos(38°)$ ii. $\cos(53.71°)$

   b. Complete the table shown. Use your calculator to find each value of $\cos(\theta)$. Give your answers correct to 2 decimal places.

θ	0°	15°	30°	45°	60°	75°	90°
$\cos(\theta)$							

   c. Summarise the trend in these values.

3. a. Calculate the following. Give your answers correct to 4 decimal places.

    i. tan(18°)   **ii.** tan(51.9°)

b. Complete the table shown. Use your calculator to find each value of tan(θ). Give your answers correct correct to 2 decimal places.

θ	0°	15°	30°	45°	60°	75°	90°
$\tan(\theta)$							

c. Determine the value of tan(89°) and tan(89.9°).

d. Comment on these results.

For questions **4–10**, use the appropriate trigonometric ratios to calculate the length of the unknown side in each of the triangles shown. Give your answers correct to 2 decimal places.

4. `WE13`

a.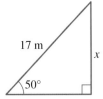

b.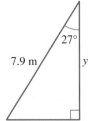

c.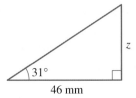

5. a.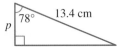

b.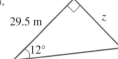

c.

6. `WE14`

a.

b.

c.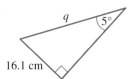

7. a.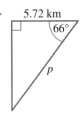

b.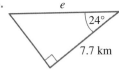

c.

8. a.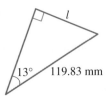

b.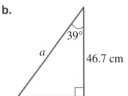

c.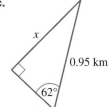

9. a.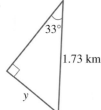

b.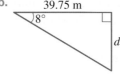

c.

10. a.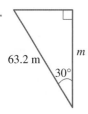

b.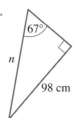

c.

11. Calculate the lengths of the unknown sides in the triangles shown. Give your answers correct to 2 decimal places.

a.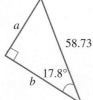

b.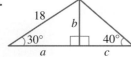

c.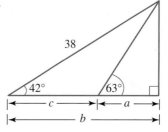

Understanding

12. **MC** Identify the value of x in the triangle shown, correct to 2 decimal places.
   A. 59.65
   B. 23.31
   C. 64.80
   D. 27.51
   E. 27.15

13. **MC** Identify the value of x in the triangle shown, correct to 2 decimal places.
   A. 99.24 mm
   C. 185.55 mm
   E. 92.55 mm
   B. 92.55 mm
   D. 198.97 mm

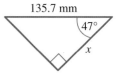

14. **MC** Identify y in the triangle shown, correct to 2 decimal places.
   A. 47.19
   C. 1.37
   E. 6.33
   B. 7.94
   D. 0.23

15. **MC** Identify the value of y in the triangle shown, correct to 2 decimal places.
   A. 0.76 km
   C. 3.83 km
   E. 3.47 m
   B. 1.79 km
   D. 3.47 km

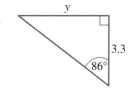

16. **WE15** A ship that was supposed to travel due north veers off course and travels N 80°E (or 080°T) for a distance of 280 km instead. The ship's path is shown in the diagram.

    a. Calculate the distance east that the ship has travelled.
    b. Determine the distance north that the ship has travelled.

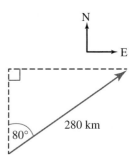

17. A rescue helicopter spots a missing surfer drifting out to sea on their damaged board. The helicopter descends vertically to a height of 19 m above sea level and drops down an emergency rope, which the surfer grabs onto.
    Due to the high winds, the rope swings at an angle of 27° to the vertical, as shown in the diagram.
    Calculate the length of the rope.

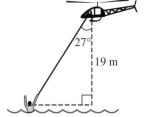

18. Walking along the coastline, Michelle (M) looks up at an angle of 55° and sees her friend Hella (H) at the lookout point on top of the cliff.
    If Michelle is 200 m from the cliff's base, determine the height of the cliff.
    (Assume Michelle and Hella are the same height.)

Reasoning

19. Using a diagram, explain why $\sin(70°) = \cos(20°)$ and $\cos(70°) = \sin(20°)$. Generally speaking, explain which cosine $\sin(\theta)$ will be equal to.

20. One method for determining the distance across a body of water is illustrated in the diagram shown.

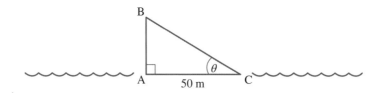

    The required distance is AB. A surveyor moves 50 m at a right angle to point C, then uses a tool called a transit to measure the angle θ (∠ACB).

    a. If $\theta = 12.3°$, show that the distance from A to B is 10.90 m.
    b. Show that a value of $\theta = 63.44°$ gives a distance from A to B of 100 m.
    c. Determine a rule that can be used to calculate the distance from A to C.

21. Explain why $\cos(0°) = 1$ and $\sin(0°) = 0$, but $\cos(90°) = 0$ and $\sin(90°) = 1$

Problem solving

22. Calculate the value of the pronumeral in each of the triangles shown in the following photos.

a.

b.

c.

23. A tile in the shape of a parallelogram has the measurements shown. Determine the tile's width, w, to the nearest mm.

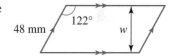

24. A pole is supported by two wires as shown. If the length of the lower wire is 4.3 m, evaluate the following to 1 decimal place.

   a. The length of the top wire
   b. The height of the pole

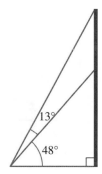

25. The frame of a kite is built from six wooden rods as shown. Evaluate the total length of wood used to make the frame of the kite. Give your answer to the nearest metre.

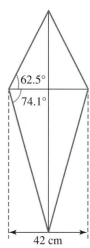

5.6 Calculating unknown angles

LEARNING INTENTION

At the end of this subtopic, you should be able to:
- use a calculator to identify an angle from a given trigonometric ratio
- determine the reference angle in a right-angled triangle when given two side lengths.

▶ 5.6.1 Inverse trigonometric ratios

eles-4760

- When we are given a reference angle, the sine, cosine and tangent functions let us determine the ratio of side lengths, but what if we only know the side lengths and we want to find the angle in a right-angled triangle?
- The inverse trigonometric functions allow us to calculate angles using sine, cosine or tangent ratios.
- We have seen that $\sin(30°) = 0.5$. This means that $30°$ is the inverse sine of 0.5. This is written as $\sin^{-1}(0.5) = 30°$.

Sine function	Inverse sine function
$\sin(30°) = 0.5$	$\sin^{-1}(0.5) = 30°$

- When reading the expression $\sin^{-1}(x)$ out loud, we say, 'the inverse sine of x'.
- When reading the expression $\cos^{-1}(x)$ out loud, we say, 'the inverse cosine of x'.
- When reading the expression $\tan^{-1}(x)$ out loud, we say, 'the inverse tangent of x'.
- You can calculate inverse trigonometric ratios using the SIN^{-1}, COS^{-1} and TAN^{-1} buttons on your calculator.

Digital technology

1. Use your calculator to determine $\sin(30°)$, then determine the inverse sine $(\sin^{-1})$ of the answer. Choose another angle and do the same thing.
2. Now determine $\cos(30°)$ and then determine the inverse cosine $(\cos^{-1})$ of the answer. Choose another angle and do the same thing.
3. Lastly, determine $\tan(45°)$ and then determine the inverse tangent $(\tan^{-1})$ of the answer. Choose another angle and do the same thing.

The fact that $\sin$ and $\sin^{-1}$ cancel each other out is useful when solving equations like the following.

Consider: $\sin(\theta) = 0.3$.

Take the inverse sine of both sides

$$\sin^{-1}(\sin(\theta)) = \sin^{-1}(0.3)$$

$$\theta = \sin^{-1}(0.3)$$

Consider: $\sin^{-1}(x) = 15°$.

Take the sine of both sides.

$$\sin(\sin^{-1}(x)) = \sin(15°)$$

$$x = \sin(15°)$$

$$\text{Similarly, } \cos(\theta) = 0.522 \text{ means that}$$

$$\theta = \cos^{-1}(0.522)$$

$$\text{and } \tan(\theta) = 1.25 \text{ means that}$$

$$\theta = \tan^{-1}(1.25).$$

WORKED EXAMPLE 16 Evaluating inverse cosine values

Evaluate $\cos^{-1}(0.3678)$, correct to the nearest degree.

THINK	WRITE
1. Set your calculator to degree mode and make the calculation.	$\cos^{-1}(0.3678) = 68.4$
2. Round the answer to the nearest whole number and include the degree symbol in your answer.	$\approx 68°$

TI | THINK	WRITE/DRAW
Ensure your calculator is set to Degree mode. In a new problem, on a Calculator page, complete the entry line as: solve $\cos^{-1}(0.3678)$ Then press ENTER.	

$\cos^{-1}(0.3678) = 68°$ (to the nearest degree)

CASIO | THINK	WRITE/DRAW
Ensure your calculator is set to Degree mode. On the Main screen, complete the entry line as: $\cos^{-1}(0.3678)$ Then press EXE.	

$\cos^{-1}(0.3678) \approx 68°$ (to the nearest degree)

WORKED EXAMPLE 17 Determining angles using inverse trigonometric ratios

Determine the size of angle θ in each of the following. Give answers correct to the nearest degree.
a. $\sin(\theta) = 0.6543$
b. $\tan(\theta) = 1.745$

THINK	WRITE
a. 1. θ is the inverse sine of 0.6543.	a. $\sin(\theta) = 0.6543$
	$\theta = \sin^{-1}(0.6543)$
2. Make the calculation using your calculator's inverse sine function and record the answer.	$= 40.8$
3. Round the answer to the nearest degree and include the degree symbol in your answer.	$\approx 41°$
b. 1. θ is the inverse tangent of 1.745.	b. $\tan(\theta) = 1.745$
2. Make the calculation using your calculator's inverse tangent function and record the answer.	$\theta = \tan^{-1}(1.745)$
	$= 60.18$
3. Round the answer to the nearest degree and include the degree symbol in your answer.	$\approx 60°$

▶ 5.6.2 Determining the angle when two sides are known

eles-4761

- If the lengths of any two sides of a right-angled triangle are known, it is possible to find an angle using inverse sine, inverse cosine or inverse tangent.

WORKED EXAMPLE 18 Determining an unknown angle when two sides are known

Determine the value of θ in the triangle shown. Give your answer correct to the nearest degree.

THINK

1. Label the given sides. These are used in CAH. Write out the ratio.

2. Substitute the given values into the cosine ratio.

3. θ is the inverse cosine of $\dfrac{12}{63}$.

4. Evaluate.

5. Round the answer to the nearest degree.

WRITE

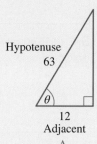

$$\cos(\theta) = \frac{A}{H}$$

$$\cos(\theta) = \frac{12}{63}$$

$$\theta = \cos^{-1}\left(\frac{12}{63}\right)$$

$$= 79.0$$

$$\approx 79°$$

WORKED EXAMPLE 19 Solving word problems using the inverse tangent

Roberta goes water skiing on the Hawkesbury River. She is going to try out a new ramp. The ramp rises 1.5 m above the water level and has a length of 6.4 m. Determine the magnitude (size) of the angle that the ramp makes with the water's surface. Give your answer correct to the nearest degree.

THINK

1. Draw a simple diagram, showing the known lengths and the angle to be found.

WRITE

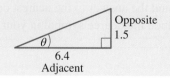

2. Label the given sides. These are used in TOA. Write out the ratio.

$$\tan(\theta) = \frac{O}{A}$$

3. Substitute the values of the pronumerals into the tangent ratio.

$$\tan(\theta) = \frac{1.5}{6.4}$$

4. θ is the inverse inverse tangent of $\frac{1.5}{6.4}$.

$$\theta = \tan^{-1}\left(\frac{1.5}{6.4}\right)$$

5. Evaluate.

$$= 13.19$$

6. Round the answer to the nearest degree.

$$\approx 13°$$

7. Write the answer in words.

The ramp makes an angle of 13° with the water's surface.

on Resources

 **eWorkbook** Topic 5 Workbook (worksheets, code puzzle and project) (ewbk-2005)

 **Digital document** SkillSHEET Rounding angles to the nearest degree (doc-10836)

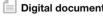 **Interactivities** Individual pathway interactivity: Calculating unknown angles (int-4500)

Finding the angle when two sides are known (int-6046)

Exercise 5.6 Calculating unknown angles

learn on

Individual pathways

■ PRACTISE	■ CONSOLIDATE	■ MASTER
1, 4, 7, 10, 11, 14, 18, 21	2, 5, 8, 12, 15, 17, 19, 22	3, 6, 9, 13, 16, 20, 23, 24

To answer questions online and to receive **immediate corrective feedback** and **fully worked solutions** for all questions, go to your learnON title at www.jacplus.com.au.

Fluency

1. **WE16** Calculate each of the following, rounding your answers to the nearest degree.
   a. $\sin^{-1}(0.6294)$
   b. $\cos^{-1}(0.3110)$
   c. $\tan^{-1}(0.7409)$

2. Calculate each of the following, rounding your answers to the nearest degree.
   a. $\tan^{-1}(1.3061)$
   b. $\sin^{-1}(0.9357)$
   c. $\cos^{-1}(0.3275)$

3. Calculate each of the following, rounding your answers to the nearest degree.
   a. $\cos^{-1}(0.1928)$
   b. $\tan^{-1}(4.1966)$
   c. $\sin^{-1}(0.2554)$

4. **WE17** Determine the size of the angle in each of the following, rounding your answers to the nearest degree.
   a. $\sin(\theta) = 0.3214$
   b. $\sin\theta = 0.6752$
   c. $\sin(\beta) = 0.8235$
   d. $\cos(\beta) = 0.9351$

5. Determine the size of the angle in each of the following, rounding your answers to the nearest degree.
   a. $\cos(\alpha) = 0.6529$
   b. $\cos(\alpha) = 0.1722$
   c. $\tan(\theta) = 0.7065$
   d. $\tan(a) = 1$

6. Determine the size of the angle in each of the following, rounding your answers to the nearest degree.
   a. $\tan(b) = 0.876$
   b. $\sin(c) = 0.3936$
   c. $\cos(\theta) = 0.5241$
   d. $\tan(\alpha) = 5.6214$

7. **WE18** Determine the value of θ in each of the following triangles, rounding your answers to the nearest degree.

a.

72
θ
49

b.

60
θ
85

c.
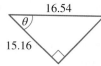
16.54
θ
15.16

8. Determine the value of θ in each of the following triangles, rounding your answers to the nearest degree.

a.

41.32
38.75
θ

b.
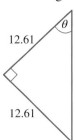
θ
12.61
12.61

c.
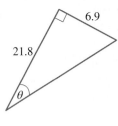
6.9
21.8
θ

9. Determine the value of θ in each of the following triangles, rounding your answers to the nearest degree.

a.
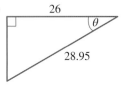
26
θ
28.95

b.

21.72
105.62
θ

c.
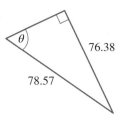
θ
76.38
78.57

10. **MC** If $\cos(\theta) = 0.8752$, identify the value of θ, correct to 2 decimal places.

   A. 61.07°   **B.** 41.19°   **C.** 25.84°   **D.** 28.93°   **E.** 29.93°

11. **MC** If $\sin(\theta) = 0.5530$, identify the value of θ, correct to 2 decimal places.

   A. 56.43°   **B.** 33.57°   **C.** 28.94°   **D.** 36.87°   **E.** 33.67°

12. **MC** Identify the value of θ in the triangle shown, correct to 2 decimal places.

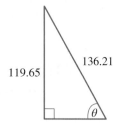

   A. 41.30°
   B. 28.55°
   C. 48.70°
   D. 61.45°
   E. 60.45°

136.21
119.65
θ

13. **MC** Identify the value of θ in the triangle shown, correct to 2 decimal places.

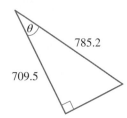

   A. 42.10°
   B. 64.63°
   C. 25.37°
   D. 47.90°
   E. 22.37°

θ
785.2
709.5

Understanding

14. A piece of fabric measuring 2.54 m by 1.5 m has been printed with parallel diagonal stripes. Determine the angle each diagonal makes with the length of the fabric.
Give your answer correct to 2 decimal places.

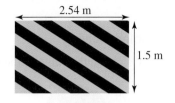

15. **WE19** Danny Dingo is perched on top of a 20-metre-high cliff. They are watching Erwin Emu, who is feeding on a bush that stands 8 m away from the base of the cliff. Danny has purchased a flying contraption that they hope will help them capture Erwin.
Calculate the angle to the cliff that Danny should follow downwards to catch their prey.
Give your answer correct to 2 decimal places.

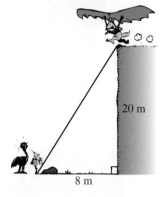

16. a. Complete the table shown.

x	0.0	0.1	0.2	0.3	0.4	0.5	0.6	0.7	0.8	0.9	1.0
$y = \cos^{-1}(x)$	90°					60°					0°

b. Plot the table from part **a** on graph paper. Alternatively you can use a spreadsheet program or a suitable calculator.

17. A zipline runs from an observation platform 400 m above sea level to a landing platform on the ground 1200 metres away.
Calculate the angle that the zipline makes with the ground at the landing platform.
Give your answer correct to 2 decimal places.

Reasoning

18. Safety guidelines for wheelchair access ramps used to state that the gradient of a ramp had to be in the ratio 1 : 20.

a. Using this ratio, show that the angle that any ramp needs to make with the horizontal is closest to 3°.
b. New regulations have changed the guidelines about the ratio of the gradient of a ramp. Now the angle the ramp makes with the horizontal must be closest to 6°.
Explain why, using this new angle size, the new ratio could be 1 : 9.5.

19. Jayani and Lee are camping with their friends Awer and Susie. Both couples have tents that are 2 m high. The top of each 2 m tent pole has to be tied with a piece of rope so that the pole stays upright. To make sure this rope doesn't trip anyone, Jayani and Lee decide that the angle between the rope and the ground should be 60°.

a. Determine the length of the rope that Jayani and Lee need to run from the top of their tent pole to the ground. Give your answer correct to 2 decimal places.
b. Awer and Susie set up their tent further into the camping ground. They want to secure their tent pole using a piece of rope that they know is somewhere between 2 and 3 metres long.
   i. Explain why Awer and Susie's rope will have to be longer than 2 m.
   ii. Show that the minimum angle Awer and Susie's rope will make with the ground will be 41.8°.

20. Use the formulas $\sin(\theta) = \dfrac{O}{H}$ and $\cos(\theta) = \dfrac{A}{H}$ to prove that $\tan(\theta) = \dfrac{\sin(\theta)}{\cos(\theta)}$.

Problem solving

21. Calculate the value of the pronumeral in each of the following, correct to 2 decimal places.

a.

5.4 cm

12 cm

θ

b.

1.2 m

θ

0.75 m

c.

0.75 m

x) 1.8 m

22. A family is building a patio extension for their house. One section of the new patio will have a gable roof. A similar structure is shown, with the planned post heights and the span of the roof given.

To allow more light in, the family wants the peak (the highest point) of the roof to be at least 5 m above ground level.

According to building regulations, the slope of the roof (the angle that the sloping edge makes with the horizontal) must be 22°.

6 m

3.2 m

a. Use trigonometry to calculate whether the roof would be high enough if the angle was 22°.
b. Using trigonometry, determine the size of the obtuse angle formed at the peak of the roof.

23. The height of a square-based prism is twice its base length. Calculate the angle the diagonal of the prism makes with the diagonal of the base.

24. A series of seven shapes are marked out as shown inside a square with a side length of 10 cm. If the dots marked on the diagram represent the midpoints of the square's sides, determine the dimensions of each of the seven smaller shapes.

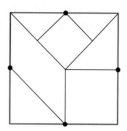

5.7 Angles of elevation and depression

LEARNING INTENTION

At the end of this subtopic, you should be able to:
- use angles of depression and elevation to determine heights and distances
- use trigonometric ratios to determine angles of depression and elevation.

⏵ 5.7.1 Angles of elevation and depression

eles-4762

- When looking up towards an object, the **angle of elevation** is the angle between the horizontal line and the line of vision.

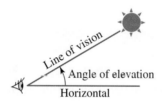

- When looking down at an object, the **angle of depression** is the angle between the horizontal line and the line of vision.

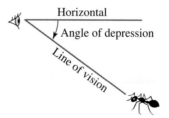

- Angles of elevation and depression are always measured from horizontal lines.
- For any two objects, A and B, the angle of elevation of B, as seen from A, is equal to the angle of depression of A, as seen from B.

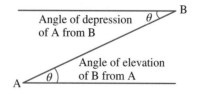

At a point 10 m from the base of a tree, the angle of elevation of the top of the tree is 38°. Calculate the height of the tree. Give your answer to the nearest centimetre.

THINK	WRITE
1. Draw a simple diagram to represent the situation. The angle of elevation is 38° from the horizontal.	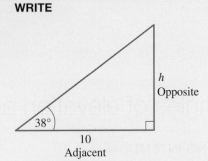
2. Label the given sides of the triangle. These sides are used in TOA. Write out the ratio.	$\tan(38°) = \dfrac{h}{10}$
3. Multiply both sides by 10.	$10\tan(38°) = h$
4. Calculate the answer, correct to 3 decimal places.	$h = 7.812$
5. Round the answer to 2 decimal places.	≈ 7.81
6. Write the answer in words.	The tree is 7.81 m tall.

A 30-metre-tall lighthouse stands on top of a cliff that is 180 m high. Determine the angle of depression (θ) of a ship from the top of the lighthouse if the ship is 3700 m from the bottom of the cliff.

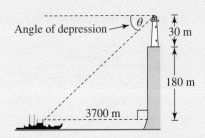

THINK	WRITE
1. Draw a simple diagram to represent the situation. The height of the triangle is $180 + 30 = 210$ m. Draw a horizontal line from the top of the triangle and mark the angle of depression, θ. Mark the alternate angle as well.	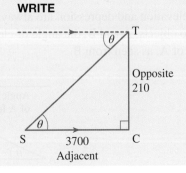

2. Label the given sides of the triangle. These sides are used in TOA. Write out the ratio.

$$\tan(\theta) = \frac{O}{A}$$

3. Substitute the given values into the ratio.

$$\tan(\theta) = \frac{210}{3700}$$

4. θ is the inverse tangent of $\frac{210}{3700}$.

$$\theta = \tan^{-1}\left(\frac{210}{3700}\right)$$

5. Calculate the answer.

$$= 3.24$$

6. Round the answer to the nearest degree.

$$\approx 3°$$

7. Write the answer in words.

The angle of depression of the ship from the top of the lighthouse is 3°.

Resources

eWorkbook   Topic 5 Workbook (worksheets, code puzzle and project) (ewbk-2005)

Digital document SkillSHEET Drawing a diagram from given directions (doc-10837)

Interactivities   Individual pathway interactivity: Angles of elevation and depression (int-4501)

                Finding the angle of elevation and angle of depression (int-6047)

Exercise 5.7 Angles of elevation and depression   learn on

Individual pathways

■ PRACTISE	■ CONSOLIDATE	■ MASTER
1, 4, 6, 9, 12	2, 5, 7, 10, 13	3, 8, 11, 14, 15, 16

To answer questions online and to receive **immediate corrective feedback** and **fully worked solutions** for all questions, go to your learnON title at www.jacplus.com.au.

Fluency

1. **WE20** A lifesaver standing on their tower 3 m above the ground spots a swimmer experiencing difficulty. The angle of depression of the swimmer from the lifesaver is 12°.

   Calculate how far the swimmer is from the lifesaver's tower. Give your answer correct to 2 decimal places.

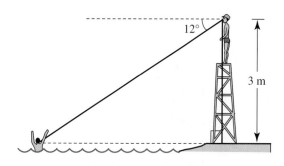

2. From the top of a 50-metre-high lookout, the angle of depression of a camp site that is level with the base of the lookout is 37°. Calculate how far the camp site is from the base of the lookout.

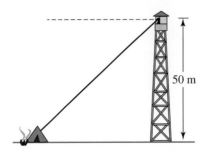

3. Building specifications require the angle of elevation of any ramp constructed for public use to be less than 3°.
   A new shopping centre is constructing its access ramps with a ratio of 7 m horizontal length to 1 m vertical height.
   Calculate the angle of elevation of these new ramps. Determine whether the new ramps meet the specifications required by ramps intended for public use.

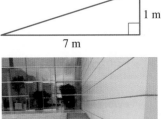

Understanding

4. From a point on the ground 60 m away from a tree, the angle of elevation of the top of the tree is 35°.

   a. Draw a labelled diagram to represent this situation.
   b. Calculate the height of the tree to the nearest metre.

5. Miriam wants to take a video of her daughter Alexandra's first attempts at crawling. When Alexandra lies on the floor and looks up at her mother, the angle of elevation is 17°.
   If Alexandra is 5.2 m away from her mother, calculate how tall Miriam is. Give your answer correct to 1 decimal place.

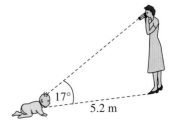

6. **WE21** Hien, who is 1.95 m tall, measures the length of the shadow he casts along the ground as 0.98 m. Determine the angle of depression of the sun's rays to the nearest degree.

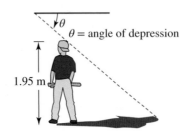

7. A ladder is 3.8 m tall and leaning against a wall.

   a. Determine the angle the ladder makes with the ground if it reaches 2.1 m up the wall. Give your answer to the nearest degree.

   b. Determine how far the foot of the ladder is from the wall. Give your answer to the nearest metre.

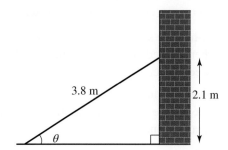

8. **MC** A lighthouse is 78 metres tall. The angle of elevation to the top of the lighthouse from point B, which is level with the base of the lighthouse, is 60°. Select the correct diagram for this information.

A.

78 m
60° B

B.

78 m
B 60°

C.

60°
B 78 m

D.

60° 78 m
B

E. None of these

Reasoning

9. Con and John are practising shots at goal. Con is 3.6 m away from the goal and John is 4.2 m away, as shown in the diagram.
If the height of the goal post is 2.44 m, determine the maximum angle of elevation that each player can kick the ball in order to score a goal. Give your answer to the nearest degree.

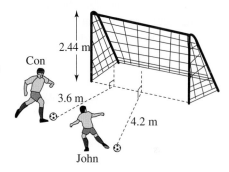

2.44 m
Con
3.6 m
4.2 m
John

10. Lifesaver Sami is sitting in a tower 10 m from the water's edge and 4 m above sea level. They spot some dolphins playing near a marker at sea directly in front of them. The marker the dolphins are swimming near is 20 m from the water's edge.

   a. Draw a diagram to represent this information.

   b. Show that the angle of depression of Sami's view of the dolphins, correct to 1 decimal place, is 7.6°.

   c. As the dolphins swim towards Sami, determine whether the angle of depression would increase or decrease. Justify your answer in terms of the tangent ratios.

11. A pair of office buildings are 100 m and 75 m high. From the top of the north side of the taller building, the angle of depression to the top of the south side of the shorter building is 20°, as shown.
Show that the horizontal distance between the north side of the taller building and the south side of the shorter building is closest to 69 m.

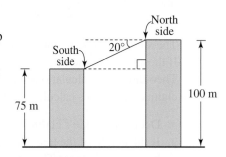

North side
South side
20°
75 m
100 m

Problem solving

12. From a rescue helicopter 80 m above the ocean, the angles of depression of two shipwreck survivors are 40° and 60° respectively. The two survivors and the helicopter are in line with each other.

    a. Draw a labelled diagram to represent the situation.
    b. Calculate the distance between the two survivors. Give your answer to the nearest metre.

13. Rouka was hiking in the mountains when she spotted an eagle sitting up in a tree. The angle of elevation of her view of the eagle was 35°. She then walked 20 m towards the tree. From her new position, her angle of elevation was 50°. The distance between the eagle and the ground was 35.5 m.

    a. Draw a labelled diagram to represent this information.
    b. If Rouka's eyes are located 9 cm below the very top of her head, calculate how tall she is. Give your answer in metres, correct to the nearest centimetre.

14. A lookout in a lighthouse tower can see two ships approaching the coast. Their angles of depression are 25° and 30°. If the ships are 100 m apart, show that the height of the lighthouse, to the nearest metre, is 242 metres.

15. As shown in the diagram, at a certain distance from an office building, the angle of elevation to the top of the building is 60°. From a distance 12 m further back, the angle of elevation to the top of the building is 45°. Show that the building is 28.4 m high.

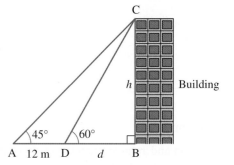

16. A gum tree stands in a courtyard in the middle of a group of office buildings. A group of three Year 9 students, Jackie, Pho and Theo, measure the angle of elevation from three different positions. They are unable to measure the distance to the base of the tree because of the steel tree guard around the base. This diagram shows the angles of elevation and the distances measured.

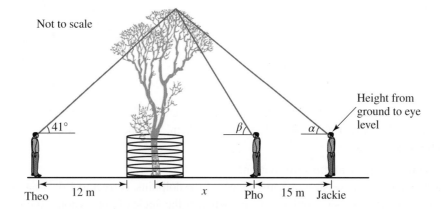

    a. Show that $x = \dfrac{15\,\tan(\alpha)}{\tan(\beta) - \tan(\alpha)}$, where x is the distance, in metres, from the base of the tree to Pho's position.
    b. The students estimate the tree to be 15 m taller than them. Pho measured the angle of elevation to be 72°. If these measurements are correct, calculate Jackie's angle of elevation, correct to the nearest degree.
    c. Theo did some calculations and determined that the tree was only about 10.4 m taller than them. Jackie claims that Theo's calculation of 10.4 m is incorrect.

       i. Decide if Jackie's claim is correct. Show how Theo calculated a height of 10.4 m.
       ii. If the height of the tree was actually 15 metres above the height of the students, determine the horizontal distance Theo should have used in his calculations. Write your answer to the nearest centimetre.

5.8 Review

5.8.1 Topic summary

Trigonometric ratios

- For a given reference angle, θ, in a right-angled triangle, the trigonometric ratios sine, cosine and tangent are constant.
  - $\sin(\theta) = \dfrac{\text{opposite (O)}}{\text{hypotenuse (H)}}$
  - $\cos(\theta) = \dfrac{\text{adjacent (A)}}{\text{hypotenuse (H)}}$ **SOH**
  - $\tan(\theta) = \dfrac{\text{opposite (O)}}{\text{adjacent (A)}}$ **CAH** **TOA**
- To remember these ratios, use the mnemonic **SOH CAH TOA** (**S**ine **O**pposite **H**ypotenuse, **C**os **A**djacent **H**ypotenuse, **T**an **O**pposite **A**djacent).

Inverse trigonometric ratios

- These ratios can be used to determine the size of a reference angle when given two side lengths.
  - $\sin^{-1}\left(\dfrac{O}{H}\right) = \theta$
  - $\cos^{-1}\left(\dfrac{A}{H}\right) = \theta$
  - $\tan^{-1}\left(\dfrac{O}{A}\right) = \theta$

PYTHAGORAS AND TRIGONOMETRY

Pythagoras' theorem

Pythagoras' theorem states the relationship between the lengths of the sides of right-angled triangles. Mathematically it is written as:

- $a^2 + b^2 = c^2$, where c is the length of the hypotenuse (the longest side) and a and b are the lengths of the other sides.
- This rule can be applied to any right-angled triangle.
- The hypotenuse is always located opposite the right angle.
- A Pythagorean triad is a group of any three whole numbers that satisfy Pythagoras' theorem.
  e.g. $\{3, 4, 5\}$ is a Pythagorean triad because $3^2 = 9$, $4^2 = 16$ and $5^2 = 25$, and $9 + 16 = 25$.

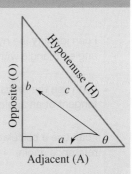

Angles of elevation and depression

- When looking upwards at an object, the **angle of elevation** is the angle between the horizontal and the upward line of sight.
- When looking downwards at an object, the **angle of depression** is the angle between the horizontal and the downward line of sight.
- The angle of elevation when looking from A to B is equal to the angle of depression when looking from B to A.

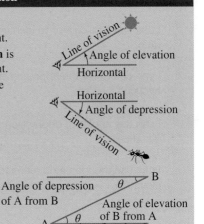

5.8.2 Success criteria

Tick the column to indicate that you have completed the subtopic and how well you have understood it using the traffic light system.

(**Green:** I understand; **Yellow:** I can do it with help; **Red:** I do not understand)

Subtopic	Success criteria	⬤	◯	◑
5.2	I can identify the hypotenuse in a right-angled triangle.			
	I can recall Pythagoras' theorem.			
	I can apply Pythagoras' theorem to find unknown side lengths.			
5.3	I can divide composite shapes into simple shapes.			
	I can calculate unknown side lengths in composite shapes and irregular triangles.			
	I can use Pythagoras' theorem to solve problems in 3D contexts.			
5.4	I can identify the hypotenuse, opposite and adjacent sides of a right-angled triangle with respect to a reference angle.			
	I can use the ratios of a triangle's sides to calculate the sine, cosine and tangent of an angle.			
5.5	I can use a calculator to calculate the trigonometric ratios for a given reference angle.			
	I can determine the lengths of unknown sides in a right-angled triangle given a reference angle and one known side length.			
5.6	I can use a calculator to identify an angle from a given trigonometric ratio.			
	I can determine the reference angle in a right-angled triangle when given two side lengths.			
5.7	I can use angles of depression and elevation to determine heights and distances.			
	I can use trigonometric ratios to determine angles of depression and elevation.			

5.8.3 Project

The Great Pyramid of Giza

The Great Pyramid of Giza, in Egypt, was built over 4500 years ago. It was constructed using approximately 2 300 000 rectangular granite blocks. This construction took over 20 years to complete. When it was finished, one side of its square base was 230 m long and its vertical height was 146.5 m.

1. Each side of the Great Pyramid of Giza has a triangular face. Use Pythagoras' theorem and the dimensions provided to calculate the height of each triangular face. Give your answer correct to 2 decimal places.
2. Special finishing blocks were added to the ends of each row of stones that were used to make the pyramid. These finishing blocks gave each triangular face a smooth, flat finish. Calculate the area of each face of the pyramid.

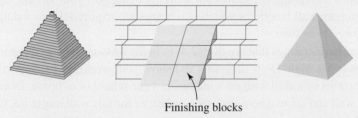

Finishing blocks

3. The edges of the pyramid join two of its faces and run all the way from the ground to the pyramid's tip. Use Pythagoras' theorem to calculate the length of these edges.

Wall braces

Builders use braces to strengthen wall frames. These braces typically run between the top and bottom horizontal sections of the wall frame.

Building industry standards require that the acute angle that a brace makes with the horizontal sections needs to be somewhere between 37° and 53°. Sometimes more than one brace may be required if the frame is particularly long, as shown in the diagram.

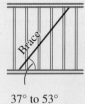 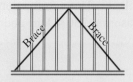 

37° to 53°

1. Cut out some thin strips of cardboard and arrange them in the shape of a rectangle to represent a wall frame. Place pins at the corners of the rectangle to hold the strips together. Notice that the frame can easily be moved out of shape. Attach a brace to this frame according to the building industry standards discussed. Write a brief comment to describe what effect this brace has on the frame.

2. Investigate what happens to the length of the brace as the acute angle it creates with the base is increased from 37° to 53°.

3. Use your findings from question 2 to calculate the angle that requires the shortest brace and the angle that requires the longest brace.

Most modern houses are constructed with a ceiling height (the height of the walls from floor to ceiling) of 2.4 m. Use this information to help with your calculations for the following questions.

4. Assume you are working with a section of wall that is 3.5 m long. Calculate the length of the longest possible brace. Draw a diagram and show your working to support your answer.

5. Calculate the minimum wall length for which two braces are required. Draw a diagram and show your working to support your answer.

6. Some older houses have ceilings that are over 2.4. m high. Answer the questions in 4 and 5 for a house with a ceiling that is 3 m high. Draw diagrams and show your workings to support your answers.

7. Take the measurements of a wall with no windows at your school or at home. Draw a scale drawing of the frame of this wall and show where the brace or braces for this wall might lie. Calculate the length and angle of each brace.

 Resources

 eWorkbook   Topic 5 Workbook (worksheets, code puzzle and project) (ewbk-2005)
 Interactivities Crossword (int-0703)
              Sudoku puzzle (int-3206)

To answer questions online and to receive **immediate corrective** feedback and **fully worked solutions** for all questions, go to your learnON title at www.jacplus.com.au.

Fluency

1. **MC** Identify the length of the third side in the triangle shown.
   - A. 34.71 m
   - B. 2.96 m
   - C. 5.89 m
   - D. 1722 cm
   - E. 58.9 cm

   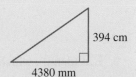

2. **MC** Select the most accurate measure for the length of the third side in the triangle shown.
   - A. 4.83 m.
   - B. 23.3 cm.
   - C. 3.94 m.
   - D. 4826 mm.
   - E. 4.83 mm.

   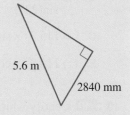

3. **MC** Select the value of x in this figure.
   - A. 5.4
   - B. 7.5
   - C. 10.1
   - D. 10.3
   - E. 10.5

   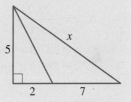

4. **MC** Identify which of the following is not a Pythagorean triad.
   - A. 3, 4, 5
   - B. 6, 8, 10
   - C. 5, 12, 13
   - D. 2, 3, 4
   - E. 12, 16, 20

5. **MC** Identify which of the following correctly names the sides and angle of the triangle shown.
   - A. $\angle C = \theta$, AB = adjacent side, AC = hypotenuse, BC = opposite side
   - B. $\angle C = \theta$, AB = opposite side, BC = hypotenuse, AC = adjacent side
   - C. $\angle A = \theta$, AB = opposite side, AC = hypotenuse, BC = adjacent side
   - D. $\angle A = \theta$, AB = adjacent side, AC = hypotenuse, BC = opposite side
   - E. $\angle A = \theta$, AB = adjacent side, BC = hypotenuse, AC = opposite side

   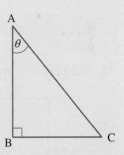

6. **MC** Select which of the following statements is correct.
   - A. $\sin(60°) = \cos(60°)$
   - B. $\cos(25°) = \cos(65°)$
   - C. $\cos(60°) = \sin(30°)$
   - D. $\sin(70°) = \cos(70°)$
   - E. $\cos(30°) = \sin(60°)$

7. **MC** Identify the value of x in the triangle shown, correct to 2 decimal places.

   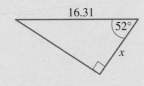

   - A. 26.49
   - B. 10.04
   - C. 12.85
   - D. 20.70
   - E. 10.06

8. **MC** Identify which of the following could be used to calculate the value of x in the triangle shown.

A. $x = \dfrac{172.1}{\cos(29°)}$ B. $x = \dfrac{172.1}{\sin(29°)}$ C. $x = 172.1 \times \sin(29°)$

D. $x = 172.1 \times \cos(29°)$ E. $x = 172.1 \times \tan(29°)$

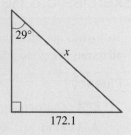

9. **MC** Identify which of the following could be used to calculate the value of x in the triangle shown.

A. $x = \dfrac{115.3}{\sin(23°)}$ B. $x = \dfrac{115.3}{\cos(67°)}$

C. $x = 115.3° \times \sin(67°)$ D. $x = \dfrac{115.3}{\cos(23°)}$

E. $x = 115.3° \times \cos(67°)$

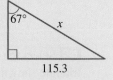

10. **MC** Identify which of the following could be used to calculate the value of x in the triangle shown.

A. $x = \dfrac{28.74}{\cos(17°)}$ B. $x = 28.74 \times \sin(17°)$

C. $x = 28.74 \times \cos(17°)$ D. $x = 28.74 \times \cos(73°)$

E. $x = 28.74 \times \sin(73°)$

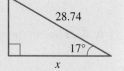

11. **MC** Select which of the following is closest to the value of $\tan^{-1}(1.8931)$.

A. 62°   B. 0.0331°   C. 1.08°   D. 69°   E. 63°

12. **MC** Select the value of θ in the triangle shown, correct to 2 decimal places.

A. 40.89°   B. 60°   C. 35.27°   D. 30°   E. 0.5°

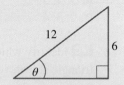

13. Calculate x, correct to 2 decimal places.

a. 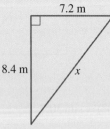   b. 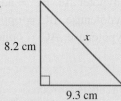   c.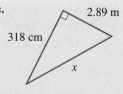

14. Calculate x, correct to 2 decimal places.

a. 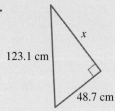   b. 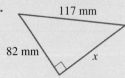   c.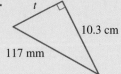

15. The top of a kitchen table measures 160 cm by 90 cm. A beetle walks diagonally across the table from one corner to the other. Calculate how far it walks. Give your answer correct to 2 decimal places.

16. A broom leans against a wall. The broom is 1.5 m long and reaches 1.2 m up the wall. Calculate how far the bottom of the broom is from the base of the wall.

17. Calculate the unknown values in the figures shown. Give your answers correct to 2 decimal places.

a.

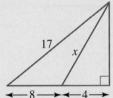

b.

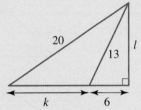

18. A beach athletics event involves three swimming legs and a run along the beach back to the start of the first swimming leg. Some of the distances of the race course are shown.
Calculate the total distance covered in the race.

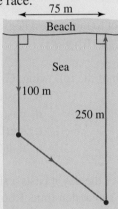

19. True or False? The following are Pythagorean triads.
a. 15, 36, 39   b. 50, 51, 10   c. 50, 48, 14

20. Label the unlabelled sides of the following right-angled triangles using the symbol θ and the words 'hypotenuse', 'adjacent' and 'opposite' where appropriate.

a.

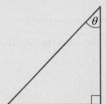

b.

c.

21. a. Write the trigonometric ratios that connect the lengths of the given sides and the size of the given angle in each of the following triangles.
b. Use these ratios to calculate the size of the angle.

i.
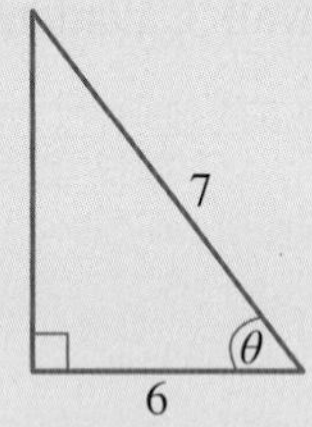

ii.
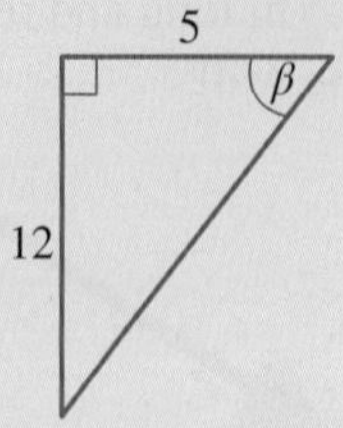

iii.
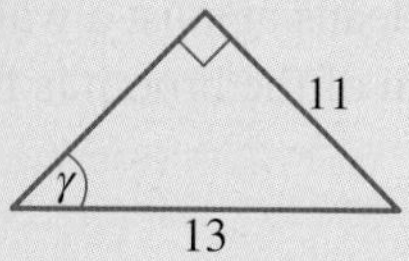

22. Use a calculator to evaluate the following trigonometric ratios, correct to 4 decimal places.

a. $\sin(54°)$ b. $\cos(39°)$ c. $\tan(12°)$

23. Calculate the values of the pronumerals in each of the following triangles. Give your answers correct to 2 decimal places.

a.
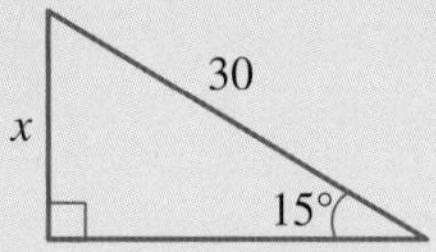

b.
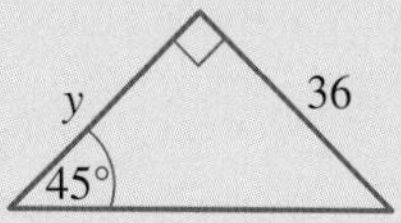

c.
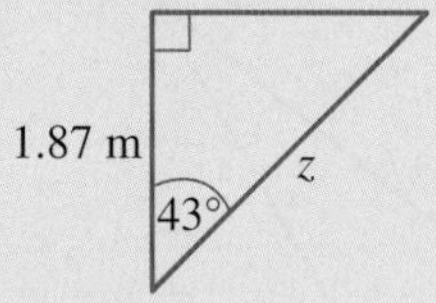

d.
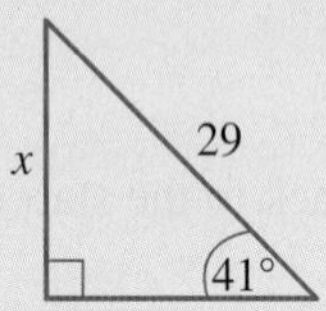

e.
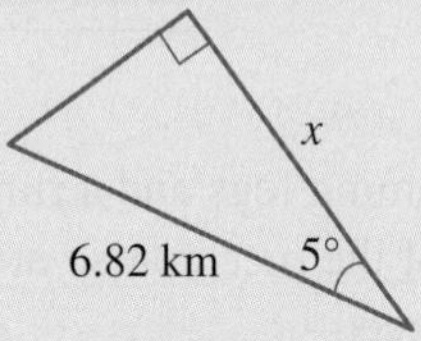

f.
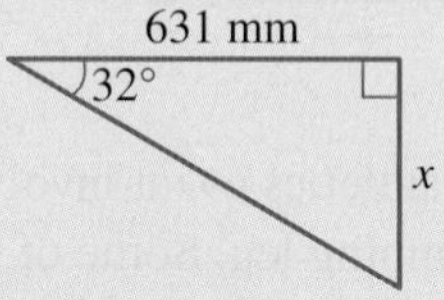

24. Evaluate each of the following, correct to the nearest degree.

a. $\sin^{-1}(0.1572)$ b. $\cos^{-1}(0.8361)$ c. $\tan^{-1}(0.5237)$

25. Calculate the size of the angle in each of the following. Give your answers correct to the nearest degree.

a. $\sin(\theta) = 0.5321$
b. $\cos(\theta) = 0.7071$
c. $\tan(\theta) = 0.8235$
d. $\cos(\alpha) = 0.3729$
e. $\tan(\alpha) = 0.5774$
f. $\sin(\beta) = 0.8660$
g. $\cos(\beta) = 0.5050$
h. $\tan(\beta) = 8.3791$

26. A tree is 6.7 m tall. At a certain time of the day it casts a shadow that is 1.87 m long. Determine the angle of depression of the rays of the sun at that time. Round your answer to 2 decimal places.

Problem solving

27. A 10-metre-high flagpole stands in the corner of a rectangular park that measures 240 m by 150 m.

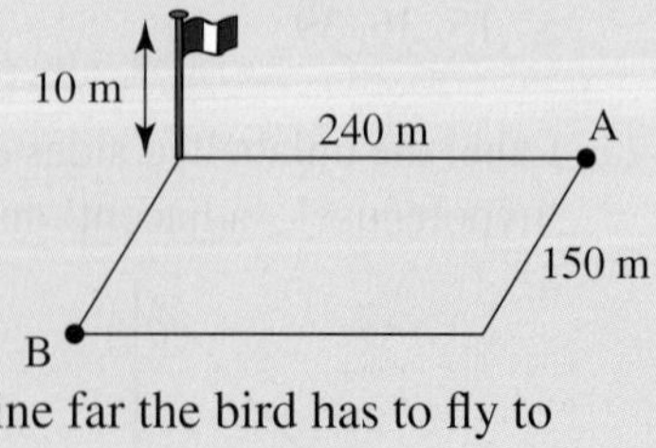

a. Determine the following, correct to 2 decimal places.
   i. The length of the diagonal of the park
   ii. The distance from A to the top of the pole
   iii. The distance from B to the top of the pole

b. A bird flies from the top of the pole to the middle of the park. Determine far the bird has to fly to reach this point.

28. Calculate the perimeters of the shapes shown.

a.

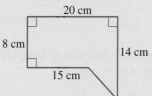

b.

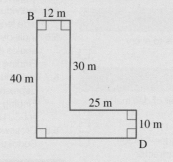

29. Evaluate the length of the shortest distance between points B and D inside the figure shown.

30. A pair of towers stand 30 m apart. From the top of tower A, the angle of depression of the base of tower B is 60° and the angle of depression of the top of tower B is 30°. Rounding your answer to the nearest metre, calculate the height of tower B.

31. Calculate the angles of a triangle whose sides can be described using the Pythagorean triad 3, 4, 5.

32. A stack of chairs is 2 m tall. The stack needs to fit through a doorway that is 1.8 m high. The maximum angle that the stack of chairs can be safely tilted is 25° to the vertical. Based on this information, determine if it is safe to try to move the stack of chairs through the doorway.

33. Evaluate the side length of the largest square that can be drawn within a circle that has a radius of r. Give your answer as a surd.

on To test your understanding and knowledge of this topic go to your learnON title at www.jacplus.com.au and complete the **post-test**.

Online Resources

 Resources

Below is a full list of **rich resources** available online for this topic. These resources are designed to bring ideas to life, to promote deep and lasting learning and to support the different learning needs of each individual.

eWorkbook

Download the workbook for this topic, which includes worksheets, a code puzzle and a project (ewbk-2005) ☐

Solutions

Download a copy of the fully worked solutions to every question in this topic (sol-0725) ☐

Digital documents

5.2 SkillSHEET Rounding to a given number of decimal places (doc-11428) ☐
SkillSHEET Rearranging formulas (doc-11429) ☐
SkillSHEET Converting units of length (doc-11430) ☐
5.4 SkillSHEET Measuring angles with a protractor (doc-10831) ☐
5.5 SkillSHEET Solving equations of the type $a = \dfrac{x}{b}$ to find x (doc-10832) ☐
SkillSHEET Solving equations of the type $a = \dfrac{b}{x}$ to find x (doc-10833) ☐
5.6 SkillSHEET Rounding angles to the nearest degree (doc-10836) ☐
5.7 SkillSHEET Drawing a diagram from given directions (doc-10837) ☐

Video eLessons

5.2 Introducing Pythagoras' theorem (eles-4750) ☐
Calculating unknown side lengths (eles-4751) ☐
Applying Pythagoras' theorem (eles-4752) ☐
Pythagorean triads (eles-4753) ☐
5.3 Using composite shapes to solve problems (eles-4754) ☐
Pythagoras' theorem in 3D (eles-4755) ☐
Pythagoras' theorem in 3 dimensions (eles-1913) ☐
5.4 What is trigonometry? (eles-4756) ☐
Trigonometric ratios (eles-4757) ☐
5.5 Values of trigonometric ratios (eles-4758) ☐
Finding side lengths using trigonometric ratio (eles-4759) ☐
5.6 Inverse trigonometric ratios (eles-4760) ☐
Finding the angle when two sides are known (eles-4671) ☐
5.7 Angles of elevation and depression (eles-4762) ☐

Interactivities

5.2 Individual pathway interactivity: Pythagoras' theorem (int-4472) ☐
Finding the hypotenuse (int-3844) ☐
Finding the shorter side (int-3845) ☐
Pythagorean triads (int-3848) ☐
5.3 Individual pathway interactivity: Applications of Pythagoras' theorem (int-4475) ☐
Composite shapes (int-3847) ☐

5.4 Individual pathway interactivity: Trigonometric ratios (int-4498) ☐
Triangles (int-3843) ☐
Trigonometric ratios (int-2577) ☐
5.5 Individual pathway interactivity: Calculating unknown side lengths (int-4499) ☐
5.6 Individual pathway interactivity: Calculating unknown angles (int-4500) ☐
Finding the angle when two sides are known (int-6046) ☐
5.7 Individual pathway interactivity: Angles of elevation and depression (int-4501) ☐
Finding the angle of elevation and angle of depression (int-6047) ☐
5.8 Crossword (int-0703) ☐
Sudoku puzzle (int-3206) ☐

Teacher resources

There are many resources available exclusively for teachers online.

To access these online resources, log on to **www.jacplus.com.au**.

Answers

Topic 5 Pythagoras and trigonometry

Exercise 5.1 Pre-test

1. 5 cm
2. 30.96 m
3. B
4. a + b
5. C
6. D
7. D
8. a. 3.47 cm   b. 5.14 mm
9. a. 0.8870   b. 0.9778   c. 3.7062
10. E
11. 23.4°
12. A
13. D
14. 7.31 m
15. 48.53 m

Exercise 5.2 Pythagoras' theorem

1. a. i. r
   ii. $r^2 = p^2 + s^2$
   b. i. x
   ii. $x = y^2 + z^2$
   c. i. k
   ii. $k^2 = m^2 + w^2$
   d. i. FU
   ii. $(FU)^2 = (VU)^2 + (VF)^2$
2. a. 7.86   b. 33.27   c. 980.95
3. a. $x = 12.49$ b. $p = 11.76\,\text{cm}$
   c. $f = 5.14\,\text{m}$ d. $c = 97.08\,\text{mm}$
4. 10.2 cm
5. a. No   b. No
   c. Yes   d. Yes
   e. No   f. Yes
6. a.

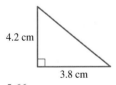

   b. 5.66 cm
7. a.

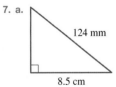

   b. 90.28 mm
8. B
9. C
10. B

11. a. 176.16 cm   b. 147.40 cm
    c. 2.62 km   d. 432.12 m
12. 2.60 m
13. The length of the diagonal can be calculated using Pythagoras' theorem.
    $$c^2 = a^2 + b^2$$
    $$c^2 = 260^2 + 480^2$$
    $$c^2 = 298\,000$$
    $$c = \sqrt{298\,000} = 546\,\text{m}$$
    The length of the triangular circuit is
    260 m + 480 m + 546 m = 1286 m.
    5 laps of this circuit is 5 × 1286 = 6430 m.
    5 laps of the regular circuit is
    5 × (260 + 260 + 480 + 480) = 7400 m.
    7400 m − 6430 m = 970 m.
14. The horizontal distance is 11.74 m, so the gradient is 0.21, which is within the limits.
15. a.

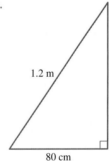

    b. 89.44 cm
    c. Yes, she can reach the hook from the top step.
16. a. i. {9, 40, 41}
    ii. {11, 60, 61}
    iii. {13, 84, 85}
    iv. {29, 420, 421}
    b. The middle number and the large number are one number apart.
17. The following figure shows the first three triangles, with the values of $\sqrt{2}$, $\sqrt{3}$ and $\sqrt{4}$.

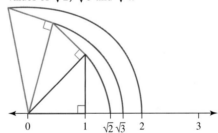

Exercise 5.3 Applications of Pythagoras' theorem

1. 12.08 cm
2. a. $k = 16.40\,\text{m}$ b. $x = 6.78\,\text{cm}$ c. $g = 4.10\,\text{km}$
3. a. $x = 4$, $y = 9.17$ b. $x = 6.93$, $y = 5.80$
   c. $x = 13$, $y = 15.20$ d. $x = 2.52$, $y = 4.32$
4. a. 30.48 cm   b. 2.61 cm   c. 47.27 cm

5. a.

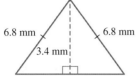

b. 11.78 mm

c. 20.02 mm^2

6. a. 0.87 m   **b.** 0.433 m^2

7. E

8. Yes. The longest umbrella that can fit in the suitcase is an umbrella that is 1.015 m long.

9. No. The longest straw that can fit in the lunchbox is a straw that is 17.72 cm long.

10. 2.42 m

11. a. $w = 0.47$ m   **b.** 0.64 m^2
   c. 12.79 m^2 **d.** $89.85

12. width $= 12$ mm, area $= 480$ mm^2

13. 17.9 cm

14. a. AD, DC, AC

   b. AD $= 4.47$ cm, DC $= 9.17$ cm, AC $= 13.64$ cm

   c. The triangle ABC is not right-angled because $(AB)^2 + (BC)^2 \neq (AC)^2$.

15. Even though a problem may be represented in 3D, right-angled triangles in 2D can often be found within the problem. This can be done by drawing a cross-section of the shape or by looking at individual faces of the shape.

16. Sample responses can be found in the worked solutions in the online resources..

17. a, b. Sample responses can be found in the worked solutions in the online resources.

18. a. 2606 mm

   b. Sample responses can be found in the worked solutions in the online resources.

19. 16.7 m

20. $w = 3.536$ m, $x = 7.071$ cm, $y = 15.909$ cm, $z = 3.536$ cm

Exercise 5.4 Trigonometric ratios

1. a. **b.** **c.**

2. a. **b.** **c.**

3. a. DE $=$ hypotenuse DF $=$ opposite $\angle E = \theta$

   b. GH $=$ hypotenuse IH $=$ adjacent $\angle H = \theta$

   c. JL $=$ hypotenuse KL $=$ opposite $\angle J = \theta$

4. D

5. $\sin(\theta) = \dfrac{4}{5}$, $\cos(\theta) = \dfrac{3}{5}$, $\tan(\theta) = \dfrac{4}{3}$

6. $\sin(\alpha) = \dfrac{i}{g}$, $\cos(\alpha) = \dfrac{h}{g}$, $\tan(\alpha) = \dfrac{i}{h}$

7. $\sin(\beta) = 0.8$, $\cos(\beta) = 0.6$, $\tan(\beta) = 1.3$

8. $\sin(\gamma) = \dfrac{24}{25}$, $\cos(\gamma) = \dfrac{7}{25}$, $\tan(\gamma) = \dfrac{24}{7}$

9. $\sin(\beta) = \dfrac{b}{c}$, $\cos(\beta) = \dfrac{a}{c}$, $\tan(\beta) = \dfrac{b}{a}$

10. $\sin(\gamma) = \dfrac{v}{u}$, $\cos(\gamma) = \dfrac{t}{u}$, $\tan(\gamma) = \dfrac{v}{t}$

11. a. $\sin(\theta) = \dfrac{12}{15} = \dfrac{4}{5}$ **b.** $\cos(\theta) = \dfrac{25}{30} = \dfrac{5}{6}$

   c. $\tan(\theta) = \dfrac{4}{5}$ **d.** $\sin(35°) = \dfrac{17}{t}$

12. a. $\sin(\alpha) = \dfrac{14.3}{17.5}$ **b.** $\sin(15°) = \dfrac{7}{x}$

   c. $\tan(\theta) = \dfrac{20}{31}$ **d.** $\cos(\alpha) = \dfrac{3.1}{9.8}$

13. D

14. B

15. D

16. Provided n is a positive value, $(m + n)$ would be the hypotenuse, because it has a greater value than both m and $(m - n)$.

17. a. Sample responses can be found in the worked solutions in the online resources.

   b. 1

18. The ratios remain constant when the angle is unchanged. The size of the triangle has no effect.

19. a. Ground   **b.** Ladder   **c.** Brick wall

20. a.

   b. **i.** $\sin(41°) = 0.65$

   ii. $\cos(41°) = 0.76$

   iii. $\tan(41°) = 0.86$

   c. $\alpha = 49°$

   d. **i.** $\sin(49°) = 0.76$

   ii. $\cos(49°) = 0.65$

   iii. $\tan(49°) = 1.16$

   e. They are equal.

   f. They are equal.

   g. The sine of an angle is equal to the cosine of its complement.

21. a. The ratio of the length of the opposite side to the length of the hypotenuse will increase.

   b. The ratio of the length of the adjacent side will decrease, and the ratio of the opposite side to the adjacent will increase.

   c. **i.** 1

   ii. 1

   iii. ∞

Exercise 5.5 Calculating unknown side lengths

1. a. i. 0.8192

   ii. 0.2011

   b.

θ	0°	15°	30°	45°	60°	75°	90°
sin (θ)	0	0.26	0.50	0.71	0.87	0.97	1.00

   c. As θ increases, so does sin(θ), starting at 0 and increasing to 1.

2. a. i. 0.7880

   ii. 0.5919

   b.

θ	0°	15°	30°	45°	60°	75°	90°
cos (θ)	1.00	0.97	0.87	0.71	0.50	0.26	0

   c. As θ increases, cos(θ) decreases, starting at 1 and decreasing to 0.

3. a. i. 0.3249

   ii. 1.2753

   b.

θ	0°	15°	30°	45°	60°	75°	90°
tan (θ)	0	0.27	0.58	1.00	1.73	3.73	Undefined

   c. tan(89°) = 57.29, tan(89.9°) = 572.96

   d. As θ increases, tan(θ) increases, starting at 0 and becoming very large. There is no value for tan (90°).

4. a. 13.02 m   b. 7.04 m   c. 27.64 mm

5. a. 2.79 cm   b. 6.27 m   c. 14.16 m

6. a. 2.95 cm   b. 25.99 cm   c. 184.73 cm

7. a. 14.06 km   b. 8.43 km   c. 31.04 m

8. a. 26.96 mm   b. 60.09 cm   c. 0.84 km

9. a. 0.94 km   b. 5.59 m   c. 41.67 m

10. a. 54.73 m   b. 106.46 cm   c. 298.54 mm

11. a. $a = 17.95, b = 55.92$

    b. $a = 15.59, b = 9.00, c = 10.73$

    c. $a = 12.96, b = 28.24, c = 15.28$

12. D

13. B

14. A

15. D

16. a. 275.75 km   b. 48.62 km

17. 21.32 m

18. 285.63 m

19. Sample responses can be found in the worked solutions in the online resources.

20. a. Sample responses can be found in the worked solutions in the online resources.

    b. Sample responses can be found in the worked solutions in the online resources.

    c. $AC = \dfrac{AB}{\tan(\theta)}$

21. Sample responses can be found in the worked solutions in the online resources.

22. a. $x = 12.87$ m   b. $h = 3.01$ m   c. $x = 2.60$ m

23. $w = 41$ mm

24. a. 5.9 m   b. 5.2 m

25. 4 m

Exercise 5.6 Calculating unknown angles

1. a. 39°   b. 72°   c. 37°

2. a. 53°   b. 69°   c. 71°

3. a. 79°   b. 77°   c. 15°

4. a. 19°   b. 42°   c. 55°   d. 21°

5. a. 49°   b. 80°   c. 35°   d. 45°

6. a. 41°   b. 23°   c. 58°   d. 80°

7. a. 47°   b. 45°   c. 24°

8. a. 43°   b. 45°   c. 18°

9. a. 26°   b. 12°   c. 76°

10. D

11. B

12. D

13. C

14. 30.56°

15. 21.80°

16. a.

x	$y = \cos^{-1}(x)$
0.0	90°
0.1	84°
0.2	78°
0.3	73°
0.4	66°
0.5	60°
0.6	53°
0.7	46°
0.8	37°
0.9	26°
1.0	0°

    b.

    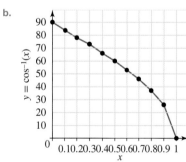

17. 18.43°

18. a. Sample responses can be found in the worked solutions in the online resources.

    b. Sample responses can be found in the worked solutions in the online resources.

19. a. 2.31 m
   b. i. Sample responses can be found in the worked solutions in the online resources.
      ii. Sample responses can be found in the worked solutions in the online resources.
20. Sample responses can be found in the worked solutions in the online resources.
21. a. $\theta = 26.01°$ b. $\theta = 32.01°$ c. $x = 22.62°$
22. a. The roof would not be high enough.
   b. 136°
23. 54.74°
24. Large square dimensions: length = 5 cm, width = 5 cm
   Large triangle dimensions: base = 5 cm, height = 5 cm
   Small square dimensions: length = $\dfrac{5\sqrt{2}}{2}$ cm,
   width $\dfrac{5\sqrt{2}}{2}$ cm
   Small triangle dimensions: base = $\dfrac{5\sqrt{2}}{2}$ cm, height $\dfrac{5\sqrt{2}}{2}$ cm
   Parallelogram dimensions: height = 5 cm, length $5\sqrt{2}$ cm

Exercise 5.7 Angles of elevation and depression

1. 14.11 m
2. 66.35 m
3. The new ramps have an angle of inclination of 8.13°. This does not meet the required specifications.
4. a.   b. 42 m
5. 1.6 m
6. 63°
7. 34°, 3 m
8. B
9. Con: 34°, John: 30°
10. a.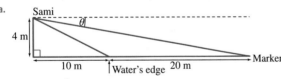
   b. $\tan(\theta) = \dfrac{3}{4}$
   $\theta = \tan^{-1}\left(\dfrac{4}{30}\right)$
   $\theta \approx 7.595$
   $\approx 7.6°$
   c.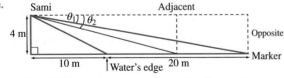
   As the dolphins swim towards Sami, the adjacent length decreases and the opposite remains unchanged.
   $\tan(\theta) = \dfrac{\text{opposite}}{\text{adjacent}}$

Therefore, θ will increase as the adjacent length decreases.
If the dolphins are at the water's edge,
$\tan(\theta) = 10$
$\theta = \tan^{-1}\left(\dfrac{4}{10}\right)$
$\theta \approx 21.801$
$\approx 21.8°$

11. Sample responses can be found in the worked solutions in the online resources.
12. a.
   b. 49 m
13. a. Sample responses can be found in the worked solutions in the online resources.
   b. 1.64 m
14. Sample responses can be found in the worked solutions in the online resources.
15. Sample responses can be found in the worked solutions in the online resources.
16. a. Sample responses can be found in the worked solutions in the online resources.
   b. 37°
   c. i. Yes
      ii. 17.26 m

Project

The Great Pyramid of Giza
1. 186.25 m
2. 21 418.75 m²
3. 218.89 m

Wall braces
1. Sample responses can be found in the worked solutions in the online resources.
2. Sample responses can be found in the worked solutions in the online resources.
3. 53° requires the shortest brace and 37° requires the longest brace.
4. 4.24 m
5. 3.62 m
6. 4.61 m, 4.52 m
7. Sample responses can be found in the worked solutions in the online resources.

Exercise 5.8 Review questions

1. C
2. D
3. D
4. D
5. D
6. C
7. B
8. B

9. D

10. C

11. A

12. D

13. a. 11.06 m

    b. 12.40 cm

    c. 429.70 cm or 4.30 m

14. a. 113.06 cm

    b. 83.46 mm

    c. 55.50 mm or 5.55 cm

15. 183.58 cm

16. 0.9 m

17. a. $x = 12.69$

    b. $l = 11.53$, $k = 10.34$

18. 593 m

19. a. True   b. False   c. True

20. a.   b.   c.

21. a. i. $\cos(\theta) = \dfrac{6}{7}$

      ii. $\tan(\beta) = \dfrac{12}{5}$

     iii. $\sin(\gamma) = \dfrac{11}{13}$

    b. i. $\theta = 31°$

      ii. $\beta = 67°$

     iii. $\gamma = 58°$

22. a. 0.8090   b. 0.7771   c. 0.2126

23. a. 7.76   b. 36.00   c. 2.56 m

    d. 19.03   e. 6.79 km   f. 394.29 mm

24. a. 9°   b. 33°   c. 28°

25. a. 32°   b. 45°   c. 39°

    d. 68°   e. 30°   f. 60°

    g. 60°   h. 83°

26. 74.41°

27. a. i. 283.02 m

      ii. 240.21 m

     iii. 150.33 m

    b. 141.86 m

28. a. 64.81 cm   b. 84.06 m

29. 59.24 m

30. 35 m

31. 90°, 53° and 37°

32. No. The stack of chairs must be tilted by 25.84° to fit through the doorway, which is more than the safe angle of 25°.

33. $\dfrac{2r}{\sqrt{2}} = \sqrt{2}r$

6 Linear and non-linear graphs

6.1 Overview

Why learn this?

We live in a world surrounded by shapes. Some of these shapes are straight, others are curved or made up of a combination of straight lines and curves. In many aspects of life, one quantity depends on another quantity, and the relationship between them can be described by equations. These equations can then be used in mathematical modelling to gain an understanding of the situation — we can draw graphs and use them to analyse, interpret and explain the relationship between the variables, and to make predictions about the future.

Scientists, engineers, health professionals and financial analysts all rely heavily on mathematical equations to model real-life situations and solve problems. Linear graphs are used extensively to represent trends, for example in the stock market or when considering the population growth of various countries.

Some of the most common curves we see each day are in the arches of bridges. The arch shape is actually a parabolic function. Because of its shape, it is very strong and stable. Architects and structural engineers use arches extensively in buildings and other structures. In space, satellites orbiting Earth follow an elliptical path, while the orbits of planets are almost circular.

Because of the many uses of linear and non-linear graphs, it is important to understand the basic concepts that you will study in this topic, such as gradient, how to calculate the distance between two points, and how to identify linear and non-linear relationships.

Where to get help

Go to your learnON title at **www.jacplus.com.au** to access the following digital resources. The Online Resources Summary at the end of this topic provides a full list of what's available to help you learn the concepts covered in this topic.

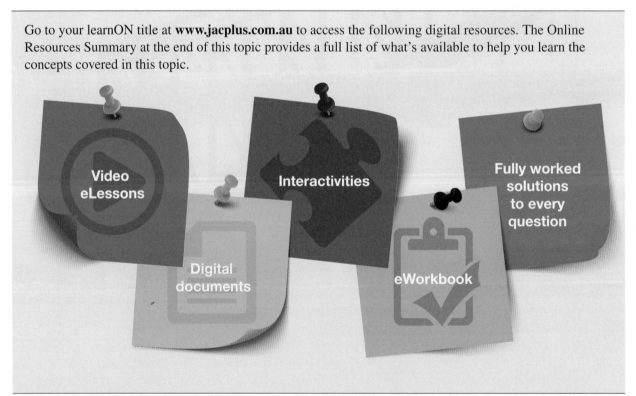

Video eLessons

Interactivities

Fully worked solutions to every question

Digital documents

eWorkbook

Exercise 6.1 Pre-test

Complete this pre-test in your learnON title at www.jacplus.com.au, and receive **automatic marks**, **immediate corrective feedback** and **fully worked solutions**.

1. State the location of the point $(-3, 0)$.

2. Using the table shown, match each point in the left-hand column with the line in the right-hand column that passes through that point.

Point	Line
a. $(3, -1)$	**A.** $y = 3 - 2x$
b. $(0, 4)$	**B.** $2y = x$
c. $(2, 1)$	**C.** $x + y = 2$
d. $(-1, 5)$	**D.** $y = x + 4$

3. **MC** Select the rule that corresponds to the table of values shown.

x	-3	-2	-1	-0	1
y	2	1	0	-1	-2

   A. $y = -x + 1$ **B.** $y = x + 1$ **C.** $y = -x - 1$ **D.** $y = x - 1$ **E.** $y = -x$

4. Calculate the gradient of the line passing through the points $\left(-3, \dfrac{1}{2}\right)$ and $\left(5, -\dfrac{7}{2}\right)$.

5. **MC** Select the gradient of the line $3x + 4y = 12$.

   A. $\dfrac{4}{3}$ **B.** 3 **C.** 4 **D.** $-\dfrac{3}{4}$ **E.** -3

6. Match the gradients and y-intercepts to the rules given by $y = mx + c$.

Gradient and y-intercept	Rule
a. $m = \dfrac{1}{2}, c = 3$	**A.** $2y - 6x = 4$
b. $m = \dfrac{1}{4}, c = 0$	**B.** $y = \dfrac{1}{2}x + 3$
c. $m = 4, c = -1$	**C.** $4y = x$
d. $m = 3, c = 2$	**D.** $y = 4x - 1$

7. **a.** Determine the rule for a straight line that passes through the origin and point $(2.4, -0.6)$.
   b. Determine the rule for a straight line that has an x-intercept $= -20$ and y-intercept $= 400$.

8. **MC** Select the equation of the linear graph shown.
   A. $y = 2x - 6$
   B. $y = 2x - 8$
   C. $y = -2x - 6$
   D. $y = -2x - 4$
   E. $y = -2x - 8$

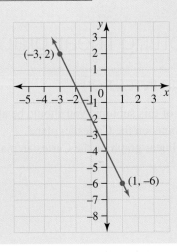

9. A student plays the game Blue Blobs. She has the following blobs on her screen, as shown in the diagram.

The student types in an equation for a line that will pass through four blobs.

a. Determine the equation of the straight line that passes through four blobs.

b. The student then types in the equation $x = -1$.
State the coordinate of the blob that both lines would hit.

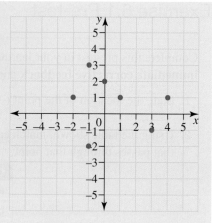

10. Determine the value of a so that the point $M\left(-\dfrac{1}{2}, -1\dfrac{1}{2}\right)$ is the midpoint of the segment joining points $A(3, 2)$ and $B(-4, a)$.

11. Calculate the exact distance between the points $(2, 5)$ and $(-3, 7)$.

12. **MC** Select the vertical asymptote for the equation $y = \dfrac{5}{x}$.

 A. $x = 0$ B. $x = 5$ C. $y = 0$ D. $y = 5$ E. $y = x$

13. **MC** Select the centre and radius for the graph $(x + 1)^2 + (y - 2)^2 = 9$.
 A. $(1, -2), r = 9$ B. $(1, -2), r = 3$ C. $(-1, 2), r = 9$ D. $(-1, 2), r = 3$ E. $(1, -2), r = 81$

14. **MC** For the graph $y = -x^2 + 4$, select the correct turning point of the parabola.
 A. a maximum at $(0, -4)$ B. a minimum at $(4, 0)$ C. a maximum at $(-4, 0)$
 D. a minimum at $(0, -4)$ E. a maximum at $(0, 4)$

15. **MC** Select the correct equation for the graph shown.
 A. $y = x^2 - 4$
 B. $y = -(x - 2)^2 - 4$
 C. $y = (x + 2)^2 + 4$
 D. $y = -(x + 2)^2$
 E. $y = -x^2 + 2$

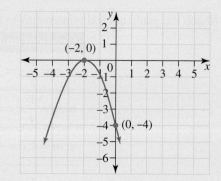

6.2 Plotting linear graphs

LEARNING INTENTION

At the end of this subtopic you should be able to:
• plot points on the Cartesian plane
• plot a linear graph on the Cartesian plane.

6.2.1 Plotting linear graphs on the Cartesian plane

eles-4779

- The **Cartesian plane** is divided into four regions (quadrants) by the x- and y-axes, as shown.
- Every point in the plane is described exactly by a pair of coordinates (x, y). The point P $(3, 2)$ is marked on the diagram.

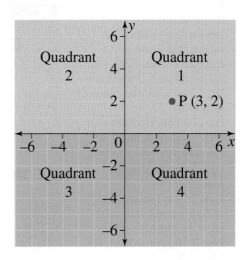

Plotting linear graphs from a rule

- A straight-line graph is called a **linear graph**.
- The rule for a linear graph can always be written in the form $y = mx + c$.
  For example, $y = 4x - 5$ or $y = x + 2$.
- A minimum of two points are needed to plot a linear graph.

Plotting a linear graph from a rule on a Cartesian plane

1. Draw a table of values for the x-values.
2. State the rule connecting x and y.
3. Substitute the x-values into the rule to find the y-values.
4. Plot the points on the Cartesian plane.
5. Label the graph.

WORKED EXAMPLE 1 Plotting a graph from a rule

Plot the graph specified by the rule $y = x + 2$ for the x-values $-3, -2, -1, 0, 1, 2, 3$.

THINK	WRITE
1. Draw a table and write in the required x-values.	<table>

x	-3	-2	-1	0	1	2	3
y							

2. Substitute each x-value into the rule $y = x + 2$ to obtain the corresponding y-value.
When $x = -3$, $y = -3 + 2 = -1$.
When $x = -2$, $y = -2 + 2 = 0$ etc.
Write the y-values into the table.

x	-3	-2	-1	0	1	2	3
y	-1	0	1	2	3	4	5

3. Plot the points from the table.

4. Join the points with a straight line and label the graph with its equation: $y = x + 2$.

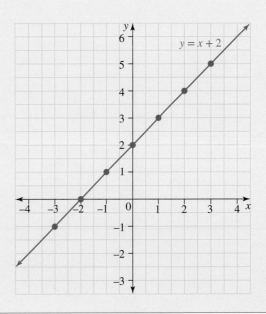

**TI	THINK**	**WRITE**	**CASIO	THINK**	**WRITE**

TI | THINK

1. In a new document, on a Lists & Spreadsheet page, label column A as *x*-values and column B as *y*-values.
Enter the *x*-values from -3 to 3 into column A. Then in cell B1, complete the entry line as:
$= a1 + 2$
Press ENTER. Hold down SHIFT and the down arrow key to fill down the *y*-values.

WRITE

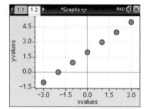

2. Open a Data & Statistics page.
Press TAB to locate the label of the horizontal axis and select the variable *x*-values. Press TAB again to locate the label of the vertical axis and select the variable *y*-values. The points will be plotted.
To change the colour of the points, select any point and press:
• CTRL
• MENU
• 3: Colour.
Choose a colour from the palette.

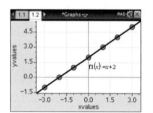

3. To join the points with a line, press:
• MENU
• 4: Analyze
• 4: Plot Function.
Complete the entry line as:
$f1(x) = x + 2$

CASIO | THINK

1. On the Spreadsheet screen, enter the *x*-values into column A. In cell B1, complete the entry line as: $= A1 + 2$
Press EXE.

WRITE

2. Highlight the required cells in column B. Press:
• Edit
• Fill
• Fill Range

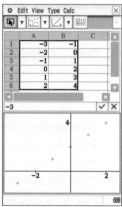

3. Highlight both columns and then press Graph. Select the scatterplot.

4. To find the equation, press:
• Calc
• Regression
• Linear Regression
The equation is $y = x + 2$. The line will appear as continuous.

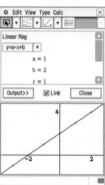

Plot two points and use them to draw the linear graph $y = 2x - 1$.

THINK

1. Choose any two x-values, for example $x = -2$ and $x = 3$.

2. Calculate y by substituting each x-value into $y = 2x - 1$.
$x = -2: y = 2 \times -2 - 1 = -5$
$x = 3: y = 2 \times 3 - 1 = 5$

3. Plot the points $(-2, -5)$ and $(3, 5)$.

4. Draw a line through the points and add a label.

WRITE

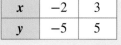

x	-2	3
y	-5	5

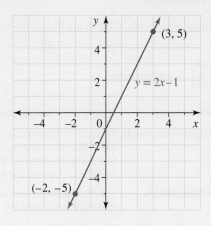

6.2.2 Points on a line

eles-4781

- Consider the line that has the rule $y = 2x + 3$ as shown in the graph
  If $x = 1$, then $y = 2(1) + 3$
  $\qquad = 5$
  so the point $(1, 5)$ lies on the line $y = 2x + 3$.
- The points $(1, -3)$ $(1, 9)$, $(1, 12)$ … are not on the line, but lie above or below it.
- The point $(-6, 4)$ is in quadrant 2.
  If $x = -6$: $\quad y = 2(-6) + 3$
  $\qquad\qquad y = -9$
  $\qquad\qquad y \neq 5$
  This shows that the point does not lie on the line.

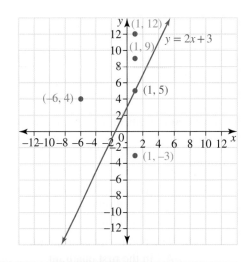

Determine whether the point $(2, 4)$ lies on the line given by:
a. $y = 3x - 2$
b. $x + y = 5$.

THINK

a. 1. Substitute $x = 2$ into the equation $y = 3x - 2$ and solve for y.

2. When $x = 2$, $y = 4$, so the point $(2, 4)$ lies on the line. Write the answer.

WRITE

a. $y = 3x - 2$
$x = 2$: $\quad y = 3(2) - 2$
$\qquad\qquad = 6 - 2$
$\qquad\qquad = 4$

The point $(2, 4)$ lies on the line $y = 3x - 2$.

b. 1. Substitute $x = 2$ into the equation $x + y = 5$ and solve for y.		**b.** $x + y = 5$ $x = 2:\ 2 + y = 5$ $y = 3$	
2. The point $(2, 3)$ lies on the line, but the point $(2, 4)$ does not. Write the answer.		The point $(2, 4)$ does not lie on the line $x + y = 5$.	

DISCUSSION

In linear equations, what does the coefficient of x determine?

 Resources

eWorkbook	Topic 6 Workbook (worksheets, code puzzle and project) (ewbk-2006)
Digital documents	SkillSHEET Plotting coordinate points (doc-6161) SkillSHEET Substituting into a rule (doc-6162) SkillSHEET Completing a table of values (doc-6163) SkillSHEET Plotting a line from a table of values (doc-6164)
Interactivities	Individual pathway interactivity: Plotting linear graphs (int-4502) Plotting linear graphs (int-3834)

Exercise 6.2 Plotting linear graphs

learn

Individual pathways

■ PRACTISE	■ CONSOLIDATE	■ MASTER
1, 2, 4, 9, 12	3, 5, 8, 11, 13	6, 7, 10, 14

To answer questions online and to receive **immediate corrective feedback** and **fully worked solutions** for all questions, go to your learnON title at www.jacplus.com.au.

Fluency

1. **MC** a. The point with coordinates $(-2, 3)$ is:
   - **A.** in quadrant 1
   - **B.** in quadrant 2
   - **C.** in quadrant 3
   - **D.** in quadrant 4
   - **E.** on the x-axis

   b. The point with coordinates $(-1, -5)$ is:
   - **A.** in the first quadrant
   - **B.** in the second quadrant
   - **C.** in the third quadrant
   - **D.** in the fourth quadrant
   - **E.** on the y-axis

   c. The point with coordinates $(0, -2)$ is:
   - **A.** in the third quadrant
   - **B.** in the fourth quadrant
   - **C.** on the x-axis
   - **D.** on the y-axis
   - **E.** in the second quadrant

2. **WE1** For each of the following rules:
   i. complete the table   ii. plot the linear graph.

x	-3	-2	-1	0	1	2	3
y							

   a. $y = x$ **b.** $y = 2x + 2$ **c.** $y = 3x - 1$ **d.** $y = -2x$

Understanding

3. **WE2** By first plotting two points, draw the linear graph given by each of the following.
   a. $y = -x$
   b. $y = \frac{1}{2}x + 4$
   c. $y = -2x + 3$
   d. $y = x - 3$

4. **WE3** Determine whether these points lie on the graph of $y = 2x - 5$.
   a. $(3, 1)$
   b. $(-1, 3)$
   c. $(0, 5)$
   d. $(5, 5)$

5. Determine whether the given point lies on the given line.
   a. $y = -x - 7, (1, -8)$
   b. $y = 3x + 5, (0, 5)$
   c. $y = x + 6, (-1, 5)$
   d. $y = 5 - x, (8, 3)$
   e. $y = -2x + 11, (5, -1)$
   f. $y = x - 4, (-4, 0)$

6. **MC** Select the line that passes through the point $(2, -1)$.
   A. $y = -2x + 5$
   B. $y = 2x - 1$
   C. $y = -2x + 1$
   D. $x + y = 1$
   E. $x + y = -1$

7. Match each point with a line passing through that point.

Point	Line
a. $(1, 1)$	A. $x + y = 4$
b. $(1, 3)$	B. $2x - y = 1$
c. $(1, 6)$	C. $y = 3x - 7$
d. $(1, -4)$	D. $y = 7 - x$

Reasoning

8. The line through $(1, 3)$ and $(0, 4)$ passes through every quadrant except one. State the quadrant through which this line does not pass. Explain your answer.

9. Answer the following questions.
   a. Determine which quadrant(s) the line $y = x + 1$ passes through.
   b. Show that the point $(1, 3)$ does not lie on the line $y = x + 1$.

10. Explain the process of how to check whether a point lies on a given line.

11. Using the coordinates $(-1, -3)$, $(0, -1)$ and $(2, 3)$, show that a rule for the linear graph is $y = 2x - 1$.

Problem solving

12. Consider the pattern of squares on the grid shown.
    Determine the coordinates of the centre of the 20th square.

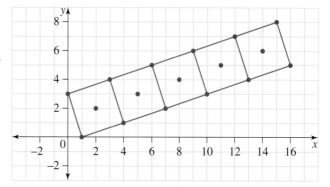

13. It is known that the mass of a certain kind of genetically modified tomato increases linearly over time. The following results were recorded.

Time, t (weeks)	1	4	6	9	16
Mass, m (grams)	6	21	31	46	81

   a. Plot the above points on a Cartesian plane.
   b. Determine the rule connecting mass with time.
   c. Show that the grams mass of one of these tomatoes is 101 grams after 20 weeks.

14. As a particular chemical reaction proceeds, the temperature increases at a constant rate. The graph represents the same chemical reaction with and without stirring. Interpret the graph and explain how stirring affects the reaction.

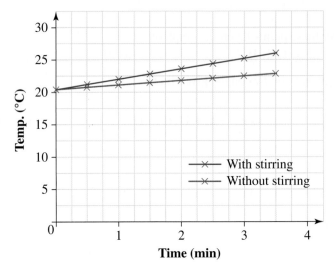

6.3 Features of linear graphs

LEARNING INTENTION

At the end of this subtopic you should be able to:
- calculate the gradient of a line or line segment
- calculate the *y*-intercept of a linear graph.

6.3.1 The gradient (*m*)

eles-4782

- The **gradient** of a line is a measure of the steepness of its slope.
- The symbol for the gradient is *m*.
- The gradient of an interval AB is defined as the rise (distance up) divided by the run (distance across).

> **Gradient formula**
>
> $$m = \frac{\text{rise}}{\text{run}}$$
>
> *(diagram of right triangle with vertices A at bottom-left, B at top, labelled rise on the vertical side and run on the horizontal side)*

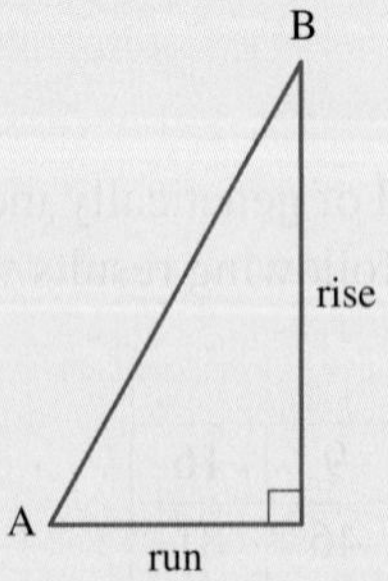

- If the line is sloping upwards (from left to right), the gradient is positive.
- If the line is sloping downwards, the gradient is negative.

Positive gradient (sloping upwards)	Negative gradient (sloping downwards)
$m = 2$	$m = -2$
$m = \dfrac{1}{2}$	$m = -\dfrac{1}{2}$

Determining the gradient of a line passing through two points

- Suppose a line passes through the points $(1, 4)$ and $(3, 8)$, as shown in the graph.
- By completing a right-angled triangle, it can be seen that:
  - the rise (difference in y-values): $8 - 4 = 4$
  - the run (difference in x-values): $3 - 1 = 2$
  - to determine the gradient:

$$
\begin{aligned}
m &= \frac{8 - 4}{3 - 1} \\
&= \frac{4}{2} \\
&= 2
\end{aligned}
$$

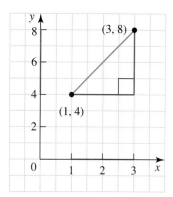

Gradient of a line

- In general, if a line passes through the points (x_1, y_1) and (x_2, y_2), then:

$$m = \frac{y_2 - y_1}{x_2 - x_1}$$

WORKED EXAMPLE 4 Determining the gradient of a line passing through two points

Calculate the gradient of the line passing through the points $(-2, 5)$ and $(1, 14)$.

THINK	WRITE
1. Let the two points be (x_1, y_1) and (x_2, y_2).	$(-2, 5) = (x_1, y_1), (1, 14) = (x_2, y_2)$
2. Write the formula for gradient.	$m = \dfrac{y_2 - y_1}{x_2 - x_1}$

3. Substitute the coordinates of the given points into the formula and evaluate.

$$m = \dfrac{14 - 5}{1 - -2}$$
$$= \dfrac{9}{1 + 2}$$
$$= \dfrac{9}{3}$$
$$= 3$$

4. Write the answer.

The gradient of the line passing through $(-2, 5)$ and $(1, 14)$ is 3.

Note: If you were to switch the order of the points and let $(x_1, y_1) = (1, 14)$ and $(x_2, y_2) = (-2, 5)$, then the gradient could be calculated as shown.

$$m = \dfrac{y_2 - y_1}{x_2 - x_1}$$
$$= \dfrac{5 - 14}{-2 - 1}$$
$$= \dfrac{-9}{-3}$$
$$= 3$$

The result is the same.

The gradient of horizontal and vertical lines

- The gradient of a horizontal line is zero, since it does not have any upwards or downwards slope.

$$m = \dfrac{\text{rise}}{\text{run}} = \dfrac{0}{\text{run}} = 0$$

- The gradient of a vertical line is undefined, as the run is zero.

$$m = \dfrac{\text{rise}}{\text{run}} = \dfrac{\text{rise}}{0} = \text{undefined}$$

WORKED EXAMPLE 5 Calculating the gradient of a straight line

Calculate the gradients of the lines shown.

a.
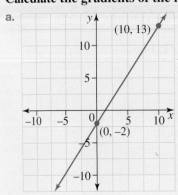

b.
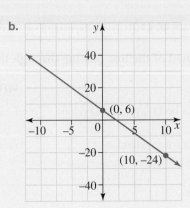

c.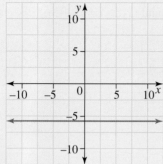

d.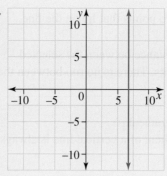

THINK	**WRITE**
a. 1. Write down two points that lie on the line.	**a.** Let $(x_1, y_1) = (0, -2)$ and $(x_2, y_2) = (10, 13)$. Rise $= y_2 - y_1 = 13 - -2 = 15$ Run $= x_2 - x_1 = 10 - 0 = 10$
2. Calculate the gradient by finding the ratio $\dfrac{\text{rise}}{\text{run}}$.	$m = \dfrac{\text{rise}}{\text{run}}$ $= \dfrac{15}{10}$ $= \dfrac{3}{2}$ or 1.5
b. 1. Write down two points that lie on the line.	**b.** Let $(x_1, y_1) = (0, 6)$ and $(x_2, y_2) = (10, -24)$. Rise $= y_2 - y_1 = -24 - 6 = -30$ Run $= x_2 - x_1 = 10 - 0 = 10$
2. Calculate the gradient.	$m = \dfrac{\text{rise}}{\text{run}}$ $= \dfrac{-30}{10}$ $= -3$
c. 1. Write down two points that lie on the line. **2.** There is no rise between the two points.	**c.** Let $(x_1, y_1) = (5, -6)$ and $(x_2, y_2) = (10, -6)$. Rise $= y_2 - y_1$ $= -6 - -6 = 0$ Run $= x_2 - x_1$ $= 10 - 5 = 5$
3. Calculate the gradient. *Note:* The gradient of a horizontal line is always zero. The line has no slope.	$m = \dfrac{\text{rise}}{\text{run}}$ $= \dfrac{0}{5}$ $= 0$

d. 1. Write down two points that lie on the line.

     d. Let $(x_1, y_1) = (7, 10)$ and $(x_2, y_2) = (7, -3)$.

2. The vertical distance between the selected points is 13 units. There is no run between the two points.

$$\text{Rise} = y_2 - y_1 = -3 - 10 = 13$$
$$\text{Run} = x_2 - x_1 = 7 - 7 = 0$$

3. Calculate the gradient.
Note: The gradient of a vertical line is always undefined.

$$m = \frac{\text{rise}}{\text{run}}$$
$$= \frac{13}{0}$$
$$= \text{undefined}$$

▶ 6.3.2 The *y*-intercept

eles-4783

- The point where a line cuts the *y*-axis is called the **y-intercept.**
- In the graph shown the line cuts the *y*-axis at $y = 2$, so the *y*-intercept is $(0, 2)$.
- The *y*-intercept of any line is easily found by substituting 0 for *x* and calculating the *y*-value.

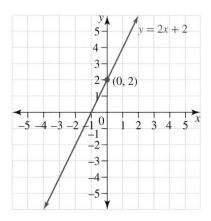

WORKED EXAMPLE 6 Determining the *y*-intercepts of lines

Determine the *y*-intercepts of the lines whose linear rules are given, and state the coordinates of the *y*-intercept.

a. $y = -4x + 7$ **b.** $5y + 2x = 10$ **c.** $y = 2x$ **d.** $y = -8$

THINK

a. 1. To calculate the *y*-intercept, substitute $x = 0$ into the equation.

2. Solve for *y*.

3. Write the coordinates of the *y*-intercept.

b. 1. To calculate the *y*-intercept, substitute $x = 0$ into the equation.

2. Solve for *y*.

3. Write the coordinates of the *y*-intercept.

WRITE

a. $y = -4x + 7$
$y = -4(0) + 7$

$y = 7$

y-intercept: $(0, 7)$

b. $5y + 2x = 10$
$5y + 2(0) = 10$

$5y = 10$
$y = 2$

y-intercept: $(0, 2)$

c. **1.** To calculate the *y*-intercept, substitute $x = 0$ into the equation.

c. $y = 2x$
$y = 2(0)$
$y = 0$

2. Write the coordinates of the *y*-intercept.

y-intercept: $(0, 0)$

d. The value of *y* is -8 regardless of the *x*-value.

d. *y*-intercept: $(0, -8)$

DISCUSSION

Why is the *y*-intercept of a graph found by substituting $x = 0$ into the equation?

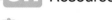 Resources

eWorkbook	Topic 6 Workbook (worksheets, code puzzle and project) (ewbk-2006)
Digital documents	SkillSHEET Transposing a linear equation to general form (doc-6165)
	SkillSHEET Finding the gradient given two points (doc-10839)
	SkillSHEET Measuring the rise and the run (doc-10840)
	SkillSHEET Finding the gradient of a line from its equation (doc-10841)
Video eLesson	Gradient (eles-1889)
Interactivities	Individual pathway interactivity: Determining linear rules (int-4506)
	The gradient (int-3836)
	Linear graphs (int-6484)

Exercise 6.3 Features of linear graphs   **learn** on

Individual pathways

■ PRACTISE	■ CONSOLIDATE	■ MASTER
1, 3, 5, 8, 11, 14	2, 4, 6, 9, 12, 15	7, 10, 13, 16, 17

To answer questions online and to receive **immediate corrective feedback** and **fully worked solutions** for all questions, go to your learnON title at www.jacplus.com.au.

Fluency

1. **WE4** Calculate the gradients of the lines passing through the following pairs of points.

   a. $(2, 10)$ and $(4, 22)$ **b.** $(1, -2)$ and $(3, -10)$
   c. $(-3, 0)$ and $(7, 0)$ **d.** $(-4, -7)$ and $(1, -1)$
   e. $(0, 4)$ and $(4, -4.8)$ **f.** $(-2, 122)$ and $(1, -13)$

2. Calculate the gradients of the lines passing through the following pairs of points.

   a. $(2, 3)$ and $(17, 3)$ **b.** $(-2, 2)$ and $(2, 2.4)$
   c. $(1, -5)$ and $(5, -15.4)$ **d.** $(-12, -7)$ and $(8.4, -7)$
   e. $(-2, -17.7)$ and $(0, 0.3)$ **f.** $(-3, 3.4)$ and $(5, 2.6)$

3. **WE6** Determine the *y*-intercepts of the lines whose rules are given below.

   a. $y = 5x + 23$ **b.** $y = 54 - 3x$ **c.** $y = 3(x - 2)$ **d.** $y = 70 - 2x$

4. Determine the *y*-intercepts of the lines whose rules are given below.

a. $y = \dfrac{1}{2}(x + 2)$ b. $y = \dfrac{x}{2} + 5.2$ c. $y = 100 - x$ d. $y = 100$

5. **WE5** Calculate the gradients of the lines shown.

a. 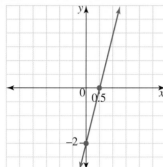 b. 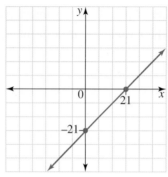 c.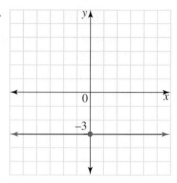

6. Calculate the gradients of the lines shown.

a. 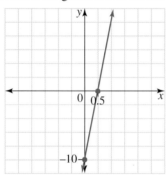 b. 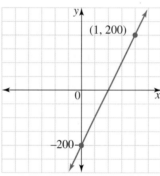 c.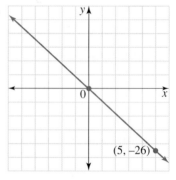

7. Calculate the gradients of the lines shown.

a. 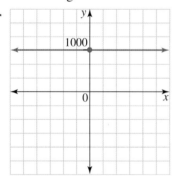 b. 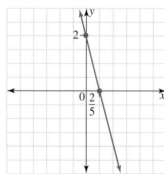 c.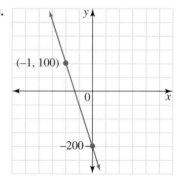

Understanding

8. **MC** Select which of the following statements about linear graphs is false.

   A. A gradient of zero means the graph is a horizontal line.
   B. A gradient can be any real number.
   C. A linear graph can have two *y*-intercepts.
   D. In the form $y = mx + c$, the *y*-intercept equals c.
   E. An undefined gradient means the graph is a vertical line.

9. Identify the *y*-intercept of the line $y = mx + c$.

10. Determine the coordinates of the *y*-intercepts of the lines with the following rules.

    a. $y = -6x - 10$ b. $3y + 3x = -12$ c. $7x - 5y + 15 = 0$
    d. $y = 7$ e. $x = 9$

Reasoning

11. Explain why the gradient of a vertical line is undefined.

12. Explain why the gradient of a horizontal line is zero.

13. Show the gradient of the line passing through the points (a, b) and (c, d) is $\dfrac{d - b}{c - a}$.

Problem solving

14. When using the gradient to draw a line, does it matter if you rise before you run or run before you rise? Explain.

15. Consider the graph shown.

    a. Determine a general formula for the gradient m in terms of x, y and c.
    b. Transpose your formula to make y the subject. Discuss what you notice about this equation.

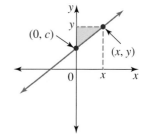

16. The price per kilogram for three different types of meat is illustrated in the graph shown.

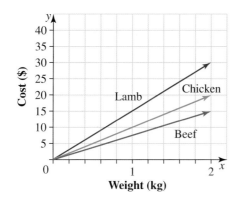

    a. Calculate the gradient (using units) for each graph.
    b. Determine the cost of 1 kg of each type of meat.
    c. Evaluate the cost of purchasing:

    i.   1 kg of lamb   ii.   0.5 kg of chicken   iii.   2 kg of beef.

    d. Calculate the total cost of the order in part **c**.
    e. Complete the table below to confirm your answer from part **d**.

Meat type	Cost per kilogram $(\$/kg)$	Weight required (kg)	Cost $= \$/kg \times kg$
Lamb		1	
Chicken		0.5	
Beef		2	
		Total cost	

17. Three right-angled triangles have been superimposed on the graph shown.

   a. Use each of these to determine the gradient of the line.
   b. Does it matter which points are chosen to determine the gradient of a line? Explain your answer.
   c. Describe the shape of the graph.

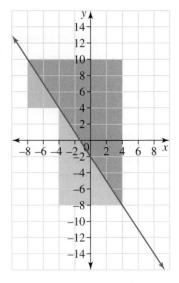

6.4 The equation of a straight line

LEARNING INTENTION

At the end of this subtopic you should be able to:
 • determine the equation of a straight line given the gradient and the y-intercept
 • determine the equation of a straight line given the gradient and one point
 • determine the equation of a straight line given two points
 • determine the equation of a straight line from a graph.

6.4.1 The general equation of a straight line

eles-4784

 • The general equation of a straight line is

$$y = mx + c$$

where m is the gradient and c is the y-intercept of the line.
 • It is very easy to state the equation of a straight line if its gradient and y-intercept are known.

WORKED EXAMPLE 7 Determining the line equation from the gradient and y-intercept

Determine the equation of the line with a gradient of −2 and y-intercept of 3.

THINK	WRITE
1. Write the equation of a straight line.	$y = mx + c$
2. Substitute the values $m = -2$, $c = 3$ to write the equation.	$m = -2$, $c = 3$ $y = -2x + 3$

▶ 6.4.2 Determining the equation of a straight line

eles-4785

- If the gradient (m) and any single point on a straight line are known, then the y-intercept can be calculated algebraically.

WORKED EXAMPLE 8 Determining the equation of a line from the gradient and a single point

Determine the equation of a straight line that goes through the point $(1, -3)$, if its gradient is -2.

THINK	WRITE
1. Write the equation of a straight line.	$y = mx + c$
2. Substitute the value $m = -2$.	$y = -2x + c$
3. Since the line passes through the point $(1, -3)$, substitute $x = 1$ and $y = -3$ into $y = -2x + c$ to calculate the value of c.	When $x = 1, y = -3$. $-3 = -2 \times 1 + c$
4. Solve for c.	$-3 = -2 + c$ $-3 + 2 = c$ $c = -1$
5. Write the rule.	$y = -2x - 1$

- If two points on a straight line are known, then the gradient (m) can be calculated using the formula $m = \dfrac{y_2 - y_1}{x_2 - x_1}$, or $m = \dfrac{\text{rise}}{\text{run}}$.
- Using the gradient and one of the points, the equation can be found using the method in Worked example 8.

WORKED EXAMPLE 9 Determining the rule of a line from two points

Determine the equation of the straight line passing through the points $(-1, 6)$ and $(3, -2)$.

THINK	WRITE
1. Write the equation of a straight line.	$y = mx + c$
2. Write the formula for calculating the gradient, m.	$m = \dfrac{y_2 - y_1}{x_2 - x_1}$
3. Let $(x_1, y_1) = (-1, 6)$ and $(x_2, x_2) = (3, -2)$. Substitute the values into the formula and determine the value of m.	$m = \dfrac{-2 - 6}{3 - -1}$ $= \dfrac{-2 - 6}{3 + 1}$ $= -\dfrac{8}{4}$ $= -2$
4. Substitute the value of m into the equation.	$y = -2x + c$
5. Select either of the two points, say $(3, -2)$, and substitute into $y = -2x + c$.	Point $(3, -2)$: $-2 = -2(3) + c$ $-2 = -6 + c$
6. Solve for c.	$c = 4$
7. Write the rule using the values $c = 4, m = -2$.	$y = -2x + 4$

TI | **THINK**	WRITE	CASIO | **THINK**	WRITE

TI \| THINK

In a new problem,
on a Calculator page,
complete the entry lines
as:

$m := \dfrac{y2 - y1}{x2 - x1} \,|\, x1 = -1$

and $y1 = 6$ and $x2 = 3$
and $y2 = -2$

$y = -2x + c$

solve

$(y = -2x + c, c) \,|\, x = 3$
and $y = -2$

$y = -2x + c \,|\, c = 4$

Press ENTER after each
entry.

WRITE

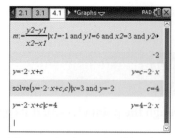

The equation is $y = 4 - 2x$.

CASIO \| THINK

1. On the Spreadsheet
   screen, complete the
   entries as:

	A	**B**
1	-1	6
2	3	-2

WRITE

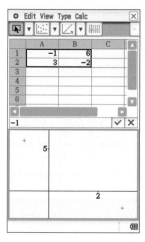

2. Highlight the data in
   both columns.
   Press Graph and select
   the scatterplot.
   The points will be
   plotted.

3. To get the equation of
   the line, press the Line
   icon. The equation will
   appear above the graph,
   and the points will be
   joined by a line.

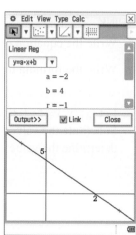

The equation of the line
passing through the points
$(-1, 6)$ and $(3, -2)$ is
$y = -2x + 4$.

- The equation of a straight line can be determined from a graph by observing the *y*-intercept and the gradient.

WORKED EXAMPLE 10 Determining the rule of a line from its graph

Determine the equation of the linear graph shown.

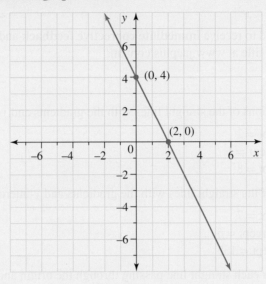

THINK

1. Read the important information from the graph. The *y*-intercept is 4 and the *x*-intercept is 2. Write the coordinates of each point.

2. Calculate the gradient using the formula $m = \dfrac{y_2 - y_1}{x_2 - x_1}$.
   Let $(x_1, y_1) = (2, 0)$ and $(x_2, y_2) = (0, 4)$.

3. From the graph, the *y*-intercept is 4.

4. Write the equation.

WRITE

The graph passes through $(0, 4)$ and $(2, 0)$.

$$m = \frac{4 - 0}{0 - 2}$$
$$= \frac{4}{-2}$$
$$= -2$$
$$y = -2x + c$$

$c = 4$

$y = -2x + 4$

 Resources

Exercise 6.4 The equation of a straight line

Individual pathways

■ PRACTISE	■ CONSOLIDATE	■ MASTER
1, 3, 5, 10, 15, 16, 19, 22, 25	2, 4, 6, 8, 11, 13, 17, 20, 23, 26, 27	7, 9, 12, 14, 18, 21, 24, 28, 29

To answer questions online and to receive **immediate corrective feedback** and **fully worked solutions** for all questions, go to your learnON title at www.jacplus.com.au.

Fluency

1. **WE7** Determine the equations of the straight lines with the gradients and y-intercepts given.

   a. Gradient $= 4$, y-intercept $= 2$
   b. Gradient $= -4$, y-intercept $= 1$
   c. Gradient $= 4$, y-intercept $= 8$

2. Determine the equations of the straight lines with the gradients and y-intercepts given.

   a. Gradient $= 6$, y-intercept $= 7$
   b. Gradient $= -2.5$, y-intercept $= 6$
   c. Gradient $= 45$, y-intercept $= 135$

3. Determine the equation for each straight line passing through the origin and with the gradient given.

   a. Gradient $= -2$
   b. Gradient $= 4$
   c. Gradient $= 10.5$

4. Determine the equation for each straight line passing through the origin and with the gradient.

   a. Gradient $= -20$
   b. Gradient $= 1.07$
   c. Gradient $= 32$

5. **WE8** Determine the equation of the straight lines with:

   a. gradient $= 1$, point $= (3, 5)$
   c. gradient $= -4$, point $= (-3, \ 4)$
   b. gradient $= -1$, point $= (3, 5)$
   d. gradient $= 2$, point $= (5, -3)$

6. Determine the equation of the straight lines with:

   a. gradient $= -5$, point $= (13, 5)$
   c. gradient $= -6$, point $= (2, -1)$
   b. gradient $= 2$, point $= (10, -3)$
   d. gradient $= -1$, point $= (-2, 0.5)$

7. Determine the equation of the straight lines with:

   a. gradient $= 6$, point $= (-6, -6)$
   c. gradient $= 1.2$, point $= (2.4, -1.2)$
   b. gradient $= -3.5$, point $= (3, 5)$
   d. gradient $= 0.2$, point $= (1.3, -1.5)$

8. Determine the equations of the straight lines with:

   a. gradient $= -4$, x-intercept $= -6$
   b. gradient $= 2$, x-intercept $= 3$
   c. gradient $= -2$, x-intercept $= 2$

9. Determine the equation of the straight lines with:

   a. gradient $= 5$, x-intercept $= -7$
   b. gradient $= 1.5$, x-intercept $= 2.5$
   c. gradient $= 0.4$, x-intercept $= 2.4$.

10. **WE9** Determine the equation for each straight line passing through the given points.

   a. $(-6, 11)$ and $(6, 23)$

   b. $(1, 2)$ and $(-5, 8)$

   c. $(4, 11)$ and $(6, 11)$

   d. $(3, 6.5)$ and $(6.5, 10)$

11. Determine the equation for each straight line passing through the given points.

   a. $(1.5, 2)$ and $(6, -2.5)$

   b. $(-7, 3)$ and $(2, 4)$

   c. $(25, -60)$ and $(10, 30)$

   d. $(5, 100)$ and $(25, 500)$

12. Determine the equation for each straight lines passing through the given points.

   a. $(1, 3)$ and $(3, 1)$

   b. $(2, 5)$ and $(-2, 6)$

   c. $(9, -2)$ and $(2, -4)$

   d. $(1, 4)$ and $(-0.5, 3)$

13. Determine the rules for the linear graphs that have the following x- and y-intercepts.

   a. x-intercept $= -3$, y-intercept $= 3$ **b.** x-intercept $= 4$, y-intercept $= 5$

   c. x-intercept $= 1$, y-intercept $= 6$ **d.** x-intercept $= -40$, y-intercept $= 35$

14. Determine the rules for the linear graphs that have the following x- and y-intercepts.

   a. x-intercept $= -8$, y-intercept $= 8$ **b.** x-intercept $= 3$, y-intercept $= 6$

   c. x-intercept $= -7$, y-intercept $= -3$ **d.** x-intercept $= -200$, y-intercept $= 50$

15. Determine the rule for each straight line passing through the origin and:

   a. the point $(4, 7)$ **b.** the point $(5, 5)$

   c. the point $(-4, 8)$ **d.** the point $(-1.2, 3.6)$

   e. the point $(-22, 48)$ **f.** the point $(-105, 35)$.

16. **WE10** Determine the equation of the line shown on each of the following graphs.

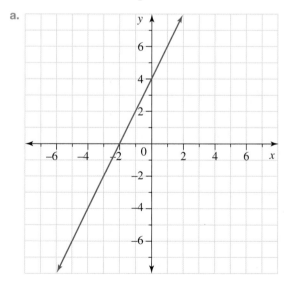

17. Determine the equation of the line shown on each of the following graphs.

a.
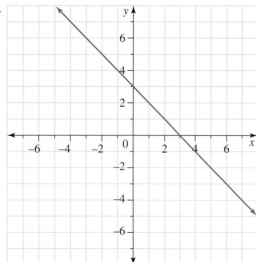

b.
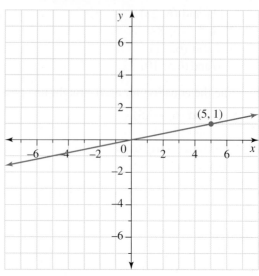

18. Determine the equation of the line shown on each of the following graphs.

a.
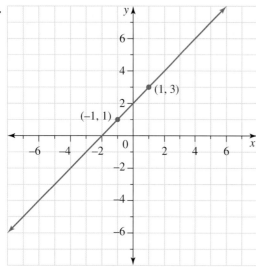

b.
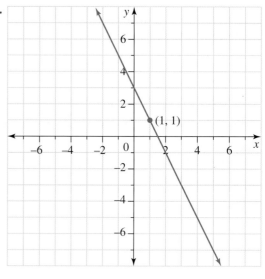

Understanding

19. **MC** a. The gradient of the straight line that passes through (3, 5) and (5, 3) is:

    A. -2 B. -1 C. 0   D. 1   E. 2

    b. A straight line with an x-intercept of 10 and a y-intercept of 20 has a gradient of:

    A. -2 B. -1 C. -0.5 D. 0   E. 0.5

    c. The rule $2y - 3x = 20$ has an x-intercept at:

    A. $-\dfrac{3}{2}$ B. $-\dfrac{2}{3}$ C. 0

    D. $\dfrac{2}{3}$ E. None of these

20. Given that the x-intercept of a straight line graph is $(-5, 0)$ and the y-intercept is $(0, -12)$:

    a. determine the equation of the straight line   b. calculate the value of y when $x = 19.3$.

21. a. Determine the equation of the straight line shown in the graph. Use the fact that when $x = 5$, $y = 7$.
b. Determine the x-intercept.

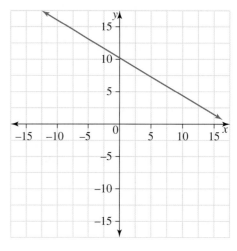

Reasoning

22. The graph shows the carbon dioxide (CO_2) concentration in the atmosphere, measured in parts per million (ppm).

   a. If the trend follows a linear pattern, determine the equation for the line.

   b. Explain why c cannot be read directly from the graph.

   c. Infer the CO_2 concentration predicted for 2020. State the assumption made when determining this value.

23. Show that the equation for the line that passes through the point $(3, 6)$ parallel to the line through the points $(0, -7)$ and $(4, -15)$ is $y = -2x + 12$.

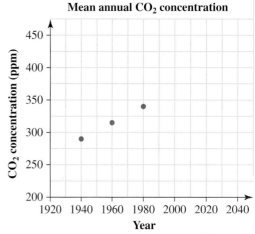

Mean annual CO_2 concentration

24. a. Determine the equations for line A and line B as shown in this graph.

   b. Write the point of intersection between line A and line B and mark it on the Cartesian plane.

   c. Show that the equation of the line that is perpendicular to line B and passes through the point $(-4, 6)$ is $y = x + 10$.

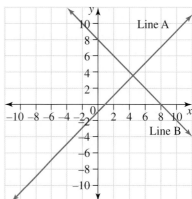

Problem solving

25. The graph shown describes the mass in kilograms of metric cups of water. Write a rule to describe the mass of water (m) relative to the number of cups (c).

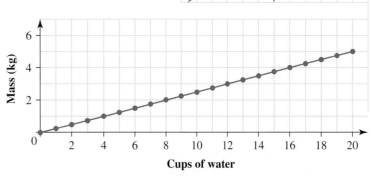

26. Harpinder plays the game *Space Galaxy* on her phone. The stars and spaceships are displayed on her screen as shown.

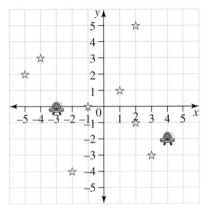

a. Copy the diagram. On the diagram, draw a straight line that will hit three stars.

b. Determine the equation of the straight line that will hit three stars.

c. Harpinder types in the equation $y = \dfrac{1}{2}x + \dfrac{1}{2}$ and manages to hit two stars. Draw the straight line on your diagram.

d. If Harpinder types in the equation from part b and the equation from part c, determine the coordinate of the star that both lines will hit.

e. If she types $y = 2$, state how many stars will she hit.

f. Give another equation of a straight line that will hit two stars.

27. The temperature of water in a kettle is $15\,°C$ before the temperature increases at a constant rate for 20 seconds to reach boiling point $(100\,°C)$. Adel argues that $T = 5t + 15$ describes the water temperature, citing the starting temperature of $15\,°C$ and that to reach $100\,°C$ in 20 seconds an increase of $5\,°C$ for every second is required.
Explain why Adel's equation is incorrect and devise another equation that correctly describes the temperature of the water.

28. A father wants to administer a children's liquid painkiller to his child. The recommended dosage is a range, $7.5 - 9\,mL$ for an average weight of $12 - 14\,kg$. The child weighs $12.8\,kg$. The father uses a linear relationship to calculate an exact dosage. Evaluate the dosage that the father calculates.

29. The graph shown displays the wages earned in three different workplaces.

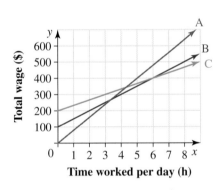

a. Identify the set allowance for each workplace.

b. Determine the hourly rates for each workplace.

c. Using your answers from parts a and b, determine linear equations that describe the wages at each workplace.

d. Match each working lifestyle below to the most appropriate workplace.

i. Working lifestyle 1: Earn the most money possible while working at most 4 hours in a day.

ii. Working lifestyle 2: Earn the most money possible while working an 8-hour day.

e. If a person works an average of 8 hours a day, determine the advantage of workplace C.

f. Calculate how much money is earned at each workplace for a:

i. 2-hour day
ii. 6-hour day.

6.5 Sketching linear graphs

LEARNING INTENTION

At the end of this subtopic you should be able to:
- sketch linear graphs using the x- and y-intercept method
- sketch linear graphs using the gradient–intercept method
- sketch vertical and horizontal lines.

- To sketch a linear graph all you need is two points on the line. Once you have found two points, the line can be drawn through those points.

▶ 6.5.1 The *x*- and *y*-intercept method

eles-4786

- One method to sketch a linear graph is to find both the **x-intercept** and the **y-intercept**.
- To determine the intercepts of a graph:
  x-intercept: let $y = 0$ and solve for x
  y-intercept: let $x = 0$ and solve for y.
- If both intercepts are zero (at the origin), another point is needed to sketch the line.

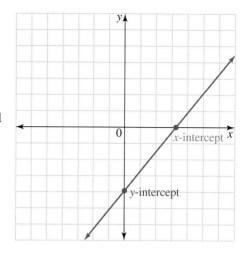

WORKED EXAMPLE 11 Sketching graphs using the *x*- and *y*-intercept method

Using the *x*- and *y*-intercept method, sketch the graphs of:

a. $2y + 3x = 6$ b. $y = \dfrac{4}{5}x + 5$ c. $y = 2x$.

THINK

a. 1. Write the rule.

   2. To calculate the *y*-intercept, let $x = 0$.
      Write the coordinates of the *y*-intercept.

   3. To calculate the *x*-intercept, let $y = 0$.
      Write the coordinates of the *x*-intercept.

   4. Plot and label the *x*- and *y*-intercepts on a set of axes and rule a straight line through them. Label the graph.

b. 1. Write the rule.

   2. The rule is in the form $y = mx + c$, so the *y*-intercept is the value of *c*.

WRITE

a. $2y + 3x = 6$

$x = 0$: $2y + 3 \times 0 = 6$
$2y = 6$
$y = 3$
y-intercept: $(0, 3)$

$y = 0$: $2 \times 0 + 3x = 6$
$3x = 6$
$x = 2$
x-intercept: $(2, 0)$

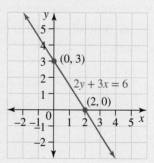

b. $y = \dfrac{4}{5}x + 5$

$c = 5$
y-intercept: $(0, 5)$

3. To calculate the x-intercept, let $y = 0$. Write the coordinates of the x-intercept.

$y = 0$:

$$y = \frac{4}{5}x + 5$$

$$0 = \frac{4}{5}x + 5$$

$$-5 = \frac{4}{5}x$$

$$x = -\frac{25}{4} \left(= -6\frac{1}{4} \right)$$

x-intercept: $\left(-\frac{25}{4}, 0 \right)$

4. Plot and label the intercepts on a set of axes and rule a straight line through them.
   Label the graph.

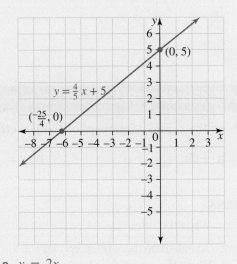

c. 1. Write the rule.

c. $y = 2x$

2. To calculate the y-intercept, let $x = 0$.
   Write the coordinates of the y-intercept.

$x = 0$:
$$y = 2 \times 0$$
$$= 0$$

y-intercept: $(0, 0)$

3. The x- and y-intercepts are the same point, $(0, 0)$, so one more point is required.
   Choose any value for x, such as $x = 3$.
   Substitute and write the coordinates of the point.

$x = 3$:
$$y = 2 \times 3$$
$$= 6$$

Another point: $(3, 6)$

4. Plot the points, then rule and label the graph.
   Label the graph.

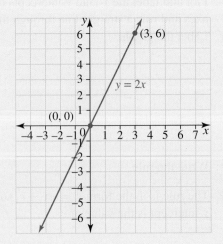

▶ 6.5.2 The gradient–intercept method

eles-4787

- To use this method, the gradient and the *y*-intercept must be known.
- The line is drawn by plotting the *y*-intercept, then drawing a line with the correct gradient through that point.

A line interval of gradient $3\left(=\dfrac{3}{1}\right)$ can be drawn with a rise of 3 and a run of 1.	A line interval of gradient $-2\left(=\dfrac{-2}{1}\right)$ can be drawn with a rise downwards of 2 and a run of 1.	A line interval of gradient $\dfrac{3}{5}$ can be shown with a rise of 3 and a run of 5.
	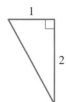	

WORKED EXAMPLE 12 Sketching graphs using the gradient–intercept method

Using the gradient–intercept method, sketch the graphs of:

a. $y = \dfrac{3}{4}x + 2$

b. $4x + 2y = 3$.

THINK

a. 1. From the equation, the *y*-intercept is 2. Plot the point $(0, 2)$.

2. From the equation, the gradient is $\dfrac{3}{4}$, so $\dfrac{\text{rise}}{\text{run}} = \dfrac{3}{4}$.
From $(0, 2)$, run 4 units and rise 3 units. Mark the point P $(4, 5)$.

3. Draw a line through $(0, 2)$ and P $(4, 5)$. Label the graph.

WRITE

a.

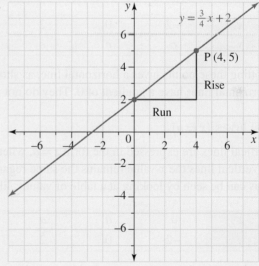

b. 1. Write the rule in gradient–intercept form: $y = mx + c$.
From the equation, $m = -2$, $c = \dfrac{3}{2}$.
Plot the point $\left(0, \dfrac{3}{2}\right)$.

b. $4x + 2y = 3$

$2y = 3 - 4x$

$y = \dfrac{3}{2} - 2x$

$y = -2x + \dfrac{3}{2}$

2. The gradient is -2, so $\dfrac{\text{rise}}{\text{run}} = \dfrac{-2}{1}$.

   From $\left(0, \dfrac{3}{2}\right)$, run 1 units and rise -2 units (i.e. go down 2 units). Mark the point $P\left(1, -\dfrac{1}{2}\right)$.

3. Draw a line through $\left(0, \dfrac{3}{2}\right)$ and $P\left(1, -\dfrac{1}{2}\right)$. Label the graph.

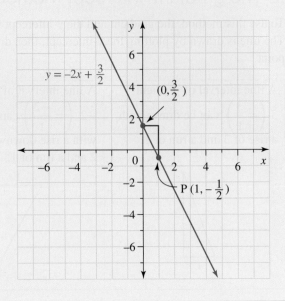

6.5.3 Horizontal and vertical lines

eles-4788

Horizontal lines $(y = c)$

- Horizontal lines are expressed in the form $y = c$, where c is the y-intercept.
- In horizontal lines the y-value remains the same regardless of the x-value.
- This can be seen by looking at a table of values, like the one shown.

x	-2	0	2	4
y	c	c	c	c

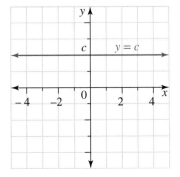

- Plotting these points gives a horizontal line, as shown in the graph.
- Horizontal lines have a gradient of 0. They do not rise or fall.

Vertical lines $(x = a)$

- Vertical lines are expressed in the form $x = a$, where a is the x-intercept.
- In vertical lines the x-value remains the same regardless of the y-value.
- This can be seen by looking at a table of values, like the one shown.

x	a	a	a	a
y	-2	0	2	4

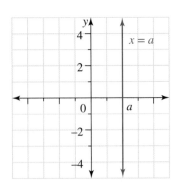

- Plotting these points gives a vertical line, as shown in the graph.
- The run of the graph is 0, so using the formula $m = \dfrac{\text{rise}}{\text{run}}$ involves dividing by zero, which cannot be done.

  The gradient is said to be **undefined**.

On a pair of axes, sketch the graphs of the following and label the point of intersection of the two lines.

a. $x = -3$

b. $y = 4$

THINK

a. 1. The line $x = -3$ is in the form $x = a$.
   This is a vertical line.

   2. Rule the vertical line where $x = -3$.
   Label the graph.

b. 1. The line $y = 4$ is in the form $y = c$.
   This is a horizontal line.

   2. Rule the horizontal line where $y = 4$.
   Label the graph.

   3. The lines intersect at $(-3, 4)$.

WRITE

a.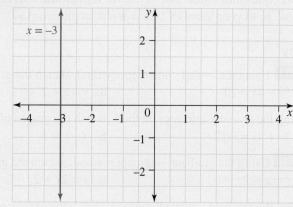

b.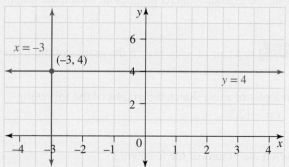

 Resources

eWorkbook Topic 6 Workbook (worksheets, code puzzle and project) (ewbk-2006)

Digital documents SkillSHEET Graphing linear equations using intercepts (doc-10842)
SkillSHEET Solving linear equations that arise when finding intercepts (doc-10843)

Video eLessons Sketching linear graphs (eles-1919)
Sketching linear graphs using the gradient–intercept method (eles-1920)

Interactivities Individual pathway interactivity: Sketching linear graphs (int-4504)
The intercept method (int-3840)
The gradient–intercept method (int-3839)
Vertical and horizontal lines (int-6049)

Exercise 6.5 Sketching linear graphs

Individual pathways

■ PRACTISE	■ CONSOLIDATE	■ MASTER
1, 3, 5, 7, 9, 11, 15, 20	4, 8, 12, 13, 16, 17, 21	2, 6, 10, 14, 18, 19, 22

To answer questions online and to receive **immediate corrective feedback** and **fully worked solutions** for all questions, go to your learnON title at www.jacplus.com.au.

Fluency

1. **WE11** Using the x- and y-intercept method, sketch the graphs of:

   a. $5y - 4x = 20$ b. $y = x + 2$ c. $y = -3x + 6$.

2. Using the x- and y-intercept method, sketch the graphs of:

   a. $3y + 4x = -12$ b. $x - y = 5$ c. $2y + 7x - 8 = 0$.

3. **WE12** Using the gradient–intercept method, sketch the graphs of:

   a. $y = x - 7$ b. $y = 2x + 2$ c. $y = -2x + 2$.

4. Using the gradient–intercept method, sketch the graphs of:

   a. $y = \dfrac{1}{2}x - 1$ b. $y = 4 - x$ c. $y = -x - 10$.

5. **WE13** On a pair of axes, sketch the graphs of the following.

   a. $y = 4$ b. $y = -3$

6. On a pair of axes, sketch the graphs of the following.

   a. $y = -12.5$ b. $y = \dfrac{4}{5}$

7. Sketch the graphs of the following.

   a. $x = 2$ b. $x = -6$

8. Sketch the graphs of the following.

   a. $x = -2.5$ b. $x = \dfrac{3}{4}$

9. Sketch the graphs of the following.

   a. $y = 3x$ b. $y = -2x$

10. Sketch the graphs of the following.

    a. $y = \dfrac{3}{4}x$ b. $y = -\dfrac{1}{3}x$

Understanding

11. **MC** Select which of the following statements about the rule $y = 4$ is false.

    A. The gradient $m = 0$.
    B. The y-intercept is at $(0, \ 4)$.
    C. The graph is parallel to the x-axis.
    D. The point $(4, \ 2)$ lies on this graph.
    E. There is no x-intercept for this graph.

12. **MC** Select which of the following statements is not true about the rule $y = -\dfrac{3}{5}x$.

 A. The graph passes through the origin.

 B. The gradient $m = -\dfrac{3}{5}$.

 C. The x-intercept is at $x = 0$.

 D. The graph can be sketched using the x- and the y-intercept method.

 E. The y-intercept is at $y = 0$.

13. $2x + 5y = 20$ is a linear equation in the form $ax + by = c$.

 a. Rearrange this equation into the form $y = mx + c$.

 b. State the gradient.

 c. State the x- and y-intercepts.

 d. Sketch this straight line.

14. A straight line has an x-intercept of -3 and a y-intercept of 5.

 a. State the gradient.

 b. Draw the graph.

 c. Write the equation in the form:

   i. $y = mx + c$ ii. $ax + by = c$.

Reasoning

15. Consider the relationship $4x - 3y = 24$.

 a. Rewrite this relationship, making y the subject.

 b. Show that the x- and y-intercepts are $(6, 0)$ and $(0, -8)$ respectively.

 c. Sketch a graph of this relationship.

16. Consider the relationship $ax + by = c$.

 a. Rewrite the relationship, making y the subject.

 b. If a, b and c are positive integer values, explain how the gradient is negative.

17. Josie accidentally spilled a drink on her work. Part of her calculations were smudged. The line $y = \dfrac{1}{2}x + \dfrac{3}{4}$ was written in the form $ax + 4y = 3$. Show that the value of $a = -2$.

18. Explain why the descriptions 'right 3 up 2', 'right 6 up 4', 'left 3 down 2', 'right $\dfrac{3}{2}$ up 1' and 'left 1 down $\dfrac{2}{3}$' all describe the same gradient for a straight line.

19. A straight line passes through the points $(3, 5)$ and $(6, 11)$.

 a. Determine the slope of the line

 b. Determine the equation of the line

 c. State the coordinates of another point that lies on the same line.

Problem solving

20. a. Match the descriptions given below with their corresponding line.

   i. Straight line with a y-intercept of $(0, 1)$ and a positive gradient

   ii. Straight line with a gradient of $1\dfrac{1}{2}$

   iii. Straight line with a gradient of -1

 b. Write a description for the unmatched graph.

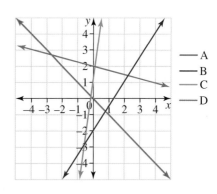

21. a. Sketch the linear equation $y = -\dfrac{5}{7}x - \dfrac{3}{4}$:

   i. using the y-intercept and the gradient
   ii. using the x- and y-intercepts
   iii. using two other points.

   b. Compare and contrast the methods and generate a list of advantages and disadvantages for each method used in part **a**. Explain which method you think is best. Give your reasons.

22. Consider these two linear graphs.
$$y - ax = b \text{ and } y - cx = d.$$
Show that if these two graphs intersect where both x and y are positive, then $a > c$ when $d > b$.

6.6 Technology and linear graphs

LEARNING INTENTION

At the end of this subtopic you should be able to:
 • produce linear graphs using digital technologies.

▶ 6.6.1 Graphing with technology

eles-4789

 • There are many digital technologies that can be used to graph linear relationships.
 • The Desmos Graphing Calculator is a free graphing tool that can be found online.
 • Other commonly used digital technologies include Microsoft Excel and graphing calculators.
 • Digital technologies can help identify important features and patterns in graphs.

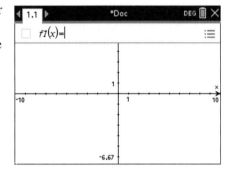

Sketching a linear graph using technology

 • Depending on the choice of digital technology used, the steps involved to produce a linear graph may vary slightly.
 • Most graphing calculators have an entry (or input) line to type in the equation of the line you wish to sketch.
 • When using the Desmos Graphing Calculator you can simply type $y = x + 1$ into the input box to produce its graph.
 • When using technologies such as a Texas Instruments or CASIO graphing calculator you will need to open a Graphs page first. The entry line for these types of technologies has a template that needs to be followed carefully. Both of these graphing calculators begin with $f1(x) =$ or $y1 =$, which is effectively saying $y =$.
 • Thus, to draw the graph of $y = x + 1$, you would simply enter '$x + 1$'.
 • The screen shows the graph of $y = x + 1$ sketched on a TI-Nspire CAS calculator.

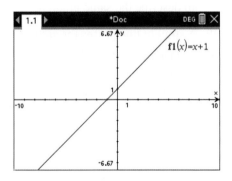

▶ 6.6.2 Graphing parallel lines

eles-4790

- Lines with the same gradient are called **parallel lines**. For a pair of parallel lines, $m_1 = m_2$.
- Digital technologies can be used to help visualise this concept. For example, $y = 3x + 1$, $y = 3x - 4$ and $y = 3x$ are all parallel lines, because $m = 3$.
- Select the digital technology of your choice and sketch these three parallel lines on the same set of axes.

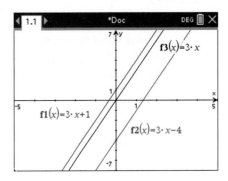

▶ 6.6.3 Graphing perpendicular lines

eles-4791

- Lines that meet at right angles are called **perpendicular lines**.
- The product of the gradients of two perpendicular lines is equal to -1.
  For a pair of perpendicular lines, $m_1 \times m_2 = -1$.
- Digital technologies can be used to help visualise this concept.
  For example, $y = 2x + 1$ and $y = -\frac{1}{2}x + 6$ are perpendicular,
  because $2 \times -\frac{1}{2} = -1$.
- Select a digital technology of your choice and sketch these two lines on the same set of axes.

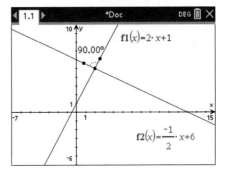

 Resources

📋 **eWorkbook**   Topic 6 Workbook (worksheets, code puzzle and project) (ewbk-2006)

🧩 **Interactivities** Individual pathway interactivity: Technology and linear graphs (int-4505)
                    Parallel lines (int-3841)

Exercise 6.6 Technology and linear graphs   **learn**on

Individual pathways

■ PRACTISE	■ CONSOLIDATE	■ MASTER
1, 2, 3, 10, 14, 15	4, 5, 6, 11, 13, 16	7, 8, 9, 12, 17

To answer questions online and to receive **immediate corrective feedback** and **fully worked solutions** for all questions, go to your learnON title at www.jacplus.com.au.

Understanding

Use technology wherever possible to answer the following questions.

1. Sketch the following graphs on the same Cartesian plane.

   a. $y = x$ b. $y = 2x$ c. $y = 3x$

     i. Describe what happens to the steepness of the graph (the gradient of the line) as the coefficient of x increases in value.

     ii. Identify where each graph cuts the x-axis (its x-intercept).

     iii. Identify where each graph cuts the y-axis (its y-intercept).

2. Sketch the following graphs on the same Cartesian plane.

   a. $y = -x$ b. $y = -2x$ c. $y = -3x$

     i. Describe what happens to the steepness of the graph as the magnitude of the coefficient of x decreases in value (becomes more negative).

     ii. Identify where each graph cuts the x-axis (its x-intercept).

     iii. Identify where each graph cuts the y-axis (its y-intercept).

3. Identify the correct word or words from the options given and rewrite the following sentences using the correct option.

   a. If the coefficient of x is **(positive/negative)**, then the graph will have an upward slope to the right. That is, the gradient of the graph is **(positive/negative)**.

   b. If the coefficient of x is negative, then the graph will have a **(downward/upward)** slope to the right. That is, the gradient of the graph is **(positive/negative)**.

   c. The bigger the magnitude of the coefficient of x (more positive or more negative), the **(bigger/smaller)** the steepness of the graph.

   d. If there is no constant term in the equation, the graph **(will/will not)** pass through the origin.

4. Sketch the graphs shown on the same Cartesian plane and answer the following questions for each graph.

   a. Is the coefficient of x the same for each graph? If so, state the coefficient.

   b. State whether the steepness (gradient) of each graph differs.

   c. Identify where each graph cuts the x-axis (its x-intercept).

   d. Identify where each graph cuts the y-axis (its y-intercept).

     i. $y = x$ ii. $y = x + 2$ iii. $y = x - 2$

5. Sketch the graphs shown on the same Cartesian plane and answer the following questions for each graph.

   a. Is the coefficient of x the same for each graph? If so, state the coefficient.

   b. State whether the steepness (gradient) of each graph differs.

   c. Identify where each graph cuts the x-axis (its x-intercept).

   d. Identify where each graph cuts the y-axis (its y-intercept).

     i. $y = -x$ ii. $y = -x + 2$ iii. $y = -x - 2$

6. Identify the correct word or words from the options given and rewrite the following sentences using the correct option.

   a. For a given set of linear graphs, if the coefficient of x is **(the same/different)**, the graphs will be parallel.

   b. The constant term in the equation is the **(y-intercept/x-intercept)** or where the graph cuts the **(y-axis/x-axis)**.

   c. The **(y-intercept/x-intercept)** can be found by substituting $x = 0$ into the equation.

   d. The **(y-intercept/x-intercept)** can be found by substituting $y = 0$ into the equation.

7. On the same Cartesian plane, sketch the following graphs.

   a. $y = x + 5$ b. $y = -x + 5$ c. $y = 3x + 5$ d. $y = -\dfrac{2}{5}x + 5$

     i. Is the coefficient of x the same for each graph? If so, state the coefficient.

     ii. State whether the steepness (gradient) of each graph differ.

     iii. Write down the gradient of each linear graph.

     iv. Identify where each graph cuts the x-axis (its x-intercept).

     v. Identify where each graph cuts the y-axis (its y-intercept).

8. Identify the correct word or words from the options given and rewrite the following sentences using the correct option.

   a. One of the general forms of the equation of a linear graph is $y = mx + c$, where m is the (**steepness/x-coordinate**) of the graph. We call the steepness of the graph the *gradient*.

   b. The value of c is the (**x-coordinate/y-coordinate**) where the graph cuts the (**x-axis/y-axis**).

   c. All linear graphs with the (**same/different**) gradient are (**parallel/perpendicular**).

   d. All linear graphs that have the same y-intercept pass through (**the same/different**) point on the y-axis.

9. For each of the following lines, identify the gradient and the y-intercept.

   a. $y = 2x$
   b. $y = x + 1$
   c. $y = -3x + 5$
   d. $y = \frac{2}{3}x - 7$

Reasoning

10. Using a CAS calculator, graph the lines $y = 3(x - 1) + 5$, $y = 2(x - 1) + 5$ and $y = -\frac{1}{2}(x - 1) + 5$. Describe what they have in common and how they differ from each other.

11. Show on a CAS calculator how $y = \frac{1}{2}x - \frac{5}{3}$ can be written as $6y = 3x - 10$.

12. A phone company charges $2.20 for international calls of 1 minute or less and $0.55 for each additional minute. Using a CAS calculator, graph the cost for calls that last for whole numbers of minutes. Explain all the important values needed to sketch the graph.

13. Shirly walks dogs after school for extra pocket money. She finds that she can use the equation $P = -15 + 10N$ to calculate her profit (in dollars) each week.

    a. Explain the real-world meaning of the numbers -15 and 10 and the variable N.
    b. What is the minimum number of dogs that Shirly must walk in order to earn a profit?
    c. Using a CAS calculator, sketch the equation.

14. Graph $y = 0.2x + 3.71$ on a CAS calculator. Explain how to use the calculator to find an approximate value when $x = 70.3$.

Problem solving

15. Plot the points $(6, 3.5)$ and $(-1, -10.5)$ using a CAS calculator and:

    a. determine the equation of the line
    b. sketch the graph, showing x- and y-intercepts
    c. calculate the value of y when $x = 8$
    d. calculate the value of x when $y = 12$.

16. A school investigating the price of a site licence for their computer network found that it would cost $1750 for 30 computers and $2500 for 60 computers.

    a. Using a CAS calculator, find a linear equation that represents the cost of a site licence in terms of the number of computers in the school.
    b. Determine is the y-intercept of the linear equation and explain how it relates to the cost of a site licence.
    c. Calculate the cost for 200 computers.
    d. Evaluate how many computers you could connect for $3000.

17. Dylan starts his exercise routine by jogging to the gym, which burns 325 calories. He then pedals a stationary bike burning 3.8 calories a minute.

    a. Graph this data using a CAS calculator.
    b. Evaluate how many calories Dylan has burnt after 15 minutes of pedalling.
    c. Evaluate how long it took Dylan to burn a total of 450 calories.

6.7 Practical applications of linear graphs

LEARNING INTENTION

At the end of this subtopic you should be able to:
- determine the linear rule from a table of values
- determine the linear rule that models a real-life problem
- interpret linear rules in real-life problems.

6.7.1 Determining a linear rule from a table of values

eles-4792

- If two **variables** are linked by a linear rule, then as one variable increases, the other increases (or decreases) at a steady rate.
- The table shown gives an example of a linear relationship.

x	0	1	2	3
y	5	8	11	14

- Each time x increases by 1, y increases by 3.
- The linear rule connecting x and y is $y = 3x + 5$.

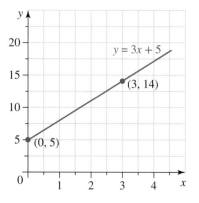

- Consider the relationship depicted in the table shown.

x	0	1	2	3
y	7	5	3	1

- Each time x increases by 1, y decreases by 2.
- The linear rule in this case is $y = -2x + 7$.

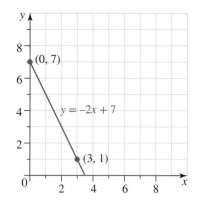

WORKED EXAMPLE 14 Determining a rule from a table of values

Determine the rule connecting x and y in each of the following value tables.

a.

x	0	1	2	3
y	-3	2	7	12

b.

x	3	4	5	6
y	12	11	10	9

THINK

a. 1. y increases at a steady rate, so this is a linear relationship. Write the rule.

   2. To calculate m: y increases by 5 each time x increases by 1. Write the value of the gradient.

   3. To determine c: From the table, when $x = 0$, $y = -3$. Write the value of the y-intercept.

   4. Write the rule.

WRITE

a. $y = mx + c$

   $m = 5$

   $c = -3$

   $y = 5x - 3$

b. 1. y decreases at a steady rate, so this is a linear relationship. Write the rule.

b. $y = mx + c$

2. To calculate m: decreases by 1 each time x increases by 1.
Write the value of the gradient.

$m = -1$

3. To determine c: From the table, when $x = 3$, $y = 12$.
To calculate the y-intercept, substitute the x- and y-values of one of the points, and solve for c.

$(3, 12):$ $\quad y = -x + c$
$12 = -(3) + c$
$12 = -3 + c$
$c = 15$

4. Write the rule.

$y = -x + 15$

▶ 6.7.2 Modelling linear relationships

eles-4793

- Relationships between real-life variables are often modelled (described) by a mathematical equation. In other words, an equation or formula is used to link the two variables.
- For example:
  - $A = l^2$ represents the relationship between the area and the side length of a square
  - $C = \pi d$ represents the relationship between the circumference and the diameter of a circle.
- If one variable changes at a constant rate compared to the other, then the two variables have a linear relationship.

WORKED EXAMPLE 15 Determining the rule to describe a relationship in a worded question

An online bookstore sells a certain textbook for \$21 and charges \$10 for delivery, regardless of the number of books being delivered.
a. Determine the rule connecting the cost (\$$C$) with the number of copies of the textbook delivered (n).
b. Use the rule to calculate the cost of delivering 35 copies of the textbook.
c. Calculate how many copies of the textbook can be delivered for \$1000.

THINK

WRITE

a. 1. Set up a table.
Cost for 1 copy $= 21 + 10$
Cost for 2 copies $= 2(21) + 10$
$= 52$

a.

n	1	2	3
C	31	52	73

2. The cost rises steadily, so there is a linear relationship. Write the rule.

$C = mn + c$

3. To calculate the gradient, use the formula $m = \dfrac{y_2 - y_1}{x_2 - x_1}$ with the points $(1, 31)$ and $(2, 52)$.

$m = \dfrac{52 - 31}{2 - 1}$
$= 21$
$C = 21n + c$

4. To determine the value of c, substitute $C = 31$ and $n = 1$.

$(1, 31):$ $31 = 21(1) + c$
$c = 10$

5. Write the rule.

$C = 21n + 10$

▶

b.	**1.** Substitute $n = 35$ and calculate the value of C.	**b.**	$C = 21(35) + 10$ $\quad = 735 + 10$ $\quad = 745$
	2. Write the answer.		The cost including delivery for 35 copies is $745.
c.	**1.** Substitute $C = 1000$ and calculate the value of n.	**c.**	$1000 = 21n + 10$ $21n = 990$ $\quad n = 47.14$
	2. You cannot buy 47.14 books, so round down. Write the answer.		For $1000, 47 copies of the textbook can be bought and delivered.

- In Worked example 15, compare the rule $C = 21n + 10$ with the original question. It is clear that the $21n$ refers to the cost of the textbooks (a variable amount, depending on the number of copies) and that 10 refers to the fixed (constant) delivery charge.
- In this case C is called the **response variable** or the **dependent variable**, because it depends on the number of copies of the textbooks (n).
- The variable n is called the **explanatory variable** or the **independent variable** because it is the variable that may explain the changes in the response variable.
- When graphing numerical data, the response variable is plotted on the vertical axis and the explanatory variable is plotted on the horizontal axis.

Exercise 6.7 Practical applications of linear graphs   **learn on**

Individual pathways

■ PRACTISE	■ CONSOLIDATE	■ MASTER
1, 3, 5, 6, 11, 16	2, 7, 9, 12, 13, 17	4, 8, 10, 14, 15, 18

To answer questions online and to receive **immediate corrective feedback** and **fully worked solutions** for all questions, go to your learnON title at www.jacplus.com.au.

Fluency

1. **WE14** Determine the linear rule linking the variables in each of the following tables.

a.
x	0	1	2	3
y	−5	1	7	13

b.
x	0	1	2	3
y	8	5	2	−1

2. Determine the linear rule linking the variables in each of the following tables.

a.
x	0	1	2	3
y	4	6	8	10

b.
x	0	1	2	3
y	1.1	2.0	2.9	3.8

3. Determine the linear rule linking the variables in each of the following tables.

a.

x	2	3	4	5
y	7	10	13	16

b.

x	5	6	7	8
y	12	11	10	9

4. Determine the linear rule linking the variables in each of the following tables.

a.

t	3	4	5	6
v	18	15	12	9

b.

d	1	2	3	4
C	11	14	17	20

5. **MC** Sasha and Fiame hire a car. They are charged a fixed fee of \$150 for hiring the car and then \$25 per day. They hire the car for d days.

Select which one of the following rules describes the number of days the car is hired and the total cost, C, they would be charged for that number of days charged.

A. $C = 25d$ **B.** $C = 150d$ **C.** $C = 175d$ **D.** $C = 25d + 150$ **E.** $C = 150d + 25$

Understanding

6. **WE15** Fady's bank balance has increased in a linear manner since he started his part-time job. After 20 weeks of work his bank balance was at \$560 and after 21 weeks of work it was at \$585.

   a. Determine the rule that relates the size of Fady's bank balance, A, and the time (in weeks) worked, t.
   b. Use the rule to calculate the amount in Fady's account after 200 weeks.
   c. Use the rule to identify the initial amount in Fady's account.

7. The cost of making a shoe increases as the size of the shoe increases. It costs \$5.30 to make a size 6 shoe, and \$6.40 to make a size 8 shoe. Assuming that a linear relationship exists:

   a. determine the rule relating cost (C) to shoe size (s)
   b. calculate much it costs to produce a size 12 shoe.

8. The number of books in a library (N) increases steadily with time (t). After 10 years there are 7200 publications in the library, and after 12 years there are 8000 publications.

   a. Determine the rule predicting the number of books in the library.
   b. Calculate how many books were there after 5.5 years.
   c. Calculate how many books will there be after 25 years.

9. A skyscraper can be built at a rate of 4.5 storeys per month.

   a. Calculate how many storeys will be built after 6 months.
   b. Calculate how many storeys will be built after 24 months.

10. The Nguyens' water tank sprang a leak and has been losing water at a steady rate. Four days after the leak occurred, the tank contained 552 L of water, and ten days later it held only 312 L.

   a. Determine the rule linking the amount of water in the tank (w) and the number of days (t) since the leak occurred.
   b. Calculate the amount of water that was initially in the tank.
   c. If water loss continues at the same rate, determine when the tank will be empty.

Reasoning

11. The pressure inside a boiler increases steadily as the temperature increases. For each $1\,°C$ increase in temperature, the pressure increases by 10 units, and at a temperature of $100\,°C$ the pressure is 1200 units. If the maximum pressure allowed is 2000 units, show that the temperature cannot exceed $180\,°C$.

12. After 11 pm a taxi company charges a $3.50 flag fall plus $2.57 for each kilometre travelled.
    a. Determine the linear rule connecting the cost, C, and the distance travelled, d.
    b. Calculate how much an 11.5 km trip will cost.
    c. If you have $22 in your pocket, calculate how far you can afford to travel, correct to 1 decimal place.

13. A certain kind of eucalyptus tree grows at a linear rate for its first 2 years of growth. If the growth rate is 5 cm per month, show that the tree will be 1.07 m tall after 21.4 months.

14. A software company claims that its staff can fix 22 bugs each month. They are working on a project to fix a program that started out with 164 bugs.
    a. Determine the linear rule connecting the number of bugs left, N, and the time in months, t, from the beginning of the project.
    b. Calculate how many bugs will be left after 2 months.
    c. Determine how long it will be until there are only 54 bugs left.
    d. Determine how long it will take to eliminate all of the bugs. Justify your answer.

15. Michael produces and sells prints of his art at a local gallery. For each print run his profit (P) is given by the equation $P = 200n - 800$, where n is the number of prints sold.
    a. Sketch the graph of this rule.
    b. Identify the y-intercept. Determine what it represents in this situation.
    c. Identify the x-intercept. Determine what it represents in this example.
    d. Identify the gradient of the graph. Determine what this means in this situation.

Problem solving

16. The cost of a taxi ride is $3.50 flag fall plus $2.14 for each kilometre travelled.
    a. Determine the linear rule connecting the cost, C, and the distance travelled, d.
    b. Calculate how much an 11.5 km trip will cost.
    c. Calculate how much a 23.1 km trip will cost.

17. Theo is going on holiday to Japan. One yen (¥) buys 0.0127 Australian dollars (A$).
    a. Write an equation that converts Australian dollars to Japanese yen (¥), where A represents amount of Australian dollars and Y represents amount of yen.
    b. Using the equation from part a, calculate how many yen Theo will receive in exchange if they have A$2500.
    c. There is a commission to be paid on exchanging currency. Theo needs to pay 2.8% for each Australian dollar they exchange into yen. Write down an equation that calculates the total amount of yen Theo will receive. Write your equation in terms of Y_T (total amount of yen) and $A (Australian dollars).

18. Burchill and Cody need to make a journey to the other branch of their store across town. The traffic is very busy at this time of the day, so Burchill catches a train that travels halfway, then walks the rest of the way. Cody travels by bike the whole way. The bike path travels along the train line and then follows the road to the other branch of their store.
    Cody's bike travels twice as fast as Burchill's walking speed, and the train travels 4 times faster than Cody's bike. Evaluate who arrives at the destination first.

6.8 Midpoint of a line segment and distance between two points

6.8.1 Calculating the midpoint of a line segment

eles-4794

- The x- and y-coordinates of the **midpoint** are halfway between those of the end points of a line segment.
- The coordinates of the midpoint, M, of a line can be found by averaging the x- and y-coordinates of the end points.

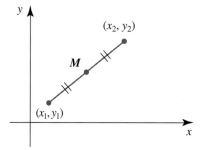

The midpoint of a line segment

The midpoint, M, of the line segment joining the points $(x_1,\ y_1)$ and $(x_2,\ y_2)$ is:

$$M = \left(\frac{x_1 + x_2}{2}, \frac{y_1 + y_2}{2} \right)$$

WORKED EXAMPLE 16 Calculating the midpoint of a line segment

Calculate the midpoint of the line segment joining $(5,\ 9)$ and $(-3, 11)$.

THINK	WRITE
1. Average the x-values: $\dfrac{x_1 + x_2}{2}$.	$x = \dfrac{5 - 3}{2}$ $= \dfrac{2}{2}$ $= 1$
2. Average the y-values: $\dfrac{y_1 + y_2}{2}$.	$y = \dfrac{9 + 11}{2}$ $= \dfrac{20}{2}$ $= 10$
3. Write the answer.	The midpoint is $(1, 10)$.

TI	THINK	WRITE	CASIO	THINK	WRITE

TI | THINK

1. On a Graphs page, set an appropriate window as shown and press:
   • MENU
   • 8: Geometry
   • Points & Lines
   • 3: Point by Coordinates

   Type the coordinates of one point on the keypad, pressing ENTER after each number.

   Repeat the above step to plot the second point.

WRITE

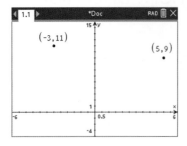

2. To determine the midpoint, press:
   • MENU
   • 8: Geometry
   • 4: Construction
   • 5: Midpoint.
   Move the cursor to one of the points and press ENTER.
   Move the cursor to the other point and press ENTER. This will place a dot on the screen representing the midpoint of the two plotted points.
   To display the midpoint's coordinates, press:
   • MENU
   • 1: Actions
   • 8: Coordinates and Equations
   Move the cursor to the midpoint and press ENTER twice.

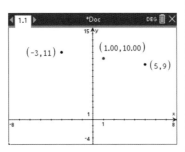

3. State the coordinates of the midpoint.

The coordinates of the midpoint are (1, 10).

CASIO | THINK

On a Geometry screen, select a pair of axes with scale and grid. This is achieved by continuous clicking on the graph icon.
To get the desired window, select:
   • View
   • Zoom out

Using the Line segment tool, place the pointer at $(-3, 11)$ and hold to place the second point at $(5, 9)$. The line segment will be automatically labelled.

WRITE

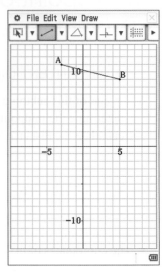

To locate the midpoint:
   • Click on the line segment.
   • Click again for the point C to be placed in the centre or midpoint.
To determine the coordinates of the midpoint C:
   • Click on the point C.
   • Use the measuring tool at the top right to display the coordinates.

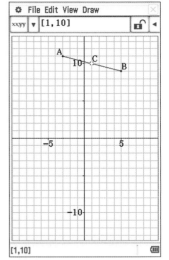

State the coordinates of the midpoint.

The coordinates of the midpoint are (1, 10).

WORKED EXAMPLE 17 Determining the coordinates of a point given the midpoint and another point of an interval

$M(7, 2)$ **is the midpoint of the line segment AB. If the coordinates of A are** $(1, -4)$**, determine the coordinates of B.**

THINK

1. Let B have the coordinates (x, y).

WRITE

$A(1, -4), B(x, y), M(7, 2)$

2. The midpoint is $(7, 2)$, so the average of the x-values is 7. Solve for x.	$\dfrac{1+x}{2} = 7$ $1 + x = 14$ $x = 13$
3. The average of the y-values is 2. Solve for y.	$\dfrac{-4+y}{2} = 2$ $-4 + y = 4$ $y = 8$
4. Write the answer.	The coordinates of point B are $(13, 8)$.

6.8.2 The distance between two points

eles-4795

- The distance between two points on the Cartesian plane is calculated using Pythagoras' theorem applied to a right-angled triangle.

Distance between two points

The distance, d, of the line segment joining the points (x_1, y_1) and (x_2, y_2) is:

$$d = \sqrt{(x_2 - x_1)^2 + (y_2 - y_1)^2}$$

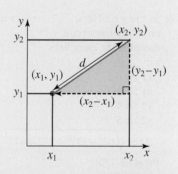

WORKED EXAMPLE 18 Calculating the distance between two points

Calculate the distance between the points $(-1, 3)$ and $(4, 5)$:
a. **exactly**   b. **correct to 3 decimal places.**

THINK

a. **1.** Draw a diagram showing the right-angled triangle (optional).

WRITE

a.

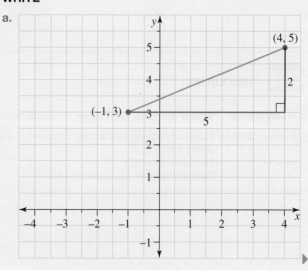

2. Write the formula for the distance between two points.

$$d = \sqrt{(x_2 - x_1)^2 + (y_2 - y_1)^2}$$

3. Let $(x_1, y_1) = (-1, 3)$ and $(x_2, y_2) = (4, 5)$. Substitute the x- and y-values into the equation.

$$d = \sqrt{(4 - -1)^2 + (5 - 3)^2}$$

4. Simplify.

$$d = \sqrt{5^2 + 2^2}$$
$$d = \sqrt{25 + 4}$$
$$d = \sqrt{29}$$

5. Write the answer.

The exact distance betwen points $(-1, 3)$ and $(4, 5)$ is $\sqrt{29}$.

b. 1. Write $\sqrt{29}$ as a decimal to 4 decimal places.   b. $\sqrt{29} = 5.3851$

2. Write the answer correct to 3 decimal places. $d \approx 5.385$

TI	THINK	WRITE	CASIO	THINK	WRITE

TI | THINK

a-b In a new problem, on a Calculator page, complete the entry lines as:

$x1 := -1$
$y1 := 3$
$x2 := 4$
$y2 := 5$
$d := \sqrt{(x2 - x1)^2 + (y2 - y1)^2}$

Press ENTER after each entry.
Press CTRL ENTER to get a decimal approximation.

WRITE

a-b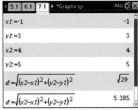

The distance is $\sqrt{29} \approx 5.385$ to 3 decimal places.

CASIO | THINK

a. Make sure the calculator is set to Standard mode. On the Main screen, complete the entry line as:

$$\sqrt{(4 - -1)^2 + (5 - 3)^2}$$

Press EXE.

b. Convert the answer to a decimal by tapping Standard and changing it to Decimal. Then press EXE.

WRITE

a. $\sqrt{29}$

b. 5.385

---

on Resources

eWorkbook   Topic 6 Workbook (worksheets, code puzzle and project) (ewbk-2006)

Interactivities Individual pathway interactivity: Midpoint of a line segment and distance between two points (int-4508)
Midpoints (int-6052)
Distance between two points (int-6051)

Exercise 6.8 Midpoint of a line segment and distance between two points

Individual pathways

■ PRACTISE	■ CONSOLIDATE	■ MASTER
1, 4, 8, 11, 13, 16, 20	2, 5, 9, 12, 14, 17, 21	3, 6, 7, 10, 15, 18, 19, 22

To answer questions online and to receive **immediate corrective feedback** and **fully worked solutions** for all questions, go to your learnON title at www.jacplus.com.au.

Fluency

1. **WE16** Calculate the midpoint of the line segment joining each of the following pairs of points.
   a. $(1, 3)$ and $(3, 5)$
   b. $(6, 4)$ and $(4, -2)$
   c. $(2, 3)$ and $(12, 1)$

2. Calculate the midpoint of the line segment joining each of the following pairs of points.
   a. $(6, 3)$ and $(10, 15)$
   b. $(4, 2)$ and $(-4, 8)$
   c. $(0, -5)$ and $(-2, 9)$

3. Calculate the midpoint of the line segment joining each of the following pairs of points.
   a. $(8, 2)$ and $(-18, -6)$
   b. $(-3, -5)$ and $(7, 11)$
   c. $(-8, -3)$ and $(8, 27)$

4. Calculate the midpoint of the segment joining each of the following pairs of points.
   a. $(7, -2)$ and $(-4, 13)$
   b. $(0, 22)$ and $(-6, -29)$
   c. $(-15, 8)$ and $(-4, 11)$
   d. $(-3, 40)$ and $(0, -27)$

5. **WE17** Determine the value of a in each series of points so that the point M is the midpoint of the line segment joining points A and B.
   a. $A(-2, a), B(-6, 5), M(-4, 5)$
   b. $A(a, 0), B(7, 3), M(8, \frac{3}{2})$
   c. $A(3, 3), B(4, a), M\left(3, \frac{1}{2}, -6\frac{1}{2}\right)$
   d. $A(-4, 4), B(a, 0), M(-2, 2)$

6. M is the midpoint of the line interval AB. Determine the coordinates of B if:
   a. $A = (0, 0)$ and $M = (2, 3)$
   b. $A = (2, 3)$ and $M = (0, 0)$
   c. $A = (-3, 2)$ and $M = (4, 2)$
   d. $A = (3, -1)$ and $M = (-2, -2)$.

7. Determine the equation of a line that has a gradient of 5 and passes through the midpoint of the line segment joining $(-1, -7)$ and $(3, 3)$.

8. **WE18** Calculate the distance between each of the following pairs of points.
   a. $(4, 5)$ and $(1, 1)$
   b. $(7, 14)$ and $(15, 8)$
   c. $(2, 4)$ and $(2, 3)$

9. Calculate the distance between each of the following pairs of points.
   a. $(12, 8)$ and $(10, 8)$
   b. $(14, 9)$ and $(2, 14)$
   c. $(5, -13)$ and $(-3, -7)$

10. Calculate the distance between each of the following pairs of points.
   a. $(-14, -9)$ and $(-10, -6)$
   b. $(0, 1)$ and $(-15, 9)$
   c. $(-4, -8)$ and $(1, 4)$

11. Calculate the distance between the following pairs of points, correct to 3 decimal places.
   a. $(-14, 10)$ and $(-8, 14)$
   b. $(6, -7)$ and $(13, 6)$
   c. $(-11, 1)$ and $(2, 2)$

12. Calculate the distance between the following pairs of points, correct to 3 decimal places
   a. $(9, 0)$ and $(5, -8)$
   b. $(2, -7)$ and $(-2, 12)$
   c. $(9, 4)$ and $(-10, 0)$

Understanding

13. Calculate the perimeter of each figure shown, giving your answers correct to 3 decimal places.

a.

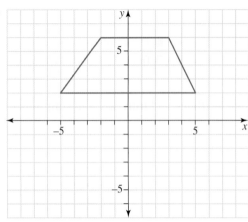

b.

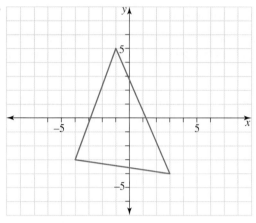

14. Calculate the perimeter of each triangle shown, giving your answers correct to 3 decimal places.

a.

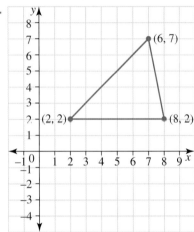

b.

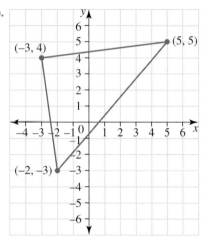

15. Two hikers are about to hike from A to B (shown on the map). Calculate the straight-line distance from A to B.

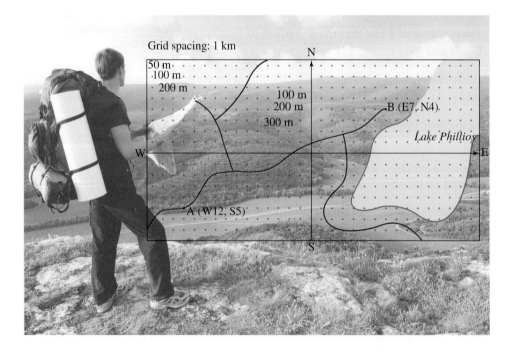

Reasoning

16. Show that the distance between the points A (2, 2) and B (6, −1) is 5.

17. The point M (−2, −4) is the midpoint of the interval AB. Show that the point B is (−9, −2), given A is (5, −6).

18. Show that the point B (6, −10) is equidistant from the points A (15, 3) and C (−7, −1).

19. Answer the following questions.
    a. Plot the following points on a Cartesian plane: A (−1, −4), B (2, 3), C (−3, 8) and D (4, −5).
    b. Show that the midpoint of the interval AC is (−2, 2).
    c. Calculate the exact distance between the points A and C.
    d. If B is the midpoint of an interval CM, determine the coordinates of point M.
    e. Show that the gradient of the line segment AB is $\frac{7}{3}$.
    f. Determine the equation of the line that passes through points B and D.

Problem solving

20. Explain what type of triangle ΔABC is if it has vertices A (−4, 1), B (2, 3) and C (0, −3).

21. Calculate the gradient of the line through the points (−1, 3) and (3 + 4t, 5 + 2t).

22. A map of a town drawn on a Cartesian plane shows the main street extending from (−4, 5) to (0, −7). There are five streetlights positioned in the street. There is one streetlight at either end, and three streetlights spaced evenly down the street. Give the position of the five lights in the street.

6.9 Non-linear relations (parabolas, hyperbolas, circles)

LEARNING INTENTION

At the end of this subtopic you should be able to:
- recognise and sketch a basic parabola from its equation
- sketch a parabola that has been translated horizontally or vertically
- recognise and sketch a hyperbola from its equation
- recognise and sketch a circle from its equation.

- There are many examples of non-linear relationships in mathematics. Some of them are the parabola, the hyperbola and the circle.

  Note: This subtopic is an introduction to non-linear relations — parabolas will be discussed in more depth in topic 12.

6.9.1 The parabola

eles-4796

- The **parabola** is a curve that is often found in nature. It is also commonly used in architecture and engineering.

- The simplest parabola has the equation $y = x^2$.
- The key features of the parabola $y = x^2$ are:
  - it has an axis of symmetry, which is the y-axis
  - the graph is concave up (it opens upwards)
  - it has a minimum **turning point**, or vertex, at $(0,0)$.
- All parabolas have the same basic shape; however, they can be wider or narrower depending on their equation.
- To sketch a parabola we can create a table of values and plot points.

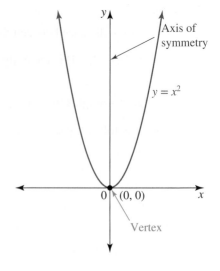

WORKED EXAMPLE 19 Graphing a parabola

Plot the graph of $y = x^2$ for values of x from -3 to 3. State the equation of the axis of symmetry and the coordinates of the turning point.

THINK	WRITE
1. Write the equation.	$y = x^2$
2. Produce a table of values using x-values from -3 to 3.	

x	-3	-2	-1	0	1	2	3
y	9	4	1	0	1	4	9

3. Draw a set of clearly labelled axes. Plot the points and join them with a smooth curve. The scale on the y-axis would be from -2 to 10 and from -4 to 4 on the x-axis.

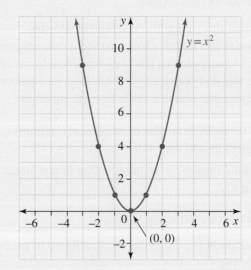

4. Write the equation of the line that divides the parabola exactly in half.

The equation of the axis of symmetry is $x = 0$.

5. Write the coordinates of the turning point.

The turning point is $(0,0)$.

| TI | THINK | WRITE | CASIO | THINK | WRITE |

1. Open a Graphs page, and press:
- MENU
- 4: Window/Zoom
- 1: Window Settings

Complete the entry lines as shown. Press OK.

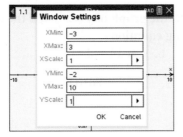

Open a Graph & Table screen and select the Window icon. Complete the entry lines as shown. Press OK.

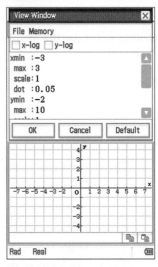

2. To sketch the graph, complete the entry line as:
$f1(x) = x^2$
Press ENTER. The graph will be displayed.

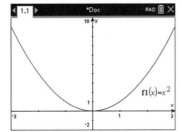

To sketch the graph, complete the entry line as:
$y1 = x^2$
Press the Graph icon. The graph will be displayed.

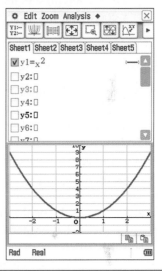

- The parabola $y = x^2$ can be transformed into other parabolas using transformations, such as:
  - vertical translations
  - horizontal translations
  - reflections.
- Other parabolas can be sketched using their key features, such as the axis of symmetry and vertex.

Vertical translations

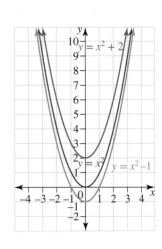

- Parabolas can be **translated** vertically by adding a number to or subtracting a number from the right-hand side of the equation.
- When a number is added to the equation the parabola is translated vertically upwards.
- When a number is subtracted from the equation the parabola is translated vertically downwards.
- The graph shown displays the parabolas $y = x^2$, $y = x^2 + 2$ and $y = x^2 - 1$ on the same set of axes.
- $y = x^2 + 2$ has been translated 2 units upwards. The turning point is now $(0, 2)$.
- $y = x^2 - 1$ has been translated 1 unit downwards. The turning point is now $(0, -1)$.

State the vertical translation (when compared with the graph of $y = x^2$) and the coordinates of the turning point for the graphs of each of the following equations.

a. $y = x^2 + 5$
b. $y = x^2 - 4$

THINK	WRITE
a. 1. Write the equation.	a. $y = x^2 + 5$
2. $+5$ means the graph is translated upwards 5 units.	Vertical translation of 5 units up
3. Translate the turning point of $y = x^2$, which is $(0, 0)$. The x-coordinate of the turning point remains 0 and the y-coordinate has 5 added to it.	The turning point becomes $(0, 5)$.
b. 1. Write the equation.	b. $y = x^2 - 4$
2. -4 means the graph is translated downwards 4 units.	Vertical translation of 4 units down
3. Translate the turning point of $y = x^2$, which is $(0, 0)$. The x-coordinate of the turning point remains 0 and the y-coordinate has 4 subtracted from it.	The turning point becomes $(0, -4)$.

Horizontal translations

- Parabolas can be translated horizontally by placing x in brackets and adding a number to it or subtracting a number from it.
- When a number is added to x, the parabola is translated to the left.
- When a number is subtracted from x, the parabola is translated to the right.
- The graphs shown displays the parabolas $y = x^2$, $y = (x + 1)^2$ and $y = (x - 2)^2$ on the same set of axes.
- $y = (x + 1)^2$ has been translated 1 unit left. The turning point is now $(-1, 0)$.
- $y = (x - 2)^2$ has been translated 2 units right. The turning point is now $(2, 0)$.

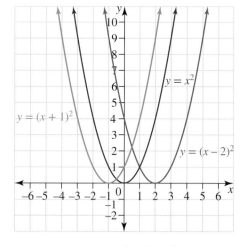

State the horizontal translation (when compared to the graph of $y = x^2$) and the coordinates of the turning point for the graphs of each of the following equations.

a. $y = (x - 3)^2$
b. $y = (x + 2)^2$

THINK	WRITE
a. 1. Write the equation.	a. $y = (x - 3)^2$
2. -3 means the graph is translated to the right 3 units.	Horizontal translation of 3 units right
3. Translate the turning point of $y = x^2$, which is $(0, 0)$. The y-coordinate of the turning point remains 0, and the x-coordinate has 3 added to it.	The turning point becomes $(3, 0)$.

b. **1.** Write the equation.

2. +2 means the graph is translated to the left 2 units.

3. Translate the turning point of $y = x^2$, which is $(0, 0)$. The y-coordinate of the turning point remains 0, and the x-coordinate has 2 subtracted from it.

b. $y = (x + 2)^2$

Horizontal translation of 2 units left

The turning point becomes $(-2, 0)$.

Reflection

- Compare the graph of $y = -x^2$ with that of $y = x^2$.
  - In each case the axis of symmetry is the line $x = 0$ and the turning point is $(0, 0)$.
  - The only difference between the equations is the $-$ sign in $y = -x^2$.
  - The difference between the graphs is that $y = x^2$ 'sits' on the x-axis and $y = -x^2$ 'hangs' from the x-axis (one is a reflection, or mirror image, of the other).
  - $y = x^2$ has a minimum turning point and $y = -x^2$ has a maximum turning point.
- Any quadratic graph for which x^2 is positive has a $\cup$ shape and is said to be concave up.
- Conversely, if x^2 is negative the graph has a $\cap$ shape and is said to be concave down.

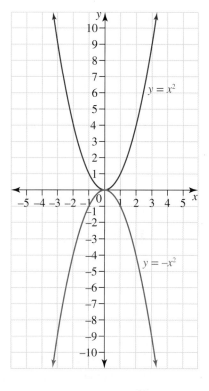

6.9.2 The hyperbola

eles-4797

- A hyperbola is a non-linear graph for which the equation is $y = \dfrac{1}{x}$ (or $xy = 1$).

WORKED EXAMPLE 22 Graphing a hyperbola

Complete the table of values below and use it to plot the graph of $y = \dfrac{1}{x}$.

x	-3	-2	-1	$-\dfrac{1}{2}$	0	$\dfrac{1}{2}$	2	3
y								

THINK

1. Substitute each x-value into the function $y = \dfrac{1}{x}$ to obtain the corresponding y-value.

WRITE

x	-3	-2	-1	$-\dfrac{1}{2}$	0	$\dfrac{1}{2}$	1	2	3
y	$-\dfrac{1}{3}$	$-\dfrac{1}{2}$	-1	-2	undefined	2	1	$\dfrac{1}{2}$	$\dfrac{1}{3}$

2. Draw a set of axes and plot the points from the table. Join them with a smooth curve.

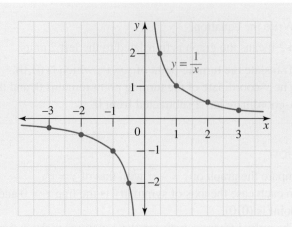

- The graph in Worked example 22 has several key features.
  - There are no values when $x = 0$ or $y = 0$.
  - The curve gets closer to the axes but never touches the x- or y-axes. We say the axes are horizontal and vertical **asymptotes**.
  - The hyperbola has two separate branches. It cannot be drawn without lifting your pen from the paper. We say it is a discontinuous graph.
  - All hyperbolas have the same basic shape but they can be wider or narrower depending on the equation.
  - Hyperbolas can be transformed in the same way as parabolas; this will be covered in later years.

▶ 6.9.3 The circle

eles-4798

- A circle is the path traced out by a point at a constant distance (the radius) from a fixed point (the centre).
- Consider the circle shown. Its centre is at the origin and its radius is r.
  - Let P (x, y) be a point on the circle.
  - By Pythagoras' theorem, $x^2 + y^2 = r^2$.
  - This relationship is true for all points on the circle.
  - The equation of a circle with centre $(0, 0)$ and radius r is: $x^2 + y^2 = r^2$.
- If the circle is translated h units to the right, parallel to the x-axis, and k units upwards, parallel to the y-axis, then the equation of a circle with centre (h, k) and radius r is: $(x - h)^2 + (y - k)^2 = r^2$.

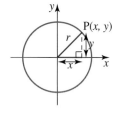

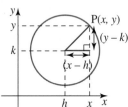

Equation of a circle with centre (h, k) and radius r

$$(x - h)^2 + (y - k)^2 = r^2$$

WORKED EXAMPLE 23 Graphing a circle centered at the origin

Sketch the graph of $x^2 + y^2 = 49$, stating the centre and radius.

THINK	WRITE
1. Write the equation.	$x^2 + y^2 = 49$

2. State the coordinates of the centre.

Centre $= (0, 0)$

3. Calculate the length of the radius by taking the square root of both sides. (Ignore the negative results.)

$r^2 = 49$
$r = 7$
Radius $= 7$ units

4. Sketch the graph.

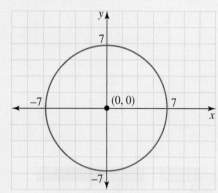

TI	THINK	WRITE

In a new problem, on a Graphs page, press:
- MENU
- 3: Graph Entry/Edit
- 2: Equation
- 3: Circle
- 1: Centre form.

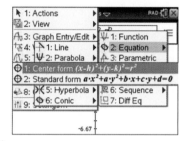

Complete the function entry line as shown. Press TAB to move between the fields. Press ENTER.

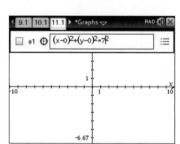

CASIO	THINK	WRITE

On the Main screen, complete the entry line as:
$x^2 + y^2 = 49$

Press the Graph icon to bring up the graphing screen.

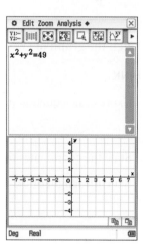

The graph of the circle is shown.

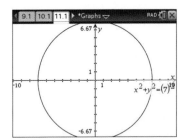

Highlight the equation and drag it down to the graphing screen.
Press the Resize icon to look at the graph on the full screen.

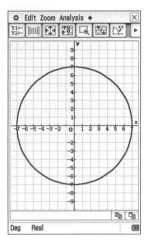

To find the centre of the circle, press:
- MENU
- 6: Analyze Graph
- 9: Analyze Conics
- 1: Centre.

Click on the circle and press ENTER. The coordinates of the centre are displayed.

To find the radius of the circle, press:
- MENU
- 6: Analyze Graph
- 9: Analyze Conics
- 7: Radius.

Click on the circle and press ENTER. The radius of the circle is displayed.

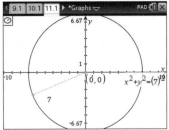

The circle $x^2 + y^2 = 49$ has its centre at $(0, 0)$ and has a radius of 7.

WORKED EXAMPLE 24 Graphing a circle

Sketch the graph of $(x - 2)^2 + (y + 3)^2 = 16$, clearly showing the centre and radius.

THINK	WRITE
1. Express the equation in general form.	$(x - h)^2 + (y - k)^2 = r^2$ $(x - 2)^2 + (y + 3)^2 = 4^2$
2. State the coordinates of the centre.	Centre $(2, -3)$
3. State the length of the radius.	$r^2 = 16$ $r = 4$ Radius $= 4$ units

4. Sketch the graph.

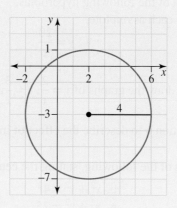

Exercise 6.9 Non-linear relations (parabolas, hyperbolas, circles)

learn on

Individual pathways

■ PRACTISE	■ CONSOLIDATE	■ MASTER
1, 2, 4, 8, 10, 14, 17	3, 5, 9, 13, 16	6, 7, 11, 12, 15, 18

To answer questions online and to receive **immediate corrective feedback** and **fully worked solutions** for all questions, go to your learnON title at www.jacplus.com.au.

Fluency

1. **WE20** State the vertical translation and the coordinates of the turning point for the graph of each of the following equations.

    a. $y = x^2 + 3$ b. $y = x^2 - 1$ c. $y = x^2 - 7$ d. $y = x^2 + \dfrac{1}{4}$

2. **WE21** State the horizontal translation and the coordinates of the turning point for the graph of each of the following equations.

    a. $y = (x - 1)^2$ b. $y = (x - 2)^2$ c. $y = (x + 10)^2$ d. $y = (x + 4)^2$

3. For each of the following graphs, give the coordinates of the turning point and state whether it is a maximum or a minimum.

    a. $y = -x^2 + 1$ b. $y = x^2 - 3$ c. $y = -(x + 2)^2$

4. **WE22** Complete the table of values shown and use it to plot the graph of $y = \dfrac{10}{x}$.

x	−5	−4	−3	−2	−1	0	1	2	3	4	5
y											

5. a. Plot the graph of each of the following hyperbolas.
   b. Write the equation of each asymptote.

   i. $y = \dfrac{5}{x}$ ii. $y = \dfrac{20}{x}$ iii. $y = \dfrac{100}{x}$

6. On the same set of axes, draw the graphs of $y = \dfrac{2}{x}, y = \dfrac{3}{x}$ and $y = \dfrac{4}{x}$.

7. Use your answer to question **6** to describe the effect of increasing the value of k on the graph of $y = \dfrac{k}{x}$.

8. **WE23** Sketch the graphs of the following, stating the centre and radius of each.

   a. $x^2 + y^2 = 49$ b. $x^2 + y^2 = 4^2$ c. $x^2 + y^2 = 36$
   d. $x^2 + y^2 = 81$ e. $x^2 + y^2 = 25$ f. $x^2 + y^2 = 100$

9. **WE24** Sketch the graphs of the following, clearly showing the centre and the radius.

   a. $(x - 1)^2 + (y - 2)^2 = 5^2$ b. $(x + 2)^2 + (y + 3)^2 = 6^2$
   c. $(x + 3)^2 + (y - 1)^2 = 49$ d. $(x - 4)^2 + (y + 5)^2 = 64$
   e. $x^2 + (y + 3)^2 = 4$ f. $(x - 5)^2 + y^2 = 100$

10. **MC** Select which of the following is the graph of $(x - 2)^2 + (y + 5)^2 = 4$.

   A.   B. 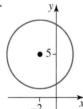   C.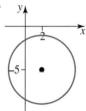

   D. 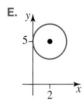   E.

Reasoning

11. Show that $y = x^2$ is the same as $y = (-x)^2$.

12. Explain the similarities and differences between $y = \dfrac{1}{x}$ and $y = -\dfrac{1}{x}$.

13. Show that the turning point for $y = (x + 3)^2 - 1$ is $(-3, -1)$. Explain whether this is a maximum or minimum turning point.

14. Show the equation of the circle that has centre $(2, 1)$ and passes through the point $(6, 1)$ is $(x - 2)^2 + (y - 1)^2 = 16$.

15. Identify the point on the circle $x^2 + y^2 - 12x - 4y = 9$ that has the lowest y-value.

Problem solving

16. At a kindergarten sports day, Edmond, Huot and Nalini ran a 20-metre race. Each child ran according to the following equations, where d is the distance, in metres, from the starting line, and t is the time, in seconds.

    Edmond: $d = 2.4 + 0.5t$

    Huot: $d = 0.1t(t - 5)$

    Nalini: $d = 0.2t(t - 5)(t - 9)$

    The following sketch shows their running paths.

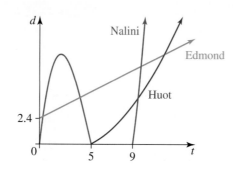

    a. Describe Nalini's race style.
    b. Determine how long it took Huot to start running.
    c. At the start of the race, determine how far Edmond was from the starting line.
    d. If the winner of the 20-metre race won the race in 10.66 seconds, determine whether Huot or Nalini won the race. Justify your answer using calculations.

17. A circle with centre (0, 2) passes through the point (3, 6). Determine the coordinates of the points where the circle crosses the y-axis.

18. Consider the relationship $\dfrac{1}{y} \doteq \dfrac{1}{x} + 1$. It appears that both variables are being raised to the power of 1. The relationship can be transposed as shown.

    $$\frac{1}{y} = \frac{1}{x} + 1$$

    $$\frac{1}{y} = \frac{1}{x} + \frac{x}{x}$$

    $$\frac{1}{y} = \frac{1 + x}{x}$$

    $$y = \frac{x}{1 + x}$$

    a. Generate a table of values for x versus y using the transposed relationship.
    b. Plot a graph to show the points contained in the table of values.
    c. Use your plot to confirm whether this relationship is linear.

6.10 Review

6.10.1 Topic summary

LINEAR AND NON-LINEAR GRAPHS

Gradient and axis intercepts

- The gradient of a straight line is given the label m. It can be determined by the following formula:

$$m = \frac{\text{rise}}{\text{run}} = \frac{y_2 - y_1}{x_2 - x_1}$$

- If the line is sloping upwards (from left to right), the gradient is positive.
- If the line is sloping downwards, the gradient is negative.
- The y-intercept of a line can be determined by letting $x = 0$, then solving for y.
- The x-intercept of a line can be determined by letting $y = 0$, then solving for x.

Equation of a straight line

- The equation of a straight line is $y = mx + c$ where:
  - m is the gradient
  - c is the y-intercept.

Sketching linear graphs

- To sketch a linear graph, all you need are two points that the line passes through.
- The x- and y-intercept method involves finding both axis intercepts, then drawing the line through them.
- The gradient and y-intercept method involves plotting the y-intercept and then one other point (usually $x = 1$) using the gradient.

Horizontal and vertical lines

- Horizontal lines are of the form $y = c$. They have a gradient of zero ($m = 0$).
- Vertical lines are of the form $x = a$. They have an undefined gradient.

Midpoint and distance

- The midpoint of two points (x_1, y_1) and (x_2, y_2) is

$$M = \left(\frac{x_2 + x_1}{2}, \frac{y_2 + y_1}{2} \right).$$

- The distance between two points, (x_1, y_1) and (x_2, y_2) is

$$d = \sqrt{(x_2 - x_1)^2 + (y_2 - y_1)^2}.$$

Non-linear graphs

- A basic **parabola** has the equation $y = x^2$.
- It has a concave upward shape and a turning point at the origin.

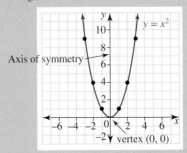

- A **hyperbola** has the equation $y = \frac{1}{x}$.
- It consists of two curved arcs that do not touch.

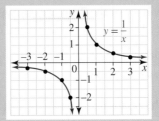

- A **circle** of radius r, when centred at the origin, has the equation $x^2 + y^2 = r^2$.

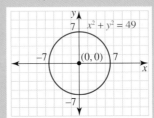

- A **circle** of radius r, when centred at the point (h, k), has the equation $(x - h)^2 + (y - k)^2 = r^2$.

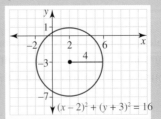

6.10.2 Success criteria

Tick the column to indicate that you have completed the subtopic and how well you have understood it using the traffic light system.

(**Green:** I understand; **Yellow:** I can do it with help; **Red:** I do not understand)

Subtopic	Success criteria	⬤	⬤	⬤
6.2	I can plot points on the Cartesian plane.			
	I can plot a linear graph on the Cartesian plane.			
6.3	I can calculate the gradient of a line or line segment.			
	I can calculate the y-intercept of a linear graph.			
6.4	I can determine the equation of a straight line given the gradient and the y-intercept.			
	I can determine the equation of a straight line given the gradient and a single point.			
	I can determine the equation of a straight line given two points.			
	I can determine the equation of a straight line from a graph.			
6.5	I can sketch linear graphs using the x- and y-intercept method.			
	I can sketch linear graphs using the gradient–intercept method.			
	I can sketch vertical and horizontal lines.			
6.6	I can produce linear graphs using digital technologies.			
6.7	I can determine the linear rule from a table of values.			
	I can determine the linear rule that models a real-life problem.			
	I can interpret linear rules in real-life problems.			
6.8	I can determine the midpoint of two points.			
	I can calculate the distance between two points.			
6.9	I can recognise and sketch a basic parabola from its equation.			
	I can sketch a parabola that has been translated horizontally or vertically.			
	I can recognise and sketch a hyperbola from its equation.			
	I can recognise and sketch a circle from its equation.			

6.10.3 Project

Path of a billiard ball

The path of a billiard ball can be mapped using mathematics. A billiard table can be represented by a rectangle, a ball by a point and its path by line segments. In this investigation, we will look at the trajectory of a single ball, unobstructed by other balls. *Note:* A billiard table has a pocket at each of its corners and one in the middle of each of its long sides.

Consider the path of a ball that is hit on its side, from the lower left-hand corner of the table, so that it travels at a 45° angle from the corner of the table. Assume that the ball continues to move, rebounding from the sides and stopping only when it comes to a pocket. Diagrams of the table drawn on grid paper are shown. Each grid square has a side length of 0.25 metres.

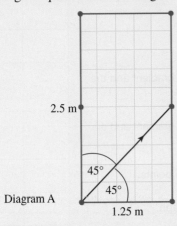

Diagram A

1.25 m

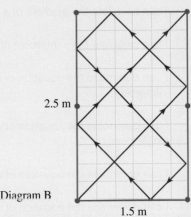

Diagram B

1.5 m

Diagram A shows the trajectory of a ball that has been hit on its side at a 45° angle, from the lower left-hand corner of a table that is 2.5 m long and 1.25 m wide. Note that, because the ball has been hit on its side at a 45° angle, it travels diagonally through each square in its path, from corner to corner. Diagram B shows the trajectory of a ball on a 2.5 m by 1.25 m table.

1. In Diagram B, how many times does the ball rebound off the sides before going into a pocket?

2. A series of eight tables of different sizes are drawn on grid paper in the following diagrams. For each table, determine the trajectory of a ball hit at 45° on its side, from the lower left-hand corner of the table. Draw the path each ball travels until it reaches a pocket.

a.

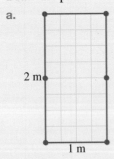

2 m

1 m

b.

2.25 m

0.75 m

c.

2.5 m

0.5 m

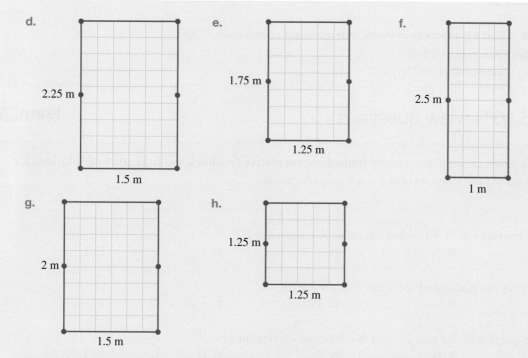

d. 2.25 m | 1.5 m

e. 1.75 m | 1.25 m

f. 2.5 m | 1 m

g. 2 m | 1.5 m

h. 1.25 m | 1.25 m

3. Determine for which tables the ball will travel through the simplest path. What is special about the shape of these tables?

4. Determine for which table the ball will travel through the most complicated path. What is special about this path? Draw another table (and path of the ball) with the same feature.

5. Determine for which tables the ball will travel through a path that does not cross itself. Draw another table (and path of the ball) with the same feature.

6. Consider the variety of table shapes. Will a ball hit on its side from the lower left-hand corner of a table at 45° always end up in a pocket (assuming it does not run out of energy)?

Simplify matters a little and consider a billiard table with no pockets in the middle of the long sides. Use a systematic way to look for patterns for tables whose dimensions are related in a special way.

7. Draw a series of billiard tables of length 3 m. Increase the width of these tables from an initial value of 0.25 m in increments of 0.25 m. Investigate the final destination (the pocket the ball lands in) of a ball hit from the lower left-hand corner. Complete the table.

	Length of table (m)	Width of table (m)	Destination pocket
a.	3	0.25	
b.	3	0.5	
c.	3	0.75	
d.	3	1	
e.	3	1.25	
f.	3	1.5	
g.	3	1.75	
h.	3	2	

8. How can you predict (without drawing a diagram) the destination pocket of a ball hit from the lower left-hand corner of a table that is 3 m long? Provide an illustration to verify your prediction.

Exercise 6.10 Review questions   learn on

To answer questions online and to receive **immediate corrective feedback** and **fully worked solutions** for all questions, go to your learnON title at www.jacplus.com.au.

Fluency

1. **MC** For the rule $y = 3x - 1$, select the value of y when $x = 2$.
   A. -1 B. 1   C. 2   D. 5   E. 7

2. **MC** Identify the gradient of the linear rule $y = 4 - 6x$.
   A. 6   B. -6 C. 4   D. -4 E. -2

3. **MC** The graph with the rule $2y - x + 6 = 0$ has an x-intercept of:
   A. -2 B. 0   C. 2   D. 6   E. 7

4. **MC** The graph with the rule $2y - x + 6 = 0$ has a y-intercept of:
   A. -6 B. -3 C. 0   D. 3   E. 6

5. **MC** Consider a linear graph that goes through the points $(6, -1)$ and $(0, 5)$. The gradient of this line is:
   A. 5   B. -5 C. 1   D. -1 E. 3

6. **MC** A straight line passes through the points $(2, 1)$ and $(5, 4)$. Its rule is:
   A. $y = x - 1$ B. $y = x + 1$ C. $y = 2x$ D. $4y = 5x$ E. $y = 3x$

7. **MC** The rule for a line whose gradient is -4 and y-intercept $= 8$ is:
   A. $y = -4x + 32$ B. $y = -4x + 8$ C. $y = 4x - 32$ D. $y = 4x - 8$ E. $-4y = x + 8$

8. **MC** Identify which of the following linear rules will *not* intersect with the straight line defined by $y = 3x$.
   A. $y = 3x + 2$ B. $y = -3x + 1$ C. $y = -3x + 2$ D. $y = -2x + 1$ E. $y = 3$

9. **MC** If $y = 2x + 1$, select from the following the point that could not be on the line.
   A. $(3, 7)$ B. $(-3, -5)$ C. $(0, 1)$ D. $(-3, 0)$ E. $(5, 11)$

10. **MC** The solution to $y = 3x + 1$ and $y = -3x + 1$ is:
    A. $(0, 1)$ B. $(1, 0)$ C. $\left(0, -\dfrac{1}{3}\right)$ D. $\left(-\dfrac{1}{3}, 0\right)$ E. $(3, 1)$

11. Write down the gradient and the y-intercept of the following linear graphs.
    a. $y = 8x - 3$ b. $y = 5 - 9x$ c. $2x + y - 6 = 0$

    d. $4x - 2y = 0$ e. $y = \dfrac{2x - 1}{3}$

12. Determine which of the following lines are parallel to the line with the equation $y = 6 - x$. There could be more than one correct answer.

    A. $y + x = 4$ **B.** $y = 13 - x$ **C.** $2y - 2x = 1$

    D. $x + 2y - 4 = 0$ **E.** $2x - 3y - 18 = 0$

13. Determine which of the following rules will yield a linear graph. There could be more than one correct answer.

    A. $3y = -5x - \dfrac{1}{2}$ **B.** $y = 3x^2 - 2\dfrac{1}{2}$ **C.** $3x + 4y + 6 = 0$ **D.** $x + y - 2xy = 0$ **E.** $2y = 4^2x + 9^2$

14. Determine the gradient of the lines shown.

    a.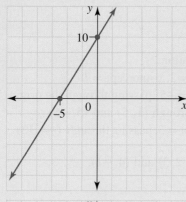
             b.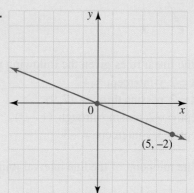

    c.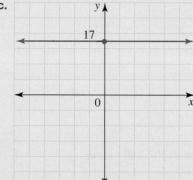

15. Calculate the gradient of the line passing through the following pairs of points.

    a. $(2, -3)$ and $(4, 1)$ **b.** $(0, -5)$ and $(4, 0)$.

16. For each of the following rules, state the gradient and the y-intercept.

    a. $y = -3x + 7$ **b.** $2y - 3x = 6$ **c.** $y = -\dfrac{2}{5}x$ **d.** $y = 4$

17. For the following rules, use the gradient–intercept method to sketch linear graphs.

    a. $y = -x + 5$ **b.** $y = 4x - 2.5$ **c.** $y = \dfrac{2}{3}x - 1$ **d.** $y = 3 - \dfrac{5}{4}x$

18. For the following rules, use the x- and y-intercept method to sketch linear graphs.

    a. $y = -6x + 25$ **b.** $y = 20x + 45$ **c.** $2y + x = -5$ **d.** $4y + x - 2.5 = 0$

19. For the following rules, use an appropriate method to sketch linear graphs.

    a. $y = -3x$ **b.** $y = \dfrac{1}{4}x$ **c.** $y = -2$ **d.** $x = 3$

20. Determine the rule of the lines shown.

a.

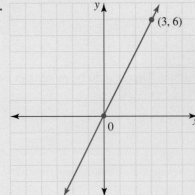

b.

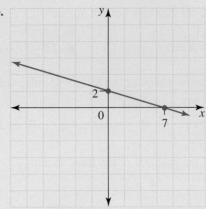

21. Determine the linear rules given the following pieces of information.

    a. Gradient $= 2$, y-intercept $= -7$
          b. Gradient $= 2$, x-intercept $= 7$

    c. Gradient $= 2$, passing through $(7, 9)$
          d. Gradient $= -5$, passing through the origin

    e. y-intercept $= -2$, passing through $(1, -3)$
          f. Passing through $(1, 5)$ and $(5, -6)$

    g. x-intercept $= 3$, y-intercept $= -3$
          h. y-intercept $= 5$, passing through $(-4, 13)$

22. Determine the midpoint of the line interval joining the points $(-2, 3)$ and $(4, -1)$.

23. Calculate the distance between the points $(1, 1)$ and $(4, 5)$.

24. Sketch each of the following, comparing it with the graph of $y = x^2$.

    a. $y = x^2 - 3$
          b. $y = (x + 2)^2$

25. a. Plot the graph of $y = \dfrac{4}{x}$.

    b. Identify what type of graph this is.

26. Sketch the circle with the equation $x^2 + y^2 = 16$. Clearly show the centre and radius.

Problem solving

27. Louise owes her friend Sula $400 and agrees to pay her back $15 per week.

    a. State a linear rule that demonstrates this debt reduction schedule and sketch the graph.

    b. Calculate how many weeks it takes Louise to repay the debt.

    c. Calculate how much she owes after 15 weeks.

    d. Evaluate how many repayments Louise needs to make to reduce her debt to $85.

28. A bushwalker is 40 km from their base camp when they decide to head back.

If they are able to walk 3.5 km each hour:

    a. determine the linear rule that describes this situation and sketch its graph

    b. calculate how long, correct to 1 decimal place, it will take them to reach base camp

    c. calculate how far they will have walked in 6.5 hours.

29. Udaya is writing test questions. She has already written 25 questions and can write a further 5 questions per hour.
   a. Represent this information as a linear equation where t hours is the time spent writing test questions and n is the number of questions written.
   b. Predict the total number of questions written after a further 8 hours assuming the same linear rule.
   c. Calculate how long, to the nearest minute, it will take Udaya to have written 53 questions.

30. Catherine earns a daily rate of $200 for working in her mother's store. She receives $5 for each necklace that she sells.
   a. Write an equation to show how much money (m) Catherine earned for the day after selling (n) necklaces.
   b. Graph the equation that you created in part a, showing the two intercepts.
   c. Explain which part of the line applies to her earnings.
   d. Explain which part of the line does not apply to her earnings.

31. Calculate the gradient of the line passing through the points $(2, 3)$ and $(6 + 4t, 5 + 2t)$. Write your answer in simplest form.

32. Determine the point on the line $y = 2x + 7$ that is also 5 units above the x-axis.

33. An experiment was conducted to collect data for two variables p and t. The data are presented in the table shown.

p	$-\dfrac{1}{2}$		$\dfrac{1}{2}$	3.6
t	$2\dfrac{1}{4}$	1.75		-5.95

   It is known that the relationship between p and t is a linear one. Determine the two values missing from the table.

34. The distance from the origin to the y-intercept of a linear graph is three times the distance from the origin to the x-intercept. The area of the triangle formed by the line and the axes is 3.375 units^2. The line has a negative gradient and a negative y-intercept.
   Determine the equation of the line.

on   To test your understanding and knowledge of this topic go to your learnON title at www.jacplus.com.au and complete the **post-test**.

Online Resources

Below is a full list of **rich resources** available online for this topic. These resources are designed to bring ideas to life, to promote deep and lasting learning and to support the different learning needs of each individual.

eWorkbook

Download the workbook for this topic, which includes worksheets, a code puzzle and a project (ewbk-2006) ☐

Solutions

Download a copy of the fully worked solutions to every question in this topic (sol-0726) ☐

Digital documents

6.2 SkillSHEET Plotting coordinate points (doc-6161) ☐
SkillSHEET Substituting into a rule (doc-6162) ☐
SkillSHEET Completing a table of values (doc-6163) ☐
SkillSHEET Plotting a line from a table of values (doc-6164) ☐
6.3 SkillSHEET Transposing a linear equation to general form (doc-6165) ☐
SkillSHEET Finding the gradient given two points (doc-10839) ☐
SkillSHEET Measuring the rise and the run (doc-10840) ☐
SkillSHEET Finding the gradient of a line from its equation (doc-10841) ☐
6.5 SkillSHEET Graphing linear equations using intercepts (doc-10842) ☐
SkillSHEET Solving linear equations that arise when finding intercepts (doc-10843) ☐

Video eLesson

6.2 Plotting linear graphs on the Cartesian plane (eles-4779) ☐
Points on a line (eles-4781) ☐
6.3 The gradient (m) (eles-4782) ☐
The y-intercept (eles-4783) ☐
Gradient (eles-1889) ☐
6.4 The general equation of a straight line (eles-4784) ☐
Determining the equation of a straight line (eles-4785) ☐
The equation of a straight line (eles-2313) ☐
6.5 The x- and y-intercept method (eles-4786) ☐
The gradient–intercept method (eles-4787) ☐
Horizontal and vertical lines (eles-4788) ☐
Sketching linear graphs (eles-1919) ☐
Sketching linear graphs using the gradient–intercept method (eles-1920) ☐
6.6 Graphing with technology (eles-4789) ☐
Graphing parallel lines (eles-4790) ☐
Graphing perpendicular lines (eles-4791) ☐
6.7 Determining a linear rule from a table of values (eles-4792) ☐
Modelling linear relationships (eles-4793) ☐
6.8 Calculating the midpoint of a line segment (eles-4794) ☐
The distance between two points (eles-4795) ☐
6.9 The parabola (eles-4796) ☐
The hyperbola (eles-4797) ☐
The circle (eles-4798) ☐

Interactivities

6.2 Individual pathway interactivity: Plotting linear graphs (int-4502) ☐
Plotting linear graphs (int-3834) ☐
6.3 Individual pathway interactivity: Determining linear rules (int-4506) ☐
The gradient (int-3836) ☐
Linear graphs (int-6484) ☐
6.4 Individual pathway interactivity: The equation of a straight line (int-4503) ☐
6.5 Individual pathway interactivity: Sketching linear graphs (int-4504) ☐
The intercept method (int-3840) ☐
The gradient–intercept method (int-3839) ☐
Vertical and horizontal lines (int-6049) ☐
6.6 Individual pathway interactivity: Technology and linear graphs (int-4505) ☐
Parallel lines (int-3841) ☐
6.7 Individual pathway interactivity: Practical applications of linear graphs (int-4507) ☐
Dependent and independent variables (int-6050) ☐
6.8 Individual pathway interactivity: Midpoint of a line segment and distance between two points (int-4508) ☐
Midpoints (int-6052) ☐
Distance between two points (int-6051) ☐
6.9 Individual pathway interactivity: Non-linear relations (parabolas, hyperbolas, circles) (int-4509) ☐
Vertical translations of parabolas (int-6055) ☐
Horizontal translations of parabolas (int-6054) ☐
Equations of circles (int-6053) ☐
6.10 Crossword (int-2688) ☐
Sudoku puzzle (int-3207) ☐

Teacher resources

There are many resources available exclusively for teachers online.

To access these online resources, log on to **www.jacplus.com.au**.

Answers

Topic 6 Linear and non-linear graphs

Exercise 6.1 Pre-test

1. On the x-axis (because $y = 0$).
2. a. C   b. D
   c. B   d. A
3. C
4. −0.5
5. D
6. a. B   b. C
   c. D   d. A
7. a. $y = -\dfrac{1}{4}x$ b. $y = 20x + 400$
8. D
9. a. $y = -x + 2$ b. $(-1, 3)$
10. $a = -5$
11. $\sqrt{29}$
12. A
13. D
14. E
15. D

Exercise 6.2 Plotting linear graphs

1. a. B   b. C   c. D

2. a. i. $y = x$

x	−3	−2	−1	0	1	2	3
y	−3	−2	−1	0	1	2	3

ii. $y = x$

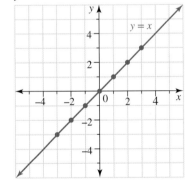

b. i. $y = 2x + 2$

x	−3	−2	−1	0	1	2	3
y	−4	−2	0	2	4	6	8

ii. $y = 2x + 2$

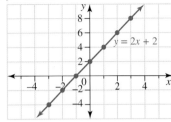

c. i. $y = 3x - 1$

x	−3	−2	−1	0	1	2	3
y	−10	−7	−4	−1	2	5	8

ii. $y = 3x - 1$

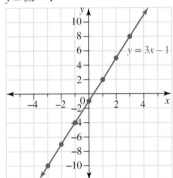

d. i. $y = -2x$

x	−3	−2	−1	0	1	2	3
y	6	4	2	0	−2	−4	−6

ii. $y = -2x$

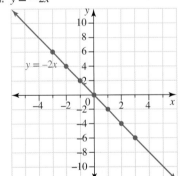

3. a. Sample points shown; $y = -x$

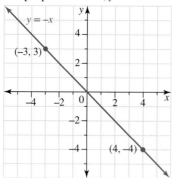

b. Sample points shown; $y = \frac{1}{2}x + 4$

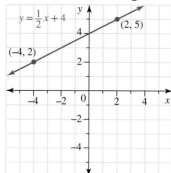

c. Sample points shown; $y = -2x + 3$

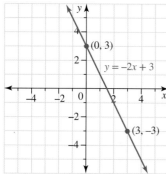

d. Sample points shown; $y = x - 3$

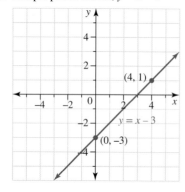

4. a. Yes   **b.** No   **c.** No   **d.** Yes

5. a. Yes   **b.** Yes   **c.** Yes
    d. No   **e.** No   **f.** No

6. D

7. a. B   **b.** A   **c.** D   **d.** C

8. By plotting the points on the Cartesian plane and joining them to make a line it can be seen that this line does not pass through the third quadrant.

9. a. The first, second and third quadrants.
   b. Answers will vary. Sample response: Substitute point $(1, 3)$ into line equation. $3 \neq 1 + 1$.

10. Answers will vary. Sample response: Substitute the point into the line equation. If the LHS equals the RHS of the equation, then the point lies on the line. Alternatively, graph the line and check the point on the Cartesian plane.

11. Answers will vary. Sample response: Plot the points on the Cartesian plane and draw a line through the points. Find the equation of the line. Check all points lie on the line by substituting into the line equation.

12. $(59, 21)$

13. a.

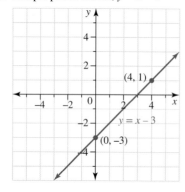

   b. $m = 5t + 1$
   c. Sample responses can be found in the worked solutions in the online resources.

14. Stirring increases the rate of reaction.

Exercise 6.3 Features of linear graphs

1. a. 6   **b.** -4 **c.** 0
   d. $\frac{6}{5}$ **e.** -2.2 **f.** -45

2. a. 0   **b.** 0.1   **c.** -2.6
   d. 0   **e.** 9   **f.** -0.1

3. a. 23   **b.** 54   **c.** -6 **d.** 70

4. a. 1   **b.** 5.2   **c.** 100   **d.** 100

5. a. 4   **b.** 1   **c.** 0

6. a. 20   **b.** 400   **c.** -5.2

7. a. 0   **b.** -5 **c.** -300

8. C

9. y-intercept $= c$

10. a. $(0, -10)$ **b.** $(0, -4)$ **c.** $(0, 3)$
   d. $(0, 7)$ **e.** No y-intercept

11. There is a rise and no run; $\dfrac{\text{rise}}{0}$ = undefined.

12. There is no rise and a run; $\dfrac{0}{\text{run}} = 0$

13. Sample responses can be found in the worked solutions in the online resources.

14. It does not matter if you rise before you run or run before you rise, as long as you take into account whether the rise or run is negative.

15. a. $m = \dfrac{y - c}{x}$ **b.** $y = mx + c$

16. a. Lamb: $m = \$15/kg$, Chicken: $m = \$10/kg$,
Beef: $m = \$7.50/kg$

b. Lamb: $\$15$, Chicken: $\$10$, Beef: $\$7.50$

c. i. $\$15$ **ii.** $\$5$ **iii.** $\$15$

d. $\$35$

e.

Meat type	Cost per kilogram ($/kg)	Weight required (kg)	Cost = $/kg × kg
Lamb	15	1	15
Chicken	10	0.5	5
Beef	7.50	2	15
		Total cost	35

17. a. $m = -\dfrac{3}{2}$

b. It does not matter which points are chosen to determine the gradient of the graph because the gradient will always remain the same.

c. Straight line with a y-intercept of $(0, -2)$ and a slope of $-\dfrac{3}{2}$

Exercise 6.4 The equation of a straight line

1. a. $y = 4x + 2$ **b.** $y = -4x + 1$ **c.** $y = 4x + 8$

2. a. $y = 6x + 7$
b. $y = -2.5x + 6$
c. $y = 45x + 135$

3. a. $y = -2x$ **b.** $y = 4x$ **c.** $y = 10.5x$

4. a. $y = -20x$ **b.** $y = 1.07x$ **c.** $y = 32x$

5. a. $y = x + 2$ **b.** $y = -x + 8$
c. $y = -4x - 8$ **d.** $y = 2x - 13$

6. a. $y = -5x + 70$ **b.** $y = 2x - 23$
c. $y = -6x + 11$ **d.** $y = -x - 1.5$

7. a. $y = 6x + 30$ **b.** $y = -3.5x + 15.5$
c. $y = 1.2x - 4.08$ **d.** $y = 0.2x - 1.76$

8. a. $y = -4x - 24$ **b.** $y = 2x - 6$
c. $y = -2x + 4$

9. a. $y = 5x + 35$ **b.** $y = 1.5x - 3.75$
c. $y = 0.4x - 0.96$

10. a. $y = x + 17$ **b.** $y = -x + 3$
c. $y = 11$ **d.** $y = x + 3.5$

11. a. $y = -x + 3.5$
b. $y = \dfrac{1}{9}x + \dfrac{34}{9}$ (or $9y = x + 34$)
c. $y = -6x + 90$
d. $y = 20x$

12. a. $y = -x + 4$
b. $y = -0.25x + 5.5$
c. $y = \dfrac{2}{7}x - \dfrac{32}{7}$ (or $7y = 2x - 32$)
d. $y = \dfrac{2}{3}x + \dfrac{10}{3}$ (or $3y = 2x + 10$)

13. a. $y = x + 3$ **b.** $y = \dfrac{-5}{4}x + 5$
c. $y = -6x + 6$ **d.** $y = \dfrac{7}{8}x + 35$

14. a. $y = x + 8$ **b.** $y = -2x + 6$
c. $y = \dfrac{-3}{7}x - 3$ **d.** $y = \dfrac{1}{4}x + 50$

15. a. $y = \dfrac{7}{4}x$ **b.** $y = x$ **c.** $y = -2x$
d. $y = -3x$ **e.** $y = \dfrac{-24}{11}x$ **f.** $y = \dfrac{-x}{3}$

16. a. $y = 2x + 4$ **b.** $y = 5x$

17. a. $y = -x + 3$ **b.** $y = \dfrac{1}{5}x$

18. a. $y = x + 2$ **b.** $y = -2x + 3$

19. a. B   **b.** A   **c.** E

20. a. $y = -2.4x - 12$ **b.** $y = -58.32$

21. a. $y = -0.6x + 10$ **b.** $\left(16\dfrac{2}{3}, 0\right)$

22. a. The equation is $y = \dfrac{5}{4}x - 2135$

b. Values of c cannot be read directly from the graph because the graph doesn't contain the origin, and since c is found at $x = 0$, we need to use the equation.

c. In 2020 the concentration of CO_2 will be 390 ppm. The assumption is that the concentration of CO_2 will continue to follow this linear pattern.

23. $m = \dfrac{-15 + 7}{4 - 0} = -2$
$y = -2x + c$
$6 = -6 + c$
$c = 12$
The equation is $y = -2x + 12$.

24. a. Line A $y = x - 1$, Line B: $y = -x + 8$

b. $(4.5, 3.5)$

c. Sample responses can be found in the worked solutions in the online resources. Possible methods include graphing on a Cartesian plane or using algebra.
Line B : $y = -x + 8, m = -1, m_\perp = 1$
$y = x + c$
$6 = -4 + c$
$c = 10$

25. $m = \dfrac{1}{4}C$

26. a, c

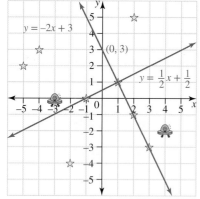

$y = -2x + 3$

$(0, 3)$

$y = \frac{1}{2}x + \frac{1}{2}$

b. $y = -2x + 3$ **c.** $(1, 1)$ **d.** 1 star

e. Sample responses can be found in the worked solutions in the online resources. Possible responses could include the following.
$y = -x - 1,\ x = 2$

27. Sample responses can be found in the worked solutions in the online resources. Responses should include the correct equation: $T = 4.25t + 15$.

28. 8.1 mL

29. a. A: $0, B: $100, C: $200

b. A: $80, B: $50 C: $33.33

c. A: $w = 80h$, B: $w = 50h + 100$, C: $w = 33.3h + 200$

d. i. C   **ii.** A

e. The advantage of location C is that it has the highest minimum pay, so you would be guaranteed at least $200 per day, regardless of how many hours you work.

f. i. A: $160, B: $200, C: $266.66

  ii. A: $480, B: $400, C: $400

Exercise 6.5 Sketching linear graphs

1. a. $5y - 4x = 20$

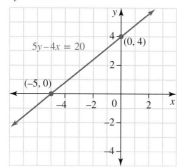

$5y - 4x = 20$

$(0, 4)$

$(-5, 0)$

b. $y = x + 2$

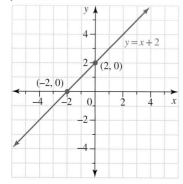

$y = x + 2$

$(2, 0)$

$(-2, 0)$

c. $y = -3x + 6$

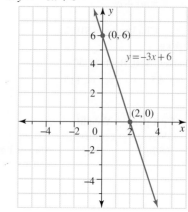

$(0, 6)$

$y = -3x + 6$

$(2, 0)$

2. a. $3y + 4x = -12$

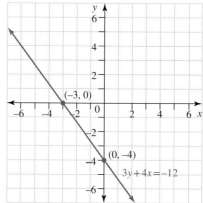

$(-3, 0)$

$(0, -4)$

$3y + 4x = -12$

b. $x - y = 5$

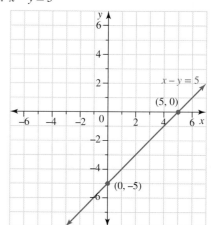

$x - y = 5$

$(5, 0)$

$(0, -5)$

c. $2y + 7x - 8 = 0$

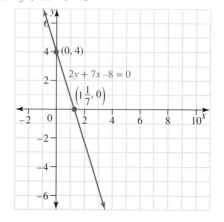

$(0, 4)$

$2y + 7x - 8 = 0$

$\left(1\frac{1}{7}, 0\right)$

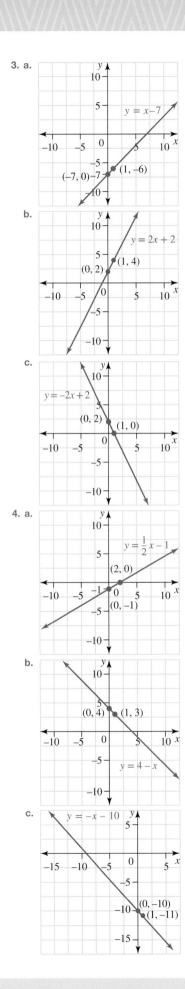

3. a. $y = x - 7$, points $(-7, 0)$, $(1, -6)$

b. $y = 2x + 2$, points $(0, 2)$, $(1, 4)$

c. $y = -2x + 2$, points $(0, 2)$, $(1, 0)$

4. a. $y = \frac{1}{2}x - 1$, points $(2, 0)$, $(0, -1)$

b. $y = 4 - x$, points $(0, 4)$, $(1, 3)$

c. $y = -x - 10$, points $(0, -10)$, $(1, -11)$

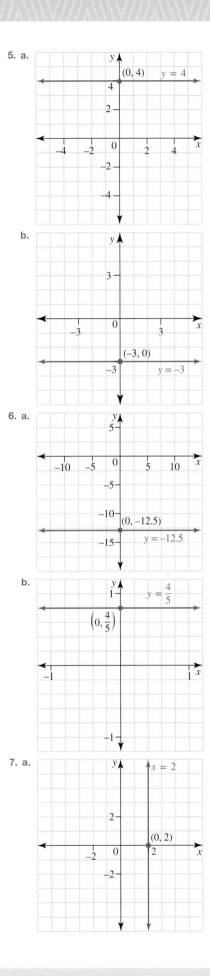

5. a. $y = 4$, point $(0, 4)$

b. $y = -3$, point $(-3, 0)$

6. a. $y = -12.5$, point $(0, -12.5)$

b. $y = \frac{4}{5}$, point $\left(0, \frac{4}{5}\right)$

7. a. $x = 2$, point $(0, 2)$

b. $x = -6$

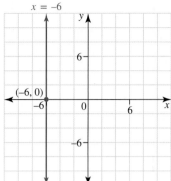

8. a. $x = -2.5$

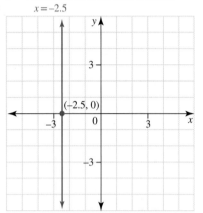

b. $x = \dfrac{3}{4}$

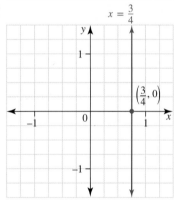

9. a.

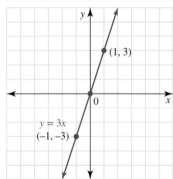

b.

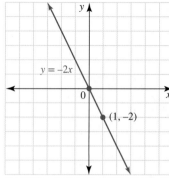

10. a.

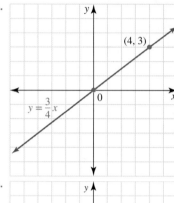

b.

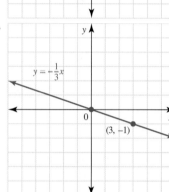

11. D

12. D

13. a. $y = -\dfrac{2x}{5} + 4$

b. $-\dfrac{2}{5}$

c. The x-intercept is 10, the y-intercept is 4.

d.

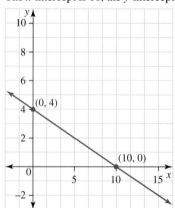

14. a. $m = \dfrac{\text{rise}}{\text{run}} = \dfrac{5}{3}$

b.

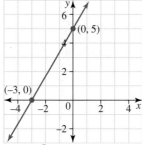

c. **i.** $y = \dfrac{5x}{3} + 5$ **ii.** $-5x + 3y = 15$

15. a. $y = \dfrac{4}{3}x - 8$

b. $y = 0,\ 4x = 24,\ x = 6$
$x = 0,\ -3y = 24,\ y = -8$

c.

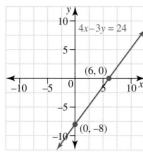

16. a. $y = \dfrac{-ax + c}{b}$

b. Sample response: The gradient is $\dfrac{-a}{b}$ substituting positive values always
results in a negative gradient.

17. Sample response:

$4y = \dfrac{4}{2}x + 3$

$4y = 2x + 3$

$-2x + 4y = 3$

18. Sample response: All descriptions use the idea that a gradient is equal to $\dfrac{\text{rise}}{\text{sun}}$, which equals $\dfrac{2}{3}$ in all of these cases.

19. a. The gradient is 2.   **b.** The equation of the line is $y = 2x - 1$.

c. One point is $(1, 1)$.

20. a. **i.** C   **ii.** B   **iii.** A

b. Straight line y-intercept of $(0, 2)$ with a slope of $-\dfrac{1}{4}$

21. a. **i.**

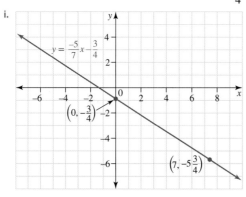

ii.

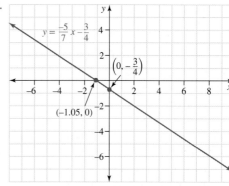

iii. Sample responses can be found in the worked solutions in the online resources.

b. Sample responses can be found in the worked solutions in the online resources.

22. Sample responses can be found in the worked solutions in the online resources.

Exercise 6.6 Technology and linear graphs

1. i. As the size of the coefficient increases, the steepness of the graph increases.

ii. Each graph cuts the x-axis at $(0, 0)$.

iii. Each graph cuts the y-axis at $(0, 0)$.

2. i. As the magnitude of the coefficient decreases, the steepness of the graph increases.

ii. Each graph cuts the x-axis at $(0, 0)$.

iii. Each graph cuts the y-axis at $(0, 0)$.

3. a. positive, positive   **b.** downward, negative
c. bigger   **d.** will

4. a. Yes, 1
b. No, the lines are parallel.
c. **i.** $(0, 0),\ x = 0$
ii. $(-2, 0),\ x = -2$
iii. $(2, 0),\ x = 2$
d. **i.** $(0, 0),\ y = 0$
ii. $(0, 2),\ y = 2$
iii. $(0, -2),\ y = -2$

5. a. Yes, -1
b. No, the lines are parallel.
c. **i.** $(0, 0),\ x = 0$
ii. $(2, 0),\ x = 2$
iii. $(-2, 0),\ x = -2$
d. **i.** $(0, 0),\ y = 0$
ii. $(0, 2),\ y = 2$
iii. $(0, -2),\ y = -2$

6. a. same   **b.** y-intercept, y-axis
c. y-intercept   **d.** x-intercept

7. i. No
ii. Yes
iii. **a.** 1   **b.** -1
c. 3   **d.** $-\dfrac{2}{5}$
iv. **a.** $(-5, 0)$, i.e. $x = -5$
b. $(5, 0)$, i.e. $x = 5$

c. $\left(-\dfrac{5}{3}, 0\right)$, i.e. $x = -\dfrac{5}{3}$

d. $\left(\dfrac{25}{2}, 0\right)$, i.e. $x = \dfrac{25}{2}$

v. a. $(0, 5)$, i.e. $y = 5$
   b. $(0, 5)$, i.e. $y = 5$
   c. $(0, 5)$, i.e. $y = 5$
   d. $(0, 5)$, i.e. $y = 5$

8. a. steepness
   b. y-coordinate, y-axis
   c. same, parallel
   d. the same

9. a. gradient = 2, y-intercept = $(0, 0)$, $y = 0$
   b. gradient = 1, y-intercept = $(0, 1)$, $y = 1$
   c. gradient = -3, y-intercept = $(0, 5)$, $y = 5$
   d. gradient = 2/3, y-intercept = $(0, -7)$, $y = -7$

10. Sample response: They all have in common the point $(1, 5)$. Their gradients are all different.

11. Sample responses can be found in the worked solutions in the online resources. Use solve on your CAS calculator.

12. Answers should show the y-intercept $(0, 2.2)$, point $(1, 2.75)$ and gradient 0.55.

13. a. N is the number of dogs walked; $-\$15$ is Shirly's starting cost and out-of-pocket expense before she walks a dog, and she earns $10 for every dog walked.

    b. She needs to walk at least two dogs a week before she can make a profit.

    c.
    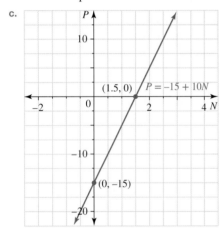

14. Sample responses can be found in the worked solutions in the online resources.

15. a. $y = 2x - 8.5$
    b.
    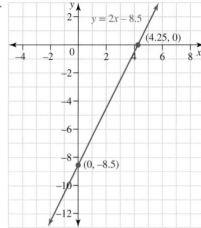

    c. $y = 7.5$
    d. $x = 10.25$

16. a. $C = 25n + 1000$
    b. $(0, 1000)$. This is the initial cost of the site licence.
    c. $6000
    d. 80

17. a. Sample response: Possible methods include plotting the points $(0, 325)$ and $(1, 328.8)$ or using the equation $c = 3.8m + 325$.

    b. 382 calories

    c. 32.9 minutes

Exercise 6.7 Practical applications of linear graphs

1. a. $y = 6x - 5$
   b. $y = -3x + 8$

2. a. $y = 2x + 4$
   b. $y = 0.9x + 1.1$

3. a. $y = 3x + 1$
   b. $y = -x + 17$

4. a. $v = -3t + 27$
   b. $C = 3d + 8$

5. D

6. a. $A = 25t + 60$
   b. $5060
   c. $60

7. a. $C = 0.55s + 2.0$
   b. $8.60

8. a. $N = 400t + 3200$
   b. 5400
   c. 13 200

9. a. 27
   b. 108

10. a. $w = -40t + 712$
    b. 712 L
    c. 18 days

11. Sample responses can be found in the worked solutions in the online resources.

12. a. $C = 2.57d + 3.50$

   b. $33.06

   c. 7.20 km

13. $5 \times 21.4 = 107 \, \text{cm} = 1.07 \, \text{m}$

14. a. $N = -22t + 164$ b. 120

   c. 5   d. 7.5 months

15. a. See the figure at the foot of the page.*

   b. $(0, -800)$, $y = -800$. Fixed costs are $800.

   c. $(4, 0)$, $x = 4$, this is the break-even amount

   d. 200, this is the sale price per print that contributes to profit

16. a. $C = 2.14d + 3.50$

   b. $28.11

   c. $52.93

17. a. $Y = 78.7A$ b. ¥196 750   c. $Y_T = 76.5A$

18. Cody arrives first.

Exercise 6.8 Midpoint of a line segment and distance between two points

1. a. $(2, 4)$ b. $(5, 1)$ c. $(7, 2)$

2. a. $(8, 9)$ b. $(0, 5)$ c. $(-1, 2)$

3. a. $(-5, -2)$ b. $(2, 3)$ c. $(0, 12)$

4. a. $\left(1\frac{1}{2}, 5\frac{1}{2}\right)$ b. $\left(-3, -3\frac{1}{2}\right)$

   c. $\left(-9\frac{1}{2}, 9\frac{1}{2}\right)$ d. $\left(-1\frac{1}{2}, 6\frac{1}{2}\right)$

5. a. 5   b. 9   c. -16 d. 0

6. a. $(4, 6)$ b. $(-2, -3)$

   c. $(11, 2)$ d. $(-7, -3)$

7. $y = 5x - 7$

8. a. 5   b. 10   c. 1

9. a. 2   b. 13   c. 10

10. a. 5   b. 17   c. 13

11. a. 7.211   b. 14.765   c. 13.038

12. a. 8.944   b. 19.416   c. 19.416

13. a. 24.472   b. 25.464

14. a. 17.788   b. 25.763

15. 21.024 km

16. $d = 5$

17. $-2 = \dfrac{x + 5}{2} \Rightarrow x = -9$

   $-4 = \dfrac{-6 + y}{2} \Rightarrow y = -2$

   The required point is $(-9, -2)$.

18. $d_{AB} = \sqrt[5]{10}$

   $d_{BC} = \sqrt[5]{10}$

   Therefore, B is equidistant from A and C.

*15. a.

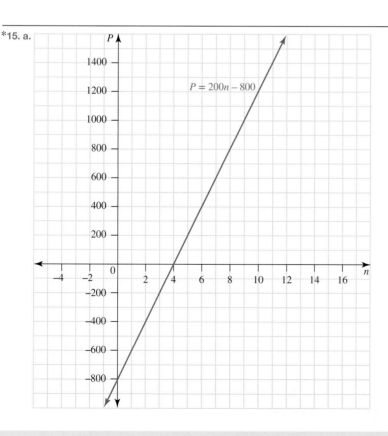

$P = 200n - 800$

19. a.

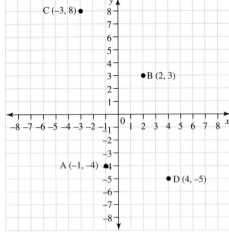

b. $M = (-2, 2)$

c. $2\sqrt{37}$

d. $(7, -2)$

e. $\dfrac{3 - (-4)}{2 - (-1)} = \dfrac{7}{3}$

f. $y = -4x + 11$

20. Sample responses can be found in the worked solutions in the online resources.

$d_{AB} = 2\sqrt{10}$

$d_{BC} = 2\sqrt{10}$

$d_{AC} = 4\sqrt{2}$

Side length AB is equal to side length BC but not equal to side length AC. Therefore, $\triangle ABC$ is an isosceles triangle.

21. $\dfrac{1}{2}$

22. $(-4, 5), (-3, 2), (-2, -1), (-1, -4), (0, -7)$

Exercise 6.9 Non-linear relations (parabolas, hyperbolas, circles)

1. a. Vertical 3 up, TP $(0, 3)$

b. Vertical 1 down, TP $(0, -1)$

c. Vertical 7 down, TP $(0, -7)$

d. Vertical $\dfrac{1}{4}$ up, TP $\left(0, \dfrac{1}{4}\right)$

2. a. Horizontal 1 right, $(1, 0)$

b. Horizontal 2 right, $(2, 0)$

c. Horizontal 10 left, $(-10, 0)$

d. Horizontal 4 left, $(-4, 0)$

3. a. $(0, 1)$, max   b. $(0, -3)$, min   c. $(-2, 0)$, max

4. See the table at the foot of the page.*

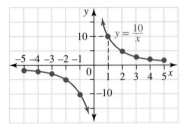

5. a. i.

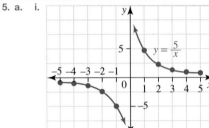

ii.

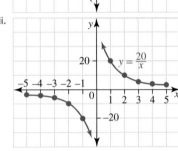

iii.

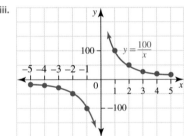

b. i. $x = 0, y = 0$

ii. $x = 0, y = 0$

iii. $x = 0, y = 0$

6.

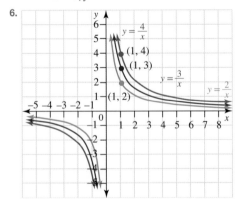

---

*4.

x	-5	-4	-3	-2	-1	0	1	2	3	4	5
y	-2	-2.5	-3.3	-5	-10	Undefined	10	5	3.3	2.5	2

7. It increases the y-values by a factor of k and hence dilates the curve by a factor of k. As k increases, the graph is further from the origin.

8. a.
Centre $(0, 0)$, radius 7

b.
Centre $(0, 0)$, radius 4

c.
Centre $(0, 0)$, radius 6

d.
Centre $(0, 0)$, radius 9

e.
Centre $(0, 0)$, radius 5

f.
Centre $(0, 0)$, radius 10

9. a.

b.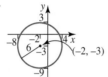

c.

d.

e.

f.

10. D

11. Sample responses can be found in the worked solutions in the online resources. Possible methods include using a table of values or graphing both equations on a Cartesian plane.

12. Similarity: they both have asymptotes of $x = 0$ and $y = 0$. Difference: one is a reflection of the other about the y-axis.

13. Sample responses can be found in the worked solutions in the online resources. Use algebra or graph the equation. The turning point is a minimum.

14. $r = 4$
The centre is $(2, 1)$. The equation is
$(x - 2)^2 + (y - 1)^2 = 16$
Sample responses can be found in the worked solutions in the online resources.

15. $(6, -5)$

16. a. Nalini started well and raced ahead of Huot and Edmond. She then turned around and went back to the starting line. She stayed at the starting line for an amount of time before turning around and sprinting to the finishing line, passing Huot and Edmond.

b. From the graph, Huot took 5 seconds to start running.

c. From the graph, at $t = 0$, Edmond is 2.4 metres from the starting line.

d. Nalini won the race.

17.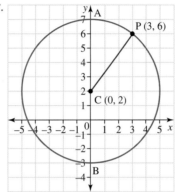

At $(0, 7)$ and $(0, -3)$

18. a.

x	-3	-2	-1	0	1	2	3
y	$\dfrac{3}{2}$	2	Undef.	0	$\dfrac{1}{2}$	$\dfrac{2}{3}$	$\dfrac{3}{4}$

b.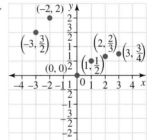

c. This relationship is not linear.

Project

1. 6

2. Sample responses can be found in the worked solutions in the online resources.

3. a and h

4. e

5. a, b, c and h

6. Yes

7.

	Length of table (m)	Width of table (m)	Destination pocket
a	3	0.25	Far left
b	3	0.5	Far left
c	3	0.75	Far left
d	3	1	Far right
e	3	1.25	Far left
f	3	1.5	Far left
g	3	1.75	Far right
h	3	2	Close right

8. Sample responses can be found in the worked solutions in the online resources.

Exercise 6.10 Review questions

1. D

2. B

3. D

4. B

5. D

6. A

7. B

8. A

9. D

10. A

11. a. gradient $= 8$, y-intercept -3

   b. gradient $= -9$, y-intercept 5

   c. gradient $= -2$, y-intercept 6

   d. gradient $= 2$, y-intercept 0

   e. gradient $= \dfrac{2}{3}$, y-intercept $-\dfrac{1}{3}$

12. A, B

13. A, C, E

14. a. 2 b. $-\dfrac{2}{5}$ c. 0

15. a. 2 b. $\dfrac{5}{4}$

16. a. $m = -3,\ c = 7$

   b. $m = \dfrac{3}{2},\ c = 3$

   c. $m = -\dfrac{2}{5},\ c = 0$

   d. $m = 0,\ c = 4$

17. a.

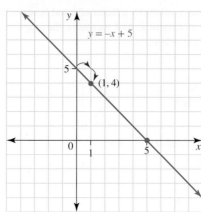

b.

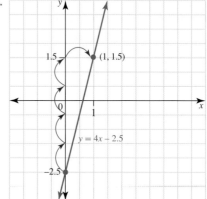

c.

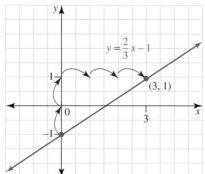

d.

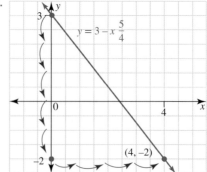

18. a.

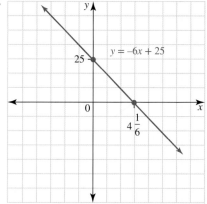

$y = -6x + 25$

b.

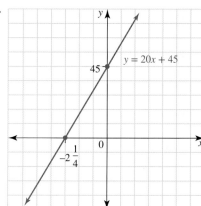

$y = 20x + 45$

c.

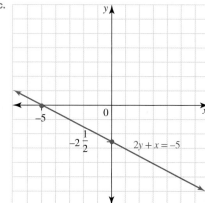

$2y + x = -5$

d.

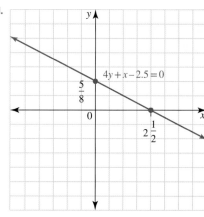

$4y + x - 2.5 = 0$

19. a.

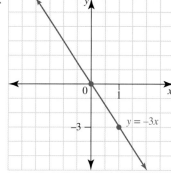

$y = -3x$

b.

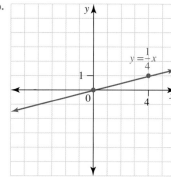

$y = \dfrac{1}{4}x$

c.

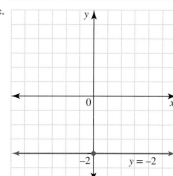

$y = -2$

d.

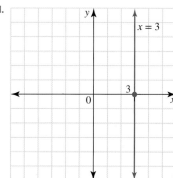

$x = 3$

20. a. $y = 2x$ **b.** $y = 2 - \dfrac{2}{7}x$

21. a. $y = 2x - 7$ **b.** $y = 2x - 14$
  c. $y = 2x - 5$ **d.** $y = -5x$
  e. $y = -x - 2$ **f.** $y = -2.75x + 7.75$
  g. $y = x - 3$ **h.** $y = -2x + 5$

22. (1.1)

23. 5 units

24. a. Vertical translation 3 units down, TP $= (0, -3)$

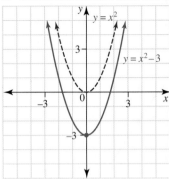

b. Horizontal translation 2 units to the left; TP $= (2, 0)$

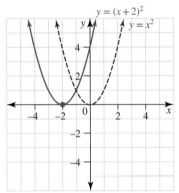

25. a.

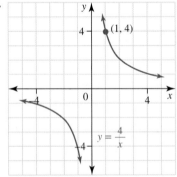

b. Hyperbola

26.

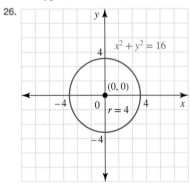

27. a. $y = 400 - 15x$

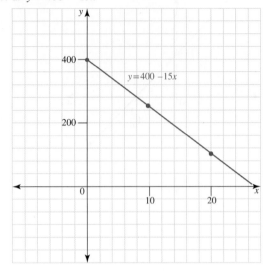

b. 26.7 or 27 weeks
c. $175
d. 21 repayments

28. a. $y = 40 - 3.5x$

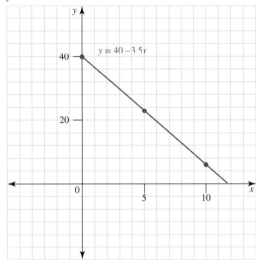

b. 11.4 hours
c. 22.75 km
29. a. $n = 5t + 25$ **b.** 65   **c.** 336 minutes
30. a. $m = 5n + 200$

b. See the figure at the foot of the page.*

c. Everything to the right of the vertical axis and including the vertical axis applies to her earnings because she can only sell 0 or more necklaces.

d. Everything to the left of the vertical axis, because she cannot sell a negative number of necklaces and because she cannot make less than her base salary of $200.

31. $\dfrac{1}{2}$

32. $(-1, 5)$

33. $p = -0.25,\ t = 0.25$

34. $y = -3x - 4.5$

*30. b.

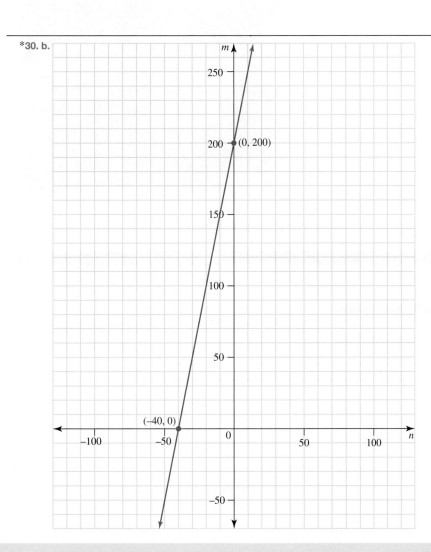

Semester review 1

The learnON platform is a powerful tool that enables students to complete revision independently and allows teachers to set mixed and spaced practice with ease.

Student self-study

Review the **Course Content** to determine which topics and subtopics you studied throughout the year. Notice the green bubbles showing which elements were covered.

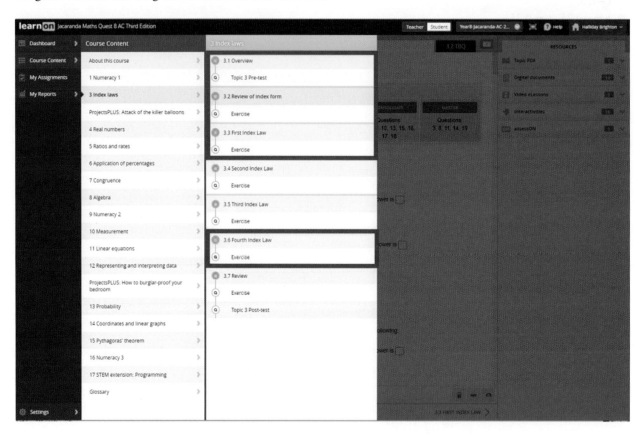

Review your results in **My Reports** and highlight the areas where you may need additional practice.

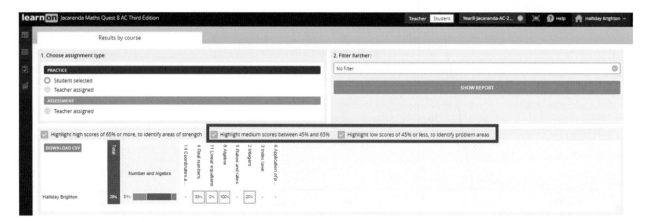

Use these and other tools to help identify areas of strengths and weakness and target those areas for improvement.

Teachers

It is possible to set questions that span multiple topics. These assignments can be given to individual students, to groups or to the whole class in a few easy steps.

Go to **Menu** and select **Assignments** and then **Create Assignment**. You can select questions from one or many topics simply by ticking the boxes as shown below.

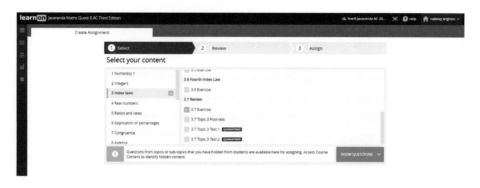

Once your selections are made, you can assign to your whole class or subsets of your class, with individualised start and finish times. You can also share with other teachers.

More instructions and helpful hints are available at www.jacplus.com.au.

7 Proportion and rates

7.1 Overview

Why learn this?

Proportion and rates are often used to compare quantities. Kilometres per hour, price per kilogram, dollars per litre and pay per hour are all examples of rates that are used in everyday life. Supermarkets give the prices of items in, for example, dollars per kilogram, so customers can compare different-sized packages of the same item and make decisions about which is the better value. Chefs use proportions when combining different ingredients from a recipe.

In geometry, the circumference of a circle is proportional to its diameter. We say, simply, that the circumference, C, is pi (π) times its diameter, d; that is, $C = \pi d$. Calculations of π date back to the time of Archimedes, but the symbol became widely used in the 1700s. The Golden Ratio, approximately 1.618, is a proportion that is found in geometry, art and architecture and has been made famous in the illustrations of Leonardo da Vinci (1452–1519).

In the construction industry, concrete is made in different proportions of gravel, sand, cement and water depending on the particular required application. Conversion rates allow businesspeople and the general population to convert Australian dollars to an equivalent amount in an overseas currency. Artists, architects and designers use the concept of proportion in many spheres of their work.

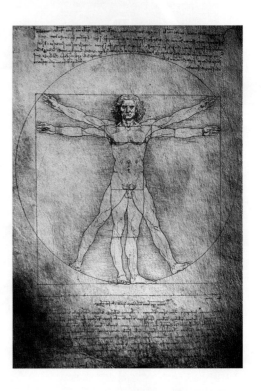

Where to get help

Go to your learnON title at **www.jacplus.com.au** to access the following digital resources. The Online Resources Summary at the end of this topic provides a full list of what's available to help you learn the concepts covered in this topic.

Video eLessons

Interactivities

Fully worked solutions to every question

Digital documents

eWorkbook

Complete this pre-test in your learnON title at www.jacplus.com.au and receive **automatic marks**, **immediate corrective feedback** and **fully worked solutions**.

1. State whether this statement is True or False.
   The expression $y \propto x$ means y is directly proportional to x.

2. **MC** Identify which of the following relationships is directly proportional.

   A.

a	0	1	2	3
b	0	2	4	6

   B.

a	0	1	2	3
b	0	1	4	9

   C.

a	1	2	3	4
b	4	3	2	1

   D.

a	0	1	2	3
b	1	3	5	7

   E.

a	1	2	3	4
b	2	3	4	5

3. Mobile phone calls are charged at $1 per minute. State if the cost of a phone bill and the number of 1-minute time periods are directly proportional.

4. State whether this statement is True or False.
   The perimeter of a square and its side length have a relationship that is directly proportional.

5. **MC** If $a \propto b$ and $a = 3$ when $b = 9$, select the rule that links a and b.

   A. $a = \dfrac{1}{3}b$ **B.** $a = 3b$ **C.** $a = \dfrac{3}{b}$ **D.** $a = 9b$ **E.** $3a = 9b$

6. The quantities x and y are related to each other by direct variation. If k is the constant of proportionality, complete the proportion statement $x = $ _____.

7. The voltage, v, that powers a mobile phone is directly proportional to the current, I amps.
   A mobile phone with consistent resistance uses a current of two amps when powered by a voltage of 5.1 volts.
   a. Determine the consistent resistance or constant of proportionality, k.
   b. Calculate the voltage required to operate a mobile phone that would use a current of 1.2 amps.

8. State whether this statement is True or False.
   The statement 'x varies directly with y' can be written as $y \propto \dfrac{1}{x}$ or $y = \dfrac{k}{x}$.

9. **MC** If y varies with the square of x and $y = 100$ when $x = 2$, then the constant of variation is:
   A. 400   **B.** 200   **C.** 100   **D.** 50   **E.** 25

10. **MC** If b is directly proportional to c^2, select the constant of variation if $b = 72$ when $c = 12$.
    A. 0.5   **B.** 2   **C.** 6   **D.** 36   **E.** 144

11. y is directly proportional to x.
    If x is tripled, select what y is.
    A. Doubled   B. Tripled   C. Reduced to one-third
    D. Reduced to one-ninth   E. Halved

12. **MC** On average, Noah kicks 2.25 goals per game of soccer. In a 16-game season, the number of goals Noah would kick is:
    A. 32   B. 36   C. 30   D. 38   E. 34

13. Tea bags in a supermarket can be bought for $1.45 per pack of 10 or for $3.85 per pack of 25. Determine the cheaper way of buying tea bags.

14. Given that $y \propto x$ and $\dfrac{y}{x} = 4$, determine the rule linking y and x.

15. Determine your average speed if you drive 320 km in five hours.

7.2 Direct linear proportion

LEARNING INTENTION

At the end of this subtopic you should be able to:
- understand the concept of direct linear proportion
- recognise and write direct linear proportion statements.

▶ 7.2.1 Direct linear proportion

eles-4818

- Proportion is described by two equal ratios.
- In **direct linear proportion**, the two variables change at the same rate.
- Suppose that ice-cream costs $3 and that you are to buy some for your friends.
- There is a relationship between the cost of the ice-cream (C) and the number of ice-creams that you buy (n).
- The relationship between the two variables (number of ice-creams and cost of the ice-cream) can be illustrated in a table or a graph.

n	0	1	2	3	4
C ($)	0	3	6	9	12

- This relationship is called direct linear proportion as it shows the following characteristics:
  - As n increases, so does C.
  - The graph of the relationship is a straight line passing through the origin as $n = 0$ and $C = 0$.

- We say that 'C is directly proportional to n' or 'C varies directly as n'.

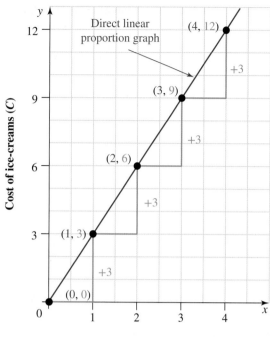

Number of ice-creams bought (n)

Expressing direct linear proportion

If the y-values increase (or decrease) directly with the x-values, then this relationship can be expressed as:

$$y \propto x$$

where $\propto$ is the symbol for proportionality — this means that y is directly proportional to x.

WORKED EXAMPLE 1 Checking for direct linear proportion

For each of the following pairs of variables, state whether direct linear proportion exists.
a. The height of a stack of photocopy paper (h) and the number of sheets (n) in the stack
b. Your Maths mark (m) and the number of hours of Maths homework you have completed (n)

THINK	WRITE
a. When n increases, so does h. When $n = 0$, $h = 0$. If graphed, the relationship would be linear.	a. $h \propto n$
b. As n increases, so does m. When $n = 0$, I may get a low mark at least, so $n \neq 0$.	b. m is not directly proportional to n.

For each of the following, determine whether direct linear proportion exists between the variables.

a.

t	0	1	2	3	4
y	0	1	3	7	15

b.

n	1	2	3	4
c	5	10	15	20

c.

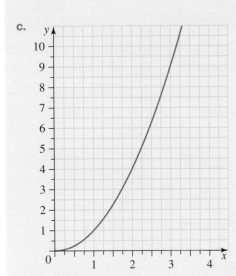

THINK

a. From the table, when t increases, so does y.
When $t = 0$, $y = 0$. The t-values increase by a constant amount but the y-values do not, so the relationship is not linear.
When t is doubled, y is not.

b. From the table, as n increases, so does c.
Extending the pattern gives $n = 0$, $c = 0$. The n-values and C-values increase by constant amounts, so the relationship is linear.

c. When x increases, so does y.
When $x = 0$, $y = 0$.
The graph is not a straight line.

WRITE

a. y is not directly proportional to t.

b. $C \alpha n$

c. y is not directly proportional to x.

Note: Not all direct proportion relationships are linear. Non-linear direct proportion will be studied in later years.

DISCUSSION

How do you know when two quantities are directly proportional?

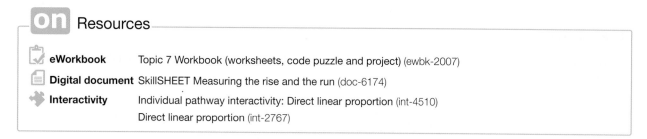
Exercise 7.2 Direct linear proportion   **learn** on

Individual pathways

■ PRACTISE	■ CONSOLIDATE	■ MASTER
1, 3, 6, 9, 13	2, 5, 7, 10, 11, 14	4, 8, 12, 15, 16

To answer questions online and to receive **immediate corrective feedback** and **fully worked solutions** for all questions, go to your learnON title at www.jacplus.com.au.

Fluency

1. **WE1** For each of the following pairs of variables, state whether direct linear proportion exists. If it does not exist, give a reason why.

   a. The distance (d) travelled in a car travelling at 60 km/h and the time taken (t)
   b. The speed of a swimmer (s) and the time the swimmer takes to complete one lap of the pool (t)
   c. The cost of a bus ticket (c) and the distance travelled (d)

2. For each of the following pairs of variables, state whether direct linear proportion exists. If it does not exist, give a reason why.

   a. The perimeter (p) of a square and the side length (l)
   b. The area of a square (A) and the side length (l)
   c. The total cost (C) of buying n boxes of pencils

3. For each of the following pairs of variables, state whether direct linear proportion exists. If it does not exist, give a reason why.

   a. The weight of an object in kilograms (k) and in pounds (p)
   b. The distance (d) travelled in a taxi and the cost (c)
   c. A person's height (h) and their age (a)

4. State whether the following statements are True or False.
   There is a direct linear relationship between:

   a. the total cost, C, of purchasing netballs and the number, n, purchased
   b. the circumference of a circle and its diameter
   c. the area of a semicircle and its radius.

5. **MC** For the table shown, if $y \propto x$, then the values of a and b would be:

x	1	2	4	8
y	3	6	a	b

   A. $a = 8$, $b = 16$ **B.** $a = 8$, $b = 24$ **C.** $a = 9$, $b = 12$
   D. $a = 12$, $b = 18$ **E.** $a = 12$, $b = 24$

Understanding

6. **WE2** For each of the following, determine whether direct linear proportion exists between the variables. If it does not, explain why.

a.

x	0	1	2	3	4
y	0	1	3	8	15

b.

a	0	1	2	3	6
M	0	8	16	24	48

c.

t	0	1	2	4	8
d	0	3	6	9	12

d.

n	0	1	2	3	4
C	10	20	30	40	50

7. For each of the following, determine whether direct proportion exists between the variables. If it does not, explain why.

a.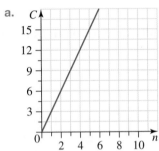

b.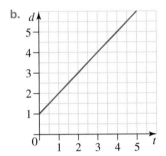

c.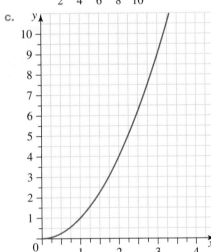

d.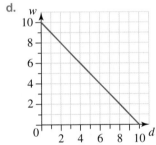

8. List five pairs of real-life variables that exhibit direct proportion.

Reasoning

9. Explain which point must always exist in a table of values if the two variables exhibit direct linear proportionality.

10. If direct linear proportion exists between two variables m and n, fill out the table and explain your reasoning.

m	0	2	5
n	0		20

11. If the variables x and y in the following table are directly proportional, find the values of a, b and c. Explain your reasoning.

x	2	4	9	a
y	8	b	c	128

12. If $y \propto x$, explain what happens to:

    a. y if x is doubled   **b.** y if x is halved   **c.** x if y is tripled.

Problem solving

13. Mobile phone calls are charged at 17 cents per 30 seconds.

    a. Does direct linear proportion exist between the cost of a phone bill and the number of 30-second time periods? Justify your answer.

    b. If a call went for 7.5 minutes, determine how much the call would cost.

14. An electrician charges \$55 for every 30 minutes or part of 30 minutes for his labour on a building site.

    a. Does direct linear proportion exist between the cost of hiring the electrician and the time spent at the building site? Justify your answer.

    b. If the electrician worked for 8.5 hours, determine how much this would cost.

15. A one-litre can of paint covers five square metres of wall.

    a. Does a direct linear proportion exist between the number of litres purchased and the area of the walls to be painted? Justify your answer.

    b. Evaluate the number of cans needed to paint a wall 5 metres long and 2.9 metres high with two coats of paint.

16. Bruce is building a pergola and needs to buy treated pine timber. He wants 4.2-metre and 5.4-metre lengths of timber. If a 4.2-metre length costs \$23.10 and a 5.4-metre length costs \$29.16, determine if direct linear proportion exists between the cost of the timber and the length of the timber per metre.

7.3 Proportionality

LEARNING INTENTION

At the end of this subtopic you should be able to:
* calculate the constant of proportionality
* express the constant of proportionality as a ratio of the two quantities.

▶ 7.3.1 Constant of proportionality

eles-4819

* Proportionality relationships ($y \propto x$) are shown by the graph with four straight lines:

$$y = \frac{1}{2}x$$

$$y = x$$

$$y = 2x$$

$$y = 3x$$

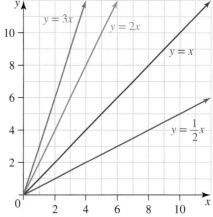

* For the equation $y = kx$:
    * k is a constant, called the **constant of proportionality**
    * y is directly proportional to x, or $y \propto x$
    * all the graphs start at $(0, 0)$
    * y is called the *dependent variable* and is normally placed on the vertical axis (y-axis)
    * x is called the *independent variable* and is normally placed on the horizontal axis (x-axis).

WORKED EXAMPLE 3 Calculating the constant of proportionality

Given that $y \propto x$ and $y = 12$ when $x = 3$, calculate the constant of proportionality and state the rule linking y and x.

THINK	WRITE
1. $y \propto x$, so write the linear rule.	$y = kx$
2. Substitute $y = 12$ and $x = 3$ into $y = kx$.	$12 = 3k$
3. Calculate the constant of proportionality by solving for k.	$k = 4$
4. Write the rule.	$y = 4x$

WORKED EXAMPLE 4 Determining the direct proportion relationship

The weight (W) of \$1 coins in a bag varies directly as the number of coins (n).
Twenty coins weigh 180 g.
a. Determine the relationship between W and n.
b. Calculate how much 57 coins weigh.
c. Determine how many coins weigh 252 g.

THINK	WRITE
Summarise the information given in a table.	$W = kn$

n	20	57	
W	180		252

	THINK		WRITE
a. 1.	Substitute $n = 20$ and $W = 180$ into $W = kn$.	a.	$180 = 20k$
			$k = 9$
2.	Solve for k.		
3.	Write the relationship between W and n.		$W = 9n$
b. 1.	State the rule.	b.	$W = 9n$
2.	Substitute 57 for n to find W.		$W = 9 \times 57$
			$= 513$
3.	Write the answer in a sentence.		Fifty-seven coins weigh 513 g.
c. 1.	State the rule.	c.	$W = 9n$
2.	Substitute 252 for W.		$252 = 9n$
3.	Solve for n.		$n = \dfrac{252}{9}$
			$= 28$
4.	Write the answer in a sentence.		Twenty-eight coins weigh 252 g.

▶ 7.3.2 Ratio

eles-4820

- If $y \propto x$, then $y = kx$, where k is constant.
  Transposing this formula gives $\dfrac{y}{x} = k$.
- The constant of proportionality, k, is the ratio of any pair of values (x, y).
- For example, this table shows that $v \propto t$.

t	1	2	3	4
v	5	10	15	20

It is clear that $\dfrac{20}{4} = \dfrac{15}{3} = \dfrac{10}{2} = \dfrac{5}{1} = 5.$

WORKED EXAMPLE 5 Applying the constant of proportionality as a ratio

Sharon works part-time and is paid at a fixed rate per hour. If she earns \$135 for 6 hours' work, calculate how much she will earn for 11 hours.

THINK	WRITE
1. Sharon's payment is directly proportional to the number of hours worked. Write the rule.	$P = kn$
2. Summarise the information given. The value of x needs to be calculated.	<table><tr><td>n</td><td>6</td><td>11</td></tr><tr><td>P</td><td>135</td><td>x</td></tr></table>
3. Since $p = kn$, $\dfrac{P}{n}$ is constant. Solve for x.	$\dfrac{135}{6} = \dfrac{x}{11}$ $x = \dfrac{135 \times 11}{6}$ $x = 247.5$
4. Write the answer in a sentence.	Sharon earns \$247.50 for 11 hours of work.

 Resources

Exercise 7.3 Proportionality

learn on

Individual pathways

■ PRACTISE	■ CONSOLIDATE	■ MASTER
1, 2, 6, 7, 12, 15	3, 4, 8, 9, 13, 16	5, 10, 11, 14, 17, 18

To answer questions online and to receive **immediate corrective feedback** and **fully worked solutions** for all questions, go to your learnON title at www.jacplus.com.au.

Fluency

1. **WE3** If a is directly proportional to b, and $a = 30$ when $b = 5$, calculate the constant of proportionality and state the rule linking a and b.

2. If $a \propto b$, and $a = 2.5$ when $b = 5$, determine the rule linking a and b.

3. If $C \propto t$, and $C = 100$ when $t = 8$, determine the rule linking C and t.

4. If $v \propto t$, and $t = 20$ when $v = 10$, determine the rule linking v and t.

5. If $F \propto a$, and $a = 40$ when $F = 100$, determine the rule linking a and F.

Understanding

6. **WE4** Springs are often used to weigh objects, because the extension of a spring (E) is directly proportional to the weight (W) of the object hanging from the spring. A 4-kg load stretches a spring by 2.5 cm.

   a. Determine the relationship between E and W.
   b. Calculate the load that will stretch the spring by 12 cm.
   c. Determine how far 7 kg will extend the spring.

7. Han finds that 40 shelled almonds weigh 52 g.

   a. Determine the relationship between the weight (W) and the number of almonds (n).
   b. Calculate how many almonds there would be in a 500 g bag.
   c. Calculate how much 250 almonds would weigh.

8. Petra knows that her bicycle wheel turns 40 times when she travels 100 m.

   a. Determine the relationship between the distance travelled (d) and the number of turns of the wheel (n).
   b. Calculate how far she goes if her wheel turns 807 times.
   c. Determine how many times her wheel turns if she travels 5 km.

9. Fiona, who operates a plant nursery, uses large quantities of potting mix. Last week she used 96 kg of potting mix to place 800 seedlings in medium-sized pots.

   a. Determine the relationship between the mass of potting mix (M) and the number of seedlings (n).
   b. Calculate how many seedlings she can pot with her remaining 54 kg of potting mix.
   c. Determine how much potting mix she will need to pot 3000 more seedlings.

10. **WE5** Tamara is paid at a fixed rate per hour. If she earns $136 for five hours of work, calculate how much will she earn for eight hours of work.

11. It costs $158 to buy 40 bags of birdseed. Calculate how much 55 bags will cost.

408 Jacaranda Maths Quest 9

Reasoning

12. If 2.5 L of lawn fertiliser will cover an area of 150 m², determine how much fertiliser is needed to cover an area of 800 m² at the same rate. Justify your answer.

13. Paul paid $68.13 for 45 L of fuel. At the same rate, determine how much he would pay for 70 L. Justify your answer.

14. Rose gold is an alloy of gold and copper that is used to make high-quality musical instruments. If it takes 45 g of gold to produce 60 g of rose gold, evaluate how much gold would be needed to make 500 g of rose gold. Justify your answer.

Problem solving

15. If Noah takes a group of friends to the movies for his birthday and it would cost $62.50 for five tickets, determine how much it would cost if there were 12 people (including Noah) in the group.

16. Sharyn enjoys quality chocolate, so she makes a trip to her favourite chocolate shop. She is able to select her favourite chocolates for $7.50 per 150 grams. Since Sharyn loves her chocolate, she decides to purchase 675 grams. Evaluate how much she spent.

17. Anthony drives to Mildura, covering an average of 75 kilometers in 45 minutes. He has to travel 610 kilometres to get to Mildura.

    a. Determine the relationship between the distance he travelled in kilometres, D, and his driving time in hours, T.
    b. Evaluate how long it will take Anthony to complete his trip.
    c. If he stops at Bendigo, 220 kilometres from his starting position, determine how long it will take him to reach Bendigo.
    d. Determine how much longer it will take him to arrive at Mildura after leaving Bendigo.

18. The volume of a bird's egg can be determined by the formula $V = kl^3$, where V is the volume in cm³, l is the length of the egg in cm and k is a constant. A typical ostrich egg is 15 cm long and has a volume of 7425 cm³. Evaluate the volume of a chicken egg that is 5 cm long.

7.4 Introduction to rates

LEARNING INTENTION

At the end of this subtopic you should be able to:
- calculate the rate connecting two quantities
- apply the rate to a worded problem.

▶ 7.4.1 Rates

eles-4821

- The word **rate** occurs commonly in news reports and conversation.
  'Home ownership rates are falling.'
  'People work at different rates.'
  'What is the current rate of inflation?'
  'Do you offer a student rate?'
  'The crime rate seems to be increasing.'

- Rate is a word often used when referring to a specific ratio.
- In general, to calculate a rate (or specific ratio), one quantity is divided by another.
- State the units for the rate, such as grams/litre (g/L) or kilometres/hour (km/h).

7.4.2 Average speed

eles-4822

- If you travel 160 km in 4 hours, your average speed is 40 km per hour.
- Speed is an example of a rate, and its unit of measurement (kilometres per hour) contains a formula.
- The word 'per' can be replaced with 'divided by', so $speed = \dfrac{distance\ (in\ km)}{time\ (in\ hours)}$.
- It is important to note the units involved.
  For example, an athlete's speed is often measured in metres per second (m/s) rather than km/h.

WORKED EXAMPLE 6 Calculating rates from statements

Calculate the rates suggested by these statements.
a. A shearer shears 1110 sheep in 5 days.
b. Eight litres of fuel costs \$12.56.
c. A cricket team scored 152 runs in 20 overs.

THINK	WRITE
a. 1. The rate suggested is 'sheep per day'.	a. $Rate = \dfrac{number\ of\ sheep}{number\ of\ days}$
2. Substitute the values for the number of sheep and the number of days.	$= \dfrac{1110}{5}$
3. Write the rate.	$= 222$ sheep per day
b. 1. The rate suggested is 'dollars per litre'.	b. $Rate = \dfrac{number\ of\ dollars}{number\ of\ litres}$
2. Substitute the values for the price per 8 litres and the number of litres.	$= \dfrac{12.56}{8}$
3. Write the rate.	$= \$1.57$ per litre
c. 1. The rate suggested is 'runs per over'.	c. $Rate = \dfrac{number\ of\ runs}{number\ of\ overs}$
2. Substitute the values for the number of runs and the number of overs.	$= \dfrac{152}{20}$
3. Write the rate.	$= 7.6$ runs per over

WORKED EXAMPLE 7 Calculating rates

The concentration of a solution is measured in g/L (grams per litre). Calculate the concentration of the solution when 10 g of salt is dissolved in 750 mL of water.

THINK	WRITE
1. Concentration is measured in g/L, which means that concentration = number of grams (mass) ÷ number of litres (volume).	Concentration = $\dfrac{\text{mass}}{\text{volume}}$
2. Substitute the values of mass and volume.	$= \dfrac{10}{0.75}$
3. Simplify.	$= 13.3$ g/L
4. Write the rate in a sentence.	The concentration of the solution is 13.3 g/L

 Resources

eWorkbook Topic 7 Workbook (worksheets, code puzzle and project) (ewbk-2007)

Interactivities Individual pathway interactivity: Introduction to rates (int-4513)
Speed (int-6457)

Exercise 7.4 Introduction to rates

learn on

Individual pathways

■ PRACTISE	■ CONSOLIDATE	■ MASTER
1, 3, 6, 9	2, 5, 7, 10	4, 8, 11

To answer questions online and to receive **immediate corrective feedback** and **fully worked solutions** for all questions, go to your learnON title at www.jacplus.com.au.

Fluency

1. Hayden drove from Hay to Bee. He covered a total distance of 96 km and took 1.5 hours for the trip. Calculate Hayden's average speed for the journey.

2. Calculate the average speed in km/h for each of the following.
   a. 90 km in 45 min
   b. 5500 km in 3 h 15 min (correct to 1 decimal place)

3. **WE6** Calculate the rates suggested by these statements, giving your answers to 2 decimal places where necessary.

   a. It costs $736 for 8 theatre tickets.
   b. Penelope decorated 72 small cakes in 3 hours.
   c. Usain Bolt has a 100-metre world sprint record of 9.58 seconds.
   d. It takes 30 hours to fill the swimming pool to a depth of 90 cm.
   e. Peter received $260 for 15 hours' work.
   f. Yan received $300 for assembling 6 air conditioners.

Understanding

4. A metal bolt of volume $25\,cm^3$ has a mass of $100\,g$. Calculate its density (mass per unit of volume).

5. **WE7** One hundred and twenty grams of sugar is dissolved in $200\,mL$ of water. Calculate the concentration of this solution in g/L. Explain your answer.

Reasoning

6. In a race between a tortoise and a hare, the hare ran at $72\,km$ per hour while the giant tortoise moved at $240\,cm$ per minute. Compare and explain the difference in the speed of the two animals.

7. A school had 300 students in 2013 and 450 students in 2015. Determine the average rate of growth in the number of students per year. Show your working.

8. The average speed of a car is determined by the distance of the journey and the time the journey takes. Explain the two ways in which the speed can be increased.

Problem solving

9. Mt Feathertop is Victoria's second-highest peak. To walk to the top involves an increase in height of $1500\,m$ over a horizontal distance of $10\,km$ ($10\,000\,m$). Determine the average gradient of the track.

10. In the 2000 Sydney Olympics, Cathy Freeman won gold in the 400-metre race. Her time was 49.11 seconds. In the 2000 Beijing Olympics, Usain Bolt set a new world record for the men's 100-metre race. His time was 9.69 seconds. Evaluate the average speed of the winner for each race in kilometres per hour.

11. Beaches are sometimes unfit for swimming if heavy rain has washed pollution into the water. A beach is declared unsafe for swimming if the concentration of bacteria is more than 5000 organisms per litre. A sample of 20 millilitres was tested and found to contain 55 organisms. Evaluate the concentration in the sample (in organisms/litre) and state whether or not the beach should be closed.

7.5 Constant and variable rates

> **LEARNING INTENTION**
>
> At the end of this subtopic you should be able to:
> - interpret a graph showing constant rates
> - draw a graph showing constant rates
> - interpret a graph showing variable rates.

7.5.1 Constant rates

eles-4823

- A **constant rate** of change means the quantity always changes by the same amount in a given time.
- Consider a car travelling along a highway at a constant speed (rate) of $90\,km/h$. After one hour it will have travelled $90\,km$, as shown in the table.

$$+1 \qquad +1 \qquad +1$$

Time (h)	0	1	2	3
Distance (km)	0	90	180	270

$$+90 \quad +90 \quad +90$$

- The distance–time graph is also shown.

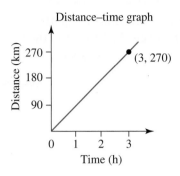

Distance–time graph

- The gradient of the graph is $\dfrac{270 - 0}{3 - 0} = 90.$
- The equation of the graph is $d = 90t$, and the gradient is equal to the speed, or rate of progress.
- A constant rate of change produces a linear (straight-line) graph.

WORKED EXAMPLE 8 Describing constant rates from a graph

Each diagram shown illustrates the distance travelled by a car over time. Describe the journey, including the speed of the car.

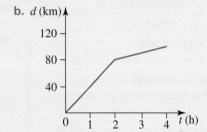

THINK

a. There are three distinct sections.

  1. In the first two hours, the car travels 80 km: $\dfrac{80}{2} = 40 \, \text{km/h.}$

  2. In the next hour, the car does not move.

  3. In the fourth hour, the car travels 40 km.

b. There are two distinct sections.

  1. In the first two hours, the car travels 80 km: $\dfrac{80}{2} = 40 \, \text{km/h.}$

  2. In the next two hours, the car travels 20 km: $\dfrac{20}{2} = 10 \, \text{km/h.}$

WRITE

a.

The car travels at a speed of 40 km/h for 2 hours.

The car stops for 1 hour.

The car then travels for 1 hour at 40 km/h.

b.

The car travels at a speed of 40 km/h for 2 hours.

The car then travels at 10 km/h for a further 2 hours.

WORKED EXAMPLE 9 Drawing a distance–time graph to illustrate constant rates

Draw a distance–time graph to illustrate the following journey.
A cyclist travels for 1 hour at a constant speed of 10 km/h, then stops for a 30-minute break before riding a further 6 km for half an hour at a constant speed.

THINK

There are three phases to the journey.

The graph starts at $(0, 0)$.

1. In the first hour, the cyclist travels 10 km. Draw a line segment from $(0, 0)$ to $(1, 10)$.

2. For the next half-hour, the cyclist is stationary, so draw a horizontal line segment from $(1, 10)$ to $(1.5, 10)$.

3. In the next half-hour, the cyclist travels 6 km. Draw a line segment from $(1.5, 10)$ to $(2, 16)$.

WRITE

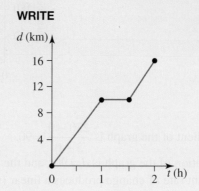

7.5.2 Variable rates

eles-4824

- In reality, a car tends not to travel at constant speed. It starts from rest and gradually picks up speed.
- Since speed $= \dfrac{\text{distance travelled}}{\text{time taken}}$, the distance–time graph will:
  - be flat, or horizontal, when the car is not moving (is stationary)
  - be fairly flat when the car is moving slowly
  - be steeper and steeper as the speed of the car increases.
- This is demonstrated in the distance–time graph shown.

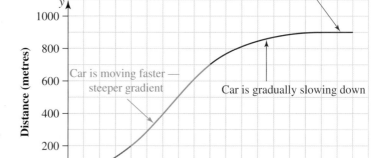

- A variable rate of change produces a non-linear graph.

WORKED EXAMPLE 10 Describing graphs with variable rates

The diagram shown illustrates the distance travelled by a car over time. Describe what is happening, in terms of speed, at each of the marked points.

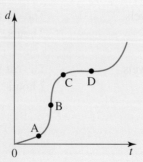

THINK	WRITE
1. At point A on the graph, the gradient is small but becoming steeper.	At point A the car is travelling slowly but accelerating.
2. At point B on the graph, the gradient is at its steepest and is not changing.	At point B the car is at its greatest speed.
3. At point C, the graph is becoming flatter.	At point C the car is slowing down.
4. At point D, the graph is horizontal.	At point D the car is stationary.

DISCUSSION

How can you tell the difference between constant and variable rates?

Exercise 7.5 Constant and variable rates

learn on

Individual pathways

■ PRACTISE	■ CONSOLIDATE	■ MASTER
1, 3, 7, 10	2, 4, 8, 11	5, 6, 9, 12, 13

To answer questions online and to receive **immediate corrective feedback** and **fully worked solutions** for all questions, go to your learnON title at www.jacplus.com.au.

Fluency

1. **WE8** Each diagram shown illustrates the distance travelled by a car over time. Describe the journey, including the speed of the car.

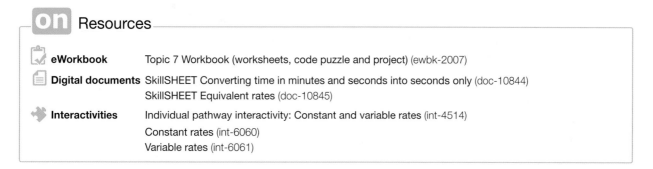

2. Two friends take part in a 24-kilometre mini-marathon. They run at constant speed. Ali takes 2 hours and Beth takes 3 hours to complete the journey.

   a. On the same diagram, draw a distance–time graph for each runner.
   b. i. Calculate the equation for each graph.
      ii. State the difference between the two graphs.

3. **WE9** Draw a distance–time graph to illustrate each of the following journeys.

   a. A cyclist rides at 40 km/h for 30 minutes, stops for a 30-minute break, and then travels another 30 km at a speed of 15 km/h.

   b. Zelko jogs at a speed of 10 km/h for one hour, and then at half the speed for another hour.

4. **WE10** The diagrams shown illustrate the distances travelled by two cars over time.

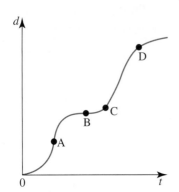 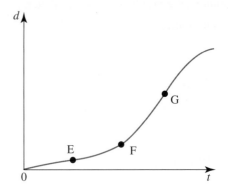

   a. Describe what is happening in terms of speed at each of the marked points.

   b. For each diagram, state at which point:

     i. the speed is the greatest

     ii. the speed is the lowest

     iii. the car is stationary.

Understanding

5. The table below shows the distance travelled, D, as a person runs for R minutes.

R (minutes)	10	20	50
D (km)	2	4	10

   a. Calculate the rate in km/minute between the time 10 minutes and 20 minutes.

   b. Determine the rate in km/minute between the distance 4 km and 10 km.

   c. Explain whether the person's speed is constant.

6. The table below shows the water used, W, after the start of a shower, where T is the time after the shower was started.

T (minutes)	0	1	2	4
W (litres)	0	20	30	100

   a. Calculate the rate of water usage in L/minute for the four stages of the shower.

   b. Explain whether the rate of water usage was constant.

Reasoning

7. Margaret and Brian left Brisbane airport at 9:00 am. They travelled separately but on the same road and in the same direction.
   Their journeys are represented by the travel graph below. Show your working.

   a. Determine the distance from the airport at which their paths crossed.
   b. Determine how far apart they were at 1:00 pm.
   c. Determine for how long each person stopped on the way.
   d. Determine the total time spent driving and the total distance for each person.
   e. Evaluate the average speed while driving for each person.

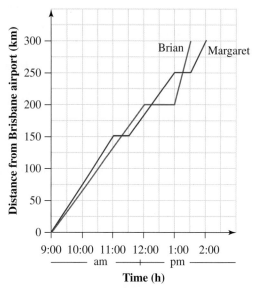

8. Hannah rode her bike along the bay one morning. She left home at 7:30 am and covered 12 km in the first hour. She felt tired and rested for half an hour. After resting she completed another 8 km in the next hour to reach her destination. Show your working.

   a. Determine how long Hannah took for the entire journey.
   b. Calculate the total distance for which she actually rode her bike.
   c. Draw a travel graph for Hannah's journey.

9. Rebecca and Joanne set off at the same time to jog 12 kilometres. Joanne ran the entire journey at constant speed and finished at the same time as Rebecca. Rebecca set out at 12 km/h, stopping after 30 minutes to let Joanne catch up, then she ran at a steady rate to complete the distance in 2 hours.

   a. Show the progress of the two runners on a distance–time graph.
   b. Determine how long Rebecca waited for Joanna to catch up. Show your working.

Problem solving

10. An internet service provider charges $30 per month plus $0.10 per megabyte downloaded. The table of monthly cost versus download amount is shown.

Download (MB)	0	100	200	300	400	500
Cost ($)	30	40	50	60	70	80

   a. Determine how much the cost increases by when the download amount increases from:

   i. 0 to 100 MB
   ii. 100 to 200 MB
   iii. 200 to 300 MB.

   b. Explain whether the cost is increasing at a constant rate.

11. Use the graph showing the volume of water in a rainwater tank to answer the following questions.

a. Determine the day(s) during which the rate of change was positive.

b. Determine the day(s) during which the rate of change was negative.

c. Determine the day(s) during which the rate of change was zero.

d. Determine the day on which the volume of water increased at the fastest rate.

e. Determine the day on which the volume of water decreased at the fastest rate.

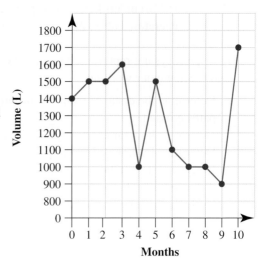

12. The graph shows the number of soft-drink cans in a vending machine at the end of each day.

a. Determine how much the number of cans changed by in the first day.

b. Determine how much the number of cans changed by in the fifth day.

c. Explain whether the number of soft-drink cans is changing at a constant or variable rate.

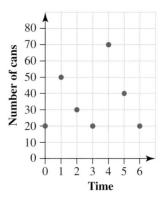

13. The table and graph show Melbourne's average daily maximum temperature over the year.

	Jan.	Feb.	Mar.	Apr.	May	June	July	Aug.	Sept.	Oct.	Nov.	Dec.
Mean maximum (°C)	25.8	25.8	23.8	20.3	16.7	14.0	13.4	14.9	17.2	19.6	21.9	24.2

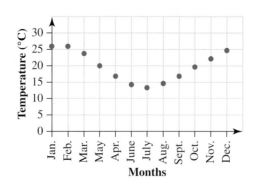

a. Determine the average maximum temperature in:
   i. February
   ii. June.

b. Calculate the change in temperature from:
   i. January to August
   ii. November to December.

c. Explain whether the temperature is changing at a constant rate.

7.6 Review

7.6.1 Topic summary

Direct linear proportion

- If y is directly proportional to x, then:
  - y varies with the value of x
  - $y \propto x$
  - as x increases (or decreases), y increases (or decreases)
  - $x = 0$ and $y = 0$ results in a straight line.

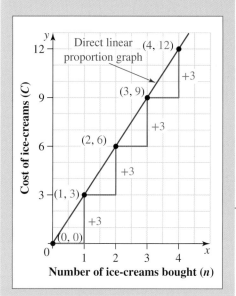

- Not all direct proportion relationships are linear.

Constant of proportionality

- If y is directly proportional to x, then:
  - $y = kx$
  - k is a constant called the **constant of proportionality** — this is the ratio of any pair of values (x, y)
  - the gradient of the straight line is k
  - all graphs start at $(0, 0)$.
- y is called the dependent variable and is normally placed on the vertical axis.
- x is called the independent variable and is normally placed on the horizontal axis.

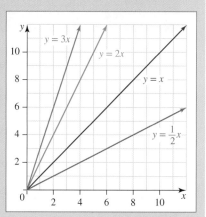

PROPORTION AND RATES

Rates

- A rate compares two quantities that have different units.
- Rates are special ratios.
- A rate is calculated by dividing one quantity with another.

Variable rates

- A variable rate means a quantity does not change by the same amount in a given time.
- Variable rates produce a non-linear graph.

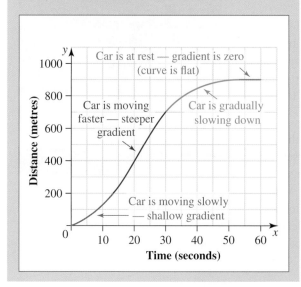

Average speed

- Average speed is an example of a rate; it is calculated by dividing total distance by total time.
- Unit: metres per second (m/s) or kilometres per hour (km/h)
- Average speed $= \dfrac{\text{distance in kilometres}}{\text{time in hours}}$

  e.g. if the distance travelled is 200 km in 4 hours, then

  average speed $= \dfrac{200 \text{ km}}{4 \text{ hours}} = 50$ km/h.

Constant rates

- A constant rate means a quantity always changes by the same amount in a given time.
- Constant rates produce a linear (straight line) graph.

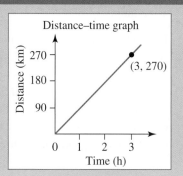

7.6.2 Success criteria

Tick a column to indicate that you have completed the subtopic and how well you think you have understood it using the traffic light system.

(**Green:** I understand; **Yellow:** I can do it with help; **Red:** I do not understand)

Subtopic	Success criteria	⬤	◯	⬤
7.2	I can understand the concept of direct linear proportion.			
	I can recognise and write direct linear proportion statements.			
7.3	I can determine the constant of proportionality.			
	I can express the constant of proportionality as a ratio of the two quantities.			
7.4	I can calculate the rate connecting two quantities.			
	I can apply the rate to a worded problem.			
7.5	I can interpret a graph showing constant rates.			
	I can draw a graph showing constant rates.			
	I can interpret a graph showing variable rates.			

7.6.3 Project

Fastest speeds

To calculate the average speed of a journey, we can use the speed formula:

$$\text{speed} = \frac{\text{distance}}{\text{time}}$$

Speed is a *rate*, because it compares two quantities of different units. Speed can be expressed in units such as km/h, km/min and m/s. The units of the quantities substituted into the numerator and denominator determine the final units of speed. In order to compare the speeds of different events, it is useful to convert them to the same units.

In 2008, the summer and winter Olympics were held. At the summer Olympics, sprinter Usain Bolt ran 100 m in 9.69 seconds. At the winter Olympics, cross-country skier Petter Northug covered 50 km in 2 hours and 5 minutes, and speed skater Mika Poutala covered 0.5 km in 34.86 seconds. Which competitor was the fastest? In order to answer this question, we need to determine the speed of each competitor. Since the information is quoted in a variety of units, we need to decide on a common unit for speed.

1. Calculate the speed of each athlete in m/s, correct to one decimal place.
2. Determine which athlete was the fastest.

Consider the speed of objects in the world around you. This task requires you to order the following objects from fastest to slowest, assuming that each object is travelling at its fastest possible speed:

electric car, submarine, diesel train, car ferry, skateboard, bowled cricket ball, windsurfer, served tennis ball, solar-powered car, motorcycle, aircraft carrier, Jaguar car, helicopter, airliner, rocket-powered car, the Concorde supersonic airliner

3. From your personal understanding and experience, order the above list from fastest to slowest.

We can make a more informed judgement if we have facts available about the movement of each of these objects. Consider the following facts:
- When travelling at its fastest speed, the Concorde could cover a distance of about 7000 km in 3 hours.
- The fastest airliner can cover a distance of 5174 km in 2 hours and a helicopter 600 km in 1.5 hours.
- It would take 9.5 hours for a rocket-powered car to travel 9652 km, 3.25 hours for an electric car to travel 1280 km and 1.5 hours for a Jaguar car to travel 525 km.
- On a sunny day, a solar-powered car can travel 39 km in half an hour.
- In 15 minutes, a car ferry can travel 26.75 km, an aircraft carrier 14 km and a submarine 18.5 km.
- It takes the fastest diesel train 0.42 hours to travel 100 km and a motorcyclist just 12 minutes to travel the same distance.
- In perfect weather conditions, a skateboarder can travel 30 km in 20 minutes and a windsurfer can travel 42 km in 30 minutes.
- A bowled cricket ball can travel 20 m in about 0.45 seconds, while a served tennis ball can travel 25 m in about 0.4 seconds.
4. Using the information above, decide on a common unit of speed and determine the speed of each of the objects.
5. Order the above objects from fastest to slowest.
6. Compare the order from question 5 with the list you made in question 3.
7. Conduct an experiment with your classmates to record the times that objects take to cover a certain distance. For example, record the time it takes to run 100 m or throw a ball 20 m. Compare the speeds with the speeds of the objects calculated during this investigation.

Exercise 7.6 Review questions

learnon

To answer questions online and to receive **immediate corrective feedback** and **fully worked solutions** for all questions, go to your learnON title at www.jacplus.com.au.

Fluency

1. **MC** y is directly proportional to x and $y = 450$ when $x = 15$. The rule relating x and y is:
   A. $y = 0.033x$ **B.** $y = 30x$ **C.** $y = 60x$ **D.** $y = 6750x$ **E.** $y = 0.3x$

2. **MC** If $y \propto x$ and $y = 10$ when $x = 50$, the constant of proportionality is:
   A. 10   **B.** 5   **C.** 1   **D.** 0.2   **E.** 0.5

3. **MC** If $y \propto x$ and $y = 10$ when $x = 50$, the value of x when $y = 12$ is:
   A. 6   **B.** 60   **C.** 40   **D.** 2.4   **E.** 24

4. In the table below, if $y \propto x$, state the rule for the relationship between x and y.

x	0	1	2	3
y	0	4	8	12

5. In the table below, if $q \propto p$, state the rule for the relationship between p and q.

p	0	3	6	9
q	0	1	2	3

6. In the graph below, y varies directly as x. State the rule for the relationship between x and y.

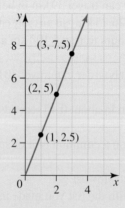

7. **MC** A speed of 60 km/h is equivalent to approximately:
   A. 17 m/ min   **B.** 1 km/s   **C.** 1 m/s
   D. 17 m/s   **E.** 60 m/s

8. **MC** A metal part has a density of 37 g/mm³. If its volume is 6 mm³, it has a mass of:

   A. 6.1 g   **B.** 6.1 kg   **C.** 222 kg

   D. 222 g   **E.** 22.2 g

9. **MC** Identify which of the following is *not* a rate.

   A. 50 km/h   **B.** 70 beats/ min   **C.** 40 kg

   D. Gradient   **E.** Slope

10. **MC** Calculate how long it takes to travel 240 km at an average speed of 60 km/h.

   A. 0.25 hours   **B.** 240 minutes   **C.** 180 hours

   D. 14 400 hours   **E.** 4 minutes

11. **MC** A plane takes two hours to travel 1600 km. Calculate how long it takes to travel 1000 km.

   A. 150 minutes   **B.** 48 minutes   **C.** 2 hours

   D. 75 minutes   **E.** 85 minutes

12. **MC** The graph shows Sandy's bank balance. Determine the length of time (in months) that Sandy has been saving.

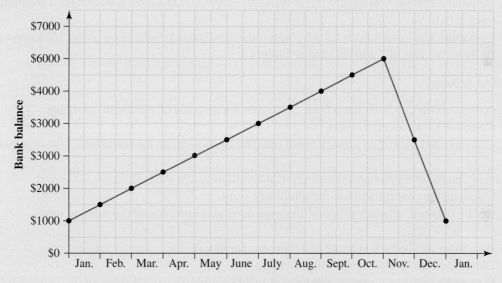

   A. 2 months   **B.** 5 months   **C.** 10 months   **D.** 12 months   **E.** 8 months

13. True or False: The following graph shows that $y \propto x$. Give your reason.

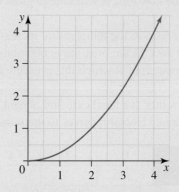

14. If $w \propto v$, and $w = 7.5$ when $v = 5$, calculate k, the constant of proportionality.

15. The wages earned by a worker for different numbers of hours are shown in the table.

Time (hours)	1	2	3	4
Wages ($)	25	50	75	100

Determine the rule relating wages, $w, to the hours worked, t, if $w \propto t$. Give your answer in hours.

16. Calculate the missing quantities in the table.

Mass	Volume	Density
500 g	20 cm³	
1500 g		50 g/cm³
	120 cm³	17 g/cm³

17. Answer the following questions.
    a. A 2000-litre water tank takes two days to fill. Express this rate in litres per hour with 2 decimal places.
    b. Determine how far you have to travel vertically to travel 600 metres horizontally if the gradient of the track is 0.3.
    c. Determine how long it takes for a 60-watt (60 joules/second) light globe to use 100 kilojoules.
    d. Evaluate the cost of 2.3 m³ of sand at $40 per m³.

Problem solving

18. Lisa drove to the city from her school. She covered a distance of 180 km in two hours.
    a. Calculate Lisa's average speed.
    b. Lisa travelled back at an average speed of 60 km/h. Determine the travel time.

19. Seventy grams of ammonium sulfate crystals are dissolved in 0.5 L of water.
    a. Calculate the concentration of the solution in g/mL.
    b. Another 500 mL of water is added. Determine the concentration of the solution now.

20. A skyscraper can be built at a rate of 4.5 storeys per month.
    a. Determine how many storeys will be built in 6 months.
    b. Determine how many storeys will be built in 24 months.

21. A certain kind of eucalyptus tree grows at a linear rate for its first two years of growth. If the growth rate is 5 cm per month, determine how long it will take to grow to a height of 1.07 m.

22. The pressure inside a boiler increases as the temperature increases. For each 1°C, the pressure increases by 10 units. At a temperature of 100 °C, the pressure is 600 units. If the boiler can withstand a pressure of 2000 units, determine the temperature at which this pressure is reached.

23. Hector has a part-time job as a waiter at a local cafe and is paid $8.50 per hour. Complete the table of values relating the amount of money Hector receives to the number of hours he works.

Number of hours	0	2	4	6	8	10
Pay						

24. Karina left home at 9:00 am. She spent some time at a friend's house, then travelled to the airport to pick up her sister. She then travelled straight back home. Her journey is shown by the travel graph.

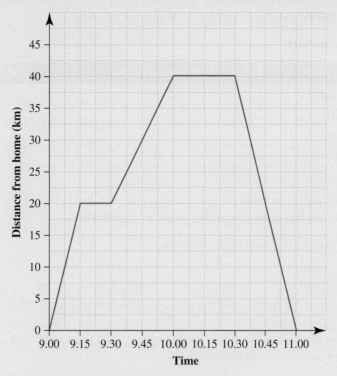

a. Determine how far Karina's house is from her friend's house.
b. Determine how much time Karina spent at her friend's place.
c. Calculate how far the airport is from Karina's house.
d. Determine how much time Karina spent at the airport.
e. Calculate how much time Karina took to drive home.
f. Calculate the average speed of Karina's journey:

   i. from her home to her friend's place
  ii. from her friend's place to the airport
 iii. from the airport to her home.

25. A commercial aircraft covers a distance of 1700 km in 2 hours 5 minutes. Evaluate the distance in metres that the aircraft travels each second.

26. A fun park charges a $10 entry fee and an additional $3 per ride. Complete the following table of values relating the total cost to the number of rides.

Rides	0	2	4	6	8	10
Cost						

Use the following information to answer questions **27–30**.

To help compare speed in different units, a length conversion chart is useful. Time is measured in the same units throughout the world, so a time conversion chart is not necessary.

Length conversion chart

Imperial unit	Conversion factor	Metric unit
Inches (in)	25.4	mm
Feet (ft)	30.5	cm
Yards (yd)	0.915	m
Miles (mi)	1.61	km

To convert an imperial length unit into its equivalent metric unit, multiply by the conversion factor. Divide by the conversion factor when converting from metric to imperial units.

27. a. Use the table to complete each of the following.

    i. 12 in = _____ mm   **ii.** 3 ft = _____ mm

    iii. 1 m = = _____ yd   **iv.** 1 km = _____ mi

  b. Determine which is the faster: a car travelling at 100 km/h or a car travelling at 100 miles/h.

28. The speed limit of 60 km/h in Australia would be equivalent to a speed limit of _____ miles/h in USA. Give your answer correct to three decimal places.

29. A launched rocket covers a distance of 17 miles in 10 seconds. Calculate its speed in km/h.

30. Here are some record speeds for moving objects.

Object	Speed
Motorcycle	149 m/sec
Train	302 miles/h
Human skiing	244 km/h
Bullet from 38-calibre revolver	4000 ft/sec

Convert these speeds to the same unit. Then place them in order from fastest to slowest.

 To test your understanding and knowledge of this topic, go to your learnON title at www.jacplus.com.au and complete the **post-test**.

Online Resources

 Resources

Below is a full list of **rich resources** available online for this topic. These resources are designed to bring ideas to life, to promote deep and lasting learning and to support the different learning needs of each individual.

eWorkbook

Download the workbook for this topic, which includes worksheets, a code puzzle and a project (ewbk-2007) ☐

Solutions

Download a copy of the fully worked solutions to every question in this topic (sol-0727) ☐

Digital documents

7.2 SkillSHEET Measuring the rise and the run (doc-6174) ☐
7.3 SkillSHEET Rounding to a given number of decimal places (doc-6175) ☐
7.4 SkillSHEET Converting time in minutes and seconds into seconds only (doc-10844) ☐
SkillSHEET Equivalent rates (doc-10845) ☐

Video eLessons

7.2 Direct linear proportion (eles-4818) ☐
7.3 Constant of proportionality (eles-4819) ☐
Ratio (eles-4820) ☐
7.4 Rates (eles-4821) ☐
Average speed (eles-4822) ☐
7.5 Constant rates (eles-4823) ☐
Variable rates (eles-4824) ☐

Interactivities

7.2 Individual pathway interactivity: Direct linear proportion (int-4510) ☐
Direct proportion (int-6056) ☐
7.3 Individual pathway interactivity: Proportionality (int-4511) ☐
Constants of proportionality (int-6057) ☐
7.4 Individual pathway interactivity: Introduction to rates (int-4513) ☐
Speed (int-6457) ☐
7.5 Individual pathway interactivity: Constant and variable rates (int-4514) ☐
Constant rates (int-6060) ☐
Variable rates (int-6061) ☐
7.6 Crossword (int-2690) ☐
Sudoku puzzle (int-3208) ☐

Teacher resources

There are many resources available exclusively for teachers online.

To access these online resources, log on to **www.jacplus.com.au**.

Answers

Topic 7 Proportion and rates

Exercise 7.1 Pre-test

1. True
2. A
3. Yes
4. True
5. A
6. $x = \dfrac{k}{y}$
7. a. $k = 2.55$ b. $v = 3.06$ amps
8. False
9. E
10. A
11. B
12. B
13. $1.45 for a pack of 10
14. $y = 4x$
15. 64 km/h

Exercise 7.2 Direct linear proportion

1. a. Yes
   b. No; as speed increases, time decreases.
   c. No; doubling distance doesn't double the cost.
2. a. Yes
   b. No; doubling the side length doesn't double the area.
   c. Yes
3. a. Yes
   b. No; doubling distance doesn't double the cost (due to the initial fee).
   c. No; doubling age doesn't double height.
4. a. True: as the number, n, increases, so does the cost, $C.
   b. True: $C = \pi D$, as C increases with D. π is a constant number.
   c. False: $A = \dfrac{1}{2}\pi r^2$, so it is not linear.
5. E
6. a. No; as the value of x increases, the value of y does not increase by a constant amount.
   b. Yes; as the value of a increases, the value of M increases by a constant amount. For every increase in a by 1, the value of M increases by 8.
   c. No; as the value of x increases, the value of y does not increase by a constant amount.
   d. No; when $n = 0$, C does not equal 0.
7. a. Yes
   b. No; when $t = 0$, d does not.
   c. No; as x doubles, y does not.
   d. No; when $d = 0$, w does not.
8. Sample responses can be found in the worked solutions in the online resources.
9. The point $(0, 0)$ must always exist in a table of values of two variables exhibiting direct proportionality.
10. The missing value is 8. The relationship between n and m can be calculated by the values 5 and 20 in the table $(n = 4m)$.
11. $a = 32, b = 16, c = 36$
12. a. y is doubled.
    b. y is halved.
    c. x is tripled.
13. a. Yes, direct proportion does exist.
    b. $2.55
14. a. Yes   b. $935
15. a. Yes, direct linear proportion does exist.
    b. 6 cans of paint
16. Direct proportion does not exist. The price per metre for the 4.2-metre length is $5.50, and the price per metre for the 5.4-metre length is $5.40.

Exercise 7.3 Proportionality

1. $k = 6, a = 6b$
2. $a = 0.5b$
3. $C = 12.5t$
4. $v = 0.5t$
5. $F = 2.5a$
6. a. $E = 0.625W$ b. 19.2 kg   c. 4.375 cm
7. a. $W = 1.3n$
   b. ≈ 385 almonds
   c. 325 g
8. a. $d = 2.5n$ b. 2017.5 m   c. 2000 turns
9. a. $M = 0.12n$ b. 450 seedlings c. 360 kg
10. $217.60
11. $217.25
12. 13.33 L
13. $105.98
14. 375 g
15. $150
16. $33.75
17. a. $D = 100T$ c. 2 hours 12 minutes
    b. 6 hours 6 minutes   d. 3 hours 54 minutes
18. 275 cm^3

Exercise 7.4 Introduction to rates

1. 64 km/h
2. a. 120 km/h
   b. 1692.3 km/h
3. a. $92 per ticket   c. 10.44 m/s   e. $17.33/h
   b. 24 cakes per hour   d. 3 cm/h   f. $50 per air conditioner
4. 4 g/cm^3
5. 600 g/L
6. The hare runs at 120 000 cm per minute; the tortoise runs at 0.144 km per hour. The hare runs 500 times faster than the tortoise.
7. 75 students/year

8. Either the distance of the journey increases and the time remains constant, or the distance of the journey remains constant and the time decreases.

9. 0.15

10. Cathy Freeman: 29.32 km/h; Usain Bolt: 37.15 km/h

11. 2750 organisms/litre. The beach should not be closed.

Exercise 7.5 Constant and variable rates

1. a. 1 hour at 60 km/h, then 1 hour at 120 km/h
   b. 1 hour at 50 km/h, a 30-minute stop, 30 minutes at 100 km/h, a 1-hour stop, then 1 hour at 100 km/h
   c. 1 hour at 20 km/h, 1 hour at 100 km/h, 1 hour at 40 km/h

2. a.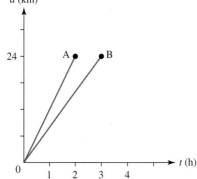
   b. i. A: $d = 12t$; B: $d = 8t$
   ii. A has a steeper gradient than B.

3. a.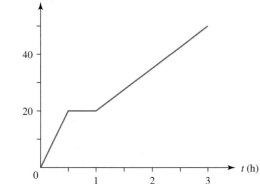
   b.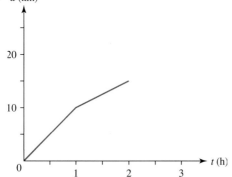

4. a. A. The car is moving with steady speed.
   B. The car is momentarily stationary.
   C. The speed is increasing.
   D. The car is slowing down.
   E. The car is moving at a slow steady speed.

F. The speed is increasing.
   G. The car is moving at a faster steady speed.
   b. i. A, G   ii. B, E   iii. B

5. a. 0.2 km/ min
   b. 0.2 km/ min
   c. Yes, both rates are the same.

6. a. 20, 10, 35, 0 L/ min
   b. No

7. a. 150 km, 200 km and 250 km
   b. 50 km
   c. Both stop for 1 hour.
   d. Brian — 300 km, 3.5 h; Margaret — 300 km, 4 h
   e. Brian — 85.7 km/h; Margaret — 75 km/h

8. a. 2.5 hours
   b. 20 km
   c.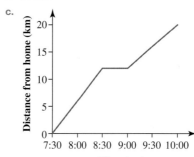

9. a.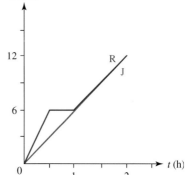
   b. 30 minutes

10. a. i. $10
    ii. $10
    iii. $10
    b. Yes

11. a. 1, 3, 5, 10   b. 4, 6, 7, 9
    c. 2, 8   d. 10
    e. 4

12. a. 30
    b. 30
    c. Variable

13. a. i. 25.8 °C
    ii. 14.0 °C
    b. i. 10.9 °C
    ii. 2.3 °C
    iii. No

Project

1. Usain Bolt = 10.3 m/s
   Petter Northug = 6.7 m/s
   Mika Poutala = 14.3 m/s

2. Mika Poutala

3. Answers will vary. Sample response: the Concorde supersonic airliner, rocket-powered car, Jaguar motorcar, helicopter, airliner, motorcycle, electric car, diesel train, served tennis ball, car ferry, skateboard, bowled cricket ball, windsurfer, submarine, aircraft carrier, solar-powered car.

4. Concorde: 2333.3 km/h
   Airliner: 2587 km/h
   Helicopter: 400 km/h
   Rocket-powered car: 1016 km/h
   Electric car: 393.8 km/h
   Jaguar car: 350 km/h
   Solar-powered car: 78 km/h
   Car ferry: 107 km/h
   Aircraft carrier: 56 km/h
   Submarine: 74 km/h
   Diesel train: 238.1 km/h
   Motorcyclist: 500 km/h
   Skateboarder: 90 km/h
   Windsurfer: 84 km/h
   Bowled cricket ball: 160 km/h
   Served tennis ball: 225 km/h

5. Airliner, Concorde, rocket-powered car, motorcyclist, helicopter, electric car, Jaguar car, diesel train, served tennis ball, bowled cricket ball, oar ferry, skateboarder, windsurfer, solar-powered car, submarine, aircraft carrier

6. Answers will vary. Concorde was the fastest and the submarine was the slowest.

7. Answers will vary. Students could measure the speed of different objects by measuring the distance covered in certain time.

Exercise 7.6 Review questions

1. B
2. D
3. B
4. $y = 4x$
5. $q = \dfrac{1}{3}p$
6. $y = 2.5x$
7. D
8. D
9. C
10. B
11. D
12. C
13. False. The graph starts at $(0, 0)$ and increases, but it is not linear. Therefore, y does not vary directly with x.
14. 1.5
15. $w = 25t$

16.

Mass	Volume	Density
500 g	20 cm³	25 g/cm³
1500 g	30 cm³	50 g/cm³
2040 g	120 cm³	17 g/cm³

17. a. 41.67 L/h b. 180 m
    c. 28 minutes d. $92

18. a. 90 km/h b. 3 h

19. a. 0.14 g/mL b. 0.07 g/mL

20. a. 27 b. 108

21. 21.4 months

22. 240 °C

23.

Number of hours	0	2	4	6	8	10
Pay	$0	$17	$34	$51	$68	$85

24. a. 20 km
    b. 15 min
    c. 40 km
    d. 30 min
    e. 30 min
    f. i. 80 km/h
       ii. 40 km/h
       iii. 80 km/h

25. $226\dfrac{2}{3}$ m/sec

26.

Rides	0	2	4	6	8	10
Cost	$10	$16	$22	$28	$34	$40

27. a. i. 304.8 mm ii. 915 mm
       iii. 1.09 yd iv. 0.62 miles
    b. The one travelling at 100 miles/h

28. 37.267 miles/h

29. 9853.2 km/h

30. The conversions may vary, but the order from fastest to slowest is:
    Bullet from 38-calibre revolver
    Motorcycle
    Train
    Human skiing

8 Financial mathematics

8.1 Overview

Why learn this?

There are many famous sayings about money. You have probably heard that 'money doesn't grow on trees', 'money can't buy happiness' and 'a fool and his money are soon parted'. Most of us have to be conscious of where our money comes from and where it goes to. Understanding the basic principles of finance is very helpful for managing everyday life. Some of you will already be earning money from a part-time or casual job. Do you know what you spend your money on, or where it goes? Do you have your own savings account? Many of you may already be saving for a new games console, your first car or your dream holiday.

In this topic you will investigate different kinds of employment as well as different investment options for saving your money. Every branch of industry and business, whether large or small, international or domestic, will have to pay their employees either an annual salary or a wage based on an hourly rate. Financial investments are another way for businesses and individuals to make money — it is important to understand how investments work to be able to decide whether an investment is a good idea or not.

A sound knowledge of financial mathematics is essential in a range of careers, including financial consultancy, accountancy, business management and pay administration.

Where to get help

Go to your learnON title at **www.jacplus.com.au** to access the following digital resources. The Online Resources Summary at the end of this topic provides a full list of what's available to help you learn the concepts covered in this topic.

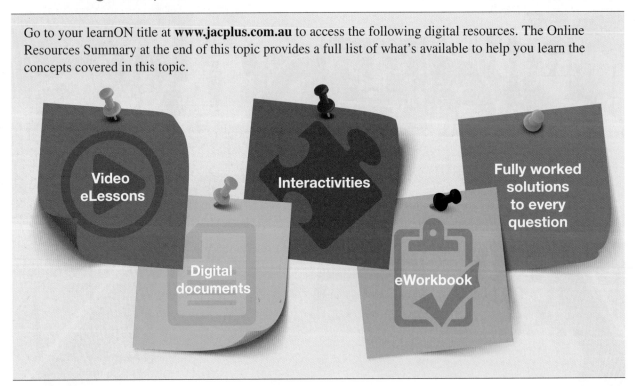

Complete this pre-test in your learnON title at www.jacplus.com.au and receive **automatic marks**, **immediate corrective feedback** and **fully worked solutions**.

1. Kira has an annual salary of $85 450.70 (without tax).
   Calculate how much she is paid:
   a. weekly   b. fortnightly   c. monthly.

2. Ishmael has a casual job at a department store. He is paid $11.50 per hour. He gets double time on a Sunday.
   Calculate his wage for a week in which he worked from 4 pm to 8 pm on Thursday and 12 pm to 6 pm on Sunday.

3. Jose is a tutor. He receives $975 for a week in which he works 30 hours. Calculate his hourly rate of pay.

4. **MC** This table shows a timesheet for Yumi, who works in an electronics store.

Day	Pay rate	Start time	Finish time
Monday	Normal	9 am	1 pm
Tuesday	Normal	9 am	5 pm
Wednesday	Normal	11 am	7 pm
Thursday	Normal	2 pm	7 pm
Friday	Normal	4 pm	8 pm
Saturday	Time-and-a-half	9 am	2 pm
Sunday	Double	12 pm	6 pm

   If Yumi's normal hourly rate is $15.30, the amount she earned for the week is:
   A. $612.00   **B.** $734.00   **C.** $818.55   **D.** $742.05   **E.** $902.70

5. Anita was offered a job with a salary of $65 000 per year, or $32.50 per hour. Assuming that she works a 40-hour week, compare the two rates to decide which is the better pay.

6. Ali earns $20.40 an hour. He needs to earn a minimum of $710 each week for his living expenses. Calculate the minimum number of hours he has to work each week. Give your answer to the nearest hour.

7. Mawluda is a casual worker in a café. She is paid time-and-a-half for working on Saturday and double time for working on Sunday. Her normal pay rate is $18.75 an hour. Over the weekend, she worked 8 hours on Saturday and 5 hours on Sunday.
   Calculate how much Mawluda earned over the weekend.

8. A bank has a special offer on a savings account of 3% p.a. simple interest.
   Peter opens an account with $800 and leaves his money there for 5 months.
   a. Calculate how much interest he earned over the 5 months.
   b. Calculate the balance of Peter's account after the 5 months.

9. **MC** If an investment of $40 000 pays 8% simple interest per year, then the value of the investment at the end of 3 years is:
   A. $3 200   B. $9 600   C. $43 200   D. $49 600   E. $50 390

10. The Liang family wants to buy a family car at a cost of $22 700. They pay a deposit of $4500 and borrow the balance at an interest rate of 11.5% p.a.
    The loan will be paid off with 48 equal monthly payments.
    a. Calculate how much interest they will pay.
    b. Calculate the total cost of the car, including the interest paid.
    c. Calculate the amount of each repayment.

11. **MC** If $10 000 is invested at a simple interest rate of 6% per annum, select the correct formula for calculating the interest earned for 6 months.
    A. $I = 10\,000 \times \dfrac{6}{100} \times 6$ B. $I = 10\,000 \times 0.6 \times 6$ C. $I = 10\,000 \times \dfrac{6}{100} \times \dfrac{1}{2}$
    D. $I = 10\,000 \times (1.6)^6$ E. $I = 10\,000 \times (1.06)^6$

12. If the simple interest charged on a loan of $6400 over 30 months was $1450, calculate the percentage rate of interest charged. Give your answer to the nearest whole percentage.

13. **MC** The amount of interest earned if $112 000 is invested for 4 years at 6.8% p.a. interest compounded annually is:
    A. $23 947.14   B. $30 464.00   C. $33 714.59   D. $142 464.00   E. $145 714.59

14. **MC** The total value of $3500 invested for 18 months at 12% p.a. compounding monthly can be calculated using:
    A. $A = 3500(1.12)^{18}$ B. $A = 3500(1.12)^{1.5}$ C. $A = 3500(1.01)^{18}$
    D. $A = 3500(1.01)^{1.5}$ E. $A = 3500 \times \dfrac{12}{100} \times 18$.

15. Alecia invests $5000 at 15% p.a. compounded annually and Cooper invests $5000 at a 15% p.a. flat rate. Compare the two investments over a period of 10 years to determine how much Alecia's investment will be worth compared to Cooper's.

8.2 Salaries and wages

8.2.1 Employees' salaries and wages

eles-4836

- Employees may be paid for their work in a variety of ways — most receive either a **wage** or a **salary**.

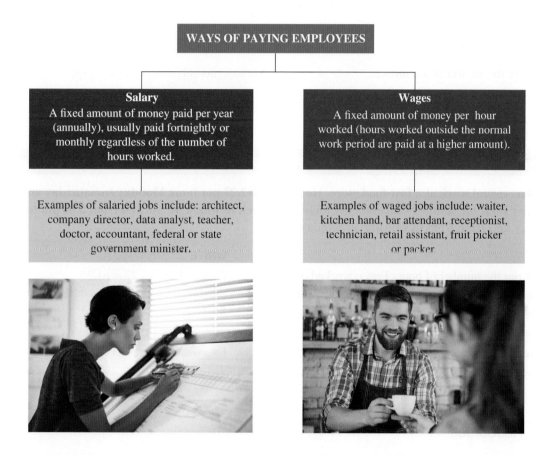

WAYS OF PAYING EMPLOYEES

Salary
A fixed amount of money paid per year (annually), usually paid fortnightly or monthly regardless of the number of hours worked.

Wages
A fixed amount of money per hour worked (hours worked outside the normal work period are paid at a higher amount).

Examples of salaried jobs include: architect, company director, data analyst, teacher, doctor, accountant, federal or state government minister.

Examples of waged jobs include: waiter, kitchen hand, bar attendant, receptionist, technician, retail assistant, fruit picker or packer.

Key points

- Normal working hours in Australia are **38 hours** per week.
- There are **52 weeks** in a year.
- There are **26 fortnights** in a year (this value is slightly different for a leap year).
- There are **12 months** in a year.

WORKED EXAMPLE 1 Calculating pay from annual salary

Susan has an annual salary of $63 048.92. Calculate how much she is paid:

a. weekly　　　　　　　**b. fortnightly**　　　　　　　**c. monthly.**

THINK

a. 1. Annual means per year, so divide the salary by 52 because there are 52 weeks in a year.

2. Write the answer in a sentence.

b. 1. There are 26 fortnights in a year, so divide the salary by 26.

2. Write the answer in a sentence.

c. 1. There are 12 months in a year, so divide the salary by 12.

2. Write the answer in a sentence.

WRITE

a. Weekly salary $= 63\,048.92 \div 52$
≈ 1212.48

Susan's weekly salary is $1212.48.

b. Fortnightly salary $= 63\,048.92 \div 26$
≈ 2424.96

Susan's fortnightly salary is $2424.96.

c. Monthly salary $= 63\,048.92 \div 12$
≈ 5254.08

Susan's monthly salary is $5254.08.

WORKED EXAMPLE 2 Calculating wage given hourly rate of pay

Frisco has casual work at a fast-food store. He is paid $12.27 per hour Monday to Saturday and $24.54 per hour on Sunday. Calculate his wage for a week in which he worked from 5.00 pm to 10 pm on Friday and from 6 pm to 9.00 pm on Sunday.

THINK

1. Work out the number of hours Frisco worked each day. He worked 5 hours on Friday and 3 hours on Sunday.

2. Calculate the total amount earned by adding the wages earned on Friday and Sunday.

3. Write the answer in a sentence.

WRITE

Friday: $5 \times 12.27 = 61.35$
Sunday: $3 \times 24.54 = 73.62$

$61.35 + 73.62 = 134.97$

Frisco's wage was $134.97.

 Resources

 eWorkbook　　Topic 8 Workbook (worksheets, code puzzle and project) (ewbk-2008)

 Digital documents SkillSHEET Converting units of time (doc-10849)
SkillSHEET Multiplying and dividing a quantity (money) by a whole number (doc-10850)
SkillSHEET Multiplying and dividing a quantity (money) by a fraction (doc-10851)
SkillSHEET Increasing a quantity by a percentage (doc-10852)
SkillSHEET Adding periods of time (doc-10853)

 Interactivities　　Individual pathway interactivity: Salaries and wages (int-4520)
Salaries (int-6067)

Exercise 8.2 Salaries and wages

Individual pathways

■ PRACTISE	■ CONSOLIDATE	■ MASTER
1, 3, 7, 10, 13, 14, 17	2, 4, 8, 11, 15, 18	5, 6, 9, 12, 16, 19, 20

To answer questions online and to receive **immediate corrective feedback** and **fully worked solutions** for all questions, go to your learnON title at www.jacplus.com.au.

Fluency

1. **WE1** Johann has an annual salary of $57 482. Calculate how much he is paid:

   a. weekly   b. fortnightly   c. monthly.

2. Nigara earns $62 300 per annum. Calculate how much she earns:

   a. weekly   b. fortnightly   c. monthly.

3. Calculate the annual salary of workers with the following weekly incomes.

   a. $368   b. $892.50   c. $1320.85

4. Calculate how much is earned per annum by people who are paid fortnightly salaries of:

   a. $995   b. $1622.46   c. $3865.31.

5. Compare the following salaries, and determine which of each pair is the higher.

   a. $3890 per month or $45 700 per annum
   b. $3200.58 per fortnight or $6700 per month

6. Calculate the hourly rate for these workers.

   a. Rahni earns $98.75 for working 5 hours.

   b. Francisco is paid $54.75 for working $4\frac{1}{2}$ hours.

   c. Nhan earns $977.74 for working a 38-hour week.

   d. Jessica works $7\frac{1}{2}$ hours a day for 5 days to earn $1464.75.

Understanding

7. Henry is a second-year apprentice motor mechanic. He receives an award wage of $12.08 per hour. Jenny, a fourth-year apprentice, earns $17.65 per hour.

   a. Calculate how much Henry earns in a 38-hour week.
   b. Determine how much more Jenny earns in the same period of time.

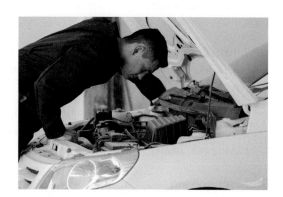

8. **WE2** Juan has casual work for which he is paid $13.17 per hour Monday to Saturday and $26.34 per hour on Sundays. Calculate his total pay for a week in which he worked from 11 am to 5 pm on Thursday and from 2 pm to 7 pm on Sunday.

9. Mimi worked the following hours in one week.

Wednesday	5 pm to 9 pm
Thursday	6 pm to 9 pm
Friday	7 pm to 11 pm

If her pay is $21.79 per hour up to 9 pm and $32.69 per hour after that, calculate her total pay for that week.

10. Decide who earns more money each week: Rhonda, who receives $38.55 an hour for 38 hours of work, or Rob, who receives $41.87 an hour for 36 hours of work.

11. Glenn is a chef. He receives $1076.92 for a week in which he works 35 hours. Calculate his hourly rate of pay.

12. Zack and Kaylah work in different department stores. Zack is paid $981.77 per week. Kaylah is paid $26.36 per hour. Calculate how many hours Kaylah must work to earn more money than Zack.

13. Calculate what pay each of the following salary earners will receive for each of the periods specified.

    a. Annual salary $83 500, paid each week
    b. Annual salary $72 509, paid each fortnight
    c. Annual salary $57 200, paid each week
    d. Annual salary $105 240, paid each month

Reasoning

14. **MC** When Jack was successful in getting a job as a trainee journalist, he was offered the following choice of four salary packages. Decide which one Jack should choose, and justify your answer.

    A. $456 per week
    B. $915 per fortnight
    C. $1980 per calendar month
    D. $23 700 per year
    E. $356 per week

15. Julie is considering two job offers for work as a receptionist. Job A pays $878.56 for a 38-hour working week. Job B pays $812.16 for a 36-hour working week. Determine which job has the higher hourly rate of pay. Justify your answer.

16. In his job as a bookkeeper, Minh works 38 hours per week and is paid $32.26 per hour. Michelle, who works 38 hours per week in a similar job, is paid a salary of $55 280 per year. Decide who has the higher-paying job. Justify your answer.

Problem solving

17. A lawyer is offered a choice between two jobs: one with a salary of $74 000 per year and a second that pays $40 per hour. Assuming that the lawyer will work 80 hours every fortnight, decide which job pays the most. Justify your answer.

18. Over the last four weeks Shahni has worked 35, 36, 34 and 41 hours. If she earns $24.45 per hour, calculate how much she earned for each of the two fortnights.

19. Jackson works a 40-hour week (8 hours a day, Monday–Friday) and earns $62 000 per annum.
    a. Calculate Jackson's hourly rate.
    b. If he works on average an extra half an hour per day from Monday to Friday and then another 4 hours over the weekend (for the same annual salary), explain how his hourly rate is affected.
    c. If Jackson was earning the hourly rate and was being paid for every hour worked, calculate his potential earnings for the year.

20. Mark saves $10 per week. Phil saves 5 cents in the first week, 10 cents the second week, and doubles the amount each week from that time on. Determine how many weeks it will take for Phil's savings to be worth more than Mark's.

8.3 Penalty rates

LEARNING INTENTION

At the end of this subtopic you should be able to:
- calculate total wages when overtime or penalty rates are included
- calculate hours paid when time-and-a-half or double time is involved.

8.3.1 Overtime and penalty rates

eles-4837

- **Overtime** is paid when a wage earner works more than their regular hours each week. These additional payments are often referred to as penalty rates.
- Penalty rates are usually paid for working on weekends, public holidays or at night.
- The extra hours are paid at a higher hourly rate, normally calculated at either time-and-a-half or double time.

Calculating overtime

The **overtime** hourly rate is usually a multiple of the regular hourly rate. Some examples of overtime rates include:
- **1.5** × regular hourly wage (**time-and-a-half**)
- **2** × regular hourly wage (**double time**)
- **2.5** × regular hourly wage (**double time-and-a-half**).

Ursula works as a waitress and earns $23.30 per hour. Last week she received the normal rate for 30 hours of work as well as time-and-a-half for 3 hours of overtime and double time for 5 hours of work on Sunday. Calculate her total pay.

THINK	WRITE
1. Calculate Ursula's normal pay for 30 hours.	Normal pay: $30 \times 23.30 = 699.00$
2. Calculate Ursula's pay for 3 hours at time-and-a-half (1.5 × regular hourly wage)	Overtime: $3 \times 1.5 \times 23.30 = 104.85$
3. Calculate Ursula's pay for 5 hours at double time (2 × regular hourly wage).	Sunday: $5 \times 2 \times 23.30 = 233.00$
4. Calculate the total amount by adding normal pay and overtime pay.	Total $= 699.00 + 104.85 + 233.00$ $= 1036.85$
5. Write the answer in a sentence.	Ursula's total pay was $1036.85.

▶ 8.3.2 Time sheets and pay slips

eles-4398

- Employers often use records to monitor the number of working hours of their employees.

FINANCIAL RECORD-KEEPING

Time sheets
Provide details of the number of hours worked by each employee, including overtime

Pay slips
Provide details of hours worked and the rate for each employee

Fiona works in a department store, and in the week before Christmas she works overtime. Her time sheet is shown. Fill in the details on her pay slip.

	Start	Finish	Normal hours	Over time 1.5
M	9:00	15:00	6	
T	9:00	17:00	8	
W	9:00	17:00	8	
T	9:00	19:00	8	2
F	9:00	19:00	8	2
S				

Pay slip for: Fiona Lee	Week ending December 21
Normal hours	
Normal rate	$17.95
Overtime hours	
Overtime rate	
Total wage	

THINK

1. Calculate the number of normal hours worked by adding the hours worked on each day of the week.

2. Calculate the number of overtime hours worked by adding the overtime hours worked on Thursday and Friday.

3. Calculate the overtime rate (1.5 × regular hourly wage).

4. Calculate the total pay by multiplying the number of normal hours by the normal rate and adding the overtime amount (calculated by multiplying the number of overtime hours by the overtime rate).

5. Fill in the amounts on the pay slip.

WRITE

Normal hours: $6 + 8 + 8 + 8 + 8 = 38$

Overtime hours: $2 + 2 = 4$

$$\text{Overtime rate} = 1.5 \times 17.95$$
$$= 26.93$$

$$\text{Total pay} = 38 \times 17.95 + 4 \times 26.93$$
$$= 789.82$$

Pay slip for: Fiona Lee	Week ending December 21
Normal hours	38
Normal rate	$17.95
Overtime hours	4
Overtime rate	$26.93
Total wage	$789.82

COLLABORATIVE TASK: Comparing pay rates and conditions

Working in small groups, collect several job advertisements from a mixture of print and digital sources. Compare the pay rates and conditions for the different positions and present a report to the class.

Exercise 8.3 Penalty rates

learnon

Individual pathways

■ PRACTISE	■ CONSOLIDATE	■ MASTER
1, 2, 6, 9, 12, 15	3, 4, 7, 10, 13, 16	5, 8, 11, 14, 17

To answer questions online and to receive **immediate corrective feedback** and **fully worked solutions** for all questions, go to your learnON title at www.jacplus.com.au.

Fluency

1. Calculate the following penalty rates:

   a. time-and-a-half when the hourly rate is $15.96
   b. double time when the hourly rate is $23.90
   c. double time-and-a-half when the hourly rate is $17.40.

2. Calculate the following total weekly wages:

   a. 38 hours at $22.10 per hour, plus 2 hours at time-and-a-half
   b. 40 hours at $17.85 per hour, plus 3 hours at time-and-a-half
   c. 37 hours at $18.32 per hour, plus 3 hours at time-and-a-half and 2 hours at double time.

3. Julio is paid $956.08 for a regular 38-hour week. Calculate:

   a. his hourly rate of pay
   b. how much he is paid for 3 hours of overtime at time-and-a-half rates
   c. his wage for a week in which he works 41 hours.

4. **WE3** Geoff is a waiter in a cafe and works 8 hours most days. Calculate what he earns for 8 hours work on the following days:

   a. a Monday, when he receives his standard rate of $21.30 per hour
   b. a Sunday, when he is paid double time
   c. a public holiday, when he is paid double time-and-a-half.

5. Albert is paid $870.58 for a 38-hour week. Determine his total wage for a week in which he works 5 extra hours on a public holiday with a double-time-and-a-half penalty rate.

Understanding

6. Jeleesa (aged 16) works at a supermarket on Thursday nights and weekends. The award rate for a 16-year-old is $7.55 per hour. Calculate what she would earn for working:

   a. 4 hours on Thursday night
   b. 6 hours on Saturday
   c. 4 hours on Sunday at double time
   d. the total of the 3 days described in parts a, b and c.

7. Jacob works in a pizza shop and is paid $13.17 per hour.

   a. Jacob is paid double time-and-a-half for public holiday work. Calculate what he earns per hour on public holidays. Give your answer to the nearest cent.

   b. Calculate Jacob's pay for working 6 hours on a public holiday.

8. If Bronte earns $7.80 on normal time, calculate how much she receives per hour:

   a. at time-and-a-half   b. at double time   c. at double time-and-a-half.

9. **WE4** a. Complete the time sheet shown. Calculate the number of hours Susan worked this week.

Day	Pay rate	Start time	Finish time	Hours worked
Monday	Normal	9:00 am	5:00 pm	
Tuesday	Normal	9:00 am	5:00 pm	
Wednesday	Normal	9:00 am	5:00 pm	
Thursday	Normal	9:00 am	5:00 pm	
Friday	Normal	9:00 am	3:00 pm	

   b. Complete Susan's pay slip for this week.

Pay slip for: Susan Jones	Week ending 17 August
Normal hours	
Normal pay rate	$25.60
Overtime hours	0
Overtime pay rate	$38.40
Total pay	

10. a. Manu works in a department store. His time sheet is shown. Complete the table.

Day	Pay rate	Start time	Finish time	Hours worked
Monday	Normal	9:00 am	5:00 pm	
Tuesday	Normal	9:00 am	5:00 pm	
Wednesday	Normal	—		
Thursday	Normal	1:00 pm	9:00 pm	
Friday	Normal	—		
Saturday	Time-and-a-half	8:00 am	12:00 pm	

   b. Complete Manu's pay slip for this week.

Pay slip for: Manu Taumata	Week ending 21 December
Normal hours	
Normal pay rate	$10.90
Overtime hours	
Overtime pay rate (time-and-a-half)	
Total pay	

11. a. Eleanor does shift work. Copy and complete their time sheet.

Day	Pay rate	Start time	Finish time	Hours worked
Monday	Normal	7:00 am	3:00 pm	
Tuesday	Normal	7:00 am	3:00 pm	
Wednesday				
Thursday				
Friday	Normal	11:00 pm	7:00 am	
Saturday	Time-and-a-half	11:00 pm	7:00 am	
Sunday	Double time	11:00 pm	7:00 am	

b. Copy and complete Eleanor's pay slip for the week.

Pay slip for: Eleanor Rigby	Week ending 15 September
Normal hours	
Normal pay rate	$16.80
Time-and-a-half hours	
Time-and-a-half pay rate	
Double time hours	
Double time pay rate	
Gross pay	

Reasoning

12. Calculate the following total weekly wages:

    a. 38 hours at $18.40 per hour, plus 2 hours at time-and-a-half

    b. 32 hours at $23.70 per hour plus 6 hours on a Sunday at double time.

13. Ruby earns $979.64 for her normal 38-hour week, but last week she also worked 6 hours of overtime at time-and-a-half rates.

    a. Calculate how much extra she earned and give a possible reason for her getting time-and-a-half rates.

    b. Calculate Ruby's total wage.

14. A standard working week is 38 hours and a worker puts in 3 hours overtime at time-and-a-half, plus 2 hours at double time. Calculate the amount of standard work hours equivalent to the total time they have worked.

Problem solving

15. Joshua's basic wage is $22 per hour. His overtime during the week is paid at time-and-a-half. Over the weekend he is paid double time. Calculate Joshua's gross wage in a week when he works his basic 40 hours, together with 1 hour overtime on Monday, 2 hours overtime on Wednesday and 4 hours overtime on Saturday.

16. Lin works 32 hours per week at $22 per hour and is paid overtime for any time worked over the 32 hours per week. In one week Lin worked 42 hours and was paid $814. Overtime is paid at 1.5 times the standard wage. Determine whether Lin was paid the correct amount. If not, then provide the correct amount.

17. The table shows the pay sheet for a small company. If a person works up to 36 hours, the regular pay is $14.50 per hour.

For hours over 36 and up to 40, the overtime is time-and-a-half.

For hours over 40, the overtime is double time.

Complete the table.

	Hours worked	Regular pay	Overtime pay	Total pay
a.	32			
b.	38.5			
c.	40.5			
d.	47.2			

8.4 Simple interest

LEARNING INTENTION

At the end of this subtopic you should be able to:
- calculate the simple interest on a loan or an investment
- apply the simple interest formula to determine the time, the rate or the principal.

8.4.1 Principal and interest

eles-4399

- When you put money in a financial institution such as a bank or credit union, the amount of money you start with is called the **principal**.
- **Interest** is the fee charged for the use of someone else's money.
- Investors are people who place money in a financial institution — they receive interest from the financial institution in return for the use of their money.
- Borrowers pay interest to financial institutions in return for being given money as a loan.
- The amount of interest is determined by the interest rate.
- Interest rates are quoted as a percentage over a given time period, usually a year.
- The total amount of money, which combines the principal (the initial amount) and the interest earned, is known as the value of the investment.

Value of an investment or loan

$$A = P + I$$

where:
- A = the total amount of money at the end of the investment or loan period
- P = the principal (the initial amount invested or borrowed)
- I = the amount of interest earned.

Sam borrowed **$450 000** from a bank for his business. **After 30 years Sam had repaid his loan, with a total of $500 000 paid to the bank. Calculate how much interest Sam paid the bank.**

THINK	WRITE
1. Write the known quantities: principal (P) and the total amount (A).	$P = \$450\,000$ $A = \$500\,000$
2. Write the formula to calculate interest paid.	$A = P + I$
3. Substitute the given values into the formula.	$500\,000 = 450\,000 + I$
4. Rearrange the formula to calculate the value of interest.	$I = 500\,000 - 450\,000$ $= 50\,000$
5. Write the answer in a sentence.	Sam paid \$50 000 in interest to the bank.

▶ 8.4.2 Simple interest

eles-4400

- **Simple interest** is the interest paid based on the principal of an investment.
- The principal remains constant and does not change from one period to the next.
- Since the amount of interest paid for each time period is based on the principal, the amount of interest paid is constant.
- Simple interest is usually used for short-term assets or loans.
  - For example, $500 is placed in an account that earns 10% p.a. simple interest every year.
  - This means $50 is paid each year (10% of $500 = $50).
  - After 4 years the total interest paid is $4 \times 50 = \$200$, so the value of the investment after 4 years is $500 + 200 = \$700$.

Time period (years)	Amount of money at the start of the year	Interest for the year
1	$500	$50
2	$550	$50
3	$600	$50
4	$650	$50

- The rate of interest on the investment or loan is i % p.a. (per annum or per year).
- The duration of the investment or loan is n years.

Formula for simple interest

$$I = Pin$$

where:
- $I =$ the amount of interest earned or paid
- $P =$ the principal
- $i =$ the interest rate as a **percentage** per annum (yearly), written as a decimal (e.g. 2% p.a. is equal to 0.02 p.a.)
- $n =$ the duration of the investment in years.

WORKED EXAMPLE 6 Calculating simple interest

Calculate the amount of simple interest earned over 4 years on an investment of $35 000 that returns 2.1% p.a.

THINK	WRITE
1. Write the simple interest formula.	$I = Pin$
2. State the known values of the variables.	$P = \$35\,000$ $i = 0.021$ $n = 4$
3. Substitute the given values into the formula and evaluate the amount of interest.	$I = 35\,000 \times 0.021 \times 4$ $I = 2940$
4. Write the answer in a sentence.	The amount of simple interest over 4 years is $2940.

• Note that the time, n, is in years. Months need to be expressed as a fraction of a year. For example, 27 months = 2 years and 3 months = 2.25 years.

WORKED EXAMPLE 7 Calculating the time of an investment

Karine invests $2000 at a simple interest rate of 4% p.a. Calculate how long she needs to invest this money to earn $120 in interest.

THINK	WRITE
1. Write the simple interest formula.	$I = Pin$
2. State the known values of the variables.	$P = \$2000$ $i = 0.04$ $n = \$120$
3. Substitute the given values into the formula and rearrange to calculate the value of the time.	$120 = 2000 \times 0.04 \times n$ $120 = 80n$ $\dfrac{120}{80} = \dfrac{80n}{80}$
4. Change the years into years and months.	$n = 1.5$ $n = 1$ year and 6 months
5. Write the answer in a sentence.	Karine will need to invest the $2000 for 1 year and 6 months

WORKED EXAMPLE 8 Calculating the total amount of a loan

Zac borrows \$3 000 for 2 years at 9% p.a. simple interest.
a. Calculate how much interest Zac is charged over the term of the loan.
b. Calculate the total amount Zac must repay.

THINK	WRITE
a. 1. Write the simple interest formula.	a. $I = Pin$
2. State the known values of the variables.	$P = \$3000$ $i = 0.09$ $n = 2$
3. Substitute the given values into the simple interest formula and evaluate the value of interest.	$I = 3000 \times 0.09 \times 2$ $I = 540$
4. Write the answer in a sentence.	Zac is charged \$540 in interest.
b. 1. State the relationship for the value of a loan.	b. $A = P + I$ $A = 3000 + 540$
2. Substitute the values for P and I and calculate the total amount.	$A = 3540$
3. Write the answer in a sentence.	To repay the loan, Zac must pay \$3540 in total.

WORKED EXAMPLE 9 Calculating simple interest on an item

The Carlon-Tozer family needs to buy a new refrigerator at a cost of
\$1679. They will pay a deposit of \$200 and borrow the balance at an
interest rate of 19.5% p.a. The loan will be paid off in 24 equal monthly
payments.

a. Calculate how much money the Carlon-Tozers need to borrow.
b. Calculate how much interest they will pay.
c. Determine the total cost of the refrigerator.
d. Calculate the amount of each payment.

THINK	WRITE
a. 1. Subtract the deposit from the cost to calculate the amount still owing.	a. $1679 - 200 = 1479$
2. Write the answer in a sentence.	They must borrow \$1479.
b. 1. State the known values of the variables and write the simple formula.	b. $P = 1479$ $i = 0.195$ $n = 2$ $I = Pin$
2. Substitute the values into the formula to calculate interest (I), by multiplying the values.	$= 1479 \times 0.195 \times 2$ $= 576.81$
3. Write the answer in a sentence.	The interest will be \$576.81.

c. 1. Add the interest to the initial cost to calculate the total cost of the refrigerator.

2. Write the answer in a sentence.

c. $1679.00 + 576.81 = 2255.81$

The total cost of the refrigerator will be $2255.81.

d. 1. Subtract the deposit from the total cost to calculate the amount to be repaid.

2. Divide the total payment into 24 equal payments to calculate each repayment.

3. Write the answer in a sentence.

d. $2255.81 - 200 = 2055.81.$

$2055.81 \div 24 = 85.66$

Each payment will be $85.66.

Technology and simple interest

- Digital technologies such as Microsoft Excel can be helpful in understanding the effects of changing P, i and n. Alternatively, you can investigate these effects using the interactivities in the Resources tab for this subtopic.

COLLABORATIVE TASK: Interesting time periods

1. In groups, complete the table below for the pair of values $P = \$100$ and $R = 6\%$, with interest calculated and paid:

   a. each year
   b. every 6 months
   c. every 4 months
   d. every 3 months
   e. every 2 months
   f. every month.

Time period	2 years	3 years	5 years	7 years	10 years	15 years
Value of the investment						

2. As a class, discuss the effect of increasing the frequency of payments (for example, by paying interest monthly) at the same time as decreasing the interest rate (for example, by dividing it by 12). Would these changes make a difference to the final investment value? Why?

Exercise 8.4 Simple interest

Individual pathways

■ PRACTISE	■ CONSOLIDATE	■ MASTER
1, 4, 6, 7, 10, 13, 18, 22	2, 5, 8, 11, 14, 17, 19, 20, 23, 24	3, 9, 12, 15, 16, 21, 25, 26

To answer questions online and to receive **immediate corrective feedback** and **fully worked solutions** for all questions, go to your learnON title at www.jacplus.com.au.

Fluency

1. **WE5** Alecia invests $400. After 3 years, she withdraws the total value of her investment, which is $496. Calculate the simple interest she earned per year.

2. **WE6** Calculate the amount of simple interest earned on an investment of $4500:

   a. returning 3.5% p.a. for 6 years
   b. returning 2.75% p.a. for 4 years
   c. returning 7.5% p.a. for 2 years
   d. returning 2.5% p.a. for 8 years.

3. Calculate the amount of simple interest earned on an investment of:

   a. $2000, returning 4.5% p.a. for 3 years and 6 months
   b. $18 200, returning 3.6% p.a. for 6 months
   c. $460, returning 2.15% p.a. for 2 years and 3 months
   d. $6700, returning 3.2% p.a. for 7 years.

4. **WE7** Calculate how long it will take an investment of $12 000 to earn the following:

   a. $1260 interest at a simple interest rate of 3.5% p.a.
   b. $2520 interest at a simple interest rate of 4.2% p.a.
   c. $405 interest at a simple interest rate of 2.25% p.a.
   d. $1485 interest at a simple interest rate of 4.5% p.a.

5. Calculate the simple interest rate per annum if:

   a. $3000 earns $270 in 3 years
   b. $480 earns $16.20 in 9 months
   c. $5500 earns $660 in 2.5 years
   d. $2750 earns $748 in 4 years.

6. **MC** If the total interest earned on a $6000 investment is $600 after 4 years, then the annual interest is:
   A. 10%   B. 7.5%   C. 5%   D. 4%   E. 2.5%

7. **WE8** Monique borrows $5000 for 3 years at 8% per annum simple interest.
   a. Calculate how much interest she is charged.   b. Calculate the total amount she must repay.

8. Calculate the simple interest earned on an investment of $15 000 at 5.2% p.a. over 30 months.

9. For each loan in the table, calculate:

   i. the simple interest
   ii. the amount repaid.

	Principal ($)	Interest rate per annum	Time
a.	1000	5%	2 years
b.	4000	16%	3 years
c.	8000	4.5%	48 months
d.	2700	3.9%	2 years 6 months
e.	15 678	9.2%	42 months

10. Calculate the final value of each of the following investments.

    a. $3000 for 2 years at 5% p.a.
    b. $5000 for 3 years at 4.3% p.a.

11. Hasim borrows $14 950 to buy a used car. The bank charges a 9.8% p.a. flat rate of interest over 60 months.

    a. Calculate the total amount he must repay.
    b. Calculate the monthly repayment.

12. Carla borrows $5200 for an overseas trip at 8.9% p.a. simple interest over 30 months. If repayment is made in equal monthly instalments, calculate the amount of each instalment.

Understanding

13. Janan invested $2000 at a simple interest rate of 4% p.a. Calculate how long he needs to invest it in order to earn $200 in interest.

14. If Jodie can invest her money at 8% p.a., calculate how much she needs to invest to earn $2000 in 2 years.

15. If the simple interest charged on a loan of $9800 over 3 years is $2352, determine the percentage rate of interest that was charged.

16. Calculate the missing quantity in each row of the table shown.

	Principal	Rate of interest p.a.	Time	Interest earned
a.	$2000	6%		$240.00
b.	$3760	5.8%		$545.20
c.		7%	3 years	$126.00
d.		4.9%	1 year 9 months	$385.88
e.	$10 000		$1\frac{1}{2}$ years	$1200.00
f.	$8500		42 months	$1041.25

17. **WE9** Mika is buying a used car for $19 998. He has a deposit of $3000 and will pay the balance in equal monthly payments over 4 years. The simple interest rate will be 12.9% p.a.

   a. Calculate how much money he needs to borrow.
   b. Calculate how much interest he will pay.
   c. Determine the total cost of the car.
   d. Determine how many payments he will make.
   e. Calculate the cost of each payment.

Reasoning

18. If a bank offers interest on its savings account of 4.2% p.a. and the investment is invested for 9 months, explain why 4.2 is not substituted into the simple interest formula as the interest rate.

19. Theresa invests $4500 at 5.72% per annum in an investment that attracts simple interest for 6 months. Show that at the end of 6 months she should expect to have $4628.70.

20. A $269 000 business is purchased on an $89 000 deposit with the balance payable over 5 years at an 8.95% p.a. flat rate.

   a. Calculate how much money is borrowed to purchase this business.
   b. Calculate how much interest is charged.
   c. Calculate the total amount that must be repaid.
   d. Determine the size of each of the equal monthly repayments.
   e. Explain two ways in which these payments could be reduced.

21. Nam has $6273 in his bank account at a simple interest rate of 4.86% per annum. After 39 days he calculates that he will have $6305.57 in his account. Did Nam calculate his interest correctly? Justify your answer by showing your calculations.

Problem solving

22. Giang is paid $79.50 in interest for an original investment of $500 for 3 years. Calculate the annual interest rate.

23. A loan is an investment in reverse: you borrow money from a bank and are charged interest. The value of a loan becomes its total cost.
   Jitto wishes to borrow $10 000 from a bank, which charges 11.5% interest per year. If the loan is over 2 years:

   a. calculate the total interest paid
   b. calculate the total cost of the loan.

24. A new sound system costs $3500, but it can be purchased for no deposit, followed by 48 equal monthly payments, at a simple interest rate of 16.2% p.a.

   a. Determine the total cost of the sound system.
   b. Under a 'no deposit, no payment for 2 years' scheme, 48 payments are still required, but the first payment isn't made for 2 years. This will stretch the loan over 6 years. Calculate how much the system will cost using this scheme.
   c. Determine the monthly payments under each of these two schemes.

25. For the following questions, assume that the interest charged on a home loan is simple interest.

    a. Tex and Molly purchase their first home and arrange for a home loan of $375 000. Their home loan interest rate rises by 0.25% per annum within the first 6 months of the loan.
    Determine the monthly increase of their repayments. Give your answer to the nearest cent.

    b. Brad and Angel's interest on their home loan is also increased by 0.25% per annum. Their monthly repayments increase by $60.
    Determine the amount of their loan. Give your answer in whole dollars.

26. Juanita sells her car for $10 984. She invests $x\%$ of the money in a bank account at a simple interest rate of 6.68% per annum for 1.5 years. She spends the remainder of the money. At the end of the investment she has exactly enough money to buy a new car for $11 002.
Determine the value of x, correct to 2 decimal places.

8.5 Compound interest

LEARNING INTENTION

At the end of this subtopic you should be able to:
- calculate the final amount received on an investment with compound interest
- calculate the interest earned on an investment.

▶ 8.5.1 Compound interest

eles-4842

- Consider $1000 invested for 3 years at 10% p.a. simple interest.
- Each year the value of the investment increases by $100, reaching a total value of $1300.
- The simple interest process can be summarised in the following table.

	Principal	Interest	Total value
Year 1	$1000	$100	$1100
Year 2	$1000	$100	$1200
Year 3	$1000	$100	$1300

Total interest = $300

- For **compound interest**, the value of the principal is increased by adding the value of the interest earned at the end of each year. In other words the principal and the interest are compounded annually.
- The compound interest process can be summarised in the following table.

	Principal	Interest	Total value
Year 1	$1000	$100	$1100
Year 2	$1100	$110	$1210
Year 3	$1210	$121	$1331

Total interest = $331

- The principal increases each year and therefore the amount of interest it earns does so as well.
- The difference between simple interest and compound interest can become enormous over many years.

Complete the table to determine the interest paid when $5000 is invested at 11% p.a. compounded annually for 3 years.

	Principal	Interest	Total value
Year 1	$5000		
Year 2			
Year 3			

Total interest =

THINK

1. Interest for year 1 = 11% of $5550

   Calculate the principal for year 2 by adding the interest to the year 1 principal.

2. Interest for year 2 = 11% of $5000
   Calculate the total value at the end of year 2. This is the principal for year 3.

3. Interest for year 3 = 11% of $6160.50.

4. Calculate the interest earned over 3 years by subtracting the year 1 principal from the final amount.

5. Complete the table.

WRITE

$11\% = \dfrac{11}{100} = 0.11$

$I = 0.11 \times 5000 = 550$
$5000 + 550 = 5550$

$0.11 \times 5550 = 610.50$
$5550 + 610.50 = 6160.50$

$0.11 \times 6160.50 = 677.66$
$6160.50 + 677.66 = 6838.16$

$6838.16 - 5000 = 1838.16$

	Principal	Interest	Total value
Year 1	$5000	$550	$5550
Year 2	$5550	$610.50	$6160.50
Year 3	$6160.50	$677.66	$6838.16

Total interest = $1838.16

- There is a quicker way of calculating the total value of the investment. Look again at Worked example 10. The investment grows by 11% each year, so its value at the end of the year is $111\% \left(\dfrac{111}{100} = 1.11\right)$ of its value at the start of the year.

$$1.11\% \text{ of } 5000$$
$$= 1.11 \times 5000$$
$$= 5550$$

- This process is repeated each year for 3 years.

$$5000 \xrightarrow{\times 1.11} 5550 \xrightarrow{\times 1.11} 6160.50 \xrightarrow{\times 1.11} 6838.16$$

$$\text{or } 5000 \times 1.11 \times 1.11 \times 1.11 = 6838.16$$

$$\text{or } 5000 \times (1.11)^3 = 6838.16$$

- After 3 years the value of the investment is $6838.16.

An investment of $2000 receives compounded interest at a rate of 8% p.a. Complete the table to calculate its value after 4 years.

Year	Start of year	End of year
Year 1	$2000	
Year 2		
Year 3		
Year 4		

THINK

1. Interest is compounded at 8%, so at the end of the first year the value is 108% of the initial value. The multiplying factor is 1.08.

2. For the value at the end of year 2, calculate 108% of the amount accumulated in year 1, so calculate 108% of 2160.

3. For the value at the end of year 3, calculate 108% of the amount accumulated in year 2, so calculate 108% of 2332.80.

4. For the value at the end of year 4, calculate 108% of the amount accumulated in year 3, so calculate 108% of 2519.424.

5. Complete the table.

WRITE

$$108\% = \frac{108}{100} = 1.08$$

$$1.08 \times 2000 = 2160$$
$$1.08 \times 2160 = 2332.80$$

$$1.08 \times 2332.80 = 2519.424$$

$$1.08 \times 2519.424 = 2720.98$$

Year	Start of year	End of year
Year 1	$2000	$2160
Year 2	$2160	$2332.80
Year 3	$2332.80	$2519.42
Year 4	$2519.42	$2720.98

TI | THINK	WRITE	CASIO | THINK	WRITE
a. In a new problem on a Calculator page type '2000'. Press ENTER.		a. On the Main screen type '2000'. Press EXE.	

TI | THINK	WRITE	CASIO | THINK	WRITE
b. Press × 1.08 and ENTER. This will multiply the previous entry of 2000 by 1.08 and display the amount of the investment after 1 year.	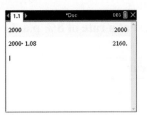	**b.** Press × 1.08 and EXE. This will multiply the previous entry of 2000 by 1.08 and display the amount of the investment after 1 year.	
c. Press ENTER to display the amount of the investment after 2 years.	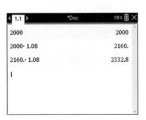	**c.** Press EXE to display the amount of the investment after 2 years.	
d. Press ENTER 2 more times (making 4 times in total) to display the amount of the investment after 4 years.	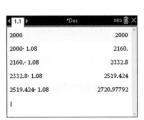	**d.** Press EXE 2 more times (making 4 times in total) to display the amount of the investment after 4 years.	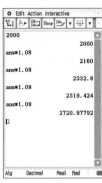

- The repeated multiplication shown in Worked example 11 can be developed into a formula for compound interest.
- In Worked example 11 the principal ($2000) was multiplied by 108% four times (because there were 4 years). The final amount, A, can be given by the formula:

$$A = 2000 \times 108\% \times 108\% \times 108\% \times 108\%$$

$$= 2000(100\% + 8\%)^4$$

$$= 2000\left(1 + \frac{8}{100}\right)^4$$

$$= 2000(1.08)^4$$

- With compound interest the final amount of an investment is calculated using the following formula.

Formula for compound interest

$$A = P(1 + i)^n$$

where:
- A = the final value of the investment
- P = the principal
- i = the interest rate as a percentage per annum (yearly), written as a decimal (e.g. 7.5% p.a. is equal to 0.075 p.a.)
- n = the duration of the investment in years.

- In future studies of financial mathematics, the duration of investment under consideration will include monthly, quarterly or weekly investments. At this stage, n will only be considered in terms of the number of years.
- To calculate the amount of compound interest earned or paid, subtract the principal from the final value of the investment.

Calculating compound interest earned

$$I = A - P$$

where:
- I = the amount of interest earned or paid
- A = the final value of the investment
- P = the principal.

WORKED EXAMPLE 12 Investing an amount earning compound interest

Peter invests $40 000 for 8 years at 7.5% p.a., compounding annually.
a. Calculate how much money Peter has in total at the end of the 8 years.
b. Calculate how much interest Peter earned on the investment.

THINK	WRITE
a. 1. Write the compound interest formula.	a. $A = P(1 + i)^n$
2. State the values of the variables P, i and n.	$P = 40\,000$ $i = 0.075$ $n = 8$
3. Substitute the values into the formula and calculate the value of the final amount.	$A = 40\,000(1 + 0.075)^8$ $= 40\,000(1.075)^8$ $= 71\,339.113\,02$
4. Write the answer in a sentence.	The final amount of Peter's investment was $71 339.11 (to the nearest cent).

b. 1. Write the relationship between the interest earned, the final amount and the principal.

b. $I = A - P$

2. State the values for A and P.

$A = 71\,339.11$
$P = 40\,000$

3. Substitute the values in the formula and Calculate the value of the interest.

c. $I = 71\,339.11 - 40\,000$
$= 31\,339.11$

4. Write the answer in a sentence.

Peter earned \$31\,339.11 in interest.

TI | THINK	WRITE	CASIO | THINK	WRITE
a. In a new problem on a Calculator page, type $40\,000\left(1 + \dfrac{7.5}{100}\right)^{8}$. Press ENTER.	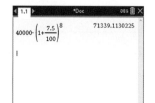	**a.** In a new problem on a Calculator page, type $40\,000\left(1 + \dfrac{7.5}{100}\right)^{8}$. Press EXE.	
b. To find the interest earned, type $-40\,000$. Press ENTER.	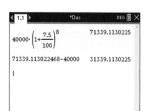	**b.** To find the interest earned, type $-40\,000$. Press EXE.	

COLLABORATIVE TASK: Compound interest conga

1. Arrange yourself into teams of 3 or more players. The number of players in each team becomes the number of years of a compound interest problem.
2. One person from your class writes a principal value and an interest rate on the board.
3. Using the interest rate and the principal of the investment written on the board, the first member of each group calculates the value of the investment after 1 year.
4. The second member of each group takes the first member's numbers and calculates the value of the investment after 2 years.
5. The rest of the members of the group take it in turns to calculate the interest for the following years.
6. The first team in each round to get the correct answer gets 3 points. The second team to get the correct answer gets 1 point. The team with the most points after the agreed-upon number of rounds is the winner.
7. Repeat the game using different compounding periods (the compounding period is the period of time after which the interest earned is added to the principal).
8. As a class, discuss the effect of decreasing compounding periods on the value of the investment.

Exercise 8.5 Compound interest

learn on

Individual pathways

■ PRACTISE	■ CONSOLIDATE	■ MASTER
1, 3, 5, 7, 11, 12, 15	2, 6, 9, 10, 13, 16	4, 8, 14, 17, 18

To answer questions online and to receive **immediate corrective feedback** and **fully worked solutions** for all questions, go to your learnON title at www.jacplus.com.au.

Fluency

1. **WE10** Complete the tables shown to determine the interest paid when:

   a. $1000 is invested at 12% p.a. compounded annually for 3 years

	Principal	**Interest**	**Total value**
Year 1	$1000		
Year 2			
Year 3			

   Total interest =

   b. $100 000 is invested at 9% p.a. compounded annually for 4 years.

	Principal	**Interest**	**Total value**
Year 1	$100 000		
Year 2			
Year 3			
Year 4			

   Total interest =

2. **WE11** Complete the tables shown to calculate the final value of each investment.

   a. $5000 invested at 12% p.a. compounded annually for 3 years

	Start of year	**End of year**
Year 1	$5000	
Year 2		
Year 3		

   b. $200 000 invested at 7% p.a. compounded annually for 3 years

	Start of year	**End of year**
Year 1	$200 000	
Year 2		
Year 3		

3. Calculate the final amount of an investment of $5000 when compound interest is earned at the rate of:

   a. 4% p.a. for 3 years

   b. 3% p.a. for 4 years

   c. 2.75% p.a. for 6 years

   d. 3.5% p.a. for 2 years.

4. Calculate the final value of the following investments when compound interest is earned:

   a. $4650 at 4.5% p.a. for 3 years

   b. $12 500 at 6.2% p.a. for 5 years

   c. $3560 at 2.4% p.a. for 4 years

   d. $25 000 at 3.2% p.a. for 10 years.

Understanding

5. **WE12** For each of the following investments, use the compound interest formula to calculate:

   i. the total value

   ii. the amount of interest paid.

   a. $8000 is invested for 8 years at 15% p.a. interest compounding annually

   b. $50 000 is invested for 4 years at 6% p.a. interest compounding annually

   c. $72 000 is invested for 3 years at 7.8% p.a. interest compounding annually

6. For each of the following investments, use the compound interest formula to calculate:

   i. the total value

   ii. the amount of interest paid.

   a. $150 000 is invested for 7 years at 6.3% p.a. interest compounding annually

   b. $3500 is invested for 20 years at 15% p.a. interest compounding annually

   c. $21 000 is invested for 10 years at 9.2% p.a. interest compounding annually

7. Peter invests $5000 for 3 years at 6% p.a. simple interest. Maria invests the same amount for 3 years at 5.8% p.a. compounding annually.

   a. Calculate the value of Peter's investment on maturity.

   b. Calculate the value of Maria's investment on maturity.

   c. Explain why Maria's investment is worth more, even though she received a lower interest rate.

8. Gianni invests $8000 at 15% p.a. compounded annually. Dylan invests $8000 at a 15% p.a. flat rate. Calculate how much more Gianni's investment is worth than Dylan's after:

   a. 1 year

   b. 2 years

   c. 5 years

   d. 10 years.

9. When Kim's granddaughter was born Kim invested $100 at the rate of 7% p.a. compounding annually. Kim plans to give it to her granddaughter on her eighteenth birthday.
   Calculate how much her granddaughter will receive when she turns 18.

10. Mai's investment account has compounded at a steady 9% for the last 10 years. If it is now worth $68 000, determine how much it was worth:

    a. last year

    b. 10 years ago.

Reasoning

11. Two investment options are available to invest $3000.
    - Invest for 5 years at 5% p.a. compounding monthly.
    - Invest for 5 years at 5% p.a. compounding weekly.

    Explain which option you would choose and why.

12. There are 3 factors that affect the value of a compound interest investment: the principal, the interest rate and the length of the investment.

    a. If the compound interest rate is 10% p.a. and the length of the investment is 2 years, calculate the value of an investment of:

       i. $1000   ii. $2000   iii. $4000.
    b. Comment on the effect that increasing the principal has on the value of the investment.

13. a. If the principal is $1000 and the compound interest rate is 10% p.a., calculate the value of an investment that lasts for:

       i. 2 years   ii. 4 years   iii. 8 years.
    b. Comment on the effect that increasing the length of the investment has on the value of the investment.

14. a. If the principal is $1000 and the length of investment is 5 years, calculate the value of an investment when the compound interest rate is:

       i. 6% p.a.   ii. 8% p.a.   iii. 10% p.a.
    b. Comment on the effect of increasing the interest rate on the value of the investment.

Problem solving

15. A bank offers a term deposit for 3 years at an interest rate of 8% p.a. with a compounding period of 6 months. Calculate the end value of a $5000 investment under these conditions.

16. A building society offers term deposits at 9% p.a., compounded annually. A credit union offers term deposits at 10% but with simple interest only.

    a. Determine which offer will result in the greatest value after 2 years.
    b. Determine which offer will result in the greatest value after 3 years.
    c. Determine how many years it will take for the compound interest offer to have the greater value.

17. Chris and Jenny each invested $10 000. Chris invested at 6.5% p.a. compounding annually. Jenny took a flat rate of interest. After 5 years, their investments had equal value.

    a. Calculate the value of Chris's investment after 5 years.
    b. Calculate Jenny's interest rate.
    c. Calculate the values of Chris's and Jenny's investments after 6 years.

18. One aspect of compound interest is of great importance to investors: how long does it take for an investment to double in value? Consider a principal of $100 and an annual interest rate of 10% (compounding annually).

    a. Determine how long it would take for this investment to be worth $200.
    b. Evaluate how long it would take for the investment to be worth $400 (to double, and then double again).

8.6 Review

8.6.1 Topic summary

Salaries

- A salary is a fixed annual (yearly) payment, regardless of the number of hours worked.
- Salaries are generally paid in weekly, fortnightly or monthly amounts.
  - Normal working hours in Australia are **38 hours** per week.
  - There are **52 weeks** in a year.
  - There are **26 fortnights** in a year (this value is slightly different for a leap year).
  - There are **12 months** in a year.

e.g. If an annual salary is $52 000:
- Weekly salary = $52 000 ÷ 52 = $1000
- Fortnightly salary = $52 000 ÷ 26 = $2000
- Monthly salary = $52 000 ÷ 12 = $4333.33.

Wages

- A wage is paid at a fixed rate per hour.
- Hours worked outside of the usual 38 or 40 hours per week are usually paid at a higher rate per hour.

e.g. If a wage's rate per hour is $24.50 and a person works 38 hours each week:
weekly pay = $24.50 × 38 = $931

Penalty rates

- Penalty rates are paid when employees work overtime or at weekends, on public holidays or at night.

e.g. If the hourly pay is $24, then:
time-and-a-half rate = $24 × 1.5
= $36/hour
double time rate = $24 × 2
= $48/hour

FINANCIAL MATHEMATICS

Simple interest

- Interest is the amount earned from investing money with a bank.
- The formula to calculate the final value of a simple interest investment is $I = Pin$, where:
  - P = the principal, or the amount initially invested
  - i = the interest rate as a percentage per annum (yearly), written as a decimal (e.g. 2% p.a. is equal to 0.02 p.a.)
  - n = the duration of the investment in years.
- The formula to calculate the value of an investment or loan is $A = P + I$, where:
  - A = the total amount of money at the end of the investment or loan period
  - P = the principal
  - I = the amount of interest earned.

Compound interest

- Compound interest is interest that is added to the principal (the amount initially invested) at the end of each year.
- Once the value of the interest is added to the principal, interest is calculated based on the principal's increased value.
- The formula to calculate the final value of a compound interest investment is $A = P(1 + i)^n$, where:
  - A = the final amount of the investment
  - P = the principal
  - i = the interest rate as a percentage per annum (yearly), written as a decimal (e.g. 7.5% p.a. is equal to 0.075 p.a.)
  - n = the duration of the investment in years.
- The formula to calculate the compound interest earned is $I = A - P$, where:
  - I = the amount of interest earned or paid
  - A = the final value of the investment
  - P = the principal.

Calculating the final amount

- The are different ways to calculate the final amount of an initial investment of $8000 for 10 years, depending on whether it is a simple interest investment or a compound interest investment.

Simple interest investments

- For $8000 over 10 years with a 3% p.a. simple interest rate (also called a flat rate):

$$I = Pin = 8000 × 0.03 × 10 = 2400$$

- Final amount = $8000 + $2400
= $10 400

Compounding interest investments

- For $8000 over 10 years with a 3% p.a. compound interest rate:

$$A = P(1 + i)^n = 8000(1 + 0.03)^{10}$$
$$= 8000(1.03)^{10}$$

- Final amount = $10 751.33

8.6.2 Success criteria

Tick the column to indicate that you have completed the subtopic and how well you think you have understood it using the traffic light system.

(**Green:** I understand; **Yellow:** I can do it with help; **Red:** I do not understand)

Subtopic	Success criteria	●	○	◐
8.2	I understand the difference between a salary and a wage.			
	I can calculate weekly, fortnightly or monthly pays from a given annual salary.			
	I can calculate the amount earned from the hours worked and the hourly rate of pay.			
8.3	I can calculate total wages when overtime or penalty rates are included.			
	I can calculate hours paid when time-and-a-half or double time is involved.			
8.4	I can calculate the simple interest on a loan or an investment.			
	I can apply the simple interest formula to determine the time, the rate or the principal.			
8.5	I can calculate the final amount received on an investment with compound interest.			
	I can calculate the interest earned on an investment.			

8.6.3 Project

Australian currency

Since decimal currency was introduced in Australia in 1966, our notes and coins have undergone many changes. Only our 5c, 10c and 20c coins are still minted as they were back then. The 1c and 2c coins are no longer in circulation, the 50c coin is a different shape, the $1 and $2 notes have been replaced by coins, and our notes have changed from paper to a special type of plastic.

Coins have two sides: an obverse side and a reverse side. The obverse side of all Australian coins depicts our reigning monarch, Queen Elizabeth II, and the year in which the coin was minted. The reverse side depicts a typical Australian feature and sometimes a special commemorative event.

1. Describe what is depicted on the reverse side of each Australian coin.

The following table includes information on Australia's coins currently in circulation. Use the table to answer questions 2 to 4.

Coin	Diameter (mm)	Mass (g)	Composition
5c	19.41	2.83	75% copper, 25% nickel
10c	23.60	5.65	75% copper, 25% nickel
20c	28.52	11.30	75% copper, 25% nickel
50c	31.51	15.55	75% copper, 25% nickel
$1	25.00	9.00	92% copper, 6% aluminium, 2% nickel
$2	20.50	6.60	92% copper 6% aluminium, 2% nickel

2. Identify the heaviest and lightest type of coin. List every type of coin in order from lightest to heaviest.
3. Identify which coin has the smaller diameter: the 5c coin or the $2 coin. Calculate the difference in size.

The following table displays information on the currency notes currently circulating in Australia. The column on the far right compares the average life of the previously used paper notes with that of the current plastic notes. Use the table to answer questions 4 to 8.

Note	Date of issue	Size (mm)	Average life of notes (months)	
			Plastic	Paper
$5	07/07/1992 24/04/1995 01/01/2001 01/09/2016	130 × 65	40	6
$10	01/11/1993 20/09/2017	137 × 65	40	8
$20	31/10/1994 09/10/2019	144 × 65	50	10
$50	04/10/1995 18/10/2018	151 × 65	About 100	24
$100	15/05/1996 29/10/2020	158 × 65	About 450	104

4. List the denominations of the notes that are available in Australian currency.
5. State the date on which Australia's first plastic note was issued and the denomination of the note.
6. The $5 note has been issued 4 times, while all other notes have only been issued twice. Suggest a reason for this.
7. Suggest a reason why each note is a different size.
8. The table clearly shows that the plastic notes last about 5 times as long as the paper notes we once used. Suggest a reason why the $50 and $100 notes last longer than the $5 and $10 notes.

Exercise 8.6 Review questions

learnon

To answer questions online and to receive **immediate corrective feedback** and **fully worked solutions** for all questions, go to your learnON title at www.jacplus.com.au.

Fluency

1. Jane earns an annual salary of $45 650. Calculate her fortnightly pay.

2. Express $638.96 per week as an annual salary.

3. Minh works as a casual shop assistant and is paid $8.20 an hour from 3.30 pm to 5.30 pm, Monday to Friday, and $9.50 an hour from 7 am to 12 noon on Saturday. Calculate his total pay for the week.

4. **MC** Below are the pay details for 5 people. Choose who receives the most money.
   A. Billy receives $18.50 per hour for a 40-hour working week.
   B. Jasmine is on an annual salary of $38 400.
   C. Omar receives $1476.90 per fortnight.
   D. Thuy receives $3205 per month.
   E. Mikke receives $735 per week.

5. Daniel earns $10 an hour for a regular 38-hour week. If he works overtime, he is paid double time. Calculate how much he would earn if he worked 42 hours in one week.

6. Bjorn earns $468.75 per week award wage for a 38-hour week.
   a. Calculate his standard hourly rate.
   b. If he is paid time-and-a-half for normal overtime, calculate his pay for a week in which he works 41 hours.

7. Phillipa is paid an annual salary of $48 800.
   Calculate how much she would be paid each time if she was paid:
   a. weekly
   b. monthly.

8. Deng works as a waiter and is paid $12.50 per hour. If he works 40 hours per week, calculate Deng's gross pay.

Problem solving

9. a. Calculate the simple interest on an investment of $4000 for 9 months at 4.9% per annum.
   b. Determine the final amount of the investment by the end of its term.

10. Calculate the monthly instalment needed to repay a loan of $12 500 over 40 months at a 9.75% p.a. flat interest rate.

11. A simple interest loan of $5000 over 4 years incurred interest of $2300. Calculate the interest rate charged.

▶

12. Complete the table shown to calculate the final value of $100 000 invested at 8.5% p.a. compounded annually for 5 years.

	Start of year	End of year
Year 1	$100 000	
Year 2		
Year 3		
Year 4		
Year 5		

13. Complete the table shown to calculate the final value of $12 000 invested at 15% p.a. compounded annually for 4 years.

	Start of year	End of year
Year 1	$12 000	
Year 2		
Year 3		
Year 4		

14. Daniela is to invest $16 000 for 2 years at 9% p.a. with interest compounded annually. Calculate:
    a. the final value of the investment
    b. the amount of interest earned.

15. Calculate the amount of interest paid on $40 000 for 5 years at 6% p.a. with interest compounded every six months.

16. Calculate 3.5% of 900.

17. Milos works in a supermarket and earns $11.70 per hour. If Milos works on Saturday he is paid at time-and-a-half. Determine the hourly rate for which Milos works on Saturday.

18. Frank invests $1000 at 5% p.a. for 3 years with interest compounded annually. Calculate the final value of Frank's investment.

19. Calculate the final value of each of the following investments if the principal is $1000.
    a. Interest rate = 8% p.a., compounding period = 1 year, time = 2 years
    b. Interest rate = 8% p.a., compounding period = 6 months, time = 2 years
    c. Interest rate = 8% p.a., compounding period = 3 months, time = 2 years

20. Jenny is given $5000 by her grandparents on her 18th birthday on the condition she invests the money until she is at least 21 years old, which she agrees to do. She finds the following investment options:
    Option 1: A local bank paying 2.5% p.a. interest compounding annually
    Option 2: A local accountant paying a 2.7% p.a. flat rate of interest
    Option 3: A building society that is offering a special deal of 2.4% p.a. interest, compounding annually with a 1% bonus added to the final value.
    Identify the option that Jenny should choose, giving a reason for your answer.

on To test your understanding and knowledge of this topic, go to your learnON title at www.jacplus.com.au and complete the **post-test**.

Online Resources

Below is a full list of **rich resources** available online for this topic. These resources are designed to bring ideas to life, to promote deep and lasting learning and to support the different learning needs of each individual.

eWorkbook

Download the workbook for this topic, which includes worksheets, a code puzzle and a project (ewbk-2008) ☐

Solutions

Download a copy of the fully worked solutions to every question in this topic (sol-0728) ☐

Digital documents

8.2   SkillSHEET Converting units of time (doc-10849) ☐
     SkillSHEET Multiplying and dividing a quantity (money) by a whole number (doc-10850) ☐
     SkillSHEET Multiplying and dividing a quantity (money) by a fraction (doc-10851) ☐
     SkillSHEET Increasing a quantity by a percentage (doc-10852) ☐
     SkillSHEET Adding periods of time (doc-10853) ☐
8.3   SkillSHEET Multiplying a quantity (money) by a decimal (doc-10854) ☐
8.5   SpreadSHEET Simple and compound interest (doc-10907) ☐

Video eLessons

8.2   Employees' salaries and wages (eles-4836) ☐
8.3   Overtime and penalty rates (eles-4837) ☐
     Time sheets and pay slips (eles-4398) ☐
8.4   Principal and interest (eles-4399) ☐
     Simple interest (eles-4400) ☐
8.5   Compound interest (eles-4842) ☐

Interactivities

8.2   Individual pathway interactivity: Salaries and wages (int-4520) ☐
     Salaries (int-6067) ☐
8.3   Individual pathway interactivity: Special rates (int-4521) ☐
     Special rates (int-6068) ☐
8.4   Individual pathway interactivity: Simple interest (int-4526) ☐
     Simple interest (int-6074) ☐
8.5   Individual pathway interactivity: Compound interest (int-4527) ☐
     Compound interest (int-6075) ☐
8.6   Crossword (int-2700) ☐
     Sudoku puzzle (int-3210) ☐

Teacher resources

There are many resources available exclusively for teachers online.

To access these online resources, log on to **www.jacplus.com.au**.

Answers

Topic 8 Financial mathematics

Exercise 8.1 Pre-test

1. a. $1643.28   b. $3286.57   c. $7120.89
2. $184
3. $32.50
4. D
5. $32.50 per hour
6. 35 hours
7. $412.50
8. a. $10   b. $810
9. D
10. a. $8372   b. $31 072   c. $553.58
11. C
12. 9% p.a
13. C
14. C
15. $7728

Exercise 8.2 Salaries and wages

1. a. $1105.42   b. $2210.85   c. $4790.17
2. a. $1198.08   b. $2396.15   c. $5191.67
3. a. $19 136   b. $46 410   c. $68 684.20
4. a. $25 870   b. $42 183.96   c. $100 498.06
5. a. $3890 per month   b. $3200.58 per fortnight
6. a. $19.75/h   b. $12.17/h
   c. $25.73/h   d. $39.06/h
7. a. $459.04   b. Jenny earns $211.66 more.
8. $210.72
9. $261.49
10. Rob earns more.
11. $30.77
12. 38 hours
13. a. $1605.77   b. $2788.81   c. $1100
    d. $8770
14. B, as the annual salary for option B is the highest.
15. Job A has a higher hourly rate.
16. Minh. Minh's annual salary is $63 745.76, which is higher than Michelle's annual salary.
17. $40 per hour. This is the option with the greatest pay.
18. $1735.95, $1833.75
19. a. $29.81   b. $25.64   c. $72 080.58
20. 12 weeks

Exercise 8.3 Penalty rates

1. a. $23.94   b. $47.80   c. $43.50
2. a. $906.10   b. $794.33   c. $833.56
3. a. $25.16   b. $113.22   c. $1069.30
4. a. $170.40   b. $340.80   c. $426.00
5. $1156.96

6. a. $30.20   b. $45.30   c. $60.40
   d. $135.90
7. a. $32.93   b. $197.55
8. a. $11.70   b. $15.60   c. $19.50
9. a. Mon: 8, Tue: 8, Wed: 8, Thu: 8, Fri: 6
   b.

Pay slip for Susan Jones	Week ending 17 August
Normal hours	38
Normal pay rate	$25.60
Overtime hours	0
Overtime pay rate	$38.40
Total pay	$972.80

10. a. Mon: 8, Tue: 8, Thu: 8, Sat: 4
    b.

Pay slip for Manu Taumata	Week ending 21 December
Normal hours	24
Normal pay rate	$10.90
Overtime hours	4
Overtime pay rate	$16.35
Total pay	$327.00

11. a. Mon: 8, Tue: 8, Fri: 8, Sat: 8, Sun: 8
    b.

Pay slip for Eleanor Rigby	Week ending 17 August
Normal hours	24
Normal pay rate	$16.80
Time-and-a half hours	8
Time-and-a half pay rate	$25.20
Double time hours	8
Double time pay rate	$33.60
Total pay	$873.60

12. a. $754.40   b. $1042.80
13. a. $232.02   b. $1211.66
14. $46\frac{1}{2}$
15. $1155
16. No. Lin should have been paid $1034.

17.

	Hours worked	Regular pay	Overtime pay	Total pay
a.	32	$464	$0	$464
b.	38.5	$522	$54.38	$576.38
c.	40.5	$522	$101.50	$623.50
d.	47.2	$522	$295.80	$817.80

Exercise 8.4 Simple interest

1. $32
2. a. $945   b. $495
   c. $675   d. $900
3. a. $315
   b. $327.60
   c. $22.25
   d. $1500.80
4. a. 3 years
   b. 5 years
   c. 1.5 years or 1 year and 6 months
   d. 2.75 years or 2 year and 9 months
5. a. 3% p.a.   b. 4.5% p.a.
   c. 4.8% p.a.   d. 6.8% p.a.
6. E
7. a. $1200   b. $6200
8. $1950
9. a. i. $100   ii. $1100
   b. i. $1920   ii. $5920
   c. i. $1440   ii. $9440
   d. i. $263.25   ii. $2963.25
   e. i. $5048.32   ii. $20 726.32
10. a. $3300   b. $5645
11. a. $22 275.50   b. $371.26
12. $211.90
13. 2.5 years
14. $12 500
15. 8%
16. a. 2 years   b. 2.5 years   c. $600
    d. $4500.06   e. 8%   f. 3.5%
17. a. $16 998   b. $8770.97   c. $28 768.97
    d. 48   e. $536.85
18. The interest rate in the simple interest formula needs to be converted from a percentage into a decimal: 4.2% = 0.042.
19. Sample responses can be found in the worked solutions in the online resources.
20. a. $180 000
    b. $80 550
    c. $260 550
    d. $4342.50
    e. Increase the size of the deposit or increase the length of time over which the loan can be repaid
21. Yes
22. 5.3%
23. a. $2300   b. $12 300
24. a. $5768   b. $6902   c. $120.17, $143.79
25. a. $78.13   b. $288 000
26. 91.04

Exercise 8.5 Compound interest

1. a.

	Principal	Interest	Total value
Year 1	$1000	$120	$1120
Year 2	$1120	$134.40	$1254.40
Year 3	$1254.40	$150.53	$1404.93

Total interest = $404.93

b.

	Principal	Interest	Total value
Year 1	$100 000	$9000	$109 000
Year 2	$109 000	$9810	$118 810
Year 3	$118 810	$10 692.90	$129 502.90
Year 4	$129 502.90	$11 655.26	$141 158.16

Total interest = $41 158.16

2. a.

	Start of year	End of year
Year 1	$5000	$5600
Year 2	$5600	$6272
Year 3	$6272	$7024.64

b.

	Start of year	End of year
Year 1	$200 000	$214 000
Year 2	$214 000	$228 980
Year 3	$228 980	$245 008.60

3. a. $5624.32   b. $5627.54
   c. $5883.84   d. $5356.13
4. a. $5306.42   b. $16 886.23
   c. $3914.26   d. $34 256.03
5. a. i. $24 472.18   ii. $16 472.18
   b. i. $63 123.85   ii. $13 123.85
   c. i. $90 196.31   ii. $18 196.31
6. a. i. $230 050.99   ii. $80 050.99
   b. i. $57 282.88   ii. $53 782.88
   c. i. $50 634.40   ii. $29 634.40
7. a. $5900
   b. $5921.44
   c. Maria's principal increases each year.
8. a. 0   b. $180
   c. $2090.86   d. $12 364.46
9. $337.99
10. a. $62 385.32   b. $28 723.93
11. The second option would be the best choice, as the shorter the time between the compounding periods, the greater the interest paid.

12. a. i. $1210 ii. $2420 iii. $4840
    b. Increasing the principal will increase the value of the investment because it will have a higher value of interest.
13. a. i. $1210 ii. $1464.10 iii. $2143.59
    b. Increasing the length of the investment will increase the value of the investment because it will have a higher value of interest.
14. a. i. $1338.23 ii. $1469.33 iii. $1610.51
    b. Increasing the interest rate will increase the value of the investment because it will have a higher value of interest.
15. $6326.60
16. a. Simple interest
    b. Simple interest
    c. 4 years
17. a. $13 700.87
    b. 7.4%
    c. Chris, $14 591.42; Jenny, $14 440
18. a. 7.27 years b. 14.55 years

Project

1. 5c coin: echidna
   10c coin: lyrebird
   20c coin: platypus
   50c coin: coat of arms
   $1 coin: five kangaroos
   $2 coin: Walpiri-Anmatyerre Aboriginal Elder Gwoya Jungarai
2. 5c, 10c, $2, $1, 20c, 50c
3. The 5c coin has a smaller diameter. It is smaller by 1.09 mm.
4. $5, $10, $20, $50, $100
5. 7 July 1992, $5
6. Sample responses can be found in the worked solutions in the online resources.
7. The different sizes allow people with vision impairment to tell the difference between each note.
8. The $50 and $100 notes are used less frequently.

Exercise 8.6 Review questions

1. $1755.77
2. $33 225.92
3. $129.50
4. A
5. $460
6. a. $12.34 b. $524.28
7. a. $938.46 b. $4066.67
8. $500
9. a. $147 b. $4147
10. $414.06
11. 11.5%

12.

	Start of year	End of year
Year 1	$100 000	$108 500
Year 2	$108 500	$117 722.50
Year 3	$117 722.50	$127 728.91
Year 4	$127 728.91	$138 585.87
Year 5	$138 585.87	$150 365.67

13.

	Start of year	End of year
Year 1	$12 000	$13 800
Year 2	$13 800	$15 870
Year 3	$15 870	$18 250.50
Year 4	$18 250.50	$20 988.08

14. a. $19 009.60 b. $3009.60
15. $13 756.66
16. 31.5
17. $17.55
18. $1157.625
19. a. $1166.40 b. $1169.86 c. $1171.66
20. The best option for Jenny would be Option 3 because it gives a larger final amount: $5422.40.

9 Measurement

9.1 Overview

Why learn this?

We live in a world surrounded by shapes and objects. Often, we ask questions such as 'how long?', 'how far?' or 'how big?'. These questions are all answered using measurement. Measurements may be in one, two or three dimensions. Examples of measurements in one dimension would be length and perimeter, temperature and time. Measurements in two dimensions include areas of shapes including the areas of curved surfaces of objects such as drink cans. Measurements in three dimensions include volumes of objects and capacity, such as the number of millilitres in a drink can.

Since objects are all around us, measurement skills are used in many real-world situations. Carpenters, builders, concreters, landscape gardeners and construction workers are just a few of the many tradespeople who need to understand various aspects of measurement when ordering or working onsite and following plans. Designers, interior decorators and architects use measurement in their drawings and calculations. Chefs measure ingredients in their cooking. Nurses and health professionals follow instructions regarding the amount of a drug to administer to a patient. To maximise profits, manufacturers need to minimise the amount of raw materials used in production. This means knowing the measurements of various parts. Understanding the basic concepts involved in measurement is beneficial in many real-world situations.

Where to get help

Go to your learnON title at **www.jacplus.com.au** to access the following digital resources. The Online Resources Summary at the end of this topic provides a full list of what's available to help you learn the concepts covered in this topic.

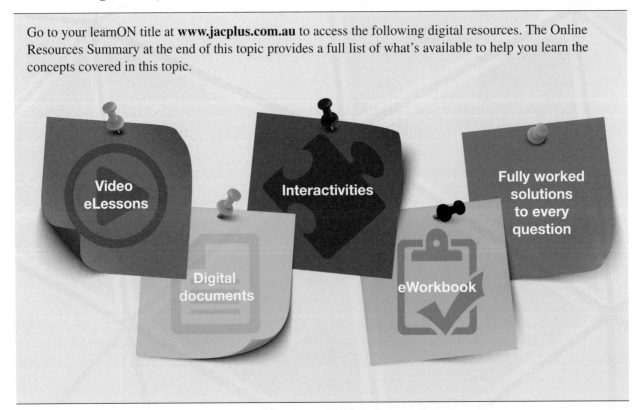

Complete this pre-test in your learnON title at www.jacplus.com.au and receive **automatic marks**, **immediate corrective feedback** and **fully worked solutions**.

1. Calculate the perimeter of this figure, giving your answer in centimetres.

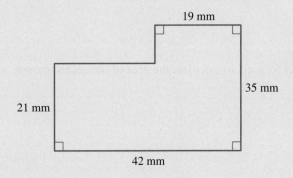

2. Calculate the area of the triangle, correct to 2 decimal places.

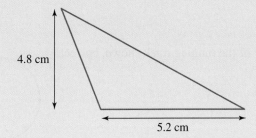

3. Calculate the perimeter of the triangle shown.

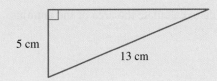

4. **MC** 0.005 m² is equivalent to:
   A. 0.000 05 cm²
   B. 0.5 cm²
   C. 50 cm²
   D. 0.005 cm²
   E. 5 cm²

5. Calculate the volume of the cuboid shown in mm³.

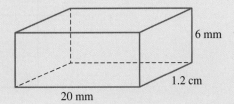

6. **MC** Select the correct formula to calculate the perimeter of the shape shown.

A. $P = \pi \times 4^2$
B. $P = 2 \times \pi \times 4^2$
C. $P = 8 \times \pi$
D. $P = \pi \times 4^2 + 8$
E. $P = 6 \times \pi + 8$

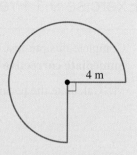

7. **MC** Select the correct formula to calculate the area of the sector shown.

A. $A = \dfrac{60}{360} \times \pi \times 5$
B. $A = \dfrac{60}{360} \times \pi \times 25$
C. $A = \dfrac{60}{300} \times \pi \times 25$
D. $A = \dfrac{60}{360} \times 2 \times \pi \times 5$
E. $A = \dfrac{60}{300} \times 2 \times \pi \times 5$

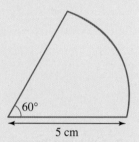

8. **MC** Select the perimeter of the running track shown, correct to 2 decimal places.

A. 94.25 m
B. 304.24 m
C. 467.12 m
D. 514.25 m
E. 916.85 m

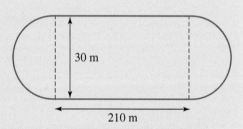

9. **MC** Select the correct formula to calculate the area of the annulus.

A. $A = \pi(7^2 - 4^2)$
B. $A = \pi(7^2 - 3^2)$
C. $A = \pi \times 4^2$
D. $A = 2 \times \pi \times (7 - 3)$
E. $A = 2 \times \pi \times (7 - 4)$

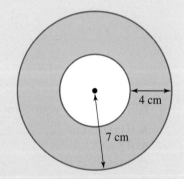

10. Calculate the surface area (SA) of the rectangular prism, giving your answer correct to 2 decimal places.

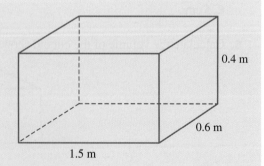

11. **MC** Select the correct formula to calculate the surface area of the triangular prism.

A. $SA = \frac{1}{2} \times 2 \times 2 \times 7$

B. $SA = \frac{1}{2} \times 2 \times 2 \times 7 \times 3$

C. $SA = \frac{1}{2} \times 2 \times 2 \times 2 + 3 \times 7 \times 2$

D. $SA = \frac{1}{2} \times 2 \times 2 + 2 \times 7 \times 2$

E. $SA = 2 \times \sqrt{3} + 3 \times 7 \times 2$

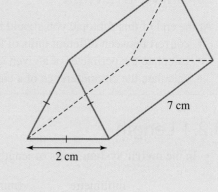

7 cm

2 cm

12. Convert 3.1 L into cm^3.

13. Calculate the capacity (in mL) of two-thirds of the soup can shown, correct to 1 decimal place.

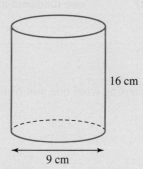

16 cm

9 cm

14. A garden bed is designed in the shape of a trapezium with the dimensions shown. If the area of the garden bed is $30\, m^2$, calculate its perimeter.

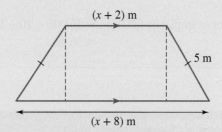

$(x + 2)$ m

5 m

$(x + 8)$ m

15. A label is made for a tin as shown in the diagram. The label is symmetrically positioned so that it is 3 cm from the top and bottom of the tin. Determine the area of paper required to make the label, correct to 1 decimal place.

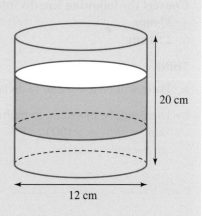

20 cm

12 cm

9.2 Measurement

LEARNING INTENTION

At the end of this subtopic you should be able to:
- convert between different units of length
- calculate the perimeter of a given shape
- calculate the circumference of a circle.

▶ 9.2.1 Length

eles-4851

- In the **metric system**, units of length are based on the metre. The following units are commonly used.

millimetre	(mm)	one-thousandth of a metre
centimetre	(cm)	one-hundredth of a metre
metre	(m)	one metre
kilometre	(km)	one thousand metres

Converting units of length

The following chart is useful when converting from one unit of length to another.

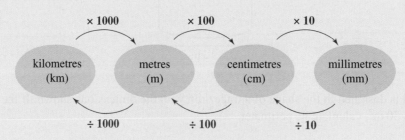

For example, $36\,000\,\text{mm} = 36\,000 \div 10 \div 100\,\text{m}$
$= 36\,\text{m}$

WORKED EXAMPLE 1 Converting units of length

Convert the following lengths into cm.
a. **37 mm**
b. **2.54 km**

THINK	WRITE
a. There will be fewer cm, so divide by 10.	a. $37 \div 10 = 3.7\,\text{cm}$
b. There will be more cm, so multiply. km → m: ×1000 m → cm: ×100	b. $2.54 \times 1000 = 2540\,\text{m}$ $2540 \times 100 = 254\,000\,\text{cm}$

9.2.2 Perimeter

eles-4852

- The **perimeter** of a plane (flat) figure is the distance around the outside of the figure.
- If the figure has straight edges, then the perimeter can be found by simply adding all the side lengths.
- Ensure that all lengths are in the same unit.

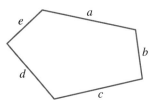

Perimeter $a + b + c + d + e$

Circumference

- **Circumference** is a special name given to the perimeter of a circle.
- Circumference is calculated using the formula:

Circumference of a circle

$C = \pi d$, where d is the **diameter** of the circle

or

$C = 2\pi r$, where r is the **radius** of the circle.

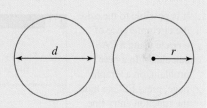

- Use the exact value of π unless otherwise directed.

WORKED EXAMPLE 2 Calculating the perimeter of a shape

Calculate the perimeter of this figure. Give your answer in centimetres.

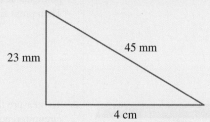

THINK	WRITE/DRAW
1. State the lengths of the sides in the same unit of length. Convert cm to mm: $4\,cm = 4 \times 10 = 40\,mm$	$23\,mm, 45\,mm, 40\,mm$
2. Add the side lengths together.	$P = 45 + 40 + 23$ $= 118\,mm$
3. As the answer is required in cm, convert mm to cm by dividing 118 by 10.	$P = \dfrac{118}{10}$ $= 11.8\,cm$
4. Write the answer in the correct unit.	The perimeter is 11.8 cm.

Calculate the circumference of the circle. Give your answer correct to 2 decimal places.

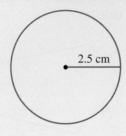

2.5 cm

THINK	WRITE
1. The radius is known, so apply the formula $C = 2\pi r$.	$C = 2\pi r$
2. Substitute $r = 2.5$ into $C = 2\pi r$.	$C = 2 \times \pi \times 2.5$
3. Calculate the circumference to 3 decimal places and then round correct to 2 decimal places.	$C \approx 15.707$ $C \approx 15.71$ cm

TI | THINK	WRITE	CASIO | THINK	WRITE
1. It is possible to measure the circumference of a circle using a CAS calculator. In a new problem, on a Graphs page, press ESC to hide the function entry line. To accurately draw a circle of radius 2.5 units with centre at the origin, we need to place a point on the *x*-axis at (2.5, 0). Press: • MENU • 8: Geometry • 1: Points & Lines • 2: Points On. Then click on the *x*-axis and press ENTER, then press ESC.	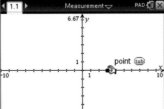	1. It is possible to measure the circumference of this circle using a CAS calculator. On the Geometry screen, press the Grid icon. Then press: • Draw • Basic object • Circle Place the centre at $(0, 0)$ and press EXE. Drag the circle so that a second point sits at $x = 2.5$. Then press EXE.	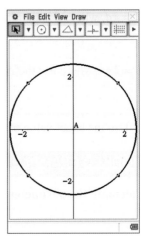
2. To see the coordinates of this point, hover over the point and press: • CTRL • MENU • 7: Coordinates and Equations. The coordinates of the point are displayed.	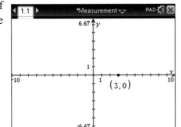	2. To measure the circumference of the circle, select the circle by placing the pointer on its circumference and then pressing EXE. Now place the pointer elsewhere on the circumference and press EXE.	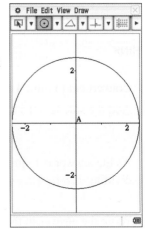

3. With the *x*-coordinate selected, click. In the text box that appears, delete the current value and type in 2.5, then press ENTER. The point will jump so that it is positioned at the correct point, $(2.5, 0)$.

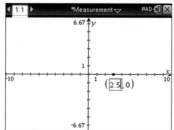

3. To measure the circumference of this circle, press:
   • Draw
   • Measurement
   • Circumference

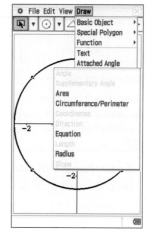

4. To see the equation of the circle, select the circle and press:
   • CTRL
   • MENU
   • 7: Coordinates and Equations.
   The equation of the circle is shown as $x^2 + y^2 = 2.5^2$.

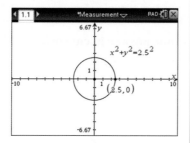

To measure the circumference of the circle, press:
   • MENU
   • 8: Geometry
   • 3: Measurement
   • 1: Length.
   Place the pointer over the circle and press ENTER. Then press ESC.

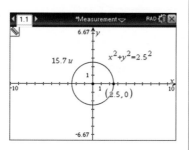

5. To check the ratio of the circumference of the circle to its radius, we will store the value of the circumference to a variable, *c*. To do this, select the text representing the value of the circumference and press:
   • CTRL
   • MENU
   • 5: Store.
   Delete the *var* and type in c, then press ENTER.

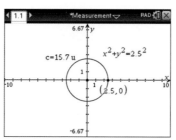

6. Open a Calculator page and store the variable r as 2.5. To do this, type: $r := 2.5$ and press ENTER.
Now find the ratio $\dfrac{c}{2r}$.
Notice that its value is the same, correct to 4 decimal places, as that of π.

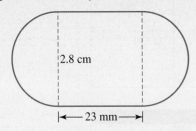

1.1	1.2	▶	*Measurement ▽	PAD ◀▮ ⊠
$r := 2.5$				2.5000
$\dfrac{c}{2 \cdot r}$				3.1416
π				π
π				3.1416
I				

- Some shapes will contain sectors of circles. To calculate the perimeter of these shapes, determine what proportion of a circle is involved, and multiply this by the perimeter of the circle.

WORKED EXAMPLE 4 Determining the perimeter of a given shape

Determine the perimeter of the shape shown. Give your answer in cm correct to 2 decimal places.

2.8 cm

|← 23 mm →|

THINK	WRITE
1. There are two straight sections and two semicircles. The two semicircles make up a full circle, the diameter of which is known, so apply the formula $C = \pi d$.	$C = \pi d$
2. Substitute $d = 2.8$ into $C = \pi d$ and give the answer to three decimal places.	$= \pi \times 2.8$ $\approx 8.796 \, \text{cm}$
3. Convert 23 mm to cm (23 mm = 2.3 cm). The perimeter is the sum of all the outside lengths.	$P = 8.796 + 2 \times 2.3$ $= 13.396$
4. Round to 2 decimal places.	$P \approx 13.40 \, \text{cm}$

---

on Resources

eWorkbook Topic 9 Workbook (worksheets, code puzzle and project) (ewbk-2009)

Digital documents SkillSHEET Converting units of length (doc-6294)
SkillSHEET Substitution into perimeter formulas (doc-10872)
SkillSHEET Perimeter of squares, rectangles, triangles and circles (doc-10873)

Interactivities Individual pathway interactivity: Measurement (int-4528)
Converting units of length (int-4011)
Perimeter (int-4013)
Circumference (int-3782)

Exercise 9.2 Measurement

Individual pathways

■ PRACTISE	■ CONSOLIDATE	■ MASTER
1, 3, 6, 7, 9, 12, 14, 20, 24	2, 4, 10, 13, 15, 17, 21, 25	5, 8, 11, 16, 18, 19, 22, 23, 26

To answer questions online and to receive **immediate corrective feedback** and **fully worked solutions** for all questions, go to your learnON title at www.jacplus.com.au.

Fluency

1. **WE1** Convert the following lengths to the units shown.
   a. $5\,cm =$ _____ mm
   b. $1.52\,m =$ _____ cm
   c. $12.5\,mm =$ _____ m
   d. $0.0322\,m =$ _____ mm

2. Convert the following lengths to the units shown.
   a. $6.57\,m =$ _____ km
   b. $64\,cm =$ _____ km
   c. $0.000\,014\,35\,km =$ _____ mm
   d. $18.35\,cm =$ _____ km

3. **WE2** Calculate the perimeter of the following figures, in millimetres.

   a.
   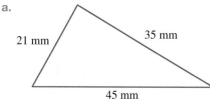

   b.
   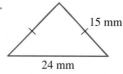

   c.
   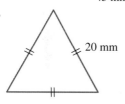

   d.
   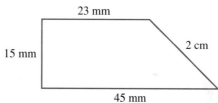

4. Calculate the perimeter of each of the following figures. Give your answers in centimetres.

   a.

   b.
   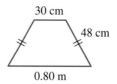

   c.

5. Calculate the perimeter of the following figures, in millimetres.

   a.
   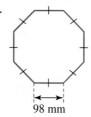

   b.
   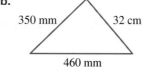

   c.
   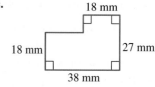

6. Calculate the perimeter of each of the squares.

a. 2.4 cm

b.
11.5 mm

c. 7.75 km

7. **WE3** Calculate the circumference of the circles. Give your answer correct to 2 decimal places.

a.
←—8 cm—→

b.
4 m

c.
22 mm

8. Calculate the circumference of the circles. Give your answer correct to 2 decimal places.

a.
7.1 cm

b.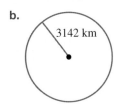
3142 km

c.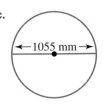
←—1055 mm—→

9. Calculate the perimeter of the rectangles shown.

a. 60 m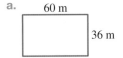
36 m

b. 500 mm
110 mm

c.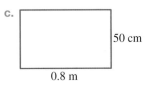
50 cm
0.8 m

10. Calculate the perimeter of the rectangles shown.

a.
9 mm
2.8 cm

b.
3 km
1.8 km

c. 100 cm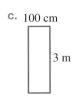
3 m

11. **MC** A circle has a radius of 34 cm. Its circumference, to the nearest centimetre, is:
   A. 102 cm   B. 214 cm   C. 107 cm   D. 3630 cm   E. 50 cm

Understanding

12. Timber is sold in standard lengths, which increase in 300-mm intervals from the smallest available length of 900 mm. (The next two standard lengths available are therefore 1200 mm and 1500 mm.)

a. Write the next four standard lengths (after 1500 mm) in mm, cm and m.
b. Calculate the number of pieces 600 mm in length that could be cut from a length of timber that is 2.4 m long.
c. If I need to cut eight pieces of timber, each 41 cm long, determine the smallest standard length I should buy.
   Note: Ignore any timber lost due to the cuts.

13. The world's longest bridge is the Akashi–Kaikyo Bridge, which links the islands of Honshu and Shikoku in Japan. Its central span covers 1.990 km.

    a. State the length of the central span, in metres.
    b. Calculate how much longer the span of the Akashi–Kaikyo Bridge is than the span of the Sydney Harbour Bridge, which is 1149 m.

14. **WE4** Determine the perimeter of the shapes shown. Give your answer correct to 2 decimal places.

a.
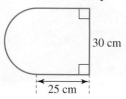

b.
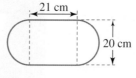

c.

15. Determine the perimeter of the shapes shown. Give your answer correct to 2 decimal places.

a.
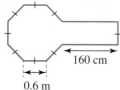

b.

c.

16. Determine the perimeter of the shapes shown. Give your answer correct to 2 decimal places.

a.

b.
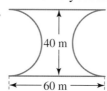

c.
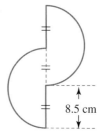

17. Determine the perimeter of the racetrack shown in the plan.

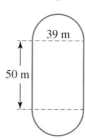

18. Yacht races are often run over a triangular course, as shown. Determine the distance covered by the yachts if they completed 3 laps of the course.

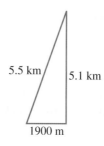

19. Use Pythagoras' theorem to calculate the length of the missing side and, hence, the perimeter of the triangular frame shown.

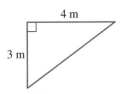

Reasoning

20. The Hubble Space Telescope is over 13 m in length. It orbits the Earth at a height of 559 km, where it can take extremely sharp images outside the distortion of the Earth's atmosphere.

    a. If the radius of the Earth is 6371 km, show that the distance travelled by the Hubble Space Telescope in one orbit, to the nearest km, is 43 542 km.

    b. If the telescope completes one orbit in 96 minutes, show that its speed is approximately 7559 m/s.

21. A bullet can travel in air at 500 m/s.

    a. Show how the bullet travels 50 000 cm in 1 second.

    b. Calculate how long it takes for the bullet to travel 1 centimetre.

    c. If a super-slow-motion camera can take 100 000 pictures each second, determine how many shots would be taken by this camera to show the bullet travelling 1 cm.

22. Edward is repainting all the lines of a netball court at the local sports stadium. The dimensions of the netball court are shown.

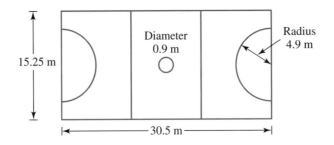

    a. Calculate the total length of lines that need to be repainted.

    Edward starts painting at 8 pm when the centre is closing, and it takes him $1\frac{1}{2}$ minutes on average to paint each metre of line.

    b. Show that it will take him 233 minutes to complete the job.

23. The radius of the Earth is accepted to be roughly 6400 km.

    a. Calculate how far, to the nearest km, you travel in one complete rotation of the Earth.

    b. As the Earth spins on its axis once every 24 hours, calculate the speed you are moving at.

    c. If the Earth is 150 000 000 km from the Sun, and it takes 365.25 days to circle around the Sun, show that the speed of the Earth's orbit around the Sun is 107 515 km/h. Give your answer to the nearest whole number.

Problem solving

24. One-fifth of an 80 cm length of jewellery wire is cut off. A further 22 cm length is then removed. Evaluate whether there is enough wire remaining to make a 40 cm necklace.

25. A church needs to repair one of its hexagonal stained glass windows. Use the information given in the diagram to calculate the width of the window.

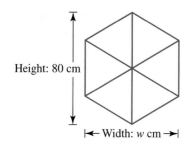

Height: 80 cm

Width: w cm

26. A spider is sitting in one top corner of a room that has dimensions 6 m by 4 m by 4 m. It needs to get to the corner of the floor that is diagonally opposite. The spider must crawl along the ceiling, then down a wall, until it reaches its destination.

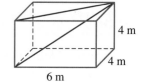

4 m

4 m

6 m

a. If the spider crawls first to the diagonally opposite corner of the ceiling, then down the wall to its destination, calculate the distance it would crawl.

b. Determine the shortest distance from the top back corner to the lower left corner.

9.3 Area

LEARNING INTENTION

At the end of this subtopic you should be able to:
- convert between units of area
- calculate areas of triangles, quadrilaterals and circles using formulas
- determine areas of composite shapes.

▶ 9.3.1 Area

eles-4853

- The diagram shows a square of side length 1 cm.
  By definition it has an area of 1 cm^2 (1 square centimetre).
  Note: This is a 'square centimetre', not a 'centimetre squared'.
- Area tells us how many squares it takes to cover a figure, so the area of the rectangle shown is 12 cm^2.
- Area is commonly measured in square millimetres $\left(\text{mm}^2\right)$, square centimetres $\left(\text{cm}^2\right)$, square metres $\left(\text{m}^2\right)$, or square kilometres $\left(\text{km}^2\right)$.

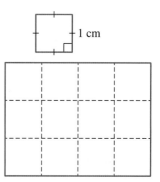

1 cm

Converting units of length

The following chart is useful when converting between units of area.

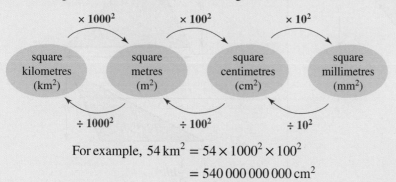

For example, $54 \text{ km}^2 = 54 \times 1000^2 \times 100^2$
$= 540\,000\,000\,000 \text{ cm}^2$

- Another common unit is the **hectare** (ha), a 100 m × 100 m square equal to 10 000 m², which is used to measure small areas of land.

> ### The hectare
>
> 1 hectare (ha) = 100 m × 100 m
>
> = 10 000 m²

WORKED EXAMPLE 5 Converting units of area

Convert 1.3 km² into:
a. **square metres**
b. **hectares.**

THINK	WRITE
a. There will be more m², so multiply by 1000².	a. $1.3 \times 1000^2 = 1\,300\,000 \text{ m}^2$
b. Divide the result of part **a** by 10 000, as 1 ha = 10 000 m².	b. $1\,300\,000 \div 10\,000 = 130 \text{ ha}$

⏵ 9.3.2 Using formulas to calculate area

eles-4854

- There are many useful formulas that can be used to find the areas of simple shapes. Some common formulas are summarised here.

Square	**Rectangle**
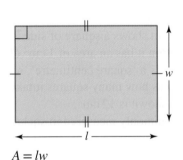 $A = l^2$	$A = lw$
Parallelogram	**Triangle**
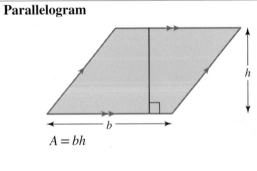 $A = bh$	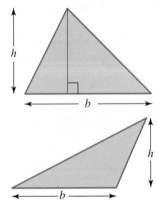 $A = \dfrac{1}{2}bh$

Trapezium	Kite
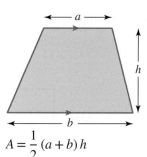 $A = \dfrac{1}{2}(a+b)\,h$	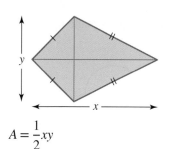 $A = \dfrac{1}{2}xy$
Circle	**Rhombus**
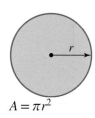 $A = \pi r^2$	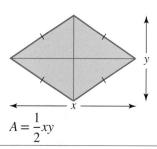 $A = \dfrac{1}{2}xy$

WORKED EXAMPLE 6 Calculating the areas of figures

Calculate the area of each of the following figures in cm², correct to 1 decimal place.

a.

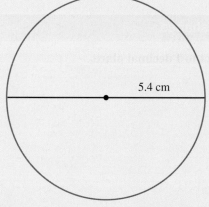

5.4 cm

b.

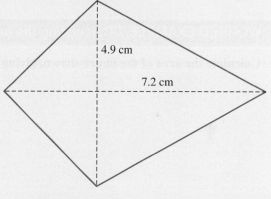

4.9 cm

7.2 cm

THINK	WRITE
a. 1. The figure is a circle. State the radius.	a. $r = 2.7\,\text{cm}$
2. Apply the formula for area of a circle: $A = \pi r^2$.	$\begin{aligned} A &= \pi r^2 \\ &= \pi \times 2.7^2 \\ &\approx 22.90 \end{aligned}$
3. Round the answer to 1 decimal place.	$A \approx 22.9\,\text{cm}^2$
b. 1. The figure is a kite. State the lengths of the diagonals.	b. $x = 4.9\,\text{cm}, \; y = 7.2\,\text{cm}$
2. Apply the formula for area of a kite: $A = \dfrac{1}{2}xy$.	$\begin{aligned} A &= \dfrac{1}{2}xy \\ &= \dfrac{1}{2} \times 4.9 \times 7.2 \\ &= 17.64 \end{aligned}$
3. Round the answer to 1 decimal place.	$A \approx 17.6\,\text{cm}^2$

ⓥ 9.3.3 Composite shapes

eles-4855

- A composite shape is made up of smaller, simpler shapes. Here are two examples.

Area = Area + Area + Area

= Area + Area

Area = Area − Area

- In the second example, the two semicircles are subtracted from the square to obtain the shaded area on the left.

WORKED EXAMPLE 7 Calculating the area of composite shapes

Calculate the area of the figure shown, giving your answer correct to 1 decimal place.

50 mm

THINK

1. Draw the diagram, divided into basic shapes.

WRITE/DRAW

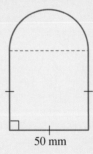

A_1

A_2

50 mm

2. A_1 is a semicircle of radius $\dfrac{50}{2} = 25$ mm.

   The area of a semicircle is half the area of a complete circle.

$A_1 = \dfrac{1}{2}\pi r^2$

3. Substitute $r = 25$ into the formula and evaluate, correct to 4 decimal places.

$$A_1 = \frac{1}{2} \times \pi \times 25^2$$
$$\approx 981.7477 \text{ mm}^2$$

4. A_2 is a square of side length 50 mm. Write the formula.

$$A_2 = l^2$$

5. Substitute $l = 50$ into the formula and evaluate A_2.

$$= 50^2$$
$$= 2500 \text{ mm}^2$$

6. Sum to calculate the total area.

$$\text{Total area} = A_1 + A_2$$
$$= 981.7477 + 2500$$
$$= 3481.7477 \text{ mm}^2$$

7. Round the final answer correct to 1 decimal place.

$$\approx 3481.7 \text{ mm}^2$$

 Resources

eWorkbook	Topic 9 Workbook (worksheets, code puzzle and project) (ewbk-2009)
Digital documents	SkillSHEET Substitution into area formulas (doc-10874)
	SkillSHEET Area of squares, rectangles, triangles and circles (doc-10875)
Video eLesson	Composite area (eles-1886)
Interactivities	Individual pathway interactivity: **Area** (int-4529)
	Conversion chart for area (int-3783)
	Area of rectangles (int-3784)
	Area of parallelograms (int-3786)
	Area of trapeziums (int-3790)
	Area of circles (int-3788)
	Area of rhombuses (int-3787)

Exercise 9.3 Area

learn on

Individual pathways

■ PRACTISE	■ CONSOLIDATE	■ MASTER
1, 2, 4, 7, 9, 13, 17, 19, 22, 27	3, 5, 8, 10, 14, 15, 18, 20, 23, 24, 28	6, 11, 12, 16, 21, 25, 26, 29, 30

To answer questions online and to receive **immediate corrective feedback** and **fully worked solutions** for all questions, go to your learnON title at www.jacplus.com.au.

Fluency

1. **MC** To convert an area measurement from square kilometres to square metres:

   A. divide by 1000 **B.** multiply by 1000 **C.** divide by 1 000 000

   D. multiply by 1 000 000 **E.** divide by 100

2. **WE5** Convert the following to the units shown.

   a. $13\,400 \text{ m}^2 =$ _____ km^2 b. $0.04 \text{ cm}^2 =$ _____ mm^2

   c. $3\,500\,000 \text{ cm}^2 =$ _____ m^2 d. $0.005 \text{ m}^2 =$ _____ cm^2

3. Convert the following to the units shown.
   a. $0.043 \text{ km}^2 =$ _____ m^2
   b. $200 \text{ mm}^2 =$ _____ cm^2
   c. $1.41 \text{ km}^2 =$ _____ ha
   d. $3800 \text{ m}^2 =$ _____ ha

4. **WE6** Calculate the area of each of the following shapes. Where appropriate, give your answer correct to 2 decimal places.

a.
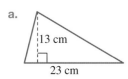
6 cm
4 cm

b.
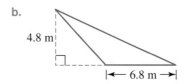
4 mm

c.
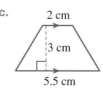
25 cm
43 cm

5. Calculate the area of each of the following shapes. Where appropriate, give your answer correct to 2 decimal places.

a.
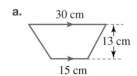
13 cm
23 cm

b.
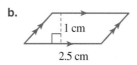
4.8 m
6.8 m

c.
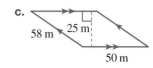
2 cm
3 cm
5.5 cm

6. Calculate the area of each of the following shapes. Where appropriate, give your answer correct to 2 decimal places.

a.
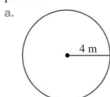
30 cm
13 cm
15 cm

b.

1 cm
2.5 cm

c.

58 m 25 m
50 m

7. Calculate the area of each of the following shapes. Where appropriate, give your answer correct to 2 decimal places.

a.
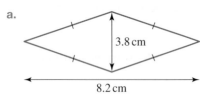
4 m

b.
2 mm

c.
3.4 m

8. Calculate the area of each of the following shapes. Where appropriate, give your answer correct to 2 decimal places.

a.
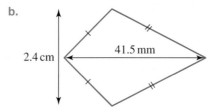
3.8 cm
8.2 cm

b.
2.4 cm
41.5 mm

c.
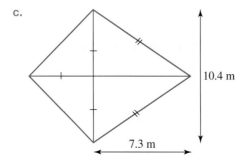
10.4 m
7.3 m

9. **WE7** Calculate the areas of the composite shapes shown. Where appropriate, express your answers correct to 1 decimal place.

a.
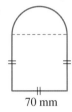
70 mm

b.
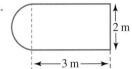
2 m
3 m

c.
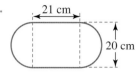
21 cm
20 cm

10. Calculate the areas of the composite shapes shown. Where appropriate, express your answers correct to 1 decimal place.

a.
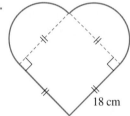
18 cm

b.

120 m
80 m

c.
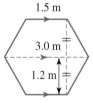
1.5 m
3.0 m
1.2 m

11. Calculate the areas of the composite shapes shown. Where appropriate, express your answers correct to 1 decimal place.

a.
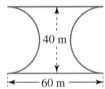
40 m
60 m

b.
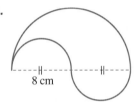
8 cm

12. Calculate the areas of the composite shapes shown. Where appropriate, express your answers correct to 1 decimal place.

a.
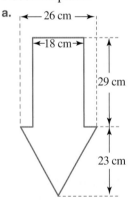
26 cm
18 cm
29 cm
23 cm

b.
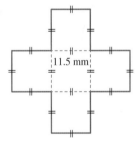
11.5 mm

Understanding

13. Calculate the area of the regular hexagon shown by dividing it into two trapeziums.

24 cm
21 cm
12 cm

14. Calculate the area of the regular octagon by dividing it into two trapeziums and a rectangle, as shown in the figure.

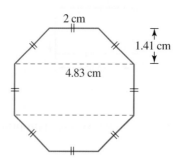

15. An annulus is a shape formed by two concentric circles (two circles with a common centre). Calculate the area of each of the annuli shown by subtracting the area of the smaller circle from the area of the larger circle. Give answers correct to 2 decimal places.

a.

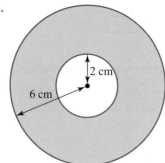

b.

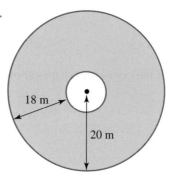

16. Calculate the area of each of the annuli shown (the shaded area). Give answers correct to 2 decimal places.

a.

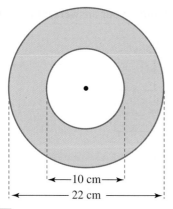

b.

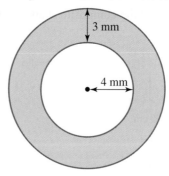

17. **MC** A pizza has a diameter of 30 cm. If your sister eats one-quarter, determine what area of pizza remains.

   A. $168.8 \, cm^2$ **B.** $530.1 \, cm^2$ **C.** $706.9 \, cm^2$ **D.** $176.7 \, cm^2$ **E.** $531.1 \, cm^2$

18. A circle has an area of $4500 \, cm^2$. Calculate its diameter to the nearest mm.

19. Determine the cost of covering the sportsground shown in the figure with turf, if the turf costs $7.50 per square metre.

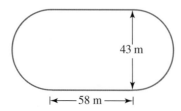

20. The Murray–Darling River Basin is Australia's largest catchment. Irrigation of farms in the Murray–Darling Basin has caused soil degradation due to rising salt levels. Studies indicate that about 500 000 hectares of the basin could be affected in the next 50 years.

   a. Convert the possible affected area to square kilometres. ($1\,km^2 = 100$ hectares)
   b. The total area of the Murray–Darling Basin is about 1 million square kilometres, about one-seventh of the continent. Calculate what percentage of this total area may be affected by salinity.

21. The plan shows two rooms, which are to be refloored.

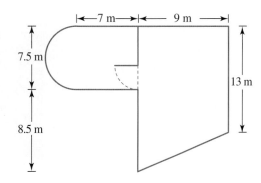

   Calculate the cost if the flooring costs $45 per square metre. Allow 10% more for wastage and round to the nearest $10.

Reasoning

22. A sheet of paper measures 29.5 cm by 21.0 cm.

   a. Calculate the area of the sheet of paper.
   b. Determine the radius of the largest circle that could be drawn on this sheet.
   c. Calculate the area of this circle.
   d. If the interior of the circle is shaded red, show that 56% of the paper is red.

23. A chessboard is made up of 8 rows and 8 columns of squares. Each square is $42\,cm^2$ in area. Show that the shortest distance from the upper right corner to the lower left corner of the chessboard is 73.32 cm.

24. Two rectangles of sides 15 cm by 10 cm and 8 cm by 5 cm overlap as shown. Show that the difference in area between the two non-overlapping sections of the rectangles is $110\,cm^2$.

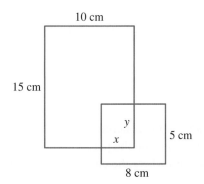

25. Answer the following questions using the information in the figure.

   a. Calculate the area of a square with side lengths 40 cm.
   b. If the midpoints of each side of the previous square are joined by straight lines to make another square, calculate the area of the smaller square.
   c. Now the midpoints of the previous square are also joined with straight lines to make another square. Calculate the area of this even smaller square.
   d. This process is repeated again to make an even smaller square. Calculate the area of this smallest square.
   e. Explain any patterns you observe.
   f. Determine the percentage of the original square's area taken up by the smallest square.
   g. Show that the area of the combined figure that is coloured pink is 1000 cm².

$\longleftarrow$ 40 cm $\longrightarrow$

26. Show that a square of perimeter $4x + 20$ has an area of $x^2 + 10x + 25$.

Problem solving

27. The area of a children's square playground is 50 m².

   a. Calculate the exact length of the playground.
   b. Pine logs 3 m long are to be laid around the playground. Determine how many logs need to be bought. (The logs can be cut into smaller pieces if required.)

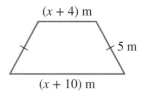

28. A sandpit is designed in the shape of a trapezium, with the dimensions shown. If the area of the sandpit is 14 m², determine its perimeter.

$(x + 4)$ m

5 m

$(x + 10)$ m

29. A rectangular classroom has a perimeter of 28 m and its length is 4 m shorter than its width. Determine the area of the classroom.

30. The playground equipment is half the length and half the width of a square kindergarten yard, as shown.

   a. Identify the fraction of the kindergarten yard that is occupied by the play equipment.
   b. During a working bee, the playground equipment area is extended 2 m in length and 1 m in width. If x represents the length of the kindergarten yard, write an expression for the area of the play equipment.
   c. Write an expression for the area of the kindergarten yard *not* taken up by the playground equipment.
   d. The kindergarten yard that is not taken up by the playground equipment is divided into 3 equal-sized sections:

      • A grassed section
      • A sandpit
      • A concrete section

   i. Write an expression for the area of the concrete section.
   ii. The children usually spend their time on the play equipment or in the sandpit. Write a simplified expression for the area of the yard where the children usually play.

9.4 Area and perimeter of a sector

LEARNING INTENTION

At the end of this subtopic you should be able to:
- identify sectors, semicircles and quadrants
- calculate the area of a sector
- calculate the length of the arc and the perimeter of a sector.

▶ 9.4.1 Sectors

eles-4856

- A **sector** is the shape created when a circle is cut by two radii.
- A circle, like a pizza, can be cut into many sectors.
- Two important sectors that have special names are the **semicircle** (half-circle) and the **quadrant** (quarter-circle).

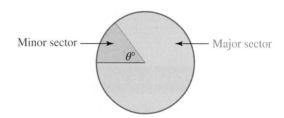
Minor sector — $\theta°$ — Major sector

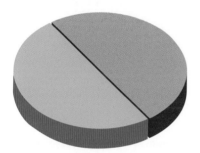
Semicircle

Quadrant

WORKED EXAMPLE 8 Calculating the area of a quadrant

Calculate the area enclosed by the figure, correct to 1 decimal place.

11 cm

THINK	WRITE
1. The figure is a quadrant or quarter-circle. Write the formula for its area.	$A = \dfrac{1}{4} \times \pi r^2$
2. Substitute $r = 11$ into the formula.	$A = \dfrac{1}{4} \times \pi \times 11^2$
3. Evaluate and round the answer correct to 1 decimal place. Include the units.	≈ 95.03 $\approx 95.0 \, \text{cm}^2$

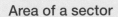 ## 9.4.2 Area of a sector

eles-4857

- Sectors are specified by the angle (θ) between the two radii.
- For example, in a quadrant, $\theta = 90°$, so a quadrant is $\dfrac{90}{360}$ or $\dfrac{1}{4}$ of a circle.

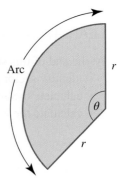

Area of a sector

- For any value of θ, the area of the sector is given by:

$$A_{\text{sector}} = \frac{\theta}{360} \times \pi r^2$$

where r is the radius of the sector.

WORKED EXAMPLE 9 Calculating the area of a sector

Calculate the area of the sector shown, correct to 1 decimal place.

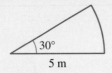

THINK	WRITE
1. This sector is $\dfrac{30}{360}$ of a circle. Write the formula for its area.	$A = \dfrac{30}{360} \pi r^2$
2. Substitute $r = 5$ into the formula.	$A = \dfrac{30}{360} \times \pi \times 5^2$ ≈ 6.54
3. Evaluate and round the answer correct to 1 decimal place. Include the units.	$\approx 6.5 \text{ m}^2$

 ## 9.4.3 Perimeter of a sector

eles-4858

- The perimeter, P, of a sector is the sum of the 2 radii and the curved section, which is called an arc of a circle.
- The length of the arc, l, will be $\dfrac{\theta}{360}$ of the circumference of the circle.

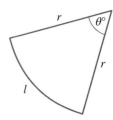

Arc length of a sector

- For any value of θ, the arc length, l, of the sector is given by:

$$l = \frac{\theta}{360} \times 2\pi r$$

where r is the radius of the sector.

- The perimeter of a sector can therefore be calculated using the following formula.

> ### Perimeter of a sector
>
> - The perimeter, P, of a sector is given by:
>
> $$P = 2r + l$$
>
> where r is the radius of the sector and l is the arc length.

WORKED EXAMPLE 10 Calculating the perimeter of a sector

Calculate the perimeter of the sector shown, correct to 1 decimal place.

THINK

1. The sector is $\dfrac{80}{360}$ of a circle. Write the formula for the length of the curved side.

2. Substitute $r = 3$ and evaluate l. Don't round off until the end.

3. Add all the sides together to calculate the perimeter.

4. Round the answer to 1 decimal place.

WRITE

$l = \dfrac{80}{360} \times 2\pi r$

$l = \dfrac{80}{360} \times 2 \times \pi \times 3$
$\approx 4.189 \text{ cm}$

$P = 4.189 + 3 + 3$
$= 10.189 \text{ cm}$

$P \approx 10.2 \text{ cm}$

  Resources

Exercise 9.4 Area and perimeter of a sector

Individual pathways

■ PRACTISE	■ CONSOLIDATE	■ MASTER
1, 2, 7, 9, 13, 15, 20, 23	3, 4, 8, 10, 14, 17, 19, 21, 24	5, 6, 11, 12, 16, 18, 22, 25

To answer questions online and to receive **immediate corrective feedback** and **fully worked solutions** for all questions, go to your learnON title at www.jacplus.com.au.

Fluency

1. Calculate the area of the semicircles shown, correct to 2 decimal places.

a. 6 cm

b. 20 cm

c. 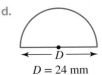 r
$r = 4.2$ cm

d. D
$D = 24$ mm

WE8 For each of the quadrants in questions **2–5**, calculate to 1 decimal place:

    i. the perimeter   **ii.** the area enclosed.

2.
4 cm

3.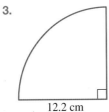
12.2 cm

4.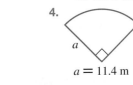
a
$a = 11.4$ m

5.
←1.5 m→

6. **MC** Select the correct formula for calculating the area of the sector shown.

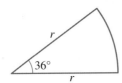
r
36°
r

  A. $A = \dfrac{1}{4}\pi r^2$ **B.** $A = \dfrac{3}{4}\pi r^2$ **C.** $A = \dfrac{1}{2}\pi r^2$

  D. $A = \dfrac{1}{100}\pi r^2$ **E.** $A = \dfrac{1}{10}\pi r^2$

WE9,10 For each of the sectors in questions **7–12**, calculate to 1 decimal place:

    i. the perimeter   **ii.** the area.

7.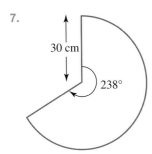
30 cm
238°

8.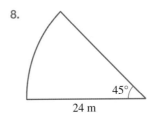
45°
24 m

9.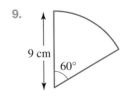
9 cm
60°

10.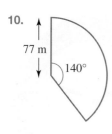

77 m

140°

11.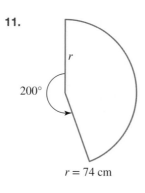

r

200°

r = 74 cm

12.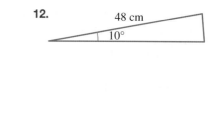

48 cm

10°

Understanding

13. A searchlight lights up the ground to a distance of 240 m. Calculate the area the searchlight illuminates if it can swing through an angle of 120°, as shown in the diagram. (Give your answer correct to 1 decimal place.)

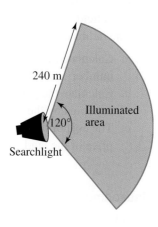

240 m

Illuminated area

120°

Searchlight

14. Calculate the perimeter, correct to 1 decimal place, of the figure shown.

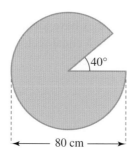

40°

80 cm

15. A goat is tethered by an 8.5 m rope to the outside of a corner post in a paddock, as shown in the diagram. Calculate the area of grass (shaded) on which the goat is able to graze. (Give your answer correct to 1 decimal place.)

16. A beam of light is projected onto a theatre stage as shown in the diagram.

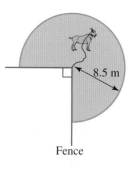

8.5 m

Fence

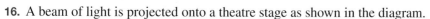

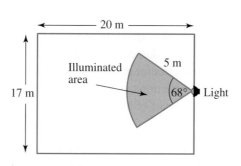

20 m

Illuminated area

5 m

17 m

68° Light

a. Calculate the illuminated area (correct to 1 decimal place) by first evaluating the area of the sector.

b. Calculate the percentage of the total stage area that is illuminated by the light beam.

17. A sector has an angle of 80° and a radius of 8 cm; another sector has an angle of 160° and a radius of 4 cm. Determine the ratio of the area of the first sector to the area of the second sector.

18. The minute hand on a vintage clock is 20 centimetres long and the hour hand is 12 centimetres long.

    a. Calculate the distance travelled by the minute hand from midday until 2 p.m.

    b. During this same time, calculate how far the hour hand travelled. Give your answers correct to the nearest centimetre.

19. Four baseball fields are to be constructed inside a rectangular piece of land. Each field is in the shape of a sector of a circle, as shown in light green. The radius of each sector is 80 m.

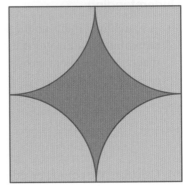

    a. Calculate the area of one baseball field, correct to the nearest whole number.

    b. Calculate the percentage, correct to 1 decimal place, of the total area occupied by the four fields.

    c. The cost of the land is $24 000 per hectare. Calculate the total purchase price of the land.

Reasoning

20. John and Jim are twins, and on their birthday they have identical birthday cakes, each with a diameter of 30 cm. Grandma Maureen cuts John's cake into 8 equal sectors. Grandma Mary cuts Jim's cake with a circle in the centre and then 6 equal portions from the rest.

John's cake

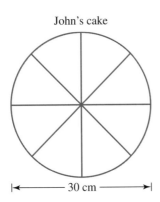

|← —— 30 cm —— →|

Jim's cake

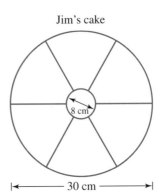

8 cm

|← —— 30 cm —— →|

    a. Show that each sector of John's cake makes an angle of 45° at the centre of the cake.

    b. Calculate the area of one slice of John's cake, correct to 1 decimal place.

    c. Calculate the area of the small central circular part of Jim's cake, correct to 1 decimal place.

    d. Calculate the area of one of the larger portions of Jim's cake, correct to 1 decimal place.

    e. If each boy eats one slice of the largest part of his own cake, does John eat more cake? Explain your answer.

21. A lighthouse has a light beam in the shape of a sector of a circle that rotates at 10 revolutions per minute and covers an angle of 40°. A person stands 200 m from the lighthouse and observes the beam. Show that the time between the end of one flash and the start of the next is approximately 5.33 seconds.

22. Answer the following questions. Where appropriate, give all answers to the nearest whole number.

   a. A donkey inside a square enclosure is tethered to a post at one of the corners. Show that a rope of length 120 m is required to prevent the donkey from eating more than half the grass in the enclosure.

   b. Suppose two donkeys are tethered at opposite corners of the square region shown. Calculate how long each rope should be so that the donkeys together can graze half of the area.

   c. This time four donkeys are tethered, one at each corner of the square region. Calculate how long each rope should be so that all the donkeys can graze only half of the area.

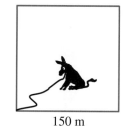
150 m

   d. Another donkey is tethered to a post inside an enclosure in the shape of an equilateral triangle. The post is at one of the vertices. Show that a rope of length 64 m is required so that the donkey eats only half of the grass in the enclosure.

   e. This time the donkey is tethered halfway along one side of the equilateral triangular region shown in the diagram. Calculate how long the rope should be so that the donkey can graze half of the area.

100 m   100 m
100 m

Problem solving

23. A new logo, divided into 6 equal sectors, has been designed based on two circles with the same centre, as shown in the diagram. The radius of the larger circle is 60 cm and the radius of the smaller circle is 20 cm. The shaded sections are to be coated with a special paint costing $145 a square metre. The edge of the shaded sections is made from a special material costing $36 a metre.
Calculate the cost to make the shaded sections. Give your answer to the nearest dollar.

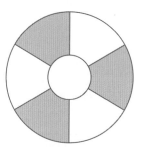

24. The metal washer shown has an inner radius of r cm and an outer radius of $(r+1)$ cm.

   a. State, in terms of r, the area of the circular piece of metal that was cut out of the washer.

   b. State, in terms of r, the area of the larger circle.

   c. Show that the area of the metal washer in terms of r is $\pi(2r+1)\,\text{cm}^2$.

   d. If r is 2 m, what is the exact area of the washer?

   e. If the area of the washer is $15\pi\,\text{cm}^2$, show that the radius would be 7 cm.

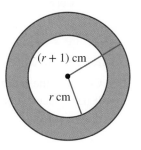
$(r+1)$ cm
r cm

25. The area of a sector of a circle is $\pi\,\text{cm}^2$, and the length of its arc is 2 cm. Determine the radius of the circle (in terms of π).

9.5 Surface area of rectangular and triangular prisms

LEARNING INTENTION

At the end of this subtopic you should be able to:
  • calculate the surface area of rectangular and triangular prisms.

▶ 9.5.1 Prisms and surface areas of prisms

eles-4859

  • A **prism** is a solid object with a uniform (unchanging) **cross-section** and all sides flat.

Triangular prism

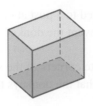

Rectangular prism (cuboid)

Hexagonal prism

- A prism can be sliced (cross-sectioned) in such a way that each 'slice' has an identical base.

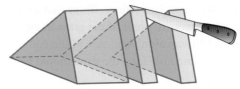

'Slicing' a prism into pieces produces congruent cross-sections.

- The following objects are not prisms because they do not have uniform cross-sections.

Cone

Sphere

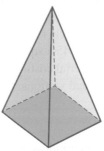

Square pyramid

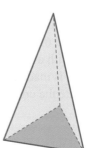

Triangular pyramid

Surface area of a prism

- Consider the **triangular prism** shown at left.
- It has 5 faces: 2 bases, which are right-angled triangles, and 3 rectangular sides. The net of the prism is drawn at right.

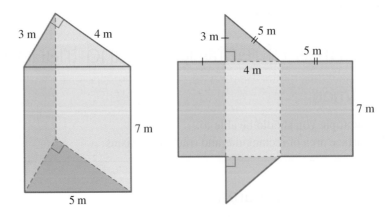

- The area of the net is the same as the total surface area (SA) of the prism.

Calculate the surface area of the rectangular prism (cuboid) shown.

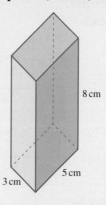

8 cm

5 cm

3 cm

THINK

1. There are 6 faces: 2 rectangular bases and 4 rectangular sides. Draw diagrams for each pair of faces and label each region.

2. Calculate the area of each rectangle by applying the formula $A = lw$, where l is the length and w is the width of the rectangle.

3. The total surface area (SA) is the sum of the area of 2 of each shape. Write the answer.

WRITE/DRAW

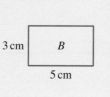

3 cm B

5 cm

R_1 8 cm

3 cm

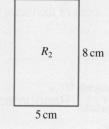

R_2 8 cm

5 cm

$B = 3 \times 5$
$\quad = 15$

$R_1 = 3 \times 8$
$\quad = 24$

$R_2 = 5 \times 8$
$\quad = 40$

$SA = 2 \times B + 2 \times R_1 + 2 \times R_2$
$\quad = 30 + 48 + 80$
$\quad = 158 \text{ cm}^2$

Calculate the surface area of the right-angled triangular prism shown.

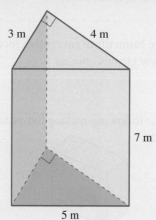

3 m   4 m

7 m

5 m

THINK	WRITE/DRAW

THINK

1. There are 5 faces: 2 triangles and 3 rectangles. Draw diagrams for each face and label each region.

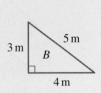

2. Calculate the area of the triangular base by applying the formula $A = \dfrac{1}{2}bh$.

$B = \dfrac{1}{2} \times 3 \times 4$

$= 6 \, \text{m}^2$

3. Calculate the area of each rectangular face by applying the formula $A = lw$.

$R_1 = 3 \times 7$ $R_2 = 4 \times 7$ $R_3 = 5 \times 7$

$= 21$ $= 28$ $= 35$

4. The total surface area is the sum of all the areas of the faces, including 2 bases.

$SA = 2 \times B + R_1 + R_2 + R_3$

$= 12 + 21 + 28 + 35$

$= 96 \, \text{m}^2$

Exercise 9.5 Surface area of rectangular and triangular prisms   **learn**

Individual pathways

■ PRACTISE	■ CONSOLIDATE	■ MASTER
1, 3, 7, 9, 12, 14, 17	2, 4, 6, 8, 10, 15, 18	5, 11, 13, 16, 19

To answer questions online and to receive **immediate corrective feedback** and **fully worked solutions** for all questions, go to your learnON title at www.jacplus.com.au.

Fluency

1. **WE11** Calculate the surface area of the following rectangular prisms (cuboids).

a.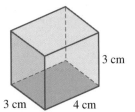

3 cm   3 cm   4 cm

b.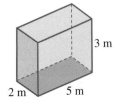

3 m   2 m   5 m

c.

1.1 m   0.8 m   1.3 m

2. Calculate the surface area of the following rectangular prisms (cuboids).

a.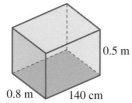

0.5 m

0.8 m 140 cm

b.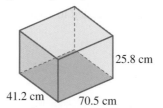

25.8 cm

41.2 cm 70.5 cm

c.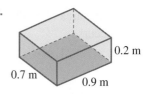

0.2 m

0.7 m 0.9 m

3. **WE12** Calculate the surface area of each of the following triangular prisms.

a.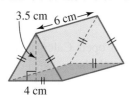

3.5 cm 6 cm

4 cm

b.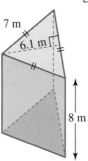

7 m

6.1 m

8 m

c.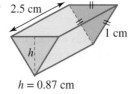

2.5 cm

1 cm

h

$h = 0.87$ cm

4. Calculate the surface area of each of the triangular prisms shown.

a.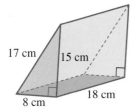

17 cm 15 cm

8 cm 18 cm

b.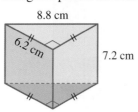

8.8 cm

6.2 cm 7.2 cm

c.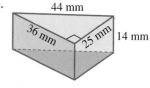

44 mm

36 mm 25 mm 14 mm

5. A shipping company is planning to buy and paint the outside surface of one of these shipping containers. Determine how many cans of paint the company should buy if the base of the container is not painted and each can of paint covers about 40 m².

6.5 m

2.8 m

3.2 m

6. The aim of the Rubik's cube puzzle is to make each face of the cube one colour. Calculate the surface area of the Rubik's cube if each small coloured square is 1.2 cm in length. Assume that there are no gaps between the squares.

1.2 cm

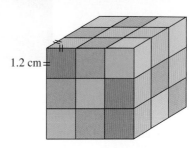

7. Determine how many square metres of iron sheet are needed to construct the water tank shown.

8. Calculate the surface area of the tank in question 7 if no top is made.

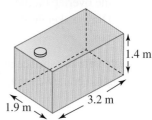

1.4 m

1.9 m 3.2 m

Understanding

9. Calculate the area of cardboard that would be needed to construct a box to pack this prism, assuming that no overlap occurs.

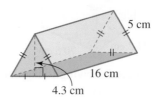

10. An aquarium is a triangular prism with the dimensions shown. The top of the tank is open. Calculate the area of glass that was required to construct the tank. Give your answer correct to 2 decimal places.

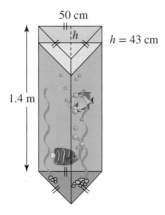

11. A tent is constructed as shown. Calculate the area of canvas needed to make the tent, including the floor.

12. a. Determine how many square centimetres of cardboard are needed to construct the shoebox shown, ignoring the overlap on the top.
    b. Draw a sketch of a net that could be used to make the box.

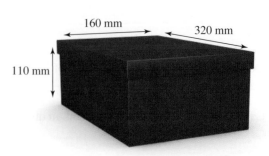

13. Calculate the surface area of a square-based prism of height 4 cm, given that the side length of its base is 6 cm.

Reasoning

14. a. Calculate the surface area of the toy block shown.

5 cm

b. If two of the blocks are placed together as shown, calculate the surface area of the prism that is formed.

5 cm

c. Calculate the surface area of the prism formed by three blocks.
d. Use the pattern to determine the surface area of a prism formed by eight blocks arranged in a line. Explain your reasoning.

15. A cube has a side length of 2 cm. Show that the least surface area of a solid formed by joining eight such cubes is 96 cm^2.

16. A prism has an equilateral triangular base with a perimeter of 12 cm. If the total surface area of the prism is 302 cm^2, show that the length of the prism is approximately 24 cm.

Problem solving

17. A wedge in the shape of a triangular prism, as shown in the diagram, is to be painted.

    a. Draw a net of the wedge so that it is easier to calculate the area to be painted.
    b. Calculate the area to be painted. (Do not include the base.)

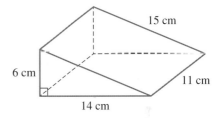

18. Ken wants to paint his son's bedroom walls blue and the ceiling white. The room measures 3 m by 4 m, with a ceiling height of 2.6 m. There is one door that measures 1 m by 2 m and one window that measures 1.8 m by 0.9 m. Each surface takes two coats of paint and 1 L of paint covers 16 m^2 on the walls and 12 m^2 on the ceiling.
    Cans of wall paint cost $33.95 for 1 L, $63.90 for 4 L, $147 for 10 L and $174 for 15 L. Ceiling paint costs $24 for 1 L and $60 for 4 L. Determine the cheapest options for Ken to paint the room.

19. A swimming pool has a length of 50 m and a width of 28 m. The shallow end of the pool has a depth of 0.80 m, which increases steadily to 3.8 m at the deep end.

    a. Calculate how much paint would be needed to paint the floor of the pool.
    b. If the pool is to be filled to the top, calculate how much water will be needed.

9.6 Surface area of a cylinder

▶ 9.6.1 Surface area of a cylinder

eles-4861

- A **cylinder** is a solid object with two identical flat circular ends and one curved side. It has a uniform cross-section.
- The net of a cylinder has two circular bases and one rectangular face. The rectangular face is the curved surface of the cylinder.
- Because the rectangle wraps around the circular base, the width of the rectangle is the same as the circumference of the circle. Therefore, the width is equal to $2\pi r$.
- The area of each base is πr^2, and the area of the rectangle is $2\pi rh$.

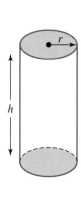

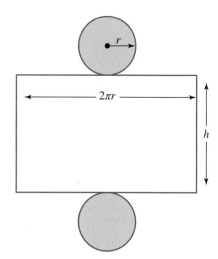

Surface area of a cylinder
$$\mathbf{SA} = 2\pi r^2 + 2\pi rh$$ $$= 2\pi r(r+h)$$

WORKED EXAMPLE 13 Calculating the surface area of a cylinder

a. Use the formula $A = 2\pi rh$ to calculate the area of the curved surface of the cylinder, correct to 1 decimal place.

b. Use the formula $SA = 2\pi rh + 2\pi r^2$ to calculate the surface area of the cylinder, correct to 1 decimal place.

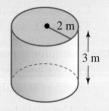

THINK	WRITE
a. 1. Write the formula for the curved surface area.	**a.** $A = 2\pi rh$
2. Identify the values of the pronumerals.	$r=2,\ h=3$
3. Substitute $r=2$ and $h=3$.	$A = 2 \times \pi \times 2 \times 3$
4. Evaluate, round to 1 decimal place and include units.	≈ 37.69 $\approx 37.7\,\text{m}^2$

b. 1. Write the formula for the surface area of a cylinder.

2. Identify the values of the pronumerals.

3. Substitute $r = 2$ and $h = 3$.

4. Evaluate, round to 1 decimal place and include units.

b. $SA = 2\pi rh + 2\pi r^2$

$r = 2, h = 3$

$SA = (2 \times \pi \times 2 \times 3) + (2 \times \pi \times 2^2)$

≈ 62.83

$\approx 62.8\, m^2$

Exercise 9.6 Surface area of a cylinder   **learn**on

Individual pathways

■ PRACTISE	■ CONSOLIDATE	■ MASTER
1, 2, 7, 9, 13, 15, 18	3, 4, 8, 10, 14, 16, 19	5, 6, 11, 12, 17, 20, 21

To answer questions online and to receive **immediate corrective feedback** and **fully worked solutions** for all questions, go to your learnON title at www.jacplus.com.au.

Fluency

WE13 For each of the cylinders in questions **1–6**, answer the following questions.

a. Use the formula $A = 2\pi rh$ to calculate the area of the curved surface of the cylinder.

b. Use the formula $SA = 2\pi rh + 2\pi r^2$ to calculate the total surface area of the cylinder.

1.

3 m

4 m

2.

3 cm

1.5 cm

3.

20 m

32 m

4.

17 cm

h

$h = 21$ cm

5.

1.4 m

1.5 m

6.

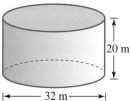

h

r

$r = 2.4$ m
$h = 1.7$ m

7. A can of energy drink has a height of 130 mm and a radius of 24 mm.
   a. Draw a net of the can.
   b. Calculate the surface area of the can, to the nearest cm^2.

8. A cylinder has a radius of 15 cm and a height of 45 mm.
   Determine its surface area correct to 1 decimal place.

Understanding

9. A cylinder has a surface area of 2000 cm^2 and a radius of 8 cm.
   Determine the cylinder's height correct to 2 decimal places.

10. A can of asparagus is 137 mm tall and has a diameter of 66 mm;
    a can of tomatoes is 102 mm tall and has a diameter of 71 mm;
    and a can of beetroot is 47 mm tall with a diameter of 84 mm.
    a. Determine which can has the largest surface area and
       which has the smallest surface area.
    b. Calculate the difference between the largest and smallest
       surface areas, correct to the nearest cm^2.

11. A 13 m-high storage tank was constructed from stainless steel
    (including the lid and the base). The diameter is 3 metres, as
    shown.
    a. Calculate the surface area of the tank. Give your answer
       correct to 2 decimal places.
    b. Determine the cost of the steel for the side of the tank if it
       comes in sheets 1 m wide that cost $60 a metre.

13 m

3 m

12. The concrete pipe shown in the diagram has the following measurements:
    $t = 30$ mm, $D = 18$ cm, $l = 27$ cm.
    Answer the following, giving your answers correct to 2 decimal places.

    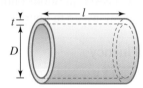

    a. Calculate the outer curved surface area.
    b. Calculate the inner curved surface area.
    c. Calculate the total surface area of both ends.
    d. Hence calculate the surface area for the entire shape.

13. Wooden mouldings are made by cutting cylindrical dowels in half, as shown.
    Calculate the surface area of the moulding.

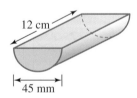

12 cm

45 mm

14. Kiara has a rectangular sheet of cardboard with dimensions 25 cm by 14 cm. She
    rolls the cardboard to form a cylinder so that the shorter side, 14 cm, is its height,
    and glues the edges together with a 1 cm overlap.

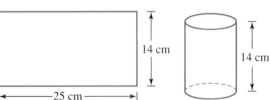

14 cm

25 cm

14 cm

   a. Determine the radius of the circle Kiara needs to construct to put at the top of her cylinder. Give your
      answer correct to 2 decimal places.
   b. Determine the total surface area of her cylinder if she also makes the top and bottom of her cylinder out
      of cardboard. Give your answer correct to 1 decimal place.

Reasoning

15. If the radius of a cylinder is twice its height, write a formula for the surface area in terms of its height only.

16. Cylinder A has a 10% greater radius and a 10% greater height than Cylinder B. Show that the ratio of their surface areas is 121 : 100.

17. Two identical cylinders (with height h cm and radius r cm) are modified slightly.
    The first cylinder's radius is increased by 10% and the second cylinder's height is increased by 20%. After these modifications have been made, the surface area of the second cylinder is greater than that of the first cylinder. Show that $h > 2.1r$.

Problem solving

18. An above-ground swimming pool has the following shape, with semicircular ends.
    Calculate how much plastic would be needed to line the base and sides of the pool.

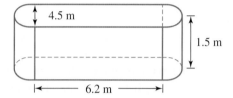

19. A over-sized solid wooden die is constructed for a children's playground. The side dimensions of the die are 50 cm. The number on each side of the die will be represented by cylindrical holes that will be drilled out of each side. Each hole will have a diameter of 10 cm and depth of 2 cm. All surfaces on the die will be painted (including the die holes). Show that the total area required to be painted is $1.63 \, \text{m}^2$.

20. The following letterbox is to be spray-painted on the outside. Calculate the total area to be spray-painted. Assume that the end of the letterbox is a semicircle above a rectangle. The letter slot is open and does not require painting.

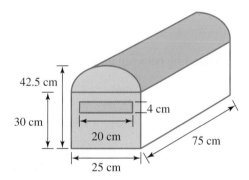

21. A timber fence is designed as shown. Evaluate how many square metres of paint are required to completely paint the fence front, back, sides and top with 2 coats of paint. Assume each paling is 2 cm in thickness and that the top of each paling is a semicircle.

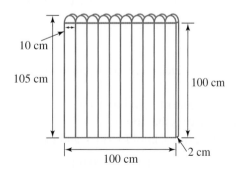

9.7 Volume of prisms and cylinders

⊙ 9.7.1 Volume and capacity

eles-4862

Volume

- The volume of a solid is the amount of space it fills or occupies.
- The diagram at far right shows a single cube of side length 1 cm. By definition, the cube has a **volume** of 1 cm³ (1 cubic centimetre). *Note:* This is a 'cubic centimetre', not a 'centimetre cubed'.
- The volume of some solids can be found by dividing them into cubes with 1 cm sides.
- Volume is commonly measured in cubic millimetres (mm³), cubic centimetres (cm³), cubic metres (m³) or cubic kilometres (km³).

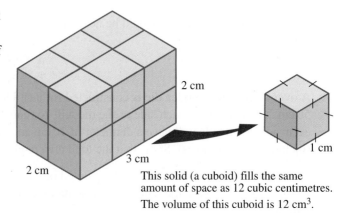

This solid (a cuboid) fills the same amount of space as 12 cubic centimetres. The volume of this cuboid is 12 cm³.

Converting units of volume

The following chart is useful when converting between units of volume.

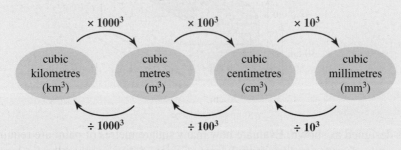

For example, $3\,m^3 = 3 \times 1000^3 \times 10^3\ mm^3$
$= 3\,000\,000\,000\ mm^3$

Capacity

- **Capacity** is another term for volume, which is usually applied to the measurement of liquids and containers.
- The capacity of a container is the volume of liquid that it can hold.
- The standard measurement for capacity is the litre (L).
- Other common units are the millilitre (mL), kilolitre (kL), and megalitre (ML), where:

$$1\,L = 1000\ mL$$
$$1\,kL = 1000\ L$$
$$1\,ML = 1\,000\,000\ L$$

Converting units of capacity

The following chart is useful when converting between units of capacity.

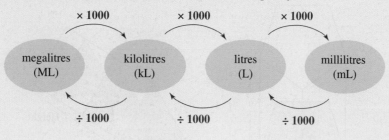

- The units of capacity and volume are related as follows:

Converting between units of volume and capacity

$$1\,cm^3 = 1\,mL$$
$$1000\,cm^3 = 1\,L$$
$$1\,m^3 = 1000\,L = 1\,kL$$

WORKED EXAMPLE 14 Converting units of volume and capacity

Convert:
a. $13.2\,L$ into cm^3
c. $0.13\,cm^3$ into mm^3
b. $3.1\,m^3$ into litres
d. $3.8\,kL$ into m^3.

THINK

a. 1. $1\,L = 1000\,mL$, so multiply by 1000.

   2. $1\,mL = 1\,cm^3$. Convert to cm^3.

b. $1\,m^3 = 1000\,L$, so multiply by 1000.

c.   There will be more mm^3, so multiply by 10^3.

d. $1\,kL = 1\,m^3$. Convert to m^3.

WRITE

a. $13.2 \times 1000 = 13\,200\,mL$

                $= 13\,200\,cm^3$

b. $3.1 \times 1000 = 3100\,L$

c. $0.13 \times 1000 = 130\,mm^3$

d. $3.8\,kL = 3.8\,m^3$

9.7.2 Volume of a prism

eles-4863

- The volume of a prism can be found by multiplying its cross-sectional area (A) by its height (h).

Volume of a prism

$$V_{prism} = A \times h$$

where:

A is the area of the base or cross-section of the prism

h is the perpendicular height of the prism.

- The cross-section (A) of a prism is often referred to as the base, even if it is not at the bottom of the prism.
- The height (h) is always measured perpendicular to the base, as shown in the following diagrams.

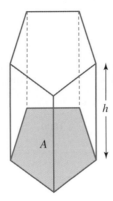

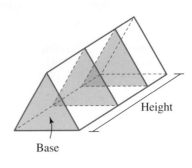

Volume of a cube or a cuboid

- A specific formula can be developed for the volumes of cubes and cuboids.

Cube	Cuboid (rectangular prism)
	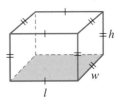
Volume = base area × height $= l^2 \times l$ $= l^3$	Volume = base area × height $= lw \times h$ $= lwh$

WORKED EXAMPLE 15 Calculating the volume of a prism

Calculate the volume of the hexagonal prism.

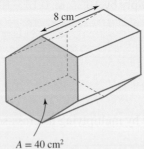

8 cm

$A = 40\ \text{cm}^2$

THINK	WRITE
1. Write the formula for the volume of a prism.	$V = A \times h$
2. Identify the values of the pronumerals.	$A = 40,\ h = 8$
3. Substitute $A = 40$ and $h = 8$ into the formula and evaluate. Include the units.	$V = 40 \times 8$ $= 320\ \text{cm}^3$

Calculate the volume of the prism.

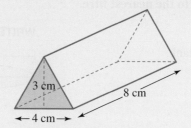

THINK	WRITE
1. The base of the prism is a triangle. Write the formula for the area of the triangle.	$A = \frac{1}{2}bh$
2. Substitute $b = 4$, $h = 3$ into the formula and evaluate.	$= \frac{1}{2} \times 4 \times 3$ $= 6 \text{ cm}^3$
3. Write the formula for the volume of a prism.	$V = A \times h$
4. State the values of A and h.	$A = 6, h = 8$
5. Substitute $A = 6$ and $h = 8$ into the formula and evaluate. Include the units.	$V = 6 \times 8$ $= 48 \text{ cm}^3$

▶ 9.7.3 Volume of a cylinder

eles-4860

- A cylinder has a circular base and uniform cross-section.
- A formula for the volume of a cylinder is shown.

Cylinder

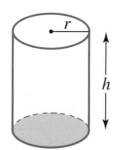

$$\text{Volume} = \text{base area} \times \text{height}$$
$$= \text{area of circle} \times \text{height}$$
$$= \pi r^2 \times h$$
$$= \pi r^2 h$$

Calculate the capacity, in litres, of a cylindrical water tank that has a diameter of 5.4 m and a height of 3 m. Give your answer correct to the nearest litre.

THINK	WRITE/DRAW
1. Draw a labelled diagram of the tank.	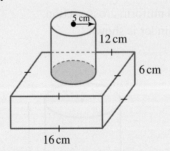
2. The base is a circle, so $A = \pi r^2$. Write the formula for the volume of a cylinder.	$V = \pi r^2 h$
3. Recognise the values of r and h. Recall that the radius, r, is half the diameter. $r = \dfrac{5.4}{2} = 2.7$ m	$r = 2.7,\ h = 3$
4. Substitute $r = 2.7$, $h = 3$ into the formula and calculate the volume.	$V = \pi \times (2.7)^2 \times 3$ $\approx 68.706\,631$ m^3
5. Convert this volume to litres (multiply by 1000).	$V = 68.706\,631 \times 1000$ $= 68\,706.631$ $\approx 68\,707$ L

Calculate the total volume of this solid.

THINK	WRITE/DRAW
1. The solid is made from two objects. Draw and label each object. Let V_1 = volume of the square prism. Let V_2 = volume of the cylinder.	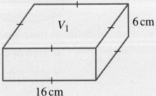 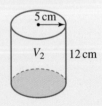
2. Calculate V_1 (square prism).	$V_1 = Ah$ $= 16^2 \times 6$ $= 1536$ cm^3

3. Calculate V_2 (cylinder).

$$V_2 = \pi r^2 h$$
$$= \pi \times 5^2 \times 12$$
$$\approx 942.478 \text{ cm}^3$$

4. To calculate the total volume, add the volumes found above.

$$V = V_1 + V_2$$
$$= 1536 + 942.478$$
$$= 2478.478$$
$$\approx 2478 \text{ cm}^3$$

Exercise 9.7 Volume of prisms and cylinders

learn on

Individual pathways

■ PRACTISE	■ CONSOLIDATE	■ MASTER
1, 4, 5, 8, 9, 12, 22, 25	2, 6, 10, 13, 14, 16, 17, 20, 23, 26	3, 7, 11, 15, 18, 19, 21, 24, 27

To answer questions online and to receive **immediate corrective feedback** and **fully worked solutions** for all questions, go to your learnON title at www.jacplus.com.au.

Fluency

1. **WE14** Convert the following units into mL.

   a. 325 cm^3 b. 2.6 m^3 c. 5.1 L d. 0.63 kL

2. Convert the following units into cm^3.

   a. 5.8 mL b. 6.1 L c. 3.2 m^3 d. 59.3 mm^3

3. Convert the following units into kL.

   a. 358 L b. 55.8 m^3 c. 8752 L d. 5.3 ML

4. Calculate the volumes of the cuboids shown. Assume that each small cube has sides of 1 cm.

a.
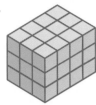

b.
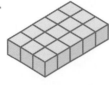

c.
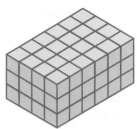

5. **WE15** Calculate the volumes of these objects.

a.
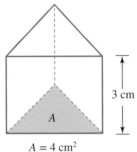

3 cm

$A = 4 \text{ cm}^2$

b.
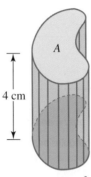

4 cm

$A = 17 \text{ cm}^2$

c.

$A = 3.2 \text{ m}^2$
$h = 3.0 \text{ m}$

6. Calculate the volumes of these objects.

a.
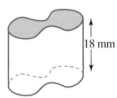

18 mm

Base area = 35 mm²

b.
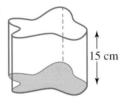

15 cm

Base area = 28 cm²

c.
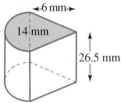

6 mm

14 mm

26.5 mm

7. Calculate the volumes of these objects.

a.
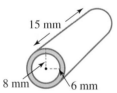

15 mm

8 mm   6 mm

b.
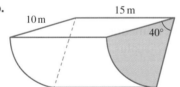

15 m

10 m

40°

c.
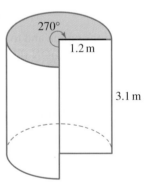

270°

1.2 m

3.1 m

8. Calculate the volumes of these rectangular prisms.

a.
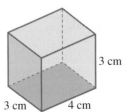

3 cm

3 cm   4 cm

b.
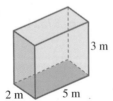

3 m

2 m   5 m

c.
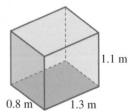

1.1 m

0.8 m   1.3 m

9. **WE16** Calculate the volumes of the prisms shown. Give your answers correct to 2 decimal places.

a.
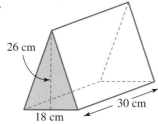
26 cm

18 cm

30 cm

b.
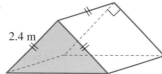
2.4 m

c.
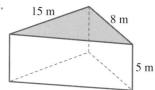
15 m

8 m

5 m

d.
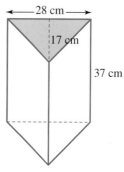
28 cm

17 cm

37 cm

10. Calculate the volume of the following cylinders. Give your answers correct to 1 decimal place.

a.
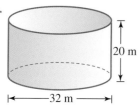
20 m

32 m

b.
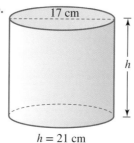
17 cm

h

$h = 21$ cm

c.
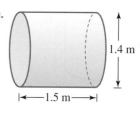
1.4 m

1.5 m

11. Calculate the volume of the following cylinders. Give your answers correct to 1 decimal place.

a.

h

r

$r = 2.4$ m
$h = 1.7$ m

b.
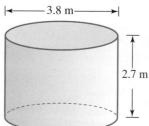
3.8 m

2.7 m

c.
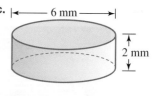
6 mm

2 mm

12. **WE17** Calculate the volume of water, in litres (L), that can fill a cylindrical water tank that has a diameter of 3.2 m and a height of 1.8 m.

13. Calculate the capacity of the Esky shown, in litres.

0.5 m

0.42 m

0.84 m

14. WE18 Calculate the volume of each solid to the nearest cm³.

a.

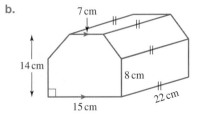

48 cm

44 cm

b.

7 cm

14 cm

8 cm

15 cm

22 cm

c.

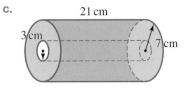

21 cm

3 cm

7 cm

15. Calculate the volume of each solid to the nearest cm³.

a.

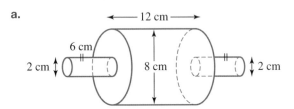

12 cm

6 cm

2 cm

8 cm

2 cm

b.

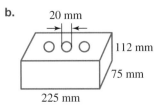

20 mm

112 mm

75 mm

225 mm

c.

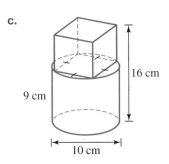

16 cm

9 cm

10 cm

Understanding

16. Calculate the capacity, in litres, of the cylindrical storage tank shown. Give your answer correct to 1 decimal place.

3.6 m   7.4 m

17. Calculate the capacity (in mL) of this cylindrical coffee plunger when it is filled to the level shown. Give your answer correct to 1 decimal place.

18. Sudhira is installing a rectangular pond in a garden. The pond is 1.5 m wide, 2.2 m long and has a uniform depth of 1.5 m.

    a. Calculate the volume of soil (in m^3) that Sudhira must remove to make the hole before installing the pond.
    b. Calculate the capacity of the pond in litres. (Ignore the thickness of the walls in this calculation.)

19. Until its closure in 2001, Fresh Kills on Staten Island outside New York City was one of the world's biggest landfill garbage dumps. (New Yorkers throw out about 100 000 tonnes of refuse weekly.) Calculate the approximate volume (in m^3) of the Fresh Kills landfill if it covers an area of 1215 hectares and is about 240 m high. (*Note:* 1 hectare = $10 000 m^2$)

20. Calculate the internal volume (in cm^3) of the wooden chest shown, correct to 2 decimal places. (Ignore the thickness of the walls.)

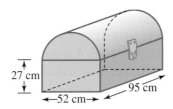

21. Answer the following questions using the given information.

    a. Calculate the volume, in m^3, of the refrigerator shown.
    b. Determine the capacity of the refrigerator if the walls are all 5 cm thick.

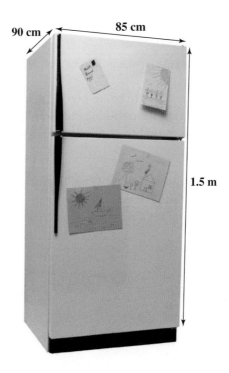

Reasoning

22. **a.** Calculate the volume of plastic needed to make the door wedge shown. Give your answer correct to 2 decimal places.

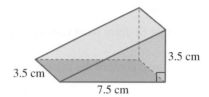

3.5 cm

3.5 cm

7.5 cm

b. The wedges can be packed snugly into cartons with dimensions 45 cm × 70 cm × 35 cm. Determine how many wedges fit into each carton.

23. A cylindrical glass is designed to hold 1.25 L. Show that its height is 132 mm if it has a diameter of 110 mm.

24. Mark is responsible for the maintenance of the Olympic (50 m) pool at an aquatic centre. The figure shows the dimensions of an Olympic pool.

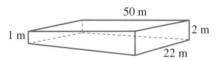

50 m

1 m

2 m

22 m

a. Identify the shape of the pool.
b. Draw the cross-section of the prism and calculate its area.
c. Show that the capacity of the pool is 1 650 000 L.
d. Mark needs to replace the water in the pool every 6 months. If the pool is drained at 45 000 L per hour and refilled at 35 000 L per hour, calculate how long it will take to:
　i. drain
　ii. refill (in hours and minutes).

Problem solving

25. A cylindrical container of water has a diameter of 16 cm and is 40 cm tall. Determine how many full cylindrical glasses can be filled from the container if the glasses have a diameter of 6 cm and are 12 cm high.

26. An internal combustion engine consists of 4 cylinders. In each cylinder a piston moves up and down. The diameter of each cylinder is called the bore and the height that the piston moves up and down within the cylinder is called the stroke (stroke = height).

a. If the bore of a cylinder is 84 mm and the stroke is 72 mm, calculate the volume (in litres) of 4 such cylinders, correct to 2 decimal places.
b. When an engine gets old, the cylinders have to be 're-bored'; that is, the bore is increased by a small amount (and new pistons put in them). If the re-boring increases the diameter by 1.1 mm, determine the increase in volume of the four cylinders (in litres). Give your answer correct to 3 decimal places.

27. Answer the following questions.

a. A square sheet of metal with dimensions 15 cm by 15 cm has a 1 cm square cut out of each corner. The remainder of the square is folded to form an open box. Calculate the volume of the box.
b. Write a general formula for calculating the volume of a box created from a metal sheet of any size with any size square cut out of the corners.

9.8 Review

9.8.1 Topic summary

MEASUREMENT

Units of measurement

- Length:

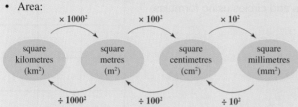

- Area:

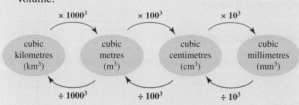

- Volume:

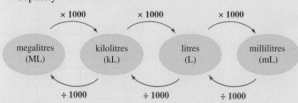

- Capacity

- When converting between units of capacity and volume:
  - $1 \text{ cm}^3 = 1 \text{ mL}$
  - $1000 \text{ cm}^3 = 1 \text{ L}$
  - $1 \text{ m}^3 = 1000 \text{ L} = 1 \text{ kL}$

Area and perimeter of a sector

- The curved part of a sector is often labeled l. Its length can be calculated by the formula $l = \dfrac{\theta}{360} \times 2\pi r$, where r is the radius and θ is the angle between the two straight sides.
- The perimeter of a sector is: $P = 2r + l$
- The area of a sector is: $A = \dfrac{\theta}{360} \times \pi r^2$

Perimeter and circumference

- The **perimeter** is the distance around a shape.
- The perimeter of a circle is called the **circumference**.
- The circumference of a circle can be calculated using either of the following formulas, where r is the radius of the circle and d is the diameter of the circle:
  - $C = 2\pi r$
  - $C = \pi d$

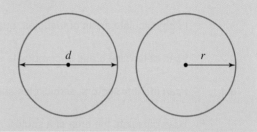

Area

- The area of a shape is the amount of space it takes up.
- The areas of common shapes can be calculated using the following formulas:
  - Square: $A = l^2$
  - Rectangle: $A = lw$
  - Parallelogram: $A = bh$
  - Triangle: $A = \dfrac{1}{2}bh$
  - Trapezium: $A = \dfrac{1}{2}(a + b)h$
  - Kite: $A = \dfrac{1}{2}xy$
  - Circle: $A = \pi r^2$
  - Rhombus: $A = \dfrac{1}{2}xy$

Surface area

- The surface area of an object can be calculated by adding the areas of all its faces.
- The formula for the surface area of a cylinder is: $SA = 2\pi r^2 + 2\pi rh = 2\pi r(r + h)$

Volume

- The volume of a prism is equal to its cross-sectional area multiplied by its height: $V_{\text{prism}} = A \times h$
- The volume of a cube is: $V_{\text{cube}} = l^3$
- The volume of a cuboid is: $V_{\text{cuboid}} = lwh$
- The volume of a cylinder is: $V_{\text{cylinder}} = \pi r^2 h$
- Capacity is another term for volume, and is usually applied to the measurement of liquids and containers.

9.8.2 Success criteria

Tick a column to indicate that you have completed the subtopic and how well you think you have understood it using the traffic light system.

(**Green:** I understand; **Yellow:** I can do it with help; **Red:** I do not understand)

Subtopic	Success criteria	⬤	⬤	⬤
9.2	I can convert between different units of length.			
	I can calculate the perimeter of a given shape.			
	I can calculate the circumference of a circle.			
9.3	I can convert between units of area.			
	I can calculate areas of triangles, quadrilaterals and circles using formulas.			
	I can determine areas of composite shapes.			
9.4	I can identify sectors, semicircles and quadrants.			
	I can calculate the area of a sector.			
	I can calculate the length of the arc and the perimeter of a sector.			
9.5	I can calculate the surface area of rectangular and triangular prisms.			
9.6	I can draw the net of a cylinder.			
	I can calculate the curved surface area of a cylinder.			
	I can calculate the total surface area of a cylinder.			
9.7	I can calculate the volume of prisms and cylinders.			
	I can convert between units of volume.			
	I can convert between units of capacity and units of volume.			

9.8.3 Project

Areas of polygons

The area of plane figures can be found using a formula. For example, if the figure is a rectangle, its area is found by multiplying its length by its width. If the figure is a triangle, the area is found by multiplying half the base length by its perpendicular height. If the figures are complex, break them down into simple shapes and determine the area of each of these shapes. The area of a complex shape is the sum of the area of its simple shapes. Drawing polygons onto grid paper is one method that can be used to determine their area.

Consider the following seven polygons drawn on 1-cm grid paper.

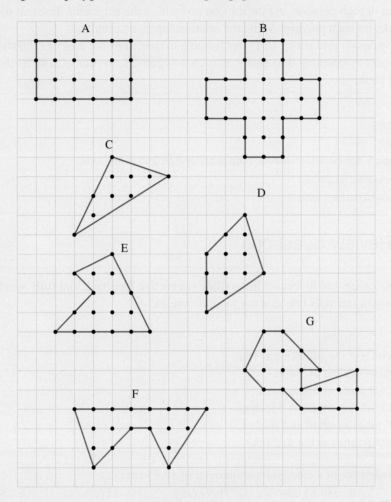

1. By counting the squares or half-squares, determine the area of each polygon in cm². In some cases, it may be necessary to divide some sections into half-rectangles in order to determine the exact area.

Formulas define a relationship between dimensions of figures. In order to search for a formula to calculate the area of polygons drawn on the grid paper, consider the next question.

2. For each of the polygons provided above, complete the following table. Count the number of dots on the perimeter of each polygon and count the number of dots that are within the perimeter of each polygon.

Polygon	Dots on perimeter (b)	Dots within perimeter (i)	Area of polygon (A)
A			
B			
C			
D			
E			
F			
G			

3. Choose a pronumeral to represent the headings in the table. Investigate and determine a relationship between the area of each polygon and the dots on and within the perimeter. Test that the relationship determined works for each polygon. Write the relationship as a formula.

4. Draw some polygons on grid paper. Use your formula to determine the area of each shape. Confirm that your formula works by counting the squares in each polygon. The results of both methods (formula and method) should be the same.

 Resources

 eWorkbook   Topic 9 Workbook (worksheets, code puzzle and project) (ewbk-2009)

Interactivities   Crossword (int-2709)

                   Sudoku puzzle (int-3211)

Exercise 9.8 Review questions

learnon

To answer questions online and to receive **immediate corrective feedback** and **fully worked solutions** for all questions, go to your learnON title at www.jacplus.com.au.

Fluency

1. **MC** Identify which of the following is true.
   A. 5 cm is 100 times as big as 5 mm.
   B. 5 metres is 100 times as big as 5 cm.
   C. 5 km is 1000 times as small as 5 metres.
   D. 5 mm is 100 times as small as 5 metres.
   E. 5 metres is 1000 times as big as 5 cm.

2. **MC** The circumference of a circle with a diameter of 12.25 cm is:
   A. 471.44 cm   B. 384.85 mm   C. 76.97 cm   D. 117.86 cm   E. 384.85 cm

3. **MC** The area of the given shape is:
   A. 216 m^2 B. 140 m^2 C. 150 m^2 D. 90 m^2 E. 141 m^2

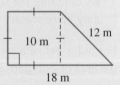

4. **MC** The area of a circle with diameter 7.5 cm is:
   A. 176.7 cm^2 B. 47.1 cm^2 C. 23.6 cm^2 D. 44.2 cm^2 E. 44.5 cm^2

5. **MC** The perimeter of the shape shown is:
   A. 1075.2 cm
   B. 55.5 cm
   C. 153.2 cm
   D. 66.1 cm
   E. 661 cm

6. **MC** The surface area and volume of a cube with side length 7 m are respectively:
   A. $294 \text{ m}^2, 343 \text{ m}^3$ B. $49 \text{ m}^2, 343 \text{ m}^3$ C. $147 \text{ m}^2, 49 \text{ m}^3$
   D. $28 \text{ m}^2, 84 \text{ m}^3$ E. $295 \text{ m}^2, 343 \text{ m}^3$

7. **MC** The surface area of a rectangular box with dimensions 7 m, 3 m, 2 m is:

    A. 41 m^2 B. 42 m^2 C. 72 m^2 D. 82 cm^2 E. 82 m^2

8. **MC** The surface area and volume of a cylinder with radius 35 cm and height 40 cm are:

    A. 16 493.36 cm^2 and 153 938 cm^3
    B. 8796.5 cm^2 and 11 246.5 cm^3
    C. 153 938 cm^2 and 11 246.5 cm^3
    D. 8796.5 cm^2 and 153 938 cm^3
    E. 16 493.36 m^2 and 153 938 cm^3

9. Convert the following lengths into the required units.

    a. 26 mm = _____ cm   b. 1385 mm = _____ cm
    c. 1.63 cm = _____ mm   d. 1.5 km = _____ m
    e. 0.077 km = _____ m   f. 2850 m = _____ km

10. Calculate the cost of 1.785 km of cable, if the cable costs $4.20 per metre.

11. Calculate the circumference of circles with the following dimensions (correct to 1 decimal place).

    a. Radius 4 cm   b. Radius 5.6 m   c. Diameter 12 cm

12. Calculate the perimeters of the following shapes (correct to 1 decimal place).

    a.
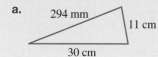

    b.

    c.
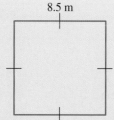

    d.
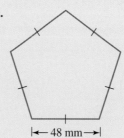

13. Calculate the perimeter of the following shapes (correct to 1 decimal place).

    a.
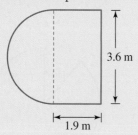

    b.
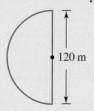

    c.
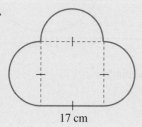

    d.
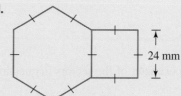

14. Calculate the area of the following shapes.

a. 5 m

b. 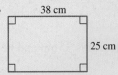 38 cm, 25 cm

c. 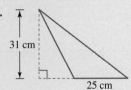 31 cm, 25 cm

d. 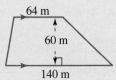 64 m, 60 m, 140 m

15. Calculate the area of the following shapes.

a. 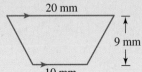 20 mm, 9 mm, 10 mm

b. 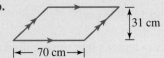 31 cm, 70 cm

c. 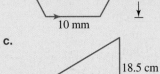 18.5 cm, 30.2 cm

d. 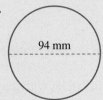 94 mm

16. Calculate the area of the following 2-dimensional shapes by dividing them into simpler shapes. (Where necessary, express your answer correct to 1 decimal place.)

a. 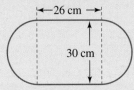 26 cm, 30 cm

b. 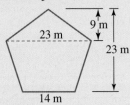 23 m, 9 m, 23 m, 14 m

c. 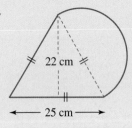 22 cm, 25 cm

d. 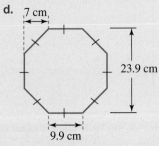 7 cm, 23.9 cm, 9.9 cm

17. Calculate the area of the sectors shown (correct to 2 decimal places).

a. 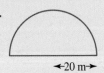 20 m

b. 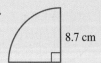 8.7 cm

c. 70°, 18 cm

d. 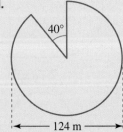 40°, 124 m

18. Calculate the perimeter of the sectors in question **17**, correct to 2 decimal places.

19. Calculate the inner surface area of the grape-collecting vat using the dimensions for length, width and depth shown.

20. Calculate the surface area of the triangular prism shown.

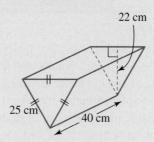

21. Calculate the volume of each of the following, giving your answers to 2 decimal places where necessary.

a.

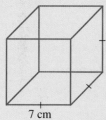

b.

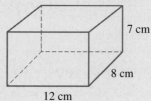

c.

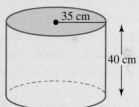

d.

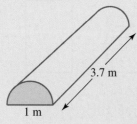

e.

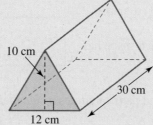

Problem solving

22. Calculate the cost of painting an outer surface (including the lid) of a cylindrical water tank that has a radius of 2.2 m and a height of 1.6 m, if the paint costs $1.90 per square metre.

23. A harvester travels at 11 km/h. The comb (harvesting section) is 8.7 m wide. Determine how many hectares per hour can be harvested with this comb.

24. Nina decides to invite some friends for a sleepover. She plans to have her guests sleep on inflatable mattresses with dimensions 60 cm by 170 cm. A plan of Nina's bedroom is shown.

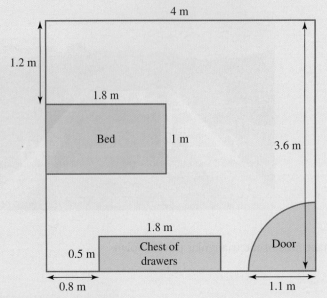

a. Calculate the area of floor space available.

b. The area of an inflatable mattress is 1.02 m². Nina's best friend suggests that, by dividing the available floor space by the area of one inflatable mattress, the number of mattresses that can fit into the bedroom can be calculated. Explain whether the friend is correct.

c. Determine how many self-inflatable mattresses can fit in the room as it is.

d. Explain how your answer to c would change if the bed could be moved within the bedroom.

25. A block of chocolate is in the shape of a triangular prism. The face of the prism is an equilateral triangle of 5 cm sides, and the length is 22 cm.

a. Determine the area of one of the triangular faces.

b. Calculate the total surface area of the block.

c. Calculate the volume of the block.

d. If it cost $0.025 per cm³ to produce the block, calculate the total cost of production.

26. An A4 sheet of paper has dimensions 210 mm × 297 mm and can be rolled two different ways (by rotating the paper) to make baseless cylinders, as illustrated. Compare the volumes to decide which shape has the greatest volume.

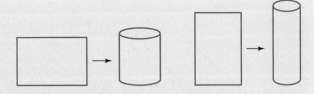

27. These congruent squares have shaded parts of circles inside them.

i. 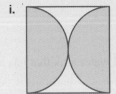   ii.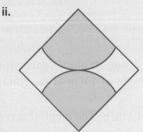

Compare the shaded area in figure **i** with that in figure **ii**.

28. The widespread manufacture of soft drinks in cans only really started in the 1960s. Today cans of soft drink are very common. They can be purchased in a variety of sizes. You can also purchase them as a single can or bulk-buy them in larger quantities. Does the arrangement of the cans in a multi-pack affect the amount of packaging used to wrap them?

For this investigation we will look at the packaging of 375 mL cans. This size of can has a radius of approximately 3.2 cm and a height of approximately 12 cm. We will consider packaging these cans in cardboard in the shape of a prism, ignoring any overlap for glueing.

   a. A 6-can pack could be packed in a single row of cans.
      i. Draw a diagram to show what this pack would look like.
      ii. Calculate the amount of cardboard required for the package.
   b. The cans could also be packaged in a 3 × 2 arrangement.
      i. Show the shape of this pack.
      ii. Calculate the amount of cardboard required for this pack.
   c. Evaluate which of the two 6-can pack arrangements is the more economical in terms of packaging.
   d. These cans can be packaged as a single layer of 4 × 3. Calculate how much cardboard this would require.
   e. They could also have a two-level arrangement of 3 × 2 each layer. Determine what packaging this would require.
   f. It is also possible to have a three-level arrangement of 2 × 2 each layer. Determine how much packaging would be needed in this case.
   g. Comment on the best way to package a 12-can pack so that it uses the minimum amount of wrapping. Provide diagrams and mathematical evidence to support your conclusion.

 To test your understanding and knowledge of this topic, go to your learnON title at www.jacplus.com.au and complete the **post-test**.

Online Resources

Below is a full list of **rich resources** available online for this topic. These resources are designed to bring ideas to life, to promote deep and lasting learning and to support the different learning needs of each individual.

eWorkbook

Download the workbook for this topic, which includes worksheets, a code puzzle and a project (ewbk-2009) ☐

Solutions

Download a copy of the fully worked solutions to every question in this topic (sol-0729) ☐

Digital documents

9.2 SkillSHEET Converting units of length (doc-6294) ☐
SkillSHEET Substitution into perimeter formulas (doc-10872) ☐
SkillSHEET Perimeter of squares, rectangles, triangles and circles (doc-10873) ☐
9.3 SkillSHEET Substitution into area formulas (doc-10874) ☐
SkillSHEET Area of squares, rectangles, triangles and circles (doc-10875) ☐
9.5 SkillSHEET Surface area of cubes and rectangular prisms (doc-10876) ☐
SkillSHEET Surface area of triangular prisms (doc-10877) ☐
9.7 SkillSHEET Volume of cubes and rectangular prisms (doc-10878) ☐
SkillSHEET Volume of triangular prisms (doc-10879) ☐
SkillSHEET Volume of cylinders (doc-10880) ☐

Video eLessons

9.2 Length (eles-4851) ☐
Perimeter (eles-4852) ☐
9.3 Area (eles-4853) ☐
Using formulas to calculate area (eles-4854) ☐
Composite shapes (eles-4855) ☐
Composite area (eles-1886) ☐
9.4 Sectors (eles-4856) ☐
Area of a sector (eles-4857) ☐
Perimeter of a sector (eles-4858) ☐
9.5 Prisms and surface area of prisms (eles-4859) ☐
Total surface area of prisms (eles-1909) ☐
9.6 Surface area of a cylinder (eles-4861) ☐
9.7 Volume and capacity (eles-4862) ☐
Volume of a prism (eles-4863) ☐
Volume of a cylinder (eles-4860) ☐

Interactivities

9.2 Individual pathway interactivity: Measurement (int-4528) ☐
Converting units of length (int-4011) ☐
Perimeter (int-4013) ☐
Circumference (int-3782) ☐

9.3 Individual pathway interactivity: Area (int-4529) ☐
Conversion chart for area (int-3783) ☐
Area of rectangles (int-3784) ☐
Area of parallelograms (int-3786) ☐
Area of trapeziums (int-3790) ☐
Area of circles (int-3788) ☐
Area of rhombuses (int-3787) ☐
9.4 Individual pathway interactivity: Area and perimeter of a sector (int-4530) ☐
Area of a sector (int-6076) ☐
Perimeter of a sector (int-6077) ☐
9.5 Individual pathway interactivity: Surface area of rectangular and triangular prisms (int-4531) ☐
Cross-sections (int-6078) ☐
Surface area of prisms (int-6079) ☐
9.6 Individual pathway interactivity: Surface area of a cylinder (int-4532) ☐
Surface area of a cylinder (int-6080) ☐
9.7 Individual pathway interactivity: Volume of prisms and cylinders (int-4533) ☐
Volume (int-3791) ☐
Prisms? (int-3792) ☐
9.8 Crossword (int-2709) ☐
Sudoku puzzle (int-3211) ☐

Teacher resources

There are many resources available exclusively for teachers online.

To access these online resources, log on to **www.jacplus.com.au**.

Answers

Topic 9 Measurement

Exercise 9.1 Pre-test

1. $15.4\,\text{cm}$
2. $12.48\,\text{cm}^2$
3. $30\,\text{cm}$
4. C
5. $1440\,\text{mm}^3$
6. E
7. B
8. D
9. B
10. $3.48\,\text{m}^2$
11. E
12. $3100\,\text{cm}^3$
13. $678.6\,\text{mL}$
14. $25\,\text{m}$
15. $527.8\,\text{cm}^2$

Exercise 9.2 Measurement

1. a. $50\,\text{mm}$ b. $152\,\text{cm}$ c. $0.0125\,\text{m}$ d. $32.2\,\text{mm}$
2. a. $0.006\,57\,\text{km}$ b. $0.000\,64\,\text{km}$
   c. $14.35\,\text{mm}$ d. $0.000\,183\,5\,\text{km}$
3. a. $101\,\text{mm}$ b. $54\,\text{mm}$ c. $60\,\text{mm}$ d. $103\,\text{mm}$
4. a. $1060\,\text{cm}$ b. $85.4\,\text{cm}$ c. $206\,\text{cm}$
5. a. $78.4\,\text{cm}$ b. $113\,\text{cm}$ c. $13\,\text{cm}$
6. a. $9.6\,\text{cm}$ b. $46\,\text{mm}$ c. $31\,\text{km}$
7. a. $25.13\,\text{cm}$ b. $25.13\,\text{m}$ c. $69.12\,\text{mm}$
8. a. $44.61\,\text{cm}$ b. $19\,741.77\,\text{km}$ c. $3314.38\,\text{mm}$
9. a. $192\,\text{m}$ b. $1220\,\text{mm}$ c. $260\,\text{cm}$
10. a. $74\,\text{mm}$ b. $9.6\,\text{km}$ c. $8\,\text{m}$
11. B
12. a. $1800\,\text{mm}$, $2100\,\text{mm}$, $2400\,\text{mm}$, $2700\,\text{mm}$, $180\,\text{cm}$, $210\,\text{cm}$, $240\,\text{cm}$, $270\,\text{cm}$, $1.8\,\text{m}$, $2.1\,\text{m}$, $2.4\,\text{m}$, $2.7\,\text{m}$
   b. 4
   c. $3300\,\text{mm}$
13. a. $1990\,\text{m}$ b. $841\,\text{m}$
14. a. $127.12\,\text{cm}$ b. $104.83\,\text{cm}$ c. $61.70\,\text{cm}$
15. a. $8\,\text{m}$ b. $480\,\text{mm}$ c. $405.35\,\text{cm}$
16. a. $125.66\,\text{cm}$ b. $245.66\,\text{m}$ c. 70.41cm
17. $222.5\,\text{m}$
18. $37.5\,\text{km}$
19. $12\,\text{m}$
20. a. $r = 6371 + 559$
    $= 6930\,\text{km}$
    $C = 2\pi r$
    $= 2 \times \pi \times 6930$
    $= 13\,860\pi$
    $= 43\,542.474$
    $\approx 43\,542\,\text{km}$

b. $96\,\text{minutes} = 96 \times 60$
   $= 5760\,\text{seconds}$
   $43\,542\,\text{km} = 43\,542\,000\,\text{m}$
   $\text{Speed} = 43\,542\,000 \div 5760$
   $= 7559.375\ldots$
   $\approx 7559\,\text{m/s}$

21. a. $500\,\text{m/s} = (500 \times 100)\,\text{cm/s}$
    $= 50\,000\,\text{cm/s}$
    b. $0.000\,02\,\text{seconds}$
    c. 2
22. a. $155.62\,\text{m}$
    b. $155.62\,\text{m} \times 1.5\,\text{minutes/metre} = 233.43\,\text{minutes}$
23. a. $40\,212\,\text{km}$
    b. $1676\,\text{km/h}$
    c. Sample responses can be found in the worked solutions in the online resources.
24. Yes, $42\,\text{cm}$ of wire remains.
25. $69.3\,\text{cm}$
26. a. $\left(2\sqrt{13} + 4\right)\,\text{m}$ b. $2\sqrt{17}\,\text{m}$

Exercise 9.3 Area

1. D
2. a. $13\,400\,\text{m}^2 = 0.0134\,\text{km}^2$
   b. $0.04\,\text{cm}^2 = 4\,\text{mm}^2$
   c. $3\,500\,000\,\text{cm}^2 = 350\,\text{m}^2$
   d. $0.005\,\text{m}^2 = 50\,\text{cm}^2$
3. a. $0.043\,\text{km}^2 = 43\,000\,\text{m}^2$
   b. $200\,\text{mm}^2 = 2\,\text{cm}^2$
   c. $1.41\,\text{km}^2 = 141\,\text{ha}$
   d. $3800\,\text{m}^2 = 0.38\,\text{ha}$
4. a. $24\,\text{cm}^2$ b. $16\,\text{mm}^2$ c. $537.5\,\text{cm}^2$
5. a. $149.5\,\text{cm}^2$ b. $16.32\,\text{m}^2$ c. $11.25\,\text{cm}^2$
6. a. $292.5\,\text{cm}^2$ b. $2.5\,\text{cm}^2$ c. $1250\,\text{m}^2$
7. a. $50.27\,\text{m}^2$ b. $3.14\,\text{mm}^2$ c. $36.32\,\text{m}^2$
8. a. $15.58\,\text{cm}^2$ b. $4.98\,\text{cm}^2$ c. $65\,\text{m}^2$
9. a. $6824.2\,\text{mm}^2$ b. $7.6\,\text{m}^2$ c. $734.2\,\text{cm}^2$
10. a. $578.5\,\text{cm}^2$ b. $7086.7\,\text{m}^2$ c. $5.4\,\text{m}^2$
11. a. $1143.4\,\text{m}^2$ b. $100.5\,\text{cm}^2$
12. a. $821\,\text{cm}^2$ b. $661.3\,\text{mm}^2$
13. $378\,\text{cm}^2$
14. $\approx 19.3\,\text{cm}^2$
15. a. $100.53\,\text{cm}^2$ b. $1244.07\,\text{m}^2$
16. a. $301.59\,\text{cm}^2$ b. $103.67\,\text{mm}^2$
17. B
18. $75.7\,\text{cm}$ or $757\,\text{mm}$
19. $\$29\,596.51$
20. a. $5000\,\text{km}^2$ b. 0.5%
21. $\$10\,150$
22. a. $619.5\,\text{cm}^2$
    b. $10.5\,\text{cm}$
    c. $346.36\,\text{cm}^2$
    d. Sample responses can be found in the worked solutions in the online resources.

23. Sample responses can be found in the worked solutions in the online resources.
24. Sample responses can be found in the worked solutions in the online resources.
25. a. $1600\,\text{cm}^2$
    b. $800\,\text{cm}^2$
    c. $400\,\text{cm}^2$
    d. $200\,\text{cm}^2$
    e. The area halves each time.
    f. 12.5%
    g. Sample responses can be found in the worked solutions in the online resources.
26. Sample responses can be found in the worked solutions in the online resources.
27. a. $\sqrt{50} = 5\sqrt{2}\ \text{m}$ b. $10\,\text{logs}$
28. $17\,\text{m}$
29. $45\,\text{m}^2$
30. a. $\dfrac{1}{4}$ b. $\dfrac{1}{4}\left(x^2 + 6x + 8\right)$

     c. $\dfrac{1}{4}\left(3x^2 - 6x - 8\right)$

     d.   i. $\dfrac{1}{12}\left(3x^2 - 6x - 8\right)$

         ii. $\dfrac{1}{6}\left(3x^2 + 6x + 8\right)$

Exercise 9.4 Area and perimeter of a sector

1. a. $14.14\,\text{cm}^2$ b. $157.08\,\text{cm}^2$
   c. $27.71\,\text{cm}^2$ d. $226.19\,\text{mm}^2$

2. i. $14.3\,\text{cm}$ ii. $12.6\,\text{cm}^2$

3. i. $43.6\,\text{cm}$ ii. $116.9\,\text{cm}^2$

4. i. $40.7\,\text{m}$ ii. $102.1\,\text{m}^2$

5. i. $5.4\,\text{m}$ ii. $1.8\,\text{m}^2$

6. E

7. i. $184.6\,\text{cm}$ ii. $1869.2\,\text{cm}^2$

8. i. $66.8\,\text{m}$ ii. $226.2\,\text{m}^2$

9. i. $27.4\,\text{cm}$ ii. $42.4\,\text{cm}^2$

10. i. $342.1\,\text{m}$ ii. $7243.6\,\text{m}^2$

11. i. $354.6\,\text{cm}$ ii. $7645.9\,\text{cm}^2$

12. i. $104.4\,\text{cm}$ ii. $201.1\,\text{cm}^2$

13. $60\,318.6\,\text{m}^2$

14. $303.4\,\text{cm}$

15. $170.2\,\text{m}^2$

16. a. $14.8\,\text{m}^2$ b. 4.4%

17. $2:1$

18. a. $251\,\text{cm}$ b. $13\,\text{cm}$

19. a. $5027\,\text{m}^2$ b. 78.5% c. $\$61\,440$

20. a. Sample responses can be found in the worked solutions in the online resources.
    b. $88.4\,\text{cm}^2$
    c. $50.3\,\text{cm}^2$

---

d. $109.4\,\text{cm}^2$
e. No, Jim eats more cake
21. Sample responses can be found in the worked solutions in the online resources.
22. a. Sample responses can be found in the worked solutions in the online resources.
    b. $85\,\text{m}$ each
    c. $60\,\text{m}$ each
    d. Sample responses can be found in the worked solutions in the online resources.
    e. $37\,\text{m}$
23. $\$250$
24. a. $\pi r^2\ \text{cm}^2$
    b. $\pi(r + 1)^2\ \text{cm}^2$
    c., e. Sample responses can be found in the worked solutions in the online resources.
    d. $5\pi\ \text{cm}^2$
25. Radius $= \pi\ \text{cm}$

Exercise 9.5 Surface area of rectangular and triangular prisms

1. a. $66\,\text{cm}^2$ b. $62\,\text{m}^2$ c. $6.7\,\text{m}^2$
2. a. $4.44\,\text{m}^2$ b. $11\,572.92\,\text{cm}^2$ c. $1.9\,\text{m}^2$
3. a. $86\,\text{cm}^2$ b. $210.7\,\text{m}^2$ c. $8.37\,\text{cm}^2$
4. a. $840\,\text{cm}^2$ b. $191.08\,\text{cm}^2$ c. $2370\,\text{mm}^2$
5. 2 cans of paint
6. $77.76\,\text{cm}^2$
7. $26.44\,\text{m}^2$
8. $20.36\,\text{m}^2$
9. $261.5\,\text{cm}^2$
10. $2.21\,\text{m}^2$
11. $15.2\,\text{m}^2$
12. a. $2080\,\text{cm}^2$
    b. Sample responses can be found in the worked solutions in the online resources.
13. $168\,\text{cm}^2$
14. a. $150\,\text{cm}^2$ b. $250\,\text{cm}^2$ c. $350\,\text{cm}^2$ d. $850\,\text{cm}^2$
15. The solid formed is a cube with side length $4\,\text{cm}$.
16. Area of base $= \dfrac{1}{2}bh$

$$= \dfrac{1}{2} \times 4 \times \sqrt{12}$$
$$= 2\sqrt{12}$$

Total surface area $= 3 \times l \times h + 2 \times \text{area of base}$

$$302 = 3 \times 4 \times h + 2 \times 2\sqrt{12}$$
$$302 = 12h + 4\sqrt{12}$$
$$302 = 12h + 13.86$$
$$12h = 288.14$$
$$h = \dfrac{288.14}{12}$$
$$h = 24.01$$
$$h \approx 24\,\text{cm}$$

17. a.

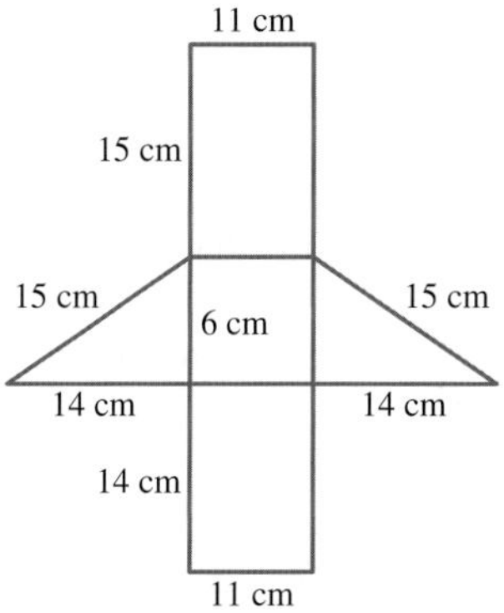

b. $315\ cm^2$

18. \$145.85 (2 L of ceiling paint, 1 L + 4 L for walls)

19. a. $1402.52\ m^2$ b. $3220\ m^3$

Exercise 9.6 Surface area of a cylinder

1. a. $75.4\ m^2$ b. $131.9\ m^2$
2. a. $28.3\ cm^2$ b. $84.8\ cm^2$
3. a. $2010.6\ m^2$ b. $3619.1\ m^2$
4. a. $1121.5\ cm^2$ b. $1575.5\ cm^2$
5. a. $6.6\ m^2$ b. $9.7\ m^2$
6. a. $25.6\ m^2$ b. $61.8\ m^2$
7. a.

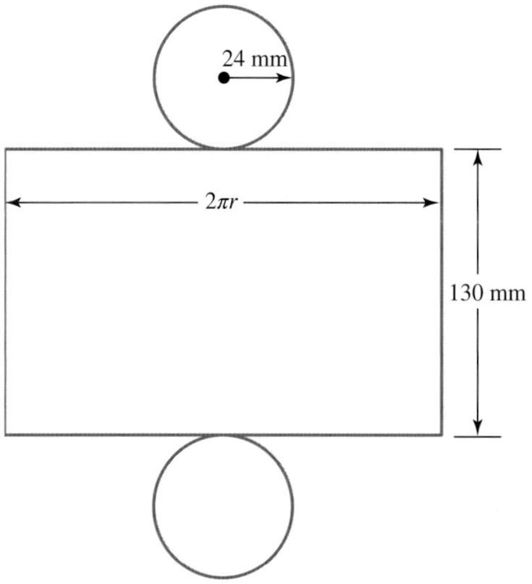

b. $232\ cm^2$

8. $\approx 1837.8\ cm^2$
9. $\approx 31.79\ cm$
10. a. Asparagus is largest; beetroot is smallest.
 b. $118\ cm^2$
11. a. $\approx 136.66\ m^2$ b. \$7351.33
12. a. $\approx 2035.75\ cm^2$ b. $\approx 1526.81\ cm^2$
 c. $\approx 395.84\ cm^2$ d. $\approx 3958.41\ cm^2$
13. $154.73\ cm^2$
14. a. $3.82\ cm$ b. $427.7\ cm^2$
15. $12\pi h^2$
16. Sample responses can be found in the worked solutions in the online resources.
17. Sample responses can be found in the worked solutions in the online resources.
18. $83.6\ m^2$
19. The area to be painted is $1.63\ m^2$.
20. $11\,231.12\ cm^2$
21. $4.34\ m^2$

Exercise 9.7 Volume of prisms and cylinders

1. a. 325 mL b. 2 600 000 mL
 c. 5100 mL d. 630 000 mL
2. a. $5.8\ cm^3$ b. $6100\ cm^3$
 c. $3\,200\,000\ cm^3$ d. $0.0593\ cm^3$
3. a. 0.358 kL b. 55.8 kL c. 8.752 kL d. 5300 kL
4. a. $36\ cm^3$ b. $15\ cm^3$ c. $72\ cm^3$
5. a. $12\ cm^3$ b. $68\ cm^3$ c. $9.6\ m^3$
6. a. $630\ mm^3$ b. $420\ cm^3$ c. $\approx 3152.7\ mm^3$
7. a. $\approx 1319.5\ mm^3$ b. $\approx 523.6\ m^3$ c. $\approx 10.5\ m^3$
8. a. $36\ cm^3$ b. $30\ m^3$ c. $1.144\ m^3$
9. a. $7020\ cm^3$ b. $6.91\ m^3$ c. $300\ m^3$ d. $8806\ cm^3$
10. a. $16\,085.0\ m^3$ b. $4766.6\ cm^3$ c. $2.3\ m^3$
11. a. $30.8\ m^3$ b. $30.6\ m^3$ c. $56.5\ mm^3$
12. 14 476.5 L
13. 176.4 L
14. a. $158\,169\ cm^3$ b. $4092\ cm^3$ c. $2639\ cm^3$
15. a. $641\ cm^3$ b. $1784\ cm^3$ c. $1057\ cm^3$
16. 75 322.8 L
17. 1145.1 mL
18. a. $4.95\ m^3$ b. 4950 L
19. $2.916 \times 10^9\ m^3$
20. $234\,256.54\ cm^3$
21. a. $1.1475\ m^3$ b. 840 L
22. a. $45.94\ cm^3$
 b. 2400 (to the nearest whole number)
23. See worked solutions in your online resources.
24. a. Prism
 b. $75\ m^2$
 c. See worked solutions in your online resources.
 d. i. 36 h 40 min ii. 47 h 9 min
25. 23
26. a. 1.60 L b. 0.042 L
27. a. $169\ cm^3$
 b. $V = x(l - 2x)(w - 2x)$

Project

1. A. $15\ cm^2$ B. $20\ cm^2$ C. $6\ cm^2$ D. $8.5\ cm^2$
 E. $10.5\ cm^2$ F. $11.5\ cm^2$ G. $12.5\ cm^2$

2.

Polygon	Dots on perimeter (b)	Dots within perimeter (i)	Area of polygon (A)
A	16	8	15
B	24	9	20
C	4	6	6
D	7	6	8.5
E	11	6	10.5
F	13	6	11.5
G	15	6	12.5

3. Sample responses can be found in the worked solutions in the online resources.

4. Sample responses can be found in the worked solutions in the online resources.

Exercise 9.8 Review questions

1. B
2. B
3. B
4. D
5. D
6. A
7. E
8. A
9. a. $26\,\text{mm} = 2.6\,\text{cm}$
   b. $1385\,\text{mm} = 138.5\,\text{cm}$
   c. $1.63\,\text{cm} = 16.3\,\text{mm}$
   d. $1.5\,\text{km} = 1500\,\text{m}$
   e. $0.077\,\text{km} = 77\,\text{m}$
   f. $2850\,\text{m} = 2.85\,\text{km}$
10. \$7497
11. a. 25.1 cm   b. 35.2 m   c. 37.7 cm
12. a. 70.4 cm   b. 30.4 mm
    c. 34 m   d. 240 mm or 24 cm
13. a. 13.1 m   b. 308.5 m
    c. 97.1 cm   d. 192 mm or 19.2 cm
14. a. $25\,\text{m}^2$ b. $950\,\text{cm}^2$
    c. $387.5\,\text{cm}^2$ d. $6120\,\text{m}^2$
15. a. $135\,\text{mm}^2$ b. $2170\,\text{cm}^2$
    c. $279.4\,\text{cm}^2$ d. $6939.8\,\text{mm}^2$
16. a. $1486.9\,\text{cm}^2$ b. $362.5\,\text{m}^2$
    c. $520.4\,\text{cm}^2$ d. $473.2\,\text{cm}^2$
17. a. $628.32\,\text{m}^2$ b. $59.45\,\text{cm}^2$
    c. $197.92\,\text{cm}^2$ d. $10\,734.47\,\text{m}^2$
18. a. 102.83 m   b. 31.07 cm
    c. 57.99 cm   d. 470.27 m
19. $48\,920\,\text{cm}^2$ or $4.892\,\text{m}^2$
20. $3550\,\text{cm}^2$
21. a. $343\,\text{cm}^3$ b. $672\,\text{cm}^3$
    c. $153\,938.04\,\text{cm}^3$ d. $1.45\,\text{m}^3$
    e. $1800\,\text{cm}^3$
22. \$99.80
23. 9.57 hectares

24. a. $10.75\,\text{m}^2$
    b. No
    c. 8
    d. 9 mattresses could fit in the room.
25. a. $10.83\,\text{cm}^2$ b. $351.65\,\text{cm}^2$
    c. $238.26\,\text{cm}^3$ d. \$5.95
26. The cylinder of height 210 mm has the greater volume.
27. The shaded parts have the same area.
28. a. i.

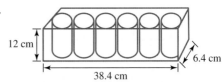

    ii. $1566.72\,\text{cm}^2$

    b. i.

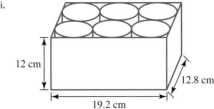

    ii. $1259.52\,\text{cm}^2$
    c. 3×2 arrangement
    d. $2058.24\,\text{cm}^2$
    e. $2027.52\,\text{cm}^2$
    f. Yes, $2170.88\,\text{cm}^2$
    g. The smallest amount of packaging required is 3×2, with two levels of packaging.

10 Probability

10.1 Overview

Why learn this?

New technologies which predict human behaviour are all based on probability. On 11 May 1997, IBM's supercomputer Deep Blue made history by defeating chess grandmaster Gary Kasparov in a six-game match under standard time controls. This was the first time that a computer had defeated the highest ranked chess player in the world. Deep Blue won by evaluating millions of possible positions each second and determining the probability of victory from each possible choice. This application of probability allowed artificial intelligence to defeat a human who had spent his life mastering the game.

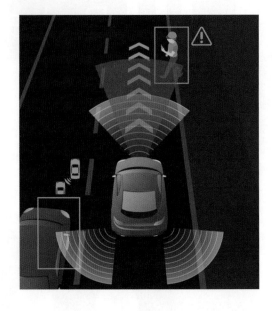

Combining advanced computing with probability is no longer used only to play games. From the daily convenience of predictive text, through to assisting doctors with cancer diagnosis, companies spend vast amounts of money on developing probability-based software. Self-driving cars use probability functions to predict the behaviour of both other cars on the road and pedestrians. The more accurate the predictive functions, the safer self-driving cars can become. Even politicians use campaign data analysis to develop models that produce predictions about individual citizens' likelihood of supporting specific candidates and issues, and the likelihood of these citizens changing their support if they're targeted with various campaign interventions.

As technology improves, so will its predictive power in determining the likelihood of certain outcomes occurring. It is important that we study and understand probability so we know how technology is being used and the impact it will have on our day-to-day life.

Where to get help

Go to your learnON title at **www.jacplus.com.au** to access the following digital resources. The Online Resources Summary at the end of this topic provides a full list of what's available to help you learn the concepts covered in this topic.

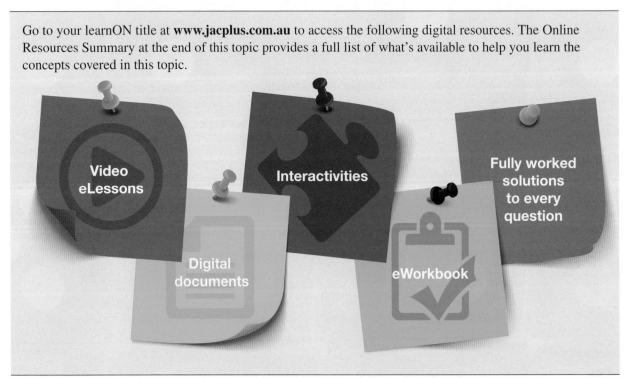

Complete this pre-test in your learnON title at www.jacplus.com.au and receive **automatic marks**, **immediate corrective feedback** and **fully worked solutions**.

1. **MC** A six-sided die is rolled, and the number uppermost is noted. The event of rolling an even number is:
   A. $\{1, 2, 3, 4, 5, 6\}$ B. $\{0, 1, 2, 3, 4, 5, 6\}$ C. $\{2, 4, 6\}$
   D. $\{1, 3, 5\}$ E. $\{2, 4, 5, 6\}$

2. The coloured spinner shown is spun once and the colour noted. Written in its simplest form, state the probability of spinning:
   a. an orange and a blue   b. an orange or a pink.

3. A coin is tossed in an experiment and the outcomes recorded.

Outcome	Heads	Tails
Frequency	72	28

   a. Identify how many trials there were.
   b. Calculate the experimental probability for tossing a Tail, giving your answer in simplest form.
   c. Calculate the theoretical probability for tossing a Tail with a fair coin, giving your answer in simplest form.
   d. In simplest form, calculate the difference in this experiment between the theoretical and experimental probabilities for tossing a Tail.

4. A random number is selected from:
   $\xi = \{1, 2, 3, 4, 5, 6, 7, 8, 9, 10, 11, 12, 13, 14, 15, 16, 17, 18, 19, 20, 21, 22, 23, 24, 25, 26, 27, 28, 29, 30\}$
   Calculate the exact probability of selecting a number that is a multiple of 3, giving your answer in simplest form.

5. Identify whether the following statement is True or False.
   If a large number of trials is conducted in an experiment, the relative (experimental) frequency of each outcome will be very close to its theoretical probability.

6. **MC** Consider the Venn diagram shown.
   If $\xi = \{$numbers between 1 and 20 inclusive$\}$, identify which of the following set B is equal to.
   A. $B = \{$multiples of 3 numbers between 1 and 20$\}$
   B. $B = \{$number 1$\}$
   C. $B = \{$even numbers between 1 and 20$\}$
   D. $B = \{$odd numbers between 1 and 20$\}$
   E. $B = \{$prime numbers between 1 and 20$\}$

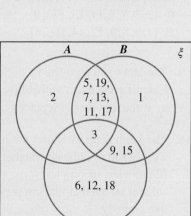

7. MC The information presented in the Venn diagram can be shown on a two-way frequency table as:

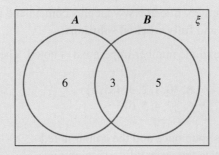

A.

	A	A'	Total
B	5	3	8
B'	0	6	6
Total	5	9	14

B.

	A	A'	Total
B	3	6	9
B'	5	0	5
Total	8	6	14

C.

	A	A'	Total
B	3	5	8
B'	6	0	6
Total	9	5	14

D.

	A	A'	Total
B	3	2	5
B'	3	3	6
Total	6	5	11

E.

	A	A'	Total
B	3	3	6
B'	2	3	5
Total	8	6	11

8. MC If $\xi=$ {numbers between 1 and 20 inclusive}, identify the complement of the event $A=$ {multiples of 5 and prime numbers}.

A. $\{1, 4, 6, 8, 9, 12, 14, 16, 18\}$
B. $\{5, 10, 15, 20\}$
C. $\{2, 3, 5, 7, 11, 13, 17, 19\}$
D. $\{2, 3, 5, 7, 10, 11, 13, 15, 17, 19, 20\}$
E. $\{4, 6, 8, 9, 12, 14, 16, 18\}$

9. The jacks and aces from a deck of cards are shuffled, then two cards are drawn. Calculate the exact probability, in simplest form, that two aces are chosen:
   a. if the first card is replaced
   b. if the first card is not replaced.

10. The Venn diagram shows the results of a survey where students were asked to indicate whether they prefer drama (*D*) or comedy (*C*) movies. In simplest form, determine the probability that a student selected at random prefers drama movies but does not like comedy.

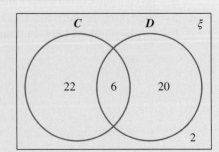

11. From a bag of mixed lollies, students can choose from three different types of lollies: red frogs, milk bottles or jelly babies. The bag contains five of each type of lolly. If Mahsa chooses two lollies without looking into the bag, calculate the probability that she will choose two different types of lollies.

12. The students in a class were asked about their sport preferences — whether they played basketball or tennis or neither. The information was recorded in a two-way frequency table.

	Basketball (*B*)	No basketball (*B'*)
Tennis (*T*)	25	20
No tennis (*T'*)	10	5

a. **MC** The student sport preferences recorded in the two-way frequency table is represented on a Venn diagram as:

A.

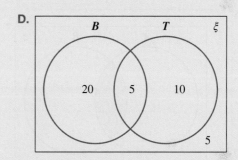

B.

C.

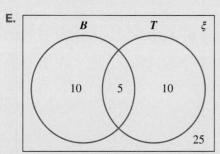

D.

E.

b. If one student is selected at random, calculate the probability that the student plays tennis only, correct to 2 decimal places.

13. Match the following Venn diagrams to the correct set notation for the shaded regions.

Venn diagram	Set notation
a.	**A.** $A \cap B$
b.	**B.** A'
c.	**C.** $A \cap B'$
d.	**D.** $A \cup B$

14. A survey of Year 8, 9 and 10 students asked the students to choose dinner options for their school camp.

Year level	Lasagna	Stir-fry
8	82	75
9	67	90
10	89	45
Total	240	210

Calculate the following probabilities.

a. The probability that a randomly selected student chose lasagne

b. The probability that a randomly selected student was in Year 10

c. The probability that a randomly selected student was in Year 8 and chose stir-fry

d. The probability that a randomly selected student who chose lasagna was in Year 9

15. **MC** A bag contains three blue balls and two red balls. A ball is taken at random from the bag and its colour noted. Then a second ball is drawn, without replacing the first one.

Identify the tree diagram that best represents this sample space.

A.

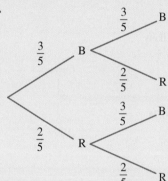

B.

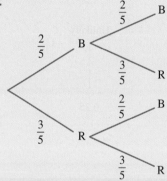

C.

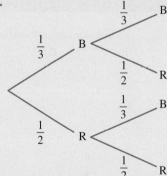

D.

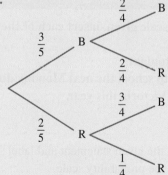

E.

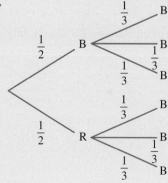

10.2 Theoretical probability

LEARNING INTENTION

At the end of this subtopic you should be able to:
- determine whether events are impossible, unlikely, likely or certain
- use key probability terminologies such as trials, outcomes, sample space and events
- calculate the theoretical probability of an event.

10.2.1 The language of probability

eles-4877

- The **probability** of an event is a measure of the likelihood that the event will take place.
- If an event is certain to occur, then it has a probability of 1.
- If an event is impossible, then it has a probability of 0.
- The probability of any other event taking place is given by a number between 0 and 1.
- An event is likely to occur if it has a probability between 0.5 and 1.
- An event is unlikely to occur if it has a probability between 0 and 0.5.

WORKED EXAMPLE 1 Placing events on a probability scale

On the probability scale given, insert each of the following events at appropriate points.

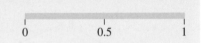

  a. **You will sleep tonight.**
  b. **You will come to school the next Monday during a school term.**
  c. **It will snow in Victoria this year.**

THINK	WRITE/DRAW
a. 1. Carefully read the given statement and label its position on the probability scale.	a. 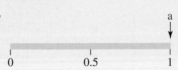
2. Write the answer and provide reasoning.	Under normal circumstances, I will certainly sleep tonight.
b. 1. Carefully read the given statement and label its position on the probability scale.	b. 

2. Write the answer and provide reasoning.

It is very likely but not certain that I will come to school on a Monday during term. Circumstances such as illness or public holidays may prevent me from coming to school on a specific Monday during a school term.

c. 1. Carefully read the given statement and label its position on the probability scale.

c.

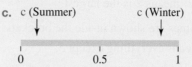

2. Write the answer and provide reasoning.

It is highly likely but not certain that it will snow in Victoria during winter. The chance of snow falling in Victoria in summer is highly unlikely but not impossible.

▶ 10.2.2 Key terms of probability

eles-4878

- The study of probability uses many special terms that must be clearly understood.
- **Chance experiment**: A chance experiment is a process, such as rolling a die, that can be repeated many times.
- **Trial**: A trial is one performance of an experiment to get a result.
  For example, each roll of the die is called a trial.
- **Outcome**: The outcome is the result obtained when the experiment is conducted.
  For example, when a normal six-sided die is rolled the outcome can be 1, 2, 3, 4, 5 or 6.
- **Sample space**: The set of all possible outcomes is called the sample space and is given the symbol ζ.
  For the example of rolling a die, $\xi = \{1, 2, 3, 4, 5, 6\}$.
- **Event**: An event is the favourable outcome of a trial and is often represented by a capital letter.
  For example, when a die is rolled, A could be the event of getting an even number; $A = \{2, 4, 6\}$.
- **Favourable outcome**: A favourable outcome for an event is any outcome that belongs to the event.
  For event A above (rolling an even number), the favourable outcomes are 2, 4 and 6.

WORKED EXAMPLE 2 Identifying sample space, events and outcomes

For the chance experiment of rolling a die:
  a. list the sample space
  b. list the events for:
    i. rolling a 4
    ii. rolling an even number
    iii. rolling at least 5
    iv. rolling at most 2

  c. list the favourable outcomes for:
    i. $\{4, 5, 6\}$
    ii. not rolling 5
    iii. rolling 3 or 4
    iv. rolling 3 and 4.

▶

THINK		WRITE	
a.	The outcomes are the numbers 1 to 6.	**a.**	$\xi = \{1, 2, 3, 4, 5, 6\}$
b. **i.**	This describes only one outcome.	**b.** **i.**	$\{4\}$
ii.	The possible even numbers are 2, 4 and 6.	**ii.**	$\{2, 4, 6\}$
iii.	'At least 5' means 5 is the smallest.	**iii.**	$\{5, 6\}$
iv.	'At most 2' means 2 is the largest.	**iv.**	$\{1, 2\}$
c. **i.**	The outcomes are shown inside the brackets.	**c.** **i.**	$4, 5, 6$
ii.	'Not 5' means everything except 5.	**ii.**	$1, 2, 3, 4, 6$
iii.	The event is $\{3, 4\}$.	**iii.**	$3, 4$
iv.	There is no number that is both 3 and 4.	**iv.**	There are no favourable outcomes.

⏵ 10.2.3 Theoretical probability

eles-4879

- When a coin is tossed, there are two possible outcomes, Heads or Tail That is, $\xi = \{H, T\}$.
- We assume the coin is unbiased, meaning each outcome is **equally likely** to occur.
- Since each outcome is equally likely, they must have the same probability.
- The total of all probabilities is 1, so the probability of an outcome is given by:
  - $\text{Pr(outcome)} = \dfrac{1}{\text{number of outcomes}}$
  - $\text{Pr(Heads)} = \dfrac{1}{2}$ and $\text{Pr(Tails)} = \dfrac{1}{2}$
- The probability of an event A is found by adding up all the probabilities of the favourable outcomes in event A.

Theoretical probability

- When determining probabilities of equally likely outcomes, use the following:
  - $\text{Pr(outcome)} = \dfrac{1}{\text{total number of outcomes}}$
  - $\text{Pr(event A)} = \dfrac{\text{number of favourable outcomes}}{\text{total number of outcomes}}$

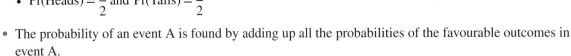

WORKED EXAMPLE 3 Calculating theoretical probability

A die is rolled and the number uppermost is noted. Determine the probability of each of the following events.

a. $A = \{1\}$ **b.** $B = \{\text{odd numbers}\}$ **c.** $C = \{4 \text{ or } 6\}$

THINK	WRITE
There are six possible outcomes.	

a. *A* has one favourable outcome.

a. $\Pr(A) = \dfrac{1}{6}$

b. *B* has three favourable outcomes: 1, 3 and 5.

b. $\Pr(B) = \dfrac{3}{6}$

$= \dfrac{1}{2}$

c. *C* has two favourable outcomes.

c. $\Pr(C) = \dfrac{2}{6}$

$= \dfrac{1}{3}$

 Resources

eWorkbook Topic 10 Workbook (worksheets, code puzzle and project) (ewbk-2010)

Digital documents SkillSHEET Probability scale (doc-6307)
SkillSHEET Understanding a deck of playing cards (doc-6308)
SkillSHEET Listing the sample space (doc-6309)
SkillSHEET Theoretical probability (doc-6310)

Interactivities Individual pathway interactivity: Theoretical probability (int-4534)
Probability scale (int-3824)
Theoretical probability (int-6081)

Exercise 10.2 Theoretical probability

learnon

Individual pathways

■ PRACTISE	■ CONSOLIDATE	■ MASTER
1, 3, 5, 8, 10, 14, 15, 19, 22	2, 4, 6, 9, 11, 16, 20, 23	7, 12, 13, 17, 18, 21, 24, 25

To answer questions online and to receive **immediate corrective feedback** and **fully worked solutions** for all questions, go to your learnON title at www.jacplus.com.au.

Fluency

1. **WE1** On the given probability scale, insert each of the following events at appropriate points. Indicate the chance of each event using one of the following terms: certain, likely, unlikely, impossible.

a. The school will have a lunch break on Friday.
b. Australia will host two consecutive Olympic Games.

2. On the given probability scale, insert each of the following events at appropriate points. Indicate the chance of each event using one of the following terms: certain, likely, unlikely, impossible.

a. At least one student in a particular class will obtain an A for Mathematics.
b. Australia will have a swimming team in the Commonwealth Games.

3. On the given probability scale, insert each of the following events at appropriate points. Indicate the chance of each event using one of the following terms: certain, likely, unlikely, impossible.

a. Mathematics will be taught in secondary schools.
b. In the future most cars will run without LPG or petrol.

4. On the given probability scale, insert each of the following events at appropriate points. Indicate the chance of each event using one of the following terms: certain, likely, unlikely, impossible.

a. Winter will be cold.
b. Bean seeds, when sown, will germinate.

5. **WE2a** For each chance experiment below, list the sample space.

a. Rolling a die
b. Tossing a coin
c. Testing a light bulb to see whether it is defective or not
d. Choosing a card from a normal deck and noting its colour
e. Choosing a card from a normal deck and noting its suit

6. **WE2b** A normal six-sided die is rolled. List each of the following events.

a. Rolling a number less than or equal to 3
b. Rolling an odd number
c. Rolling an even number or 1
d. Not rolling a 1 or 2
e. Rolling at most a 4
f. Rolling at least a 5

7. **WE2c** A normal six-sided die is rolled. List the favourable outcomes for each of the following events.

a. $A = \{3, 5\}$
b. $B = \{1, 2\}$
c. $C =$ 'rolling a number greater than 5'
d. $D =$ 'not rolling a 3 or a 4'
e. $E =$ 'rolling an odd number or a 2'
f. $F =$ 'rolling an odd number and a 2'
g. $G =$ 'rolling an odd number and a 3'

8. A card is selected from a normal deck of 52 cards and its suit is noted.

a. List the sample space.
b. List each of the following events.
    i. Drawing a black card
    ii. Drawing a red card
    iii. Not drawing a heart
    iv. Drawing a black or a red card

9. Determine the number of outcomes there are for:
a. rolling a die
b. tossing a coin
c. drawing a card from a standard deck
d. drawing a card and noting its suit
e. noting the remainder when a number is divided by 5.

10. A card is drawn at random from a standard deck of 52 cards.
    Note: 'At random' means that every card has the same chance of being selected.
    Calculate the probability of selecting:

    a. an ace
    b. a king
    c. the 2 of spades
    d. a diamond.

11. **WE3** A card is drawn at random from a deck of 52. Determine the probability of each event below.

    a. $A = \{5 \text{ of clubs}\}$
    b. $B = \{\text{black card}\}$
    c. $C = \{5 \text{ of clubs or queen of diamonds}\}$
    d. $D = \{\text{hearts}\}$
    e. $E = \{\text{hearts or clubs}\}$

12. A card is drawn at random from a deck of 52. Determine the probability of each event below.

    a. $F = \{\text{hearts and } 5\}$ b. $G = \{\text{hearts or } 5\}$ c. $H = \{\text{aces or kings}\}$
    d. $I = \{\text{aces and kings}\}$ e. $J = \{\text{not a } 7\}$

13. A letter is chosen at random from the letters in the word PROBABILITY. Determine the probability that the letter is:

    a. B
    b. not B
    c. a vowel
    d. not a vowel.

14. The following coloured spinner is spun and the colour is noted. Determine the probability of each of the events given below.

    a. $A = \{\text{blue}\}$
    b. $B = \{\text{orange}\}$
    c. $C = \{\text{orange or pink}\}$
    d. $D = \{\text{orange and pink}\}$
    e. $E = \{\text{not blue}\}$

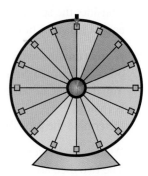

Understanding

15. A bag contains four purple balls and two green balls.

    a. If a ball is drawn at random, then calculate the probability that it will be:

       i. purple
       ii. green.

    b. Design an experiment like the one in part **a** but where the probability of drawing a purple ball is 3 times that of drawing a green ball.

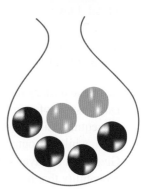

16. Design spinners (see question 14) using red, white and blue sections so that:

    a. each colour has the same probability of being spun
    b. red is twice as likely to be spun as either of the other two colours
    c. red is twice as likely to be spun as white and three times as likely to be spun as blue.

17. A bag contains red, green and blue marbles. Calculate how many marbles there must be in the bag for the following to be true when a single marble is selected at random from the bag.

   a. Each colour is equally likely to be selected and there are at least six red marbles in the bag.
   b. Blue is twice as likely to be selected as the other colours and there are at least five green marbles in the bag.
   c. There is a $\frac{1}{2}$ chance of selecting red, a $\frac{1}{3}$ chance of selecting green and a $\frac{1}{6}$ chance of selecting blue from the bag when there is between 30 and 40 marbles in the bag.

18. a. A bag contains seven gold and three silver coins. If a coin is drawn at random from the bag, calculate the probability that it will be:

   i. gold
   ii. silver.

   b. After a gold coin is taken out of the bag, a second coin is then selected at random. Assuming the first coin was not returned to the bag, calculate the probability that the second coin will be:

   i. gold
   ii. silver.

Reasoning

19. Do you think that the probability of tossing Heads is the same as the probability of tossing Tails if your friend tosses the coin? Suggest some reasons that it might not be.

20. If the following four probabilities were given to you, explain which two you would say were not correct.

$$0.725, -0.5, 0.005, 1.05$$

21. A coin is going to be tossed five times in a row. During the first four flips the coin comes up Heads each time. What is the probability that the coin will come up Heads again on the fifth flip? Justify your answer.

Problem solving

22. Consider the spinner shown. Discuss whether the spinner has an equal chance of falling on each of the colours.

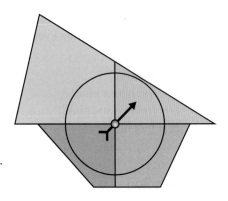

23. A box contains two coins. One is a double-headed coin, and the other is a normal coin with Heads on one side and Tails on the other. You draw one of the coins from a box and look at one of the sides. It is Heads.
    Determine the probability that the other side also shows Heads.

24. 'Unders and Overs' is a game played with two normal six-sided dice. The two dice are rolled, and the numbers uppermost added to give a total. Players bet on the outcome being 'under 7', 'equal to 7' or 'over 7'.
    If you had to choose one of these outcomes, which would you choose? Explain why.

25. Justine and Mary have designed a new darts game for their Year 9 Fete Day. Instead of a circular dart board, their dart board is in the shape of two equilateral triangles. The inner triangle (bullseye) has a side length of 3 cm, while the outer triangle has side length 10 cm.
    Given that a player's dart falls in one of the triangles, determine the probability that it lands in the bullseye. Write your answer correct to 2 decimal places.

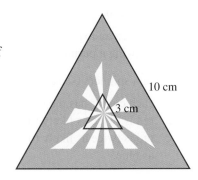

10.3 Experimental probability

▶ 10.3.1 Relative frequency

eles-4880

- A die is rolled 12 times and the outcomes are recorded in the table shown.

Outcome	1	2	3	4	5	6
Frequency	3	1	1	2	2	3

In this chance experiment there were 12 trials.
The table shows that the number 1 was rolled 3 times out of 12.

- So the **relative frequency** of 1 is 3 out of 12, or $\dfrac{3}{12} = \dfrac{1}{4}$.

As a decimal, the relative frequency of 1 is equal to 0.25.

Relative frequency

- The relative frequency of an outcome is given by:

$$\text{Relative frequency} = \frac{\text{frequency of an outcome}}{\text{total number of trials}}$$

- As the number of trials becomes larger, the relative frequency of each outcome will become very close to the theoretical probability.

WORKED EXAMPLE 4 Calculating relative frequency

For the chance experiment of rolling a die, the following outcomes were noted.

Outcome	1	2	3	4	5	6
Frequency	3	1	4	6	3	3

a. **Calculate the number of trials.**
b. **Identify how many threes were rolled.**
c. **Calculate the relative frequency for each number written as a decimal.**

THINK	WRITE
a. Adding the frequencies will give the number of trials.	a. $1 + 3 + 4 + 6 + 3 + 3 = 20$ trials
b. The frequency of 3 is 4.	b. 4 threes were rolled.

▶

c. Add a relative frequency row to the table and complete it. The relative frequency is calculated by dividing the frequency of the outcome by the total number of trials.

c.

Outcome	1	2	3	4	5	6
Frequency	3	1	4	6	3	3
Relative frequency	$\dfrac{3}{20} =$ 0.15	$\dfrac{1}{20} =$ 0.05	$\dfrac{4}{20} =$ 0.2	$\dfrac{6}{20} =$ 0.3	$\dfrac{3}{20} =$ 0.15	$\dfrac{3}{20} =$ 0.15

▶ 10.3.2 Experimental probability

eles-4881

- When it is not possible to calculate the theoretical probability of an outcome, carrying out simulations involving repeated trials can be used to determine the **experimental probability**.
- The relative frequency of an outcome is the experimental probability.

Experimental probability

- The experimental probability of an outcome is given by:

$$\text{Experimental probability} = \frac{\text{frequency of an outcome}}{\text{total number of trials}}$$

- The spinner shown is not symmetrical, and the probability of each outcome cannot be determined theoretically.
- The experimental probability of each outcome can be found by using the spinner many times and recording the outcomes. As more trials are conducted, the experimental probability will become more accurate and closer to the true probability of each section.

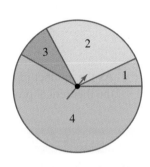

WORKED EXAMPLE 5 Calculating experimental probabilities

The spinner shown above was spun 100 times and the following results were achieved.

Outcome	1	2	3	4
Frequency	7	26	9	58

a. Calculate the number of trials.
b. Calculate the experimental probability of each outcome.
c. Calculate or recognise the sum of the four probabilities.

THINK

a. Adding the frequencies will determine the number of trials.

WRITE

a. $7 + 26 + 9 + 58 = 100$ trials

b. The experimental probability equals the relative frequency. This is calculated by dividing the frequency of the outcome by the total number of trials.

b. $\Pr(1) = \dfrac{7}{100}$

$= 0.07$

$\Pr(2) = \dfrac{26}{100}$

$= 0.26$

$\Pr(3) = \dfrac{9}{100}$

$= 0.09$

$\Pr(4) = \dfrac{58}{100}$

$= 0.58$

c. Add the probabilities (they should equal 1).

c. $0.07 + 0.26 + 0.09 + 0.58 = 1$

10.3.3 Expected number of results

- If we tossed a coin 100 times, we would expect there to be 50 Heads, since $\Pr(\text{Heads}) = \dfrac{1}{2}$.

Expected number of results

- The expected number of favourable outcomes from a series of trials is found from:

Expected number = probability of outcome × number of trials

- The probability of an outcome can be the theoretical probability or an experimental probability.

WORKED EXAMPLE 6 Expected number of outcomes

Calculate the expected number of results in the following situations:
a. The number of Tails after flipping a coin 250 times
b. The number of times a one comes up after rolling a dice 120 times
c. The number of times a royal card is picked from a deck that is reshuffled with the card replaced 650 times

THINK

WRITE

a. The probability of getting a Tail is $\Pr(\text{Tails}) = \dfrac{1}{2}$.

a. Number of Tails $= \Pr(\text{Tails}) \times 250$

$= \dfrac{1}{2} \times 250$

$= 125$ times

b. The probability of getting a one is $\Pr(\text{one}) = \dfrac{1}{6}$.

b.
$$\begin{aligned} \text{Number of ones} &= \Pr(\text{one}) \times 120 \\ &= \frac{1}{6} \times 120 \\ &= 20 \text{ times} \end{aligned}$$

c. A deck of cards has 52 cards and 12 of them are royal cards. This means the probability of getting a royal is $\Pr(\text{royal}) = \dfrac{12}{52} = \dfrac{3}{13}$.

c.
$$\begin{aligned} \text{Number of royals} &= \Pr(\text{royal}) \times 650 \\ &= \frac{3}{13} \times 650 \\ &= 150 \text{ times} \end{aligned}$$

Digital technology

Scientific calculators have a random function that allows the user to generate integers at random. This can be used to create a random set of data for probability or statistics questions.

The ⬤ button brings up the probability menu.

The functions nPr, nCr and ! will be used in Year 10.

Press across to select the **RAND** submenu and then **randint(** to generate random whole numbers.

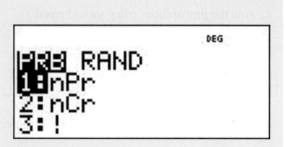

randint(1, 10) will generate a random integer between 1 and 10.

The comma button is found by pressing the ⬤ button and then the decimal button ⬚.

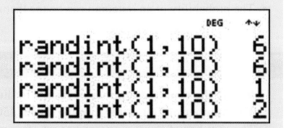

COLLABORATIVE TASK

Construct an irregular spinner using cardboard and a toothpick. By carrying out a number of trials, estimate the probability of each outcome.

Exercise 10.3 Experimental probability   **learn**on

Individual pathways

■ PRACTISE	■ CONSOLIDATE	■ MASTER
1, 3, 8, 10, 11, 15, 17, 19, 22	2, 4, 6, 9, 12, 18, 20, 23, 24	5, 7, 13, 14, 16, 21, 25, 26, 27

To answer questions online and to receive **immediate corrective feedback** and **fully worked solutions** for all questions, go to your learnON title at www.jacplus.com.au.

Fluency

1. **WE4** Each of the two tables shown contains the results of a chance experiment (rolling a die).
   For each table, calculate:
   i. the number of trials held
   ii. the number of fives rolled
   iii. the relative frequency for each outcome, correct to 2 decimal places
   iv. the sum of the relative frequencies.

   a.
Number	1	2	3	4	5	6
Frequency	3	1	5	2	4	1

   b.
Number	1	2	3	4	5	6
Frequency	52	38	45	49	40	46

2. A coin is tossed in two chance experiments. The outcomes are recorded in the tables shown. For each experiment, calculate:
   i. the relative frequency of both outcomes
   ii. the sum of the relative frequencies.

   a.
Outcome	H	T
Frequency	22	28

   b.
Outcome	H	T
Frequency	31	19

3. **WE5** An unbalanced die was rolled 200 times and the following outcomes were recorded.

Number	1	2	3	4	5	6
Frequency	18	32	25	29	23	73

   Using these results, calculate:

   a. Pr(6)   b. Pr(odd number)   c. Pr(at most 2)   d. Pr(not 3).

4. A box of matches claims on its cover to contain 100 matches.
   A survey of 200 boxes established the following results.

Number of matches	95	96	97	98	99	100	101	102	103	104
Frequency	1	13	14	17	27	55	30	16	13	14

If you were to purchase a box of these matches, calculate the probability that:

a. the box would contain 100 matches
b. the box would contain at least 100 matches
c. the box would contain more than 100 matches
d. the box would contain no more than 100 matches.

5. A packet of chips is labelled as weighing 170 grams. This is not always the case and there will be some variation in the weight of each packet. A packet of chips is considered underweight if its weight is below 168 grams. Chips are made in batches of 1000 at a time. A sample of the weights from a particular batch are shown below.

Weight (grams)	166	167	168	169	170	171	172	173	174
Frequency	2	3	13	17	27	18	9	5	1

Calculate the probability that:

a. a packet of chips is its advertised weight
b. a packet of chips is above its advertised weight
c. a packet of chips is underweight.
A batch of chips is rejected if more than 50 packets in a batch are classed as underweight.
d. Explain if the batch that this sample of chips is taken from should be rejected.

Understanding

6. Here is a series of statements based on experimental probability. If a statement is not reasonable, give a reason why.

   a. I tossed a coin five times and there were four Heads, so Pr(Heads) = 0.8.
   b. Sydney Roosters have won 1064 matches out of the 2045 that they have played, so Pr(Sydney will win their next game) = 0.52.
   c. Pr(the sun will rise tomorrow) = 1
   d. At a factory, a test of 10 000 light globes showed that 7 were faulty. Therefore, Pr(faulty light globe) = 0.0007.
   e. In Sydney it rains an average of 143.7 days each year, so Pr(it will rain in Sydney on the 17th of next month) = 0.39.

7. At a birthday party, some cans of soft drink were put in a container of ice. There were 16 cans of Coke, 20 cans of Sprite, 13 cans of Fanta, 8 cans of Sunkist and 15 cans of Pepsi.
   If a can was picked at random, calculate the probability that it was:

   a. a can of Pepsi
   b. not a can of Fanta.

8. **WE6** Calculate the expected number of Tails if a fair coin is tossed 400 times.

9. Calculate the expected number of threes if a fair die is rolled 120 times.

10. In Tattslotto, six numbers are drawn from the numbers 1, 2, 3, ... 45. The number of different combinations of six numbers is 8 145 060. If you buy one ticket, what is the probability that you will win the draw?

A. $\dfrac{1}{8\,145\,060}$ B. $\dfrac{1}{45}$

C. $\dfrac{45}{8\,145\,060}$ D. $\dfrac{1}{6}$

E. $\dfrac{6}{8\,145\,060}$

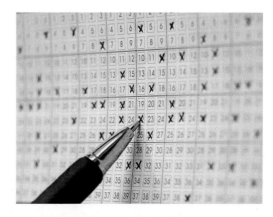

11. **MC** A survey of high school students asked 'Should Saturday be a normal school day?' 350 students voted yes, and 450 voted no. From the following, recognise the probability that a student chosen at random said no.

A. $\dfrac{7}{16}$ B. $\dfrac{9}{16}$ C. $\dfrac{7}{9}$ D. $\dfrac{9}{14}$ E. $\dfrac{1}{350}$

12. In a poll of 200 people, 110 supported party M, 60 supported party N and 30 were undecided. If a person is chosen at random from this group of people, calculate the probability that he or she:

a. supports party M
c. supports a party

b. supports party N
d. is not sure what party to support.

13. A random number is picked from $N = \{1, 2, 3, \ldots 100\}$. Calculate the probability of picking a number that is:

a. a multiple of 3
b. a multiple of 4 or 5
c. a multiple of 5 and 6.

14. The numbers 3, 5 and 6 are combined to form a three-digit number such that no digit may be repeated.

a. i. Recognise how many numbers can be formed.
   ii. List them.
b. Determine Pr(the number is odd).
c. Determine Pr(the number is even).
d. Determine Pr(the number is a multiple of 5).

15. **MC** In a batch of batteries, 2 out of every 10 in a large sample were faulty. At this rate, calculate how many batteries are expected to be faulty in a batch of 1500.

A. 2   B. 150   C. 200
D. 300   E. 750

16. Svetlana, Sarah, Leonie and Trang are volleyball players. The probabilities that they will score a point on serve are 0.6, 0.4, 0.3 and 0.2 respectively. Calculate how many points on serve are expected from each player if they serve 10 times each.

17. **MC** A survey of the favourite leisure activity of 200 Year 9 students produced the following results.

Activity	Playing sport	Fishing	Watching TV	Video games	Surfing
Number of students	58	26	28	38	50

The probability (given as a percentage) that a student selected at random from this group will have surfing as their favourite leisure activity is:

A. 50%   B. 100%   C. 25%   D. 0%   E. 29%

18. The numbers 1, 2 and 5 are combined to form a three-digit number, allowing for any digit to be repeated up to three times.

 a. Recognise how many different numbers can be formed.
 b. List the numbers.
 c. Determine Pr(the number is even).
 d. Determine Pr(the number is odd).
 e. Determine Pr(the number is a multiple of 3).

Reasoning

19. John has a 12-sided die numbered 1 to 12 and Lisa has a 20-sided die numbered 1 to 20. They are playing a game where the first person to get the number 10 wins.
They are rolling their dice individually.

 a. Calculate Pr(John gets a 10).
 b. Calculate Pr(Lisa gets a 10).
 c. Explain whether the game is fair.

20. At a supermarket checkout, the scanners have temporarily broken down and the cashiers must enter in the bar codes manually. One particular cashier overcharged 7 of the last 10 customers she served by entering the incorrect bar code.

 a. Based on the cashier's record, determine the probability of making a mistake with the next customer.
 b. Explain if another customer should have any objections with being served by this cashier.

21. If you flip a coin six times, determine how many of the possible outcomes could include a Tail on the second toss.

Problem solving

22. In a jar, there are 600 red balls, 400 green balls, and an unknown number of yellow balls. If the probability of selecting a green ball is $\dfrac{1}{5}$, determine how many yellow balls are in the jar.

23. In a jar there are an unknown number of balls, N, with 20 of them green. The other colours contained in the jar are red, yellow and blue, with Pr(red or yellow) $= \dfrac{1}{2}$, Pr(red or green) $= \dfrac{1}{4}$ and Pr(blue) $= \dfrac{1}{3}$.
Determine the number of red, yellow and blue balls in the jar.

24. The biological sex of babies in a set of triplets is simulated by flipping three coins. If a coin lands Tails up, the baby is male. If a coin lands Heads up, the baby is female. In the simulation, the trial is repeated 40 times.
The following results show the number of Heads obtained in each trial:

 0, 3, 2, 1, 1, 0, 1, 2, 1, 0, 1, 0, 2, 0, 1, 0, 1, 2, 3, 2, 1, 3, 0, 2, 1, 2, 0, 3, 1, 3, 0, 1, 0, 1, 3, 2, 2, 1, 2, 1

 a. Calculate the probability that exactly one of the babies in a set of triplets is female.
 b. Calculate the probability that more than one of the babies in the set of triplets is female.

25. Use your calculator to generate three random sets of numbers between 1 and 6 to simulate rolling a dice. The number of trials in each set will be:

 a. 10
 b. 25
 c. 50.

Calculate the relative frequency of each outcome (1 to 6) and comment on what you notice about experimental and theoretical probability.

26. A survey of the favourite foods of Year 9 students is recorded, with the following results.

Meal	Tally
Hamburger	45
Fish and chips	31
Macaroni and cheese	30
Lamb souvlaki	25
BBQ pork ribs	21
Cornflakes	17
T-bone steak	14
Banana split	12
Corn-on-the-cob	9
Hot dogs	8
Garden salad	8
Veggie burger	7
Smoked salmon	6
Muesli	5
Fruit salad	3

a. Estimate the probability that macaroni and cheese is the favourite food of a randomly selected Year 9 student.
b. Estimate the probability that a vegetarian dish is a randomly selected student's favourite food.
c. Estimate the probability that a beef dish is a randomly selected student's favourite food.

27. A spinner has six sections of different sizes. Steven conducts an experiment and finds the following results:

Section	1	2	3	4	5	6
Frequency	6	24	15	30	48	12

Determine the angle size of each section of the spinner using the results above.

10.4 Venn diagrams and two-way tables

LEARNING INTENTION

At the end of this subtopic you should be able to:
- identify the complement of an event A
- create and interpret Venn diagrams and two-way tables
- use a Venn diagram or two-way table to determine $A \cap B$
- use a Venn diagram or two-way table to determine $A \cup B$.

▶ 10.4.1 The complement of an event

eles-4883
- Suppose that a die is rolled. The sample space is given by: $\xi = \{1, 2, 3, 4, 5, 6\}$.
- If A is the event 'rolling an odd number', then $A = \{1, 3, 5\}$.
- There is another event called 'the **complement** of A', or 'not A'. This event contains all the outcomes that do not belong to A. It is given the symbol A'.
- In this case $A' = \{2, 4, 6\}$.

- A and A' can be shown on a **Venn diagram**.

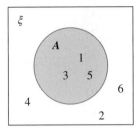

A is shaded.

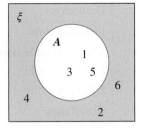

A' (not A) is shaded.

Complementary events

- For the event A, the complement is denoted A', and the two are related by the following:

$$Pr(A) + Pr(A') = 1$$

WORKED EXAMPLE 7 Determining the complement

For the sample space $\xi = \{1, 2, 3, 4, 5\}$, list the complement of each of the following events.
a. $A = \{\text{multiples of } 3\}$
b. $B = \{\text{square numbers}\}$
c. $C = \{1, 2, 3, 5\}$

THINK	WRITE
a. The only multiple of 3 in the set is 3. Therefore $A = \{3\}$. A' is every other element of the set.	a. $A' = \{1, 2, 4, 5\}$
b. The only square numbers are 1 and 4. Therefore $B = \{1, 4\}$. B' is every other element of the set.	b. $B' = \{2, 3, 5\}$
c. $C = \{1, 2, 3, 5\}$. C' is every other element of the set.	c. $C' = \{4\}$

10.4.2 Venn diagrams: the intersection of events

eles-4884

- A Venn diagram consists of a rectangle and one or more circles.
- A Venn diagram contains all possible outcomes in the sample space and will have the ξ symbol in the top left corner.
- A Venn diagram is used to illustrate the relationship between sets of objects or numbers.
- All outcomes for a given event will be contained within a specific circle.
- Outcomes that belong to multiple events will be found in the overlapping region of two or more circles.
- The overlapping region of two circles is called the intersection of the two events and is represented using the $\cap$ symbol.

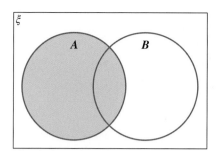

The circle on the left contains all outcomes in event A.

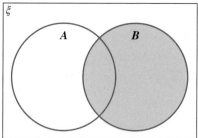

The circle on the right contains all outcomes in event B.

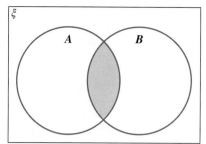

The overlap or intersection of the two circles contains the outcomes that are in event A 'and' in event B. This is denoted by $A \cap B$.

- A Venn diagram for two events A and B has four distinct regions.
- $A \cap B'$ contains the outcomes in event A and not in event B.
- $A \cap B$ contains the outcomes in event A and in event B.
- $A' \cap B$ contains the outcomes not in event A and in event B.
- $A' \cap B'$ contains the outcomes not in event A and not in event B.

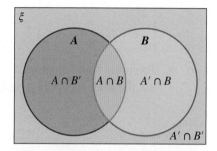

WORKED EXAMPLE 8 Setting up a Venn diagram

In a class of 20 students, 5 study Art, 9 study Biology and 2 students study both.
Let A = {students who study Art} and B = {students who study Biology}.
a. Create a Venn diagram to represent this information.
b. Identify the number of students represented by the following and state what these regions represent:
   i. $A \cap B$ ii. $A \cap B'$ iii. $A' \cap B$ iv. $A' \cap B'$

THINK

a. Draw a sample space with events A and B.

   Place a 2 in the intersection of both circles since we know 2 students take both subjects.

   Since 5 study Art and there are 2 already in the middle, place a 3 in the remaining section of circle A.

   Since 9 study Biology and there are 2 already in the middle, place a seven in the remaining section of circle B.

   The total number inside the three circles is $3 + 2 + 7 = 12$. This means there must be $20 - 12 = 8$ outside of the two circles. Place an 8 outside the circles, within the rectangle.

WRITE/DRAW

a.

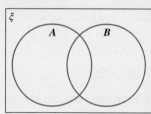

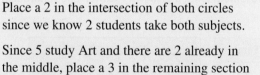

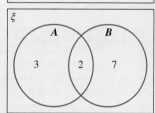

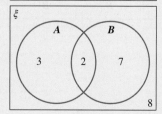

b. i. From the Venn diagram, $A \cap B = 2$. These are the students in both A and B.

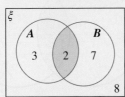

b. i. There are 2 students who study Art and Biology.

b. ii. From the Venn diagram, $A \cap B' = 3$. These are the students in A and not in B.

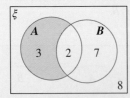

b. ii. There are 3 students who study Art and not Biology, i.e. 3 students study Art only.

b. iii. From the Venn diagram, $A' \cap B = 7$. These are the students not in A and in B.

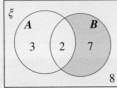

b. iii. There are 7 students who do not study Art and do study Biology, i.e. 7 students study Biology only.

b. iv. From the Venn diagram, $A' \cap B' = 8$. These are the students not in A and not in B.

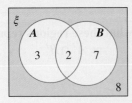

b. iv. There are 8 students who do not study Art and do not study Biology.

⊙ 10.4.3 Two-way tables

eles-4885

- The information in a Venn diagram can also be represented using a **two-way table**. The relationship between the two is shown below.

	Event B	**Event B′**	**Total**
Event A	$A \cap B$	$A \cap B'$	A
Event A′	$A' \cap B$	$A' \cap B'$	A'
Total	B	B'	

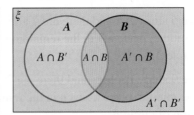

WORKED EXAMPLE 9 Creating a two-way table

In a class of 20 students, 5 study Art, 9 study Biology and 2 students study both. Create a two-way table to represent this information.

THINK	WRITE
1. Create an empty two-way table.	

	Biology	Not Biology	Total
Art			
Not Art			
Total			

2. Fill the table in with the information provided in the question.
   - 2 students study both subjects
   - 5 in total take Art
   - 9 in total take Biology
   - 20 students in the class

	Biology	Not Biology	Total
Art	2		5
Not Art			
Total	9		20

3. Use the totals of the rows and columns to fill in the gaps in the table.
   - $9 - 2 = 7$ Biology and not Art
   - $5 - 2 = 3$ Art and not Biology
   - $20 - 9 = 11$ not Biology total
   - $20 - 5 = 15$ not Art total

	Biology	Not Biology	Total
Art	2	3	5
Not Art	7	8	15
Total	9	11	20

4. The last value, not Biology and not Art, can be found from either of the following:
   - $11 - 3 = 8$
   - $15 - 7 = 8$

⊳ 10.4.4 Number of outcomes

eles-4886

- If event A contains seven outcomes or members, this is written as $n(A) = 7$.
- So $n(A \cap B') = 3$ means that there are three outcomes that are in event A and not in event B.

WORKED EXAMPLE 10 Determining the number of outcomes in an event

For the Venn diagram shown, write down the number of outcomes in each of the following.

a. M b. M' c. $M \cap N$ d. $M \cap N'$ e. $M' \cap N'$

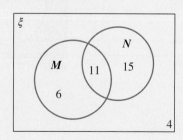

THINK	**WRITE/DRAW**

a. Identify the regions showing M and add the outcomes.

a.

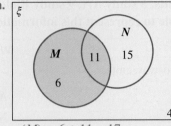

$n(M) = 6 + 11 = 17$

b. Identify the regions showing M' and add the outcomes.

b.

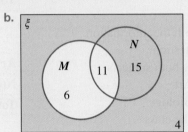

$n(M') = 4 + 15 = 19$

c. $M \cap N$ means 'M and N'. Identify the region.

c.

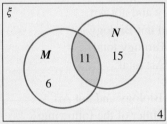

$n(M \cap N) = 11$

d. $M \cap N'$ means 'M and not N'. Identify the region.

d.

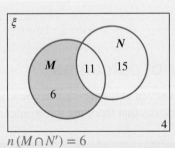

$n(M \cap N') = 6$

e. $M' \cap N'$ means 'not M and not N'. Identify the region.

e.

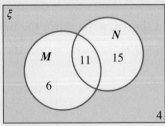

$n(M' \cap N') = 4$

Show the information from the Venn diagram on a two-way table.

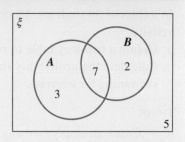

THINK

1. Draw a 2×2 table and add the labels A, A', B and B'.

2. There are 7 elements in A and B.
   There are 3 elements in A and 'not B'.
   There are 2 elements in 'not A' and B.
   There are 5 elements in 'not A' and 'not B'.

3. Add in a column and a row to show the totals.

WRITE

	A	A'
B		
B'		

	A	A'
B	7	2
B'	3	5

	A	A'	Total
B	7	2	9
B'	3	5	8
Total	10	7	17

Show the information from the two-way table on a Venn diagram.

	Left-handed	Right-handed
Blue eyes	7	20
Not blue eyes	17	48

THINK

Draw a Venn diagram that includes a sample space and events L for left-handedness and B for blue eyes. (Right-handedness $= L'$)
$n(L \cap B) = 7$
$n(L \cap B') = 17$
$n(L' \cap B) = 20$
$n(L' \cap B') = 48$

DRAW

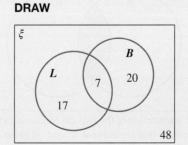

In a class of 30 students, 15 swim for exercise and 20 run for exercise and 5 participate in neither activity.

a. Create a two-way table to represent this information.

b. Calculate the probability that a randomly chosen student from this class does running and swimming for exercise.

THINK

a. 1. Create an empty two-way table.

2. Fill the table in with the information provided in the question.
   - 5 students do neither activity.
   - 15 in total swim.
   - 20 in total run.
   - 30 students are in the class.

3. Use the totals of the rows and columns to fill in the gaps in the table.

b. The probability a student runs and swims is given by:

$$\Pr(\text{run} \cap \text{swim}) = \frac{\text{number in run} \cap \text{swim}}{\text{total in the class}}$$

WRITE

a.

	Swim	Not swim	Total
Run			
Not run			
Total			

	Swim	Not swim	Total
Run			20
Not run		5	
Total	15		30

	Swim	Not swim	Total
Run	10	10	20
Not run	5	5	10
Total	15	15	30

b.
$$\Pr(\text{run} \cap \text{swim}) = \frac{\text{number in run} \cap \text{swim}}{\text{total in the class}}$$

$$= \frac{10}{30}$$

$$= \frac{1}{3}$$

10.4.5 Venn diagrams: the union of events

eles-4887

- The **intersection** of two events $(A \cap B)$ is all outcomes in event A 'and' in event B.
- The **union** of two events $(A \cup B)$ is all outcomes in events A 'or' in event B.

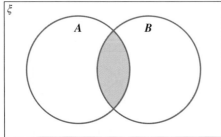

The intersection of the two circles contains the outcomes that are in event A 'and' in event B. This is denoted by $A \cap B$.

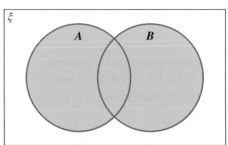

Everything contained within the two circles is an outcome that is in event A 'or' in event B. This is denoted by $A \cup B$.

Use the Venn diagram shown to calculate the value of the
following:
 a. $n(A)$
 b. $\Pr(B)$
 c. $n(A \cup B)$
 d. $\Pr(A \cap B)$

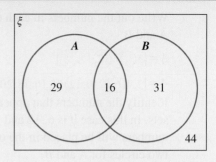

THINK

a. Identify the number of outcomes in the
A circle.

b. Identify the number of outcomes in the
B circle.
$$\Pr(B) = \frac{\text{number of favourable outcomes in } B}{\text{total number of outcomes}}$$

c. Identify the number of outcomes in either
of the two circles.

d. Identify the number of outcomes in the
intersection of the two circles.
$$\Pr(A \cap B) = \frac{\text{favourable outcomes in } A \cap B}{\text{total number of outcomes}}$$

WRITE

a. $n(A) = 29 + 16 = 45$

b. $\Pr(B) = \dfrac{\text{favourable outcomes in } B}{\text{total number of outcomes}}$
$$= \frac{n(B)}{\text{total number of outcomes}}$$
$$= \frac{16 + 31}{29 + 16 + 31 + 44}$$
$$= \frac{47}{120}$$

c. $n(A \cup B) = 29 + 16 + 31$
$$= 76$$

d. $\Pr(A \cap B) = \dfrac{\text{number of favourable outcomes in } A \cap B}{\text{total number of outcomes}}$
$$= \frac{n(A \cap B)}{\text{total number of outcomes}}$$
$$= \frac{16}{29 + 16 + 31 + 44}$$
$$= \frac{16}{120}$$
$$= \frac{2}{15}$$

a. **Place the elements of the following sets of numbers in their correct position in a single Venn diagram.**

$$\xi = \{\text{Number 1 to 20 inclusive}\}$$
$$A = \{\text{Multiples of 3 from 1 to 20 inclusive}\}$$
$$B = \{\text{Multiples of 2 from 1 to 20 inclusive}\}$$

b. **Use this Venn diagram to determine the following:**
 i. $A \cap B$ ii. $A \cup B$ iii. $A \cap B'$ iv. $A' \cup B'$

THINK	WRITE/DRAW

a. Write out the numbers in each event A and B:

$A = \{3, 6, 9, 12, 15, 18\}$

$A = \{2, 4, 6, 8, 10, 12, 14, 16, 18, 20\}$

Identify the numbers that appear in both sets. In this case it is 6, 12 and 18. These numbers will be placed in the overlap of the two circles for A and B.

All numbers not in A or B are placed outside the two circles.

After placing the numbers in the Venn diagram, check that all numbers from 1 to 20 are written down.

a.

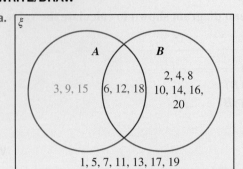

b.

i. $A \cap B$ are the numbers in A 'and' in B.

ii. $A \cup B$ are the numbers in A 'or' in B.

iii. $A \cap B'$ are the number in A and not in B. Refer to the four sections of the Venn diagram to locate this region.

iv. $A' \cup B'$ is any number that is not in A 'or' not in B. This ends up being any number not in $A \cap B$.

b.

i. $A \cap B = \{6, 12, 18\}$

ii. $A \cup B = \{2, 3, 4, 6, 8, 9, 10, 12, 14, 15, 16, 18, 20\}$

iii. $A \cap B' = \{3, 9, 15\}$

iv. $A' \cup B' = \left\{ \begin{array}{l} 1, 2, 3, 4, 5, 7, 8, 9, 10, 11, 13, \\ 14, 15, 16, 17, 19, 20 \end{array} \right\}$

WORKED EXAMPLE 16 Calculating the probability of the union of two events

In a class of 24 students, 11 students play basketball, 7 play tennis, and 4 play both sports.
a. Show the information on a Venn diagram.
b. If one student is selected at random, then calculate the probability that:
    i. the student plays basketball
    ii. the student plays tennis or basketball
    iii. the student plays tennis or basketball but not both.

THINK	WRITE/DRAW

a. 1. Draw a sample space with events B and T.

a.

2. $n(B \cap T) = 4$
    $n(B \cap T') = 11 - 4 = 7$
    $n(T \cap B') = 7 - 4 = 3$
    So far, 14 students out of 24 have been placed.
    $n(B' \cap T') = 24 - 14 = 10$

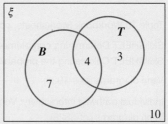

b. i. Identify the number of students who play basketball.

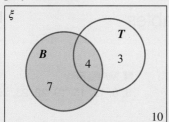

$$Pr(B) = \frac{\text{number of favourable outcomes}}{\text{total number of outcomes}}$$

ii. Identify the number of students who play tennis or basketball.

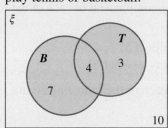

iii. Identify the number of students who play tennis or basketball but not both.

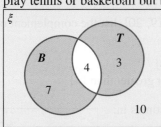

b. i. $Pr(B) = \dfrac{\text{number of students who play basketball}}{\text{total number of students}}$

$$= \frac{n(B)}{24}$$

$$= \frac{11}{24}$$

ii. $Pr(T \cup B) = \dfrac{n(T \cup B)}{24}$

$$= \frac{14}{24}$$

$$= \frac{7}{12}$$

iii. $n(B \cap T') + n(B' \cap T) = 3 + 7$

$$= 10$$

Pr(tennis or basketball but not both)

$$= \frac{10}{24}$$

$$= \frac{5}{12}$$

DISCUSSION

How will you remember the difference between when one event and another occurs and when one event or another occurs?

Exercise 10.4 Venn diagrams and two-way tables   **learn** on

Individual pathways

■ PRACTISE	■ CONSOLIDATE	■ MASTER
1, 4, 5, 10, 13, 16, 17, 20, 24	2, 6, 7, 9, 11, 14, 18, 21, 25	3, 8, 12, 15, 19, 22, 23, 26, 27

To answer questions online and to receive **immediate corrective feedback** and **fully worked solutions** for all questions, go to your learnON title at www.jacplus.com.au.

Fluency

1. **WE7** For the sample space $\xi = \{1, 2, 3, 4, 5, 6, 7, 8, 9, 10\}$, list the complement of each of the following events.

   a. $A = \{\text{evens}\}$ b. $B = \{\text{multiples of 5}\}$
   c. $C = \{\text{squares}\}$ d. $D = \{\text{numbers less than 8}\}$

2. If $\xi = \{11, 12, 13, 14, 15, 16, 17, 18, 19, 20\}$, list the complement of each of the following events.

   a. $A = \{\text{multiples of 3}\}$
   b. $B = \{\text{numbers less than 20}\}$
   c. $C = \{\text{prime numbers}\}$
   d. $D = \{\text{odd numbers or numbers greater than 16}\}$

3. If $\xi = \{1, 2, 3, 4, 5, 6, 7, 8, 9, 10, 11, 12, 13, 14, 15, 16, 17, 18, 19, 20\}$, list the complement of each of the following events.

   a. $A = \{\text{multiples of 4}\}$ b. $B = \{\text{primes}\}$
   c. $C = \{\text{even and less than 13}\}$ d. $D = \{\text{even or greater than 13}\}$

4. **WE10** For the Venn diagram shown, write down the number of outcomes in:

   a. ξ b. S
   c. T d. $T \cap S$
   e. $T \cap S'$ f. $S' \cap T'$.

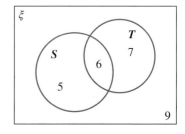

5. **WE11** Show the information from question 4 on a two-way table.

6. **WE12** Show the information from this two-way table on a Venn diagram.

	S	S'
V	21	7
V'	2	10

7. For each of the following Venn diagrams, use set notation to write the name of the region coloured in:
   i. blue
   ii. pink.

   a.
   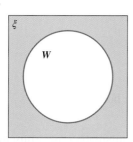

   b.
   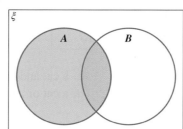

   c.
   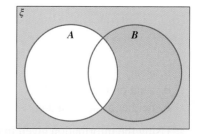

8. **WE9, 13** The membership of a tennis club consists of 55 men and 45 women. There are 27 left-handed people, including 15 men.

   a. Show the information on a two-way table.
   b. Show the information on a Venn diagram.
   c. If one member is chosen at random, calculate the probability that the person is:

   i. right-handed
   ii. a right-handed man
   iii. a left-handed woman.

9. **WE14** Using the information given in the Venn diagram, if one outcome is chosen at random, determine:
   a. $Pr(L)$
   b. $Pr(L')$
   c. $Pr(L \cap M)$
   d. $Pr(L \cap M')$.

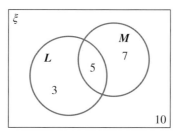

10. **WE15** Place the elements of the following sets of numbers in their correct position in a single Venn diagram.
    $\xi = \{$numbers between 1 to 10 inclusive$\}$
    $A = \{$odd numbers from 1 to 10$\}$
    $B = \{$squared numbers between 1 to 10 inclusive$\}$

11. Place the elements of the following sets of numbers in their correct position in a single Venn diagram.
    $\xi = \{$numbers between 1 to 25 inclusive$\}$
    $A = \{$multiples of 3 from 1 to 25$\}$
    $B = \{$numbers that are odd or over 17 from 1 to 25 inclusive$\}$

12. Place the elements of the following sets of numbers in their correct position in a single Venn diagram.

$A = \{$prime numbers from 1 to 20$\}$

$B = \{$even numbers from 1 to 20$\}$

$C = \{$multiples of 5 from 1 to 20$\}$

$\xi = \{$numbers between 1 and 20 inclusive$\}$

13. Using the information given in the table, if one family is chosen at random, calculate the probability that they own:

Pets owned by families		
	Cat	No cat
Dog	4	11
No dog	16	9

a. a cat

b. a cat and a dog

c. a cat or a dog or both

d. a cat or a dog but not both

e. neither a cat nor a dog.

14. Using the information given in the table, if a person is chosen at random, calculate the probability that for exercise, this person:

Type of exercise		
	Cycling	No cycling
Running	12	19
No running	13	6

a. cycles

b. cycles and runs

c. cycles or runs

d. cycles or runs but not both.

15. A barista decides to record what the first 45 customers order on a particular morning. The information is partially filled out in the table shown.

	Croissant	No croissant	Total
Coffee	27		
No coffee		7	
Total	36		

a. Fill in the missing information in the table.

b. Using the table, calculate the probability that the next customer:

   i. orders a coffee and a croissant

  ii. order a coffee or a croissant

 iii. orders a coffee or a croissant but not both.

c. The barista serves an average of 315 customers a day.

If the café is open six days a week, calculate how many coffees the barista will make each week.

Understanding

16. A group of athletes was surveyed and the results were shown on a Venn diagram.

    $S = \{\text{sprinters}\}$ and $L = \{\text{long jumpers}\}$.

    a. Write down how many athletes were included in the survey.
    b. If one of the athletes is chosen at random, calculate the probability that the athlete competes in:
        i. long jump
        ii. long jump and sprints
        iii. long jump or sprints
        iv. long jump or sprints but not both.

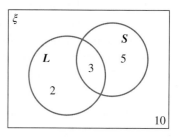

17. **WE8a,16** In a class of 40 students, 26 take a train to school, 19 take a bus and 8 take neither of these. Let $T = \{\text{takes the train}\}$ and $B = \{\text{takes the bus}\}$.

    a. Show the information on a Venn diagram.
    b. If one student is selected at random, calculate the probability that:
        i. the student takes the bus
        ii. the student takes the train or the bus
        iii. the student takes the train or the bus but not both.

18. **WE8b** If $\xi = \{\text{children}\}$, $S = \{\text{swimmers}\}$ and $R = \{\text{runners}\}$, describe in words each of the following.

    a. S'
    b. $S \cap R$
    c. $R' \cap S'$
    d. $R \cup S$

19. A group of 12 students was asked whether they liked hip hop (H) and whether they liked classical music (C). The results are shown in the table.

	C	H
Ali	✓	✓
Anu		
Chris		✓
George		✓
Imogen		✓
Jen	✓	✓
Luke	✓	✓
Pam	✓	
Petra		
Roger	✓	
Seedevi		✓
Tomas		

a. Show the results on:
    i. a Venn diagram
    ii. a two-way table.

b. If one student is selected at random, calculate:
    i. $\Pr(H)$
    ii. $\Pr(H \cup C)$
    iii. $\Pr(H \cap C)$
    iv. $\Pr(\text{student likes classical or hip hop but not both})$.

Reasoning

20. One hundred Year 9 Maths students were asked to indicate their favourite topic in mathematics. Sixty chose Probability, 50 chose Measurement and 43 chose Algebra. Some students chose two topics: 15 chose Probability and Algebra, 18 chose Measurement and Algebra, and 25 chose Probability and Measurement. Five students chose all three topics.

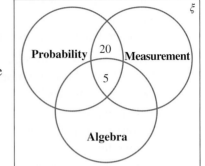

a. Copy and complete the Venn diagram shown.
b. Calculate how many students chose Probability only.
c. Calculate how many students chose Algebra only.
d. Calculate how many students chose Measurement only.
e. Calculate how many students chose any two of the three topics.

A student is selected at random from this group. Calculate the probability that this student has chosen:

f. Probability
g. Algebra
h. Algebra and Measurement
i. Algebra and Measurement but not Probability
j. all of the topics.

21. Create a Venn diagram using two circles to accurately describe the relationships between the following quadrilaterals: rectangle, square and rhombus.

22. Use the Venn diagram shown to write the numbers of the correct regions for each of the following problems.

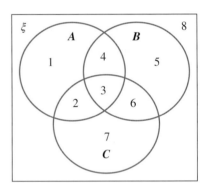

a. $A' \cup (B' \cap C)$
b. $A \cap (B \cap C')$
c. $A' \cap (B' \cap C')$
d. $A \cup (B \cap C)'$

23. A recent survey taken at a cinema asked 90 teenagers what they thought about three different movies. In total, 47 liked 'Hairy Potter', 25 liked 'Stuporman' and 52 liked 'There's Something About Fred'.
16 liked 'Hairy Potter' only.
4 liked 'Stuporman' only.
27 liked 'There's Something About Fred' only.
There were 11 who liked all three films and 10 who liked none of them.

a. Construct a Venn diagram showing the results of the survey.
b. Calculate the probability that a teenager chosen at random liked 'Hairy Potter' and 'Stuporman' but not 'There's Something About Fred'.

Problem solving

24. 120 children attended a school holiday program during September. They were asked to select their favourite board game from Cluedo, Monopoly and Scrabble. They all selected at least one game, and four children chose all three games.
In total, 70 chose Monopoly and 55 chose Scrabble.
Some children selected exactly two games — 12 chose Cluedo and Scrabble, 15 chose Monopoly and Scrabble, and 20 chose Cluedo and Monopoly.

a. Draw a Venn diagram to represent the children's selections.
b. Calculate the probability that a child selected at random did not choose Cluedo as a favourite game.

25. Valleyview High School offers three sports at Year 9: baseball, volleyball and soccer. There are 65 students in Year 9.

2 have been given permission not to play sport due to injuries and medical conditions.

30 students play soccer.

9 students play both soccer and volleyball but not baseball.

9 students play both baseball and soccer (including those who do and don't play volleyball).

4 students play all three sports.

12 students play both baseball and volleyball (including those who do and don't play soccer).

The total number of players who play baseball is one more than the total of students who play volleyball.

a. Determine the number of students who play volleyball.

b. If a student was selected at random, calculate the probability that this student plays soccer and baseball only.

26. A Venn diagram consists of overlapping ovals which are used to show the relationships between sets. Consider the numbers 156 and 520.

Show how a Venn diagram could be used to determine their:

a. HCF   b. LCM.

27. A group of 200 shoppers was asked which type of fruit they had bought in the last week. The results are shown in the table.

Fruit	Number of shoppers
Apples (A) only	45
Bananas (B) only	34
Cherries (C) only	12
A and B	32
A and C	15
B and C	26
A and B and C	11

a. Display this information in a Venn diagram.

b. Calculate $n(A \cap B' \cap C)$.

c. Determine how many shoppers purchased apples and bananas but not cherries.

d. Calculate the relative frequency of shoppers who purchased:

   i. apples

   ii. bananas or cherries.

e. Estimate the probability that a shopper purchased cherries only.

10.5 Two-step experiments

▶ 10.5.1 The sample space of two-step experiments

eles-4888

- A two-step experiment involves two separate actions. It may be the same action repeated (tossing a coin twice) or two separate actions (tossing a coin and rolling a dice).
- Imagine two bags (that are not transparent) that contain coloured counters. The first bag has a mixture of black and white counters, and the second bag holds blue, magenta (pink) and purple counters. In a probability experiment, one counter is to be selected at random from each bag and its colour noted.

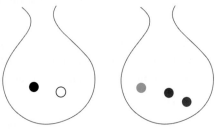

Bag 1   **Bag 2**

- The sample space for this experiment can be found using a table called an **array** that systematically displays all the outcomes.

			Second action		
			Bag 2		
			B	**M**	**P**
First action	**Bag 1**	**B**	B B	B M	B P
		W	W B	W M	W P

The sample space, $\xi = \{BB, BM, BP, WB, WM, WP\}$.
- The sample space can also be found using a **tree diagram**.

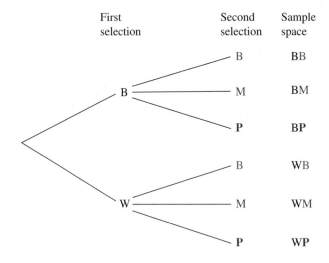

Two dice are rolled and the numbers uppermost are noted.
List the sample space in an array.
a. Recognise how many outcomes there are.
b. Calculate how many outcomes contain at least one 5.
c. Calculate Pr(at least one 5).

THINK

Draw an array (table) showing all the possible outcomes.

WRITE/DRAW

		Second die					
		1	2	3	4	5	6
First die	1	1, 1	1, 2	1, 3	1, 4	1, 5	1, 6
	2	2, 1	2, 2	2, 3	2, 4	2, 5	2, 6
	3	3, 1	3, 2	3, 3	3, 4	3, 5	3, 6
	4	4, 1	4, 2	4, 3	4, 4	4, 5	4, 6
	5	5, 1	5, 2	5, 3	5, 4	5, 5	5, 6
	6	6, 1	6, 2	6, 3	6, 4	6, 5	6, 6

a. The table shows 36 outcomes.

b. Count the outcomes that contain 5. The cells are shaded in the table.

c. There are 11 favourable outcomes and 36 in total.

a. There are 36 outcomes.

b. Eleven outcomes include 5.

c. $\text{Pr(at least one 5)} = \dfrac{11}{36}$

Two coins are tossed and the outcomes are noted.
Show the sample space on a tree diagram.
a. Recognise how many outcomes there are.
b. Calculate the probability of tossing at least one Head.

THINK

1. Draw a tree representing the outcomes for the toss of the first coin.

WRITE/DRAW

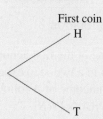

First coin

H

T

2. For the second coin the tree looks like this:

Second coin

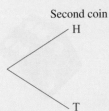

Add this tree to both ends of the first tree.

First coin   Second coin   Sample space

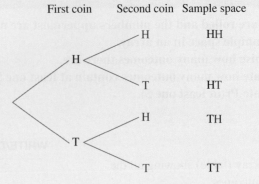

3. List the outcomes.

   a. Count the outcomes in the sample space.

   b. Three outcomes have at least one Head.

   a. There are four outcomes (HH, HT, TH, TT).

   b. $\text{Pr(at least one Head)} = \dfrac{3}{4}$

10.5.2 Two-step experiments

eles-4889

- When a coin is tossed, $\text{Pr}(H) = \dfrac{1}{2}$, and when a die is rolled, $\text{Pr}(3) = \dfrac{1}{6}$.

- If a coin is tossed and a die is rolled, what is the probability of getting a Head *and* a 3?
- Consider the sample space.

	1	2	3	4	5	6
H	H, 1	H, 2	**H, 3**	H, 4	H, 5	H, 6
T	T, 1	T, 2	T, 3	T, 4	T, 5	T, 6

There are 12 outcomes, and $\text{Pr}(\text{Head and 3}) = \dfrac{1}{12}$.

- In this case, $\text{Pr}(\text{Head and 3}) = \text{Pr}(H) \times \text{Pr}(3)$; that is, $\dfrac{1}{12} = \dfrac{1}{2} \times \dfrac{1}{6}$.

Two-step probabilities

If A is the outcome of the first action and B is the outcome of
the second action in a two-step experiment, then:
- $A \cap B$ is the outcome of A followed by B
- $\text{Pr}(A \cap B) = \text{Pr}(A) \times \text{Pr}(B)$.

WORKED EXAMPLE 19 Calculating two-step probabilities

In one cupboard Joe has two black T-shirts and one yellow one. In his drawer there are three pairs of white socks and one black pair. If he selects his clothes at random, calculate the probability that his socks and T-shirt will be the same colour.

THINK

If they are the same colour then they must be black.

$Pr(\text{black T-shirt}) = Pr(B_t) = \dfrac{2}{3}$

$Pr(\text{black socks}) = Pr(B_s) = \dfrac{1}{4}$

WRITE

$$Pr(B_t \cap B_s) = Pr(B_t) \times Pr(B_s)$$

$$= \dfrac{2}{3} \times \dfrac{1}{4}$$

$$= \dfrac{1}{6}$$

10.5.3 Experiments with replacement

eles-4890

- If a two-step experiment requires an object to be selected, say from a bag, the person doing the selecting has two options after the first selection.
  - Place the object bag back in the bag, meaning the experiment is being carried out **with replacement**. In this case, the number of objects in the bag remains constant.
  - Permanently remove the object from the bag, meaning the experiment is being carried out **without replacement**. In this situation, the number of the objects in the bag is reduced by 1 after every selection.

WORKED EXAMPLE 20 Calculating two-step experiments with replacement

A bag contains three pink and two blue counters. A counter is taken at random from the bag, its colour is noted, then it is returned to the bag and a second counter is chosen.
a. Show the outcomes on a tree diagram.
b. Calculate the probability of each outcome.
c. Calculate the sum of the probabilities.

THINK

a. 1. Draw a tree for the first trial. Write the probability on the branch.
   Note: The probabilities should sum to 1.

a. 2. For the second trial the tree is the same. Add this tree to both ends of the first tree, then list the outcomes.

WRITE/DRAW

a.

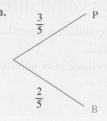

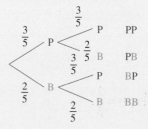

b. For both draws $\Pr(P) = \dfrac{3}{5}$ and $\Pr(B) = \dfrac{2}{5}$.

Use the rule $\Pr(A \cap B) = \Pr(A) \times \Pr(B)$ to determine the probabilities.

b.
$$\Pr(P \cap P) = \Pr(P) \times \Pr(P)$$
$$= \dfrac{3}{5} \times \dfrac{3}{5}$$
$$= \dfrac{9}{25}$$

$$\Pr(P \cap B) = \Pr(P) \times \Pr(B)$$
$$= \dfrac{3}{5} \times \dfrac{2}{5}$$
$$= \dfrac{6}{25}$$

$$\Pr(B \cap P) = \Pr(B) \times \Pr(P)$$
$$= \dfrac{2}{5} \times \dfrac{3}{5}$$
$$= \dfrac{6}{25}$$

$$\Pr(B \cap B) = \Pr(B) \times \Pr(B)$$
$$= \dfrac{2}{5} \times \dfrac{2}{5}$$
$$= \dfrac{4}{25}$$

c. Add the probabilities.

c. $\dfrac{9}{25} + \dfrac{6}{25} + \dfrac{6}{25} + \dfrac{4}{25} = 1$

- In Worked example 20, $\Pr(P) = \dfrac{3}{5}$ and $\Pr(B) = \dfrac{2}{5}$ for both trials.
  This would not be true if a counter is selected *but not replaced.*

10.5.4 Experiments without replacement

eles-4891

- Let us consider again the situation described in Worked example 20, and consider what happens if the first marble is not replaced.
- Initially the bag contains three pink and two blue counters, and either a pink counter or a blue counter will be chosen.

- $\Pr(P) = \dfrac{3}{5}$ and $\Pr(B) = \dfrac{2}{5}$.

- If the counter is not replaced, then the sample space is affected as follows:

If the first counter randomly selected is pink, then the sample space for the second draw looks like this:

So $\Pr(P) = \dfrac{2}{4}$ and $\Pr(B) = \dfrac{2}{4}$.

If the first counter randomly selected is blue, then the sample space for the second draw looks like this:

So $\Pr(P) = \dfrac{3}{4}$ and $\Pr(B) = \dfrac{1}{4}$.

WORKED EXAMPLE 21 Calculating two-step experiments without replacement

A bag contains three pink and two blue counters. A counter is taken at random from the bag and its colour is noted, then a second counter is drawn, without replacing the first one.

a. Show the outcomes on a tree diagram.
b. Calculate the probability of each outcome.
c. Calculate the sum of the probabilities.

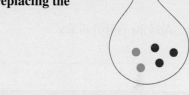

THINK	WRITE/DRAW
a. Draw a tree diagram, listing the probabilities.	a. 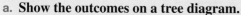

b. Use the rule $\Pr(A \cap B) = \Pr(A) \times \Pr(B)$ to determine the probabilities.

b. $\Pr(P \cap P) = \Pr(P) \times \Pr(P)$

$$= \frac{3}{5} \times \frac{2}{4}$$

$$= \frac{6}{20}$$

$$= \frac{3}{10}$$

$\Pr(P \cap B) = \Pr(P) \times \Pr(B)$

$$= \frac{3}{5} \times \frac{2}{4}$$

$$= \frac{6}{20}$$

$$= \frac{3}{10}$$

$$Pr(B \cap P) = Pr(B) \times Pr(P)$$

$$= \frac{2}{5} \times \frac{3}{4}$$

$$= \frac{6}{20}$$

$$= \frac{3}{10}$$

$$Pr(B \cap B) = Pr(B) \times Pr(B)$$

$$= \frac{2}{5} \times \frac{1}{4}$$

$$= \frac{2}{20}$$

$$= \frac{1}{10}$$

c. Add the probabilities.

c. $\dfrac{3}{10} + \dfrac{3}{10} + \dfrac{3}{10} + \dfrac{1}{10} = 1$

WORKED EXAMPLE 22 Language of two-step experiments

Consider the situation presented in Worked example 21. Use the tree diagram to calculate the following probabilities:
a. **Pr(a pink counter and a blue counter)**
b. **Pr(a pink counter then a blue counter)**
c. **Pr(a matching pair)**
d. **Pr(different colours)**

THINK	WRITE
a. Think of the outcomes that have a pink and a blue counter. There are two: PB and BP. Add their probabilities together to find the answer.	a. Pr(a pink counter and a blue counter) $= Pr(PB) + Pr(BP)$ $= \dfrac{3}{10} + \dfrac{3}{10}$ $= \dfrac{3}{5}$
b. In this case, we have to take the order the counters are selected into account. The only outcome that has a pink then a blue is PB.	b. Pr(a pink counter then a blue counter) $= Pr(PB)$ $= \dfrac{3}{10}$
c. A matching pair implies both counters are the same colour. These outcomes are BB and PP.	c. Pr(a matching pair) $= Pr(BB) + Pr(PP)$ $= \dfrac{1}{10} + \dfrac{3}{10}$ $= \dfrac{2}{5}$

d. One way to approach this is to think that if the colours are different, this is all outcomes not counted in part **c**. Thus, we can find the answer by subtracting the answer for part **c** from 1.

d. Pr(different colours)
$$= 1 - \text{Pr(a matching pair)}$$
$$= 1 - \frac{2}{5}$$
$$= \frac{3}{5}$$

DISCUSSION

How does replacement affect the probability of an event occurring?

 Resources

 eWorkbook   Topic 10 Workbook (worksheets, code puzzle and project) (ewbk-2010)

 Video eLesson Tree diagrams (eles-1894)

 Interactivities Individual pathway interactivity: Two-step experiments (int-4537)

Two-step experiments (int-6083)

Exercise 10.5 Two-step experiments   **learn**on

Individual pathways

■ PRACTISE	■ CONSOLIDATE	■ MASTER
1, 2, 4, 5, 8, 9, 12, 15, 18, 21, 22	3, 7, 10, 13, 16, 19, 23, 24	6, 11, 14, 17, 20, 25, 26, 27

To answer questions online and to receive **immediate corrective feedback** and **fully worked solutions** for all questions, go to your learnON title at www.jacplus.com.au.

Fluency

1. **WE17** If two dice are rolled and their *sum* is noted, complete the array below to show the sample space.

		Die 1					
		1	**2**	**3**	**4**	**5**	**6**
Die 2	**1**	2					
	2					7	
	3						
	4						
	5						
	6			9			

a. Calculate Pr(rolling a total of 5).
b. Calculate Pr(rolling a total of 1).
c. Recognise the most probable outcome.

2. In her cupboard Rosa has three scarves (red, blue and pink) and two beanies (brown and purple). If she randomly chooses one scarf and one beanie, show the sample space in an array.

3. A ten-sided die is rolled and then a coin is flipped.

   a. Use an array to determine all the outcomes.
   b. Calculate:

      i. Pr(even and a Head)
      ii. Pr(even or a Head)
      iii. Pr(a Tail and a number greater than 7)
      iv. Pr(a Tail or a number greater than 7).

4. One box contains red and blue pencils, and a second box contains red, blue and green pencils. If one pencil is chosen at random from each box and the colours are noted, draw a tree diagram to show the sample space.

5. **WE18** A bag contains three discs labelled 1, 3 and 5, and another bag contains two discs, labelled 2 and 4, as shown below. A disc is taken from each bag and the larger number is recorded.

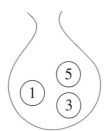 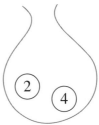

   a. Complete the tree diagram below to list the sample space.

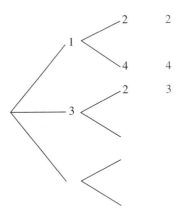

   b. Calculate:
      i. Pr(2)   ii. Pr(1)   iii. Pr(odd number).

6. Two dice are rolled and the difference between the two numbers is found.

   a. Use an array to find all the outcomes.
   b. Calculate:

      i. Pr(odd number)   ii. Pr(0)
      iii. Pr(a number more than 2)   iv. Pr(a number no more than 2).

7. **WE19** A die is rolled twice. Calculate the probability of rolling:
   a. a 6 on the first roll
   b. a double 6
   c. an even number on both dice
   d. a total of 12.

8. A coin is tossed twice.
   a. Show the outcomes on a tree diagram.
   b. Calculate:
      i. Pr(2 Tails)
      ii. Pr(at least 1 Tail).

9. **WE20** A bag contains three red counters and one blue counter. A counter is chosen at random. A second counter is drawn with replacement.
   a. Show the outcomes and probabilities on a tree diagram.
   b. Calculate the probability of choosing:
      i. a red counter then a blue counter
      ii. two blue counters.

10. A bag contains five red, six green and four blue counters. A counter is chosen at random. A second counter is drawn with replacement. Use a tree diagram to calculate the probability of choosing:
    a. a red and a green counter
    b. a red counter then a green counter
    c. a red or green counter
    d. a matching pair of the same colour.

11. A bag contains eight black, nine white and three red counters. A counter is chosen at random. A second counter is drawn with replacement. Use a tree diagram to calculate the probability of choosing:
    a. at least one black counter
    b. a black counter and a white counter
    c. a black counter or a white counter
    d. a matching pair of the same colour
    e. different colours.

12. **WE21** A bag contains three black balls and two red balls. If two balls are selected, randomly, without replacement:
    a. show the outcomes and their probabilities on a tree diagram
    b. calculate Pr(2 red balls).

13. A bag contains five red, three blue and two green balls. If two balls are randomly selected without replacement, calculate:
    a. Pr(two red balls)
    b. Pr(a red ball then a green ball)
    c. Pr(a red ball and a green ball)
    d. Pr(the same colour selected twice).

14. A bag contains five black, one white and six red balls. If two balls are randomly selected without replacement, calculate:
    a. Pr(two red balls)
    b. Pr(a red ball then a white ball)
    c. Pr(a red ball and a black ball)
    d. Pr(the same colour selected twice).

Understanding

15. The kings and queens from a deck of cards are shuffled, then two cards are chosen. Calculate the probability that two kings are chosen:

    a. if the first card is replaced
    b. if the first card is not replaced.

16. The 12 royal cards from a deck of cards are shuffled together, then two cards are chosen. Determine the probability that:

    a. two queens are chosen if the first card is replaced
    b. a matching pair (same value, not suit) is chosen if the first card is not replaced
    c. two cards from different suits are chosen if the first card is not replaced.

17. Each week John and Paul play two sets of tennis against each other. They each have an equal chance of winning the first set. When John wins the first set, his probability of winning the second set rises to 0.6, but if he loses the first set, he has only a 0.3 chance of winning the second set.

    a. Show the possible outcomes on a tree diagram.
    b. Calculate:
       i. Pr(John wins both sets)
       ii. Pr(Paul wins both sets)
       iii. Pr(they win one set each).

Reasoning

18. A bag contains four red and six yellow balls. If the first ball drawn is yellow, explain the difference in the probability of drawing the second ball if the first ball was replaced compared to not being replaced.

19. James has six different ties, five different shirts and three different suits that he can choose from when getting ready for work.

    a. Determine how many days he can go without repeating an item of clothing.
    b. Determine how many possible combinations of clothing there are.
    c. James receives a new shirt and tie for his birthday. Determine how many more combinations of clothing are now possible.

20. Three dice are tossed and the total is recorded.

    a. State the smallest and largest possible totals.
    b. Calculate the probabilities for all possible totals.

Problem solving

21. You draw two cards, one after the other without replacement, from a deck of 52 cards. Calculate the probability of:

    a. drawing two aces
    b. drawing two face cards (J, Q, K)
    c. getting a 'pair' (22, 33, 44 ... QQ, KK, AA).

22. A chance experiment involves flipping a coin and rolling two dice. Determine the probability of obtaining Tails and two numbers whose sum is greater than 4.

23. In a jar there are 10 red balls and 6 green balls. Jacob takes out two balls, one at a time, without replacing them. Calculate the probability that both balls are the same colour.

24. A coin is being tossed repeatedly. Determine how many possible outcomes there will be if it is tossed:

 a. 3 times
 b. 5 times
 c. 4 times
 d. x times.

25. A box of chocolates contains milk chocolates and dark chocolates. The probability of selecting a dark chocolate from a full box is $\frac{1}{6}$. Once a dark chocolate has been taken from the box, the chance of selecting a second dark chocolate drops to $\frac{1}{7}$.

 a. Calculate how many chocolates are in the box altogether.
 b. If two chocolates were randomly selected from the box, calculate the probability of getting two of the same type of chocolate.

26. Claire's maths teacher decided to surprise the class with a four-question multiple-choice quiz. If each question has four possible options, calculate the probability the Claire passes the test given she guesses every question.

27. In the game of 'Texas Hold'Em' poker, five cards are progressively placed face up in the centre of the table for all players to use.
 At one point in the game there are three face-up cards (two hearts and one diamond). You have two diamonds in your hand for a total of three diamonds. Five diamonds make a flush.
 Given that there are 47 cards left, determine the probability that the next two face-up cards are both diamonds.

10.6 Review

10.6.1 Topic summary

Probability of an outcome

- The probability of any outcome always falls between 0 and 1.
- An outcome that is **certain** has a probability of 1.
- An outcome that is **impossible** has a probability of 0.
- An outcome that is **likely** has a probability between 0.5 and 1.
- An outcome that is **unlikely** has a probability between 0 and 0.5.

PROBABILITY

Key terms

- **Trial:** A single performance of an experiment to produce a result, such as rolling a die
- **Sample space:** The set of all possible outcomes, represented by the symbol ξ. When rolling a 6-sided die, $\xi = \{1, 2, 3, 4, 5, 6\}$.
- **Event:** A set of favourable outcomes
  For example, A can represent the event of rolling an even number on a 6-sided die: $A = \{2, 4, 6\}$.
- **Complement:** All outcomes that are not part of an event. The complement of event A above is denoted by A'. $A' = \{1, 3, 5\}$.

Experimental probability

- **Experimental probability** is used when it is difficult or impossible to determine the theoretical probability.
- **Experimental probability** = Relative frequency
- **Relative frequency** = $\dfrac{\text{frequency of outcome}}{\text{total number of trials}}$
- The more trials conducted, the more accurate the experimental probability will be.
- **Expected number** = probability of outcome × number of trials

Venn diagrams and two-way tables

- These are two different methods that can be used to represent a sample space and visualise the interaction of different events.
- For example, in a class of 20 students, 5 study Art (event A), 9 study Biology (event B) and 2 study both subjects.
  - Venn diagram:

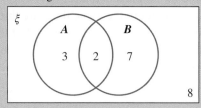

  - Two-way table:

	B	B'	Total
A	2	3	5
A'	7	8	15
Total	9	11	20

- Using these we can see:
  - $n(A' \cap B) = 7$
  - $P(A \cap B') = \dfrac{3}{20}$

Two-step experiments

- Two-step experiments involve two separate actions performed one after the other.
- The sample space of a two-step experiment can be represented with an array or a tree diagram.
- An experiment can be conducted with replacement or without replacement.
- If an experiment is performed without replacement, the probabilities will change for the second action.

		Bag 2		
		B	**M**	**P**
Bag 1	**B**	BB	BM	BP
	W	WB	WM	WP

First selection	Second selection	Sample space
B	B	BB
	M	BM
	P	BP
W	B	WB
	M	WM
	P	WP

Sections of a Venn diagram

- A Venn diagram can be split into four distinct sections, as shown.

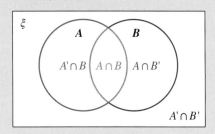

Intersection and union

- The intersection of two events A and B is written $A \cap B$. These are the outcomes that are both in A and in B.

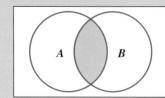

- The union of two events A and B is written $A \cup B$. These are the outcomes that are either in A or in B.

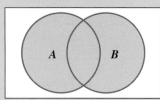

10.6.2 Success criteria

Tick a column to indicate that you have completed the subtopic and how well you think you have understood it using the traffic light system.

(**Green:** I understand; **Yellow:** I can do it with help; **Red:** I do not understand)

Subtopic	Success criteria	⬤	◯	⬤
10.2	I can determine whether an event is likely or unlikely.			
	I am familiar with key terms such as *trials*, *outcomes*, *events* and *sample space*.			
	I can calculate theoretical probability of an event.			
10.3	I can calculate the relative frequency and experimental probability of an outcome.			
	I can explain the difference between theoretical and experimental probability.			
10.4	I can identify the complement of an event A.			
	I can create and interpret a Venn diagram.			
	I can create and interpret a two-way table.			
	I can use a Venn diagram or two-way table to find $A \cap B$.			
	I can use a Venn diagram or two-way table to find $A \cup B$.			
10.5	I can display the sample space for a two-step experiment using an array and a tree diagram.			
	I can solve two-step probabilities for experiments with replacement.			
	I can solve two-step probabilities for experiments without replacement.			

10.6.3 Project

Sand-rings

A class of students and their teacher spent a day at the beach as part of a school excursion.

Part of the day was devoted to activities involving puzzles in the sand. One of the popular — and most challenging — puzzles was 'sand-rings'. Sand-rings involves drawing three rings in the sand, as shown in the diagram.

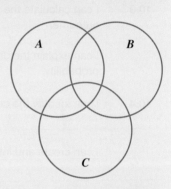

The first sand-rings puzzle requires eight shells to be arranged inside the circles, so that four shells appear inside circle A, five shells appear inside circle B and six shells appear inside circle C. The overlapping of the circles shows that the shells can be counted in two or three circles. One possible arrangement is shown below. Use this diagram to answer questions **1** to **4**.

1. How many shells appear inside circle A, but not circle B?
2. How many shells appear in circles B and C, but not circle A?

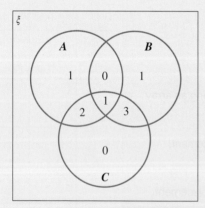

A shell is selected at random from the sand.

3. Calculate the probability it came from circle A.
4. Calculate the probability it was not in circle C.
5. The class was challenged to find the rest of the arrangements of the eight shells. (*Remember:* four shells need to appear in circle A, five in circle B and six in circle C.)

After completing the first puzzle, the students are given new rules. The number of shells to be arranged in the circles is reduced from eight to six. However, the number of shells to be in each circle remains the same; that is, four shells in circle A, five shells in circle B and six shells in circle C.

6. Using six shells, calculate how many ways the shells can be arranged so that there are four, five and six shells in the three circles.

7. Explain the system or method you used to determine your answer to question 6 above. Draw diagrams to show the different arrangements.

8. Using seven shells, calculate how many ways the shells can be arranged so that there are four, five and six shells in the three circles.

9. Again, explain the system or method you used to determine your answer to question 8. Draw diagrams to show the different arrangements.

10. Determine the minimum number of shells required to play sand-rings, so that there are four, five and six shells in the three circles.

11. Modify the rules of this game so that different totals are required for the three circles. Challenge your classmates to find all possible solutions to your modified game.

---

on Resources

 eWorkbook   Topic 10 Workbook (worksheets, code puzzle and project) (ewbk-2010)

 Interactivities Crossword (int-2712)

Sudoku puzzle (int-3212)

Exercise 10.6 Review questions

learn on

To answer questions online and to receive **immediate corrective feedback** and **fully worked solutions** for all questions, go to your learnON title at www.jacplus.com.au.

Fluency

1. **MC** In a trial, it was found that a drug cures $\dfrac{2}{5}$ of those treated by it. If 700 sufferers are treated with the drug, calculate how many of them are not expected to be cured.

   A. 280   B. 420   C. 140   D. 350   E. 100

2. **MC** If a die is rolled and a coin tossed, recognise the probability of a 6–Heads result.

   A. $\dfrac{1}{6}$ B. $\dfrac{1}{2}$ C. $\dfrac{1}{8}$ D. $\dfrac{1}{12}$ E. $\dfrac{1}{3}$

▶

3. **MC** Twelve nuts are taken from a jar containing macadamias and cashews. If three macadamias are obtained, the experimental probability of obtaining a cashew is:

A. $\dfrac{1}{12}$ B. $\dfrac{1}{4}$ C. $\dfrac{1}{3}$ D. $\dfrac{3}{4}$ E. $\dfrac{11}{12}$

4. **MC** From a normal pack of 52 playing cards, one card is randomly drawn and replaced. If this is done 208 times, the number of red or picture cards expected to turn up is:

A. 150   B. 130   C. 128   D. 120   E. 144

5. **MC** A cubic die with faces numbered 2, 3, 4, 5, 6 and 6 is rolled. The probability of rolling an even number is:

A. $\dfrac{1}{3}$ B. $\dfrac{2}{3}$ C. $\dfrac{1}{6}$ D. $\dfrac{1}{2}$ E. 1

6. **MC** The probability of rolling an odd number or a multiple of 2 using the die in question **5** is:

A. 1   B. $\dfrac{1}{3}$ C. $\dfrac{1}{4}$ D. $\dfrac{3}{4}$ E. $\dfrac{4}{3}$

The following information should be used to answer questions **7** and **8**.

Students in a Year 9 class chose the following activities for a recreation day.

Activity	Tennis	Fishing	Golf	Bushwalking
Number of students	8	15	5	7

7. **MC** If a student is selected at random from the class, the probability that the student chose fishing is:

A. $\dfrac{1}{7}$ B. $\dfrac{2}{7}$ C. $\dfrac{3}{7}$ D. $\dfrac{4}{7}$ E. $\dfrac{5}{7}$

8. **MC** If a student is selected at random, the probability that the student did not choose bushwalking is:

A. $\dfrac{1}{35}$ B. $\dfrac{2}{5}$ C. $\dfrac{3}{5}$ D. $\dfrac{4}{5}$ E. $\dfrac{4}{35}$

9. The mass of 40 students in a Year 9 Maths class was recorded in a table.

Mass (kg)	Less than 50	50−<55	55−<60	60−<65	65 and over
Number of students	4	6	10	15	5

Calculate the experimental probability of selecting a student who has:
a. a mass of 55 kg or more, but less than 60 kg
b. a mass less than 50 kg
c. a mass of 65 kg or greater.

10. Calculate the following expected values.
    a. The number of Heads in 80 tosses of a coin
    b. The number of sixes in 200 rolls of a die
    c. The number of hearts if a card is picked from a reshuffled pack and replaced 100 times

11. A normal six-sided die is rolled. Calculate the probability of getting an odd number or a multiple of 4.

12. A card is drawn from a pack of 52 cards. Calculate the probability that the card is a heart or a club.

13. Insert each of the letters **a** to **d** to represent the following events at appropriate places on the probability scale shown.

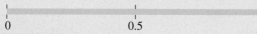

   0   0.5   1

   a. You will go to school on Christmas Day.
   b. All Year 9 students can go to university without doing Year 10.
   c. Year 9 students will study Maths.
   d. An Australian TV channel will telecast the news at 6:00 pm.

14. Indicate the set that each of the shaded regions represents.

   a. Subject preference

   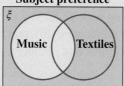

   b. Leisure activity

   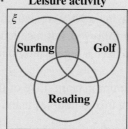

   c. Favourite drinks

   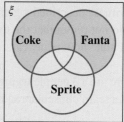

15. An equal number of red (R), black (B) and yellow (Y) counters makes up a total of 30 in a bag.
   a. List the sample space for picking a counter from the bag.
   b. Event *A* is 'draw a yellow counter, then randomly draw another counter from the bag'. List the sample space of event *A*.
   c. Explain whether 'choosing a green counter' is an outcome.

Problem solving

16. Teachers at a school opted for the choice of morning recess refreshments shown in the Venn diagram.
   a. Recognise how many teachers are in the set 'cake ∩ coffee'.
   b. State the total number of teachers surveyed.
   c. If a teacher is selected at random, calculate the probability that the teacher:
      i. chose tea
      ii. chose coffee only
      iii. chose milk
      iv. did not choose tea, coffee, cake or milk
      v. did not choose coffee.

   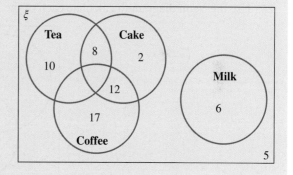

17. Thirty-two students ordered fried rice, chicken wings and dim sims for lunch. Four students ordered all three, two ordered fried rice and chicken wings only, three ordered fried rice and dim sims only, and three ordered chicken wings and dim sims only.
   When the waiter organised the orders, he found that 16 students ordered fried rice and 18 students ordered dim sims.
   a. Show this information on a Venn diagram.
   b. Calculate how many students ordered chicken wings only.
   c. If a student is picked at random, calculate the probability that the student has:
      i. ordered chicken wings and dim sims
      ii. ordered fried rice
      iii. not ordered dim sims.

18. The following are options for dorm rooms at a university. You are required to choose one option from each of the four categories. How many different combinations of rooms are there to choose from?
    - Upstairs or downstairs rooms
    - Single or double rooms
    - Male or female roommates
    - A choice of 10 different locations on campus

    Calculate how many different combinations of rooms there are to choose from.

19. A future king is the oldest male child. The future king of Mainland has two siblings. Determine the probability that he has an older sister.

20. A witness described a getaway car as having a NSW registration plate starting with TLK. The witness could not remember the three digits that followed, but recalled that all three digits were different. Calculate how many cars in NSW could have a registration plate with these letters and numbers.

21. There are 12 people trying out for a badminton team. Five of them are girls. Calculate the probability that a team chosen at random to play is a mixed doubles team.

22. An ace is chosen from a deck of standard cards and not replaced. A king is then chosen from the deck. Calculate the probability of choosing an ace and a king in this order.

23. If you randomly select one number from 1 to 500 (inclusive), calculate the probability that the selected number will have at least one 4 in the digits.

24. If you flip a coin eight times, calculate how many of the possible outcomes would you expect to have a Head on the second toss.

25. Use the Venn diagram to calculate each of the following.

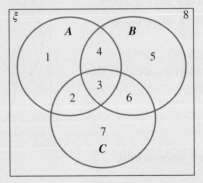

    a. $A' \cup (B' \cap C')$
    c. $A \cap (B' \cup C')$
    b. $A' \cap (B \cap C')$
    d. $(A \cup B \cup C)' \cap (A \cup B \cup C)$

on To test your understanding and knowledge of this topic go to your learnON title at www.jacplus.com.au and complete the **post-test**.

Online Resources

Below is a full list of **rich resources** available online for this topic. These resources are designed to bring ideas to life, to promote deep and lasting learning and to support the different learning needs of each individual.

eWorkbook

Download the workbook for this topic, which includes worksheets, a code puzzle and a project (ewbk-2010) ☐

Solutions

Download a copy of the fully worked solutions to every question in this topic (sol-0730) ☐

Digital documents

10.2 SkillSHEET Probability scale (doc-6307) ☐
SkillSHEET Understanding a deck of playing cards (doc-6308) ☐
SkillSHEET Listing the sample space (doc-6309) ☐
SkillSHEET Theoretical probability (doc-6310) ☐
10.3 SkillSHEET Determining complementary events (doc-6311) ☐
SkillSHEET Calculating the probability of a complementary event (doc-6312) ☐

Video eLessons

10.2 The language of probability (eles-4877) ☐
Key terms of probability (eles-4878) ☐
Theoretical probability (eles-4879) ☐
10.3 Relative frequency (eles-4880) ☐
Experimental probability (eles-4881) ☐
Expected number of results (eles-4882) ☐
10.4 The complement of an event (eles-4883) ☐
Venn diagrams: the intersection of events (eles-4884) ☐
Two-way tables (eles-4885) ☐
Number of outcomes (eles-4886) ☐
Venn diagrams: the union of events (eles-4887) ☐
Venn diagrams (eles-1934) ☐
10.5 The sample space of two-step experiments (eles-4888) ☐
Two-step experiments (eles-4889) ☐
Experiments with replacement (eles-4890) ☐
Experiments without replacement (eles-4891) ☐
Tree diagrams (eles-1894) ☐

Interactivities

10.2 Individual pathway interactivity: Theoretical probability (int-4534) ☐
Probability scale (int-3824) ☐
Theoretical probability (int-6081) ☐
10.3 Individual pathway interactivity: Experimental probability (int-4535) ☐
Experimental probability (int-3825) ☐

10.4 Individual pathway interactivity: Venn diagrams and two-way tables (int-4536) ☐
Venn diagrams (int-3828) ☐
Two-way tables (int-6082) ☐
10.5 Individual pathway interactivity: Two-step experiments (int-4537) ☐
Two-step experiments (int-6083) ☐
10.6 Crossword (int-2712) ☐
Sudoku puzzle (int-3212) ☐

Teacher resources

There are many resources available exclusively for teachers online.

To access these online resources, log on to **www.jacplus.com.au**.

Answers

Topic 10 Probability

Exercise 10.1 Pre-test

1. C

2. a. Zero   b. $\dfrac{3}{8}$ or 0.375

3. a. 100   b. $\dfrac{7}{25}$ c. $\dfrac{1}{2}$ d. $\dfrac{11}{50}$

4. $\dfrac{1}{3}$

5. True

6. D

7. C

8. A

9. a. $\dfrac{1}{4}$ b. $\dfrac{3}{14}$

10. Pr(drama not comedy) $= \dfrac{2}{5}$

11. $\dfrac{5}{7}$

12. a. B   b. 0.33

13. a. C   b. D   c. A   d. B

14. a. $\dfrac{8}{15}$ b. $\dfrac{67}{225}$ c. $\dfrac{1}{6}$ d. $\dfrac{67}{240}$

15. D

Exercise 10.2 Theoretical probability

1.

a. Certain   b. Unlikely

2.

a. Likely   b. Certain

3.

a. Certain   b. Likely

4.

a. Likely   b. Likely

5. a. $\{1, 2, 3, 4, 5, 6\}$
   b. $\{H, T\}$
   c. $\{$defective, not defective$\}$
   d. $\{$red, black$\}$
   e. $\{$hearts, clubs, diamonds, spades$\}$

6. a. $\{1, 2, 3\}$ b. $\{1, 3, 5\}$
   c. $\{1, 2, 4, 6\}$ d. $\{3, 4, 5, 6\}$
   e. $\{1, 2, 3, 4\}$ f. $\{5, 6\}$

7. a. 3, 5
   b. 1, 2
   c. 6
   d. 1, 2, 5, 6
   e. 1, 2, 3, 5
   f. No favourable outcomes
   g. 3

8. a. $\{$hearts, clubs, diamonds, spades$\}$
   b.   i. $\{$clubs, spades$\}$
       ii. $\{$hearts, diamonds$\}$
       iii. $\{$clubs, diamonds, spades$\}$
       iv. $\{$hearts, clubs, diamonds, spades$\}$

9. a. 6   b. 2   c. 52
   d. 4   e. 5

10. a. $\dfrac{1}{13}$ b. $\dfrac{1}{13}$ c. $\dfrac{1}{52}$ d. $\dfrac{1}{4}$

11. a. $\dfrac{1}{52}$ b. $\dfrac{1}{2}$ c. $\dfrac{1}{26}$ d. $\dfrac{1}{4}$ e. $\dfrac{1}{2}$

12. a. $\dfrac{1}{52}$ b. $\dfrac{4}{13}$ c. $\dfrac{2}{13}$ d. 0   e. $\dfrac{12}{13}$

13. a. $\dfrac{2}{11}$ b. $\dfrac{9}{11}$ c. $\dfrac{4}{11}$ d. $\dfrac{7}{11}$

14. a. $\dfrac{1}{4}$ b. $\dfrac{1}{8}$ c. $\dfrac{5}{16}$ d. 0   e. $\dfrac{3}{4}$

15. a. i. $\dfrac{2}{3}$ ii. $\dfrac{1}{3}$
   b. Sample responses can be found in the worked solutions in the online resources.

16. Sample responses can be found in the worked solutions in the online resources.

17. a. 18   b. 20   c. 36

18. a. i. $\dfrac{7}{10}$ ii. $\dfrac{3}{10}$
   b. i. $\dfrac{2}{3}$ ii. $\dfrac{1}{3}$

19. Sample responses can be found in the worked solutions in the online resources.

20. Probabilities must be between 0 and 1, so -0.5 and 1.05 can't be probabilities.

21. $\dfrac{1}{2}$

22. The coloured portions outside the arc of the spinner shown are of no consequence. The four colours within the arc of the spinner are of equal area (each $\dfrac{1}{4}$ circle), so there is equal chance of falling on each of the colours.

23. $\dfrac{2}{3}$

24. There are 36 outcomes, 15 under 7, 6 equal to 7 and 15 over 7. So, you would have a greater chance of winning if you chose 'under 7' or 'over 7' rather than 'equal to 7'.

25. 0.09

Exercise 10.3 Experimental probability

1. a. i. 16
   ii. 4
   iii.

Outcome	1	2	3	4	5	6
Relative frequency	0.19	0.06	0.31	0.13	0.25	0.06

   iv. 1

   b. i. 270
   ii. 40
   iii.

Outcome	1	2	3	4	5	6
Relative frequency	0.19	0.14	0.17	0.18	0.15	0.17

   iv. 1

2. a. i. r.f. (H) = 0.44, r.f. (T) = 0.56
   ii. 1
   b. i. r.f. (H) = 0.62, r.f. (T) = 0.38
   ii. 1

3. a. 0.365   b. 0.33   c. 0.25   d. 0.875

4. a. 0.275   b. 0.64   c. 0.365   d. 0.635

5. a. $\dfrac{27}{95}$ b. $\dfrac{33}{95}$

   c. $\dfrac{1}{19}$ d. Yes, reject this batch.

6. a. Not reasonable; not enough trials were held.
   b. Not reasonable; the conditions are different under each trial.
   c. Reasonable; the Sun rises every morning, regardless of the weather or season.
   d. Reasonable; enough trials were performed under the same conditions.
   e. Not reasonable; monthly rainfall in Sydney is not consistent throughout the year.

7. a. $\dfrac{5}{24}$ b. $\dfrac{59}{72}$

8. 200

9. 20

10. A

11. B

12. a. $\dfrac{11}{20}$ b. $\dfrac{3}{10}$ c. $\dfrac{17}{20}$ d. $\dfrac{3}{20}$

13. a. $\dfrac{33}{100}$ b. $\dfrac{40}{100} = \dfrac{2}{5}$ c. $\dfrac{3}{100}$

14. a. i. 6
    ii. $\{356, 365, 536, 563, 635, 653\}$
    b. $\dfrac{2}{3}$
    c. $\dfrac{1}{3}$
    d. $\dfrac{1}{3}$

15. D

16. Svetlana 6, Sarah 4, Leonie 3, Trang 2

17. C

18. a. 27
    b. $\left\{ \begin{array}{l} 111, 112, 115, 121, 122, 125, 151, 152, 155, 211, 212, \\ 215, 221, 222, 225, 251, 252, 255, 511, 512, 515, 521, \\ 522, 525, 551, 552, 555 \end{array} \right\}$
    c. $\dfrac{1}{3}$
    d. $\dfrac{2}{3}$
    e. $\dfrac{1}{3}$

19. a. $\dfrac{1}{12}$
    b. $\dfrac{1}{20}$
    c. No, because John has a higher probability of winning.

20. a. $\dfrac{7}{10}$
    b. Yes, far too many mistakes

21. 32

22. 1000 balls

23. Red = 10, yellow = 50, blue = 40

24. a. $\dfrac{7}{20}$ b. $\dfrac{2}{5}$

25. Sample responses can be found in the worked solutions in the online resources.

26. a. $\dfrac{30}{241}$ b. $\dfrac{91}{241}$ c. $\dfrac{59}{241}$

27. 16, 64, 40, 80, 128 and 32 degrees.

Exercise 10.4 Venn diagrams and two-way tables

1. a. $A' = \{1, 3, 5, 7, 9\}$
   b. $B' = \{1, 2, 3, 4, 6, 7, 8, 9\}$
   c. $C' = \{2, 3, 5, 6, 7, 8, 10\}$
   d. $D' = \{8, 9, 10\}$

2. a. $A' = \{11, 13, 14, 16, 17, 19, 20\}$
   b. $B' = \{20\}$
   c. $C' = \{12, 14, 15, 16, 18, 20\}$
   d. $D' = \{12, 14, 16\}$

3. a. $A' = \{1, 2, 3, 5, 6, 7, 9, 10, 11, 13, 14, 15, 17, 18, 19\}$
   b. $B' = \{1, 4, 6, 8, 9, 10, 12, 14, 15, 16, 18, 20\}$
   c. $C' = \{1, 3, 5, 7, 9, 11, 13, 14, 15, 16, 17, 18, 19, 20\}$
   d. $D' = \{1, 3, 5, 7, 9, 11, 13\}$

4. a. 27   b. 11   c. 13   d. 6
   e. 7   f. 9

5.

	T	T'
S	6	5
S'	7	9

6.
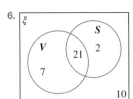

7. a. i. W' **ii.** None
  b. i. $A \cap B'$ **ii.** $A \cap B$
  c. i. $A' \cap B'$ **ii.** $B \cap A'$

8. a.

	Left-handed	Right-handed
Male	15	40
Female	12	33

  b.
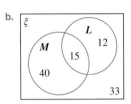

  c. i. $\dfrac{73}{100}$ **ii.** $\dfrac{2}{5}$ **iii.** $\dfrac{3}{25}$

9. a. $\dfrac{8}{25}$ **b.** $\dfrac{17}{25}$ **c.** $\dfrac{1}{5}$ **d.** $\dfrac{3}{25}$

10.
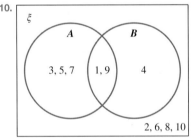

11.
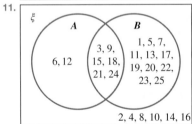

12.
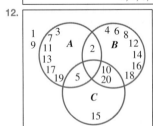

13. a. $\dfrac{1}{2}$ **b.** $\dfrac{1}{10}$ **c.** $\dfrac{31}{40}$
  d. $\dfrac{27}{40}$ **e.** $\dfrac{9}{40}$

14. a. $\dfrac{1}{2}$ **b.** $\dfrac{6}{25}$
  c. $\dfrac{22}{25}$ **d.** $\dfrac{16}{25}$

15. a.

	Croissant	No croissant	Total
Coffee	27	2	29
No coffee	9	7	16
Total	36	9	45

  b. i. $\dfrac{3}{5}$ **ii.** $\dfrac{38}{45}$ **iii.** $\dfrac{11}{45}$
  c. 1218

16. a. 16
  b. i. $\dfrac{5}{16}$ **ii.** $\dfrac{3}{16}$ **iii.** $\dfrac{5}{8}$ **iv.** $\dfrac{7}{16}$

17. a.
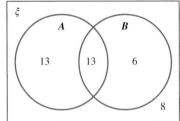

  b. i. $\dfrac{29}{40}$ **ii.** $\dfrac{4}{5}$ **iii.** $\dfrac{19}{40}$

18. a. Children who are not swimmers
  b. Children who are swimmers and runners
  c. Children who neither swim nor run
  d. Children who swim or run or both

19. a. i.
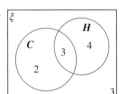

  ii.

	H	H'
C	3	2
C'	4	3

  b. i. $\dfrac{7}{12}$ **ii.** $\dfrac{3}{4}$ **iii.** $\dfrac{1}{4}$ **iv.** $\dfrac{1}{2}$

20. a.
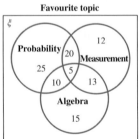

  b. 25 **c.** 15 **d.** 12

e. 43 f. $\dfrac{3}{5}$ g. $\dfrac{43}{100}$

h. $\dfrac{9}{50}$ i. $\dfrac{13}{100}$ j. $\dfrac{1}{20}$

21.
Quadrilaterals

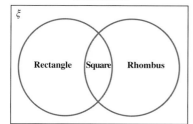

22. a. 2, 5, 6, 7, 8 b. 4
  c. 8 d. 1, 2, 3, 4, 5, 7, 8

23. a.

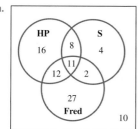

b. $\dfrac{8}{90} = \dfrac{4}{45}$

24. a.

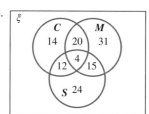

b. $\dfrac{7}{12}$

25. a. 31 students b. $\dfrac{1}{13}$

26.

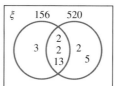

a. HCF $= 2 \times 2 \times 13 = 52$
b. LCM $= 3 \times 2 \times 2 \times 13 \times 2 \times 5 = 1560$

27. a.

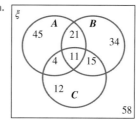

b. 4
c. 21
d. i. $\dfrac{81}{200}$ ii. $\dfrac{97}{200}$

e. $\dfrac{3}{50}$

Exercise 10.5 Two-step experiments

1. See table at the foot of the page.*
  a. $\dfrac{1}{9}$ b. 0 c. A total of 7

2.

		Scarves		
		R	**Bl**	**Pi**
Beanies	**Br**	Br, R	Br, Bl	Br, Pi
	Pu	Pu, R	Pu, Bl	Pu, Pi

3. a. See table at the foot of the page.*
  b. i. $\dfrac{1}{4}$ ii. $\dfrac{3}{4}$ iii. $\dfrac{3}{20}$ iv. $\dfrac{13}{20}$

*1.

		Die 1					
		1	**2**	**3**	**4**	**5**	**6**
Die 2	**1**	2	3	4	5	6	7
	2	3	4	5	6	7	8
	3	4	5	6	7	8	9
	4	5	6	7	8	9	10
	5	6	7	8	9	10	11
	6	7	8	9	10	11	12

*3. a.

		Dice roll									
		1	**2**	**3**	**4**	**5**	**6**	**7**	**8**	**9**	**10**
Coin toss	**H**	1, H	2, H	3, H	4, H	5, H	6, H	7, H	8, H	9, H	10, H
	T	1, T	2, T	3, T	4, T	5, T	6, T	7, T	8, T	9, T	10, T

4.

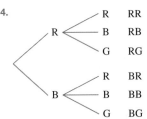

5. a.

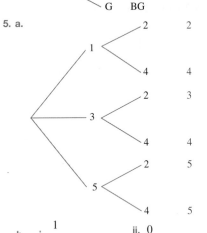

b. i. $\dfrac{1}{6}$ ii. 0 iii. $\dfrac{1}{2}$

6. a. See table at the foot of the page.*

b. i. $\dfrac{1}{2}$ ii. $\dfrac{1}{6}$ iii. $\dfrac{1}{3}$ iv. $\dfrac{2}{3}$

7. a. $\dfrac{1}{6}$ b. $\dfrac{1}{36}$ c. $\dfrac{1}{4}$ d. $\dfrac{1}{36}$

8. a.

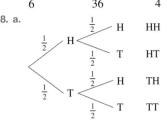

b. i. $\dfrac{1}{4}$ ii. $\dfrac{3}{4}$

9. a.

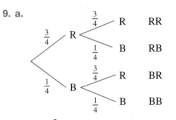

b. i. $\dfrac{3}{16}$ ii. $\dfrac{1}{16}$

10. a. $\dfrac{4}{15}$ b. $\dfrac{2}{15}$ c. $\dfrac{209}{225}$ d. $\dfrac{77}{225}$

11. a. $\dfrac{16}{25}$ b. $\dfrac{9}{25}$ c. $\dfrac{391}{400}$

d. $\dfrac{77}{200}$ e. $\dfrac{123}{200}$

12. a.

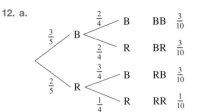

b. $\dfrac{1}{10}$

13. a. $\dfrac{2}{9}$ b. $\dfrac{1}{9}$ c. $\dfrac{2}{9}$ d. $\dfrac{14}{45}$

14. a. $\dfrac{5}{22}$ b. $\dfrac{1}{22}$ c. $\dfrac{5}{11}$ d. $\dfrac{25}{66}$

15. a. $\dfrac{1}{4}$ b. $\dfrac{3}{14}$

16. a. $\dfrac{1}{9}$ b. $\dfrac{3}{11}$ c. $\dfrac{8}{11}$

17. a.

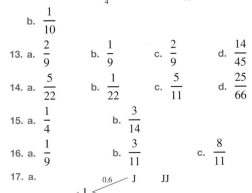

b. i. 0.3   ii. 0.35   iii. 0.35

18. If the first ball is replaced, the probability of drawing a yellow ball stays the same on the second draw, i.e. $\left(\dfrac{3}{5}\right)$.

---

*6. a.

		Die 1					
		1	**2**	**3**	**4**	**5**	**6**
	1	0	1	2	3	4	5
	2	1	0	1	2	3	4
Die 2	**3**	2	1	0	1	2	3
	4	3	2	1	0	1	2
	5	4	3	2	1	0	1
	6	5	4	3	2	1	0

If the first ball isn't replaced, the probability of drawing a yellow ball on the second draw decreases, i.e. $\left(\dfrac{5}{9}\right)$.

19. a. 3 days   b. 90 combinations
    c. 36 new combinations

20. a. Smallest total: 3, largest total: 18
    b.

Total	Probability
3	$\dfrac{1}{216}$
4	$\dfrac{3}{216}$
5	$\dfrac{6}{216}$
6	$\dfrac{10}{216}$
7	$\dfrac{15}{216}$
8	$\dfrac{21}{216}$
9	$\dfrac{25}{216}$
10	$\dfrac{27}{216}$
11	$\dfrac{27}{216}$
12	$\dfrac{25}{216}$
13	$\dfrac{21}{216}$
14	$\dfrac{15}{216}$
15	$\dfrac{10}{216}$
16	$\dfrac{6}{216}$
17	$\dfrac{3}{216}$
18	$\dfrac{1}{216}$

21. a. $\dfrac{1}{221}$ b. $\dfrac{11}{221}$ c. $\dfrac{1}{17}$

22. $\dfrac{5}{12}$

23. $\dfrac{1}{2}$

24. a. 8   b. 32   c. 16   d. 2^x

25. a. 36   b. $\dfrac{5}{7}$

26. $\dfrac{67}{256}$

27. $\dfrac{45}{1081}$

Project

1. 3
2. 4
3. $\dfrac{1}{2}$
4. $\dfrac{1}{4}$
5. 18
6. 2
7. Sample responses can be found in the worked solutions in the online resources.
8. 8
9. Sample responses can be found in the worked solutions in the online resources.
10. 6

Exercise 10.6 Review questions

1. B
2. D
3. D
4. C
5. B
6. A
7. C
8. D
9. a. $\dfrac{1}{4}$ b. $\dfrac{1}{10}$ c. $\dfrac{1}{8}$
10. a. 40   b. 33   c. 25
11. $\dfrac{2}{3}$
12. $\dfrac{1}{2}$
13.
14. a. Students who do not like music
    b. People who like surfing and golf as leisure activities, but not reading
    c. People who like Coke or Fanta or both but not Sprite
15. a. $\{R, B, Y\}$
    b. $\{(YR), (YY), (YB)\}$
    c. No, there is no green counter.
16. a. 12
    b. 60
    c. i. $\dfrac{3}{10}$ ii. $\dfrac{17}{60}$ iii. $\dfrac{1}{10}$
      iv. $\dfrac{1}{12}$ v. $\dfrac{31}{60}$

17. a.

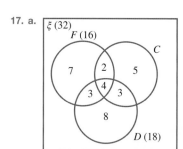

b. 5

c. i. $\dfrac{7}{32}$ ii. $\dfrac{1}{2}$ iii. $\dfrac{7}{16}$

18. 80

19. $\dfrac{3}{7}$

20. 720

21. $\dfrac{35}{66}$

22. $\dfrac{4}{663}$

23. $\dfrac{176}{500}$

24. 128

25. a. $\{1, 5, 6, 7, 8\}$ **b.** $\{5\}$

    c. $\{1, 2, 4\}$ **d.** ϕ

11 Statistics

Learning sequence

11.1 Overview

Why learn this?

The science of statistics is concerned with understanding the world around us. It presents a way of processing data about people, places and things in order to make that data useful and easier to understand.

One example of statistics is the data that social media companies collect about how you use their platforms. This data is statistically analysed and used to adapt these platforms to the way you use them (and to help the company maximise profits). Another example of statistics is the Australian Bureau of Meteorology — they collect data about the weather, analyse the statistics from that data and then use that knowledge to help predict what the weather will be like tomorrow, next week and even next year.

A good understanding of statistics can help you develop critical analysis skills so that you are not tricked by bad or misleading statistics. It is also important because statistics are a valuable tool used in a broad range of careers including product design, science, sports coaching and finance. Next time you are reading an article or watching a news report, keep your eye out for statistics and you'll start to notice that they are everywhere!

Where to get help

Go to your learnON title at **www.jacplus.com.au** to access the following digital resources. The Online Resources Summary at the end of this topic provides a full list of what's available to help you learn the concepts covered in this topic.

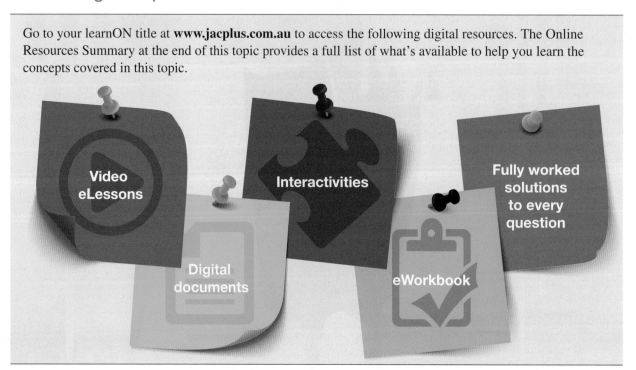

Complete this pre-test in your learnON title at **www.jacplus.com.au**, and receive **automatic marks**, **immediate corrective feedback** and **fully worked solutions**.

1. Match and classify the following different types of data.

Example	Data type
a. Attendance at a concert	**i.** nominal
b. Mass of bag of lollies	**ii.** ordinal
c. Eye colour	**iii.** discrete
d. The quality of food at a restaurant	**iv.** continuous

2. Thirty Year 9 students from were surveyed on the number of hours they spend each week using social media on their phones. State whether this type of data is a population or a sample.

3. Fifty members of a tennis club were selected to participate in a survey. The club has 80 male adult, 70 female adult, 55 male adolescent and 45 female adolescent members.
State how many members are required from each of these four groups to perform a stratified sample survey.

4. The stem and leaf plot shown represents the speeds of cars recorded by a roadside camera outside a school where the speed limit is: 40 km/h.

Key: $1 \,|\, 9 = 19$ km/h

Stem	Leaf
1	9
2	5 7 7 9 9
3	0 0 4 5 6 7 8 8
4	0 0 0 0 2 4 5 5
5	2 3
6	0

a. State how many cars had their speed recorded.
b. Determine how many cars were travelling at a speed over 40 km/h.

5. The dot plot shown displays the results of a maths test for a Year 9 class.
The results show the students' marks out of 20.

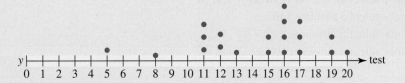

a. State how many students are in the class.
b. Determine how many students scored more than 15 out of 20 for the test.

6. Match the different types of data shown in the table with the most appropriate type of graph.

Data type	Type of graph
a. Categorical data	**i.** Box plot
b. Numerical continuous data	**ii.** Stem plot
c. Numerical discrete data	**iii.** Bar chart

7. **MC** Based on the pie chart shown, the percentage of students who catch public transport to school is approximately:

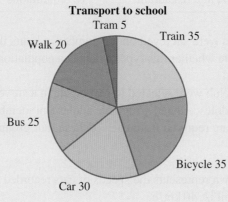

Transport to school

Tram 5
Train 35
Walk 20
Bus 25
Bicycle 35
Car 30

A. 40%   B. 43%   C. 52%   D. 60%   E. 65%

8. The histogram shows the weights of passengers' luggage for a flight to Perth.
   a. State the type of data shown the histogram.
   b. State how many passengers exceeded the maximum limit of 20 kg.
   c. Explain whether the data distribution is skewed or symmetrical.

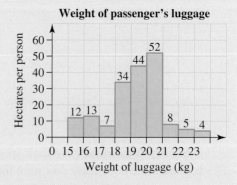

Weight of passenger's luggage

Hectares per person

60, 50, 40, 30, 20, 10, 0

12 13 7 34 44 52 8 5 4

0 15 16 17 18 19 20 21 22 23
Weight of luggage (kg)

9. **MC** Identify the correct meaning of the spreadsheet function =RANDBETWEEN(1, 5).
   A. Randomly select a real number between 1 and 5.
   B. Randomly select an integer between 1 and 5.
   C. Randomly select a number 5 times.
   D. Randomly select 5 numbers.
   E. Randomly select a number between 2 and 4.

10. **MC** Megan wanted to select a random number in the range 10 to 20. Select the command she used to generate the number in her spreadsheet.
   A. = range (10, 20)
   B. = ran (10, 20)
   C. = randbetween (10, 20)
   D. = rand (10, 20)
   E. = random (10, 20)

11. Calculate the median for the frequency table shown.

Score	Frequency
0	3
1	8
2	10
3	14
4	5
5	2

12. Examine the grouped frequency table shown and answer the following questions.

Score	Frequency
0– < 20	3
20– < 40	4
40– < 60	8
60– < 80	15
80– < 100	7

   a. Calculate the mean for the grouped frequency table, rounded to 1 decimal place.
   b. **MC** Select the modal class for the grouped frequency table.
      A. 0– < 20   B. 20– < 40   C. 40– < 60
      D. 60– < 80   E. 80– < 100

13. Determine the missing value from the set of numbers {8, 8, 10, 12, 15} if the mean is 10, the mode is 8 and the median is 9.

14. **MC** Select the range and interquartile range for the stem plot shown, which presents the weight of school bags.
   A. Range = 36 and interquartile range = 10
   B. Range = 36 and interquartile range = 17
   C. Range = 34 and interquartile range = 9
   D. Range = 34 and interquartile range = 10
   E. Range = 34 and interquartile range = 11

Key: 1|3 = 13 kg

Stem	Leaf
0	2 3 4 5 8
1	0 0 1 1 2 2 2 3
1	5 5 5 6 7 7 8 8 9 9 9
2	0 0 1 1 2 2 4
2	5 5 8
3	1 3
3	6

15. **MC** Select the correct statement to describe symmetrical distributions.
   A. The data on the right have more low values.
   B. Distribution has two peaks.
   C. Mean has a smaller value than median.
   D. Mean has a larger value than median.
   E. Mean, median and mode have similar values.

11.2 Sampling

LEARNING INTENTION

At the end of this subtopic you should be able to:
- identify different types of data
- determine if a data set is a sample or a population
- generate random numbers.

11.2.1 Types of data

eles-4909

- Data can be **categorical** or **numerical**.
- There are two types of categorical data:
  - **nominal** data
  - **ordinal** data.
- There are two types of numerical data:
  - **discrete** data
  - **continuous** data.

Determining types of data

- The information in the flowchart shown can be used to determine the type of data being considered.

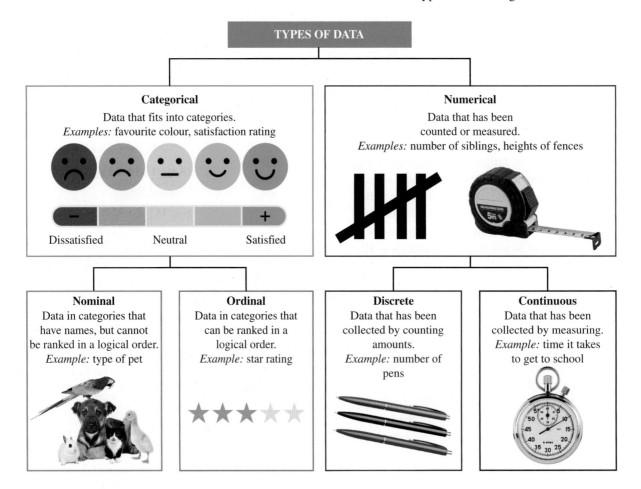

WORKED EXAMPLE 1 Classifying data as categorical or numerical

Classify each of the following data as categorical data or numerical data.
a. Favourite sport watched on TV
b. Quantity of books carried in a school bag by 20 Year 9 students
c. Cars passing your home each hour in a day
d. Eye colour of 100 preschool children

THINK

a. Decide how the data would initially be recorded, as either numbers or categories. Examples of sports watched on TV could include football, netball, soccer or tennis.

b. Decide how the data would be recorded, as either numbers or categories. The quantity of books carried in a school bag by 20 Year 9 students will be collected by counting the number of books.

c. Decide how the data would be recorded, as numbers or categories. Data for cars passing your home each hour of a day would be collected by counting the number of cars.

d. Decide how the data would be recorded, as numbers or categories. Eye colour of 100 preschool children could include blue, brown, green or hazel.

WRITE

a. The sports watched on TV are collected into categories so this is categorical data.

b. Because the recorded data would be collected by counting the number of books in a school bag, this is numerical data.

c. Because the recorded data would be collected by counting the number of cars, this is numerical data.

d. Because the eye colour of 100 preschool children would be collected into categories, this is categorical data.

WORKED EXAMPLE 2 Classifying types of data

Classify each of the following types of data into the correct groups, stating whether it is categorical or numerical. If it is categorical, state whether it is nominal or ordinal. If it is numerical, state whether it is discrete or continuous.
a. Incomes of senior players from the Australian Women's Cricket team
b. Type of transport used to go to work
c. Daily attendance at a zoo
d. Hair colour of 100 Year 9 students
e. Mass of an individual Easter egg
f. A student's behavioural report

THINK

a. 1. The incomes of the senior players from the Australian women's cricket team cover a wide range and are not able to be precisely given. They will be grouped into categories such as $50 000–$99 999, $100 000–$149 999 and > $150 000.

WRITE

a. These incomes are collected into categories, therefore, they are categorical data.

▶

2. For categorical data, decide whether the categories can be ranked in a logical order (categorical ordinal) or not (categorical nominal).

b. 1. Types of transport used to go to work could include car, bicycle, tram, bus or train.
   2. For categorical data, decide whether the categories can be ranked in a logical order (categorical ordinal) or not (categorical nominal).

c. 1. Visitors at a zoo each day would be counted and recorded as a whole number (integer).

   2. For numerical data, decide whether it is collected by counting (numerical discrete data) or by measuring (numerical continuous data).

d. 1. Hair colour of 100 Year 9 students could include black, brown, blonde, brunette or redhead.
   2. For categorical data, decide if the categories can be ranked in a logical order (categorical ordinal) or not (categorical nominal).

e. 1. Mass of an individual Easter egg would be measured on a scale.

   2. For numerical data, decide whether it is collected by counting (numerical discrete data) or by measuring (numerical continuous data).

f. 1. A student's behavioural report could be described as excellent, good, satisfactory, improvement needed or unsatisfactory.
   2. For categorical data, decide whether the categories can be ranked in a logical order (categorical ordinal) or not (categorical nominal).

Because the incomes could be ranked in a logical order, the data can be classified as categorical ordinal data.

b. The possible transport modes are categories; therefore, they are categorical data.

   Because types of transport cannot be logically ranked, the data can be classified as categorical nominal data.

c. The recorded data would be the number of visitors at a zoo; therefore, this is numerical data.
   The visitors would be counted; therefore, it is classified as numerical discrete data.

d. The hair colours are categories; therefore, this is categorical data.

   Because hair colours cannot be logically ranked, the data can be classified as categorical nominal data.

e. Because the recorded data would be the mass of an individual egg, this is numerical data.
   The mass of each egg can be recorded as a whole number (for example, 55 g), or more accurately as a decimal number (for example, 55.23 g). Therefore, it is numerical continuous data.

f. A student's behavioural report uses categories; therefore, it is categorical data.

   Because the student's behavioural report can be logically ranked, it can be classified as categorical ordinal data.

eles-4910

• The **size** of the data collected can be classified as a **population** or a **sample**.

Determining the size of a data set

• Data is classified as a *population* if data is collected from an entire group.
  Examples of population data include surveying all students at a school or measuring the size of all frogs in a pond.

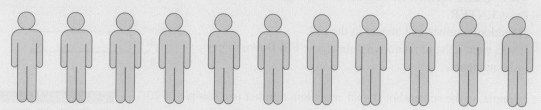

In this example, 11 out of 11 people in the group were surveyed.

• Data is classified as a *sample* if data is collected from parts of a group.
  Examples of sample data include surveying only some students at a school or measuring only some of the frogs in a pond.

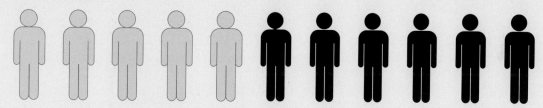

In this example, 5 out of 11 people in the group were surveyed.

WORKED EXAMPLE 3 Classifying the data collection methods

For the following data collection activities, define whether the amount of data collected is a sample or a population.

a. **Sixty Year 7 students collected data on the amount of money they contributed to a charity, then used this data to calculate the average donation from the 1523 students at their school.**

b. **A data set of the force (measured in newtons) needed to break wooden ice-block sticks**

c. **The average age of the members of your own family**

THINK

a. Only 60 Year 7 students were surveyed out of a total of 1523.

b. An unknown number of wooden ice-block sticks are required.

c. Every family member is surveyed to find the average age of the group.

WRITE

a. This is a sample, because only 60 of the 1523 students were surveyed.

b. This is a sample, because there is an unknown number of wooden ice-block sticks. It is impossible to collect data on the entire population (which would mean every wooden ice-block stick in the world!).

c. This is a population, because every member of the group has been surveyed.

⊙ 11.2.3 Generating random numbers

• Digital technology, such as spreadsheet software, often features random number generators that can be used to create random data.

Digital technology

• In a spreadsheet, click on a cell and type **=RANDBETWEEN**(<lowest number>,<highest number>).
• Press **ENTER**.
• A random number will appear in the cell.
• If you want a list of random numbers, use **Fill Down** to get more numbers.

For example, in the screenshot shown, a random number in a sample of 20 is generated by:
• clicking on a cell
• typing **=RANDBETWEEN**(1, 20)
• pressing **ENTER**.

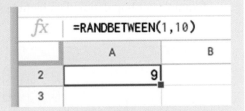

WORKED EXAMPLE 4 Selecting a sample using random number generator

Use a random number generator on a spreadsheet to select a sample for the following surveys.
a. **Randomly select 4 days from an entire year.**
b. **Randomly select 6 months that fall between January 2022 and December 2023.**

THINK	WRITE
a. 1. Use the Julian date system, which associates each day of the year with the numbers 1 to 365.	a.
2. Click on a cell. Type =RANDBETWEEN(1, 365). Click on the bottom right corner of the cell. Fill down by dragging your cursor down four cells. *Note:* If any numbers are repeated, fill down as required to generate another number to replace the repeated number.	*fx* =RANDBETWEEN(1, 365)
3. Use the Julian calendar to convert each number to a calendar date.	Day 206 is 25 July Day 188 is 7 July Day 214 is 2 August Day 114 is 24 April

b. 1. Convert the months to numbers by creating a numbered list.

b. Label the months numerically as:
1. January 2022
2. February 2022

. . .

23. November 2023
24. December 2023.

2. Click on a cell.
Type =RANDBETWEEN (1, 24)
Press ENTER.
Click on the bottom right corner of the cell.
Fill down by dragging your cursor down six cells.
Note: If any numbers are repeated, fill down an extra cell to replace the repeated numbers.

fx	=RANDBETWEEN(1,24)	
	A	B
3	15	
4	14	
5	6	
6	21	
7	22	
8	12	
9		

3. Convert the numbers back to months to write the answer.

15 is March 2023
14 is February 2023
6 is June 2022
21 is September 2023
22 is October 2023
12 is December 2022

COLLABORATIVE TASK: Designing a survey

1. Working in pairs, think of a topic on which you would like to gather your classmates' opinions. Some suggestions for things to ask them about include pets, favourite movies, favourite social platform or school subjects.
2. Think of five questions you would like to ask your classmates about this topic, ensuring that you collect both categorical and numerical data.
3. When coming up with your questions, you should consider what sorts of answers you want. Do you want yes and no answers, or would you like questions that require people to make a judgement on a scale? Your questions should be easy to understand and must be relevant to your topic.
4. Write down how you intend to collect the data, the type of data you expect to collect, and how you think it is best to represent this data. Think about the possible responses you may get. It is important to consider the methods used to collect data, because they could possibly bias the results of the analysis, which would lead to incorrect conclusions.
5. As a class, listen to and offer constructive advice on the questions that each pair has developed. Your feedback might include comments about the type of responses you might expect, how you could collect the data, whether the questions make sense (or if they could give a biased result), and whether they are relevant to the topic.
6. Once you have received your feedback, rewrite some of your questions and rethink the way you will collect the data.
7. Collect your data and keep it ready to use again during this topic.

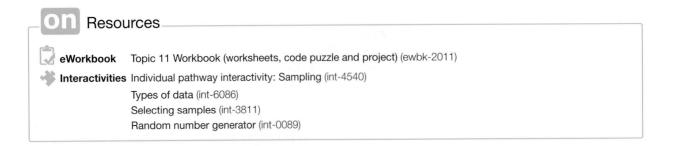
Exercise 11.2 Sampling   learn on

Individual pathways

■ PRACTISE	■ CONSOLIDATE	■ MASTER
1, 4, 7, 8, 13, 16	2, 5, 10, 11, 14, 17	3, 6, 9, 12, 15, 18

To answer questions online and to receive **immediate corrective feedback** and **fully worked solutions** for all questions, go to your learnON title at www.jacplus.com.au.

Fluency

1. **WE1** Classify each of the following data sets according to whether they are numerical or categorical data.

   a. Favourite TV programs of each student in your year level
   b. Shoe size of the top ten international models

2. Classify each of the following data sets according to whether they are numerical or categorical data.

   a. Hours of games played at home each week by each student in your class
   b. Birth weight of each baby in a mothers' group

3. **WE2** Classify each of the following numerical data sets according to whether they are discrete or continuous.

   a. Length of a stride of each golfer in a local golf club
   b. Distance travelled on one full tank from the top ten most popular vehicles

4. Classify each of the following numerical data sets according to whether they are discrete or continuous.

   a. Rainfall recorded for each day during spring
   b. Shoe size of each student in your class

5. Classify each of the following categorical data sets as nominal or ordinal.

   a. Satisfaction levels of customers (Satisfied, Indifferent, Dissatisfied)
   b. Favourite magazine of each student in your year level

6. Classify each of the following categorical data sets as nominal or ordinal.

   a. Sales ranking of the top ten magazines in Australia
   b. Age group of the readers of the top ten magazines in Australia

7. Classify each of the following categorical data sets as nominal or ordinal.

   a. Types of fish caught while fishing off a pier
   b. Genres of movies on a streaming service

8. **WE3** Classify each of the following data collections according to whether they are samples or populations.

   a. Favourite TV programs of secondary school students, determined by surveying students in Year 8
   b. Size of shoes to be stocked in a department store, determined by the shoe sizes of the top ten international models
   c. Number of hours your classmates spend playing games at home, determined by surveying each student in your class

9. Classify each of the following data collections according to whether they are samples or populations.

   a. Survey of first 40 customers at the opening of a new store to gauge customer satisfaction
   b. The most popular duco colours for cars sold in the last month in Victoria, taken from the registration board database of all 2310 cars registered last month

Understanding

10. Identify the missing word from each of the following sentences.

    a. When writing questions for a survey, it is important to consider which type of _____ will be useful for you to collect.
    b. If you _____ an entire population, this data can be used for the purpose of a _____.
    c. If the data you collect is grouped numerical data, it is best to use a _____ to visually represent the data.

11. **WE4** Complete the following.

    a. Use a random number generator to select six days from the month of March.
    b. Use a random number generator to select five students from your mathematics class.
    c. State whether your selections from a. and b. are samples or populations.

12. a. Randomly select five days of a non-leap year.
    b. State whether this data is a sample or population.

Reasoning

13. The following are 2020 statistics for motorcycle and car fatalities in Australia.

    Motorcycle fatalities: 190
    Car fatalities: 732
    Registered motorcycles: 880 881
    Registered cars: 14 679 249
    Kilometres travelled by motorcycles (millions): 1683
    Kilometres travelled by cars (millions): 162 983
    Population of Australia: 25 693 059

    a. State whether motorcycles or cars are the safer mode of transport, according to the data. Explain your answer.
    b. Calculate the ratios or percentages that could be extracted from this data to reinforce your conclusion.
    c. State the information that you would include if you wanted to argue for the safety of riding a motorcycle. State the information that you would leave out in order to make this argument.
    d. Write a sentence to convince your readers that a motorcycle is a safer mode of transport.
    e. Write a sentence to convince your readers that a car is a safer mode of transport.

14. Explain how data is classified into different types, giving examples for each type.

15. Identify and justify two sets of information a person might want to collect from each of the following sources.

   a. A maternity hospital
   b. An airport
   c. Shoppers at a supermarket

Problem solving

16. Questions are being written for a survey about students' views on sports. Select the questions from the following list that you would use. Explain your choices.

   a. What do you think about sport?
   b. List these sports in order from your most favourite to least favourite: football; tennis; netball; golf.
   c. How often do you watch sports on television? Choose from these options: once a week; once a day; once a month; once a year; never.
   d. Why do you like sports?
   e. How many live sporting events do you attend each year?
   f. Can you play a sport?
   g. Which sports do you watch on the television? Choose from these options: football; tennis; netball; golf; all of these; none of these.

17. Businesses collect and analyse all sorts of data.

   a. Suggest some types of data that businesses might want to collect.
   b. Suggest some ways in which businesses might collect data.
   c. Provide some examples of how businesses might use the data that they collect.

18. A company involved in constructing a desalination plant in the Wonthaggi region of Victoria wishes to gather some data on public opinion about the presence of the desalination plant in the area and its possible effects on the environment. The company has decided to conduct a survey and has asked you to plan and implement the survey and then analyse the data and interpret the results.

   a. Suggest three questions you could include in the survey.
   b. Explain what you are trying to find out by asking these questions.
   c. State the types of data you would collect.
   d. Explain how this data could be displayed visually.
   e. Suggest what the company could use this data for.

11.3 Collecting data

> **LEARNING INTENTION**
>
> At the end of this subtopic you should be able to:
> - display data in frequency tables
> - display data in dot plots
> - display data in stem-and-leaf plots.

⏵ 11.3.1 Frequency tables

eles-4912

- Recording and organising data allows graphs and tables to be drawn, which makes it possible to analyse the data.
- Data needs to be recorded and organised so that graphs and tables can be drawn and analysis conducted.

- A **frequency table** can be used to record and organise data.
- Data in frequency tables can be two sets of numerical data or one set of categorical and one set of numerical data.
- A frequency table looks like this:

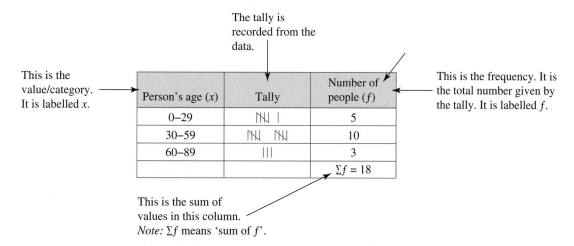

The tally is recorded from the data.

This is the value/category. It is labelled x.

This is the frequency. It is the total number given by the tally. It is labelled f.

Person's age (x)	Tally	Number of people (f)
0–29	⁙	5
30–59	⁙ ⁙	10
60–89	|||	3
		$\Sigma f = 18$

This is the sum of values in this column.
Note: Σf means 'sum of f'.

WORKED EXAMPLE 5 Drawing frequency tables with values

The following is a result of a survey of 30 students showing the number of people in their family living at home. Draw a frequency table to summarise the data.

2, 4, 5, 6, 4, 3, 4, 3, 2, 5, 6, 6, 5, 4, 5,
3, 3, 4, 5, 8, 7, 4, 3, 4, 2, 5, 4, 6, 4, 5

THINK

1. Identify the data as numerical discrete data.

2. There are only seven different scores so the data can be left ungrouped.

3. Use a frequency table of two (or three) columns. The first column is for the possible scores (x) and can be labelled 'Number of members in the family'.
The last column is for the frequency (f) and can be labelled 'Number of families'.
An optional column between these two can be used to keep a tally for each category.

4. Add up the frequency column $\left(\sum f\right)$ to confirm that 30 results have been recorded.

WRITE

Number of members in the family (x)	Tally	Number of families (f)
2					3
3	⁙	5
4	⁙ ||||	9
5	⁙ ||	7
6						4
7			1
8			1
		$\sum f = 30$

▶ 11.3.2 Dot plots

eles-4913

- A **dot plot** is another way of recording and organising data.
- Data can be two sets of numerical data or one set of categorical and one set of numerical data.
- A dot plot looks like this:

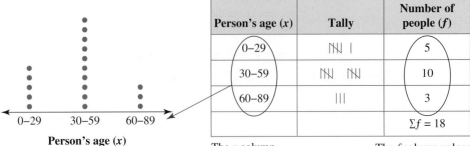

The *x* column categories are listed along the base of the plot.

The *f* column values indicate the number of dots.

WORKED EXAMPLE 6 Drawing a dot plot

A survey was conducted, asking students about the type of vehicle driven by their parents. Their responses were collected and recorded as shown. Display the data as a frequency table and as a dot plot.

Sedan	4WD	Sedan	Sedan
4WD	SUV	SUV	Station wagon
SUV	Convertible	Station wagon	4WD
Sedan	Sports car	Convertible	Station wagon
Sedan	Station wagon	SUV	Station wagon
SUV	SUV	Sedan	Sedan

THINK

1. Identify this data as categorical ordinal data.

2. There are six different scores (or categorical labels).

3. Use a frequency table of two (or three) columns.
   The first column is for the possible scores (*x*) and can be labelled as 'Type of vehicle'.
   The last column is for the frequency (*f*) and can be labelled 'Number of vehicles'.
   An optional middle column can be used to keep a tally for each category.

4. Add up the frequency column $\left(\sum f\right)$ to confirm that 24 scores have been recorded.

WRITE

Type of vehicle (*x*)	Tally	Number of vehicles (*f*)
4WD	|||	3
SUV	ℍℕ I	6
Sedan	ℍℕ II	7
Station wagon	ℍℕ	5
Sports car			1
Convertible	||	2
		$\sum f = 24$

5. Place a dot for each vehicle type in the appropriate column.

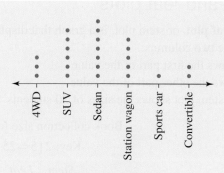

Type of vehicle

WORKED EXAMPLE 7 Drawing a frequency table with categories

The final scores for 30 golfers were recorded as follows. Summarise the data as a frequency table.

73, 69, 75, 79, 68, 85, 78, 76, 72, 73,
71, 70, 81, 73, 74, 78, 68, 75, 76, 72,
63, 72, 74, 71, 70, 75, 66, 82, 87, 78

THINK

1. Identify the data as numerical discrete data.

2. The scores range from 63 to 87. This is a range of 24 shots.

3. For this example the data will be grouped by in ranges of 2 shots.
   This will give $24 \div 2 = 12$ groups.
   Note that the groups have been made to start with even numbers, giving us 13 groups.

4. Add up the frequency column to check that all 30 scores have been recorded.

WRITE

Golf scores (x)	Tally	Number of players (f)
62–63	|	1
64–65		0
66–67	|	1
68–69	|||	3
70–71	||||	4
72–73	⊬|	6
74–75	⊬	5
76–77	||	2
78–79	|||	4
80–81	|	1
82–83	|	1
84–85	|	1
86–87	|	1
		$\sum f = 30$

▶ 11.3.3 Stem-and-leaf plots

eles-4914

- A **stem-and-leaf plot**, or stem plot, is a graph that displays numerical data.
- Stem plots have two columns:
  - the stem shows the first part of the value
  - the leaf shows the other part of the value.
- The following stem plot shows the sizes of 20 students' book collections.

Book collection size for 20 students

Key: $2\,|\,5 = 25$ books

Stem	Leaf
0	4
1	46
2	2358
3	0446779
4	1359
5	46

- From the stem plot shown, the scores collected were 4, 14, 16, 22, 23, 25, 28, 30, 34, 34, 36, 37, 37, 39, 41, 43, 45, 49, 54, 56.

WORKED EXAMPLE 8 Summarising data in a stem plot

The following scores represent the distance (in km) travelled by a group of people over a particular weekend. Summarise the data in a stem plot.

48, 67, 87, 2, 34, 105, 34, 45, 63, 98, 12, 23, 35, 54, 65, 41, 34, 23, 12, 38, 18, 58, 53, 44, 39, 29

THINK

1. Identify the data as numerical discrete data. Write a title and a key.

2. The scores range from 2 to 105. Grouping them in ranges of 10 will give 11 categories.
   The stem will hold the tens place value and the leaf will hold the unit/ones place value (the last digit) for each score.

WRITE/DRAW

Distance travelled over one weekend

Key: $4\,|\,7 = 47\,\text{km}$

Stem	Leaf
0	2
1	228
2	339
3	044589
4	1458
5	348
6	357
7	
8	7
9	8
10	5

3. Transfer each score to the stem plot by recording its last digit in the leaf column alongside the row of the stem column that corresponds to its tens place digit.

4. Rearrange each row in the leaf column so that the digits are in ascending order from left to right.

- Sometimes the leaves of a stem plot can become too long. This can be fixed by breaking the stems into smaller intervals. The second part of the interval is marked with an asterisk (*).

WORKED EXAMPLE 9 Representing data in a stem plot with smaller stem intervals

The heights of 30 students are measured (in cm) and recorded as follows.

125, 143, 119, 136, 127, 131, 139, 122, 140, 118, 120, 123, 132, 134, 127,
129, 124, 131, 138, 133, 122, 128, 130, 135, 141, 139, 121, 138, 131, 126

Represent the data in a stem plot.

THINK	WRITE/DRAW
1. Write a title and key, then draw up the stem plot with the numbers in each row of the leaf column arranged in ascending order from the stem.	Heights of 30 students (in cm) Key: 11|8 = 118 cm Stem | Leaf 11 | 8 9 12 | 0 1 2 2 3 4 5 6 7 7 8 9 13 | 0 1 1 1 2 3 4 5 6 8 8 9 9 14 | 0 1 3
2. The leaves of the two middle stem values are long. They would be easier to interpret with the stem broken into smaller intervals, for example intervals of 5. The stem of 12 would then include the numbers from 120 to 124 while the stem of 12* would include the numbers from 125 to 129. Reconstruct the stem plot according to this new set of intervals.	Heights of 30 students (in cm) Key: 12*|5 = 125 cm Stem | Leaf 11 | 8 9 12 | 0 1 2 2 3 4 12* | 5 6 7 7 8 9 13 | 0 1 1 1 2 3 4 13* | 5 6 8 8 9 9 14 | 0 1 3

 Resources

 eWorkbook　　Topic 11 Workbook (worksheets, code puzzle and a project) (ewbk-2011)

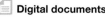 **Digital documents** SkillSHEET Presenting data in a frequency table (doc-6317)
　　　　　　　　　SkillSHEET Arranging a set of data in ascending order (doc-10954)

 Interactivities　Individual pathway interactivity: Collecting data (int-4541)
　　　　　　　　　Collecting data (int-3807)
　　　　　　　　　Stem plots (int-6242)

Exercise 11.3 Collecting data

learnon

Individual pathways

■ PRACTISE	■ CONSOLIDATE	■ MASTER
1, 4, 8, 9, 12, 15	2, 5, 6, 10, 13, 16	3, 7, 11, 14, 17, 18

To answer questions online and to receive **immediate corrective feedback** and **fully worked solutions** for all questions, go to your learnON title at www.jacplus.com.au.

Fluency

1. **WE5,6** The following set of data represents the scores achieved in a multiple choice Maths test.

   4, 6, 8, 3, 6, 9, 1, 3, 5, 6, 4, 7, 5, 9, 3, 2, 7, 8, 9, 6, 5, 4, 6, 5, 3, 5, 7, 6, 7, 7

   a. Present the data in a frequency table.
   b. Present the data as a dot plot.

2. The following set of data shows the types of pets kept by 18 people.
   cat, dog, fish, cat, cat, cat, dog, dog, bird, turtle, dog, cat, turtle, cat, dog, dog, turtle, bird

   a. Present the data in a frequency table.
   b. Present the data as a dot plot.
   c. Explain whether this data can be presented in a stem plot.

3. a. **WE7** Present the data shown in the table as a frequency table by grouping the data in intervals of 10 (0–9, 10–19, ...).

   b. **WE8** Present the data as a stem plot.

| State | Death by motor vehicle accidents: Males aged 15–24 years (rates per 100 000 people) | | | |
	2008	2009	2010	2011
NSW	14	13	18	9
Vic.	19	25	14	12
Qld	25	17	26	15
SA	23	34	22	26
WA	22	24	36	27
Tas.	42	18	36	42
NT	43	60	71	45
ACT	0	11	14	21

4. A sample of students surveyed by the Australian Bureau of Statistics were asked the colour of their eyes. The results were as follows:

Brown	Hazel	Blue	Brown	Brown	Green	Green	Brown	Brown	Hazel
Brown	Brown	Brown	Brown	Blue	Brown	Green	Blue	Blue	Blue
Brown	Brown	Brown	Brown	Hazel	Blue	Hazel	Blue	Brown	Blue
Brown	Blue	Green	Hazel	Hazel	Hazel	Green	Blue	Blue	Brown

   a. Set up a frequency table for the data.
   b. Construct a dot plot for the data.
   c. Comment on the proportion of people who have brown eyes.

5. Draw a stem-and-leaf plot for each of the following sets of data. Comment on each distribution.

   a. 18, 22, 20, 19, 20, 21, 19, 20, 21
   b. 24, 19, 31, 43, 20, 36, 26, 19, 27, 24, 31, 42, 29, 25, 38
   c. 346, 353, 349, 368, 371, 336, 346, 350, 359, 362

6. **WE9** Redraw the stem plots in question 5 so that the stems are arranged in intervals of 5 rather than intervals of 10.

7. The following times (to the nearest minute) were achieved by 40 students during a running race.

23	45	25	48	21	56	33	34	63	43	42	41	26	44
45	41	40	39	37	53	26	55	48	39	29	52	57	33
31	32	71	60	49	52	32	28	47	42	37	33		

   a. Construct a stem plot with groupings of 5 minutes.
   b. Construct a frequency table. Use groupings of 5 minutes.
   c. If the top 10 runners were chosen for the representative team, identify the qualifying time as given by:
      i. the stem plot
      ii. the frequency table.
   d. Identify the number of times that the finish time of 33 minutes was recorded. Explain which summary (stem plot or frequency table) was able to give this information.

8. The ages of participants in a weights class at a gym are listed below:

   $$17, 21, 36, 38, 23, 45, 32, 53, 18, 25, 14, 29, 42, 26, 18, 27, 37, 19, 34, 20, 35$$

   Display the data as a stem plot using intervals of 5.

9. **MC** Consider the stem plot shown.

   Key $7|6 = 7.6$

Stem	Leaf
7	8
8	0 8 9
9	1 6 7 8
10	3 5 8
11	2

   Identify which of the following data sets matches the stem plot.

   A. 78, 80, 88, 89, 91, 96, 97, 98, 103, 105, 108, 112
   B. 8, 0, 8, 9, 1, 6, 7, 8, 3, 5, 8, 2
   C. 7.8, 8.0, 8.8, 8.9, 9.1, 9.6, 9.7, 9.8, 1.03, 1.05, 1.08, 1.12
   D. 7.8, 8.0, 8.8, 8.9, 9.1, 9.6, 9.7, 9.8, 10.3, 10.5, 10.8, 11.2
   E. 78, 8089, 91 678, 10 358, 112

Understanding

10. The following data represent the life expectancy (in years) of Australians from 40 different age groups.

83.7	84.5	84.7	85.2	84.6	84.5	85.9	84.9	88.3	86.5
84.1	84.5	84.8	84.4	84.7	84.5	86.0	85.0	88.3	86.6
84.1	84.5	84.8	84.4	84.7	85.7	87.3	85.0	88.8	86.6
84.1	84.5	84.9	84.5	84.7	85.8	87.6	85.1	88.8	86.9

   a. Present the data as a frequency table.
   b. Present the data as a stem plot.

11. Choose the most appropriate technique for summarising the data in the table. Justify your choice.

Median age of Australian females and males from 2002 to 2009		
Year	Females	Males
2002	36.6	35.2
2003	36.9	35.3
2004	37.1	35.5
2005	37.3	35.7
2006	37.4	35.9
2007	37.6	36.1
2008	37.6	36.1
2009	37.7	36.1

Reasoning

12. The dot plot shown is a Year 9 student's attempt to display the values:

8, 10, 10, 11, 11, 12, 12, 15, 15, 15, 15, 17, 18, 19

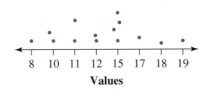

a. List the mistakes that the student made in drawing the dot plot.
b. Redraw the dot plot so that it correctly displays all of the data given.

13. a. Sometimes you are asked to represent two sets of data on separate stem-and-leaf plots. Suggest how you could represent two sets of data on a single stem plot while keeping them separate. Demonstrate your method using the data sets below that show the heights (in cm) of 12 boys and 12 girls in a Year 8 class:

Boys	154	156	156	158	159	160	162	164	168	171	172	180
Girls	145	148	152	152	155	156	160	161	161	163	165	168

b. Consider the stem-and-leaf plot below. Explain how you could make this plot easier to read and use.

```
Stem │ Leaf
   1 │ 0 2 2 2 5 7 8 9
   2 │ 0 0 0 0 1 1 2 2 2 2 2 4 4 7 7 7 7 8 8 8 8 9 9
   3 │ 0 2 2 3 3 3 5 5 5 7 8 9
```

14. The dot plots below display Maths test results for two Year 9 classes. The results show the marks out of 20.

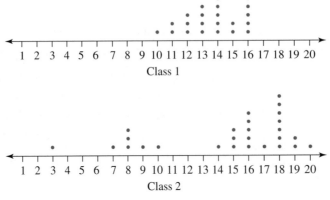

a. Determine the number of students in each class.
b. For each class, determine the number of students who scored 15 out of 20 on the test.
c. For each class, determine the number of students who scored more than 10 on the test.
d. Use the dot plots to describe the performance of each class on the test.

Problem solving

15. The following data shows the speeds of 30 cars recorded by a roadside camera outside a school, where the speed limit is 40 km/h.

    20, 27, 30, 36, 45, 39, 15, 22, 29, 30, 30, 38, 40, 40, 40,
    40, 42, 44, 20, 45, 29, 30, 34, 37, 40, 45, 60, 38, 35, 32

    a. Present the data as an ordered stem-and-leaf plot.
    b. Write a paragraph for the school newsletter about the speed of cars outside the school.

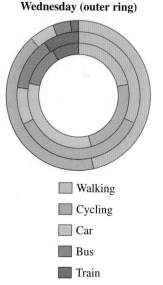

16. A Year 9 Science class recorderd the masses of mice (measured in grams) as shown below.

    25.1, 24.8, 25.1, 27.3, 25.3, 29.5, 24.5, 26.7, 24.0, 26.3, 25.4,
    26.3, 23.9, 25.8, 25.4, 24.6, 25.1, 23.9, 33.2, 24.5, 28.1, 27.3

    a. Draw an ordered stem-and-leaf plot for this data.
    b. Calculate the range of this data.
    c. Draw an ordered stem-and-leaf plot for the masses of the mice rounded to the nearest gram. Discuss whether a stem-and-leaf plot is suitable for the rounded data. Compare the stem-and-leaf plot with the original data. Explain your conclusions.

17. A doughnut graph can be used to compare two or more sets of data using the basic principles of a pie graph. A doughnut graph comparing students' methods of travel to school on Monday, Tuesday and Wednesday is shown.

    Method of travel to school
    Monday (inner ring)
    Tuesday (middle ring)
    Wednesday (outer ring)

    a. List the similarities and differences between a doughnut graph and a pie graph.
    b. Write a series of instructions for drawing a doughnut graph.
    c. Test your instructions on the following data by creating a doughnut graph to represent it.

Method of travel	Monday	Tuesday	Wednesday
Walking	5	7	25
Cycling	6	8	23
Car	8	2	3
Bus	3	3	2
Train	2	2	1

☐ Walking
☐ Cycling
☐ Car
☐ Bus
☐ Train

d. Compare your doughnut chart with the one shown.
e. Suggest some advantages and disadvantages of using a doughnut graph instead of a multiple column graph to compare two or more sets of data.

18. Analysis of written English prose reveals the proportion of vowels and consonants to be roughly the same in every large sample — approximately 7.11. The letter *e* accounts for one-third of the occurrence of vowels. In a passage consisting of 10 000 letters, approximately how many times would you expect the letter *e* to occur?

11.4 Displaying data

11.4.1 Bar charts and pie charts

eles-4915

- **Bar charts** and **pie charts** are used to display categorical data.

Displaying data in bar charts and pie charts

- A bar chart for the following data is shown as an example.

Items purchased from canteen	
Item purchased	**Number (f)**
Sandwich	35
Juice box	10
Chocolate bar	12
Fruit salad	5

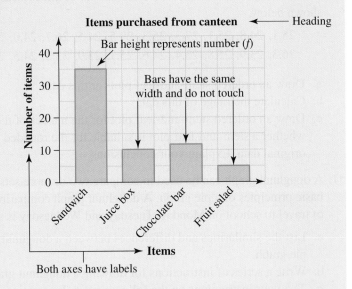

- A pie chart for the same data is shown as an example.

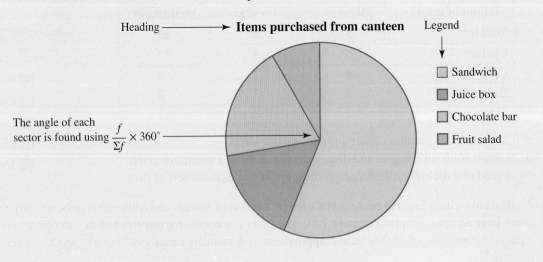

The angle of each sector is found using $\frac{f}{\Sigma f} \times 360°$

The Chalemet family is having a major reunion. The age groups of the family members who are attending have been recorded in the table below.

Age group	Number
Baby	20
Toddler	24
Child	34
Teenager	36
Adult	50
Pensioner	36

Display the frequency table as a:

a. bar chart

b. pie chart.

THINK

a. 1. Give the bar chart a suitable title.

2. Identify the category names and use them to label the horizontal axis.
It is the convention to leave a space between each bar on a bar chart.

3. Label the vertical axis with a suitable scale and title.

4. Draw each bar so that its height matches the frequency of the category it represents.

WRITE

a.

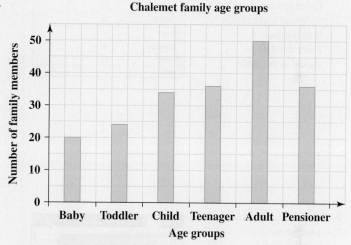

Chalemet family age groups

b. 1. Add one extra column to the frequency table to calculate the size of the sector.
The size of each sector is determined by its angle. To calculate this angle, use the use the rule angle = $\dfrac{f}{\sum f} \times 360°$.

b.

Age group	Number (f)	Size of sector
Baby	20	$\dfrac{20}{200} \times 360° = 36°$
Toddler	24	$\dfrac{24}{200} \times 360° = 43.2° \approx 43°$
Child	34	$\dfrac{34}{200} \times 360° = 61.2° \approx 61°$
Teenager	36	$\dfrac{36}{200} \times 360° = 64.8° \approx 65°$
Adult	50	$\dfrac{50}{200} \times 360° = 90°$
Pensioner	36	$\dfrac{36}{200} \times 360° = 64.8° \approx 65°$
$\sum$	$\sum f = 200$	$360°$

2. Give the pie chart a suitable title.
3. Create a legend with a suitable colour for each category.
4. Draw a circle.
5. Using a protractor, measure and draw the correct angle of each sector.
6. Colour the sectors using the legend.

b. **Chalemet family age groups**

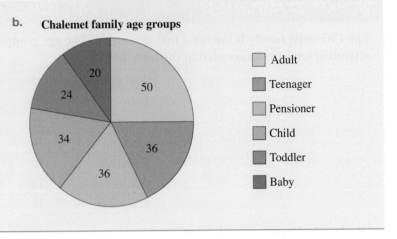

- Adult
- Teenager
- Pensioner
- Child
- Toddler
- Baby

TI | **THINK**	**WRITE**	CASIO | **THINK**	**WRITE**

a. 1. In a new problem, open a Lists & Spreadsheet page. Label column A as agegroup and column B as number. Enter the data as shown in the table.

a.

a. 1. Open a new Spreadsheet screen. Enter the data into columns A and B.

a.

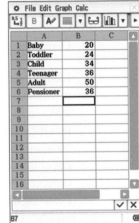

2. To create a summary plot, press:
- MENU
- 3: Data
- 8: Summary Plot.
Set the X List (category) as agegroup and the Summary List as number.
You have the option of displaying the summary plot in a split screen or in a new page.
Choose New Page for Display On.
Use TAB to move to OK.
Press ENTER.
Hover the pointer over each bar to see the percentage frequency.

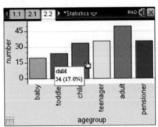

2. Highlight columns A and B. Press:
- Graph
- Bar
- Stacked.
Note: This is a frequency bar graph. To draw a percentage frequency bar graph, change the entered data to percentages.

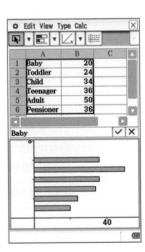

	TI	THINK		WRITE		CASIO	THINK		WRITE

TI | THINK   **WRITE**

b. To draw the pie chart, press:
  • MENU
  • 1: Plot Type
  • 9: Pie Chart.
  Hover the pointer over each bar to see the percentage frequency.

b.

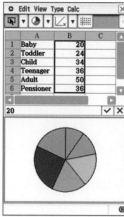

CASIO | THINK   **WRITE**

b. To draw the pie chart, highlight the numerical data only.
  Then press:
  • Graph
  • Pie.
  Hover the pointer over each sector to see the percentage frequency.

b.

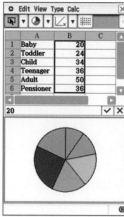

▶ 11.4.2 Histograms

eles-4916

- **Histograms** are used to display numerical data.
- They are used most commonly to display frequency data.
- The horizontal axis of a histogram can be labelled with:
  • values
  • ranges
  • mid points of ranges.

Displaying data in histograms

Histograms with values along the horizontal axis

- The x-axis of this type of histogram shows the values or results of the data collected.
- The y-axis of this type of histogram shows the frequency of each result.

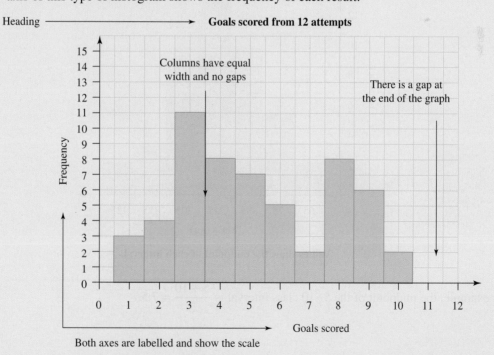

Histograms with ranges along the horizontal axis

- Range is the difference between the maximum and minimum value of a data set.

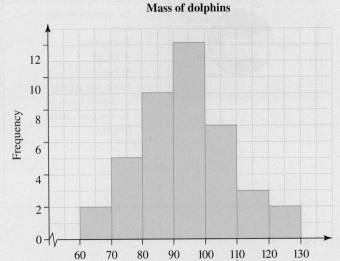

Mass of dolphins

Values show the range of each interval.

- In this type of histogram, each column has the range of each interval shown on each of its edges.
- See subtopic 11.6 for more information about ranges.

Displaying data with midpoints along the horizontal axis

- The midpoint of a class interval is calculated by taking the average of the two extreme values of that class interval.

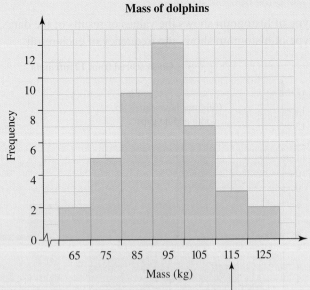

Mass of dolphins

Values show the midpoint of each interval.

- For example, the midpoint of the 5−10 class interval is $\dfrac{5+10}{2} = 7.5$.

A sample of 40 people was surveyed regarding the number of hours per week they spent on their phones per week. The results are listed below.

12.9, 18.3, 9.9, 17.1, 20, 7.8, 24.2, 16.7, 9.1, 27, 7.2, 16, 26.5, 15, 7.4, 28, 11.4, 20, 9, 11, 23.7, 19.8, 29, 12.6, 19, 12.5, 16, 21, 8.3, 5, 16.4, 20.1, 17.5, 10, 24, 21, 5.9, 13.8, 29, 25

a. Calculate the range for the data.
b. Organise the data into five class intervals and create a frequency distribution table that displays class intervals, midpoints and frequencies.
c. Construct a histogram to represent the data.

THINK

a. Determine the range of the number of hours spent on phones.

b. 1. Determine the size of the class intervals. Class intervals have been recorded as 5–<10, 10–<15 and so on, to accommodate the continuous data.

2. Rule a table with three columns, headed 'Hours on phone each week' (class interval), 'Midpoint' (class centre) and 'Frequency'.

3. Tally the scores and enter the information into the frequency column.

4. Calculate the total of the frequency column.

c. 1. Rule a set of axes on graph paper. Label the horizontal axis 'Number of hours on phone' and the vertical axis 'Frequency'.
2. Draw the first column so that it starts and finishes halfway between class intervals and reaches a vertical height of 9 people. Make sure you leave space between the y-axis and the first column.
3. Draw the columns for each of the other scores.

WRITE/DRAW

a. Range = largest value – smallest value
  = 29 – 5
  = 24

b. Class intervals of 5 hours will create 5 groups, and cover a range of 25 numbers.

Hours on phone each week	Midpoint	Frequency (f)
5– < 10	7.5	9
10– < 15	12.5	7
15– < 20	17.5	10
20– < 25	22.5	8
25– < 30	27.5	6
	Total	40

c.

Hours spent on phone

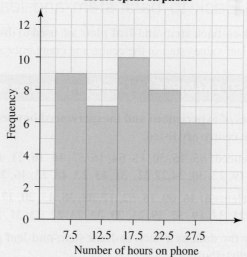

A stem plot was used to display the weight of
airline luggage for flights to the Gold Coast.
Convert the stem plot into a histogram.

Weights of airline luggage
Key: $21 \mid 3 = 21.3$ kg

Stem	Leaf
18	589
19	2367
20	578899
21	0269
22	019
23	2

THINK

1. Identify the data as numerical continuous data
   with six groups or class intervals. Label the
   x-axis as 'Weight of airline luggage'. Leave a
   half-interval gap at the beginning and use an
   appropriate scale for the x-axis.

2. Label the y-axis as 'Number of luggage pieces'.

3. Draw the first column on the x-axis to a height
   of its frequency.

4. Repeat for all class intervals.

DRAW

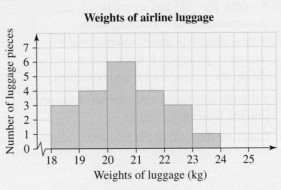

Weights of airline luggage

11.4.3 Back-to-back stem-and-leaf plots

eles-4917

- **Back-to-back stem-and-leaf plots** are used to display two related data sets.
- The leaf values start in the centre and count outwards.

The ages of experienced and inexperienced bowlers using a ten-pin
bowling centre are listed.

Experienced: 65, 15, 50, 15, 54, 16, 57, 16, 16, 21, 17, 28,17, 27, 17,
35, 18, 19, 22, 30, 34,22,22, 31,43, 23, 48, 23, 46, 25, 30, 21.

Inexperienced: 16, 60, 16, 52, 17, 38, 38, 43, 20, 17, 45, 18,
45, 36, 21, 34, 19, 32, 29, 21, 23, 32, 23, 22, 23, 31, 25, 28.

Display the data as a back-to-back stem-and-leaf plot and comment
on the distribution.

| | | THINK | | | | | WRITE | |

THINK

1. Rule three columns, headed Leaf (Inexperienced), Stem and Leaf (Experienced).

2. Make a note of the smallest and largest values of both sets of data (15 and 65). List the stems in ascending order in the middle column.

3. Beginning with the experienced bowlers, work through the given data and enter the leaf (unit component) of each value in a row beside the appropriate stem.

4. Repeat step 3 for the set of data for inexperienced bowlers.

5. Include a key to the plot that informs the reader of the meaning of each entry.

6. Redraw the stem-and-leaf plot so that the numbers in each row of the leaf columns are in ascending order.
Note: The smallest values should be closest to the stem column and increase as they move away from the stem.

7. Comment on any interesting features.

WRITE

Leaf (Inexperienced)	Stem	Leaf (Experienced)
987766	1	5566677789
8532331910	2	1872223351
1224688	3	50410
553	4	386
2	5	047
0	6	5

Key: 1 | 5 = 15

Leaf (Inexperienced)	Stem	Leaf (Experienced)
987766	1	5566677789
9853332110	2	1122233578
8864221	3	00145
553	4	368
2	5	047
0	6	5

The youngest experienced bowler attending the ten-pin bowling centre is 15 and the oldest is 65. The youngest and oldest inexperienced bowlers attending the ten-pin bowling centre are 16 and 60 respectively. There were an equal amount of experienced and inexperienced bowlers in their 20s. The next largest number of experienced bowlers were in their teens, while the next largest number of inexperienced bowlers were in their 30s.

 Resources

 eWorkbook Topic 11 Workbook (worksheets, code puzzle and project) (ewbk-2011)

Interactivity Individual pathway interactivity: Displaying data (int-4542)

Exercise 11.4 Displaying data

Individual pathways

■ PRACTISE	■ CONSOLIDATE	■ MASTER
1, 2, 5, 8, 12, 16	3, 6, 7, 9, 11, 13, 17	4, 10, 14, 15, 18

To answer questions online and to receive **immediate corrective feedback** and **fully worked solutions** for all questions, go to your learnON title at www.jacplus.com.au.

Fluency

1. **WE10** The number of participants in various classes at the local gym is given as:

   Aerobics — 120 members   Pilates — 60 members   Aqua aerobics — 30 members

   Boxing — 15 members   Spin — 45 members   Step — 90 members.

   Represent this information as a:

   a. bar chart   b. pie chart.

2. **WE11** In a survey, 40 people were asked about the number of hours a week they spent on their phone. The results are listed below.

   10.3, 13, 7.1, 12.4, 16, 11, 6, 14, 6, 11.1, 5.8, 14, 12.9, 8.5, 27.5, 17.3, 13.5, 8, 14, 10.2, 13, 7.9, 15, 10.7, 16, 8.4, 18.4, 14, 21, 28, 9, 12, 11.5, 13, 9, 13, 29, 5, 24, 11

   a. State whether this data is continuous or discrete.
   b. Determine the range of the data.
   c. Organise the data into class intervals and create a frequency distribution table that displays the class intervals, midpoints and frequencies.
   d. Construct a histogram.

3. The number of phone calls made per week in a sample of 56 people is listed below.

   21, 50, 8, 64, 33, 58, 35, 61, 3, 51, 5, 62, 16, 44, 56, 17, 59, 23, 34, 57, 49, 2, 24, 50, 27, 33, 55, 7
   52, 17, 54, 78, 69, 53, 2, 42, 52, 25, 48, 63, 12, 72, 36, 66, 15, 28, 67, 13, 23, 10, 72, 72, 89, 80

   a. Organise the data into a grouped frequency distribution table using a suitable class interval.
   b. Display the data as a combined histogram.

4. For the following data, construct:

   a. a frequency distribution table
   b. a histogram.

Class sizes at Jacaranda Secondary College									
38	24	20	23	27	27	22	17	30	26
25	16	29	26	15	26	19	22	13	25
21	19	23	18	30	20	23	16	24	18
12	26	22	25	14	21	25	21	31	25

5. **WE12** The number of bananas sold over a 5-week period at a high-school canteen is shown in the following stem-and-leaf plot. Represent this information as a histogram.

Key: $1|0 = 10$

Stem	Leaf
0	2 4 5
1	0 01 3
2	0 0 1 4
3	1

6. The distances 300 students have to travel to attend their school are summarised in the table below.

Distance km	Number of students
$0-<2$	112
$2-<4$	65
$4-<6$	56
$6-<8$	44
$8-<10$	15
$10-<12$	8

Represent this information a histogram.

7. **MC** Examine the bar chart shown and state the number of students who catch public transport to school.

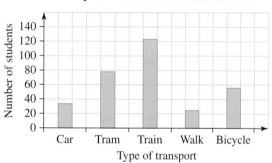

Transport mode of 300 students to school

A. 80 **B.** 100 **C.** 120 **D.** 200 **E.** 310

Understanding

8. An investigation into transport needs for an outer suburb community recorded the number of passengers boarding a bus during each of its journeys.

12, 43, 76, 24, 21, 46, 24, 46, 24, 21, 46, 54, 109, 87, 23,
78, 37, 22, 139, 65, 78, 89, 52, 23, 30, 54, 56, 32, 66, 49

Construct a histogram using class intervals of 20.

9. A drama teacher is organising his DVD collection into categories based on the genre of the movie or TV show. He recorded the following for the first 20 DVDs he categorised.
musical, musical, drama, comedy, horror, musical, comedy, action, drama, comedy, drama, drama, musical, drama, comedy, romance, drama, musical, romance, drama

a. Show the data set as a bar chart.

He continued with the task, and the classification for his collection of 200 DVDs is summarised in the frequency table shown below.

Genre	Number of DVDs	Size of sector for a pie chart
Musical	52	
Drama	98	
Comedy	30	
Horror	10	
Action	5	
Romance	5	
Total	200	360°

b. Copy and complete the frequency table.
c. Construct a pie chart for the whole 200-DVD collection.

10. The height, in centimetres, of 32 students in Year 9 was recorded as follows.
167, 162.9, 157, 166.7, 146.5, 163, 156.1, 168, 159.4, 170.1,
152.1, 152.1, 174.5, 156, 163, 157.8, 161, 178, 151, 148, 166.7,
154.6, 150.3, 166, 160, 155.9, 164, 157, 171.1, 168, 158, 162

a. State whether the data is continuous or discrete.
b. Organise the data into seven classes, starting at 145 cm, and create a frequency distribution table.
c. Display the data as a histogram.
d. Explain whether it is possible to obtain the range of heights from the frequency distribution table without being given any other information.
e. **MC** Determine how many students stood at least 160 cm tall.

A. 15   B. 16   C. 17   D. 7   E. 8

f. **MC** Calculate what percentage of students were under the minimum height of 165 cm required to qualify for the basketball team.

A. 21%   B. 30%   C. 87%   D. 70%   E. 13%

g. Reorganise the data into class intervals of 4 cm (for example, 145−148 cm).
h. Draw a new histogram and compare it to your histogram from part c. Discuss any advantages or disadvantages of having a smaller class interval.

11. The following histogram illustrates the number of days that members of a tennis team practised in a week.

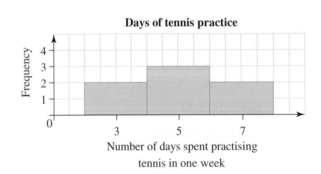

Days of tennis practice

Number of days spent practising tennis in one week

a. Describe what the frequency in this problem represents.
b. Calculate the average number of days that the team members practised in a week.

Reasoning

12. **WE13** The number of goals scored in football matches by Mitch and Yani were recorded as follows:

Mitch	0	3	1	0	1	2	1	0	0	1
Yani	1	2	0	1	0	1	2	2	1	1

   a. Display the data as a back-to-back stem-and-leaf plot and comment on the distribution.
   b. Calculate how many times Mitch and Yani scored more than a single goal.
   c. Determine who scored the most goals in any single match.
   d. Determine who scored the greatesst number of goals overall.
   e. State who is the most consistent performer. Explain your answer.

13. Percentages in a Mathematics exam for two classes were as follows:

9A	32	65	60	54	85	73	67	65	49	96	57	68
9B	46	74	62	78	55	73	60	75	73	77	68	81

   a. Construct a back-to-back stem-and-leaf plot of the data.
   b. Calculate the percentage of each group that scored above 50. Give your answers to the nearest whole number.
   c. Identify the group with more scores over 80.
   d. Compare the clustering for each group.
   e. Comment on extreme values.
   f. Calculate the average percentage for each group.

14. **MC** The back-to-back stem-and-leaf plot displays the heights of a group of Year 9 students.
   a. The total number of Year 9 students is:

      A. 13
      B. 17
      C. 30
      D. 36
      E. 171

      Key: $13\,|\,7 = 137\,$cm

Leaf (males)	Stem	Leaf (females)
9 8	13	7 8
9 8 8 7 6	14	3 5 6
9 8 8	15	1 2 3 7
7 6 6 5	16	3 5 6
8 7 6	17	1

   b. The tallest male and shortest female heights respectively are:

      A. 186 cm and 137 cm
      B. 171 cm and 148 cm
      C. 137 cm and 188 cm
      D. 178 cm and 137 cm
      E. 178 cm and 171 cm

   c. Compare the heights of boys and girls.

15. Suggest two reasons stem plots rather than frequency tables are preferred by statisticians.

Problem solving

16. When reading the menu at the local Chinese restaurant, you notice that the dishes are divided into sections. The sections are labelled chicken, beef, duck, vegetarian and seafood.

    a. Identify the type of data that this menu presents.
    b. Suggest the best way to represent this data.

17. You are working in the marketing department of a company that makes sunscreen and your boss wants to know if there are any unusual results from the market research you completed recently.

    You completed a survey of people who use your sunscreen, asking them how many times a week they used it. Their answers are given in the data set shown.
    Analyse the data set to work out what you will say to your boss, and try to explain any unusual variables.

    7, 2, 5, 4, 7, 5, 7, 2, 5, 4, 3, 5, 7, 7, 4, 5, 21, 5, 5, 2

18. The pie chart shown displays data collected about the types of roses in the local botanical gardens.

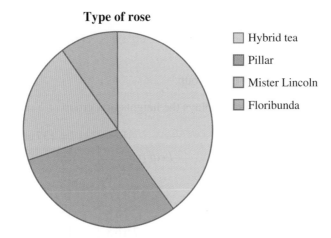

Type of rose

- Hybrid tea
- Pillar
- Mister Lincoln
- Floribunda

    a. Carefully measure the angle of each section and record the results in a table.
    b. Determine the smallest possible total number of rose bushes in the original data.
    c. Create a bar chart displaying the data, assuming the total number of bushes is your answer from **b**.

11.5 Measures of centre

⏵ 11.5.1 Mean, median and mode

eles-4918

- There are three values commonly used to measure the centre of data. These are the **mean**, **median** and **mode**.
  - The mean is also known as the average.
  - The median is the middle of the data.
  - The mode is the most common value in the data.

Calculating the mean

For raw data, calculate the mean using the formula:

$$\bar{x} = \frac{x}{n} = \frac{\textbf{sum of all data values}}{\textbf{number of all data values}}$$

For example, for the data $2, 2, 2, 3, 3, 7, 12, 13, 19$:

$$\text{sum of data values} = 1 + 2 + 2 + 2 + 3 + 3 + 7 + 13 + 13 = 63$$

$$\text{number of data values} = 9$$

$$\text{mean} = \frac{63}{9} = 7$$

For data in frequency tables, calculate the mean using the formula:

$$\bar{x} = \frac{xf}{f} = \frac{\textbf{sum of (data values} \times \textbf{frequency)}}{\textbf{sum of frequencies}}$$

Calculating the median

Start with data in order from smallest to largest.

The median is the middle of the data.

The middle value will be the $\frac{n+1}{2}$th score, where n is the number of scores or pieces of data.

For example, for the data $2, 2, 2, 3, 3, 7, 12, 13, 19$:

there are 9 scores, so the middle value will be the 5th score (because $\frac{9+1}{2} = 5$)

$$2 \; 2 \; 2 \; 3 \; 3 \; 7 \; 12 \; 13 \; 19$$

$$\text{median} = 3$$

Note: If n is even, find the median by averaging the two middle scores.

Calculating the mode

To calculate the mode, determine the most common value in the data.

For example, for the data 2, 2, 2, 3, 3, 7, 12, 13, 19:

> the most common score is 2
>
> mode $= 2$

Note: It is possible to have more than one mode.

WORKED EXAMPLE 14 Calculating the median, mean and mode

Consider the data set shown.

$$3, 7, 4, 5, 6, 2, 8, 3, 3, 5$$

Calculate the:
a. **median**
b. **mean**
c. **mode.**

THINK	WRITE
a. 1. The median is the middle of the data. Organise the data from smallest to largest.	a. 2 3 3 3 4 5 5 6 7 8
2. The median, or middle score, is found by first counting how many scores there are. If there is an even number of scores, then the median is the average of the two middle scores. Use the rule: median score $= \dfrac{n+1}{2}$th score where n is the number of scores.	Median score $= \dfrac{n+1}{2}$th score $= \dfrac{10+1}{2}$ $= 5.5$th score The median score is between the 5th and 6th scores, which is between 4 and 5. $\therefore$ Median $= 4.5$
b. 1. The mean is the average of the data values Add up all the data values.	b. Mean $= \bar{x}$ $= \dfrac{x}{n} = \dfrac{\text{sum of all data values}}{\text{number of data values}}$
2. Divide the result by the total number of scores.	$\bar{x} = \dfrac{3+7+4+5+6+2+8+3+3+5}{10}$ $\bar{x} = \dfrac{46}{10}$ $= 4.6$
c. The mode is the score that occurs the most frequently in the list.	c. Mode $= 3$

The combined number of goals scored in each of the 48 first round games of this year's World Cup is summarised in the frequency table below.

Total combined goals scored	Number of matches
0	6
1	13
2	12
3	9
4	5
5	2
6	0
7	1
$\sum f$	48

Calculate the:

a. median number of goals scored
b. mean number of goals scored
c. modal number of goals scored.

THINK

a. 1. The median is the middle value of the data set.
The sum of the frequencies is the number of scores, n. If there is an even number of scores, then the median is the average of the 2 middle scores.

2. Determine where the 24th and 25th scores lie.

b. 1. The mean is the average value. Set up a third column on the frequency table and multiply the score and its frequency.

2. Add up all of the products of xf.

WRITE

a.
$$\text{Median score} = \frac{n+1}{2} \text{th match}$$
$$= \frac{48+1}{2}$$
$$= 24.5 \text{th match}$$

This score is between the 24th match and the 25th match.

The 24th score is 2 combined goals.
The 25th score is 2 combined goals.
Median = 2 combined goals

b.

Combined goals (X)	Number of matches (f)	$x \times f$
0	6	0
1	13	13
2	12	24
3	9	27
4	5	20
5	2	10
6	0	0
7	1	7
$\sum f$	48	$\sum xf = 101$

3. Divide the result by the sum of the frequencies, f.

$$\bar{x} = \frac{xf}{f} = \frac{\text{sum of (scores} \times \text{frequency)}}{\text{sum of frequencies}}$$

$$= \frac{101}{48}$$

$$= 2.1$$

c. The mode is the score with the highest frequency.   c. Mode = 1 combined goal scored

---

- Mean, median and mode can be calculated from plots.

WORKED EXAMPLE 16 Calculating the median from a plot

Calculate the median of each data set.

a.   Key : $1|8 = 18$

Stem	Leaf
1	8 9
2	2 2 5 7 7 8
3	0 1 4 6 7
4	0 5

b.

16 17 18 19 20 21 22 23 24

THINK

a. 1. Check that the scores are arranged in ascending order.

2. Locate the position of the median using the rule $\frac{n+1}{2}$, where $n = 15$.

Note: The position of the median is $\frac{15+1}{2} = 8$ (the eighth score).

3. Write the answer in a sentence.

b. 1. Locate the position of the median using the rule $\frac{n+1}{2}$, where $n = 12$.

Note: The position of the median is $\frac{12+1}{2} = 6.5$. This means it is between the sixth and seventh scores.

3. Calculate the average of the two middle scores.

4. Write the answer in a sentence.

WRITE

a. The scores are arranged in ascending order.

Key : $1|8 = 18$

Stem	Leaf
1	8 9
2	2 2 5 7 7 **8**
3	0 1 4 6 7
4	0 5

The median of the set of data is 18.

b. The 6th score is 19.
The 7th score is 19.

$$\text{Median} = \frac{19+19}{2}$$

$$= \frac{38}{2}$$

$$= 19$$

The median of the set of data is 19.

▶ 11.5.2 Grouped data

- **Grouped data** is given when a frequency table shows a range instead of a value.
- To calculate the mean of a grouped data set, multiply the midpoint of each class interval by the corresponding frequency.
- The median of a grouped data set is the middle value when the values are in order.
- The mode of a grouped data set is the most common class interval. This is known as the **modal class**. For example, let us calculate the mean, median and mode for the following data.

Interval	Frequency (f)
$10-<15$	3
$15-<20$	8
$20-<25$	6
$\sum f$	17

- To calculate the mean, add a column to the data table showing midpoints, and another column showing the product of frequency and midpoint.

Interval	Frequency (f)	Midpoint (x_{mid})	$f \times x_{mid}$
$10-<15$	3 $\times$	12.5	$3 \times 12.5 = 37.5$
$15-<20$	8 $\times$	17.5	$8 \times 17.5 = 140$
$20-<25$	6 $\times$	22.5	$6 \times 22.5 = 135$
Σf	17	$\Sigma f \times x_{min}$	$37.5 + 140 + 135 = \mathbf{312.5}$

- Mean $= \dfrac{312.5}{17} = 18.38$ (rounded to 2 decimal places)
- To calculate the median: since there are 17 data values, the middle value will be the 9th value (because $\dfrac{17+1}{2} = 9$).

Interval	Frequency (f)	
$10-<15$	3	← First 3 values — need 6 more to get to 9
$15-<20$	8	← The median is the 6th value in this group of 8
$20-<25$	6	
Σf	17	

- To calculate the median, look at the interval group that contains the median and use the rule:
  - median $= 15 + \dfrac{6}{8} \times 5 = 18.75$
- To calculate the modal class, determine the interval with the highest frequency.
  - Modal class $= 15-<20$.

The actual diagonal lengths of 45 televisions in a store are presented as a frequency table.
Calculate the:

a. **median (to the nearest inch)**   b. **mean (to the nearest inch)**   c. **modal class.**

Diagonal length (inches)	Number of TVs (f)
10– < 20	1
20– < 30	3
30– < 40	8
40– < 50	19
50– < 60	9
60– < 70	3
70– < 80	2
f	45

THINK

a. 1. The median is the middle value. The sum of the frequencies is the number of values, n. If there is an odd number of values, then the median is the middle value.

2. Determine where the 23rd value lies in the table by looking at the cumulative frequencies.

< 20 inches $= 1$ TV
< 30 inches $= 4$ TVs
< 40 inches $= 12$ TVs
< 50 inches $= 31$ TVs

The 23rd value is in the $40 - < 50$ inches group. It is the 11th value in that group of 19 TVs.

b. 1. The mean is the average value. Set up a third and fourth column in the frequency table.

2. Record the midpoint as the middle value of each interval, for example $\dfrac{10 + 20}{2} = 15$.

3. Multiply the midpoint of each interval, x_{mid} and the interval's frequency, f.

4. Add up all the products of $x_{mid} \times f$.

WRITE/DISPLAY

a. Median value $= \dfrac{n + 1}{2}$th value

$= \dfrac{45 + 1}{2}$

$= 23$rd value

The 23rd value is in the $40 - < 50$ inches class.
It is the 11th value in a group of 19 values.
$\dfrac{11}{19} \times 10$ inches $= 5.79$ inches

Median $= 40 + 5.79$ inches
$= 45.79$ inches
≈ 46 inches

b.

Diagonal length (inches)	Number of TVs (f)	Class midpoint x_{mid} (inches)	Product of x_{mid} and f
10– < 20	1	15	$1 \times 15 = 15$
20– < 30	3	25	$3 \times 25 = 75$
30– < 40	8	35	$8 \times 35 = 280$
40– < 50	19	45	$19 \times 45 = 855$
50– < 60	9	55	$9 \times 55 = 495$
60– < 70	3	65	$3 \times 65 = 195$
70– < 80	2	75	$2 \times 75 = 150$
Σf	$= 45$	$\Sigma x_{mid} \times f$	$= 2065$

5. Divide $x_{mid} \times f$ by the sum of the frequencies, f.

$$\bar{x} = \frac{xf}{f} = \frac{\text{data values (score} \times \text{frequency)}}{\text{sum of frequencies}}$$

$$= \frac{2065}{45}$$

$$= 45.89 \approx 46 \text{ inches}$$

6. Write the answer in a sentence.

The mean TV size is 46 inches.

c. The modal class is the class interval with the highest frequency.

c. The modal class is 40 to less than 50 inches.

DISCUSSION

It is important to understand what the numbers on an exam report mean, as these are used to make decisions about student learning. A student in a Year 9 exam receives a mean, median and range of scores. In pairs, discuss and write down what each of these measures suggests about the score of an individual student in comparison to the class.

---

on Resources

eWorkbook   Topic 11 Workbook (worksheets, code puzzle and a project) (ewbk-2011)

Digital documents SkillSHEET Finding the mean (doc-10955)
                     SkillSHEET Finding the middle score (doc-10956)
                     SkillSHEET Finding the middle score for data arranged in a dot plot (doc 10957)
                     SkillSHEET Finding the most frequent score (doc-10958)

Video eLesson   Mean and median (eles-1905)

Interactivities Individual pathway interactivity: Measures of central tendency (int-4543)
                     Mean (int-3818)
                     Median (int-3819)
                     Mode (int-3820)

Exercise 11.5 Measures of centre   learn on

Individual pathways

■ PRACTISE	■ CONSOLIDATE	■ MASTER
1, 2, 3, 6, 10, 13, 14, 19	4, 7, 8, 12, 15, 16, 20	5, 9, 11, 17, 18, 21, 22

To answer questions online and to receive **immediate corrective feedback** and **fully worked solutions** for all questions, go to your learnON title at www.jacplus.com.au.

Fluency

1. **WE14** Calculate the median, mean and mode for the data sets given. Confirm your answers for the median and mean using a graphics calculator. Where necessary, give your answers to 2 decimal places.

   a. 27, 32, 45, 48, 53, 55, 55, 57, 59, 61, 75, 81
   b. 8, 8, 2, 8, 8, 8, 8, 8, 9, 9, 1, 9, 5, 9, 7, 10, 0

2. **WE15** Calculate the median, mean (to 2 decimal places) and mode for the ungrouped frequency table shown.

Score	Frequency
0	2
1	5
2	12
3	10
4	7
5	2

3. **WE16a** Calculate the median of each of the following data sets.

a. Key: $1|0 = 10$

Stem	Leaf
1	0 2
2	1 3 3 5
3	
4	4

b. **WE16b**

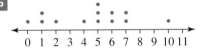

4. Calculate the median of each of the following data sets.

a. Key: $10|0 = 100$

Stem	Leaf
10	0
11	0 2 2 2
12	0 4 6 6
13	3

b.

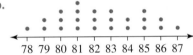

5. Calculate the median of each of the following data sets.

a. Key: $6.1|8 = 6.18$

Stem	Leaf
6.1	8 8 9
6.2	0 5 6 8
6.3	0 1 2 4 4 4

b.

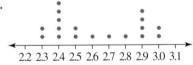

6. For the stem plot given below, calculate:

a. the median
b. the mean (to 2 decimal places)
c. the mode.

Key: $3^*|7 = 37$

Stem	Leaf
1	3 4
1*	5 6 7 9
2	0 2 3 3 4
2*	5 5 5 6 6 7 8
3	1 2 3 4
3*	5 7
4	1

7. **WE17** For the following grouped frequency tables, calculate the:

   i. median
   ii. mean
   iii. modal class.

   Where necessary, give your answers to 1 decimal place.

a.

Score x	Frequency f
120–129	2
130–139	8
140–149	10
150–159	5
160–169	1

b.

Score	Frequency
120–129	2
130–139	8
140–149	10
150–159	16
160–169	5
170–179	1

Hint: The tables above show grouped discrete data, for which the midpoint of the first class interval is $\dfrac{120 + 129}{2} = 124.5$.

8. **MC** Select the correct mean, median and mode for the following set of data.

$$1, 5, 2, 8, 7, 3, 5, 7, 2, 8, 7, 6, 4, 5, 5$$

   A. $5, 6, 5$ B. $6, 3, 5$ C. $5, 5, 5$ D. $5, 5, 7$ E. $6, 5, 6$

9. **MC** Select the median number of members per family surveyed as presented in this frequency table.

   A. 3   B. 4   C. 4.5   D. 4.6   E. 53

Number of family members	Frequency
3	19
4	53
5	21
6	4
7	2
8	1

10. **MC** Select the best estimate of the mean score for the grouped frequency table shown.

   A. 100   B. 150   C. 125   D. 130   E. 145

Score	Frequency
0 – <50	1
50 – <100	4
100 – <150	10
150 – <200	3
200 – <250	1
250 – <300	1

Understanding

11. The frequency table shown is a summary of 100 long-jump attempts by a group of primary school students during a recent school sports carnival. For the data given, determine the:

    a. type of data
    b. median
    c. mean
    d. mode.

    Where necessary, round your answers to the nearest whole number.

Long jump distance (cm)	Number of jumps
$120 - <140$	22
$140 - <160$	44
$160 - <180$	17
$180 - <200$	14
$200 - <220$	2
$220 - <240$	0
$240 - <260$	0
$260 - <280$	1

12. The grouped frequency table shown is a summary of 40 golf scores recorded at the local 9-hole golf course. For the data given, determine the:

    a. type of data
    b. median
    c. mean
    d. modal class.

Golf score	Number of games
30–39	4
40–49	32
50–59	2
60–69	1
70–79	1

13. A survey of the number of people living in each house on a residential street produced the following data:

    $$2, 5, 1, 6, 2, 3, 2, 1, 4, 3, 4, 3, 1, 2, 2, 0, 2, 4$$

    a. Prepare a frequency distribution table with an $f \times x$ column and use it to find the average (mean) number of people per household. Give your answer to 1 decimal place.
    b. Draw a dot plot of the data and use it to find the median number of people per household.
    c. Calculate the modal number of people per household.
    d. Explain which of the measures would be most useful to:
       i. a real-estate agent renting out houses
       ii. a government population survey
       iii. an ice-cream truck owner.

Reasoning

14. If the mean $= 11$, mode $= 4$, and the median$= 9$ and the missing values are positive integers:
    a. determine the missing two values in the following set of numbers:
       4, ___, 19, ____, 21, 4
    b. calculate two different values that meet the stated criteria
    c. explain whether you can substitute 9 for both missing values.

15. A class of 26 students had a median mark of 54 in Mathematics. In reality, though, no-one actually obtained this result.
    a. Explain how this is possible.
    b. Explain how many students must have scored below 54.

16. A tyre manufacturer selects 48 tyres at random from the production line for testing. The total distance travelled during the safe life of each tyre is shown in the table.

Distance in km ('000)	82	78	56	52	50	46
Number of tyres	2	4	10	16	12	4

    a. Calculate the mean, median and mode.
    b. Explain which measure best describes average tyre life.
    c. Recalculate the mean with the 6 longest-lasting tyres removed. By how much is it lowered? Give your answer to the nearest km.
    d. Determine what tyre life has the highest probability of being selected if you selected a tyre at random.
    e. In a production run of 10 000 tyres, determine how many could be expected to last for a maximum of 50 000 km.
    f. As the manufacturer, state the distance for which you would be prepared to guarantee your tyres. Explain your answer.

17. A small business pays the following wages (in thousands of dollars) to its employees:
    18, 18, 18, 18, 26, 26, 26, 35, 80 (boss)
    a. **MC** Select the wage earned by most workers at this business.
       A. $29 400   B. $18 000   C. $29 000   D. $26 000   E. $19 000
    b. Calculate the average wage at this business.
    c. **MC** Select the median of this distribution of wages.
       A. $26 000   B. $26 500   C. $18 000   D. $18 500   E. $26 400
    d. Explain which measure might be used in wage negotiations by:
       i. the union, representing the employees (other than the boss)
       ii. the boss.

18. a. Create a data set of 8 numbers that meet the following criteria.
       • Data values are positive integers
       • Mean $= 9.5$
       • Median $= 7$
       • Mode $= 6$
       • Maximum value in data set $= 21$
       • Minimum value in data set $= 1$
    b. Explain whether there is only one data set that can meet this criteria

Problem solving

19. Answer the following questions.

    a. Create a set of five values with a mean of 4 and a median of 5.
       Demonstrate that your data set meets the given criteria.
    b. Change only one of the numbers in your set so that the mean is now 5 and the median remains 5.
       Demonstrate that your modified set meets the new criteria.

20. Seven students completed a challenging Science test. The test was scored out of 15. Six of the students got the following marks: 4, 4, 5, 5, 5, 6. Determine the mark that the 7th student obtained if:

    a. the mean score is the same as the median score
    b. the mean score is greater than the median score
    c. the mean score is equal to the modal score
    d. the mean score is greater than 6.

21. Calculate the mean for this data set.

Score	4	8	12	16	20	24
Frequency	$4a$	$3a$	$2a$	a	$\dfrac{a}{2}$	$\dfrac{a}{4}$

22. On a piece of paper is a list of six 2-digit prime numbers. They have a mean and median of 39. Their mode is 31 and the smallest number is 13. What are the six numbers? Explain how you got your answer.

11.6 Measures of spread

LEARNING INTENTION

At the end of this subtopic you should be able to:
  • calculate range and interquartile range
  • determine whether a distribution is symmetrical, skewed or bimodal.

11.6.1 Range and interquartile range

eles-4920

  • The **range** is the difference between the maximum and minimum values in a data set.
  • To calculate interquartile range, the data is first organised into four sections called quartiles, each containing 25% of the data.
  • The word 'quartiles' comes from the word 'quarter'.
  • The **interquartile range (IQR)** is the difference between the first **quartile** and the third quartile.

Range

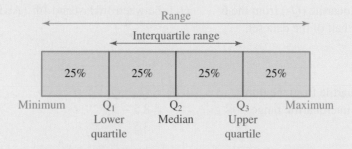

Range

Interquartile range

| 25% | 25% | 25% | 25% |

Minimum Q_1 Q_2 Q_3 Maximum

Lower   Median   Upper
quartile   quartile

- To calculate range, use the following formula:

$$\text{Range} = \textbf{maximum value} - \textbf{minimum value}$$

Interquartile range (IQR)

- To calculate interquartile range (IQR), use the following formula:

$$IQR = Q_3 - Q_1$$

where:

Q_1 is the median of the first half of the data

Q_3 is the median of the second half of the data.

WORKED EXAMPLE 18 Calculating the range and interquartile range of a data set

Calculate the range and interquartile range of the following data set.

$$5, 6, 8, 4, 12, 5, 7, 8, 4, 10, 9, 6, 5$$

THINK	WRITE
1. Arrange data from minimum to maximum value.	4 4 5 5 5 6 6 7 8 8 9 10 12
2. Calculate the range.	Range = maximum value − minimum value $= 12 - 4$ $= 8$
3. Calculate the median by using the rule: median $= \dfrac{n+1}{2}$th data value.	Median $= \dfrac{n+1}{2}$th data value $= \dfrac{13+1}{2}$ $= $ 7th data value Median is the 7th data value. $\therefore$ Median $= 6$
4. Calculate the first quartile (Q_1) from the 6 values in the lower half of the data set.	4 4 5 5 5 6 **6** 7 8 8 9 10 12 Q_1 is between 3rd and 4th data values $Q_1 = 5$

5. Calculate the third quartile (Q_3) from the 6 values in the upper half of the data set.

Q_3 is between 3rd last and 4th last data values.

$$Q_3 = \frac{8 + 9}{2}$$

$$= 8.5$$

6. Subtract the first quartile from the third quartile to find the interquartile range.

$$IQR = Q_3 - Q_1$$
$$= 8.5 - 5$$
$$= 3.5$$

TI	THINK	WRITE/DISPLAY

TI | THINK

1. In a new problem, open a Lists & Spreadsheet page.
Label column A as x values.
Enter the data.

2. Press:
   • MENU
   • 4: Statistics
   • 1: Stat Calculations
   • 1: One-Variable Statistics.
Set Number of Lists = 1.
Press OK.
Complete the table as shown.
Press ENTER.

The summary statistics will appear as shown. Scroll down to find the measures required.

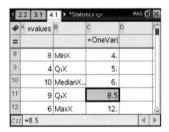

$$Range = MaxX - MinX$$
$$= 12 - 4$$
$$= 8$$

$$IQR = Q_3X - Q_1X$$
$$= 8.5 - 5$$
$$= 3.5$$

CASIO | THINK

1. Open a new Statistics screen.
Enter the data into List1.

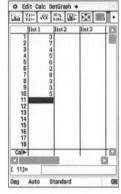

2. Press:
   • Calc
   • One-Variable
Fill in the Set Calculation table as shown:
   • Xlist: list1
   • Freq: 1
Press OK.

The summary statistics will appear as shown. Scroll down to find the measures required.

$$Range = Max_x - Min_x$$
$$= 12 - 4$$
$$= 8$$

$$IQR = Q_3 - Q_1$$
$$= 8.5 - 5$$
$$= 3.5$$

For the stem plot shown, determine:
a. the range
b. the interquartile range.

Key: $4 \mid 7 = 4.7$ kg

Stem	Leaf
1	9
2	2 7
3	0 1 3 7
4	2 4 5 7 8
5	1 2 8
6	0

THINK

1. Calculate the range by finding the difference between the minimum and maximum value.

2. Count the number of data values, n.

3. Calculate the median by using the rule median $= \dfrac{n+1}{2}$ th data value.

4. Calculate the first quartile, Q_1 (1st quartile) from the 8 values in the lower half of the data set.

5. Calculate the third quartile, Q_3 (3rd quartile from the 8 values in the upper half of the data set.

6. Determine the interquartile range by subtracting the first quartile from the third quartile.

WRITE

Range = maximum value − minimun value
$$= 6.0 - 1.9$$
$$= 4.1 \text{ kg}$$

$n = 16$

Median $= \dfrac{n+1}{2}$ th data value

$$= \dfrac{16+1}{2}$$

$$= 8.5 \text{th data value}$$

The median is between the 8th and 9th data values, i.e. between 4.2 and 4.4.
∴ Median $= 4.3$ kg

Stem	Leaf
1	9
2	2 7
3	0 1 3 7
4	2 4 5 7 8
5	1 2 8
6	0

Q_1 is between the 4th value (3.0) and the 5th value (3.1).

$$\therefore Q_1 = \dfrac{3.0 + 3.1}{2}$$

$$= 3.05 \text{ kg}$$

Q_3 is between the 4th last value (5.1) and the 5th last value (4.8), that is 4.95 kg.

$$\therefore Q_3 = \dfrac{4.8 + 5.1}{2}$$

$$= 4.95 \text{ kg}$$

IQR $= Q_3 - Q_1$
$$= 4.95 - 3.05$$
$$= 1.9 \text{ kg}$$

⏵ 11.6.2 Types of distributions

- The type of distribution describes the shape of the data.
- **Symmetrical distributions** have a peak in the centre and evenly balanced data above and below the centre.
  - In symmetrical distributions, the mean, median and mode have similar values.

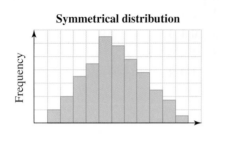

Stem	Leaf
0	7
1	2 3
2	2 4 5 7 9
3	0 2 3 6 8 8
4	4 7 8 9 9
5	2 7 8
6	1 3

- Skewed distributions are not symmetrical.
  - **Positively skewed distributions** (or right-skewed distributions) have more low values.
    - In positively skewed distributions, the mean has a larger value than the median.
  - **Negatively skewed distributions** (or left-skewed distributions) have more high values.
    - In negatively skewed distributions, the mean has a smaller value than the median.

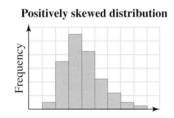

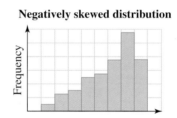

- **Bimodal distributions** have two peaks (the prefix *bi-* means *two*).

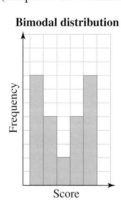

Types of distribution and measures of centre

Type of distribution	Most appropriate measure of centre
Symmetrical	Mean
Skewed	Median

For each of the distributions shown, describe the type of distribution, determining its measure of centre using the most appropriate measures.

a.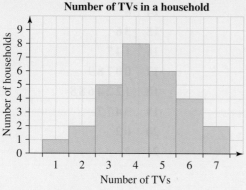

b.

THINK

a. 1. Describe the shape of the distribution.

2. Select an appropriate measure of centre.

3. Determine the mean.

4. Write the answer in a sentence.

b. 1. Describe the shape of the distribution.

2. Select the appropriate measures of centre.

3. Determine the median by first determining the number of data values.
$25 + 45 + 30 + 10 + 5 + 1 + 1 + 5 = 121$

4. Write the answer in a sentence.

WRITE

a. The graph is roughly symmetrical.

Measure of centre: mean

$$\text{Mean} = \bar{x} = \frac{\sum x}{n} = \frac{\text{sum of all data values}}{\text{number of data values}}$$

$$= \frac{120}{28}$$

$$= 4.3$$

The mean number of TVs per household is 4.3.

b. The graph is positively skewed.

Measure of centre: median

$$\text{Median} = \frac{n+1}{2}\text{th data value}$$

$$= \frac{122+1}{2}$$

$$= 61.5$$

The median is the average of the 61st and 62nd data values.

$$\text{Median} = \frac{2+2}{2}$$

$$= 2$$

The median number of members in a family is 2.

Determine the best measures of centre for the heights of 25 people shown in the stem plot. Give your answer to the nearest centimetre.

Key: 17|8 = 178 cm

Stem	Leaf
14	8
15	3 6
16	1 4 7
17	0 0 7 8 9
18	1 2 2 4 6 8 8
19	3 4 7 7
20	0 2
21	2

THINK	WRITE
1. Describe the distribution.	The distribution is symmetrical (or normal).
2. State the most appropriate measure of centre.	The most appropriate measure of centre is the mean.
3. Calculate the mean.	$\text{Mean} = \bar{x} = \dfrac{x}{n} = \dfrac{\text{sum of all data values}}{\text{number of data values}}$
	$= \dfrac{4509}{25}$
	$= 180.36$
4. Write the answer in a sentence.	The mean height of the 25 people is 180 cm.

 Resources

 eWorkbook   Topic 11 Workbook (worksheets, code puzzle and project) (ewbk-2011)

 Interactivities Individual pathway interactivity: Measures of spread (int-4622)
Range (int-3822)
The interquartile range (int-4813)
Skewness (int-3823)

Exercise 11.6 Measures of spread

learnon

Individual pathways

■ PRACTISE	■ CONSOLIDATE	■ MASTER
1, 4, 5, 8, 11, 14	3, 6, 10, 12, 15	2, 7, 9, 13, 16

To answer questions online and to receive **immediate corrective feedback** and **fully worked solutions** for all questions, go to your learnON title at www.jacplus.com.au.

Fluency

1. **WE18** Calculate the range and interquartile range for the following raw data sets.

   a. 18, 22, 25, 27, 29, 31, 35, 37, 40, 42, 44
   b. 120, 124, 125, 135, 135, 137, 141, 145, 151

2. **WE19** This stem plot gives the spread of 25 cars caught by a roadside speed camera.
   Determine:

   a. the range
   b. the interquartile range.

   Key: $8 \mid 2 = 82 \, \text{km/h}$
   $8^* \mid 6 = 86 \, \text{km/h}$

Stem	Leaf
8	2 2 4 4 4 4
8*	5 5 6 6 7 9 9 9
9	0 1 1 2 4
9*	5 6 9
10	0 2
10*	
11	4

3. For the following raw data sets, calculate the range and interquartile range.

   a. 20, 45, 73, 65, 85, 127, 101
   b. 1.22, 1.28, 1.48, 1.52, 1.62, 1.72, 1.85, 1.95, 2.03

4. For the ungrouped frequency table shown, calculate:

   a. the spread — the range and interquartile range
   b. the measures of centre — the mode, the median and mean (correct to 2 decimal places).

Score	Frequency
0	24
1	17
2	12
3	7
4	2
5	1

5. For each of the following histograms, state whether the distribution is symmetrical, positively skewed or negatively skewed.

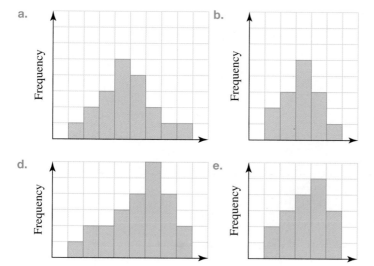

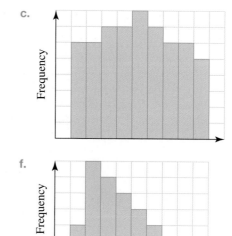

6. State the type of distribution shown by each of the following stem plots.

a. Key: 14 | 6 = 14.6

Stem	Leaf
12	0
13	8 9
14	0 8 9 9
15	1 5 6 6 6 7 8
16	3 5 7 8
17	0 2
18	3 5

b. Key: 14 | 6 = 146

Stem	Leaf
13	1 2 4 6 7 8 9
14	0 4 5 6 8 9 9
15	1 5 6 6 7 8
16	3 7 8
17	0 2
18	3

c. Key: 14 | 6 = 1.46

Stem	Leaf
12	7 8
13	6 7 8 9
14	0 9
15	1 5 6
16	3 7 8 9 9
17	0 2
18	3

d. Key: 14 | 6 = 1.46

Stem	Leaf
1	1
2	
3	6 9
4	0 4 9
5	1 3 5 6 8
6	2 3 7 8 9 9
7	0 4 6 7 7 8

Understanding

7. **WE20** This histogram represents the results of data collected from households in a suburban street. Describe the type of distribution, determining its centre using the most appropriate measure.

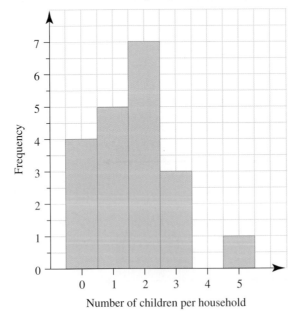

Children per household

8. **WE21** The number of customers served at a cafe each hour for the past 24-hour period is summarised in the stem plot shown.

   a. State the type of data and type of distribution.
   b. Choose the most appropriate measure of centre.
   c. Calculate the centre chosen in b.
   d. Calculate the interquartile range.

Key: 2 | 4 = 24 customers

Stem	Leaf
0	0 0 1 4
0*	5 5 8 8 9 9 9
1	1 2 3 4 4 4
1*	5 5 5 7
2	0 2
2*	
3	4

9. **MC** State the most likely type of distribution for the frequency table shown.

Score	0	1	2	3	4	5	6	7
Frequency	3	4	5	7	10	22	34	45

   A. negatively skewed distribution
   B. positively skewed distribution
   C. bimodal distribution
   D. normal distribution
   E. negatively skewed distribution with an outlier

10. Consider the distribution shown.

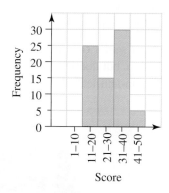

   a. State whether the data is symmetrical.
   b. State the modal class.
   c. Explain whether the mean and median can be seen from the graph. If they can, state their values.

Reasoning

11. A Geography teacher with a class of 20 students commented that test results for their class out of a score of 50 were negatively skewed. Show and explain a data set for the possible results obtained by the 20 students.

12. Show a data set for which the range and interquartile range are the same. Explain what factors cause this.

13. Explain the effect on the range of a data set when a set of scores has:

   a. each score increased by 5
   b. each score doubled
   c. each score increased by 25%.

Problem solving

14. A symmetrical distribution is one in which the mean and median are close to equal. A uniform distribution (one where the elements are all the same) is symmetrical. Give an example of a non-uniform distribution that is symmetrical.

15. This back-to-back stem-and-leaf plot displays the speeds of 25 cars caught by a speed camera on a major highway on Easter Sunday and Christmas Day.

Key: 6 | 3 = 63 km/h

Easter Sunday		Christmas Day
8 6 3	6	2 3 6 9 9
8 7 4 1	7	1 2 2 4 5 5 7 8
6 6 5 4 2 1	8	0 0 3 4 6 7
9 8 7 7 5 2 1	9	1 2 3 3
7 7 5 2 1	10	2 4

The police commissioner was heard to comment that he was surprised to find drivers behaved better on Christmas Day than they did on Easter Sunday.

a. The mean speed for Easter Sunday was 87.68 km/h and the mean speed for Christmas Day was 79.92 km/h. Explain what this tells us about the drivers on these different days.
b. The interquartile range on Easter Sunday was 21 and on Christmas Day it was 17.5. Explain what this tells us about drivers on these different days.
c. Do your results from part a and b support the police commissioner's statement? Explain your answer.

16. Waterway Secondary has one junior and one senior rowing team. Each team has eight rowers and one cox (the cox steers the boat and keeps the rhythm, but doesn't row). The heights of all the rowing team members in centimetres is shown in the following stem-and-leaf plot.

Key:16 | 4 = 164

Stem	Leaf
13	5 9
14	
15	9
16	2 4 5 5 8
17	0 1
18	0 1 2 2 4 5
19	0 4

a. This data set has two outliers. Identify these outliers and state their value.
b. Excluding the two outliers, calculate the mean of the data set.
c. Excluding the two outliers, identify the data set's distribution type.
d. Assuming that the junior rowers are shorter than the senior rowers, separate the data into junior team and senior team heights. Present the data in two stem-and-leaf plots.
e. Calculate the mean height of the:
   i. junior team
   ii. senior team.
f. Using your answers to b and e, explain why it is better to separate bimodal data before finding the mean.

11.7 Review

11.7.1 Topic summary

STATISTICS

Types of data

- Categorical data is expressed using labelled categories.
  - Nominal data cannot be ranked in a logical order; e.g. favourite colours, pet types.
  - Ordinal data can be ranked in a logical order; e.g. AFL players' salaries.
- Numerical data is expressed using numbers.
  - Discrete data is collected by counting; e.g. number of pens, number of days.
  - Continuous data is collected by measuring; e.g. the mass of eggs, the distance between school and each student's home.

Size of data set

- Population: Data collected from an entire group; e.g. when all members of a family are surveyed.
- Sample: Data collected from only parts of a group; e.g. when only Year 9 students are surveyed at a school.

Random number generator

- To create random data on a spreadsheet, use the command =RANDBETWEEN()
- Then enter the lowest and highest desired value within the brackets, separated by a comma.

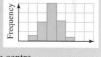

Collecting data

- Methods of collecting data include:
  - frequency tables

Person's age (x)	Tally	Number of people (f)
0–29	Ⅳ I	5
30–59	Ⅳ Ⅳ	10
60–89	III	3
		$\Sigma f = 18$

  - dot plots

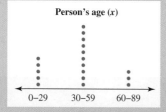

Person's age (x)

0–29   30–59   60–89

  - stem-and-leaf plots.

Heights of 30 students (in cm)
Key: 11|8 = 118 cm

Stem	Leaf
11	8 9
12	0 1 2 2 3 4 5 6 7 7 8 9
13	0 1 1 1 2 3 4 5 6 8 8 9 9
14	0 1 3

Displaying data

- Methods of displaying data include:
  - bar charts

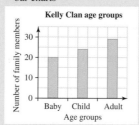

  - pie charts

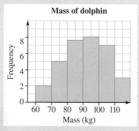

  - histograms

(histogram: Mass of dolphin)

  - back-to-back stem-and-leaf plots.

Leaf (female)	Stem	Leaf (male)
987766	1	55667789
53231910	2	18722351
1224688	3	50410
553	4	386
2	5	047

Measure of centre

- The **mean** is also known as the average.
  - For raw data:
  $$\bar{x} = \frac{\Sigma x}{n} = \frac{\text{sum of all data values}}{\text{number of data values}}$$
  - For data in a frequency table:
  $$\bar{x} = \frac{\Sigma xf}{\Sigma f} = \frac{\text{sum of (data values} \times \text{frequency)}}{\text{sum of frequencies}}$$
- The **median** is the middle value of the data.
  $\frac{n+1}{2}$ th data value, where n is the number of pieces of data or scores.
- The **mode** is the most common value in the data.

Spread of data

- Range = maximum value − minimum value
- Interquartile range (IQR) = $Q_3 - Q_1$

Types of distribution

Types of distribution include:

- symmetrical distribution, which has a peak in the centre and evenly balanced data above and below the centre.

- skewed data, which is not evenly balanced.
  - For positively skewed data, the mean has a larger value than the median.
  - For negatively skewed data, the mean has a smaller value than the median.

- bimodal data, which has a distribution pattern with two peaks.

11.7.2 Success criteria

Tick the column to indicate that you have completed the subtopic and how well you think you have understood it using the traffic light system.

(**Green:** I understand; **Yellow:** I can do it with help; **Red:** I do not understand)

Subtopic	Success criteria			
11.2	I can identify different types of data.			
	I can determine if a data set is a sample or a population.			
	I can generate random numbers.			
11.3	I can display data in frequency tables.			
	I can display data in dot plots.			
	I can display data in stem-and-leaf plots.			
11.4	I can display data in bar charts, pie charts and histograms.			
	I can present data as a back-to-back stem-and-leaf plots.			
11.5	I can calculate the mean, median and mode of raw data.			
	I can calculate the mean, median and mode of data in tables and plots.			
	I can calculate the mean, median and mode of grouped data.			
11.6	I can calculate range and interquartile range.			
	I can determine whether a distribution is symmetrical, skewed or bimodal.			

11.7.3 Project

Graphical displays of data

Statistical data can be communicated using words, tables or graphs.

When data is presented in words, it can be difficult to identify and compare trends.

When data is presented in table form, additional calculations are often needed to gain a complete understanding.

When data is presented as a graph it has an immediate visual impact, but the impression that graphs give can often be misleading, particularly if the graph is not looked at carefully. It is easy to manipulate graphs to produce a desired effect.

Consider the two graphs shown.

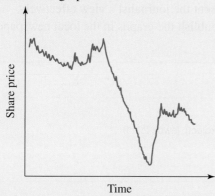

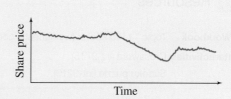

These graphs both display the same data (the value of a company's shares over the same period of time), but they look quite different. If the objective of the graph is to create the impression that the value of the shares has dropped quite dramatically, the first graph would serve this purpose. On the other hand, if we are to be convinced that the shares have not changed very much in value, the second graph reflects this point of view.

1. What technique is used to create two quite opposing impressions in the two graphs?

The following table shows the times for the men's and women's 100-m sprints in the Summer Olympic Games. A sports journalist wants to convince readers that, since 1928, women have been more successful in reducing their time than men.

Year	100 m men	Men's time (seconds)	100 m women	Women's time (seconds)
1928	Percy Williams, CAN	10.80	Betty Robinson, USA	12.20
1932	Eddie Tolan, USA	10.30	Stella Walsh, POL	11.90
1936	Jesse Owens, USA	10.30	Helen Stephens, USA	11.50
1948	Harrison Dillard, USA	10.30	Fanny Blankers-Koen, HOL	11.90
1952	Lindy Remigino, USA	10.40	Marjorie Jackson, AUS	11.50
1956	Bobby Morrow, USA	10.50	Betty Cuthbert, AUS	11.50
1960	Armin Hary, GER	10.20	Wilma Rudolph, USA	11.00
1964	Bob Hayes, USA	10.00	Wyomia Tyus, USA	11.40
1968	Jim Hines, USA	9.95	Wyomia Tyus, USA	11.08
1972	Valery Borzov, URS	10.14	Renate Stecher, GDR	11.07
1976	Hasely Crawford, TRI	10.06	Annegret Richter, FRG	11.08
1980	Allan Wells, GBR	10.25	Lyudmila Kondratyeva, URS	11.06
1984	Carl Lewis, USA	9.99	Evelyn Ashford, USA	10.97
1988	Carl Lewis, USA	9.92	Florence Griffith Joyner, USA	10.54
1992	Linford Christie, GBR	9.96	Gail Devers, USA	10.82
1996	Donovan Bailey, CAN	9.84	Gail Devers, USA	10.94
2000	Maurice Greene, USA	9.87	Marion Jones, USA	10.75
2004	Justin Gatlin, USA	9.85	Yuliya Nesterenko, BLR	10.93
2008	Usain Bolt, JAM	9.69	Shelley-Ann Fraser, JAM	10.78
2012	Usain Bolt, JAM	9.63	Shelly-Ann Fraser-Pryce, JAM	10.75
2016	Usain Bolt, JAM	9.81	Elaine Thompson, JAM	10.71

Note: The Summer Olympic Games were not held in 1940 and 1944.

2. Using graph paper, draw two graphs to support the sports journalist's view.
3. Describe the features of your graphs that helped you to present the journalist's view effectively.
4. Imagine you are the sports journalist and you are about to publish the graphs in the local newspaper. Write a short article to accompany the graphs.

---

on Resources

eWorkbook   Topic 11 Workbook (worksheets, code puzzle and a project) (ewbk-2011)

Interactivities Crossword (int-2715)
                  Sudoku puzzle (int-3213)

Exercise 11.7 Review questions

learn on

To answer questions online and to receive **immediate corrective** feedback and **fully worked solutions** for all questions, go to your learnON title at www.jacplus.com.au.

Fluency

1. **MC** Recognise which one of the following is an example of numerical discrete data.
   A. Your favourite weeknight television show
   B. The speed of a car recorded on a speed camera
   C. The number of home runs by a baseball player
   D. Placing in a field of eight swimmers of a 100 m freestyle race
   E. The heights of different animals at a zoo

The following information relates to questions **2** and **3**.

Number of bedrooms	1	2	3	4	5	6
Number of homes	4	9	32	36	7	2

2. **MC** For the data given, select the most likely type of distribution.
   A. A negatively skewed distribution
   B. A positively skewed distribution
   C. Skewed to the left
   D. A symmetrical distribution
   E. A bimodal distribution

3. **MC** Select the mode for the data collected.
   A. 36   B. 5   C. 4   D. 3   E. 90

4. **MC** The results of a Science test marked out of 60 are represented by the stem plot shown. The median, mean and mode, respectively, are:
   A. 60, 42, 52
   B. 40, 45, 60
   C. 45, 60, 45
   D. 45, 44, 60
   E. 42, 44, 45

Key: $2 \mid 3 = 23$

Stem	Leaf
0	8
1	8
2	3 5
3	0 3 6
4	2 4 5 5 5 8
5	4 7 8 8 9
6	0 0 0 0

5. **MC** Select the best estimate of the mean score for the grouped frequency table shown.

Score	Frequency
$0 - <50$	12
$50 - <100$	24
$100 - <150$	23
$150 - <200$	4

A. 63
B. 90
C. 92
D. 100
E. 85

6. **MC** Identify the range and IQR for the following set of data: $46, 46, 49, 53, 61, 63, 67, 72, 81, 84, 93$.
A. 46 and 34
B. 46 and 65
C. 47 and 63
D. 47 and 32
E. 93 and 63

7. For the following surveys or studies, decide what type of data would be collected and how it should be displayed.
a. The blood type of all workers at a shoe factory
b. A survey of 500 Australians' views on bringing back corporal punishment (strongly agree, agree, disagree, strongly disagree)
c. Number of phones in the households of the students in your class
d. The top 20 fastest recorded times for the women's 100 metres butterfly

8. Classify each of the following data collections according to whether they are samples or populations.
a. A survey of 1000 travellers on the quality of train service provided.
b. A survey of all the students at a secondary college about the state government's rules on underage alcohol laws
c. Performance in Mathematics exams by this year's Year 9 cohort

9. The following set of data shows the types of phones owned by the first 30 shoppers surveyed.
iPhone, Samsung, Motorola, iPhone, iPhone, iPhone, Samsung, Samsung, Nokia, Motorola, Samsung, iPhone, Motorola, iPhone, Samsung, Samsung, Motorola, iPhone, Samsung, Nokia, Samsung, Motorola, iPhone, iPhone, iPhone, iPhone, Samsung, Samsung, Nokia, Motorola
a. Suggest two appropriate ways to display this data.
b. Present this data in the ways suggested in part a

10. The following data represent the number of songs on people's favourite Spotify playlists. An initial analysis of 25 playlists resulted in the following data being collected.

$$47, 103, 45, 111, 67, 54, 59, 44, 31, 89, 34, 2, 43, 96, 70, 90, 66, 48, 67, 82, 74, 56, 35, 9, 77$$

a. Present the data as a stem-and-leaf plot.
b. Comment on the shape of distribution.
c. Present the data as a frequency table grouped in intervals of 20.
d. State the type of data and type of distribution, and give the most appropriate measure of centre.

11. The following table shows the salaries for AFL players in 2019.

AFL players' salaries	Number of players
0–$100 000	146
$100 001–$200 000	188
$200 001–$300 000	107
$300 001–$400 000	57
$400 001–$500 000	24
$500 001–$600 000	12
$600 001–$700 000	4
$700 001–$800 000	0
$800 001–$900 000	4
Total number of players	542

a. Draw a histogram to display the data.
b. Describe the type of distribution the histogram displays.
c. Calculate the median salary of these football players (to the nearest thousand).
d. Calculate the modal salary for AFL footballers in 2019.

Problem solving

12. The workers in an office are trying to obtain a salary rise. In the previous year, all ten people who work in the office each received a 2% salary rise, while the company CEO received a 42% salary rise.
    a. Calculate the mean salary rise received last year. Give your answer to 1 decimal place.
    b. Calculate the median percentage salary rise received last year.
    c. Calculate the modal percentage salary rise received last year.
    d. The company is trying to avoid paying the salary rise. Explain the statistics you think they would quote about last year's rise. Give your reason for using these statistics.
    e. Explain the statistics that you think the trade union would quote about salary rises. Give your reason for using these statistics.
    f. Identify which statistic you think is the most 'honest' reflection of last year's salary rises. Explain your reasoning.

13. A data set with the values 8, ___, 25, ___, 21, 8 has a mean of 15, mode of 8 and a median of 14. If the missing values are positive integers:
    a. calculate two values for the set
    b. calculate two different values that meet the given criteria.

14. Create a data set of 9 numbers that meet the following criteria.
    - Data values are positive integers
    - Mean = 8
    - Median = 7
    - Mode = 10
    - Maximum value in data set = 14
    - Minimum value in data set = 3

on To test your understanding and knowledge of this topic go to your learnON title at www.jacplus.com.au and complete the **post-test**.

Online Resources

 **Resources**

Below is a full list of **rich resources** available online for this topic. These resources are designed to bring ideas to life, to promote deep and lasting learning and to support the different learning needs of each individual.

eWorkbook

Download the workbook for this topic, which includes worksheets, a code puzzle and a project (ewbk-2011) ☐

Solutions

Download a copy of the fully worked solutions to every question in this topic (sol-0731) ☐

Digital documents

11.2 SkillSHEET Presenting data in a frequency table (doc-6317) ☐
SkillSHEET Arranging a set of data in ascending order (doc-10954) ☐
11.3 SkillSHEET Finding the mean (doc-10955) ☐
SkillSHEET Finding the middle score (doc-10956) ☐
SkillSHEET Finding the middle score for data arranged in a dot plot (doc-10957) ☐
SkillSHEET Finding the most frequent score (doc-10958) ☐

Video eLessons

11.2 Types of data (eles-4909) ☐
Data set size (eles-4910) ☐
Generating random numbers (eles-4911) ☐
11.3 Frequency tables (eles-4912) ☐
Dot plots (eles-4913) ☐
Stem-and-leaf plots (eles-4914) ☐
11.4 Bar charts and pie charts (eles-4915) ☐
Histograms (eles-4916) ☐
Back-to-back stem-and-leaf plots (eles-4917) ☐
11.5 Mean, median and mode (eles-4918) ☐
Grouped data (eles-4919) ☐
Mean and median (eles-1905) ☐
11.6 Range and interquartile range (eles-4920) ☐
Types of distributions (eles-4921) ☐

Interactivities

11.2 Individual pathway interactivity: Sampling (int-4540) ☐
Types of data (int-6086) ☐
Selecting samples (int-3811) ☐
Random number generator (int-0089) ☐
11.3 Individual pathway interactivity: Collecting data (int-4541) ☐
Collecting data (int-3807) ☐
Stem plots (int-6242) ☐

11.4 Individual pathway interactivity: Displaying data (int-4542) ☐
Back-to-back stem plot (int-2773) ☐
11.5 Individual pathway interactivity: Measures of central tendency (int-4543) ☐
Mean (int-3818) ☐
Median (int-3819) ☐
Mode (int-3820) ☐
11.6 Individual pathway interactivity: Measures of spread (int-4622) ☐
Range (int-3822) ☐
The interquartile range (int-4813) ☐
Skewness (int-3823) ☐
11.7 Crossword (int-2715) ☐
Sudoku puzzle (int-3213) ☐

Teacher resources

There are many resources available exclusively for teachers online.

To access these online resources, log on to **www.jacplus.com.au**.

Answers

Topic 11 Statistics

Exercise 11.1 Pre-test

1. a. iii   b. iv   c. i   d. ii
2. Sample
3. 16 male adult, 14 female adult, 11 male adolescent, 9 female adolescent
4. a. 25   b. 7
5. a. 20   b. 10
6. a. ii   b. iii   c. i
7. B
8. a. Discrete   b. 17   c. Skewed
9. B
10. C
11. 2.5
12. a. 60.3   b. D
13. 7
14. D
15. E

Exercise 11.2 Sampling

1. a. Categorical   b. Numerical
2. a. Numerical   b. Numerical
3. a. Continuous   b. Continuous
4. a. Continuous   b. Discrete
5. a. Ordinal   b. Nominal
6. a. Ordinal   b. Ordinal
7. a. Nominal   b. Nominal
8. a. Sample   b. Sample   c. Population
9. a. Sample   b. Population
10. a. data   b. survey, census c. graph
11. a. RANDBETWEEN $(1, 31)$
    b. RANDBETWEEN $(1, \text{class size})$
    c. Samples
12. a. RANDBETWEEN $(1, 365)$
    b. Sample
13. a. Car
    b. Sample responses can be found in the worked solutions in the online resources.
    c. Sample responses can be found in the worked solutions in the online resources. Answers could include the lower number of motor cycle fatalities as compared to the number of car fatalities.
    d. According to the data, there were 190 motorcycle fatalities in 2020, compared to 732 car fatalities.
    e. Sample responses can be found in the worked solutions in the online resources.
14. Sample responses can be found in the worked solutions in the online resources. Answers will vary but should include the following groups.

Categorical data can be:
- nominal — arranged in categories, for example by colour or size
- ordinal — arranged in categories that have an order, for example, karate ability, which ranges from white belt (beginner) to black belt (accomplished).

Numerical data can be:
- discrete — data can have only particular values, for example amounts of money or shoe sizes
- continuous — data can have any value within a range, for example the time it takes a person to run around an oval.

15. a. Sample responses can be found in the worked solutions in the online resources. Answers could include:
    Data that might be collected from a maternity hospital:
    - birth weights of babies
    - length of time in labour.
    b. Sample responses can be found in the worked solutions in the online resources. Answers could include:
    Data that might be collected from an airport:
    - number of delays
    - time spent waiting for flights.
    c. Sample responses can be found in the worked solutions in the online resources. Answers could include:
    Data that might be collected from shoppers in a supermarket:
    - shoppers' favourite brands of detergent
    - the amount of money spent on an average weekly shop.
16. a. Sample responses can be found in the worked solutions in the online resources. Answers could include:
    'What do you think of sport?' — This question should not be used because it is too broad.
    b. Sample responses can be found in the worked solutions in the online resources. Answers could include:
    'List these sports in order from your most favourite to your least favourite: football, tennis, netball, golf.'
    c. Sample responses can be found in the worked solutions in the online resources.
    d. Sample responses can be found in the worked solutions in the online resources. Answers could include:
    'Why do you like sports?' — This question should not be used because it is too broad.
    e. Sample responses can be found in the worked solutions in the online resources.
    f. Sample responses can be found in the worked solutions in the online resources. Answers could include
    'Can you play a sport?' — This question should not be used because it is too broad. It would be better to compile a list of sports and ask about those specifically.
    g. Sample responses can be found in the worked solutions in the online resources.
    'Which of these sports do you watch on the television? Football, tennis, netball, golf, all of these, none of these.' — This is a good question, though it must be noted that results may vary depending on the selection of the sample population.
17. a. Sample responses can be found in the worked solutions in the online resources.

b. Sample responses can be found in the worked solutions in the online resources. Answers could include:
Data can be collected manually, or by using technology to capture data as specific events occur. For example, a computer system might be set up to automatically record the number of sales made per month.

c. Sample responses can be found in the worked solutions in the online resources. Answers could include:
Data is used for forward planning and budgeting purposes. For example, it can be used to identify the most or least popular items that the company sells; this data can then be used to see which items require updating or which items should no longer be sold.

18. a. Sample responses can be found in the worked solutions in the online resources.

b. Sample response: We are trying to find out whether the population approves or disapproves of the desalination plant and we are also trying to capture data about their reasons.

c. Sample response: You are collecting categorical data in both instances.

d. Sample response: You could use pie charts, column graphs or bar graphs.

e. Sample responses can be found in the worked solutions in the online resources.

Exercise 11.3 Collecting data

1. a. **Frequency of Maths test results**

Score x	Frequency f
1	1
2	1
3	4
4	3
5	5
6	6
7	5
8	2
9	3

b. **Multiple choice Maths test results**

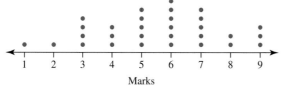

Marks

2. **Frequency of type of pet**

a.

Type of pet x	Frequency f
Cat	6
Dog	6
Fish	1
Bird	2
Turtle	3

b. **Type of pet**

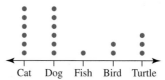

Cat   Dog   Fish   Bird   Turtle

c. No, only numeric data can be presented in a stem plot.

3. a.

Deaths by motor vehicle accidents (rates per 100 000 people)	Frequency f
0– < 10	2
10– < 20	11
20– < 30	10
30– < 40	3
40– < 50	4
50– < 60	0
60– < 70	1
70– < 80	1
	$f = 32$

b. Deaths by motor vehicle accidents (rates per 100 000 people)
Key: 7 | 3 = 73 deaths per 100 000 people

Stem	Leaf
0	0 9
1	1 2 3 4 4 4 5 7 8 8 9
2	1 2 2 3 4 5 5 6 6 7
3	4 6 6
4	2 2 3 5
5	
6	0
7	1

4. a. **Frequency of eye colour in a sample of 40 people**

Score x	Frequency f
Brown	17
Blue	11
Hazel	7
Green	5
	$f = 40$

b. Eye colour in a sample of 40 people

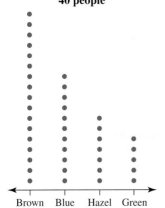

Brown Blue Hazel Green

c. 17 out of 40 people (or 42.5% of the people) have brown eyes.

5. a. Key: $1\,|\,8 = 18$

Stem	Leaf
1	8 9 9
2	0 0 0 1 1 2

b. Key: $1\,|\,9 = 19$

Stem	Leaf
1	9 9
2	0 4 4 5 6 7 9
3	1 1 6 8
4	2 3

c. Key: $33\,|\,6 = 336$

Stem	Leaf
33	6
34	6 6 9
35	0 3 9
36	2 8
37	1

6. a. Key: $1^*|\,8 = 18\;\;2\,|\,0 = 20$

Stem	Leaf
1^*	8 9 9
2	0 0 0 1 1 2

b. Key: $1^*|\,9 = 19\;\;2\,|\,0 = 20$

Stem	Leaf
1^*	9 9
2	0 4 4
2^*	5 6 7 9
3	1 1
3^*	6 8
4	2 3

c. Key: $33^*|\,6 = 336\;\;35\,|\,0 = 350$

Stem	Leaf
33^*	6
34	
34^*	6 6 9
35	0 3
35^*	9
36	2
36^*	8
37	1

7. a. Running times for 40 students

Key: $4\,|\,3 = 43$ minutes, $4^*|\,8 = 48$ minutes

Stem	Leaf
2	1 3
2^*	5 6 6 8 9
3	1 2 2 3 3 3 4
3^*	7 7 9 9
4	0 1 1 2 2 3 4
4^*	5 5 7 8 8 9
5	2 2 3
5^*	5 6 7
6	0 3
6^*	
7	1

b.

Run times (minutes) x	Number of runners f
$20 - <25$	2
$25 - <30$	5
$30 - <35$	7
$35 - <40$	4
$40 - <45$	7
$45 - <50$	6
$50 - <55$	3
$55 - <60$	3
$60 - <65$	2
$65 - <70$	0
$70 - <75$	1
$f =$	40

c. i. 32 min
ii. 30 min $- < 35$ min
d. 3, stem plot

8. Key: $1 \mid 4 = 141^* \mid 7 = 17$

Stem	Leaf
1	4
1*	7 8 8 9
2	0 1 3
2*	5 6 7 9
3	2 4
3*	5 6 7 8
4	2
4*	5
5	3

9. D

10. a.

Life expectancy (years) x	Number of age groups f
83 − < 84	1
84 − < 85	21
85 − < 86	7
86 − < 87	5
87 − < 88	2
88 − < 89	4
89 − < 90	0
$f =$	40

b. Life expectancy of Australians across 40 different age groups.
Key: $86 \mid 1 = 86.1$ years

Stem	Leaf
83	7
84	1 1 1 4 4 5 5 5 5 5 5 5 6 7 7 7 7 8 8 9 9
85	0 0 1 2 7 8 9
86	0 5 6 8 9
87	3 6
88	3 3 8 8
89	

11. Median age of Australian females and males from 2002 to 2009
Key: $36 \mid 1 = 36.1$ years

Stem	Leaf
35	2 3
35*	5 7 9
36	1 1 1
36*	6 9
37	1 3 4
37*	6 6 7

Note: The alternative method — showing categories for 35, 36 and 37 years — is not appropriate because it only allows for 3 groups, which is below the 5 minimum groups allowed.

Score x	Frequency f
35 − < 36	5
36 − < 37	5
37 − < 38	6
	16

b.

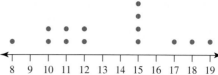

13. a. Key: $15 \mid 4 = 154$ cm

Leaf (boys)	Stem	Leaf (girls)
	14	5 8
9 8 6 6 4	15	2 2 5 6
8 4 2 0	16	0 1 1 3 5 8
2 1	17	
0	18	

b. The data could be grouped into smaller intervals, such as 5, instead of 10. For example, the stem value of 2 could be broken into two groups, encompassing values 20–24 and 25–29.

14. a. Class 1 : 25, class 2 : 27

b. Class 1 : 2, class 2 : 3

c. Class 1 : 24, class 2 : 20

d. A total of 11 students in class 2 scored better than class 1, but 6 students scored worse than class 1. Class 2's results are more widely spread than class 1's results.

15. a. Key: $2 \mid 1 = 21$ km/h

Stem	Leaf
1	5
2	0 0 2 7 9 9
3	0 0 0 0 2 4 5 6 7 8 8 9
4	0 0 0 0 0 2 4 5 5 5
5	
6	0

b. Sample responses can be found in the worked solutions in the online resources. A possible answer could include that there were a high number of speeds ending in 0 (possibly indicating a machine malfunction), and most of the speeds (24 out of 30) were within the legal speed limit.

16. a. Key: $25 \mid 1 = 25.1$ g

Stem	Leaf
23	9 9
24	0 5 5 6 8
25	1 1 1 3 4 4 8
26	3 3 7
27	3 3
28	1
29	5
30	
31	
32	
33	2

b. The range is $33.2 - 23.9 = 9.3$.

c. Sample responses can be found in the worked solutions in the online resources. Answers could mention that the plot for rounded values provides less detail.

Key: $2\,|\,4 = 24\,g$

$2^*|\,6 = 26\,g$

Stem	Leaf
2	4 4
2*	5 5 5 5 5 5 5 5 5 6 6 6 7 7 7 8
3	0 3

17. a. Sample responses can be found in the worked solutions in the online resources. Answers could mention that a doughnut graph and a pie graph are both circular, and both use colour to emphasise different categories.

b. Draw circles of different size with the same centre. Then use the same instructions as for creating a pie chart, but mark only the lines between the two circles.

c. **Methods of travel to school**

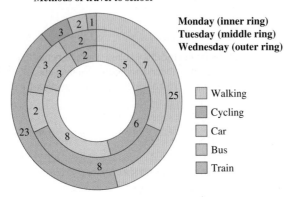

Monday (inner ring)
Tuesday (middle ring)
Wednesday (outer ring)

d. They are the same graph.

e. Sample responses can be found in the worked solutions in the online resources. Answers could include that a disadvantage of using a doughnut graph could be that you are restricted in the number of circles you can use, so this type of graph would not be suitable for large data sets.

18. 1296

Exercise 11.4 Displaying data

1. a. See the figure at the foot of the page.*

b. **Pie chart of gym class participation**

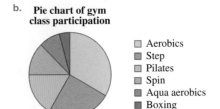

- Aerobics
- Step
- Pilates
- Spin
- Aqua aerobics
- Boxing

2. a. Continuous
b. 24

c.

Class interval	Midpoint (class centre)	Tally	Frequency
5 − < 10	7.5	⊔Ⱶ ⊔Ⱶ |	11
10 − < 15	12.5	⊔Ⱶ ⊔Ⱶ ⊔Ⱶ ||||	19
15 − < 20	17.5	⊔Ⱶ	5
20 − < 25	22.5	||	2
25 − < 30	27.5	|||	3
		Total	40

d.

Hours of phone use per week

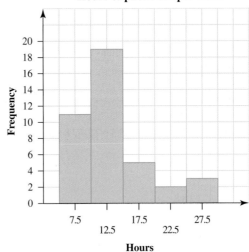

*1. a.

Gym class participation

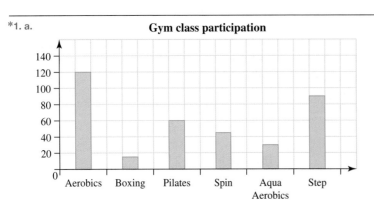

3. a.

Class interval	Midpoint (class centre)	Tally	Frequency
0 – 9	4.5	卌I	6
10 – 19	14.5	卌II	7
20 – 29	24.5	卌III	8
30 – 39	34.5	卌	5
40 – 49	44.5	IIII	4
50 – 59	54.5	卌 卌II	12
60 – 69	64.5	卌III	8
70 – 79	74.5	IIII	4
80 – 89	84.5	II	2
		Total	56

b. See the figure at the foot of the page.*

4. a.

Class interval	Midpoint (class centre)	Tally	Frequency
12 – 15	13.5	IIII	4
16 – 19	17.5	卌 II	7
20 – 23	21.5	卌 卌 I	11
24 – 27	25.5	卌 卌 III	13
28 – 31	29.5	IIII	4
32 – 35	33.5	____	0
36 – 39	37.5	I	1
		Total	40

b. See the figure at the foot of the page.*

---

*3. b.

Number of phone calls made per week

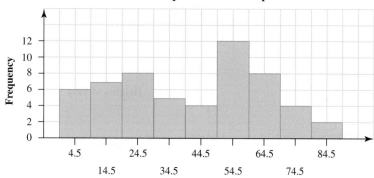

Number of phone calls

*4. b.

Class sizes at Jacaranda SC

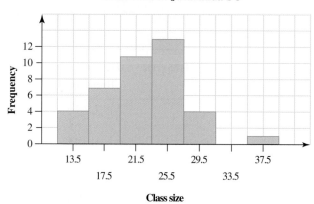

Class size

5.

Number of bananas sold at recess

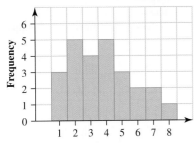

6.

Distances travelled

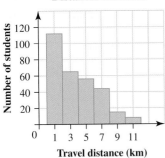

7. D

8.

Number of passengers on bus journeys

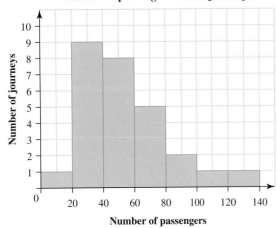

9. a.

Bar chart of DVD genres (20 DVDs)

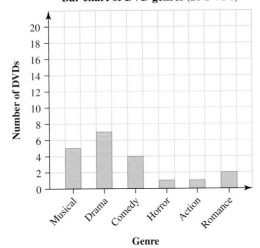

b.

Genre	Number of DVDs	Size of sector for a pie chart
Musical	52	93.6°
Drama	98	176.4°
Comedy	30	54°
Horror	10	18°
Action	5	9°
Romance	5	9°
Total	200	360°

c.

DVD genres (sample size 200)

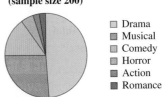

☐ Drama
☐ Musical
☐ Comedy
☐ Horror
☐ Action
☐ Romance

10. a. Continuous

b.

Class interval	Midpoint (class centre)	Tally	Frequency
145 − < 150	147.5	||	2
150 − < 155	152.5	⊥⊥⊥⊥	5
155 − < 160	157.5	⊥⊥⊥⊥ |||	8
160 − < 165	162.5	⊥⊥⊥⊥ ||	7
165 − < 170	167.5	⊥⊥⊥⊥ |	6
170 − < 175	172.5	|||	3
175 − < 180	177.5	|	1
		Total	32

c.

Heights of Year 9 students

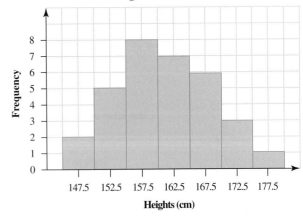

d. No. Individual scores are lost when data are grouped into class intervals.

e. C

f. D

g.

Class interval	Midpoint (class centre)	Tally	Frequency
145 – < 148	146.5	|	1
148 – < 151	149.5	||	2
151 – < 154	152.5	|||	3
154 – < 157	155.5	||||	4
157 – < 160	158.5	JHT	5
160 – < 163	161.5	||||	4
163 – < 166	164.5	|||	3
166 – < 169	167.5	JHT |	6
169 – < 172	170.5	||	2
172 – < 175	173.5	|	1
175 – < 178	176.5	____	0
178 – < 181	179.5	|	1
		Total	32

h. See the figure at the foot of the page.*

11. a. Number of players   b. 5

12. a. Key: 0 | 0 = 0

Leaf (Mitch)	Stem	Leaf (Yani)
3 2 1 1 1 1 0 0 0 0	0	0 0 1 1 1 1 1 2 2 2

Mitch scored between 0 and 3 goals inclusive. Yani scored between 0 and 2 goals inclusive.

b. Mitch: 2, Yani: 3

c. Mitch

d. Yani: 11

e. Yani, because he scored goals in more games.

13. a. Key: 3 | 2 = 32

Leaf (9A)	Stem	Leaf (9B)
2	3	
9	4	6
7 4	5	5
8 7 5 5 0	6	0 2 8
3	7	3 3 4 5 7 8
5	8	1
6	9	

b. 9A: 83%, 9B: 92%

c. 9A: 2

d. Sample responses can be found in the worked solutions in the online resources.

e. Sample responses can be found in the worked solutions in the online resources.

f. 9A: 64.25%, 9B: 68.5%

14. a. C

b. D

c. In general the boys are slightly taller than the girls, with the tallest 3 students all being boys.

15. Sample responses can be found in the worked solutions in the online resources.

16. a. Categorical

b. A column graph or a pie chart

17. The average amount of times per week people used the sunscreen was 5. The minimum amount of times was 2 and the maximum amount was 21, although this value was much higher than the rest of the data. Fifty per cent of the people surveyed used the sunscreen between 4 and 7 times per week.

18. a. Floribunda: 36°, Mister Lincon: 72°, Pillar: 108°, Hybrid tea: 144°

b. 10

*10. h.

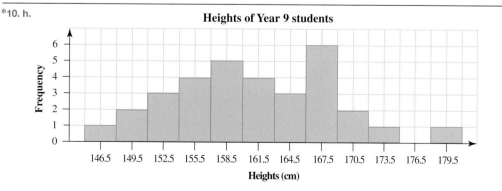

c.

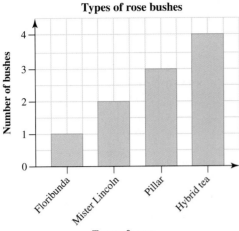

Types of rose bushes

(Bar graph: y-axis "Number of bushes" from 0 to 4; x-axis "Types of roses" with categories Floribunda = 1, Mister Lincoln = 2, Pillar = 3, Hybrid tea = 4)

Exercise 11.5 Measures of centre

1. a. 55, 54, 55 b. 8, 6.88, 8
2. 2.5, 2.55, 2
3. a. 23 b. 5
4. a. 116 b. 82
5. a. 6.28 b. 2.55
6. a. 25 b. 25.24 c. 25
7. a. 143.5, 142.6, 140–149
   b. 150.9, 148.5, 150–159
8. C
9. B
10. D
11. a. Numerical continuous
    b. 153
    c. 157
    d. 140− < 160
12. a. Numerical discrete
    b. 45.2
    c. 45.25
    d. 40 − 49
13. a.

Score (x)	Frequency (f)	Freq ×score (f×x)
0	1	0
1	3	3
2	6	12
3	3	9
4	3	12
5	1	5
6	1	6
	n = 18	fx = 47

$\bar{x} = 2.6$

b.

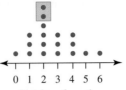

(Dot plot: Number of people in household, 0 to 6)

c. 2
d. i. Median ii. Mean iii. Mode

14. a. Sample responses can be found in the worked solutions in the online resources. Possible answers include 7, 11 and 8, 10.
    b. Sample responses can be found in the worked solutions in the online resources. Possible answers include 7, 11 and 8, 10.
    c. No — 4 is the mode.
15. a. The median was calculated by taking the average of the 2 middle scores.
    b. 13
16. a. 55 250 km, 52 000 km, 52 000 km
    b. Discuss in class.
    c. 51 810 km (it is reduced by 3440 km)
    d. 52 000 km
    e. 3333
    f. 50 000 km (92% of tyres last that distance or more)
17. a. B
    b. $29 444
    c. A
    d. Sample responses can be found in the worked solutions in the online resources.
18. a. Sample responses can be found in the worked solutions in the online resources.
    b. No
19. Sample responses can be found in the worked solutions in the online resources. Possible answers include:
    a. 1, 2, 5, 6, 6 $\dfrac{1+2+5+6+6}{5} = \dfrac{20}{5} = 4$
    b. 1, 2, 5, 6, 11 $\dfrac{1+2+5+6+11}{5} = \dfrac{25}{5} = 5$
20. a. 6 b. > 6 c. 6 d. > 13
21. 8.93
22. 13, 31, 31, 47, 53 and 59

Exercise 11.6 Measures of spread

1. a. 26, 15 b. 31, 18.5
2. a. 32 km/h b. 10 km/h
3. a. 107, 56 b. 0.81, 0.52
4. a. 5, 2 b. 0, 1, 1.19

5. a. Symmetrical
   c. Symmetrical
   e. Negatively skewed
   b. Symmetrical
   d. Negatively skewed
   f. Positively skewed

6. a. Symmetrical
   c. Bimodal
   b. Positively skewed
   d. Negatively skewed with an outlier

7. The data is positively skewed with an outlier. This indicates that the mean will not be a suitable measure of centre. The median, with a value of 2, is an appropriate measure of centre because it is not affected by the presence of an outlier.

8. a. Numerical discrete data with a positively skewed distribution
   b. Median is the best measure of centre for a skewed distribution.
   c. Median $= 11.5$ customers
   d. IQR $= 8.5$ customers

9. A

10. a. No
    b. 31–40
    c. The mean and median cannot be seen directly from the graph itself. You would need to calculate the mean and the median using a frequency distribution table, which can be constructed from the distribution shown.

11. Sample responses can be found in the worked solutions in the online resources. Possible answers could include that for data to be negatively skewed, the data needs to be a set with few low scores spread over a large range to the left. The most common scores are at the right. Assuming that the horizontal axis has test results in increasing order along the horizontal axis, it could be assumed that many of his students achieved good results.

12. Answers will vary. Possible data sets will have the highest score $= Q_3$ and the lowest score $= Q_1$.

13. a. No change
    b. Range will double
    c. Range will increase by 25%

14. Sample responses can be found in the worked solutions in the online resources. Possible answers include $\{1, 2, 2, 3, 3, 3, 4, 4, 5\}$

15. a. Sample responses can be found in the worked solutions in the online resources.
    b. Sample responses can be found in the worked solutions in the online resources.
    c. The Police Commissioner's claim is not supported, because all statistical measures were lower on Christmas Day. Sample responses can be found in the worked solutions in the online resources.

16. a. 135 cm and 139 cm
    b. 175 cm
    c. Bimodal
    d. Heights of junior rowing team

    Key: $15 \,|\, 9 = 159$ cm

Stem	Leaf
15	9
16	2 4 5 5 8
17	0 1

    Heights of senior rowing team

    Key: $18 \,|\, 0 = 180$ cm

Stem	Leaf
18	0 1 2 2 4 5
19	0 4

    e. Junior team: 165.5 cm, senior team: 184.75 cm,
    f. Sample responses can be found in the worked solutions in the online resources.

Project

1. Stretching the axes vertically gives the impression of a fall in the value of the shares. Stretching the axes horizontally gives the impression the value of the shares has not changed very much.

2. See figure at the foot of the page.*

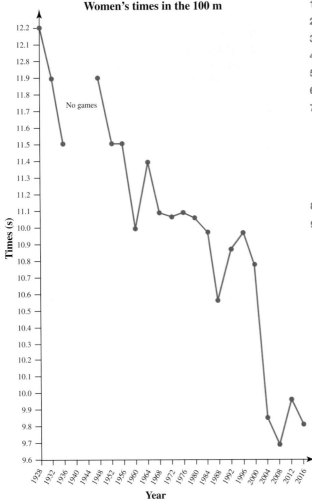

Women's times in the 100 m

Times (s) (vertical axis)
12.2, 12.1, 12.8, 11.9, 11.8, 11.7, 11.6, 11.5, 11.4, 11.3, 11.2, 11.1, 10.0, 10.9, 10.8, 10.7, 10.6, 10.5, 10.4, 10.3, 10.2, 10.1, 10.0, 9.9, 9.8, 9.7, 9.6

No games

Year (horizontal axis)
1928, 1932, 1936, 1940, 1944, 1948, 1952, 1956, 1960, 1964, 1968, 1972, 1976, 1980, 1984, 1988, 1992, 1996, 2000, 2004, 2008, 2012, 2016

3. For the women's graph, stretching the vertical axis and reducing the distance between the values on the horizontal axis gives the impression women's times have fallen dramatically. To show that the men's times have not fallen greatly, you can reduce the distance between the values on the vertical axis and stretch the values on the horizontal axis.

4. Sample responses can be found in the worked solutions in the online resources.

Exercise 11.7 Review questions

1. C
2. D
3. C
4. D
5. B
6. D
7. a. Categorical nominal; frequency table, dot plot, bar chart or pie chart
   b. Categorical ordinal; frequency table, dot plot, bar chart or pie chart
   c. Numerical discrete; stem plot
   d. Numerical continuous: box plot
8. a. Sample   b. Sample   c. Population
9. a. This is categorical nominal data. It could be displayed as a frequency table, dot plot, bar chart or pie chart.
   b. Examples of a dot plot and frequency table are shown. Pie or bar charts are also acceptable.

Mobiles owned by 30 shoppers

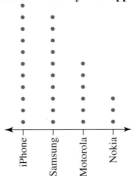

Type of mobile

iPhone, Samsung, Motorola, Nokia

Mobiles owned by 30 shoppers	
Brand of mobile	**Number of mobiles**
iPhone	11
Samsung	10
Motorola	6
Nokia	3

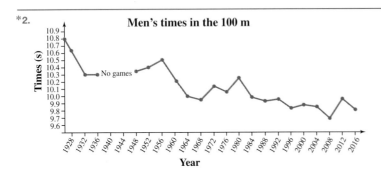

*2.
Men's times in the 100 m

Times (s) (vertical axis)
10.9, 10.8, 10.7, 10.6, 10.5, 10.4, 10.3, 10.2, 10.1, 10.0, 9.9, 9.8, 9.7, 9.6

No games

Year (horizontal axis)
1928, 1932, 1936, 1940, 1944, 1948, 1952, 1956, 1960, 1964, 1968, 1972, 1976, 1980, 1984, 1988, 1992, 1996, 2000, 2004, 2008, 2012, 2016

10. a. Key: $1 \mid 2 = 12$ songs

Stem	Leaf
0	2 9
1	
2	
3	1 4 5
4	3 4 5 7 8
5	4 6 9
6	6 7 7
7	0 4 7
8	2 9
9	0 6
10	3
11	1

b. Positively skewed with outliers

c. **Number of songs on favourite Spotify playlist**

Number of songs	Number of playlists
0-19	2
20-39	3
40-59	8
60-79	6
80-99	4
100-119	2

d. Numerical discrete, symmetrical distribution, mean $= 60$ (rounded to the nearest whole number) and standard deviation $= 27.3$

11. a. See figure at the foot of the page.*

b. Positively skewed

c. The median is $166\,755$, which is $167\,000$ to the nearest thousand.

d. The modal class is $100\,001$–$200\,000$

12. a. 5.6%

b. 2%

c. 2%

d. In order to give the impression that the pay rise is generous, the company would quote the mean.

e. The trade union would quote the median or the mode, because they would be arguing that the wage rise is too low.

f. The mode is probably the most 'honest' reflection of last year's wage rise, because it represents the situation of most of the workers.

13. a. 8, 20   b. 9, 19

14. Sample response: 3, 4, 5, 6, 7, 10, 10, 13, 14

*11. a.

2019 AFL player salaries

12 Quadratic equations and graphs

LEARNING SEQUENCE

12.1 Overview

Why learn this?

Galileo did a lot of important things in his lifetime. One of these was showing that objects launched into the air at an angle follow a parabolic path as they fall back to the ground. You can see this for yourself every time you throw a basketball, kick a football or hit a golf ball. As a ball moves through the air, gravity acts on it, causing it to follow a parabolic curved path. The parabolic curve that it follows can be described mathematically by quadratic equations. Every ball that is thrown, every arrow shot from a bow and every rocket launched into space has a quadratic equation that describes that object's position in space and also in time.

Quadratic equations also describe the parabolic shape of radio telescope dishes like the one at Parkes in New South Wales. The curved shape of these telescope dishes allows radio waves from space objects like quasars, galaxies and nebulae to be focused at a receiver to get a stronger signal. It was even used to relay transmissions between NASA and Apollo 11 during the first moon landing. On a much smaller scale, parabolic satellite dishes allow us to get clear internet and television. The bend in a banana, the loop of a rollercoaster, the curve of some suspension bridges or the path of a droplet of water from a fountain — these are all examples of the parabolas and quadratic equations that are all around us.

Where to get help

Go to your learnON title at **www.jacplus.com.au** to access the following digital resources. The Online Resources Summary at the end of this topic provides a full list of what's available to help you learn the concepts covered in this topic.

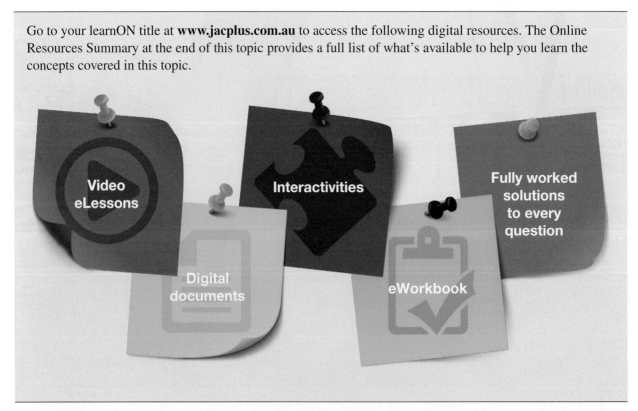

Exercise 12.1 Pre-test

Complete this pre-test in your learnON title at www.jacplus.com.au and receive **automatic marks**, **immediate corrective feedback** and **fully worked solutions**.

1. **MC** The expression $(a-4)^2 - 25$ factorises to:
   A. $(a-9)(a+1)$
   B. $(a+9)(a-1)$
   C. $(a-7)(a+3)$
   D. $(a-4)(a+4) - 25$
   E. $(a-4)(a+4) - 5$

2. State whether the following is True or False.
   A monic quadratic trinomial is an expression in the form $ax^2 + bx + c$, where a, b and c are constants.

3. Factorise $m^2 + 2m - 8$.

4. Factorise the equation $2w^2 - 10w + 12$ in its simplest form.

5. **MC** Select the solutions to the quadratic equation $-(x-3)^2 + 4 = 0$.
   A. $x = 3$ and $x = -4$
   B. $x = -3$ and $x = 4$
   C. $x = 3$ and $x = 4$
   D. $x = -5$ and $x = -1$
   E. $x = 5$ and $x = 1$

6. Identify the coordinate of the turning point for the graph shown.

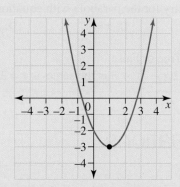

7. **MC** Identify the x-intercepts of the graph shown.
   A. 0 and 1
   B. 0 and 2
   C. 0 and 4
   D. 1 and 2
   E. 1 and 4

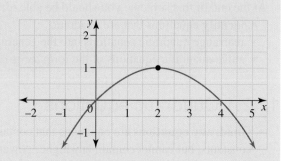

8. **MC** Select the equation that matches this table of values.

x	-3	-2	-1	0	1	2
y	-8	-3	0	1	0	-3

   A. $y = (1-x)^2$ B. $y = x^2 + x$ C. $y = x^2 - 1$ D. $y = 1 - x^2$ E. $y = x^2 + 1$

9. **MC** The graph of $y = 2x^2$ is:
   A. wider than $y = x^2$
   B. the same as $y = x^2$
   C. narrower than $y = x^2$
   D. a reflection of $y = x^2$ in the x-axis
   E. a reflection of $y = x^2$ in the y-axis.

10. **MC** Select the equation for a parabola that has a maximum turning point and is narrower than $y = x^2$.
    A. $y = 2x^2$ B. $y = \frac{1}{2}x^2$ C. $y = (x-2)^2 + 1$ D. $y = -2x^2$ E. $y = -\frac{1}{2}x^2$

11. **MC** Select the equation of a quadratic relation in the form $y = ax^2$ which passes through $(-2, 1)$.
    A. $y = 4x^2$ B. $y = 2x^2$ C. $y = -x^2$ D. $y = -2x^2$ E. $y = \frac{1}{4}x^2$

12. For the graph of the equation $y = -9 - x^2$, state the coordinate of the turning point.

13. **MC** The graph of $y = x^2$ has been translated 3 units to the right and 2 units down. The new equation is:
    A. $y = (x+3)^2 + 2$ B. $y = (x-3)^2 - 2$ C. $y = (x-2)^2 + 3$
    D. $y = (x+2)^2 - 3$ E. $y = (x+2)^2 + 3$

14. **MC** Identify the axis of symmetry for the parabola with equation $y = (x-4)(x+2)$.
    A. $x = -2$ B. $x = -1$ C. $x = 1$ D. $x = 4$ E. $y = -8$

15. State whether the following is True or False.
    The equation for the graph $y = 2(x-1)^2 + 3$ is the same as $y = 2x^2 - 4x + 5$.

12.2 Quadratic equations

LEARNING INTENTION

At the end of this subtopic you should be able to:
- identify equations that are quadratic
- rearrange quadratic equations into the general form $y = ax^2 + bx + c$
- solve quadratic equations of the form $ax^2 + c = 0$.

12.2.1 Quadratic equations

eles-4929

- **Quadratic equations** are equations that contain an x^2 term. They may also contain the variable x and constants, but cannot contain any higher powers of x such as x^3.
- Examples of quadratic equations include $x^2 + 2x - 7 = 0$, $2x^2 = 18$ and $x^2 = 5x$.
- The general form of a quadratic equation is $y = ax^2 + bx + c$.
- Any quadratic equation can be expressed in general form by rearranging and combining like terms.

Rearrange the following quadratic equations so that they are in general form and state the values of a, b and c.

a. $5x^2 - 2x + 3 = 2x^2 + 4x - 12$

b. $\dfrac{x^2}{2} - \dfrac{1}{6} = x\left(\dfrac{x}{3}\right) - 4$

c. $x(3 - 2x) = 4(x - 6)$

THINK	WRITE
a. 1. Write the equation.	a. $5x^2 - 2x + 3 = 2x^2 + 4x - 12$
2. Subtract $2x^2$ from both sides of the equation.	$3x^2 - 2x + 3 = 4x - 12$
3. Subtract $4x$ from both sides of the equation.	$3x - 6x + 3 = -12$
4. Add 12 to both sides of the equation. The equation is now in the general form $ax^2 + bx + c = 0$.	$3x^2 - 6x + 15 = 0$
5. Write the values of a, b and c.	$a = 3, b = -6, c = 15$
b. 1. Write the equation.	b. $\dfrac{x^2}{2} - \dfrac{1}{6} = x\left(\dfrac{x}{3}\right) - 4$
2. Expand the bracket. Lowest common denominator (LCD) = 6.	$\dfrac{x^2}{2} - \dfrac{1}{6} = \dfrac{x^2}{3} - 4$
3. Multiply through by 6.	$\dfrac{x^2}{2} \times \dfrac{6}{1} - \dfrac{1}{6} = \dfrac{x^2}{3} \times \dfrac{6}{1} - 4 \times 6$
4. Simplify each fraction.	$3x^2 - 1 = 2x^2 - 24$
5. Collect like terms on the left-hand side. The equation is now in the general form $ax^2 + bx + c = 0$.	$x^2 - 1 = -24$ $x^2 + 23 = 0$
6. Write the values of a, b and c.	$a = 1, b = 0, c = 23$
c. 1. Write the equation.	c. $x(3 - 2x) = 4(x - 6)$
2. Expand the brackets.	$3x - 2x^2 = 4x - 24$
3. To collect like terms on the left-hand side of the equation, subtract $4x$ from both sides.	$-2x^2 - x = -24$
4. Add 24 to both sides.	$-2x - x + 24 = 0$
5. Multiply all terms by -1 to make the x^2 term positive. The equation is now in the general form $ax^2 + bx + c = 0$.	$2x^2 + x - 24 = 0$
6. Write the values of a, b and c.	$a = 2, b = 1, c = -24$

▶ 12.2.2 Solving equations of the form $ax^2 + c = 0$

eles-4930

- Solving a quadratic equation means finding the values of x that satisfy the equation.
- Some quadratic equations have two solutions, some have only one solution, and some have no solutions.
- Quadratic equations of the form $ax^2 + c = 0$ can be solved using square roots. To solve such an equation, rearrange it so that the x^2 term is isolated on one side of the equation. Then take the square root of both sides.

WORKED EXAMPLE 2 Solving equations of the form $ax^2 + c = 0$

Solve the following equations.

a. $2x^2 - 18 = 0$ b. $x^2 + 9 = 0$ c. $3x^2 + 4 = 4$

THINK	WRITE
a. 1. Write the equation. The aim is to make x the subject.	**a.** $2x^2 - 18 = 0$
2. Add 18 to both sides.	$2x^2 = 18$
3. Divide both sides by 2.	$x^2 = 9$
4. Take the square root of both sides.	$x = 3, -3$
5. Write the solutions.	Two solutions: $x = 3$, $x = -3$
b. 1. Write the equation. The aim is to make x the subject.	**b.** $x^2 + 9 = 0$
2. Take 9 from both sides.	$x^2 = -9$
3. Take the square root of both sides.	-9 has no square root.
4. Write the solutions.	The equation has no solution.
c. 1. Write the equation. The aim is to make x the subject.	**c.** $3x^2 + 4 = 4$
2. Take 4 from both sides.	$3x^2 = 0$
3. Divide both sides by 3.	$x^2 = 0$
4. Take the square root of both sides. Zero has only one square root.	$x = 0$ One solution: $x = 0$
5. Write the solutions.	

TI	THINK	DISPLAY/WRITE	CASIO	THINK	DISPLAY/WRITE
a–c. On a Calculator page, press: • Menu • 3. Algebra • 1. Solve Complete the entry lines as: solve $(2x^2 - 18 = 0, x)$ solve $(x^2 + 9 = 0, x)$ solve $(3x^2 + 4 = 4, x)$ Press ENTER after each entry.	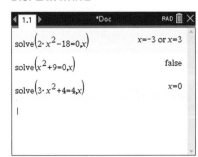	On a Main screen, complete the entry lines as: solve $(2x^2 - 18 = 0)$ solve $(x^2 + 9 = 0)$ solve $(3x^2 + 4 = 4)$ Press EXE after each entry.	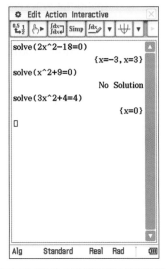

Confirming the solutions to quadratic equations

- Solutions to quadratic equations can be checked by substitution.
- Substitute the solution(s) into the equation, and if the RHS of the equation is equal to the LHS, the solutions are correct.

WORKED EXAMPLE 3 Confirming solutions to quadratic equations

Determine whether any of the following are solutions to the quadratic equation $x^2 = 4x - 3$.

a. $x = 1$ b. $x = 2$ c. $x = 3$

THINK	WRITE
a. 1. Substitute $x = 1$ into both sides of the equation.	a. $x^2 = 4x - 3$
2. Evaluate the left-hand side.	$\begin{aligned} \text{LHS} &= (1)^2 \\ &= 1 \end{aligned}$
3. Evaluate the right-hand side.	$\begin{aligned} \text{RHS} &= 4(1) - 3 \\ &= 4 - 3 \\ &= 1 \end{aligned}$
4. Since the left-hand side equals the right-hand side, $x = 1$ is a solution. Write the answer.	$x = 1$ is a solution to the equation.
b. 1. Substitute $x = 2$ into both sides of the equation.	b. $x^2 = 4x - 3$
2. Evaluate the left-hand side.	$\begin{aligned} \text{LHS} &= (2)^2 \\ &= 4 \end{aligned}$
3. Evaluate the right-hand side.	$\begin{aligned} \text{RHS} &= 4(2) - 3 \\ &= 8 - 3 \\ &= 5 \end{aligned}$
4. The left-hand side does not equal the right-hand side, so $x = 2$ is not a solution. Write the answer.	$x = 2$ is not a solution to the equation.
c. 1. Substitute $x = 3$ into both sides of the equation.	c. $x^2 = 4x - 3$
2. Evaluate the left-hand side.	$\begin{aligned} \text{LHS} &= (3)^2 \\ &= 9 \end{aligned}$
3. Evaluate the right-hand side.	$\begin{aligned} \text{RHS} &= 4(3) - 3 \\ &= 12 - 3 \\ &= 9 \end{aligned}$
4. The left-hand side equals the right-hand side, so $x = 3$ is a solution. Write the answer.	$x = 3$ is a solution to the equation.

---

on Resources

eWorkbook Topic 12 workbook (worksheets, code puzzle and a project) (ewbk-2012)

Interactivity Individual pathway interactivity: Quadratic equations (int-8349)

Exercise 12.2 Quadratic equations

Individual pathways

■ PRACTISE	■ CONSOLIDATE	■ MASTER
1, 5, 7, 11, 14	2, 3, 8, 10, 12, 15	4, 6, 9, 13, 16

To answer questions online and to receive **immediate corrective feedback** and **fully worked solutions** for all questions, go to your learnON title at www.jacplus.com.au.

Fluency

1. **WE1** Rearrange each of the following quadratic equations so that they are in the form $ax^2 + bx + c = 0$.

   a. $3x - x^2 + 1 = 5x$
   b. $5(x - 2) = x(4 - x)$
   c. $x(5 - 2x) = 6(5 - x^2)$

2. Rearrange each of the following quadratic equations so that they are in the form $ax^2 + bx + c = 0$.

   a. $4x^2 - 5x + 2 = 6x - 4x^2$
   b. $5(x - 12) = x(4 - 2x)$
   c. $x^2 - 4x - 16 = 1 - 4x$

3. **MC** Select which of the following is a quadratic equation.

   A. $2x - 1 = 0$
   B. $2^x - 1 = 0$
   C. $x^2 - x = 1 + x^2$
   D. $x^2 - x = 1 - x^2$
   E. $x^3 + 2x = 0$

4. **MC** Select which of the following is not a quadratic equation.

   A. $x(x - 1) = 2x - 1$
   B. $-3x^2 + 2x = 1$
   C. $3(x + 2) + 5(x + 3) = 2(x + 1)$
   D. $2(x - 1) + 3x = x(2x - 3)$
   E. $2x^3 + x = 2 - 2x^2$

5. **WE2** Solve the following quadratic equations.

   a. $x^2 - 16 = 0$
   b. $2x^2 + 18 = 0$
   c. $3x^2 + 2x = x(x + 2)$

6. Solve the following quadratic equations.

   a. $2(x^2 + 7) = 16$
   b. $x^2 + 17 = 13$
   c. $-3x^2 + 17 = 5$

Understanding

7. **WE3** Determine whether $x = 4$ is a solution to the following equations.

   a. $x^2 = x + 12$
   b. $x^2 = 3x + 1$
   c. $x^2 = 4x$

8. Determine whether $x = -3$ is a solution to the following equations.

   a. $x^2 = x + 12$
   b. $x^2 = 3x + 1$
   c. $x^2 = 4x$

9. Determine whether $x = 0$ is a solution to the following equations.

   a. $x^2 = x + 12$
   b. $x^2 = 3x + 1$
   c. $x^2 = 4x$

10. Determine whether the following are solutions to the equation $5(x - 1)^2 + 7 = 27$.

    a. $x = -1$
    b. $x = 1$
    c. $x = 0$

Reasoning

11. Is $(x + 4)^2$ equal to $x^2 + 16$? Explain using a numerical example.

12. a. Show that the shaded area can be represented by $A = x^2(4 - \pi)$.
    b. If the shaded area $= 10\,\text{cm}^2$, find the value of x correct to 3 significant figures.
    c. Why can't x be equal to -3.41?

13. Explain why $x^2 + 7x + 4 = 7x$ has no solutions.

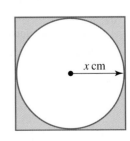

x cm

Problem solving

14. Five squares of increasing size and five area formulas are given below.

    a. Use factorisation to find the side length that correlates to each area formula.
    b. Using the area given and side lengths found, match the squares below with the appropriate algebraic expression for their area.

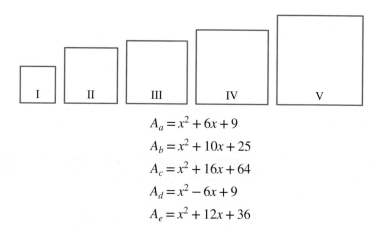

$$A_a = x^2 + 6x + 9$$
$$A_b = x^2 + 10x + 25$$
$$A_c = x^2 + 16x + 64$$
$$A_d = x^2 - 6x + 9$$
$$A_e = x^2 + 12x + 36$$

    c. If $x = 5$ cm, use the formula given to calculate the area of each square.

15. The sum of 5 and the square of a number is 41.

    a. Determine the number.
    b. Is there more than one possible number? If so, determine the other number and explain why there is more than one.

16. Two less than a number is squared and its result is tripled. The difference between this and 7 is 41.

    a. Write the algebraic equation.
    b. Determine both possible answers that satisfy the criteria.

12.3 The Null Factor Law

LEARNING INTENTION

At the end of this subtopic you should be able to:
- recall the Null Factor Law
- solve quadratic equations using the Null Factor Law.

12.3.1 Using the Null Factor Law to solve equations

eles-4931

- The **Null Factor Law** states that if the product of two or more factors is zero, then at least one of the factors must be zero.

> **The Null Factor Law**
>
> If $a \times b = 0$ then:
>
> $a = 0$ or $b = 0$ or both a and b equal 0.

- For example, if $(x-5)(x-2)=0$ then:

$$(x-5)=0 \text{ or } (x-2)=0 \text{ or both } x-5=x-2=0.$$
$$\text{If } x-5=0 \quad \text{If } x-2=0$$
$$x=5 \qquad x=2$$

Both $x=5$ and $x=2$ make the equation true.
- This can be checked by substituting the values into the original equation.
  For example:

If $x=5$, then:
$$(x-5)(x-2) = (5-5)(x-2)$$
$$= 0(x-2)$$
$$= 0$$

If $x=2$, then:
$$(x-5)(x-2) = (x-5)(2-2)$$
$$= (x-5)0$$
$$= 0$$

- *Note:* Quadratic equations can have a maximum of two solutions. This is known as the **Fundamental Theorem of Algebra**.

WORKED EXAMPLE 4 Using the Null Factor Law to solve quadratics

Solve each of the following quadratic equations.

a. $(x-2)(2x+1)=0$

b. $(4-3x)(6+11x)=0$

c. $x(x-3)=0$

d. $(x-1)^2=0$

THINK	WRITE
a. 1. The product of 2 factors is 0, so apply the Null Factor Law.	a. $(x-2)(2x+1)=0$
2. One of the factors must equal zero.	Either $x-2=0$ or $2x+1=0$
3. Solve the equations.	$x=0 \qquad 2x=-1$ $=-\dfrac{1}{2}$
4. Write the solutions.	$x=-\dfrac{1}{2}, x=2$
b. 1. The product of two factors is 0, so apply the Null Factor Law.	b. $(4-3x)(6+11x)=0$
2. One of the factors must equal zero.	Either $4-3x=0$ or $6+11x=0$
3. Solve the equations.	$-3x=-4 \qquad 11x=-6$ $x=\dfrac{4}{3} \qquad x=-\dfrac{6}{11}$
4. Write the solutions.	$x=\dfrac{4}{3}, -\dfrac{6}{11}$
c. 1. The product of two factors is 0, so apply the Null Factor Law.	c. $x(x-3)=0$
2. One of the factors must equal zero.	Either $x=0$ or $x-3=0$
3. Solve the equations.	$x=3$
4. Write the solutions.	$x=0, 3$

d. 1. Write the equation.

2. Rewrite the squared factors as the product of two factors.

3. Apply the Null Factor Law.

4. Solve the equations.
The two solutions are the same.

5. Write the solution.

d. $(x-1)^2 = 0$

$(x-1)(x-1) = 0$

Either $x - 1 = 0$ or $x - 1 = 0$
$x = 1$ $x = 1$

$x = 1$

TI | THINK

a-c On a Calculator page, press:
- Menu
- 3. Algebra
- 1. Solve

Complete the entry lines as:
solve
$((x-2)(2x+1) = 0, x)$
solve
$((4-3x)(6+11x) = 0, x)$
solve $((x-1)^2 = 0, x)$
Press ENTER after each entry.

DISPLAY/WRITE

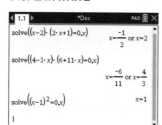

CASIO | THINK

On a Main screen, complete the entry lines as:
solve
$((x-2)(2x+1) = 0)$
solve
$((4-3x)(6+11x) = 0)$
solve $((x-1)^2 = 0)$
Press EXE after each entry.

DISPLAY/WRITE

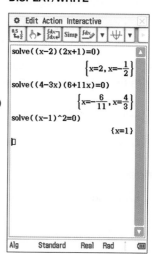

   Resources

eWorkbook	Topic 12 Workbook (worksheets, code puzzle and project) (ewbk-2012)
Digital document	SkillSHEET Solving linear equations (doc-10984)
Video eLesson	The Null Factor Law (eles-2312)
Interactivities	Individual pathway interactivity: The Null Factor Law (int-8350)
	The Null Factor Law (int-6095)

Exercise 12.3 The Null Factor Law

learn

Individual pathways

■ PRACTISE	■ CONSOLIDATE	■ MASTER
1, 6, 9, 12	2, 4, 7, 10, 13	3, 5, 8, 11, 14

To answer questions online and to receive **immediate corrective feedback** and **fully worked solutions** for all questions, go to your learnON title at www.jacplus.com.au.

Fluency

1. **WE4** Solve each of the following quadratic equations.

   a. $(x-2)(x+3) = 0$

   b. $(2x+4)(x-3) = 0$

   c. $(x+2)(x-3) = 0$

   d. $(2x+5)(4x+3) = 0$

   e. $(x+4)(2x+1) = 0$

2. Solve each of the following quadratic equations.

a. $(2x-1)(x+30)=0$ b. $(2x+1)(3-x)=0$ c. $(1-x)(3x-1)=0$

d. $x(x-2)=0$ e. $\left(x+\dfrac{1}{3}\right)\left(2x-\dfrac{1}{2}\right)=0$

3. Solve each of the following quadratic equations.

a. $(5x-1.5)(x+2.3)=0$ b. $\left(2x+\dfrac{1}{3}\right)\left(2x-\dfrac{1}{3}\right)=0$ c. $(x-2)^2=0$

d. $x(4x-15)=0$ e. $(x+4)^2=0$

4. The Null Factor Law can be extended to products of more than two factors. Use this to find all the solutions to the following equations.

a. $(x-2)(x+2)(x+3)=0$ b. $(x+2)(x+2)(2x-5)=0$
c. $(x+2)(x+2)(x+4)=0$ d. $x(x+2)(3x+12)=0$

5. The Null Factor Law can be extended to products of more than two factors. Use this to find all the solutions to the following equations.

a. $(2x-2.2)(x+2.4)(x+2.6)=0$ b. $(2x+6)\left(x+\dfrac{1}{2}\right)(9x-15)=0$

c. $3(x-3)^2=0$ d. $(x+1)(x-2)^2=0$

Understanding

6. **MC** Select the solutions to $(2x-4)(x+7)=0$.

A. $x=4$, $x=7$ B. $x=4, x=-7$ C. $x=2$, $x=7$
D. $x=2, x=-7$ E. $x=-4, x=7$

7. **MC** The Null Factor Law cannot be applied to the equation $x\left(x+\dfrac{1}{2}\right)\left(\dfrac{x}{2}-1\right)=1$ because:

A. there are more than two factors
B. the right-hand side equals 1
C. the first factor is a simple x-term
D. the third term has x in a fraction
E. there are three terms on the left-hand side

8. Rewrite the following equations so that the Null Factor Law can be used. Then solve the resulting equation.

a. $x^2+10x=0$ b. $2x^2-14x=0$ c. $25x^2-40x=0$

Reasoning

9. Explain why $x^2-x-6=0$ has the same solution as $-2x^2+2x+12=0$.

10. a. Explain why $x=2$ is not a solution of the equation $(2x-2)(x+2)=0$.
    b. Determine the solutions to the equation.

11. Explain what is the maximum number of solutions a quadratic such as $x^2-x-56=0$ can have.

Problem solving

12. A bridge is constructed with a supporting structure in the shape of a parabola, as shown in the diagram. The origin $(0, 0)$ is at the left-hand edge of the bridge, which is 100 m long.

    a. Identify the maximum height of the bridge support.
    b. If the equation of the support is $y = ax(b - x)$, determine the values of a and b. (*Hint:* Let $y = 0$.)
    c. Calculate the height of the support when $x = 62$.

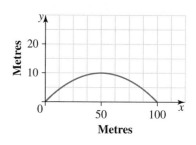

13. Consider a ball thrown upwards so that it reaches a height h of metres after t seconds. The expression $20t - 4t^2$ represents the height of the ball, in metres, after t seconds.

$h = 20t - 4t^2$

    a. Factorise the expression for height.
    b. Calculate the height of the ball after:
       i. 1 second
       ii. 5 seconds.
    c. Explain if factorising the expression made it easier to evaluate.

14. The product of 7 more than a number and 14 more than that same number is 330. Determine the possible values of the number.

12.4 Solving quadratic equations with two terms

LEARNING INTENTION

At the end of this subtopic you should be able to:
- solve quadratic equations with two terms using the Null Factor Law.

12.4.1 Solving quadratic equations of the form $ax^2 + c = 0$

eles-4932

- In section 12.2.2, we saw how to solve equations in the form $ax^2 + c = 0$ using algebra. In most cases, we can use the Null Factor Law to solve them instead.
- To use the Null Factor Law the quadratic needs to be factorised and the RHS of the equation must be 0.

WORKED EXAMPLE 5 Using the Null Factor Law to solve $ax^2 + c = 0$

Solve each of the following quadratic equations.

a. $x^2 - 1 = 0$

b. $2x^2 - 18 = 0$

THINK	WRITE
a. 1. Write the equation: $x^2 - 1$ is the difference of two squares.	a. $x^2 - 1 = 0$
2. Factorise the left-hand side using the difference of two squares rule.	$(x + 1)(x - 1) = 0$
3. Apply the Null Factor Law.	$x + 1 = 0 \qquad x - 1 = 0$ $x = -1 \qquad x = 1$
4. State the solutions.	$x = 1, x = -1$ (This can be abbreviated to $x = \pm 1$.)
b. 1. Write the equation.	b. $2x^2 - 18 = 0$
2. Take out the common factor.	$2(x^2 - 9) = 0$
3. Divide both sides of the equation by 2.	$x^2 - 9 = 0$
4. Factorise the left-hand side using the difference of two squares rule.	$(x + 3)(x - 3) = 0$
5. Apply the Null Factor Law.	$x + 3 = 0 \qquad x - 3 = 0$ $x = -3 \qquad x = 3$
6. State the solutions.	$x = \pm 3$

▶ 12.4.2 Solving quadratic equations of the form $ax^2 + bx = 0$

eles-4933

- The equation $ax^2 + bx = 0$ can be solved easily using the Null Factor Law.
- To solve an equation of this form, factorise the LHS to $x(ax + b) = 0$ and then apply the Null Factor Law.
- Since both terms on the left-hand side involve the variable x, one solution will always be $x = 0$.

WORKED EXAMPLE 6 Solving equations in the form $ax^2 + bx = 0$

Solve each of the following equations.

a. $x^2 + 4x = 0$

b. $-2x^2 - 4x = 0$

THINK	WRITE
a. 1. Write the equation.	a. $x^2 + 4x = 0$
2. Factorise by taking out a common factor of x.	$x(x + 4) = 0$
3. Apply the Null Factor Law.	$x = 0 \qquad x + 4 = 0$ $x = 0 \qquad x = -4$
4. Write the solutions.	$x = -4, \qquad x = 0$
b. 1. Write the equation.	b. $-2x^2 - 4x = 0$
2. Factorise by taking out the common factor of $-2x$.	$-2x(x + 2) = 0$

3. Apply the Null Factor Law.

$$-2x = 0 \qquad x + 2 = 0$$
$$x = 0 \qquad x = -2$$

4. Write the solutions.

$$x = -2, \qquad x = 0$$

WORKED EXAMPLE 7 Solving word problems using Null Factor Law

If the square of a number is multiplied by 5, the answer is 45. Calculate the number.

THINK	WRITE
1. Define the number.	Let x be the number.
2. Write an equation that can be used to find the number.	$5x^2 = 45$
3. Transpose to make the right-hand side equal to zero. Solve the equation.	$5x^2 - 45 = 0$ $5(x^2 - 9) = 0$ $x^2 - 9 = 0$ $x^2 - 3^2 = 0$ $(x + 3)(x - 3) = 0$ $x + 3 = 0 \quad$ or $\quad x - 3 = 0$ $x = -3 \quad$ or $\quad x = 3$
4. Write the answer in a sentence.	The number is either 3 or -3.

 Resources

Exercise 12.4 Solving quadratic equations with two terms   **learn** on

Individual pathways

■ PRACTISE	■ CONSOLIDATE	■ MASTER
1, 4, 7, 9, 12, 16	2, 5, 8, 10, 13, 17	3, 6, 11, 14, 15, 18

To answer questions online and to receive **immediate corrective feedback** and **fully worked solutions** for all questions, go to your learnON title at www.jacplus.com.au.

Fluency

1. **WE5** Solve each of the following quadratic equations using the Null Factor Law.

   a. $x^2 - 9 = 0$ b. $x^2 - 16 = 0$ c. $2x^2 - 18 = 0$
   d. $2x^2 - 50 = 0$ e. $100 - x^2 = 0$

2. Solve each of the following quadratic equations using the Null Factor Law.

   a. $49 - x^2 = 0$ b. $3x^2 - 27 = 0$ c. $5x^2 - 20 = 0$
   d. $x^2 + 6 = 0$ e. $2x^2 + 18 = 0$

3. Solve each of the following quadratic equations using the Null Factor Law.

   a. $-x^2 + 9 = 0$ b. $-3x^2 + 48 = 0$ c. $-4x^2 + 100 = 0$
   d. $x^2 = 0$ e. $-x^2 = 0$

4. **WE6** Solve each of the following equations.

   a. $x^2 + 6x = 0$ b. $x^2 - 8x = 0$ c. $x^2 + 9x = 0$
   d. $x^2 - 11x = 0$ e. $2x^2 - 12x = 0$

5. Solve each of the following equations.

   a. $2x^2 - 15x = 0$ b. $3x^2 - 2x = 0$ c. $4x^2 + 7x = 0$
   d. $2x^2 - 5x = 0$ e. $x^2 + x = 0$

6. Solve each of the following equations.

   a. $4x^2 - x = 0$ b. $-x^2 - 5x = 0$ c. $-2x^2 - 24x = 0$
   d. $-x^2 + 18x = 0$ e. $x^2 - 2.5x = 0$

7. **MC** Select the solutions to $4x^2 - 36 = 0$.

   A. $x = 3$ and $x = -3$ B. $x = 9$ and $x = -9$ C. $x = 1$ and $x = -1$
   D. $x = 2$ and $x = -2$ E. $x = 6$ and $x = -6$

8. **MC** Select the solutions to $x^2 - 5x = 0$.

   A. $x = 1$ and $x = 5$ B. $x = 0$ and $x = -5$ C. $x = 0$ and $x = 5$
   D. $x = -1$ and $x = 5$ E. $x = -5$ and $x = 5$

Understanding

9. **WE7** If the square of a number is multiplied by 2, the answer is 32. Calculate the number.

10. A garden has two vegetable plots. One plot is a square; the other plot is a rectangle with one side 3 m shorter than the side of the square and the other side 4 m longer than the side of the square. Both plots have the same area. Sketch a diagram and determine the dimensions of each plot.

11. The square of a number is equal to 10 times the same number. Calculate the number.

Reasoning

12. Explain why the equation $x^2 + 9 = 0$ cannot be solved using the Null Factor Law.

13. Solve the equation $m^2x^2 - n^2 = 0$ using the Null Factor Law.

14. Explain why the equation $x^2 + bx = 0$ can only be solved using the Null Factor Law.

15. A quadratic expression may be written as $ax^2 + bx + c$. Using examples, explain why b and c may take any values, but a cannot equal zero.

Problem solving

16. The square of a number which is then tripled, is equal to 24 times the same number. Calculate the number.

17. The sum of 6 times the square of a number and 72 times the same number is equal to zero. Calculate the number.

18. A rectangular horse paddock has a width of x and a length that is 30 metres longer than its width. The area of the paddock is $50x$ square metres.

   a. Write the algebraic equation for the area described.
   b. Factorise the equation.
   c. Solve the equation.
   d. Can you use all your solutions? Explain why or why not.

12.5 Factorising and solving monic quadratic equations

LEARNING INTENTION

At the end of this subtopic you should be able to:
- factorise monic quadratic equations
- solve monic quadratic equations using factorisation and the Null Factor Law.

▶ 12.5.1 Quadratic trinomials

eles-4934

- A **quadratic trinomial** is an expression of the form $ax^2 + bx + c$, where a, b and c are constants.
  - For the time being we will look only at quadratic trinomials for which $a = 1$; that is, trinomials of the form $x^2 + bx + c$.
- The area model of binomial expansion can be reversed to find a pattern for factorising a general quadratic expression.

  For example:

$$(x+f)(x+h) = x^2 + fx + hx + fh$$
$$= x^2 + (f+h)x + fh$$

$$(x+4)(x+3) = x^2 + 4x + 3x + 12$$
$$= x^2 + 7x + 12$$

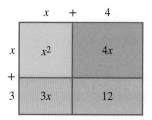

Factorising monic quadratics

- To factorise a general quadratic, look for factors of c that sum to b.

$$x^2 + bx + c = (x+f)\,(x+h)$$

Factors of c that sum to b

For example, $x^2 + 7x + 12 = (x+3)(x+4)$.
$$3 + 4 = 7$$
$$3 \times 4 = 12$$

WORKED EXAMPLE 8 Factorising quadratic expressions

Factorise the following quadratic expressions.

a. $x^2 + 5x + 6$

b. $x^2 + 10x + 24$

THINK		WRITE

a. 1. The general quadratic expression has the pattern $x^2 + 5x + 6 = (x + f)(x + h)$. f and h are a factor pair of 6 that add to 5.
Calculate the sums of factor pairs of 6. The factors of 6 that add to 5 are 2 and 3, as shown in blue.

a.

Factors of 6	Sum of factors
1 and 6	7
2 and 3	5

2. Substitute the values of f and h into the expression in its factorised form.

$(x^2 + 5x + 6) = (x + 2)(x + 3)$

b. 1. The general quadratic expression has the pattern $x^2 + 10x + 24 = (x + f)(x + h)$, where f and h are a factor pair of 24 that add to 10.
Calculate the sums of factor pairs of 24. The factors of 24 that add to 10 are 4 and 6, as shown in blue.

b.

Factors of 24	Sum of factors
1 and 24	25
2 and 12	14
3 and 8	11
4 and 6	10

2. Substitute the values of f and h into the expression in its factorised form.

$x^2 + 10x + 24 = (x + 4)(x + 6)$

WORKED EXAMPLE 9 Factorising quadratic expressions with negative terms

Factorise the following expressions.

a. $x^2 - 9x + 18$

b. $x^2 + 6x - 16$

THINK		WRITE

a. 1. The general quadratic expression has the pattern $x^2 - 9x + 18 = (x + f)(x + h)$, where f and h are a factor pair of 18 that add to -9.
Calculate the sums of factor pairs of 18. As shown in blue, -3 and -6 are factors of 18 that add to -9.

a.

Factors of 18	Sum of factors
1 and 18	19
-1 and -18	-19
2 and 9	11
-2 and -9	-11
3 and 6	9
-3 and -6	-9

2. Substitute the values of f and h into the expression in its factorised form.

$x^2 - 9x + 18 = (x - 3)(x - 6)$

b. 1. The general quadratic expression has the pattern $x^2 + 6x - 16 = (x + f)(x + h)$, where f and h are a factor pair of -16 that add to 6.
Calculate the sums of factor pairs of -16.
As shown in blue, -2 and 8 are factors of -16 that add to 6.

b.

Factors of -16	Sum of factors
1 and -16	-15
-1 and 16	15
2 and -8	-6
-2 and 8	6

2. Substitute the values of f and h into the expression in its factorised form.

$x^2 - 6x + 16 = (x - 2)(x + 8)$

12.5.2 Solving quadratic equations of the form $x^2 + bx + c = 0$

eles-4935

- If the expression $x^2 + bx + c$ can be factorised, then the quadratic equation $x^2 + bx + c = 0$ can be solved using the Null Factor Law.

WORKED EXAMPLE 10 Solving monic quadratic equations

Solve the following quadratic equations.

a. $x^2 - 5x - 6 = 0$

b. $x^2 + 14x = 15$

THINK	WRITE
a. 1. Write the equation.	a. $x^2 - 5x - 6 = 0$
2. Factorise $x^2 - 5x - 6$.	$(x - 6)(x + 1) = 0$
3. Apply the Null Factor Law.	$x - 6 = 0 \quad x + 1 = 0$
4. Solve the equations.	$x = 6 \qquad x = -1$
5. Write the solutions.	$x = -1, \quad x = 6$
b. 1. Write the equation and transpose so that the right-hand side equals 0.	b. $x^2 + 14x = 15$ $x^2 + 14x - 15 = 0$
2. Factorise $x^2 - 14x - 15$.	$(x + 15)(x - 1) = 0$
3. Apply the Null Factor Law.	$x + 15 = 0 \quad x - 1 = 0$ $x = -15 \qquad x = 1$
4. Solve the linear equations and write the solutions.	$x = -15, \quad x = 1$

TI | THINK	DISPLAY/WRITE	CASIO | THINK	DISPLAY/WRITE
a–b In a new problem on a Calculator page, complete the entry lines as: solve $(x^2 - 5x - 6 = 0, x)$ solve $(x^2 + 14x = 15, x)$ Press ENTER after each entry. Press ENTER after each entry.	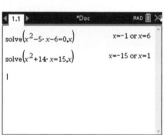 a. $x^2 - 5x - 6 = 0$ $\Rightarrow x = -1, x = 6$ b. $x^2 + 14x = 15$ $\Rightarrow x = -15, x = 1$	On the Main screen, complete the entry lines as: solve $(x^2 - 5x - 6 = 0)$ solve $(x^2 + 14x = 15)$ Press EXE after each entry.	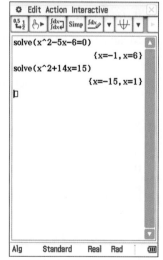 a. $x^2 - 5x - 6 = 0$ $\Rightarrow x = -1, x = 6$ b. $x^2 + 14x = 15$ $\Rightarrow x = -15, x = 1$

Exercise 12.5 Factorising and solving monic quadratic equations

learn on

Individual pathways

■ PRACTISE	■ CONSOLIDATE	■ MASTER
1, 4, 7, 10, 13, 18, 22, 25	2, 5, 8, 11, 14, 16, 19, 20, 23, 26	3, 6, 9, 12, 15, 17, 21, 24, 27, 28

To answer questions online and to receive **immediate corrective feedback** and **fully worked solutions** for all questions, go to your learnON title at www.jacplus.com.au.

Fluency

1. Expand each of the following.

   a. $(x+4)(x+2)$
   b. $(x-2)(x-4)$
   c. $(x-4)(x-5)$
   d. $(x+4)(x+5)$
   e. $(m+1)(m+5)$
   f. $(m-1)(m-5)$

2. Expand each of the following.

   a. $(t+8)(t+11)$
   b. $(t-10)(t-20)$
   c. $(x+2)(x-3)$
   d. $(x+3)(x-2)$
   e. $(v+5)(v-8)$
   f. $(v-5)(v+8)$

3. Expand each of the following.

   a. $(x+7)(x-2)$
   b. $(t+1)(t-12)$
   c. $(n+15)(n-2)$
   d. $(a+3)(a-6)$
   e. $(z+4)(z+4)$
   f. $(z+11)(z-8)$

4. Make a complete systematic list of pairs of positive integers that add up to:

   a. 5
   b. 6
   c. 12
   d. 20.

5. Make a complete systematic list of pairs of negative integers that add up to:

   a. -5
   b. -7
   c. -8
   d. -11.

6. Make a systematic list of 5 pairs of integers with opposite signs whose sum is:

   a. 6
   b. -6
   c. 3
   d. -3
   e. -8
   f. 14.

7. **WE8** Factorise each of the following.

   a. x^2+4x+3
   b. x^2-4x+3
   c. $x^2+12x+11$
   d. $x^2-12x+11$
   e. a^2+6a+5

8. Factorise each of the following.

   a. a^2-6a+5
   b. $x^2-7x+12$
   c. $x^2-7x+10$
   d. $n^2+8n+16$
   e. $n^2+10n+16$

9. Factorise each of the following.

   a. $y^2-12y+27$
   b. $x^2-13x+42$
   c. $t^2-8t+12$
   d. $t^2+11t+18$
   e. u^2+5u+6

10. **WE9** Factorise each of the following.

   a. $x^2 + 3x - 18$
   b. $x^2 - 3x - 18$
   c. $x^2 - 2x - 15$
   d. $x^2 + 2x - 15$
   e. $n^2 - 13n - 14$

11. Factorise each of the following.

   a. $n^2 + 2n - 35$
   b. $v^2 + 5v - 6$
   c. $v^2 - 5v - 6$
   d. $t^2 + 4t - 12$
   e. $t^2 - 5t - 14$

12. Factorise each of the following.

   a. $x^2 - x - 20$
   b. $x^2 + x - 20$
   c. $n^2 + n - 90$
   d. $n^2 - 3n - 70$
   e. $x^2 - 4x - 5$

13. **WE10** Solve each of the following quadratic equations.

   a. $x^2 - 6x + 8 = 0$
   b. $x^2 + 6x + 8 = 0$
   c. $x^2 + 6x + 5 = 0$
   d. $x^2 + x - 6 = 0$
   e. $x^2 + 2x - 15 = 0$

14. Solve each of the following quadratic equations.

   a. $x^2 + 4x + 4 = 0$
   b. $x^2 + 2x - 24 = 0$
   c. $x^2 - 5x - 24 = 0$
   d. $x^2 - x - 12 = 0$
   e. $x^2 + 13x + 12 = 0$

15. Solve each of the following quadratic equations.

   a. $x^2 - 10x = 11$
   b. $x^2 + x = 20$
   c. $x^2 + 29x = -100$
   d. $x^2 - 15x = -50$
   e. $0 = x^2 - 2x - 8$

16. **MC** Select the solutions to the quadratic equation $x^2 + 2x - 8 = 0$.

   A. $x = 4$ and $x = -2$
   B. $x = -4$ and $x = 2$
   C. $x = -4$ and $x = 4$
   D. $x = -2$ and $x = 2$
   E. $x = -4$ and $x = -2$

17. **MC** Select the solutions to the quadratic equation $x^2 - 7x - 8 = 0$.

   A. $x = 1$ and $x = 8$
   B. $x = -1$ and $x = -8$
   C. $x = -8$ and $x = 1$
   D. $x = -1$ and $x = 8$
   E. $x = -1$ and $x = 1$

Understanding

18. Consider the quadratic trinomial $x^2 + 7x + c$ where c is a positive integer.

   a. Factorise the expression if $c = 6$.
   b. Determine what other positive whole number values c can take if the expression is to be factorised.
      Factorise the expression for each of these values.

19. Consider the quadratic trinomial $x^2 + 3x + c$ where c is a negative integer.

   a. Factorise the expression if $c = -4$.
   b. Determine three more values of c for which the expression can be factorised, and factorise each one.

20. **MC** When factorised, $x^2 - 3x - 18$ is equal to:

   A. $(x - 3)(x - 6)$
   B. $(x - 3)(x + 6)$
   C. $(x + 3)(x + 6)$
   D. $(x + 3)(x - 6)$
   E. $(x + 2)(x - 9)$.

21. The rectangle shown has an area of 45 cm^2. By solving a quadratic equation, determine the dimensions of the rectangle.

Reasoning

22. Show that $x^2 + 8x + 10$ has no factors, if only whole numbers can be used.

23. Show that $x^2 + 12x + 24$ has no factors, if only whole numbers can be used.

24. The number of diagonals in a polygon is given by the formula $D = \frac{n}{2}(n-3)$, where n is the number of sides in the polygon.

    a. Determine how many diagonals there are in:
        i. a triangle
        ii. a square
        iii. a decagon.
    b. Determine what type of polygon has:
        i. 20 diagonals
        ii. 170 diagonals.

Problem solving

25. a. Write algebraic expressions for each of the parts below. Your friend picks a number, n.
        i. He adds 1 to it.
        ii. He adds 7 to it.
        iii. He multiplies these two new numbers together.
        iv. He then divides the product in part iii by the product of 2 more than the original number, n, and 1 more than the original number, n.
    b. The result of the division is 2. What number, n, did he begin with?

26. Rectangular floor mats have an area of $x^2 + 2x - 15$.

    a. If the length of the mat is $(x + 5)$ cm, develop an expression for the width.
    b. If the length of the mat is 70 cm, calculate the width.
    c. If the width of the mat is 1 m, calculate the length.

27. A girl, her brother and her teacher are shown below. The product of the teacher's and brother's ages is 256.

    a. Write an algebraic expression in terms of g to represent the teacher's age.
    b. Use the algebraic expressions to write an expression for the product of the brother's and teacher's ages and expand this product.
    c. Write an equation using your answer to part b for the product of their ages. *Hint:* An equation has an 'is equal to' part to it.
    d. Rearrange your answer to part c by collecting like trms and making it equal to zero.
    e. Determine the value of g by solving the equation in part d using the Null Factor Law.
        Note: Discard any value that does not make sense.
    f. State the ages of the girl, her brother and the teacher.

28. A particular rectangle has an area of $120\,\text{m}^2$.

   a. Generate a list of possible whole number dimensions for the rectangle.

   b. The area of the rectangle can also be expressed as $A = x^2 + 2x - 48$. Determine algebraic expressions for the length and the width of the rectangle in terms of x.

   c. Use your list of possible dimensions and your algebraic expressions for width and length to determine the dimensions of the rectangle. *Hint:* The value of x in the length and width must be the same.

12.6 Graphs of quadratic functions

12.6.1 Graphs of quadratic functions

eles-4936

- When quadratic functions are graphed, they produce curved lines called parabolas.
- The graph shown is a typical parabola with features as listed below.

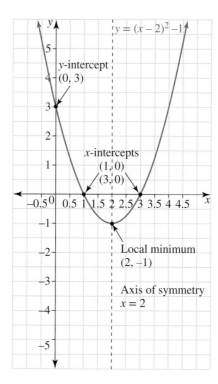

- The dotted line is the **axis of symmetry**; the parabola is the same on either side of this line.
- The **turning point** is the point where the graph changes direction. It is located on the axis of symmetry.
- The turning point is a local minimum if it is the lowest point on the graph, and a local maximum if it is the highest point on the graph.
- The x-intercept(s) is where the graph crosses (or sometimes just touches) the x-axis. Not all parabolas have x-intercepts.
- The y-intercept is where the graph crosses the y-axis. All parabolas have one y-intercept.

For each of the following graphs:
  i. state the equation of the axis of symmetry
  ii. state the coordinates of the turning point
  iii. indicate whether it is a maximum or a minimum turning point.

a.

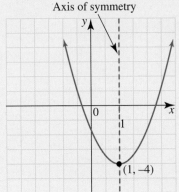

b.

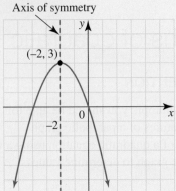

THINK

a.
  i. State the equation of the vertical line that cuts the parabola in half.

  ii. State the turning point.

  iii. Determine the nature of the turning point by observing whether it is the highest or lowest point of the graph.

b.
  i. State the equation of the vertical line that cuts the parabola in half.

  ii. State the turning point.

  iii. Determine the nature of the turning point by observing whether it is the highest or lowest point of the graph.

WRITE

a.
  i. Axis of symmetry is $x = 1$.

  ii. The turning point is at $(1, -4)$.

  iii. Minimum turning point

b.
  i. Axis of symmetry is $x = -2$.

  ii. The turning point is at $(-2, 3)$.

  iii. Maximum turning point

The x- and y-intercepts

- The **x-intercept** is where the graph crosses (or just touches) the x-axis.
- The **y-intercept** is where the graph crosses the y-axis. All parabolas have one y-intercept.
- When sketching a parabola, the x-intercepts (if any) and the y-intercept should always be marked on the graph, with their respective coordinates.

For each of the following graphs, state:
 i. the equation of the axis of symmetry
 ii. the coordinates of the turning point and whether the point is a maximum or a minimum
 iii. the x- and y-intercepts.

a.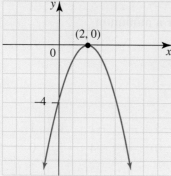

b.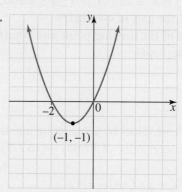

THINK

a. i. State the equation of the vertical line that cuts the parabola in half.

   ii. State the turning point and its nature; that is, determine whether it is the highest or lowest point of the graph.

   iii. Observe where the parabola crosses the x-axis. In this case, the graph touches the x-axis when $x = 2$, so there is only one x-intercept.

   Observe where the parabola crosses the y-axis.

b. i. State the equation of the vertical line that cuts the parabola in half.

   ii. State the turning point and its nature; that is, determine whether it is the highest or lowest point of the graph.

   iii. Observe where the parabola crosses the x-axis.

   Observe where the parabola crosses the y-axis.

WRITE

a. i. Axis of symmetry is $x = 2$.

   ii. Maximum turning point is at $(2, 0)$.

   iii. The x-intercept is 2. It occurs at the point $(2, 0)$.

   The y-intercept is -4. It occurs at the point $(0, -4)$.

b. i. Axis of symmetry is $x = -1$.

   ii. Minimum turning point is at $(-1, -1)$.

   iii. The x-intercepts are -2 and 0. They occur at the points $(-2, 0)$ and $(0, 0)$.

   The y-intercept is 0. It occurs at the point $(0, 0)$.

▶ 12.6.2 Plotting points to graph quadratic functions

eles-4937

- If there is a rule connecting y and x, a table of values can be used to determine actual coordinates.
- When drawing straight line graphs, a minimum of two points is required. For parabolas there is *no minimum* number of points, but between 6 and 12 points is a reasonable number.
- The more points used, the 'smoother' the parabola will appear. The points should be joined with a smooth curve, not ruled.
- Ensure that points plotted include (or are near) the main features of the parabola, namely the axis of symmetry, the turning point and the x- and y-intercepts.

- Occasionally a list of x-values will be provided and the corresponding y-values can be calculated. In the following example, the set of x-values is specified as $-4 \le x \le 2$.

WORKED EXAMPLE 13 Plotting points to graph quadratic functions

Plot the graph of $y = x^2 + 2x - 3$, $-4 \le x \le 2$ and, hence, state:
a. **the equation of the axis of symmetry**
b. **the coordinates of the turning point and whether it is a maximum or a minimum**
c. **the x- and y-intercepts.**

THINK

WRITE/DRAW

1. Write the equation.

$y = x^2 + 2x - 3$

2. Complete a table of values by substituting into the equation each integer value of x from -4 to 2. For example, when $x = -4$, $y = (-4)^2 + 2 \times (-4) - 3 = 5$.

x	-4	-3	-2	-1	0	1	2
y	5	0	-3	-4	-3	0	5

3. List the coordinates of the points.

$(-4, 5), (-3, 0), (-2, -3), (-1, -4), (0, -3),$
$(1, 0), (2, 5)$

4. Draw and label a set of axes, plot the points listed and join the points to form a smooth curve.

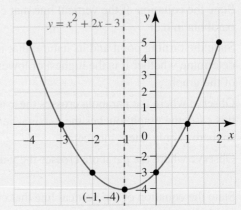

a. Find the equation of the line that divides the parabola exactly into two halves.

a. Axis of symmetry is $x = -1$.

b. Find the point where the graph turns or changes direction, and decide whether it is the highest or lowest point of the graph. State the coordinates of this point.

b. Minimum turning point is at $(-1, -4)$.

c. 1. State the x-coordinates of the points where the graph crosses the x-axis.

c. The x-intercepts are at -3 and 1. They occur at the points $(-3, 0)$ and $(1, 0)$.

2. State the y-coordinate of the point where the graph crosses the y-axis.

The y-intercept is at -3. It occurs at the point $(0, -3)$.

TI	THINK	DISPLAY/WRITE	CASIO	THINK	DISPLAY/WRITE

TI | THINK

a In a new problem, on a Graphs page, complete the function entry line as:
$f1(x) = x^2 + 2x - 3|$
$-4 \leq x \leq 2$
Then press ENTER.

DISPLAY/WRITE

a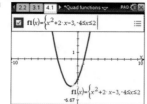

CASIO | THINK

a 1 On the Graph & Table screen, complete the function entry line as:
$y1 = x^2 + 2x - 3|$
$-4 \leq x \leq 2$
Then press EXE. Press the icon.

DISPLAY/WRITE

a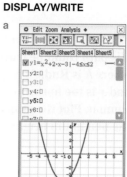

2 To determine the equation of the axis of symmetry, we need to know the x-coordinate of the turning point.
Press the icon.

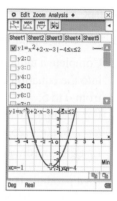

The axis of symmetry is $x = -1$

b To determine the equation of the axis of symmetry, we need to know the x-coordinate of the turning point.
Press:
• MENU
• 6: Analyze Graph
• 2: Minimum.
Press ENTER to fix the lower bound, then drag the vertical line to the right to locate an upper bound and press ENTER. State the coordinates of the turning point and whether it is a maximum or a minimum.

b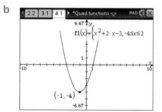

The axis of symmetry is $x = -1$.
The turning point occurs at $(-1, -4)$ and it is a mimimum point.

b State the coordinates of the turning point and whether it is a maximum or a minimum.

b The turning point occurs at $(-1, -4)$ and it is a minimum point.

• A rule connecting x and y will be occasionally provided. From this rule, pairs of x- and y-values can be calculated. In the following example, the rule is given as $h = -5x^2 + 25x$.
• Graphs can be drawn using a graphics calculator, a graphing program on your computer or by hand.

WORKED EXAMPLE 14 Using graphs of quadratic functions

Rudie, the cannonball chicken, was fired out of a cannon. His path could be traced by the equation $h = -5x^2 + 25x$, where h is Rudie's height, in metres, above the ground and x is the horizontal distance, in metres, from the cannon. Plot the graph for $0 \le x \le 5$ and use it to find the maximum height of Rudie's path.

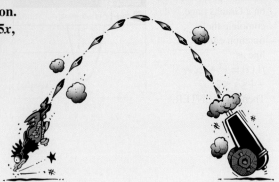

THINK	WRITE
1. Write the equation.	$h = -5x^2 + 25x$
2. Complete a table of values by substituting into the equation each integer value of x from 0 to 5. For example, when $x = 0$, $h = -5 \times 0^2 + 25 \times 0 = 0$.	
3. List the coordinates of the points.	$(0, 0), (1, 20), (2, 30), (3, 30), (4, 20), (5, 0)$
4. As a parabola is symmetrical, the greatest value of h must be greater than 30 and occurs when x lies between 2 and 3, so find the value of h when $x = 2.5$.	When $x = 2.5, h = 5 \times (2.5)^2 + 25 \times 2.5$ $= 31.25$
5. Draw and label a set of axes, plot the points from the table and join the points to form a smooth curve.	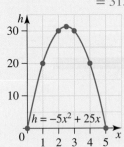
6. The maximum height is the value of h at the highest point of the graph.	$h = 31.25$
7. Answer the question in a sentence.	The maximum height of Rudie's path is 31.25 metres.

Table of values:

x	1	2	3	4	5
h	0	20	30	30	20

---

on Resources

Exercise 12.6 Graphs of quadratic functions

learn on

Individual pathways

■ PRACTISE	■ CONSOLIDATE	■ MASTER
1, 4, 5, 9, 10, 11, 16, 19	2, 6, 8, 12, 14, 17, 20	3, 7, 13, 15, 18, 21, 22

To answer questions online and to receive **immediate corrective feedback** and **fully worked solutions** for all questions, go to your learnON title at www.jacplus.com.au.

Fluency

1. **WE11** For each of the graphs below:

   i. state the equation of the axis of symmetry
   ii. give the coordinates of the turning point
   iii. indicate whether it is a minimum or maximum turning point.

   a. 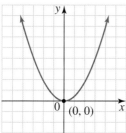 b. 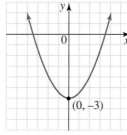 c.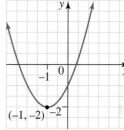

   d. 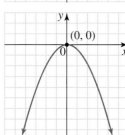 e. 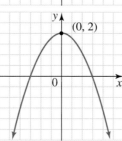 f.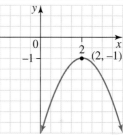

2. For each of the following graphs, state:

   i. the equation of the axis of symmetry
   ii. the coordinates of the turning point
   iii. whether the turning point is a maximum or minimum.

   a. 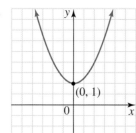 b. 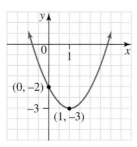 c.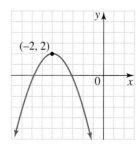

   d. 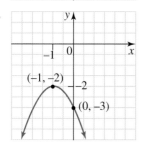 e. 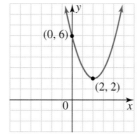 f.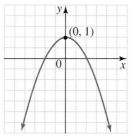

3. **WE12** For each of the following graphs, state:

   i. the equation of the axis of symmetry
   ii. the coordinates of the turning point and whether the point is a maximum or a minimum
   iii. the x- and y-intercepts.

   a.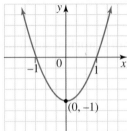

   b.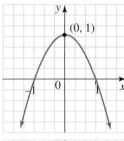

   c.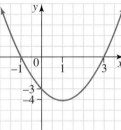

   d.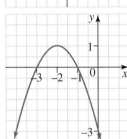

   e.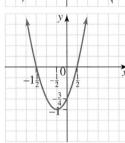

   f.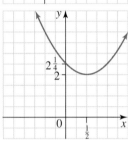

4. **MC** a. Identify the axis of symmetry for the graph shown.

   A. $x = 0$ B. $x = -2$ C. $x = -4$
   D. $y = 0$ E. the y-axis

   b. Identify the coordinates of the turning point for the graph.
   A. $(0, 0)$ B. $(-2, 4)$ C. $(-2, -4)$
   D. $(2, -4)$ E. $(2, 4)$

   c. Identify the y-intercept.
   A. 0   B. -2 C. -4
   D. 2   E. 4

   d. Identify the x-intercepts.
   A. 0 and 4   B. 0 and 2   C. 0 and -2
   D. 0 and -4 E. -2 and -4

   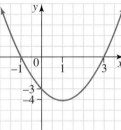

For questions **5** to **7**, plot the graph of the quadratic equation and hence state:
 i. the equation of the axis of symmetry
 ii. the coordinates of the turning point and whether it is a maximum or a minimum
 iii. the x- and y-intercepts.

5. **WE13** a. $y = x^2 + 8x + 15, -7 \le x \le 0$ b. $y = x^2 - 1, -3 \le x \le 3$

6. a. $y = x^2 - 4x, -1 \le x \le 5$ b. $y = x^2 - 2x + 3, -2 \le x \le 4$

7. a. $y = x^2 + 12x + 35, -9 \le x \le 0$ b. $y = -x^2 + 4x + 5, -2 \le x \le 6$

8. Consider the equations for $-3 \le x \le 3$:

   i. $y = x^2 + 2$
   ii. $y = x^2 + 3$

      a. Make a table of values and plot the points on the same set of axes.
      b. State the equation of the axis of symmetry for each equation.
      c. State the x-intercepts for each equation.

Understanding

9. Consider the table of values below.

x	−4	−3	−2	−1	0	1	2	3	4
y	12	5	0	−3	−4	−3	0	5	12

   a. Plot these points on graph paper. State the shape of the graph.
   b. Locate the axis of symmetry.
   c. Locate the y-intercept.
   d. Locate the x-intercept(s).

10. Consider the function $y = x^2 + x$. Complete this table of values for the function.

x	−6	−4	−2	0	2	4	6	8
y				0				

11. **MC** Select which of the following rules is not a parabola.

   A. $y = -2x^2$ **B.** $y = 2x^2 - x$ **C.** $y = -2 \div x^2$ **D.** $y = -2 + x^2$ **E.** $y = 3x^2$

12. Consider the graph of $y = -x^2$.

   a. State the turning point of this graph.
   b. State whether the turning point is a maximum or a minimum.

13. Given the following information, make a sketch of the graph involved.

   a. Maximum turning point $= (-2, -2)$, y-intercept $= (0, -6)$
   b. Minimum turning point $= (-3, -2)$, x-intercept $(1, 0)$, and $(-7, 0)$

14. Sketch a graph where the turning point and the x-intercept are the same. Suggest a possible equation.

15. Sketch a graph where the turning point and the y-intercept are the same. Suggest a possible equation.

Reasoning

16. On a set of axes, sketch a parabola that has no x-intercepts and has an axis of symmetry $x = -2$. Explain if the parabola has a maximum or minimum turning point, or both.

17. If the axis of symmetry of a parabola is $x = -4$ and one of the x-intercepts is $(10, 0)$, show the other x-intercept is $(-18, 0)$.

18. If the x-intercepts of a parabola are $(-2, 0)$ and $(5, 0)$, show that the axis of symmetry is $x = 1.5$.

Problem solving

19. Consider the parabola given by the rule $y = x^2$ and the straight line given by $y = x$. Show that the two graphs meet at $(0, 0)$ and $(1, 1)$.

20. a. The axis of symmetry of a parabola is $x = -4$. If one x-intercept is -10, determine the other x-intercept.
   b. Suggest a possible equation for the parabola.

21. **WE14** A missile was fired from a boat during a test. The missile's path could be traced by the equation $h = -\dfrac{1}{2}x^2 + x$, where h is the missile's height above the ground, in kilometres, and x is the horizontal distance from the boat, in kilometres.
   Plot the graph for $0 \le x \le 2$ and use it to find the maximum height of the missile's path, in metres.

22. SpaceCorp sent a lander to Mars to measure the temperature change over a period of time. The results were plotted on a set of axes, as shown. From the graph it can be seen that the temperature change follows the quadratic rule $T = -h^2 + 22h - 21$, where T is the temperature in degrees Celsius, and the time elapsed, h, is in hours.

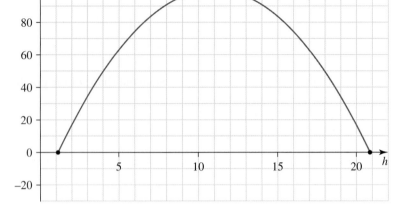

a. State the initial recorded temperature on Mars.
b. State when the temperature measured $0\,^\circ C$.
c. Identify when the highest temperature was recorded.
d. State the highest temperature recorded.

12.7 Sketching parabolas of the form $y = ax^2$

12.7.1 The graph of the quadratic function $y = x^2$

eles-4938

- The simplest parabola, $y = x^2$, is shown.

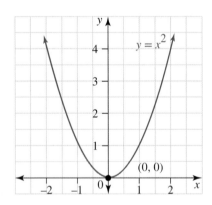

- Both the x- and y-axes are clearly indicated, along with their scales.
- The turning point $(0, 0)$ is indicated.
- The x- and y-intercepts are indicated. For this graph they are all $(0, 0)$.
- This is an example of a parabola that just touches (does not cross) the x-axis at $(0, 0)$.

▶ 12.7.2 Parabolas of the form $y = ax^2$, where $a > 0$

eles-4939

- A coefficient in front of the x^2 term affects the **dilation** of the graph, making it wider or narrower than the graph of $y = x^2$.
- If $a > 1$ then the graph becomes narrower, whereas if $0 < a < 1$, the graph becomes wider.
- The graph shows the effect of varying the coefficient a. Notice that $y = 3x^2$ is narrower than $y = x^2$ and $y = \frac{1}{4}x^2$ is wider than $y = x^2$.

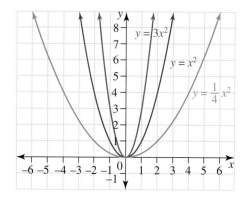

Parabolas of the form $y = ax^2$, where $a > 0$

For all parabolas of the form $y = ax^2$, where $a > 0$:
- the axis of symmetry is $x = 0$
- the turning point is $(0, 0)$
- the x-intercept is $(0, 0)$
- the y-intercept is $(0, 0)$
- the shape of the parabola is always upright, or a U shape (∪).

WORKED EXAMPLE 15 Sketching dilation of parabolas with $a > 0$

On the same set of axes, sketch the graph of $y = x^2$ and $y = 3x^2$, marking the coordinates of the turning point and the intercepts. State which graph is narrower.

THINK	WRITE/DRAW
1. Write the equation of the first graph.	$y = x^2$
2. State its axis of symmetry.	The axis of symmetry is $x = 0$.
3. State the coordinates of the turning point.	The turning point is $(0, 0)$.
4. State the intercepts.	The x-intercept is $(0, 0)$ and the y-intercept is also $(0, 0)$.
5. Calculate the coordinates of one other point.	When $x = 1$, $y = 1$; $(1, 1)$
6. Write the equation of the second graph.	$y = 3x^2$
7. State its axis of symmetry.	The axis of symmetry is $x = 0$.
8. State the coordinates of the turning point.	The turning point is $(0, 0)$.
9. State the intercepts.	The x-intercept is 0 and the y-intercept is 0.
10. Calculate the coordinates of one other point.	When $x = 1$, $y = 3$; $(1, 3)$

▶

11. Sketch the graphs, labelling the turning point.

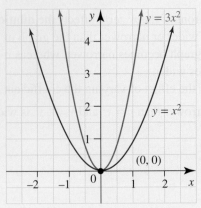

12. State which graph is narrower.

The graph of $y = 3x^2$ is narrower.

12.7.3 Parabolas of the form $y = ax^2$, where $a < 0$

- When $a < 0$, the graph is inverted; that is, it is $\cap$ shaped.
- A coefficient in front of the x^2 term affects the dilation of the graph, making it wider or narrower than the graph of $y = x^2$.
- If $-1 < a < 0$, the graph is wider than $y = -x^2$.
- If $a < -1$, the graph is narrower than $y = -x^2$.
  - The graph shows the effect of varying the coefficient a.

  Notice that $y = -3x^2$ is narrower than $y = -x^2$ and $y = -\dfrac{1}{4}x^2$ is wider than $y = -x^2$.

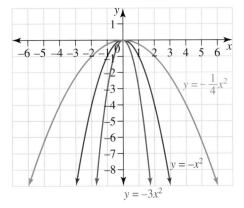

Parabolas of the form $y = ax^2$, where $a < 0$

For all parabolas of the form $y = ax^2$, where $a < 0$:
- the axis of symmetry is $x = 0$
- the turning point is $(0, 0)$
- the x-intercept is $(0, 0)$
- the y-intercept is $(0, 0)$
- the shape of the parabola is always inverted or an upside-down U shape ($\cap$).

WORKED EXAMPLE 16 Sketching dilation of parabolas with $a < 0$

On the same set of axes sketch the graphs of $y = -x^2$ and $y = -2x^2$, marking the coordinates of the turning point and the intercept. State which graph is narrower.

THINK	WRITE/DRAW
1. Write the equation of the first graph.	$y = -x^2$
2. State its axis of symmetry.	The axis of symmetry is $x = 0$.
3. State the coordinates of the turning point.	The turning point is $(0, 0)$.
4. State the intercepts.	The x-intercept is $(0, 0)$ and the y-intercept is also $(0, 0)$.

5. Calculate the coordinates of one other point.   When $x=1$, $y=-1$; $(1,-1)$

6. Write the equation of the second graph. $y=-2x^2$

7. State its axis of symmetry.   The axis of symmetry is $x=0$.

8. State the coordinates of the turning point.   The turning point is $(0,0)$.

9. State the intercepts.   The x-intercept is $(0,0)$ and the y-intercept is also $(0,0)$.

10. Calculate the coordinates of one other point.   When $x=1$; $y=-2$; $(1,-2)$

11. Sketch the two graphs on a single set of axes, labelling the turning point (as well as intercepts and maximum).

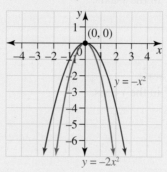

12. State which graph is narrower.   The graph of $y=-2x^2$ is narrower.

Exercise 12.7 Sketching parabolas of the form $y=ax^2$ **learn**on

Individual pathways

■ PRACTISE	■ CONSOLIDATE	■ MASTER
1, 5, 7, 11, 14	2, 4, 8, 12, 15	3, 6, 9, 10, 13, 16

To answer questions online and to receive **immediate corrective feedback** and **fully worked solutions** for all questions, go to your learnON title at www.jacplus.com.au.

Fluency

1. **WE15** On the same set of axes, sketch the graph of $y=x^2$ and $y=4x^2$, marking the coordinates of the turning point and the intercepts. State which graph is narrower.

2. On the same set of axes, sketch the graph of $y=x^2$ and $y=\dfrac{1}{2}x^2$, marking the coordinates of the turning point and the intercepts. State which graph is narrower.

3. Sketch the graph of the following table. State the equation of the graph.

x	-4	-3	-2	-1	0	1	2	3	4
y	4	2.25	1	0.25	0	0.25	1	2.25	4

4. **a.** **WE16** Using the same set of axes, sketch the graphs of $y = -x^2$ and $y = -0.5x^2$, marking the coordinates of the turning point and the intercepts. State which graph is narrower.

   b. Using the same set of axes, sketch the graph of $y = -5x^2$, marking the coordinates of the turning point and the intercepts. State which graph is the narrowest.

5. **MC** **a.** The graph of $y = -3x^2$ is:

   A. wider than $y = x^2$ **B.** narrower than $y = x^2$
   C. the same width as $y = x^2$ **D.** a reflection of $y = x^2$ in the x-axis.
   E. a reflection of $y = x^2$ in the y-axis.

   b. The graph of $y = \frac{1}{3}x^2$ is:

   A. wider than $y = x^2$ **B.** narrower than $y = x^2$
   C. the same width as $y = x^2$ **D.** a reflection of $y = x^2$ in the x-axis.
   E. a reflection of $y = x^2$ in the y-axis.

   c. The graph of $y = \frac{1}{2}x^2$ is:

   A. wider than $y = \frac{1}{4}x^2$ **B.** narrower than $y = \frac{1}{4}x^2$

   C. the same width as $y = \frac{1}{4}x^2$ **D.** a reflection of $y = \frac{1}{4}x^2$ in the x-axis.

   E. a reflection of $y = x^2$ in the y-axis.

6. Match each of the following parabolas with the appropriate equation from the list.

   i. $y = 3x^2$
   ii. $y = -x^2$
   iii. $y = 4x^2$
   iv. $y = \frac{1}{2}x^2$
   v. $y = -4x^2$
   vi. $y = -2x^2$

   a. 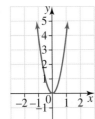   **b.** 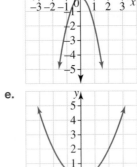   **c.**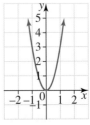

   d. 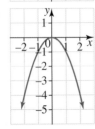   **e.**   **f.**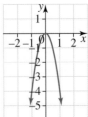

Understanding

7. Write an equation for a parabola that has a minimum turning point and is narrower than $y = x^2$.

8. Write an equation for a parabola that has a maximum turning point and is wider than $y = x^2$.

9. Determine the equation of a quadratic relation if it has an equation of the form $y = ax^2$ and passes through:

   a. $(1, 3)$ **b.** $(-1, -1)$.

10. Consider the equation $y = -3.5x^2$. Calculate the values of y when x is:

    a. 10   b. -10 c. -3 d. 1.5   e. -2.2.

Reasoning

11. a. Sketch the following graphs on the same axes: $y = x^2$, $y = 2x^2$ and $y = -3x^2$. Shade the area between the two graphs above the x-axis and the area inside the graph below the x-axis. Describe the shape that has been shaded.

    b. Sketch the following graphs on the same axes: $y = x^2$, $y = \dfrac{1}{3}x^2$ and $y = -4x^2$. Shade the area inside the graphs of $y = x^2$ and $y = -4x^2$. Also shade the area between the graph of $y = \dfrac{1}{3}x^2$ and the x-axis. Describe the shape that you have drawn.

12. The photograph shows the parabolic shape of a skate ramp. The rule of the form $y = ax^2$ describes the shape of the ramp. If the top of the ramp has coordinates $(3, 6)$, find a possible equation that describes the shape. Justify your answer.

13. The total sales of a fast food franchise vary as the square of the number of franchises in a given city. Let S be the total sales (in millions of dollars per month) and f be the number of franchises. If sales $= \$25$ million when $f = 4$, then:

    a. show that the equation relating S and f is $S = 1\,562\,500\,f^2$

    b. determine the number of franchises needed to (at least) double the sales from $\$25\,000\,000$.

Problem solving

14. The amount of power (watts) in an electric circuit varies as the square of the current (amperes). If the power is 100 watts when the current is 2 amperes, calculate:

    a. the power when the current is 4 amperes

    b. the power when the current is 5 amperes.

15. Xanthe and Carly were comparing parabolas on their CAS calculators. Xanthe graphed $y = 0.001x^2$ with window settings $-1000 \le x \le 1000$ and $0 \le y \le 1000$. Carly graphed $y = x^2$ with window settings $-k \le x \le k$ and $0 \le y \le k$. Except for the scale markings, the graphs looked exactly the same. What is the value of k?

16. The parabola $y = x^2$ is rotated $90°$ clockwise about the origin. Determine the equation.

12.8 Sketching parabolas of the form $y = (x-h)^2 + k$

LEARNING INTENTION

At the end of this subtopic you should be able to:

- find the axes of symmetry, y-intercepts and turning points for graphs of the form $y = ax^2 + c$ and $y = (x-h)^2 + k$
- sketch graphs of the form $y = ax^2 + c$, $y = (x-h)^2$ and $y = (x-h)^2 + k$
- describe the transformations that transform the graph of $y = x^2$ into $y = (x-h)^2 + k$.

12.8.1 Sketching parabolas of the form $y = ax^2 + c$

eles-4941

- A parabola with the equation $y = x^2 + c$ is simply the graph of $y = x^2$ moved vertically by c units.
- If $c > 0$, the graph of $y = x^2$ moves up by c units.
- If $c < 0$, the graph of $y = x^2$ moves down by c units.
- The turning point of such a parabola is $(0, c)$.

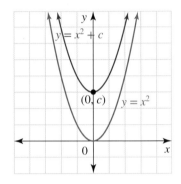

WORKED EXAMPLE 17 Sketching parabolas of the form $y = x^2 + c$ and $y = -x^2 + c$

For each part of the question, sketch the graph of $y = x^2$, then, on the same axes, sketch the given graph, clearly labelling the turning point.

a. $y = x^2 + 2$　　　　　　　　　b. $y = -x^2 - 3$

THINK

a. 1. Sketch the graph of $y = x^2$ by drawing a set of labelled axes, marking the turning point $(0, 0)$ and noting that it is symmetrical about the y-axis.

2. Determine the turning point of $y = x^2 + 2$ by adding 2 to the y-coordinate of the turning point of $y = x^2$.

WRITE/DRAW

a.

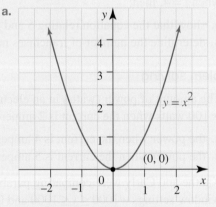

The turning point of $y = x^2 + 2$ is $(0, 2)$.

3. Using the same axes as for the graph of $y = x^2$, sketch the graph of $y = x^2 + 2$, marking the turning point and making sure that it is the same width as the graph of $y = x^2$. (The coefficient of x^2 is the same for both graphs.)

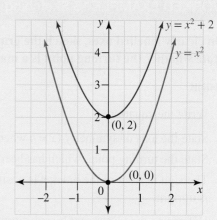

b. 1. Sketch the graph of $y = x^2$ by drawing a set of labelled axes, marking the turning point $(0, 0)$ and noting that it is symmetrical about the y-axis.

b.

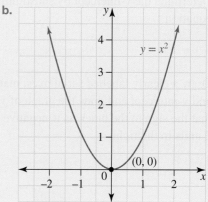

2. Determine the turning point of $y = -x^2 - 3$ by subtracting 3 from the y-coordinate of the turning point of $y = x^2$.

The turning point of
$y = -x^2 - 3$ is $(-0, -3)$.

3. Using the same axes as for the graph of $y = x^2$, sketch the graph of $y = -x^2 - 3$, marking the turning point, inverting the graph and making sure that the graph is the same width as the graph of $y = x^2$.

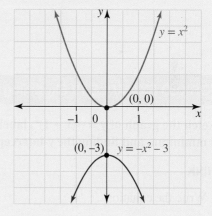

12.8.2 Sketching parabolas of the form $y = (x - h)^2$

eles-4942

- Subtracting a constant value h from x in $y = x^2$ translates the parabola h units to the right. The parabola's shape is otherwise unchanged.
  - The equation is $y = (x - h)^2$, where h is a positive quantity.
  - The x-intercept occurs when $x = h$.
- Adding a constant value h to x in $y = x^2$ translates the parabola h units to the left; otherwise the parabola's shape is unchanged.
  - The equation is $y = (x + h)^2$, where h is a positive quantity.
  - The x-intercept occurs when $x = -h$.
- The y-intercept occurs when $x = 0$, and is always at $y = h^2$.

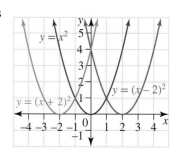

WORKED EXAMPLE 18 Sketching parabolas of the form $y = (x - h)^2$

On clearly labelled axes, sketch the graph of $y = (x - 2)^2$, marking the turning point and y-intercept. State whether the turning point is a maximum or a minimum.

THINK	WRITE/DRAW
1. Write the equation.	$y = (x - 2)^2$
2. State the axis of symmetry ($x = h$ where h is 2).	The axis of symmetry is $x = 2$.
3. State the turning point, which has been moved to the right.	The turning point is $(2, 0)$.
4. The sign in front of the bracket is positive so the parabola is upright.	Minimum turning point
5. Determine the y-intercept by substituting $x = 0$ into the equation.	y-intercept: when $x = 0$, $$y = (0 - 2)^2$$ $$= 4$$ The y-intercept is 4.
6. Draw a clearly labelled set of axes, mark the turning point and y-intercept and draw the graph of $y = (x - 2)^2$. Note that the sign in the brackets is negative so the graph moves 2 units to the right.	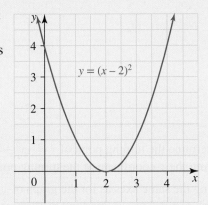

WORKED EXAMPLE 19 Sketching parabolas of the form $y = -(x + h)^2$

Sketch the graph of $y = -(x + 1)^2$, labelling the turning point, stating whether it is a maximum or minimum, and finding the y-intercept and the equation of the axis of symmetry.

THINK	WRITE/DRAW
1. Write the equation.	$y = -(x + 1)^2$
2. State the turning point, which has been moved to the left.	The turning point is $(-1, 0)$.
3. The sign in front of the bracket is negative so the parabola is inverted.	Maximum turning point
4. Determine the y-intercept by substituting $x = 0$ into the equation.	y-intercept: when $x = 0$, $$y = -(0 + 1)^2$$ $$= -1$$ The y-intercept is -1.
5. State the axis of symmetry ($x = h$, where h is -1).	The axis of symmetry is $x = -1$.

6. Draw a clearly labelled set of axes, mark the turning point and y-intercept and draw the graph of $y = -(x + 1)^2$.

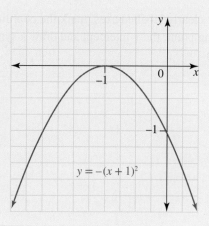

$y = -(x + 1)^2$

12.8.3 Sketching parabolas of the form $y = (x - h)^2 + k$

eles-4943

- The equation $y = (x - h)^2 + k$ combines a vertical translation of k and a horizontal translation of h together.
- The equation $y = (x - h)^2 + k$ is called **turning point form** because the turning point is given by the coordinates (h, k).

Sketching the graph of $y = (x - h)^2 + k$

1. Identify the values of h and k.
2. Mark the following:
   - the turning point at (h, k)
   - y-intercept at $(0, h^2 + k)$
   - axis of symmetry at $x = h$.
3. Calculate the positions of the x-intercepts by setting $y = 0$ and solving for x. Mark their positions on the x-axis.
4. Connect the points you have marked with a single neat curve to form a parabola.

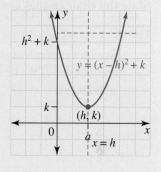

WORKED EXAMPLE 20 Sketching graphs of the form $y = (x + h)^2 - k$

Sketch the graph of $y = (x + 2)^2 - 1$, marking the turning point and the y-intercept, and indicate the type of turning point.

THINK	WRITE/DRAW
1. Write the equation.	$y = (x + 2)^2 - 1$
2. State the turning point. As the equation is in the form $y = (x - h)^2 + k$, the turning point is (h, k).	The turning point is $(-2, -1)$.
3. There is no sign outside the brackets, so the parabola is upright.	Minimum turning point
4. Determine the y-intercept by substituting $x = 0$ into the equation.	y-intercept: when $x = 0$, $$y = (0 + 2)^2 - 1$$ $$= 3$$

5. Draw clearly labelled axes, mark the coordinates of the turning point, the y-intercept and draw a symmetrical graph.

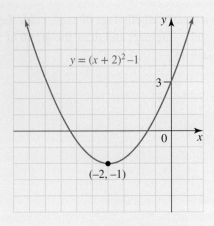

$y = (x + 2)^2 - 1$

3

0

$(-2, -1)$

- Note that in the previous example the graph has shifted 2 units left ($h = 2$) and 1 unit down ($k = -1$).
- The graph has the same shape as $y = x^2$.

WORKED EXAMPLE 21 Sketching variations of the $y = x^2$ graph

On the same set of axes, sketch the graphs of each of the following, clearly marking the coordinates of the turning point and the y-intercept:

a. $y = x^2$ b. $y = (x - 2)^2$ c. $y = (x - 2)^2 + 1$.

State the changes that are made from a to b and from a to c.

THINK	WRITE/DRAW
a. Sketch the graph of $y = x^2$, marking the coordinates of the turning point and the y-intercept.	a. 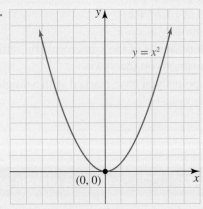 $y = x^2$ $(0, 0)$
b. 1. Write the equation.	b. $y = (x - 2)^2$
2. Determine the coordinates of the turning point.	The turning point is $(2, 0)$.
3. Calculate the y-intercept.	y-intercept: when $x = 0$, $$y = (0 - 2)^2$$ $$= 4$$

4. On the same set of axes, sketch the graph of $y = (x - 2)^2$, marking the coordinates of the turning point and the y-intercept.

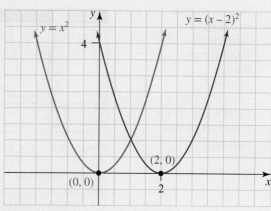

c. 1. Write the equation.

c. $y = (x - 2)^2 + 1$

2. Determine the coordinates of the turning point.

The turning point is $(2, 1)$.

3. Calculate the y-intercept.

y-intercept: when $x = 0$,

$$y = (0 - 2)^2 + 1$$
$$= 4 + 1$$
$$= 5$$

4. On the same set of axes, sketch the graph of $y = (x - 2)^2 + 1$, marking the coordinates of the turning point and the y-intercept.

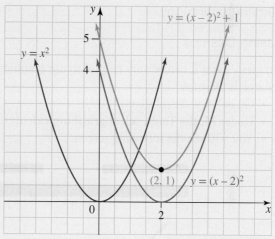

5. State how $y = x^2$ is changed to form $y = (x - 2)^2$.
State how $y = x^2$ is changed to form $y = (x - 2)^2 + 1$.

If $y = x^2$ is moved 2 units to the right, the resulting graph is $y = (x - 2)^2$.
If $y = x^2$ is moved 2 units to the right and 1 unit up, the resulting graph is $y = (x - 2)^2 + 1$.

TI | THINK	DISPLAY/WRITE	CASIO | THINK	DISPLAY/WRITE

a–c In a new problem, on a Graphs page, complete the function entry lines as:
$$f1(x) = x^2$$
$$f2(x) = (x-2)^2$$
$$f3(x) = (x-2)^2 + 1$$
Press ENTER after each entry.

a–c

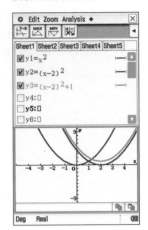

The graph of $y = x^2$ has a turning point at $(0, 0)$. If this graph is moved 2 units to the right, it is the graph of $y = (x-2)^2$, which has a turning point at $(2, 0)$. If this graph is moved 1 unit up, it is the graph of $y = (x-2)^2 + 1$, which has a turning point at $(2, 1)$.

On the Graph & Table screen, complete the function entry lines as:
$$y1 = x^2$$
$$y2 = (x-2)^2$$
$$y3 = (x-2)^2 + 1$$
Press EXE after each entry.
Press the Graph icon to view the graph.

If $y = (x-h)^2 + k$, then h controls the horizontal translation of the graphs and k controls the vertical translations.

Digital technology

Using Desmos

In many cases it can be very helpful to use graphing software. One program that can be accessed free of charge is Desmos.
- Go to www.desmos.com.
- Under **Maths Tools, select graphing calculator**.
- You will see the following screen:

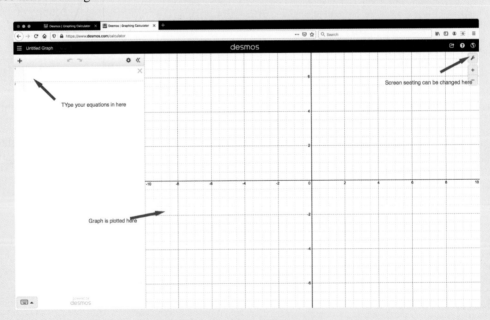

- Enter one of following equations into the box at top left, then click the + icon to create a new box, and repeat for the remaining equations: $y = x^2$; $y = 2x^2$; $y = -2x^2$.

- You will see the three parabolas appear. Note that the colours of the parabolas match the coloured circles in front of each of your equations in the panel.

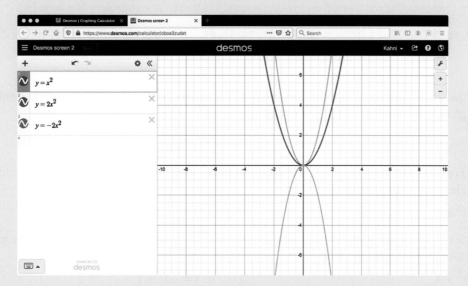

- Clear parabolas from the graph space by clicking ✕ at right of each equation box.
- Now try the equation $y = (x - 3)^2 - 2$.
- Clicking on the grey circles at intersection points on the graph will display their coordinates. Below we can see the coordinates for the turning point, the x-intercepts and the y-intercept.

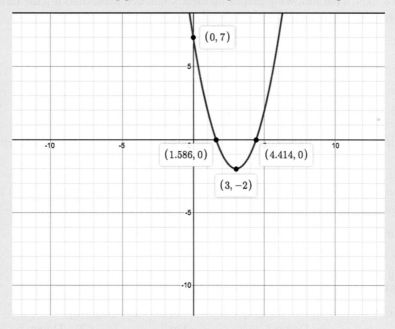

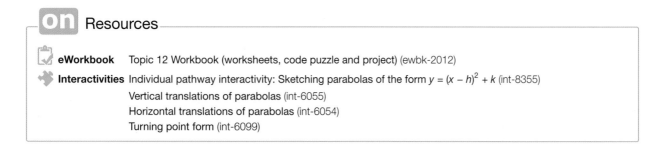

Resources

eWorkbook   Topic 12 Workbook (worksheets, code puzzle and project) (ewbk-2012)

Interactivities Individual pathway interactivity: Sketching parabolas of the form $y = (x - h)^2 + k$ (int-8355)
   Vertical translations of parabolas (int-6055)
   Horizontal translations of parabolas (int-6054)
   Turning point form (int-6099)

Exercise 12.8 Sketching parabolas of the form $y = (x - h)^2 + k$ learn on

Individual pathways

■ PRACTISE	■ CONSOLIDATE	■ MASTER
1, 4, 9, 10, 13, 16, 17, 20, 24, 27, 31, 36	2, 5, 7, 11, 14, 18, 21, 23, 26, 29, 32, 34, 37, 38	3, 6, 8, 12, 15, 19, 22, 25, 28, 30, 33, 35, 39, 40

To answer questions online and to receive **immediate corrective feedback** and **fully worked solutions** for all questions, go to your learnON title at www.jacplus.com.au.

Fluency

1. **WE17** For each part of the question, sketch the graph of $y = x^2$; then, on the same axes, sketch the given graph, clearly labelling the turning point.

   a. $y = x^2 + 1$ b. $y = x^2 - 4$

2. For each part of the question, sketch the graph of $y = x^2$; then, on the same axes, sketch the given graph, clearly labelling the turning point.

   a. $y = x^2 + 4$ b. $y = -x^2 - 1$

3. For each part of the question, sketch the graph of $y = x^2$; then, on the same axes, sketch the given graph, clearly labelling the turning point.

   a. $y = x^2 - 1$ b. $y = -x^2 + 1$

For questions 4 to 6, sketch the following graphs on clearly labelled axes, marking the turning point.
State whether the turning point is a maximum or a minimum.

4. a. $y = x^2 + 2$ b. $y = -x^2 + 3$

5. a. $y = x^2 - 5$ b. $y = -x^2 + 4$

6. a. $y = -x^2 - 3$ b. $y = x^2 - \dfrac{1}{2}$

7. Sketch the following graphs, indicating the turning point and estimating the x-intercepts.

   a. $y = 4 - x^2$ b. $y = -4 - x^2$ c. $y = 1 - x^2$

8. a. Does the turning point change if there is a negative number in front of the x^2 term in the equation $y = x^2$?
   b. Explain how a negative coefficient of x^2 affects the graph.

9. **MC** a. The turning point for the graph of the equation $y = -x^2 + 8$ is:

        A. $(0, 0)$ **B.** $(-1, 8)$ **C.** $(0, 8)$ **D.** $(0, -8)$ **E.** $(1, 8)$

   b. The turning point of the graph of the equation $y = x^2 - 16$ is:

        A. $(1, -16)$ **B.** $(-16, 1)$ **C.** $(0, 0)$ **D.** $(0, -16)$ **E.** $(-1, 16)$

   c. The graph of $y = x^2 - 7$ moves the graph of $y = x^2$ in the following way:

        A. Up 1   **B.** Down 1   **C.** Up 7   **D.** Down 7   **E.** Left 7

   d. The y-intercept of the graph of $y = -x^2 - 6$ is:

        A. 1   **B.** -1 **C.** 6   **D.** -6 **E.** -7

For questions 10–12, sketch the graphs of each of the following on clearly labelled axes, marking the turning point and y-intercept.

State whether the turning point is a maximum or a minimum.

10. a. $y = (x - 1)^2$ b. $y = (x + 2)^2$

11. a. $y = (x - 3)^2$ b. $y = (x + 4)^2$

12. a. $y = (x - 5)^2$ b. $y = (x + 6)^2$

WE19 For questions 13–15 sketch the graphs of each of the following on clearly labelled axes, marking the turning point and y-intercept.

State whether the turning point is a maximum or a minimum.

13. a. $y = -(x - 1)^2$ b. $y = -(x + 2)^2$

14. a. $y = -(x - 3)^2$ b. $y = -(x + 4)^2$

15. a. $y = -(x - 5)^2$ b. $y = -(x + 6)^2$

16. **MC** a. The axis of symmetry for the graph $y = (x + 3)^2$ is:

     A. $y = 0$ B. $x = 3$ C. $x = -3$ D. $x = 2$ E. $x = 0$

    b. The turning point of the graph $y = (x + 3)^2$ is:

     A. $(0, 0)$ B. $(1, 3)$ C. $(1, -3)$ D. $(-3, 0)$ E. $(3, 0)$

WE20 For questions 17–19 sketch the graph of each of the following, marking the turning point, the type of turning point and the y-intercept.

17. a. $y = (x - 1)^2 + 1$ b. $y = (x + 2)^2 - 1$

18. a. $y = (x - 3)^2 + 2$ b. $y = (x + 3)^2 - 2$

19. a. $y = -(x + 2)^2 - 1$ b. $y = -(x - 1)^2 + 2$

20. **WE21** On the same set of axes, sketch the following graphs, clearly marking the coordinates of the turning points and the y-intercepts.

State the changes that are made from **a** to **b** and from **a** to **c**.

    a. $y = x^2$ b. $y = (x - 1)^2$ c. $y = (x - 1)^2 + 3$

21. On the same set of axes, sketch the following graphs, clearly marking the coordinates of the turning points and the y-intercepts.

State the changes that are made from **a** to **b** and from **a** to **c**.

    a. $y = x^2$ b. $y = (x + 3)^2$ c. $y = (x + 3)^2 + 2$

22. On the same set of axes, sketch the following graphs, clearly marking the coordinates of the turning points and the y-intercepts.

State the changes that are made from **a** to **b** and from **a** to **c**.

    a. $y = x^2$ b. $y = (x - 4)^2$ c. $y = (x - 1)^2 - 1$

23. **MC** a. For the graph of $y = (x + 5)^2 - 2$, the coordinates of the turning point are:

     A. $(5, -2)$ B. $(-2, 5)$ C. $(-2, -5)$ D. $(-5, -2)$ E. $(-5, 2)$

    b. For the graph of $y = -(x - 3)^2 + 7$, the axis of symmetry is:

     A. $x = 3$ B. $x = 7$ C. $x = -3$ D. $y = 3$ E. $x = -7$

    c. For the graph of $y = (x - 1)^2 - 4$, the y-intercept is:

     A. 1 B. -1 C. -3 D. 4 E. -4

Understanding

24. Match each of the following parabolas with the appropriate equation from the list.

i. $y = x^2 - 3$

ii. $y = x^2 + 3$

iii. $y = 3 - x^2$

iv. $y = x^2 + 2$

v. $y = -x^2 + 2$

vi. $y = -x^2 - 2$

a.
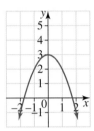

b.
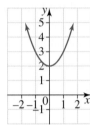

c.
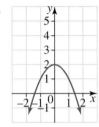

d.
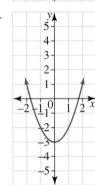

e.
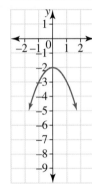

f.
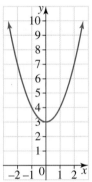

25. The vertical cross-section through the top of the mountain called the Devil's Tower can be approximated by the graph $y = -x^2 + 5$. Sketch the graph. If the x-axis represents the sea level, and both x and y are in kilometres, find the maximum height of the mountain.

26. Match each of the following parabolas with the appropriate equation from the list.

i. $y = (x - 2)^2$

ii. $y = x^2$

iii. $y = (x + 3)^2$

iv. $y = -(x + 2)^2$

v. $y = (x - 3)^2$

vi. $y = -(x - 2)^2$

a.
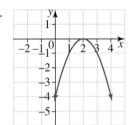

b.
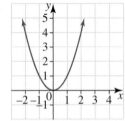

c.
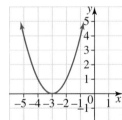

d. 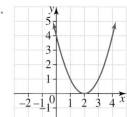   **e.** 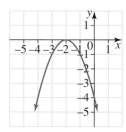   **f.**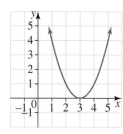

27. Determine the equation of the parabola that is of the form $y = (x - h)^2$ and passes through:

   a. $(3, 1)$
   b. $(-1, 9)$.

28. State the differences, if any, between the following pairs of parabolas.

   a. $y = (1 - x)^2$ and $y = (x - 1)^2$
   b. $y = (x - 2)^2$ and $y = (x + 2)^2$
   c. $y = -(x - 4)^2$ and $y = (4 - x)^2$

29. Write the equations for the following transformations on $y = x^2$.

   a. Reflection in the x-axis and translation of 2 units to the left
   b. Translation of 1 unit down and 3 units to the right
   c. Translation of 1 unit left and 2 units up
   d. Reflection in the x-axis, translation of 4 units to the right and 6 units up

30. **MC** A parabola has an equation of $y = x^2 - 6x + 14$. Its turning point form is:

   A. $y = (x - 6)^2 + 14$
   B. $y = (x - 14)^2 - 6$
   C. $y = (x - 3)^2 + 5$
   D. $y = (x - 5)^2 + 3$
   E. $y = (x + 3)^2 + 14$

Reasoning

31. Explain why the equation presented in the form $y = (x - h)^2 + k$ is known as 'turning point form'.

32. Show that the equation of the parabola that is of the form $y = x^2 + c$ and passes through:

   a. $(2, 1)$ is $y = x^2 - 3$
   b. $(-3, -1)$ is $y = x^2 - 10$.

33. The figure below shows the span of the Gateshead Millennium Bridge in England.

A set of axes has been superimposed onto the photo. Use the coordinates $(200, 0)$ and $(0, -75)$ to show that a possible equation is $y = -0.001\,875\,(x - 200)^2$.

34. Sketch $y = x^2 - 2$ and $y = -x^2 + 2$ on the same set of axes. Use algebra to explain where they intersect.

35. **a.** Describe how to transform the parabola $y = (x-3)^2$ to obtain the parabolas $y = (x-3)^2 - 2$ and $y = (x-3)^2 + 1$.
    b. Another parabola is created by moving $y = (x-3)^2$ so that its turning point is $(5, -4)$. Write an equation for this parabola.

Problem solving

36. Nikki wanted to keep a carnivorous plant, so after school she recorded the temperature on her windowsill for 8 hours every day for several months. One summer evening the temperature followed the relationship $t = (h-5)^2 + 15$, where t is the temperature in degrees Celsius, h hours after 4 pm.

    a. Calculate the temperature on the windowsill at 4 pm.
    b. Show that the minimum temperature reached during the 8-hour period is $15\,^\circ\text{C}$.
    c. Calculate the number of hours it took for the windowsill to reach the minimum temperature.
    d. Sketch a graph of the relationship between the temperature and the number of hours after recording began. Mark the turning point and the t-intercept on the graph.

37. For the equation $y = x^2 - 5$, calculate the exact value of the y-coordinate when the x-coordinate is $2 + \sqrt{5}$.

38. A rocket is shot in the air from a given point $(0, 0)$. The rocket follows a parabolic trajectory. It reaches a maximum height of 400 m and lands 300 m away from the launching point.

    a. Calculate how far horizontally from the launching point the rocket is when it reaches its maximum height.
    b. State the equation of the path of the rocket.

39. The path of a ball rolling off the end of a table follows a parabolic curve and can be modelled by the equation $y = ax^2 + c$. A student rolls a ball off a tabletop that is 128 cm above the floor, and the ball lands 80 cm horizontally away from the desk.
    If the student sets a cup 78 cm above the floor to catch the ball during its fall, determine where the cup should be placed.

40. Use the three points $(-2, 1)$, $(3, 1)$ and $(0, 7)$ to determine a, h and k in the equation $y = a(x-h)^2 + k$.

12.9 Sketching parabolas of the form $y = (x+a)(x+b)$

12.9.1 Sketching parabolas of the form $y = (x+a)(x+b)$

eles-4944

- The equation $y = (x+a)(x+b)$ consists of a pair of linear factors $(x+a)$ and $(x+b)$ multiplied together.

Determining important points to sketch the $y = (x + a)(x + b)$ parabola

- The x-intercepts are found by setting each factor to 0:

$$(x + a) = 0, \ x = -a$$

$$(x + b) = 0, \ x = -b$$

- The y-intercept can be found by letting $x = 0$ in the original equation:

$$y = (0 + a)(0 + b) = ab$$

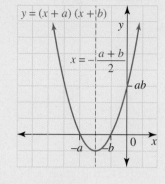

- The axis of symmetry is halfway between the x-intercepts $-a$ and $-b$; that is: $x = -\dfrac{a + b}{2}$.
- Substitute the x-value of the axis of symmetry into $y = (x + a)(x + b)$ to find the y-coordinate of the turning point.

WORKED EXAMPLE 22 Sketching parabolas in intercept form

Sketch the graph of $y = (x - 4)(x + 2)$ by first finding the x- and y-intercepts and then the turning point.

THINK	WRITE/DRAW
1. Write the equation.	$y = (x - 4)(x + 2)$
2. Determine the x-intercepts.	x-intercepts: when $y = 0$, $(x - 4)(x + 2) = 0$ $x - 4 = 0$ or $x + 2 = 0$ $\qquad x = 4 \qquad\qquad x = -2$ The x-intercepts are -2 and 4.
3. Determine the y-intercept.	y-intercept: when $x = 0$, $y = (0 - 4)(0 + 2)$ $\quad = -4 \times 2$ $\quad = -8$ The y-intercept is -8.
4. Determine the x-value of the turning point by averaging the values of the two x-intercepts.	At the turning point, $x = \dfrac{4 + -2}{2} = 1$.
5. Determine the y-value of the turning point by substituting the x-value of the turning point into the equation of the graph.	When $x = 1$, $y = (1 - 4)(1 + 2)$ $\quad = -3 \times 3$ $\quad = -9$
6. State the coordinates of the turning point.	The turning point is $(1, -9)$.

7. Sketch the graph.

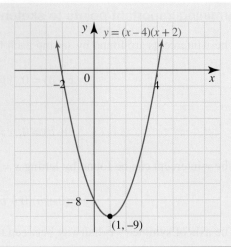

- Sometimes, it is possible to rearrange equations in the general form ($y = x^2 + bx + c$) so that they are in intercept or factor form $y = (x + a)(x + b)$.

WORKED EXAMPLE 23 Sketching parabolas of the form $y = x^2 + bx + c$

Sketch the graph of $y = x^2 + 6x + 8$.

THINK	WRITE/DRAW
1. Write the equation.	$y = x^2 + 6x + 8$
2. Factorise the expression on the right-hand side of the equation, $x^2 + 6x + 8$. Remember that to factorise an equation, you should find the factors of c that add up to b.	$= (x + 2)(x + 4)$
3. Determine the two x-intercepts.	x-intercepts: when $y = 0$, $(x + 2)(x + 4) = 0$ $x + 2 = 0 \quad$ or $x + 4 = 0$ $\quad x = -2 \qquad\quad x = -4$ The x-intercepts are -4 and -2.
4. Determine the y-intercept.	y-intercept: when $x = 0$, $y = 0 + 0 + 8$ $\quad = 8$ The y-intercept is 8.
5. Determine the x-value of the turning point.	At the turning point, $x = \dfrac{-2 + -4}{2}$ $\quad = -3$
6. Determine the y-value of the turning point.	When $x = -3$ $y = (-3 + 2)(-3 + 4)$ $\quad = -1 \times 1$ $\quad = -1$
7. State the turning point.	The turning point is $(-3, -1)$.

8. Sketch the graph.

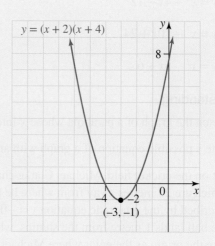

$y = (x + 2)(x + 4)$

$(-3, -1)$

on Resources

eWorkbook   Topic 12 Workbook (worksheets, code puzzle and project) (ewbk-2012)

Digital documents SkillSHEET Solving quadratic equations of the form $(x + a)(x + b) = 0$ (doc-10992)

                          SkillSHEET Factorising quadratic trinomials of the form $ax^2 + bx + c$ where $a = 1$ (doc-11011)

                          SkillSHEET Solving quadratic trinomials of the type $ax^2 + bx + c = 0$ where $a = 1$ (doc-11012)

Interactivity   Individual pathway interactivity: Sketching parabolas of the form $y = (x + a)(x + b)$ (int-8356)

                    Parabolas of the form $y = (x + a)(x + b)$ (int-6100)

Exercise 12.9 Sketching parabolas of the form $y = (x + a)(x + b)$ learn on

Individual pathways

■ PRACTISE	■ CONSOLIDATE	■ MASTER
1, 4, 7, 10, 11, 15, 18	2, 5, 8, 12, 14, 16, 19	3, 6, 9, 13, 17, 20

To answer questions online and to receive **immediate corrective feedback** and **fully worked solutions** for all questions, go to your learnON title at www.jacplus.com.au.

Fluency

WE22 For questions **1–3**, sketch the graph of each of the following by first finding the x- and y-intercepts and then the turning point.

1. **a.** $y = (x + 2)(x + 6)$ **b.** $y = (x - 3)(x - 5)$

2. **a.** $y = (x - 3)(x + 1)$ **b.** $y = (x + 4)(x - 6)$

3. **a.** $y = (x - 5)(x + 1)$ **b.** $y = (x + 1)(x - 2)$

WE23 For questions **4–9**, sketch the graph of each of the following.

4. **a.** $y = x^2 + 8x + 12$ **b.** $y = x^2 + 8x + 15$

5. **a.** $y = x^2 + 6x + 5$ **b.** $y = x^2 + 10x + 16$

6. **a.** $y = x^2 + 2x - 8$ **b.** $y = x^2 + 4x - 5$

7. a. $y = x^2 - 4x - 12$
   b. $y = x^2 - 6x - 7$

8. a. $y = x^2 - 10x + 24$
   b. $y = x^2 - 6x + 5$

9. a. $y = x^2 - 5x + 6$
   b. $y = x^2 - 3x - 10$

Understanding

10. Sketch the graph of a parabola whose x-intercepts are at $x = -4$ and $x = 6$ and whose y-intercept is at $y = 2$.

11. State a possible equation of a graph with an axis of symmetry at $x = 4$ and one intercept at $x = 6$.

12. Out on the cricket field, Michael Clarke chases the ball. He picks it up, runs 4 metres towards the stumps (which are 25 metres away from where he picks up the ball). He then throws the ball, which follows the path described by the quadratic equation $y = -0.1(x - m)(x - n)$, where m and n are positive integers. The ball lands 1 metre from the stumps, where another player quickly scoops it up and removes the bails. Take the origin as the point where Michael picked up the ball initially.

    a. Calculate how far the ball has travelled horizontally while in flight.
    b. Calculate the values of m and n.
    c. Calculate the x-intercepts.
    d. Calculate the axis of symmetry.
    e. Calculate the turning point.
    f. Identify the highest point reached by the ball.
    g. Determine how far horizontally the ball has travelled when it reaches its highest point.
    h. Sketch the flight of the ball, showing all of the relevant details on the graph.

13. McDonald's golden arches were designed by Jim Schindler in 1962. The two arches each approximate the shape of a parabola.
    A large McDonald's sign stands on the roof of a shopping centre. The shape of one of the parabolas can be modelled by the quadratic function $y = -x^2 + 8x - 7$, where both x and y are measured in metres and the bottom of the outer edges of the arches touches the x-axis. The complete sign is supported by a beam underneath the arches.
    Determine the minimum length required for this beam.

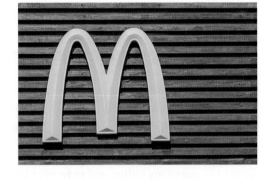

14. For the equation $y = (x - 3)(x - 9)$:

    a. determine the turning point
    b. rewrite the equation in turning point form
    c. expand both forms of the equation, showing that they are equivalent
    d. sketch the equation showing the x-intercepts and turning point.

Reasoning

15. The dimensions of a rectangular backyard can be given by $(x + 2)$ m and $(x - 4)$ m. Determine the value of x if the yard's area is 91 m^2.

16. The height of an object, $h(t)$, thrown into the air is determined by the formula $h(t) = -8t^2 + 128t$, where t is time in seconds and h is height in metres.

    a. Explain if the graph of the formula has a maximum or a minimum turning point.
    b. Calculate the maximum height of the object and the time that this height is reached.

17. Daniel is in a car on a roller-coaster ride as shown in the following graph, where the height, h, is in metres above the ground and time for the ride, t, is in minutes.

    a. The ride can be represented by three separate equations. Show that the first section of the ride is $h = 2t$ from $t = 0$ to $t = 6$. The middle section is a parabola and the final section is a straight line. Determine the other two equations for the ride.

    b. State the required domain (set of possible t-values) for each section of the ride.

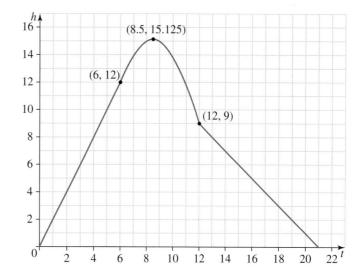

Problem solving

18. On a set of axes, sketch several parabolas of the form $y = -x^2 + bx$. What do you notice about the turning points of each graph sketched?

19. Determine a quadratic equation given the x-intercepts of the parabola are $-\dfrac{1}{2}$ and $\dfrac{3}{4}$. Explain if there is more than one possible equation.

20. Determine the equation of the parabola that passes through the points $(-1, -2)$, $(1, -4)$ and $(3, 10)$.

12.10 Applications

LEARNING INTENTION

At the end of this subtopic you should be able to:
- determine solutions to practical problems using your knowledge of quadratic equations.

⊙ 12.10.1 Applications of quadratic equations

eles-4945

- Quadratic graphs and equations can be used to solve practical problems in science and engineering.
- They are often involved in problems where a maximum or minimum needs to be found. This will usually occur at the turning point of the parabola.
- When working with physical phenomena, ensure that the solution satisfies any physical constraints of the problem. For example, if measurements of length or time are involved, there can be no negative solutions.

WORKED EXAMPLE 24 Applying quadratic equations to word problems

A flare is fired from a yacht in distress off the coast of Brisbane. The flare's height, h metres above the horizon t seconds after firing, is given by $h = -2t^2 + 18t + 20$.

a. Calculate when the flare will fall into the ocean.
b. Calculate the height of the flare after 2 seconds.
c. Determine at what other time the flare will be at the same height.
d. Calculate how long the flare is above the 'lowest visible height' of 56 m.

▶

THINK	WRITE
a. Substitute $h = 0$ into the equation and solve for t using the Null Factor Law. Since t cannot equal -1 seconds, select the appropriate solution.	**a.** $\begin{aligned} 0 &= -2t^2 + 18t + 20 \\ &= -2(t^2 - 9t - 10) \\ &= -2(t - 10)(t + 1) \\ t &= 10 \text{ or } -1 \\ t &= 10 \text{ seconds} \end{aligned}$
b. 1. Substitute $t = 2$ into the equation and solve for h.	**b.** $\begin{aligned} h &= -2(2)^2 + 18(2) + 20 \\ &= -8 + 36 + 20 \\ &= 48 \end{aligned}$
2. Write the answer in a sentence.	The flare is 48 metres high after 2 seconds.
c. Substitute $h = 48$ into the equation, rearrange, factorise and solve for t.	**c.** $\begin{aligned} 48 &= -2t^2 + 18t + 20 \\ 0 &= -2t^2 + 18t - 28 \\ &= -2(t^2 - 9t + 14) \\ &= -2(t - 7)(t - 2) \\ t &= 7 \text{ seconds} \end{aligned}$
d. 1. Substitute $h = 56$ into the equation, rearrange, factorise and solve for t.	**d.** $\begin{aligned} 56 &= -2t^2 + 18t + 20 \\ &= -2t^2 + 18t - 36 \\ &= -2(t^2 - 9t + 18) \\ &= -2(t - 6)(t - 3) \\ t &= 3 \text{ seconds and } 6 \text{ seconds} \end{aligned}$
2. Write the answer in a sentence.	The flare is above 56 m for 3 seconds (between 3 and 6 seconds).

TI	THINK	DISPLAY/WRITE	CASIO	THINK	DISPLAY/WRITE
a–d	On a Calculator page, complete the entry lines as: solve $\left(0 = -2t^2 + 18t + 20, t\right)$ $-2t^2 + 18t + 20 \mid t = 2$ solve $\left(48 = -2t^2 + 18t + 20, t\right)$ solve $\left(56 = -2t^2 + 18t + 20, t\right)$ Press ENTER after each entry.	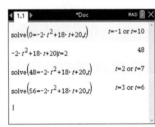	On a Main screen, complete the entry lines as: solve $\left(0 = 2t^2 + 18t + 20, t\right)$ $2t^2 + 18t + 20 \mid t = 2$ solve $\left(48 = 2t^2 + 18t + 20, t\right)$ solve $\left(56 = 2t^2 + 18t + 20, t\right)$ Press EXE after each entry.	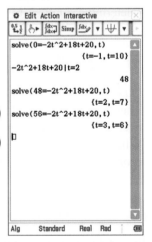

 Resources

Exercise 12.10 Applications

Individual pathways

■ PRACTISE	■ CONSOLIDATE	■ MASTER
1, 4, 7, 9, 12	2, 5, 8, 10, 13	3, 6, 11, 14, 15

To answer questions online and to receive **immediate corrective feedback** and **fully worked solutions** for all questions, go to your learnON title at www.jacplus.com.au.

Fluency

1. A spurt of water emerging from an outlet just below the surface of an ornamental fountain follows a parabolic path described by the equation $h = -x^2 + 8x$, where h is the height of the water and x is the horizontal distance from the outlet in metres.

   a. Sketch the graph of $h = -x^2 + 8x$.
   b. State the maximum height of the water above the surface of the fountain.

2. The position d metres below the starting point of a ball when it is dropped from a great height is given by the equation $d = 250 - 4.9t^2$. If it falls for 1 second, it drops 4.9 m, giving a value of $d = 245.1$ m.

   a. Calculate how far it has dropped after:
      i. 2 seconds
      ii. 7 seconds.

   b. Sketch a graph of this relation.

3. A car travels along a highway for a number of minutes according to the relationship $P = 20t^2 + 20t - 120$, where P is the distance from home in metres and t is time in minutes.

   a. State the distance from home when $t = 0$.
   b. Calculate how long it takes the car to reach home.

Understanding

4. **WE24** A rocket fired from Earth travels in a parabolic path. The equation for the path is $h = -0.05d^2 + 4d$, where h is the height in km above the surface of the earth and d is the horizontal distance travelled in km.

   a. Calculate the height of the rocket after:
      i. 30 km
      ii. 60 km.

   b. Calculate how far away the rocket lands.
   c. Determine the maximum height of the rocket and how far it travelled before reaching this height.
   d. Sketch the path of the rocket.

5. The height of a golf ball hit from the top of a hill is given by the quadratic rule $h = -t^2 + 5t + 14$, where h is in metres and t in seconds.

   a. Identify the height the golf ball was hit from.
   b. Calculate the height of the ball after 2 seconds.
   c. Calculate when the golf ball hits the ground.
   d. Calculate the maximum height of the ball.
   e. Sketch the graph of the flight of the ball.

6. Cave Ltd manufactures teddy bears. The daily profit, P, is given by the rule $P = -n^2 + 70n - 1200$, where n is the number of teddy bears produced.

   a. If they produce 40 bears a day, calculate the profit.
   b. Sketch the graph of P for appropriate values of n.
   c. Determine how many bears they need to produce before they start making a profit.
   d. Determine the maximum profit and how many teddy bears they need to manufacture to make this amount.

7. A school playground is to be mulched and the grounds keeper needs to mark out a rectangular perimeter of 80 m.

   a. Write an expression for the perimeter of the playground in terms of x and y.
   b. Show that $y = 40 - x$.
   c. Write an expression for the area to be mulched, A, in terms of x only.
   d. Sketch the graph of A against x for suitable values of x.
   e. Determine the maximum area of the playground.
   f. Determine the dimensions of the playground for this maximum area.

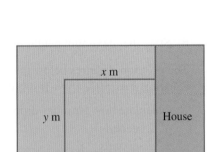

8. A woman wished to build a fence around part of her backyard, as shown in the diagram. The fence will have one side abutting the wall of the house. She has enough fencing material for 55 m of fence.

   a. Show that $y = 55 - 2x$.
   b. Write an expression for the area enclosed by the fence in terms of x alone.
   c. Determine the dimensions of the fence such that the area is a maximum and calculate that area.

Reasoning

9. A basketball thrown from the edge of the court to the goal shooter is described by the formula $h = -t^2 + 6t + 1$, where h is the height of the basketball in metres after t seconds.

   a. Identify the height the ball was thrown from.
   b. Calculate the height of the basketball after 2 s.
   c. Show that the ball was first at a height of 6 m above the ground at 1 second.
   d. Determine the time interval during which the ball was above a height of 9 m.
   e. Plot the path of the basketball.
   f. Calculate the maximum height of the basketball during its flight.
   g. Determine how long the ball was in flight if it was caught at a height of 1 m above the ground on its downward path.

10. Jack and his dog were playing outside. Jack was throwing a stick in the air for his dog to catch. The height of the stick (in metres) followed the equation $h = 20t - 5t^2$.

  a. The graph of h is a parabola. State if the parabola is upright or inverted.
  b. Factorise the expression on the right-hand side of the equation.
  c. Find the t-intercepts. These will be the two points at which the stick is on the ground, once at take-off and once at landing. Calculate how long the stick remained in the air. (For simplicity, assume that Jack throws the stick from ground level.)
  d. Justify whether the parabola has a maximum or minimum turning point, and determine its coordinates.
  e. Calculate the maximum height reached by the stick.
  f. Use the information you have found to produce a sketch of the path of the stick.

11. An engineer wishes to build a footbridge in the shape of an inverted parabola across a 40-m wide river. The bridge will be symmetrical and the greatest difference between the lowest part and highest part will be 4 m. Taking the origin as one side, show that the equation that models this footbridge is $y = -0.01x(x - 40)$. Explain what x and y are, and state the domain (the possible values that x can take).

Problem solving

12. A car engine spark plug produces a spark of electricity. The size of the spark depends on how far apart the terminals are. The percentage performance, Z, of a certain brand is thought to be $Z = -400(g - 0.5)^2 + 100$, where g is the distance between the terminals.

  a. Sketch the graph of Z.
  b. Identify when the performance is greatest.
  c. From your graph, determine the values of g for which the percentage performance is greater than 50%.

13. The monthly profit or loss, p (in thousands of dollars), for a new brand of soft drink is given by $p = -2(x - 7.5)^2 + 40.5$, where x is the number (integer) of months after its introduction (when $x = 0$).

  a. Determine the month when the greatest profit was made.
  b. Determine the months when the company made a profit.

14. NASA uses a parabolic flight path to simulate zero gravity and the gravity experienced on the moon. Use the internet to investigate the flight path and create a mathematical model to represent the flight path.

15. An arch bridge is modelled in the shape of a parabolic arch. The arch span is 50 m wide and the maximum height of the arch above water level is 5.5 m. A floating platform 35-m wide is towed under the bridge. Determine the greatest height of the deck above water level if the platform is to be towed under the bridge with at least a 35-cm clearance on either side.

12.11 Review

12.11.1 Topic summary

QUADRATIC EQUATIONS AND GRAPHS

Quadratic equations

- Quadratic equations are equations in which the largest power of x is 2. They contain an x^2 term.
- The general form of a quadratic equation is $y = ax^2 + bx + c$, where $a \neq 0$.
- Quadratic equations for which $a = 1$ are called monic quadratic equations.
- The largest number of solutions a quadratic equation can have is 2.

Axis of symmetry

- Every parabola has an axis of symmetry. This is the vertical line that passes through the parabola's turning point.

Turning point

- The turning point is the point at which a parabola changes direction.
- If $a > 0$ the turning point is the lowest point on the graph. This point is called the local minimum.
- If $a < 0$ the turning point is the highest point on the graph. This point is called the local maximum.

Null Factor Law

- If $a \times b = 0$, then:
  $a = 0$ or $b = 0$ or $a = b = 0$

Turning point form

- The turning point form of a parabola is $y = (x - h)^2 + k$.
- The turning point of a parabola is (h, k).
- A basic parabola is translated h units horizontally and k units vertically.

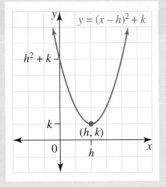

Axis intercepts

- All parabolas have a single y-intercept. The location of this y-intercept can be found by letting $x = 0$.
- Parabolas can have 0, 1 or 2 x-intercepts.
- x-intercepts can be found by letting $y = 0$ and solving for x. If the equation can't be solved, there is no x-intercept.

Parabolas of the form $y = ax^2$

- Parabolas of the form $y = ax^2$ have a turning point at the origin.
- If $a > 0$ the graph is ∪-shaped and the turning point is a local minimum.
- If $a < 0$ the graph is ∩-shaped and the turning point is a local maximum.
- Larger values of a create narrower graphs. Smaller values of a create wider graphs.

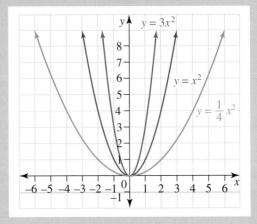

Factor form

- Parabolas written in the form $y = (x + a)(x + b)$ have x-intercepts at $x = -a$ and $x = -b$.
- The y-intercept is $y = ab$.
- The axis of symmetry will be halfway between the x-intercepts.

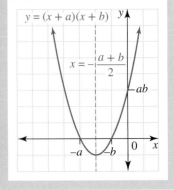

12.11.2 Success criteria

Tick a column to indicate that you have completed the subtopic and how well you think you have understood it using the traffic light system.

(**Green:** I understand; **Yellow:** I can do it with help; **Red:** I do not understand)

Subtopic	Success criteria	⬤	◯	⬤
12.2	I can identify equations that are quadratic.			
	I can rearrange quadratic equations into the general form $y = ax^2 + bx + c$.			
	I can solve quadratic equations in the form $ax^2 + c = 0$.			
12.3	I can recall the Null Factor Law.			
	I can solve quadratic equations using the Null Factor Law.			
12.4	I can solve quadratic equations with two terms using the Null Factor Law.			
12.5	I can factorise monic quadratic equations.			
	I can solve monic quadratic equations using factorisation and the Null Factor Law.			
12.6	I can identify the features of a parabolic graph.			
	I can plot the graph of a quadratic equation using a table of values.			
	I can use a parabolic graph to determine values.			
12.7	I can sketch parabolas for quadratic equations of the form $y = ax^2$.			
	I can describe the features of a parabola of the form $y = ax^2$.			
	I can describe the effect of changing the value of a on parabolas of the form $y = ax^2$.			
12.8	I can find the axes of symmetry, y-intercepts and turning points for graphs of the forms $y = ax^2 + c$ and $y = (x - h)^2 + k$.			
	I can sketch graphs of the form $y = ax^2 + c$, $y = (x + h)^2$ and $y = (x + h)^2 + k$.			
	I can describe the transformations that transform the graph of $y = x^2$ into $y = (x - h)^2 + k$.			
12.9	I can identify the x- and y-intercepts, the axis of symmetry and the turning point of a parabolas of the form $y = (x + a)(x + b)$.			
	I can factorise quadratic equations of the form $y = x^2 + bx + c$ to sketch them using their x-intercepts.			
12.10	I can determine solutions to practical problems using my knowledge of quadratic equations.			

12.11.3 Project

Constructing a parabola

The word *parabola* comes from the Greek language and means 'thrown', because it is the path followed by a projectile in flight. Notice that the water streams shown in the photo are moving in the path of a parabola. This investigation explores the technique of folding paper to display the shape of a parabola. The instructions are given below. Take care with each step to ensure your finished product is a well-constructed parabola that can be used in later parts of this investigation.

Forming a parabola by folding paper

- Take a sheet of A4 paper. Cut it into two pieces by dividing the longer side into two. Only one of the halves is required for this investigation.
- Along one of the longer sides of your piece of paper, mark points that are equally spaced 1 cm apart. Start with the first point being on the very edge of the paper.
- Turn over the piece of paper and mark a point, X, 3 cm above the centre of the edge that has the markings on the reverse side.
- Fold the paper so that the first point you marked on the edge touches point X. Make a sharp crease and open the paper flat.
- Fold the paper again so that the second mark touches the point X. Crease and unfold again.
- Repeat this process until all the marks have been folded to touch point X.
- With the paper flat and the point X facing up, you should notice the shape of a parabola appearing in the creases.

1. Trace the curve with a pencil.
   The point X is called the focus of the parabola. Consider the parabola to represent a mirror. Rays of light from the focus would hit the mirror (parabola) and be reflected. The angle at which each ray hits the mirror is the same size as the angle at which it is reflected.

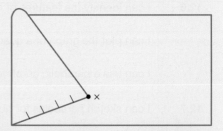

2. Using your curve traced from your folding activity, accurately draw a series of lines to represent rays of light from the point X to the parabola (mirror). Use a protractor to carefully measure the angle each line makes with the mirror and draw the path of these rays after reflection in the mirror.

3. Draw a diagram to describe your finding from question 2. Provide a brief comment on your description.

4. Retrace your parabola onto another sheet of paper. Take a point other than the focus and repeat the process of reflection of rays of light from this point by the parabolic mirror.

5. Draw a diagram to describe your finding from question 4. Provide a brief comment on your description.

6. Give examples of where these systems could be used in society.

Exercise 12.11 Review questions

learn on

To answer questions online and to receive **immediate corrective feedback** and **fully worked solutions** for all questions, go to your learnON title at www.jacplus.com.au.

Fluency

1. **MC** Identify which of the following is *not* a quadratic equation.

   A. $x^2 - 1 = 0$ B. $x^2 - 1 + 2x = 0$ C. $x^2 - \dfrac{1}{2} = 0$ D. $x^2 - \dfrac{1}{x} = 0$ E. $x^2 + 4 = 0$

2. **MC** Identify which of the following is in general form.

   A. $x^2 + 1 = 0$ B. $x^2 - 1 = 2$ C. $2x^2 = x^2 + 3$ D. $x^2 = x^2 + 1$ E. $x^2 + 5 = x$

3. **MC** The Null Factor Law cannot be applied to the equation $x(x + 3)(x - 2) = 1$ because:

   A. the first factor is a simple x term.   B. there are more than two factors.
   C. the right-hand side equals 1.   D. the second term is positive.
   E. there is no x^2 term.

4. **MC** If the solutions to the quadratic equation $(x - 3)(x - b) = 0$ are 3 and -5, then b is equal to:

   A. 5   B. -5 C. 3   D. -3 E. 1

5. **MC** Select the solutions to $3x^2 - 27 = 0$.

   A. $x = 3$ and $x = -3$ B. $x = 9$ and $x = -9$ C. $x = 1$ and $x = -1$
   D. $x = 2$ and $x = -2$ E. There are no solutions.

6. Calculate the solutions to the quadratic equation $(x - 3)(2x + 8) = 0$.

7. Calculate the solutions to the equation $(4 - x)(2x - 7) = 0$.

8. Factorise each of the following quadratic trinomials.

   a. $c^2 + 5c + 4$ b. $p^2 + 10p - 24$ c. $y^2 - 10y + 24$
   d. $x^2 + 3x + 2$ e. $m^2 - 7m + 10$ f. $m^2 + 24m + 44$

9. Solve the following equations, identifying those with no real solutions.

   a. $x^2 + 11x + 10 = 0$ b. $3x^2 + 6x = 0$ c. $-2x^2 - 1 = 0$

10. Determine the solutions to the following equations.

    a. $(x + 2)^2 - 16 = 0$ b. $4(x - 3)^2 - 36 = 0$ c. $(x + 1)^2 = 25$

11. **MC** a. The graph of $y = -4x^2$ is:
    A. wider than $y = x^2$.   B. narrower than $y = x^2$.
    C. the same width as $y = x^2$.   D. a reflection of $y = x^2$ in the x-axis.
    E. a reflection of $y = x^2$ in the y-axis.

b. The graph of $y = \dfrac{1}{2}x^2$ is:

    A. wider than $y = x^2$.
    B. narrower than $y = x^2$.

    C. the same width as $y = x^2$.
    D. a reflection of $y = x^2$ in the x-axis.

    E. a reflection of $y = x^2$ in the y-axis.

c. Compared to the graph of $y = x^2$, the graph of $y = -2x^2$ is:

    A. half as wide.
    B. twice as wide.

    C. moved 2 units to the right.
    D. moved 2 units up.

    E. moved 2 units to the left.

12. **MC** **a.** The turning point for the graph of the equation $y = -x^2 + 6$ is:

    A. $(0, 0)$ **B.** $(-1, 6)$ **C.** $(0, 6)$ **D.** $(0, -6)$ **E.** $(1, -6)$

b. The turning point of the graph of the equation $y = x^2 - 12$ is:

    A. $(1, -12)$ **B.** $(-12, 1)$ **C.** $(0, 12)$ **D.** $(0, -12)$ **E.** $(-1, 12)$

c. The graph of $y = x^2 - 4$ moves the graph of $y = x^2$ in the following way:

    A. Up 4   **B.** Down 4   **C.** No change   **D.** Left 4 units   **E.** Right 4 units

d. The y-intercept of the graph of $y = -x^2 - 3$ is:

    A. 1   **B.** -1 **C.** 3   **D.** -3 **E.** -4

13. **MC** **a.** The axis of symmetry for the graph $y = (x + 1)^2$ is:

    A. $y = 0$ **B.** $x = 1$ **C.** $x = -1$ **D.** $x = 2$ **E.** $y = -1$

b. The turning point of the graph $y = (x + 1)^2$ is:

    A. $(0, 0)$ **B.** $(1, 1)$ **C.** $(1, 1)$ **D.** $(-1, 0)$ **E.** $(1, 0)$

c. Compared to the graph of $y = x^2$, the graph of $y = (x - 2)^2$ is:

    A. twice as wide.
    B. moved 2 units to the left.
    C. moved 2 units to the right.

    D. moved 2 units up.
    E. moved 2 units down.

14. **MC** **a.** For the graph of $y = (x + 6)^2 - 1$, the coordinates of the turning point are:

    A. $(6, -1)$ **B.** $(-1, 6)$ **C.** $(-1, -6)$ **D.** $(-6, -1)$ **E.** $(-6, 1)$

b. For the graph of $y = -(x - 4)^2 + 5$, the axis of symmetry is:

    A. $x = 4$ **B.** $x = 5$ **C.** $x = -4$ **D.** $y = -5$ **E.** $y = 5$

c. For the graph of $y = (x - 2)^2 - 3$, the y-intercept is:

    A. -5 **B.** 1   **C.** -4 **D.** -3 **E.** -1

15. **MC** To change the graph of $y = x^2$ to make the graph of $y = 2(x - 1)^2 - 3$, we need to:

A. move it 1 unit to the right and 3 units down.

B. make it narrower by a factor of 3, move it 1 unit to the left and 3 units down.

C. make it wider by a factor of 2, move it 1 unit to the right and 3 units down.

D. make it narrower by a factor of 2, move it 1 unit to the right and 3 units down.

E. reflect the graph in the x-axis, and move it 1 unit to the right and 3 units up.

16. **MC** Human cannonball Stephanie Smith was fired from a cannon. Her flight path is a parabola. If c is her height above the net and b is her horizontal distance from the cannon, identify the rule that describes her flight.

    A. $y = x^2 + c$
    B. $y = bx^2 + c$
    C. $y = (x - b)^2 + c$

    D. $y = -(x - b)^2 + c$
    E. $y = -bx^2 + c$

17. 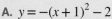 Identify the rule for the curve shown.

   A. $y = -(x+1)^2 - 2$
   B. $y = (x+1)^2 + 2$
   C. $y = (x+1)^2 - 2$
   D. $y = -(x+2)^2 - 1$
   E. $y = -(x+1)^2 + 2$

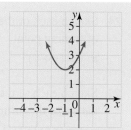

18. MC Identify the equation describing the sketch in the diagram shown.

   A. $y = (x-1)(x-3)$ B. $y = (x+1)(x+3)$ C. $y = (x-1)(x+3)$
   D. $y = (x+1)(x-3)$ E. $y = (x-3)^2$

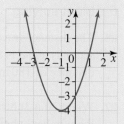

19. MC Select the x-intercepts of the graph $y = x^2 - x - 6$.

   A. -2 and 3 B. -3 and 2 C. 2 and 3
   D. -2 and -3 E. -2 and 2

20. For each of the following graphs, state the equation of the axis of symmetry, the coordinates of the turning point, whether the point is a maximum or a minimum, and the x- and y-intercepts.

   a.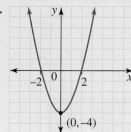

   b.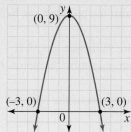

   c.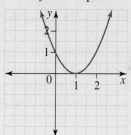

   d.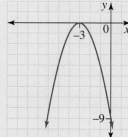

   e.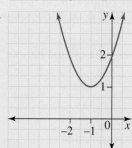

   f.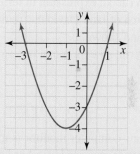

21. Draw a table of values, use it to plot the graph of each of the following equations, and hence state the equation of the axis of symmetry, the coordinates of the turning point and whether it is a maximum or a minimum, and the x- and y-intercepts.

   a. $y = x^2 - 4x, -2 \le x \le 6$
   b. $y = -x^2 - 2x + 8, -5 \le x \le 3$
   c. $y = 2x^2 - 4x + 4, -2 \le x \le 3$
   d. $y = -x^2 + 6x - 5, 0 \le x \le 6$

22. Sketch each of the following graphs, labelling the axis of symmetry, the turning point and the intercepts, and stating the type of turning point.

   a. $y = 2x^2$
   b. $y = \dfrac{1}{2}x^2$
   c. $y = -4x^2$
   d. $y = -\dfrac{1}{3}x^2$

23. Sketch each of the following graphs, labelling the axis of symmetry, the turning point and the y-intercept, and stating the nature of the turning point.

   a. $y = x^2 + 2$
   b. $y = x^2 - 4$
   c. $y = x^2 + 5$
   d. $y = x^2 - 3$

24. Sketch each of the following graphs, labelling the turning point and the *y*-intercept, and stating the type of turning point.

   a. $y = (x+1)^2$　　　**b.** $y = -(x-2)^2$　　　**c.** $y = (x-5)^2$　　　**d.** $y = -(x+2)^2$

25. Sketch each of the following graphs, labelling the turning point and the intercepts, and stating the nature of the turning point.

   a. $y = (x-1)^2 + 1$　　　**b.** $y = (x+2)^2 + 3$　　　**c.** $y = (x-3)^2 - 2$　　　**d.** $y = -(x+1)^2 - 4$

Problem solving

26. The distance travelled by a motorbike is given by the formula $d = 18t + 2t^2$, where *t* is the time in seconds and *d* is the distance in metres. Calculate how long it would take the motorbike to travel a distance of $180\,\text{m}$.

27. When 20 is subtracted from the square of a certain number, the result is 8 times that number. Determine the number, which is negative.

28. A ball is thrown from the balcony of an apartment building. The ball is *h* metres above the ground when it is a horizontal distance of *x* metres from the building. The path of the ball follows the rule $h = -x^2 + 3x + 28$. Determine how far from the building the ball will land.

29. David owned land in the shape of a square with side length *p*. He decided to sell part of this land by reducing it by 50 metres in one direction and 90 metres in the other.

   a. Write an expression in terms of *p* for the area of the original land.
   b. Write an expression in terms of *p*, in its simplest form, for the area of land David has remaining after he sells the section of land shown.
   c. Write an expression in terms of *p* for the area of land he sold.

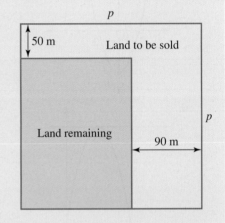

30. A new computer monitor is made up of a rectangular screen surrounded by a hard plastic frame in which speakers can be inserted.

   a. Write expressions for the length and width of the screen in terms of *x*.
   b. Write an expression for the area of the screen, using expanded form.
   c. Calculate the area of the screen if $x = 30\,\text{cm}$.

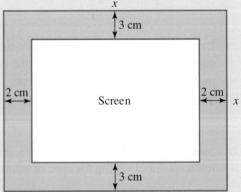

31. An astronaut needs to find a positive number such that twice the number plus its square gives 35.

   a. Write this as an equation.
   b. Solve the equation to find the answer for her.

32. A window cleaner was 3 metres off the ground, cleaning the windows of a high-rise building, when he dropped his bucket. The height, h m, of the bucket above the ground t seconds after it is dropped is given by the equation $h = 3 - t^2 - 2t$.
    a. Identify the value of h when the bucket hits the ground.
    b. Rewrite the equation, replacing h with the value it takes when the bucket hits the ground.
    c. Solve the equation.
    d. Calculate the time it takes for the bucket to reach the ground.

33. A railway bridge has an arch below it that can be modelled by the equation $H = -(x + 1)^2 + 16$, where H is the height of the bridge, in metres. Sketch a graph of the bridge and, hence, calculate its maximum height.

34. Sketch the graph of each of the following by finding the x- and y-intercepts and the turning point.
    a. $y = (x + 2)(x - 2)$
    b. $y = (x + 3)(x - 7)$
    c. $y = x^2 - 2x - 15$
    d. $y = x^2 + 3x + 2$

35. A valley in the countryside has a river running through the centre of it. A vertical cross-section of the valley can be modelled by the equation $H = \dfrac{1}{10}x^2 - 6x + 100$, where H is the height above sea level, in metres, and x is the horizontal distance, in metres. If two villages are situated in the valley, one a horizontal distance of 10 metres from the river and the other 20 metres from the river, determine the height of each village above sea level.

36. Children playing on a cricket pitch throw a ball that follows the path $y = \dfrac{1}{120}(x - 16)^2$, where y is the height, in yards, above the ground and x is the horizontal distance from the stumps at the end from which the ball is thrown. If the stumps are 22 yards away and 0.5 yards high, explain if the ball will hit the stumps. Determine by sketching the graph. (As a cricket pitch is 22 yards long, a yard is a convenient unit to use in this example. A yard is about 91.5 cm.)

37. A toy rocket is thrown up into the air from a balcony. The path the rocket takes is given by the equation $h = 49 - (t - 2)^2$ where h is the height above the ground in metres and t is the time in seconds.
    a. Calculate the height of the balcony.
    b. Calculate when the rocket lands on the ground.
    c. Calculate the maximum height of the rocket.
    d. Sketch a graph of the rockets path for $0 \le t \le 9$.

38. A piece of wire 1 m long is bent into the shape of a rectangle.
    a. If w cm is the width and l cm is the length, write a rule connecting w and l.
    b. If the area of the rectangle is A cm^2, write an equation connecting l in terms of A.
    c. Sketch the graph of A versus l.
    d. Calculate the maximum area enclosed by the wire.

39. The organising committee of the Greensborough Tennis Club found that the profit, P, from a Saturday morning barbeque depended on the number of sausages sold, n. The estimated profit is given by $P = -n^2 + 50n - 400$.

a. Calculate the profit if the number of sausages sold is 35.

b. Sketch the graph of P versus n for $0 \le n \le 50$.

c. Calculate the maximum profit possible and the number of sausages that need to be sold to make this amount.

d. Determine the minimum number of sausages that would need to be sold to avoid a loss.

40. In a demonstration using the width of a street, a skateboarder performs an ollie over a distance of 2 m. He remains on the ground for $\dfrac{1}{4}$ m before he begins the ollie, the path of which can be described by the quadratic equation $y = -2(x - p)(x - q)$ (where all dimensions are in metres), and lands with 35 cm to spare before hitting the opposite edge of the curb. Take the origin as the start of his run.

a. Calculate how far he has travelled horizontally.

b. Determine the values of p and q.

c. State the axis of symmetry of the ollie.

d. Determine the highest point reached by the skateboarder.

e. Sketch the path of the skateboarder, showing all of the relevant details on the graph.

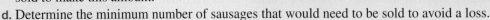 To test your understanding and knowledge of this topic, go to your learnON title at www.jacplus.com.au and complete the **post-test**.

Online Resources

 **Resources**

Below is a full list of **rich resources** available online for this topic. These resources are designed to bring ideas to life, to promote deep and lasting learning and to support the different learning needs of each individual.

📋 eWorkbook

Download the Workbook for this topic, which includes worksheets, a code puzzle and a project (ewbk-2012) ☐

📋 Solutions

Download a copy of the fully worked solutions to every question in this topic (sol-0722) ☐

📄 Digital documents

12.3 SkillSHEET Solving linear equations (doc-10984) ☐
12.4 SkillSHEET Factorising expressions of the type $ax^2 + bx$ (doc-10985) ☐
12.5 SkillSHEET Factorising quadratic trinomials (doc-10981) ☐
   SkillSHEET Finding the factor pair that adds to a given number (doc-10982) ☐
12.6 SkillSHEET Equation of a vertical line (doc-10989) ☐
   SkillSHEET Substitution into quadratic equations (doc-10990) ☐
   SkillSHEET Plotting coordinate points (doc-10991) ☐
12.7 SkillSHEET Solving quadratic equations of the form $(x + a)(x + b) = 0$ (doc-10992) ☐
   SkillSHEET Factorising quadratic trinomials of the form $ax^2 + bx + c$ where $a = 1$ (doc-11011) ☐
   SkillSHEET Solving quadratic trinomials of the type $ax^2 + bx = c = 0$ where $a = 0$ (doc-11012) ☐

▶ Video eLessons

12.2 Quadratic equations (eles-4929) ☐
   Solving equations of the form $ax^2 + c = 0$ (eles-4930) ☐
12.3 Using the Null Factor Law to solve equations (eles-4931) ☐
   The Null Factor Law (eles-2312) ☐
12.4 Solving quadratic equations of the form $ax^2 + c = 0$ (eles-4932) ☐
   Solving quadratic equations of the form $ax^2 + bx = 0$ (eles-4933) ☐
12.5 Quadratic trinomials (eles-4934) ☐
   Solving quadratic equations of the form $ax^2 + bx + c = 0$ (eles-4935) ☐
12.6 Graphs of quadratic functions (eles-4936) ☐
   Plotting points to graph quadratic functions (eles-4937) ☐
12.7 The graph of the quadratic function $y = x^2$ (eles-4938) ☐
   Parabolas of the form $y = ax^2$, where $a > 0$ (eles-4939) ☐
   Parabolas of the form $y = ax^2$, where $a < 0$ (eles-4940) ☐
12.8 Sketching parabolas of the form $y = ax^2 + c$ (eles-4941) ☐
   Sketching parabolas of the form $y = (x - h)^2$ (eles-4942) ☐
   Sketching parabolas of the form $y = (x - h)^2 + k$ (eles-4943) ☐

12.9 Sketching parabolas of the form $y = (x+a)(x+b)$ (eles-4944) ☐
12.10 Applications of quadratic equations (eles-4945) ☐

🐾 Interactivities

12.2 Individual pathway interactivity: Quadratic equations (int-8349) ☐
12.3 Individual pathway interactivity: The Null Factor Law (int-8350) ☐
   The Null Factor Law (int-6095) ☐
12.4 Individual pathway interactivity: Solving quadratic equations with two terms (int-8351) ☐
12.5 Individual pathway interactivity: Factorising and solving monic quadratics (int-8352) ☐
   Factorising monic quadratics (int-6092) ☐
12.6 Individual pathway interactivity: Graphs of quadratic functions (int-8353) ☐
   The y-intercept (int-3837) ☐
12.7 Individual pathway interactivity: Sketching parabolas of the form $y = ax^2$ (int-8354) ☐
   Dilation of parabolas (int-6096) ☐
12.8 Individual pathway interactivity: Sketching parabolas of the form $y = (x - h)^2 + k$ (int-8355) ☐
   Vertical translations of parabolas (int-6055) ☐
   Horizontal translations of parabolas (int-6054) ☐
12.9 Individual pathway interactivity: Sketching parabolas of the form $y = (x + a)(x + b)$ (int-8356) ☐
   Parabolas of the form $y = (x + a) (x + b)$ (int-6100) ☐
12.10 Individual pathway interactivity: Applications (int-8357) ☐
12.11 Crossword (int-0708) ☐
   Sudoku puzzle (int-3215) ☐

Teacher resources

There are many resources available exclusively for teachers online.

To access these online resources, log on to **www.jacplus.com.au**.

Answers

Topic 12 Quadratic equations and graphs

Exercise 12.1 Pre-test

1. A
2. False
3. $(m-2)(m+4)$
4. $2(w-3)(w-2)$
5. E
6. $(1,-3)$
7. C
8. D
9. C
10. D
11. E
12. $(0,-9)$
13. B
14. C
15. True

Exercise 12.2 Quadratic equations

1. a. $a=1, b=2, c=-1$
   b. $a=1, b=1, c=-10$
   c. $a=4, b=5, c=-30$
2. a. $a=8, b=-11, c=2$
   b. $a=2, b=1, c=-60$
   c. $a=1, b=0, c=-17$
3. D
4. C
5. a. $x=4, x=-4$
   b. There is no solution.
   c. $x=0$
6. a. $x=1, x=-1$
   b. There is no solution.
   c. $x=2, x=-2$
7. a. Y   b. N   c. Y
8. a. Y   b. N   c. N
9. a. N   b. N   c. Y
10. a. Y   b. N
11. When $x=2$:
    $$\begin{aligned}(x+4)^2 &= x^2+8x+16\\ &= 2^2+8\times2+16\\ &= 4+16+16\\ &= 36\end{aligned}$$
    $$\begin{aligned}x^2+16 &= 2^2+16\\ &= 4+16\\ &= 20\end{aligned}$$
    The two expressions are not equal.

12. a. Area of square $= (2x)^2 = 4x^2$
    Area of circle $= \pi x^2$
    $$\begin{aligned}A &= \text{Area of square} - \text{Area of circle}\\ A &= 4x^2 - \pi x^2\\ A &= x^2(4-\pi)\end{aligned}$$
    b. $10 = x^2(4-\pi)$
    $$\frac{10}{4-\pi} = x^2$$
    $$x = \sqrt{\frac{10}{4-\pi}} \approx 3.41\,\text{cm}$$
    c. x is a length, and lengths do not have negative values.
13. If $x^2+7x+4=7x$, then $x^2+4=0$, which means that $x^2=-4$ and, finally, $x=\sqrt{-4}$.
    As the square root of a negative number is not a real number, x has no solutions in the real number range.
14. a. Side length_a $= x+3$
    Side length_b $= x+5$
    Side length_c $= x+8$
    Side length_d $= x-3$
    Side length_e $= x+6$
    b. I $= (x-3)^2$
    II $= (x+3)^2$
    III $= (x+5)^2$
    IV $= (x+6)^2$
    V $= (x+8)^2$
15. a. 6 or -6
    b. There are two possible numbers, as the square root of a positive number, in this case 36, has two possible solutions, one positive and one negative.
16. a. $3(x-2)^2 - 7 = 41$
    $7 - 3(x-2)^2 = 41$ has no solution.
    b. -2 and 6

Exercise 12.3 The Null Factor Law

1. a. $x=-3, x=2$ b. $x=-2, x=3$
   c. $x=-2, x=3$ d. $x=-2\frac{1}{2}, x=-\frac{3}{4}$
   e. $x=-4, x=-\frac{1}{2}$
2. a. $x=\frac{1}{2}, x=-30$ b. $x=-\frac{1}{2}, x=3$
   c. $x=1, x=\frac{1}{3}$ d. $x=0, x=2$
   e. $x=-\frac{1}{3}, x=\frac{1}{4}$
3. a. $x=-2.3, x=0.3$ b. $x=-\frac{1}{6}, x=\frac{1}{6}$
   c. $x=2$ d. $x=0, x=\frac{15}{4}$
   e. $x=-4$
4. a. $x=2, x=-2, x=-3$
   b. $x=-2, x=2.5$
   c. $x=-2, x=-4$
   d. $x=0, x=-2, x=-4$

5. a. $x = 1.1, x = -2.4, x = -2.6$

   b. $x = -3, x = -\dfrac{1}{2}, x = 1\dfrac{2}{3}$

   c. $x = 3$

   d. $x = -1, x = 2$

6. D

7. B

8. a. $x(x + 10) = 0; \ x = 0 \textbf{ or } x = -10$

   b. $2x(x - 7) = 0; \ x = 0 \textbf{ or } x = 7$

   c. $5x(5x - 8) = 0; \ x = 0 \textbf{ or } x = \dfrac{8}{5}$

9. By dividing both sides of the equation $-2x^2 + 2x + 12 = 0$ by -2, we get $(x - 3)(x + 2) = 0$. Since the equations are equivalent, they have the same solution(s).

10. a. When $x = 2$, the first bracket equals 2 and the second bracket equals 4; therefore, the product is 8.

   b. -2 and 1

11. A quadratic can have a maximum of two solutions, because a quadratic can at most be factorised into two separate pairs of brackets, each of which represents one solution.

12. a. 10 metres

   b. $y = \dfrac{1}{250}x(100 - x); a = \dfrac{1}{250}, b = 100$

   c. 9.424 m

13. a. $h = -4t(t - 5)$

   b. i. 16 m

     ii. 0 m

   c. Sample responses can be found in the worked solutions in the online resources.

14. -29 or 8

Exercise 12.4 Solving quadratic equations with two terms

1. a. $x = -3, x = 3$ b. $x = -4, x = 4$
   c. $x = -3, x = 3$ d. $x = -5, x = 5$
   e. $x = -10, x = 10$

2. a. $x = -7, x = 7$ b. $x = -3, x = 3$
   c. $x = -2, x = 2$ d. No real solutions
   e. No real solutions

3. a. $x = -3, x = 3$ b. $x = -4, x = 4$
   c. $x = -5, x = 5$ d. $x = 0$
   e. $x = 0$

4. a. $x = 0, x = -6$ b. $x = 0, x = 8$
   c. $x = 0, x = -9$ d. $x = 0, x = 11$
   e. $x = 0, x = 6$

5. a. $x = 0, x = 7.5$ b. $x = 0, x = \dfrac{2}{3}$
   c. $x = 0, x = -1\dfrac{3}{4}$ d. $x = 0, x = 2\dfrac{1}{2}$
   e. $x = 0, x = -1$

6. a. $x = 0, x = \dfrac{1}{4}$ b. $x = 0, x = -5$
   c. $x = 0, x = -12$ d. $x = 0, x = 18$
   e. $x = 0, x = 2.5$

7. A

8. C

9. 4 or -4

10. The square plot is 12 m $\times$ 12 m; the rectangular plot is 16 m $\times$ 9 m.

11. The number is 0 or 10.

12. $x^2 + 9$ cannot be factorised.

13. $x = \pm\dfrac{n}{m}$

14. x cannot be isolated, so the only way to solve the equation will be to factorise it and use the Null Factor Law.

15. If $a = 0$, then the expression is not quadratic.

16. 0 or 8

17. -12 or 0

18. a. $x(x + 30) = 50x$

   b. $x(x - 20) = 0$

   c. $x = 0$ or $x = 20$

   d. No, x cannot be 0, because the width has to have a positive value.

Exercise 12.5 Factorising and solving monic quadratics equations

1. a. $x^2 + 6x + 8$ b. $x^2 - 6x + 8$
   c. $x^2 - 9x + 20$ d. $x^2 + 9x + 20$
   e. $m^2 + 6m + 5$ f. $m^2 - 6m + 5$

2. a. $t^2 + 19t + 88$ b. $t^2 - 30t + 200$
   c. $x^2 - x - 6$ d. $x^2 + x - 6$
   e. $v^2 - 3v - 40$ f. $v^2 - 13v - 40$

3. a. $x^2 + 5x - 14$ b. $t^2 - 11t - 12$
   c. $n^2 + 13n - 30$ d. $a^2 - 3a - 18$
   e. $z^2 + 8z + 16$ f. $z^2 + 3z - 88$

4. a. 1, 4; 2, 3   b. 1, 5; 2, 4; 3, 3
   c. 1, 11; 2, 10 ... 6, 6   d. 1, 19; 2, 18 ... 10, 10

5. a. $-1, -4; -2, -3$

   b. $-1, -6; -2, -5; -3, -4$

   c. $-1, -7; -2, -6; -3, -5; -4, -4$

   d. $-1, -10; -2, -9... -5, -6$

6. a. 7 and -1; 8 and -2; 9 and -3; 10 and -4; 11 and -5

   b. 1 and -7; 2 and -8; 3 and -9; 4 and -10; 5 and -11

   c. 4 and -1; 5 and -2; 6 and -3; 7 and -4; 8 and -5

   d. 1 and -4; 2 and -5; 3 and -6; 4 and -7; 5 and -8

   e. 1 and -9; 2 and -10; 3 and -11; 4 and -12; 5 and -13

   f. 15 and -1; 16 and -2; 17 and -3; 18 and -4; 19 and -5

7. a. $(x + 1)(x + 3)$ b. $(x - 1)(x - 3)$
   c. $(x + 1)(x + 11)$ d. $(x - 1)(x - 11)$
   e. $(a + 5)(a + 1)$

8. a. $(a - 5)(a - 1)$ b. $(x - 4)(x - 3)$
   c. $(x - 5)(x - 2)$ d. $(n + 4)^2$
   e. $(n + 8)(n + 2)$

9. a. $(y - 9)(y - 3)$ b. $(x - 7)(x - 6)$
   c. $(t - 6)(t - 2)$ d. $(t + 9)(t + 2)$
   e. $(u + 2)(u + 3)$

10. a. $(x+6)(x-3)$ b. $(x-6)(x+3)$
    c. $(x-5)(x+3)$ d. $(x+5)(x-3)$
    e. $(n-14)(n+1)$

11. a. $(n+7)(n-5)$ b. $(v+6)(v-1)$
    c. $(v-6)(v+1)$ d. $(t+6)(t-2)$
    e. $(t-7)(t+2)$

12. a. $(x-5)(x+4)$ b. $(x-4)(x+5)$
    c. $(n+10)(n-9)$ d. $(n-10)(n+7)$
    e. $(x-5)(x+1)$

13. a. $x=2, x=4$ b. $x=-2, x=-4$
    c. $x=-1, x=-5$ d. $x=2, x=-3$
    e. $x=3, x=-5$

14. a. $x=-2$ b. $x=4, x=-6$
    c. $x=8, x=-3$ d. $x=-3, x=4$
    e. $x=-12, x=-1$

15. a. $x=11, x=-1$ b. $x=4, x=-5$
    c. $x=-25, x=-4$ d. $x=5, x=10$
    e. $x=4, x=-2$

16. B

17. D

18. a. $(x+1)(x+6)$
    b. $x^2+7x+10=(x+2)(x+5)$,
      $x^2+7x+12=(x+3)(x+4)$

19. a. $(x-1)(x+4)$
    b. Sample responses can be found in the worked solutions in the online resources.

20. D

21. 9 m by 5 m

22. Sample responses can be found in the worked solutions in the online resources.

23. Sample responses can be found in the worked solutions in the online resources.

24. a   i. 0
     ii. 2
     iii. 35
   b   i. Octagon
     ii. Icosahedron

25. a.   i. $n+1$ ii. $n+7$
     iii. $(n+1)(n+7)$ iv. $\dfrac{(n+7)}{(n+2)}$
   b. 3

26. a. $w=(x-3)$ cm
    b. $w=62$ cm
    c. $l=108$ cm

27. a. $g+18$
    b. $g^2+12g-108$
    c. $g^2+12g-108=256$
    d. $g^2+12g-364=0$
    e. $g=14$
    f. The girl is 14 years old, her brother is 8 years old and her teacher is 32 years old.

28. a. 1×120 m
     2×60 m
     3×40 m
     4×30 m
     5×24 m
     6×20 m
     8×15 m
     10×12 m
    b. $l=x+8, w=x-6$
    c. 6×20 m

Exercise 12.6 Graphs of quadratic functions

1. a. Axis of symmetry: $x=0$, turning point $(0,0)$, minimum
    b. Axis of symmetry: $x=0$, turning point $(0,-3)$, minimum
    c. Axis of symmetry: $x=-1$, turning point $(-1,-2)$, minimum
    d. Axis of symmetry: $x=0$, turning point $(0,0)$, maximum
    e. Axis of symmetry: $x=0$, turning point $(0,2)$, maximum
    f. Axis of symmetry: $x=2$, turning point $(2,-1)$, maximum

2. a. $x=0$, $(0,1)$, minimum
    b. $x=1$, $(1,-3)$, minimum
    c. $x=-2$, $(-2,2)$, maximum
    d. $x=-1$, $(-1,-2)$, maximum
    e. $x=2$, $(2,2)$, minimum
    f. $x=0$, $(0,1)$, maximum

3. a. $x=0$; TP $(0,-1)$, minimum; x-intercepts are -1 and 1, y-intercept is -1.
    b. $x=0$; TP $(0,1)$, maximum; x-intercepts are -1 and 1, y-intercept is 1.
    c. $x=1$; TP $(1,-4)$, minimum; x-intercepts are -1 a and 3, y-intercept is -3.
    d. $x=-2$; TP $(-2,1)$, maximum; x-intercepts are -3 and -1, y-intercept is -3.
    e. $x=-\dfrac{1}{2}$; TP $\left(-\dfrac{1}{2},-1\right)$, minimum; x-intercepts are $-1\dfrac{1}{2}$ and $\dfrac{1}{2}$, y-intercept is $-\dfrac{3}{4}$.
    f. $x=\dfrac{1}{2}$; TP $\left(\dfrac{1}{2},2\right)$, minimum; no x-intercepts, y-intercept is $2\dfrac{1}{4}$.

4. a. B   b. C   c. A   d. D

5. a. $y=x^2+8x+15$, $-7\le x\le 0$

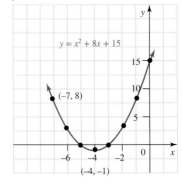

i. $x = -4$

   ii. $(-4, -1)$, minimum

   iii. x-intercepts are -5 and -3, y-intercept is 15.

b. $y = x^2 - 1$, $-3 \leq x \leq 3$

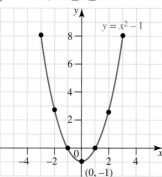

   i. $x = 0$

   ii. $(0, -1)$, minimum

   iii. x-intercepts are -1 and 1, y-intercept is -1.

6. a. $y = x^2 - 4x$, $-1 \leq x \leq 5$

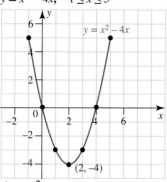

   i. $x = 2$

   ii. $(2, -4)$, minimum

   iii. x-intercepts are 0 and 4, y-intercept is 0.

b. $y = x^2 - 2x + 3$, $-2 \leq x \leq 4$

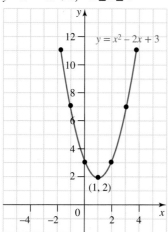

   i. $x = 1$

   ii. $(1, 2)$, minimum

   iii. No x-intercepts, y-intercept is 3.

7. a. $y = x^2 + 12x + 35$, $-9 \leq x \leq 0$

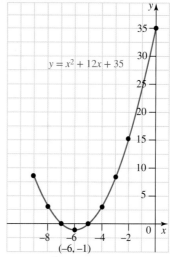

   i. $x = -6$

   ii. $(-6, -1)$, minimum

   iii. x-intercepts are -7 and -5, y-intercept is 35.

b. $y = -x^2 + 4x + 5$, $-2 \leq x \leq 6$

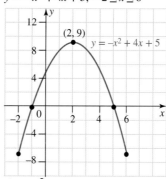

   i. $x = 2$

   ii. $(2, 9)$, maximum

   iii. x-intercepts are -1 and 5, y-intercept is 5.

8.

x	-3	-2	-1	0	1	2	3
$y = x^2 + 2$	11	6	3	2	3	6	12
$y = x^2 + 3$	12	7	4	3	4	7	12

a.

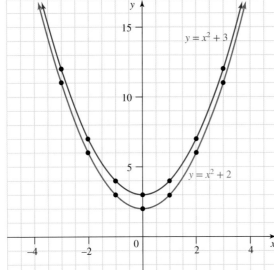

$y = x^2 + 3$

$y = x^2 + 2$

b. Axis of symmetry: $x = 0$ for both equations.

c. No x-intercepts for either equation.

9. a.

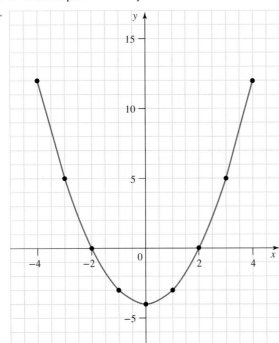

The shape is a parabola.

b. $x = 0$

c. $y = -4$

d. $x = -2$, $x = 2$

10.

x	-6	-4	-2	0	2	4	6	8
y	30	12	2	0	6	20	42	72

11. C

12. a. $(0, 0)$ **b.** Maximum

13. a.

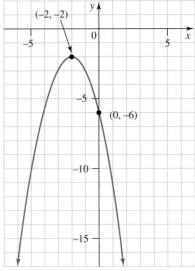

$(-2, -2)$

$(0, -6)$

b.

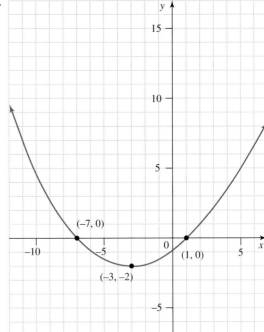

$(-7, 0)$

$(1, 0)$

$(-3, -2)$

14. Answers will vary but must be of the form $y = a(x - h)^2$.

15. Answers will vary but must be of the form $y = ax^2 + c$.

16. Answers will vary. If the parabola is upright, it has a minimum turning point. If the parabola is inverted, it has a maximum turning point.

17. Axis of symmetry crosses halfway between the x-intercepts. Let the unknown x-intercept be at p. Then,
$$-4 = \frac{10 + p}{2};$$
$$-8 = 10 + p;$$
$$p = -18$$
Therefore the other x-intercept is at $(-18, 0)$.

18. Let $(a, 0)$ be the point at which the axis of symmetry crosses the x-axis.

As a will lie halfway between the x-intercepts $(-2, 0)$ and $(5, 0)$;

$$a = \frac{-2 + 5}{2} = \frac{3}{2} = 1.5$$

Therefore, the axis of symmetry has the equation $x = 1.5$.

19.

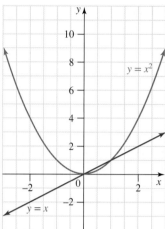

By inspection or algebra, the graphs meet at $(0, 0)$ and again at $(1, 1)$.

20. a. $\dfrac{x_2 + 10}{2} = -4$

$x_2 = 2$

b. Answers will vary but must be of the form $y = a(x - 2)(x + 10)$.

21.

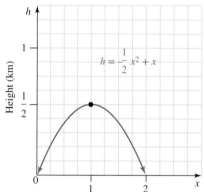

The maximum height is 500 m.

22. a. $-21\,°C$ **b.** 1 hour and at 21 hours
c. $t = 11$ hours **d.** $100\,°C$

1.

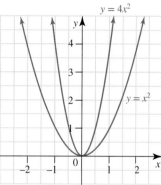

$y = 4x^2$ is narrower.
Turning point for each is at $(0, 0)$; x-int and y-int is 0 for both.

2.

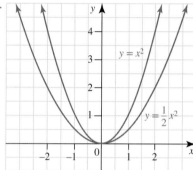

$y = x^2$ is narrower.
Turning point for each is at $(0, 0)$; x-int and y-int is 0 for both.

3.

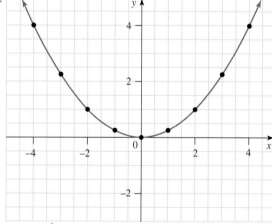

$y = 0.25x^2$

4. a, b. See the figure at the foot of the page.*

5. a. B   b. A   c. B

6. a. iii   b. vi   c. i
   d. ii   e. iv   f. v

7. Sample responses can be found in the worked solutions in the online resources; for example, $y = 2x^2$.

8. Sample responses can be found in the worked solutions in the online resources; for example, $y = -0.5x^2$.

9. a. $y = 3x^2$ b. $y = -x^2$

10. a. -350 b. -350 c. -31.5
    d. -7.875 e. -16.94

11. a.

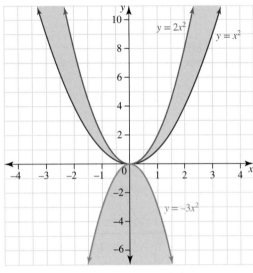

b.

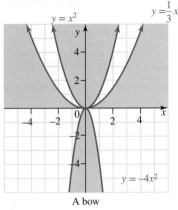

A bow

12. $y = \dfrac{2}{3}x^2$

13. a. Sample responses can be found in the worked solutions in the online resources.

   b. 6

14. a. 400 watts   b. 625 watts

15. $k = 1$

16. $x = y^2$

*4. a, b.

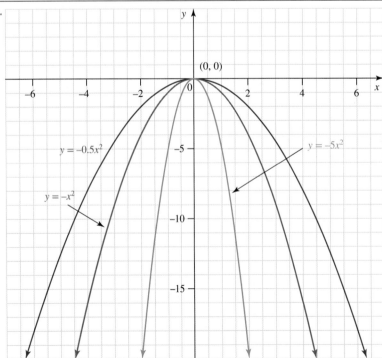

Exercise 12.8 Sketching parabolas of the form $y = (x - h)^2 + k$

1. a.

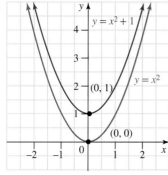

b.

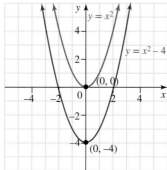

2. a.

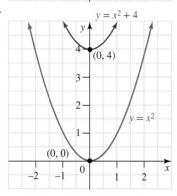

b.

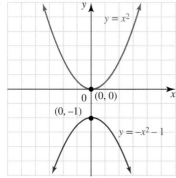

3. a.

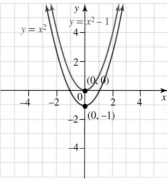

b.

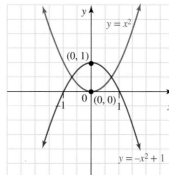

4. a.

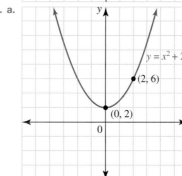

Minimum

b.

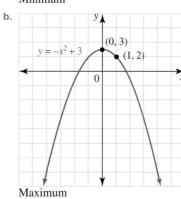

Maximum

5. a.

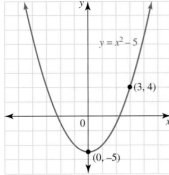

$y = x^2 - 5$

(3, 4)

0

(0, −5)

Minimum

b.

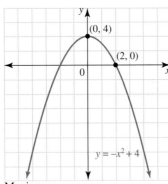

(0, 4)

(2, 0)

0

$y = -x^2 + 4$

Maximum

6. a.

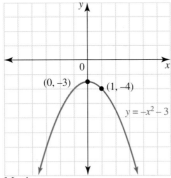

0

(0, −3) (1, −4)

$y = -x^2 - 3$

Maximum

b.

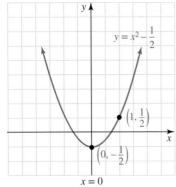

$y = x^2 - \dfrac{1}{2}$

$\left(1, \dfrac{1}{2}\right)$

$\left(0, -\dfrac{1}{2}\right)$

$x = 0$

Minimum

7. a. *x*-intercepts: (−2, 0), (2, 0)

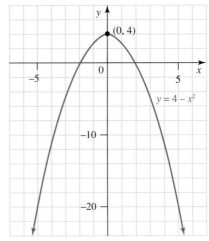

(0, 4)

−5

0

5

$y = 4 - x^2$

−10

−20

b. No *x*-intercepts

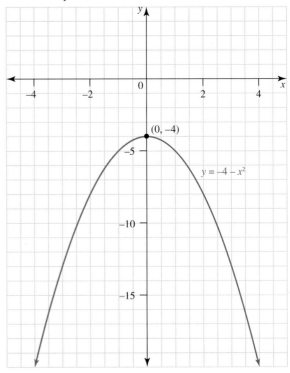

−4

−2

0

2

4

(0, −4)

−5

$y = -4 - x^2$

−10

−15

c. *x*-intercepts: $(-2, 0)$, $(2, 0)$
See the figure at the foot of the page.*

8. a. No

   b. A negative sign inverts the graph.

9. a. C   b. D   c. D   d. D

10. a. Minimum turning point $(1, 0)$

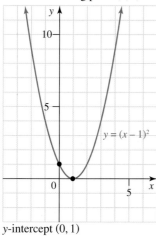

y-intercept $(0, 1)$

b. Minimum turning point $(-2, 0)$

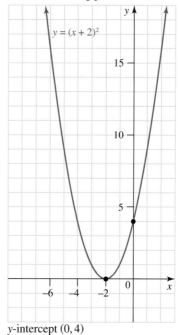

y-intercept $(0, 4)$

---

*7. c.

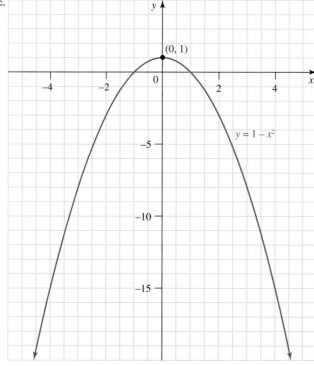

11. a. Minimum turning point $(3, 0)$

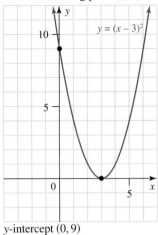

y-intercept $(0, 9)$

b. Minimum turning point $(-4, 0)$

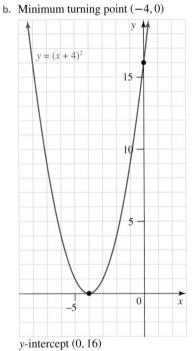

y-intercept $(0, 16)$

12. a. Minimum turning point $(5, 0)$

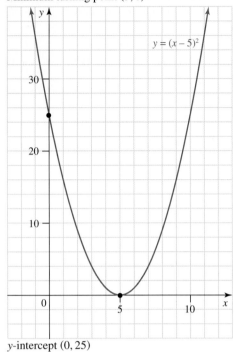

y-intercept $(0, 25)$

b. Minimum turning point $(-6, 0)$

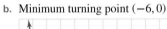

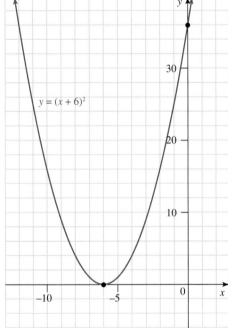

y-intercept $(0, 36)$

13. a. Maximum turning point $(1, 0)$

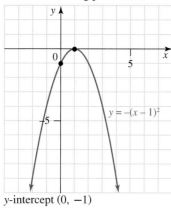

$y = -(x - 1)^2$

y-intercept $(0, -1)$

b. Maximum turning point $(-2, 0)$

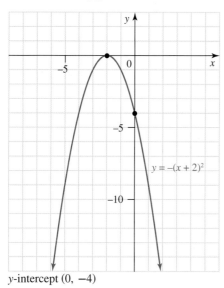

$y = -(x + 2)^2$

y-intercept $(0, -4)$

14. a. Maximum turning point $(3, 0)$

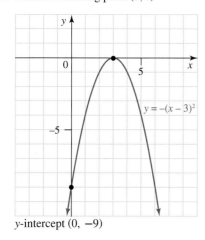

$y = -(x - 3)^2$

y-intercept $(0, -9)$

b. Maximum turning point $(-4, 0)$

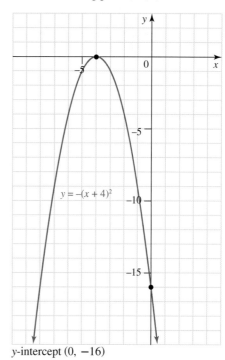

$y = -(x + 4)^2$

y-intercept $(0, -16)$

15. a. Maximum turning point $(5, 0)$

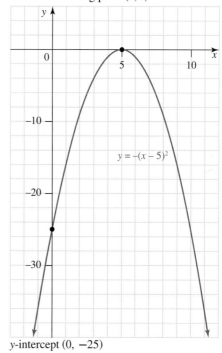

$y = -(x - 5)^2$

y-intercept $(0, -25)$

b. Maximum turning point $(-6, 0)$

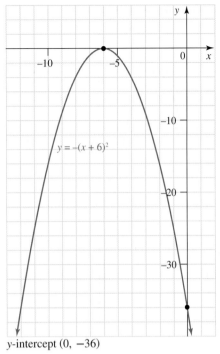

$y = -(x + 6)^2$

y-intercept (0, −36)

16. a. C **b.** D

17. a.

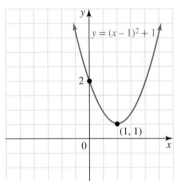

$y = (x - 1)^2 + 1$

(1, 1)

Minimum

b.

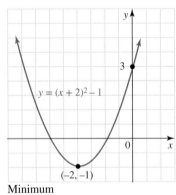

$y = (x + 2)^2 - 1$

(−2, −1)

Minimum

18. a.

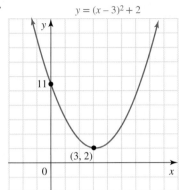

$y = (x - 3)^2 + 2$

(3, 2)

Minimum

b.

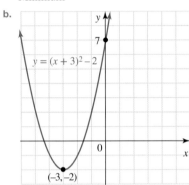

$y = (x + 3)^2 - 2$

(−3, −2)

Minimum

19. a.

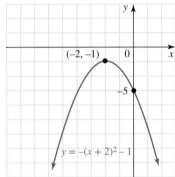

(−2, −1)

−5

$y = -(x + 2)^2 - 1$

Maximum

b.

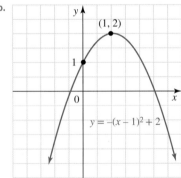

(1, 2)

1

$y = -(x - 1)^2 + 2$

Maximum

20. a.

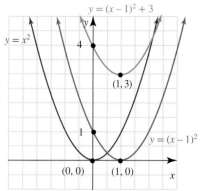

$y = x^2$
$y = (x-1)^2 + 3$
4
(1, 3)
1
$y = (x-1)^2$
(0, 0) (1, 0)

b. 1 right

c. 1 right, 3 up

21. a.

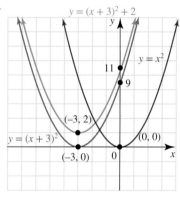

$y = (x+3)^2 + 2$
$y = x^2$
11
9
(-3, 2)
$y = (x+3)^2$
(0, 0)
(-3, 0)

b. 3 left

c. 3 left, 2 up

22. a.

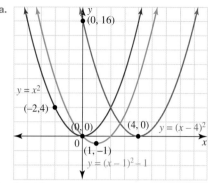

(0, 16)
$y = x^2$
(-2, 4)
(0, 0) (4, 0) $y = (x-4)^2$
(1, -1)
$y = (x-1)^2 - 1$

b. 4 right

c. 4 right, 1 down

23. a. D   **b.** A   **c.** C

24. a. iii

   b. iv

   c. v

   d. i

   e. vi

   f. ii

25. 5 km above sea level.
      See figure at the foot of the page.*

---

*25.

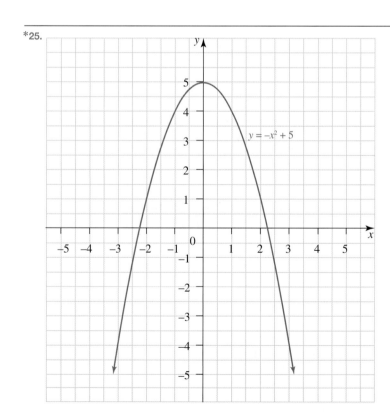

$y = -x^2 + 5$

26. a. vi   b. ii   c. iii

    d. i   e. iv   f. v

27. a. $y = (x - 2)^2$ or $y = (x - 4)^2$

    b. $y = (x - 2)^2$ or $y = (x + 4)^2$

28. a. No difference

    b. 2nd graph is translated 4 units to the left of the 1st graph.

    c. 2nd graph is 'inverted' and is the mirror image of the 1st graph in the x-axis.

29. a. $y = -(x + 2)^2$ b. $y = (x - 3)^2 - 1$

    c. $y = (x + 1)^2 + 2$ d. $y = -(x - 4)^2 + 6$

30. C

31. The graph is a basic parabola that has been translated h units horizontally and k units vertically. Therefore the turning point becomes (h, k). This form allows the turning point to be determined very quickly and easily.

32. Students should substitute the points into the given equations and show that the equations are true.

33. Students should substitute the points into the given equations and show that the equations are true.

34. Intersections at $\left(-\sqrt{2}, 0\right)$ and $\left(\sqrt{2}, 0\right)$

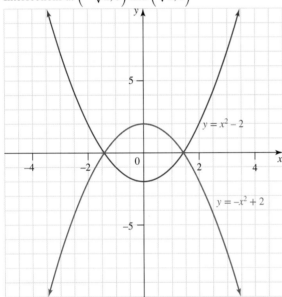

35. a. $y = (x - 3)^2 - 2$ is obtained by translating $y = (x - 3)^2$ vertically 2 units down.

    $y = (x - 3)^2 + 1$ is obtained by translating $y = (x - 3)^2$ vertically 1 unit up.

    b. $y = (x - 5)^2 - 4$

36. a. 40 °C

    b. Sample responses can be found in the worked solutions in the online resources.

    c. 5 hours

d.

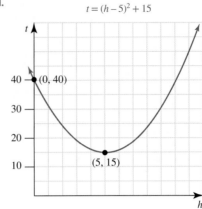

$t = (h - 5)^2 + 15$

37. $y = 4 + 4\sqrt{5}$

38. a. 150 metres

    b. $y = \dfrac{-4}{225}(x - 150)^2 + 400$

39. 50 cm horizontally away from the desk.

40. $y = -\left(x - \dfrac{1}{2}\right)^2 + \dfrac{29}{4}$

Exercise 12.9 Sketching parabolas of the form $y = (x + a)(x + b)$

1. a.

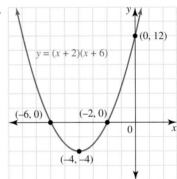

b.

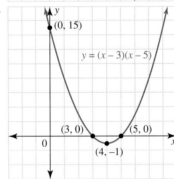

2. a.

$y = (x-3)(x+1)$

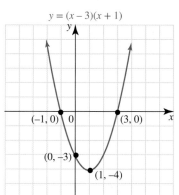

b.

$y = (x+4)(x-6)$

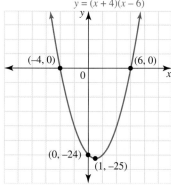

3. a.

$y = (x-5)(x+1)$

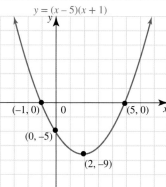

b.

$y = (x+1)(x-2)$

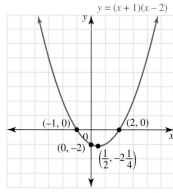

4. a.

$y = x^2 + 8x + 12$

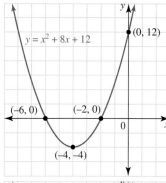

b.

$y = x^2 + 8x + 15$

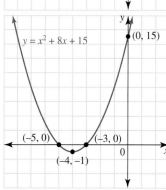

5. a.

$y = x^2 + 6x + 5$

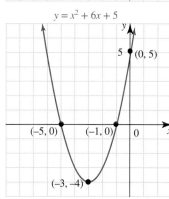

b.

$y = x^2 + 10x + 16$

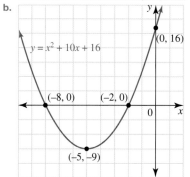

6. a.

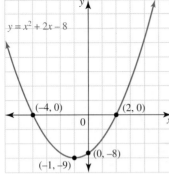

$y = x^2 + 2x - 8$

$(-4, 0)$ $(2, 0)$ $(0, -8)$ $(-1, -9)$

b.

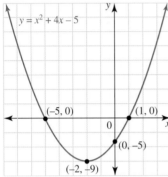

$y = x^2 + 4x - 5$

$(-5, 0)$ $(1, 0)$ $(0, -5)$ $(-2, -9)$

7. a.

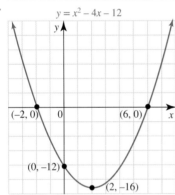

$y = x^2 - 4x - 12$

$(-2, 0)$ $(6, 0)$ $(0, -12)$ $(2, -16)$

b.

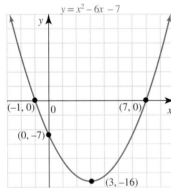

$y = x^2 - 6x - 7$

$(-1, 0)$ $(7, 0)$ $(0, -7)$ $(3, -16)$

8. a.

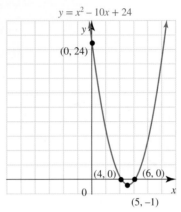

$y = x^2 - 10x + 24$

$(0, 24)$ $(4, 0)$ $(6, 0)$ $(5, -1)$

b.

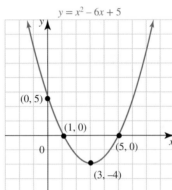

$y = x^2 - 6x + 5$

$(0, 5)$ $(1, 0)$ $(5, 0)$ $(3, -4)$

9. a.

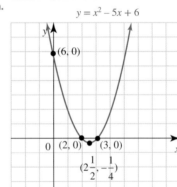

$y = x^2 - 5x + 6$

$(6, 0)$ $(2, 0)$ $(3, 0)$ $\left(2\frac{1}{2}, -\frac{1}{4}\right)$

b.

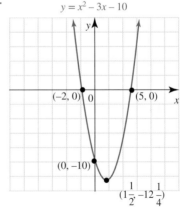

$y = x^2 - 3x - 10$

$(-2, 0)$ $(5, 0)$ $(0, -10)$ $\left(1\frac{1}{2}, -12\frac{1}{4}\right)$

10.

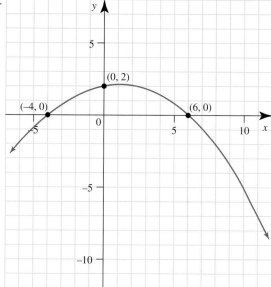

11. $y = (x-2)(x-6)$

12. a. 20 m

  b. 4, 24

  c. $(4, 0)$, $(24, 0)$

  d. $x = 14$

  e. $(14, 10)$

  f. 10 m

  g. 10 m

  h.

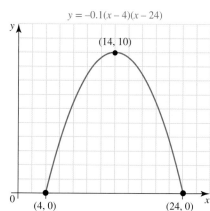

$y = -0.1(x-4)(x-24)$

13. 12 m

14. a. $(6, -9)$

  b. $y = (x-6)^2 - 9$

  c. Intercept form expanded:
    $y = x^2 - 12x + 27$
    Turning point form expanded:
    $y = x^2 - 12x + 36 - 9 = x^2 - 12x + 27$

d.

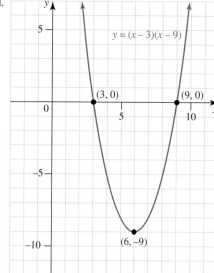

15. $x = 11$

16. a. Since the coefficient of t^2 w is negative, the turning point is a maximum.

  b. Maximum height $= 512$ m at $t = 8$ seconds

17. a. From $t = 6$ to $t = 12$; $h = -0.5(t - 8.5)^2 + 15.125$
    From $t = 12$ to $t = 21$; $h = -t + 21$

  b. $0 \leq t \leq 6, 6 \leq t \leq 12, 12 \leq t \leq 21$

18. The general turning point is $\left(\dfrac{b}{2}, \dfrac{b^2}{4} \right)$.

19. Answers will vary but must be of the general form
$$y = a\left(x + \frac{1}{2}\right) - \left(x - \frac{3}{4}\right).$$

20. $y = 2x^2 - x - 5$

Exercise 12.10 Applications

1. a.   **b.** 16 m

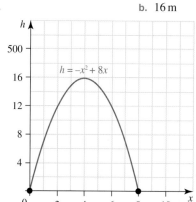

2. a. i. 19.6 m

    ii. 240.1 m

b.

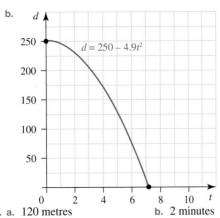

3. **a.** 120 metres   **b.** 2 minutes

4. **a.** **i.** 75 km

     ii. 60 km

   b. 80 km

   c. 80 km high after travelling 40 km horizontally

   d.

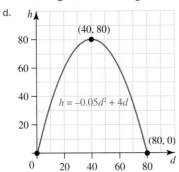

5. **a.** 14 m

   b. 20 m

   c. 7 s

   d. 20.25 m

   e.

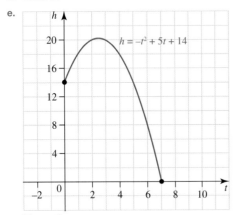

6. **a.** No profit

b.

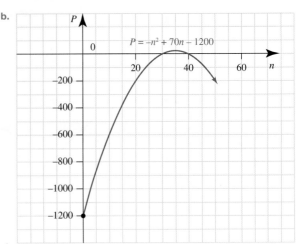

   c. 30 bears

   d. \$25; 35 bears

7. **a.** $2(x + y) = 80$

   b. Sample responses can be found in the worked solutions in the online resources.

   c. $A = x(40 - x)$

   d.

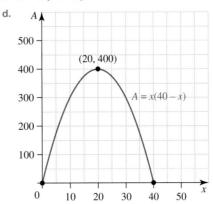

   e. $400 \, \text{m}^2$

   f. $x = 20 \, \text{m}; y = 20 \, \text{m}$

8. a. Perimeter $-2x + y = 55$ m, so $y = 55 - 2x$

b. $55x - 2x^2$

c. $x = 13.75$ m, $y = 27.5$ m, area $= 378.125$ m^2.

9. a. 1 m

b. 9 m

c. Sample responses can be found in the worked solutions in the online resources.

d. From 2 to 4 seconds

e.

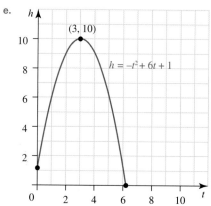

$h = -t^2 + 6t + 1$

f. 10 m

g. 6 s

10. a. Inverted

b. $h = 5t(4 - t)$

c. $t = 0, t = 4$; 4 seconds

d. Maximum turning point at $(2, 20)$

e. 20 m

f.

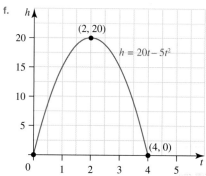

$h = 20t - 5t^2$

11. Let $x = $ horizontal distance, $y = $ vertical distance across the bridge $(0 \le x \le 40)$

$y = -0.01x(x - 40)$

12. a.

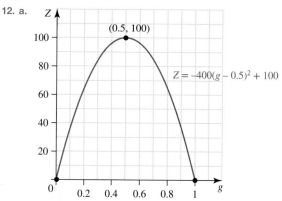

$Z = -400(g - 0.5)^2 + 100$

b. $g = 0.5$

c. $0.15 \le g \le 0.85$

13. a. 7th to 8th month

b. 3rd and 12th months

14. Sample responses can be found in the worked solutions in the online resources.

15. The height of the platform deck above the water is 2.7 m.

Project

1-5. Students need to construct a parabola by folding paper with the directions given in this project. Draw incident and reflected rays of light and use a protractor to carefully measure the angle each line makes with the mirror. Also comment on the diagram.

6. Parabolas could be seen in real life in, for example, water shot by the fountain in a parabolic path, a ball thrown into the air, bridges, headlights, satellite dishes or telescopes.

Exercise 12.11 Review questions

1. D

2. A

3. C

4. B

5. A

6. $x = -4, x = 3$

7. $x = 3.5, x = 4$

8. a. $(c + 1)(c + 4)$ **b.** $(p - 2)(p + 12)$

c. $(y - 6)(y - 4)$ **d.** $(x + 1)(x + 2)$

e. $(m - 2)(m - 5)$ **f.** $(m + 22)(m + 2)$

9. a. $x = -10, x = -1$

b. $x = -2, x = 0$

c. No real solutions

10. a. $x = -6, x = 2$

b. $x = 0, x = 6$

c. $x = -6, x = 4$

11. a. B   **b.** A   **c.** A

12. a. C   **b.** D   **c.** B   **d.** D

13. a. C   **b.** D   **c.** C

14. a. D   **b.** A   **c.** B

15. D

16. D

17. *B*

18. C

19. C

20. a. $x = 0$, TP $(0, -4)$, minimum, x-intercepts are -2 and 2, y-intercept is -4.

b. $x = 0$, TP $(0, 9)$, maximum, x-intercepts are 3 and -3, y-intercept is 9.

c. $x = 1$, TP $(1, 0)$, minimum, x-intercepts is 1, y-intercept is 1.

d. $x = -3$, TP $(-3, 0)$, maximum, x-intercepts is -3, y-intercept is -9.

e. $x = -1$, TP $(-1, 1)$, minimum, no real x-intercepts, y-intercept is 2.

f. $x = -1$, TP $(-1, -4)$, minimum, x-intercepts are 1 and -3, y-intercept is -3.

21. a. $y = x^2 - 4x, \; -2 \leq x \leq 6$

x	−2	−1	0	1	2	3	4	5	6
y	12	5	0	−3	−4	−3	0	5	12

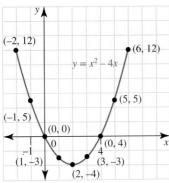

$x = 2$, TP $(2, -4)$, minimum, x-intercepts are 0 and 4, y-intercept is 0.

b. $y = -x^2 - 2x + 8, \; -5$

x	−5	−4	−3	−2	−1	0	1	2	3
y	−7	0	5	8	9	8	5	0	−7

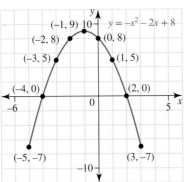

$x = -1$, TP $(-1, 9)$, maximum, x-intercepts are −4 and 2,
y-intercept is 8.

c. $y = 2x^2 - 4x + 4, \; -2 \leq x \leq 3$

x	−2	−1	0	1	2	3
y	20	10	4	2	4	10

See the figure at the foot of the page.*
$x = 1$, TP $(1, 2)$, minimum, no x-intercepts, y-intercept is 4.

d. $y = -x^2 + 6x - 5, \; 0 \leq x \leq 6$

x	0	1	2	3	4	5	6
y	−5	0	3	4	3	0	−5

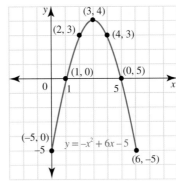

$x = 3$, TP $(3, 4)$, maximum, x-intercepts are 1 and 5,
y-intercept is −5.

22. a.

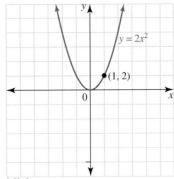

Minimum

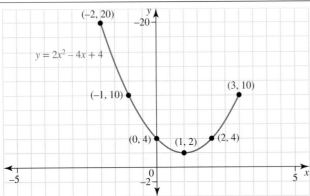

b.

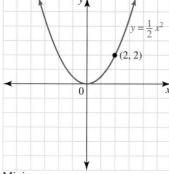

$y = \frac{1}{2}x^2$

(2, 2)

Minimum

c.

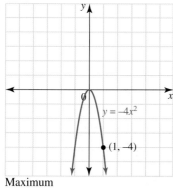

$y = -4x^2$

(1, −4)

Maximum

d.

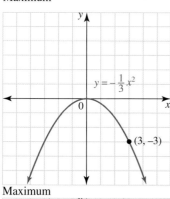

$y = -\frac{1}{3}x^2$

(3, −3)

Maximum

23. a.

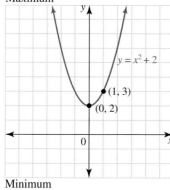

$y = x^2 + 2$

(1, 3)

(0, 2)

Minimum

b.

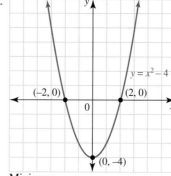

$y = x^2 - 4$

(−2, 0)   (2, 0)

(0, −4)

Minimum

c.

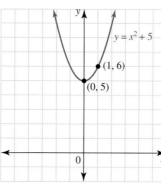

$y = x^2 + 5$

(1, 6)

(0, 5)

Minimum

d.

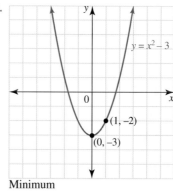

$y = x^2 - 3$

(1, −2)

(0, −3)

Minimum

24. a.

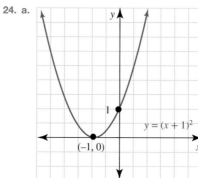

1

$y = (x + 1)^2$

(−1, 0)

Minimum

b.

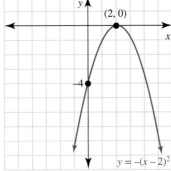

Maximum

c.

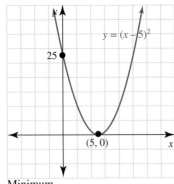

Minimum

d.

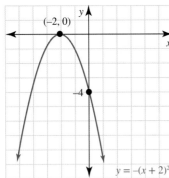

Maximum

25. a.

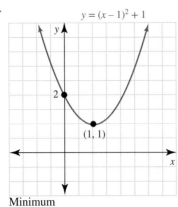

Minimum

b.

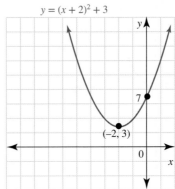

Minimum

c.

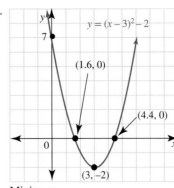

Minimum

d.

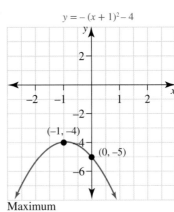

Maximum

26. 6 s

27. $x = -2$

28. 7 metres

29. a. p^2

   b. $p^2 - 140p + 4500$

   c. $140p - 4500$

30. a. Length $= (x - 4)$ cm, width $= (x - 6)$ cm

   b. $\left(x^2 - 10x + 24\right)$ cm^2

   c. 624 cm^2

31. a. $x^2 + 2x = 35$ **b.** 5

32. a. $h = 0$ **b.** $0 = 3 - t^2 - 2t$

   c. $t = -3$ or $t = 1$ **d.** 1 second

33.

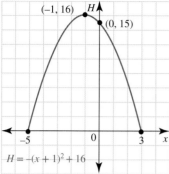

$H = -(x+1)^2 + 16$

34. a.

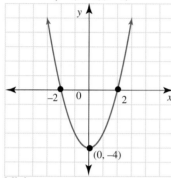

$y = (x+2)(x-2)$

Minimum

b.

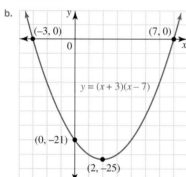

$y = (x+3)(x-7)$

Minimum

c.

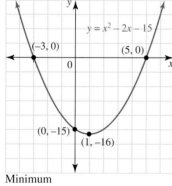

$y = x^2 - 2x - 15$

Minimum

d.

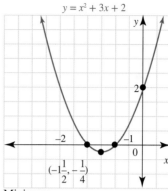

$y = x^2 + 3x + 2$

Minimum

35. 50 metres, 20 metres

36.

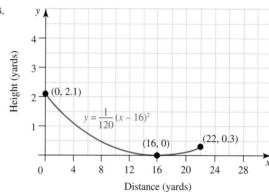

$y = \frac{1}{120}(x-16)^2$

Distance (yards)

Yes, the ball will hit the stumps because they are higher than 0.3 yards.

37. a. 45 m

b. 9 s

c. 49 m

d.

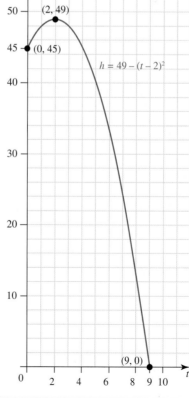

$h = 49 - (t-2)^2$

38. a. $w + l = 50$
   b. $A = l(50 - l)$
   c.

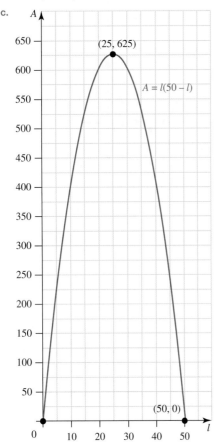

   d. 625 cm^2

39. a. $125

   b.

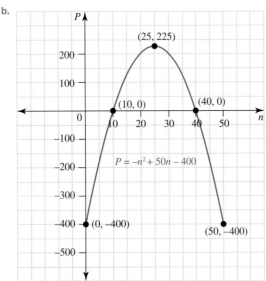

   c. $225 for 25 sausages
   d. 10 or 40 sausages

40. a. 2.6 m
   b. $p = 0.25$, $q = 2.25$
   c. $x = 1.25$
   d. 2 m
   e.

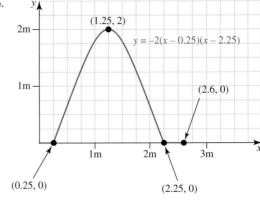

Semester review 2

The learnON platform is a powerful tool that enables students to complete revision independently and allows teachers to set mixed and spaced practice with ease.

Student self-study

Review the **Course Content** to determine which topics and subtopics you studied throughout the year. Notice the green bubbles showing which elements were covered.

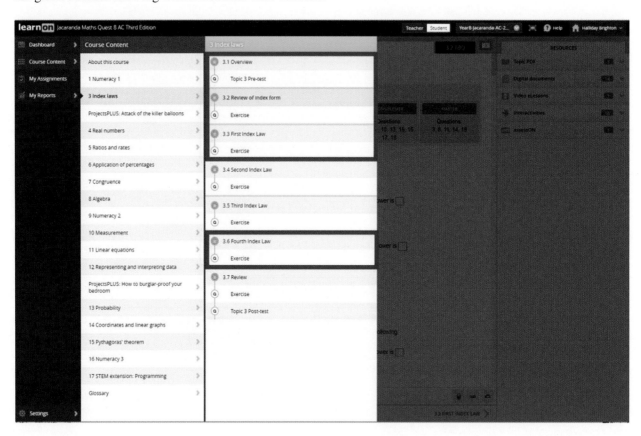

Review your results in **My Reports** and highlight the areas where you may need additional practice.

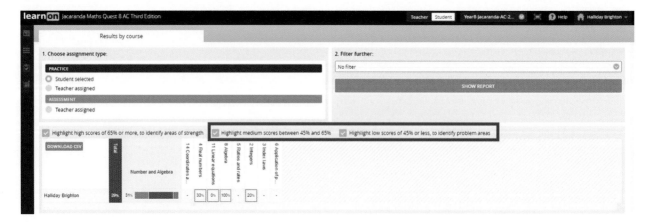

Use these and other tools to help identify areas of strengths and weakness and target those areas for improvement.

Teachers

It is possible to set questions that span multiple topics. These assignments can be given to individual students, to groups or to the whole class in a few easy steps.

Go to **Menu** and select **Assignments** and then **Create Assignment**. You can select questions from one or many topics simply by ticking the boxes as shown below.

Once your selections are made, you can assign to your whole class or subsets of your class, with individualised start and finish times. You can also share with other teachers.

More instructions and helpful hints are available at www.jacplus.com.au.

GLOSSARY

adjacent side the side next to the reference angle in a right-angled triangle

algorithm a step-by-step set of tasks to solve a particular problem. A program is an implementation of an algorithm.

angle of depression angle measured down from a horizontal line (through the observation point) to the line of vision

angle of elevation angle measured up from the horizontal line (through the observation point) to the line of vision

area amount of surface enclosed by a shape; measured in square units, such as square metres (m^2) and square kilometres (km^2)

array (probability) a table that systematically displays all the outcomes of an experiment; (programming) a list of values. A JavaScript array takes the form [x0, x1, x2, ...], where x0, x1, x2 etc. are the different values of the array.

asymptote a line that a graph approaches but never touches

axis of symmetry the straight line that sits midway between two halves of a symmetrical graph, or between an object and its image. An object is reflected along an axis of symmetry (mirror).

back-to-back stem-and-leaf plot a method for comparing two data distributions by attaching two sets of 'leaves' to the same 'stem' in a stem-and-leaf plot; for example, to compare the pulse rate before and after exercise

bar chart (graph) a graph drawn in a similar way to a column graph, with horizontal bars instead of vertical columns, with categories graphed on the vertical axis and the frequencies (numbers) on the horizontal axis

bimodal distribution a distribution of data that has two modes

binomial an expression containing two terms, for example $x + 3$ or $2y - z^2$

Booleans a JavaScript data type with two possible values: true or false. JavaScript Booleans are used to make logical decisions.

bubble sort a simple sorting algorithm. The algorithm scans an array, comparing adjacent values. If any adjacent values are out of place, they are swapped. This scanning process is repeated until no swaps are made during a complete scan.

capacity the maximum amount of fluid contained in an object, usually applied to the measurement of liquids and measured in units such as millilitres (mL), litres (L) and kilolitres (kL)

Cartesian coordinates (x, y) pairs of numbers that give the position of a point on the 2-dimensional flat plane known as the Cartesian plane. The first number is the x-coordinate and the second number is the y-coordinate.

Cartesian plane an area that consists of two number lines at right angles to each other and includes all the space between the two number lines

categorical data data that can be organised into categories; for example, satisfaction rating or favourite colour. Categorical data can be either numerical or nominal.

chance experiment a process, such as rolling a die, that can be repeated many times

circumference distance around the outside of a circle, given by the rule $2\pi r$ or πD, where r is the radius and D is the diameter of the circle

coefficient the number part of a term

complement (of a set) the complement of a set, A, written A'; the set of elements that are in ξ but not in A

compound interest interest that is paid on the sum of the principal plus preceding interest over time

congruent figures identical figures with exactly the same shape and size

console a special region in a web browser for monitoring the running of JavaScript programs

constant a term with a fixed value

constant of proportionality the number multiplying x in the equation $y = kx$, which is equal to the gradient of the corresponding graph; also called the constant of variation

constant rate rate of change that changes a quantity by the same amount for a given unit of time

continuous data data that has been collected by measuring; for example, the time it takes to get to school

cosine ratio of the adjacent side to the hypotenuse in a right-angled triangle; $\cos\theta = \dfrac{\text{adjacent}}{\text{hypotenuse}}$ or $\cos\theta = \dfrac{A}{H}$

cross-section identical 'slice' produced when cuts are made across a prism parallel to its ends

cube root the inverse of cubing (raising to the power 3); a cube root is equivalent to raising a number to an index of $\dfrac{1}{3}$

cylinder a solid object with two identical flat circular ends and one curved side

dependent variable a variable with a value that changes in response to changes in the value of other variables

diameter a straight line passing through the centre of a circle from one side of the circumference to the other

difference of two squares an expression in which one perfect square is subtracted from another, for example $x^2 - 100$; this can be factorised by $(a + b)(a - b) = a^2 - b^2$

direct linear proportion describes a particular relationship between two variables or quantities; as one variable increases, so does the other variable

discrete data data that has been collected by counting amounts; for example, number of pens

Distributive Law the product of one number with the sum of two others equals the sum of the products of the first number with each of the others; for example, $4(6 + 2) = 4 \times 6 + 4 \times 2$. It is also applicable to algebra; for example, $3x(x + 4) = 3x^2 + 12x$.

dot plot a way of recording and organising data that displays each member of a sample as an individual dot and assigns each dot a value along an x-axis

equally likely describes outcomes that have the same probability of occurring in an experiment

equation a mathematical statement that shows two equal expressions

event set of favourable outcomes in each trial of a probability experiment

expanding using the Distributive Law to remove the brackets from an expression

experimental probability probability determined by observing an experiment and gathering data, expressed by the formula: $\text{Pr(event)} = \dfrac{\text{number of ways the event can occur}}{\text{total number of equally likely outcomes}}$

expression group of terms separated by $+$ or $-$ signs, but which do not contain equals signs.

factorising writing a number or term as the product of a pair of its factors

favourable outcome any outcome of a trial that belongs to the event; for example, if A is the event of getting an even number when rolling a die, the favourable outcomes are 2, 4 and 6

Fifth Index Law to remove brackets containing a product, raise every part of the product to the index outside the bracket; therefore, $(ab)^m = a^m b^m$

First Index Law when terms with the same base are multiplied, the indices are added; therefore, $a^m \times a^n = a^{m+n}$

FOIL diagrammatic method of expanding a pair of brackets where the letters in **FOIL** represent the order of the expansion; **F**irst, **O**uter, **I**nner and **L**ast

formula a mathematical rule, usually written using pronumerals

Fourth Index Law to remove brackets, multiply the indices inside the brackets by the index outside the brackets. Where no index is shown, assume that it is 1; therefore, $(a^m)^n = a^{mn}$

frequency table a kind of table that is used to record and organise data

Fundamental Theorem of Algebra theorem that states that quadratic equations can have a maximum of two solutions

gradient (slope) (m) measure of how steep something is; the gradient of a straight line is given by $m = \dfrac{\text{rise}}{\text{run}}$ or $m = \dfrac{y_2 - y_1}{x_2 - x_1}$ and is constant anywhere along that line

grouped data data presented using a frequency table to show a range instead of a value

hectare (ha) a unit of area equal to the space enclosed by a square with side lengths of 100 m ($1 \text{ ha} = 10\,000 \text{ m}^2$); often used to measure land area. There are 100 hectares in 1 square kilometre.

highest common factor (HCF) the largest of the set of factors common to two or more numbers; for example, the HCF of 16 and 24 is 8

histogram column graph with no gaps between the columns and each column 'straddles' an x-axis score, starting and finishing halfway between scores. The x-axis scale is continuous and usually a half-interval is left before the first column and after the last column.

hyperbola the shape of the graph formed by an inverse variation relationship

hypotenuse longest side of a right-angled triangle, always opposite the right angle

image an enlarged or reduced copy of a shape (the original shape is called the object)

independent variable variable that does not change value in response to changes in the value of other variables

index (programming) an integer that points to a particular item in an array

index notation a short way of writing a number or variable when it is multiplied by itself repeatedly (for example, $6 \times 6 \times 6 \times 6 = 6^4$)

infinite loop when a program gets into a state where it will not finish without external intervention. The problem happens in the loop when it is impossible to change the terminating condition.

integer (Z) any number from the set of numbers that includes positive whole numbers, negative whole numbers and zero

interest a fee charged for the use of someone else's money, normally as a percentage of the amount borrowed

interquartile range (IQR) difference between the upper quartile, Q_U and the lower quartile, Q_L, expressed as $IQR = Q_U - Q_L$. It is the range of the middle half of the data.

intersection in probability, an intersection of two events (written as $A \cap B$) describes all outcomes in event A and in event B

inverse operation an operation that reverses the action of another; for example, $+$ is the inverse of $-$, and $\times$ is the inverse of $\div$

irrational number (I) a number that cannot be written as a fraction. Examples of irrational numbers include surds, π and non-terminating, non-recurring decimals.

like terms terms containing exactly the same pronumeral part, including the power; for example, $3ab$ and $7ab$ are like terms, but $5a$ is not

linear equations equations in which the variable has an index (power) of 1

linear graph a relationship between two variables where a straight line is formed when the points are plotted on a graph

literal equation formulas are defined as literal equations because rearranging formulas does not require finding a value for the pronumeral (or pronumerals)

mean measure of the centre (average) of a set of data, given by $\text{mean} = \dfrac{\text{sum of all scores}}{\text{number of scores}}$ or $\bar{x} = \dfrac{\Sigma x}{n}$. When data are presented in a frequency distribution table, $\bar{x} = \dfrac{\Sigma(f \times x)}{n}$.

median measure of the centre (middle) of a set of data. The median is the middle score for an odd number of scores arranged in numerical order (ordered). For an even number of scores, the median is the average of the two middle scores when they are ordered. Its location is determined by the rule $\dfrac{n+1}{2}$.

metric system the system of measurement based on the metre

midpoint halfway point between the end points of a line segment

modal class term used when analysing grouped data, given by the class interval with the highest frequency

mode measure of the centre of a set of data; the score that occurs most often. There may be no mode (for example, when all scores occur only once), one mode, or more than one mode (for example, when two or more scores occur equally frequently).

monomial an expression with one term

Monte Carlo method a method that approximates the solution to a problem using many random samples. The more random samples generated, the higher the accuracy of the final result.

multistep equation an equation that performs more than two operations on a pronumeral

natural numbers (N) positive whole numbers; 1, 2, 3, 4, …

negatively skewed distribution describes a plot that has larger amounts of data as the values of the data increase

nominal data data in categories that have names, but cannot be ranked in a logical order; for example, brands of car or types of pet

Null Factor Law (NFL) law used to solve quadratic equations, which states that if $a \times b = 0$, then either $a = 0$ or $b = 0$ or both $a = 0$ and $b = 0$

numbers (programming) a JavaScript data type that represents numerical values

numerical data data that has been counted or measured; for example, number of siblings or heights of fences. Numerical data can be either discrete and continuous.

object the original shape before it is copied and the copy is enlarged or reduced (the changed shape is called the image)

opposite side the side opposite the reference angle in a right-angled triangle

ordinal data data in categories that can be ranked in a logical order; for example, star ratings or race results

outcome result obtained from a probability experiment

overtime the time in which a worker is working hours in addition to, or outside, their regular hours of employment

parabola graph of a quadratic function, for example $y = x^2$ or $y = x^2 + 8$

parallel lines lines with the same gradient

perfect square a number or expression that is the result of the square of a whole number

perimeter distance around the outside of a shape

perpendicular lines lines that intersect at right angles to each other

pi (π) represents the ratio of the circumference of any circle to its diameter. The number represented by π is irrational, with an approximate value of $\frac{22}{7}$ and a decimal value of $\pi = 3.141\ 59\ ....$

pie chart (graph) a type of graph mostly used to represent categorical data. A circle is used to represent all the data, with each category being represented by a sector of the circle. The size of each sector is proportional to the size of that category compared to the total.

population the whole group from which a sample is drawn

positively skewed distribution describes a plot that has larger amounts of data as the values of the data decrease

principal the amount of money you start with when you put money into a financial institution such as a bank or credit union

prism solid shape with identical opposite ends joined by straight edges; a 3-dimensional figure with uniform cross-section

probability likelihood or chance of a particular event (result) occurring, given by the formula:
$$\text{Pr(event)} = \frac{\text{number of favourable outcomes}}{\text{number of possible outcomes}}.$$ The probability of an event occurring ranges from 0 (impossible) to 1 (certain) inclusive.

pronumeral a letter or symbol that stands for a number

proportion equality of two or more ratios

pseudo-random appearing to be random. A pseudo-random number or sequence of numbers is generated by a mathematical algorithm. The algorithm is seeded with a different value each time so the same sequence is not generated again.

Pythagoras' theorem theorem stating that in any right-angled triangle, the square on the hypotenuse is equal to the sum of the squares on the other two sides. This is often expressed as $c^2 = a^2 + b^2$.

Pythagorean triad a group of any three whole numbers that satisfy Pythagoras' theorem; for example, $\{5, 12, 13\}$ and $\{7, 24, 25\}$ are Pythagorean triads ($5^2 + 12^2 = 13^2$)

quadrant a quarter of a circle; one of the four regions of a Cartesian plane produced by the intersection of the x- and y-axes

quadratic equation an equation in which the term with the highest power is a squared term; for example, $x^2 + 2x - 7 = 0$

quadratic trinomial an expression of the form $ax^2 + bx + c$, where a, b and c are constants

quartile one of the quarters of a data sample that has been divided into four equal parts

radius (plural: radii) the distance from the centre of a circle to its circumference

range the difference between the maximum and minimum values in a data set

rate a ratio that compares the size of one quantity with that of another quantity; a method of measuring the change in one quantity in response to another

ratio comparison of two or more quantities of the same kind

rational number (Q) a number that can be expressed as a fraction with whole numbers both above and below the dividing sign

real number (R) any number from the set of all rational and irrational numbers

recurring decimal a decimal number with one or more digits repeated continuously; for example, 0.999 ... A rational number can be expressed exactly by placing a dot or horizontal line over the repeating digits; for example, $8.343\,434 ... = 8.\dot{3}\dot{4}$ or $8.\overline{34}$.

relative frequency frequency of a particular score divided by the total sum of the frequencies; given by the rule:

$$\text{Relative frequency of a score} = \frac{\text{frequency of the score}}{\text{total sum of frequencies}}$$

right-angled triangle a triangle with one 90° angle

salary a fixed annual amount of payment to an employee, usually paid in fortnightly or monthly instalments

sample part of a whole population

sample space the set of all possible outcomes from a probability experiment, written as ξ or S. A list of every possible outcome is written within a pair of curled brackets {}.

scale factor the factor by which an object is enlarged or reduced compared to the image

scientific notation (standard form) the format used to express very large or very small numbers. To express a number in standard form, write it as a number between 1 and 10 multiplied by a power of 10; for example, $64\,350\,000$ can be written as 6.435×10^7 in standard form.

Second Index Law when terms with the same base are divided, the indices are subtracted; therefore, $a^m \div a^n = a^{m-n}$.

sector the shape created when a circle is cut by two radii

seed the starting number used in a pseudo-random number generator

semicircle half of a circle

similar figures identical shapes with different sizes. Their corresponding angles are equal in size and their corresponding sides are in the same ratio, called a scale factor.

simple interest the interest accumulated when the interest payment in each period is a fixed fraction of the principal. The formula used to calculate simple interest is $I = \dfrac{Pin}{100}$, where I is the interest earned (in $) when a principal of $$P$ is invested at an interest rate of $i\%$ p.a. for a period of n years.

sine ratio of the opposite side to the hypotenuse in a right-angled triangle; $\sin\theta = \dfrac{\text{opposite}}{\text{hypotenuse}}$ or $\sin\theta = \dfrac{O}{H}$

Sixth Index Law to remove brackets containing a fraction, multiply the indices of both the numerator and denominator by the index outside the brackets; therefore, $\left(\dfrac{a}{b}\right)^m = \dfrac{a^m}{b^m}$

selection sort a simple sorting algorithm. The selection sort algorithm swaps the minimum value with the item at the start of the array. This minimum value is locked in place and the process is repeated on the rest of the array until the array is sorted.

square root the inverse of squaring (raising to the power 2); a square root of a number is equivalent to raising that number to an index of $\dfrac{1}{2}$

standard form *see* scientific notation

stem-and-leaf plot format used to display organised data. Each piece of data in a stem-and-leaf plot is made up of two components, a 'stem' and a 'leaf'.

stratified selection selection of a sample such that groups within a population have a similar representation in the sample

strings a JavaScript data type that represents text

subject (of a formula) the pronumeral or variable that is written by itself, usually on the left-hand side of the equation

surd any nth root of a number that results in an irrational number

symmetrical distribution a distribution that has a clear centre and an even spread on either side; also referred to as a bell-shaped or normal distribution

tangent ratio of the opposite side to the adjacent side in a right-angled triangle; $\tan\theta = \dfrac{\text{opposite}}{\text{adjacent}}$ or $\tan\theta = \dfrac{O}{A}$

term part of an expression. Terms may contain one or more pronumerals, such as $6x$ or $3xy$, or they may consist of a number only.

terminate to end; terminating decimals are decimals with a fixed number of decimal places, for example 0.6 and 2.54

Third Index Law any term (excluding 0) with an index of 0 is equal to 1; therefore, $a^0 = 1$

translate (quadratics) to move a parabola horizontally (left/right) or vertically (up/down)

tree diagram diagram used to list all possible outcomes of two or more events; the branches show the possible links between one outcome and the next outcome

trial one performance of an experiment to get a result; for example, each roll of a die is a trial

triangular prism a prism with a triangular base

trigonometry the branch of mathematics that deals with the relationship between the sides and the angles of a triangle

trinomial an algebraic expression containing three terms

turning point point at which the graph of a quadratic function (parabola) changes direction (either up or down)

turning point form the form of the equation of a graph that gives the coordinates of the turning point; for a parabola with the equation $y = (x - h)^2 + k$, the turning point is (h, k)

two-step equation an equation in which two operations have been performed on the pronumeral; for example, $2y + 4 = 12, 6 - x = 8, \dfrac{x}{3} - 4 = 2$

two-step experiment an experiment involving two separate actions — this could be the same action repeated (tossing a coin twice) or two separate actions (tossing a coin and rolling a dice)

two-way table a table listing all the possible outcomes of a probability experiment in a logical manner

undefined a numerical value that cannot be calculated

union in probability, the union of two events (written as $A \cup B$) is all possible outcomes in events A or in event B

variable (algebra) a letter or symbol in an equation or expression that may take many different values; (programming) a named container or memory location that holds a value

Venn diagram a series of circles, representing sets, within a rectangle, representing the universal set. They show the relationships between the sets.

vertex (geometry) point where two rays or arms of an angle meet; (quadratics) *see* turning point

volume amount of space a 3-dimensional object occupies; measured in cubic units, such as cubic centimetres (cm^3) and cubic metres (m^3)

wage a payment to an employee based on a fixed rate per hour

with replacement describes an experiment carried out without each event reducing the number of possible events; for example, taking an object out of a bag and then placing it back in the bag before taking another object out of the bag. In this case, the number of objects in the bag remains constant.

without replacement describes an experiment carried out with each event reducing the number of possible events; for example, taking an object permanently out of a bag before taking another object out of the bag. In this case, the number of the objects in the bag is reduced by one after every selection.

***x*-intercept** point where a graph cuts the *x*-axis

***y*-intercept** point where a graph cuts the *y*-axis. In the equation of a straight line, $y = mx + c$, the constant term, c, represents the *y*-intercept of that line.

INDEX

non-linear relations
  circle 364–7
  hyperbola 363–4
  parabola 359–63
non-right-angled triangles 259
Null Factor Law 689–90
  word problems using 695
number line, surds on 41
numbers
  cube roots 38
  irrational 7
  rational 4–5
  real *see* real numbers
  rounding *see* rounding numbers
  square roots 38
  surds 41–6
numerical data 608

O

object 191, 230
octahedron 207
  nets of 207
Ohm's Law 74
opposite side 265
ordinal data 608
outcomes 545
  expected number of 553
  experimental probability of 552
  number of 563–6
overtime 439–40
  calculating 439
  definition 439
  and penalty rates 439–40
  wages with 440

P

parabolas 359–63, 703
  constructing 742
  dilation of 713
  features of 704
  form $y = ax^2$ 713–4
  horizontal translations 362–3
  reflection 363
  sketching *see* sketching parabolas
  vertical translations 361–2
parallel lines
  definition 345
  graphing 345
pay slips 440–1
penalty rates 439–41
  definition 439
  overtime and 439–40
  time sheets and pay slips 440–1
perfect squares 101
perimeter 477–80
  circumference 477–80
  of sector 495–7

shape 477
perpendicular lines
  definition 345
  graphing 345
pi (π)
  special number 8
  symbol 8
  value 2
pie charts 626–9
  displaying data in 626
  frequency table as 627
plotting linear graphs 314–7
  on Cartesian plane 315–7
  points on line 317–8
  from rule 315–7
polygons, area of 524
polyhedron 206–7
population 611
positive indices
  converting scientific notation
    with 34
  products of powers with 29
  quotients of powers with 29
positively skewed distributions 655
prime factors 15
  product of 15
principal 445–6
prisms 501–3
  base 514
  cross-section 514
  and cylinders 512–6
  definition 501
  rectangular 503
  surface area of 502–3
  triangular 502, 503
  volume of 513–5
probability 538
  definition 544
  expected number of results 553–4
  experimental 551–3, 559
  key terms of 545–6
  language of 544–5
  scale 544
  theoretical 544–6
  two-step experiments 576–82
  two-way tables 559–69
  union of two events 568
  Venn diagrams 559–69
problem solving, with linear equations
    162–5
pronumerals 70–4, 168
  on both sides, equations with 158
  congruent triangles to 200
  definition 70
  in denominator, algebraic fractions
    with 149–50

expressions 70–1
  substitution 71–3
proportion 190, 398
  direct linear 400–2
  value in 191
proportionality 405–7
  constant of 405–7
  ratio 407
Pythagoras 246
Pythagoras' theorem 250–4, 355
  in 3D 260
  angle of depression 289–90
  angle of elevation 289–90
  applications of 252–3, 258–60
    composite shapes to solve
      problems 258–60
  definition 250
  introducing 250–1
  to non-right-angled triangles 259
  problems using 253
  Pythagorean triads 253–4
  trigonometric ratios 265–8
  unknown angles 282–4
  unknown side lengths 251–2,
    273–6
Pythagorean triads 253–4

Q

quadrant 495
  area of 495
quadratic equations 682
  applications of 735–6
  definition 684
  factorising 697–9
  of form $ax^2 + bx = 0$ 694
  of form $ax^2 + c = 0$ 686–7, 693–4
  of form $x2 + bx + c = 0$ 699
  general 685
  graphs of quadratic functions
    703–8
  monic quadratic equations 697–9
  Null Factor Law 689–90
  sketching parabolas *see* sketching
    parabolas
  solutions to 687
  with two terms 693–4
  word problems 735
quadratic functions, graphs of 703–8
quadratic trinomials 697–9
quartile 650
quilt squares 128

R

radii 495
raising powers 22–4
  more index laws 22–4

writing (*cont.*)